T5-AGA-247

Other McGraw-Hill Books of Interest

AWWA • *Water Quality and Treatment*

BRATER, KING, LINDELL & WEI • *Handbook of Hydraulics*

DODSON • *Storm Water Pollution Control: Industry and Construction NPDES Compliances*

GRIGG • *Water Resources Management*

MAIDMENT • *Handbook of Hydrology*

MAYS • *Water Resources Handbook*

PARMLEY • *Hyudraulics Field Manual*

ZIPPARO & HASEN • *Davis' Handbook of Applied Hydraulics*

Water Resources

Environmental Planning, Management, and Development

Asit K. Biswas Editor-in-Chief
President
International Society for Ecological Modelling
Oxford, England

McGraw-Hill
New York San Francisco Washington, D.C. Auckland Bogotá
Caracas Lisbon London Madrid Mexico City Milan
Montreal New Delhi San Juan Singapore
Sydney Tokyo Toronto

Library of Congress Cataloging-in-Publication Data

Water resources : environmental planning, management, and development / Asit K. Biswas, editor-in-chief.
p. cm.
Includes index.
ISBN 0-07-005483-5 (hardcover: acid-free paper)
1. Water quality management. 2. Water-supply—Management.
I. Biswas, Asit K.
TD365.W4 1996
333.91′15—dc21 96-38133
CIP

McGraw-Hill

A Division of The ***McGraw-Hill*** *Companies*

1 2 3 4 5 6 7 8 9 0 DOC/DOC 9 0 2 1 0 9 8 7

ISBN 0-07-005483-5

The sponsoring editor for this book was Larry S. Hager, the editing supervisor was Stephen M. Smith, and the production supervisor was Donald F. Schmidt. It was set in Century Schoolbook by Donald A. Feldman of McGraw-Hill's Professional Book Group composition unit.

Printed and bound by R. R. Donnelley & Sons Company.

Contents

Contributors vii
Preface ix

Chapter 1. Water Development and Environment
Asit K. Biswas 1

Chapter 2. Land Use, Erosion, and Water Resources
David Pimentel, C. Harvey, P. Resosudarmo, K. Sinclair, D. Kurz, M. McNair, S. Crist, L. Shpritz, L. Fitton, R. Saffouri, and R. Blair 37

Chapter 3. Sedimentation
Everett V. Richardson 73

Chapter 4. Waterlogging and Salinity
Safwat Abdel-Dayem 99

Chapter 5. Groundwater and the Environment
Ashim Das Gupta 117

Chapter 6. Water-Quality Monitoring
Deborah V. Chapman 209

Chapter 7. Water-Quality Prediction and Management
Christian Jokiel and Daniel P. Loucks 249

Chapter 8. Eutrophication
Mitsuru Sakamoto 297

Chapter 9. Wastewater Reuse
Takashi Asano 381

Chapter 10. Inland Fisheries
U. Barg, I. G. Dunn, T. Petr, and R. L. Welcomme 439

Chapter 11. Aquatic Weeds
P. R. F. Barrett 477

Chapter 12. Institutional Principles for Sound Management of Water and Related Environmental Resources
Harald D. Frederiksen 529

Chapter 13. Economic Mechanisms for Managing Water Resources: Pricing, Permits, and Markets
K. William Easter, Nir Becker, and Yacov Tsur 579

Chapter 14. Social Impacts
Thayer Scudder 623

Chapter 15. Resettlement
Thayer Scudder 667

Index 711

Contributors

Safwat Abdel-Dayem *Drainage Research Institute, National Water Research Center, Cairo, Egypt* (Chap. 4)

Takashi Asano *Department of Civil and Environmental Engineering, University of California at Davis, Davis, California* (Chap. 9)

U. Barg *Food and Agriculture Organization of the United Nations, Fisheries Department, Rome, Italy* (Chap. 10)

P. R. F. Barrett *Consultant in Aquatic Plant Management, Oxford, England* (Chap. 11)

Nir Becker *Department of Applied Economics, University of Minnesota, St. Paul, Minnesota* (Chap. 13)

Asit K. Biswas *President, International Society for Ecological Modelling, Oxford, England* (Chap. 1)

R. Blair *College of Agriculture and Life Sciences, Cornell University, Ithaca, New York* (Chap. 2)

Deborah V. Chapman *Environment Consultant, Kinsale, County Cork, Ireland* (Chap. 6)

S. Crist *College of Agriculture and Life Sciences, Cornell University, Ithaca, New York* (Chap. 2)

Ashim Das Gupta *Professor, Water Resources Engineering, School of Civil Engineering, Asian Institute of Technology, Bangkok, Thailand* (Chap. 5)

I. G. Dunn *Aquatic Biological Consultancy Services Ltd., Chatham, Kent, England* (Chap. 10)

K. William Easter *Department of Applied Economics, University of Minnesota, St. Paul, Minnesota* (Chap. 13)

L. Fitton *College of Agriculture and Life Sciences, Cornell University, Ithaca, New York* (Chap. 2)

Harald D. Frederiksen *World Bank, Washington, D.C.* (Chap. 12)

C. Harvey *College of Agriculture and Life Sciences, Cornell University, Ithaca, New York* (Chap. 2)

Christian Jokiel *IWW, University of Technology, Aachen, Germany* (Chap. 7)

D. Kurz *College of Agriculture and Life Sciences, Cornell University, Ithaca, New York* (Chap. 2)

Daniel P. Loucks *Cornell University, Ithaca, New York* (Chap. 7)

M. McNair *College of Agriculture and Life Sciences, Cornell University, Ithaca, New York* (Chap. 2)

T. Petr *Food and Agriculture Organization of the United Nations, Fisheries Department, Rome, Italy* (Chap. 10)

David Pimentel *College of Agriculture and Life Sciences, Cornell University, Ithaca, New York* (Chap. 2)

P. Resosudarmo *College of Agriculture and Life Sciences, Cornell University, Ithaca, New York* (Chap. 2)

Everett V. Richardson *Senior Associate, Ayres Associates, and Professor Emeritus, Civil Engineering, Colorado State University, Fort Collins, Colorado* (Chap. 3)

R. Saffouri *College of Agriculture and Life Sciences, Cornell University, Ithaca, New York* (Chap. 2)

Mitsuru Sakamoto *Professor, School of Environmental Science, University of Shiga Prefecture, Hikone, Japan* (Chap. 8)

Thayer Scudder *California Institute of Technology, Pasadena, California, and Institute for Development Anthropology, Binghamton, New York* (Chaps. 14, 15)

L. Shpritz *College of Agriculture and Life Sciences, Cornell University, Ithaca, New York* (Chap. 2)

K. Sinclair *College of Agriculture and Life Sciences, Cornell University, Ithaca, New York* (Chap. 2)

Yacov Tsur *Department of Applied Economics, University of Minnesota, St. Paul, Minnesota* (Chap. 13)

R. L. Welcomme *Food and Agriculture Organization of the United Nations, Fisheries Department, Rome, Italy* (Chap. 10)

Preface

"Water," said the eminent Greek philosopher Pindar, "is the best of all things." As the twenty-first century dawns, the importance of this statement is becoming evident in many parts of the world, even though Pindar lived some two-and-a-half millennia ago.

Increasing population and higher levels of human activities, including effluent disposals to surface and groundwater sources, have made sustainable management of water resources a very complex task throughout the world. In addition, per capita demand for water in most countries is steadily increasing as more and more people achieve higher standards of living and as lifestyles are changing steadily. For example, even in an industralized nation like Japan, per capita water demand exactly doubled during the period 1965 to 1991. The situation is more critical for developing countries, where the rates of population growth are the highest and, also, per capita demand for water from steadily increasing numbers of affluent people are putting a stress on the management of water sources.

Environmentally sound management of water resources in individual countries has become increasingly difficult during the past 25 years, and all the current trends indicate that the situation is likely to become more complex in the coming decades. Accordingly, sustainable planning and management of water resources has become a priority consideration for the future welfare of humankind.

The problem is further complicated by the fact that in the field of sustainable water management, there are often no universal solutions, both geographically and in terms of time. For example, solutions that work in the United States may not be the most appropriate for China, or what may work well in China may not prove to be an efficient solution for India, Turkey, or Mexico, primarily because of differing technical, economic, social, institutional, political, and environmental conditions. Similarly, solutions that were successfully used in a country two decades ago may need to be carefully reviewed and then perhaps modified for use at present.

In addition, there are many areas where our existing knowledge base may be somewhat limited, and needs to be substantially improved. For example, there are very few (certainly less than ten) water projects in the world where detailed environmental impact assessments were carried out a few years after the projects began operation and then were compared with what was forecasted prior to construction. In the absence of such field investigations, our knowledge of forecasting environmental impacts reliably has made only limited advance during the past two decades.

In this book, we have made a determined attempt to analyze and review the various environmental issues associated with water resources planning, management, and development from an interdisciplinary perspective, as well as to analyze global situations. The contributors are all leading international authorities in their fields, and together they provide experiences from most parts of the developed and developing world.

I am indeed most grateful to the contributors, all of whom promptly agreed to write the chapters when I first approached them. Without their active collaboration and support, this book simply would not have been possible. I would also like to express my very special appreciation to Larry Hager of McGraw-Hill, who consistantly encouraged me from the time we initially discussed the idea for this book in Washington, D.C., and throughout its preparation.

Asit K. Biswas

Chapter

1

Water Development and Environment

Asit K. Biswas
President
International Society for Ecological Modelling
Oxford, England

Introduction

Throughout history water has been considered a natural resource critical to human survival. From the earliest evolution of hominid species around the lake shores of northern Kenya to the development of the main civilizations on the banks of certain major rivers, human history can generally be considered to be water-centered. The early important civilizations developed and flourished on the banks of major rivers such as the Nile, Euphrates, Tigris, and Indus. Human history can, in fact, be written in terms of interactions and interrelations between humans and water (Biswas, 1970).

It is not difficult to realize why civilizations and habitats often developed along the banks of several strategically important rivers. Easy availability of water for drinking, farming, and transportation was an important requirement for survival. Human survival and welfare generally depended on regular availability and control of water. Floods and droughts inflicted major pains, often contributing to deaths of human beings and livestock. Because water played a very important role, when Rishi Narada of India, probably the earliest leading authority on politics who lived many centuries before the Christian era, met the great Pandava king, Yuddhistira, his greeting was water-centered because of its importance: "I hope your realm has reservoirs that are large and full of water, located in different parts of the land, so that agriculture does not depend on the caprice of the Rain God." Proper water control meant that the ravages due to droughts and subsequent famines could be significantly reduced.

Somewhat later, again in India, fiscal policies were sometimes linked with water. Because of the critical importance of water availability to ensure a good agricultural harvest in a semiarid country, the eminent Indian statesman Kautilya discussed the importance of rainfall for the economic and social well-being of the nation. In his epic *Arthasastra* (science of politics and administration), which was probably written toward the end of the fourth century B.C., Kautilya discussed the organization of a network of rain gauges throughout the country. The network was considered important, and thus essential, for two very good reasons. First, land taxes were based on the amounts of rainfall received each year, since rainfalls were considered to be proxies for agricultural production, and hence the incomes of the farmers. Second, good information on rainfall was essential to farmers for planting crops, and thus maximizing agricultural production on which national security and well-being depended.

Similarly, in Egypt, a country which the historian Herodotus considered to be a "gift" of the River Nile, the flood levels of the river have been noted for approximately 5000 years. Agricultural production, and thus the survival of the ancient Egyptians, depended on the annual inundation of the Nile, and hence its flood levels were considered to be an important indicator of the next agricultural harvest and thus their future welfare.

The fact that water control and management received such emphasis in countries as far away as India or Egypt approximately three to five millennia ago clearly indicates that the importance of this resource in the development process of the arid and semiarid regions was clearly recognized from very ancient times in different parts of the world. Despite very significant technological development, this situation has not changed much over the past several millennia, especially for arid and semiarid countries. On the contrary, the importance and relevance of water in such regions has, if anything, increased in recent decades, and it is likely to remain so in the foreseeable future.

Total Global Water Use

The total global water use has steadily increased throughout recorded history, and the trends observed in the twentieth century have been no exception. However, a closer and detailed analysis of the total global water use in recent decades would indicate two significant differences worth noting, compared to the trends observed in earlier times.

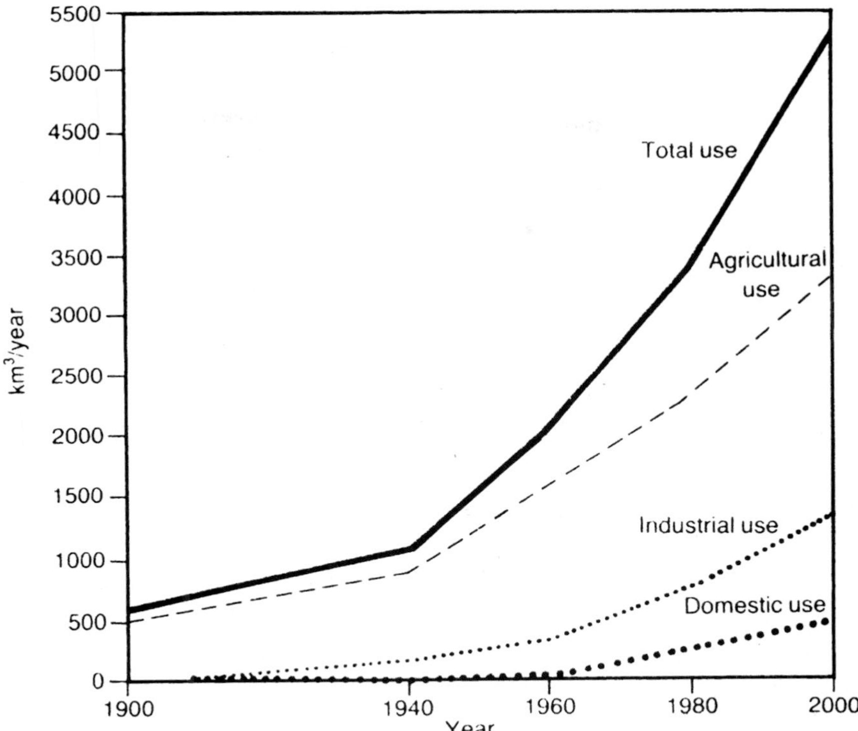

Figure 1.1 Expected increase in total water use during the twentieth century.

1. The rate of increase in total global water use accelerated remarkably in the twentieth century compared to earlier periods, particularly after 1940 (Fig. 1.1). Currently there are no visible indications that this rate is likely to level off in the near future. Since water is a limited resource, clearly such a high rate cannot be sustained indefinitely in the future. At a certain stage, and most likely within the next one to four decades, total global water use is likely to level off as a result of physical, technical, economic, and environmental constraints, first in certain individual countries and then globally.

2. At nearly 5000 km^3 per annum, the current level of global water use is extraordinarily high compared to past consumption, and the total demand is still increasing rapidly.

Figure 1.1 also indicates that total global water use is likely to have increased about tenfold from 1900 to 1999. It should, however, be noted that the estimates of global water use, as shown in Fig. 1.1, should be taken as indicative rather than definitive because of the paucity of reliable data on total water use at national levels, even for developed industrialized countries such as the United States or Germany. The situation is even worse for major water-consuming developing countries such as China or India, where the quality of data on total national water use leaves much to be desired. Thus, all data available at present on the total global water use should be considered to be gross estimates, and thus only indicative at best.

It is often generally assumed, somewhat erroneously, that the increase in global water use has paralleled population growth. While population growth is unquestionably an important factor in increasing the total global water requirements, there are other contributory factors as well, some of which are discussed later in the present chapter. Suffice it to note at present the fact that the total water-use growth rate has been significantly higher than the population growth rate in the present (twentieth) century. In fact, if the recent decades are considered, total global water use has grown at almost three times faster than the population growth rate. If these trends continue well into the future, doubling of the world's population would mean a sixfold increase in the total global water requirements (Falkenmark and Biswas, 1995). This would most likely be an unsustainable situation on a long-term basis.

The very high rate of increase in total global water use witnessed in the recent past does not axiomatically mean that this rate will continue indefinitely well into the far future. It is likely that as water requirements escalate, and as the prices charged for water for all uses gradually increase so as to approach the real cost of obtaining and managing that resource, water-use practices will become increasingly more efficient in all sectors in most parts of the world. This trend will, in all probability, decrease the demand growth rates by encouraging water-conservation practices. Accordingly, there is a high probability that during the earlier part of the twenty-first century, certain fundamental structural changes will significantly alter the growth trends witnessed in the past. Similar structural changes were witnessed in the early 1970s in the energy sector, when higher energy prices significantly altered energy demand and use patterns fundamentally on a long-term basis by making the sector increasingly more efficient in terms of production and use. There is no reason to believe that the water sector would behave any differently compared to the energy sector, if water prices increase perceptibly in the future to reflect their true cost.

Equally, it is highly unlikely that the general pattern of various types of water use will remain somewhat similar to what they are at present. Past experience indicates that as the structure of economic activities changes with time in different countries, and as it becomes more expensive and difficult to develop new sources of water and thus continually increase the supply of water available, continuing tradeoffs occur between the various uses. This pattern is most likely to continue in the future. For example, on a global basis, agriculture accounted for nearly 90 percent of all water used in 1900. By the year 2000, its share is expected to decline to around 62 percent, that is, a 28 percent reduction in 100 years. An analysis of all the current trends indicates that the percentage share of agricultural water use is likely to continue to erode steadily well into the twenty-first century, even though more and more food has to be produced to feed an ever-expanding global population.

In contrast, for the industrial water-use sector, where the value-added factor is much higher than in agriculture, and as the process of industrialization started to accelerate significantly in many parts of the world after 1940, the percentage share of the total industrial water use in the current century is expected to increase nearly fourfold, from 6 to about 24 percent. In all probability, this trend is likely to continue well into the twenty-first century.

These figures are global averages. This means that there are often significant differences in water-use patterns between different countries, depending on their respective levels of economic development, prevailing physical and climatic conditions, social norms, environmental requirements, and other relevant factors. For example, in major agrarian developing countries such as China or India, agriculture still accounts for more than 85 percent of all water used. Corresponding figures for industrialized countries such as Japan, the United States, and the United Kingdom are approximately 60, 42, and 3 percent, respectively.

As water requirements have increased with time, countries have progressively increased its supply by steadily increasing the extent of utilization of their available resources. Figure 1.2 shows the overall macro picture of the extent of water available that has been exploited in the various continents over time during the present century. Current estimates indicate that the ratio of water consumption to water availability in Asia, which has the highest volume of water requirements of all the continents at present, is likely to reach 22 percent by the year 2000. The growth rate in water requirements in Asia during recent decades has been phenomenal, since as recently as in 1960 the corresponding ratio was only about 6 percent. Furthermore, at a ratio of 22 percent, the extent of available water utilization in Asia would be the highest among all the continents, and would be nearly twice the global average (Biswas, 1994).

Figure 1.2 also indicates that the extent of water utilization in Europe has closely shadowed that of Asia since 1940. However, the total volume of water used in Europe is significantly less than that of

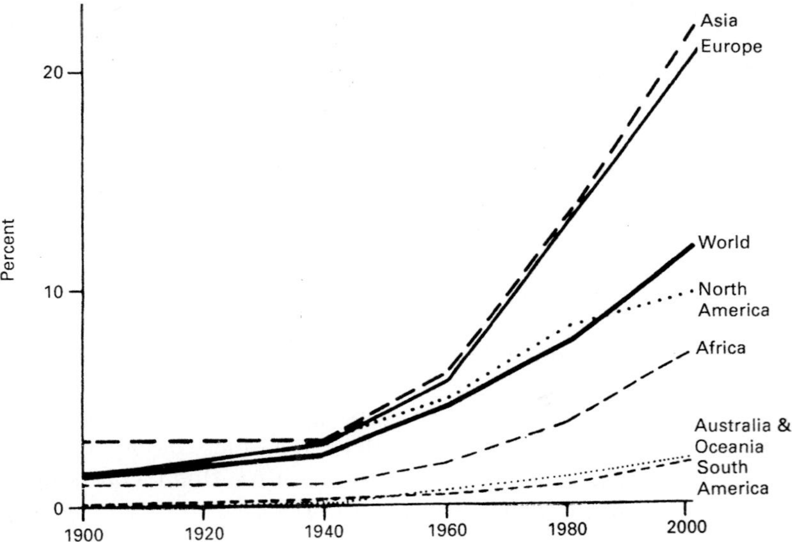

Figure 1.2 Dynamics of the ratio of water consumption to water resources.

Asia, and thus the problem faced in Asia is unquestionably that of a higher order of magnitude. The fundamental question that cannot yet be answered is to what extent the water available in a country could be developed economically with appropriate social and environmental safeguards. At present even appropriate methodologies to carry out such analyses with a reasonable degree of accuracy are still missing. This is an area that will need increasing attention in the future.

As demands for all types of traditional water use (domestic, industrial, and agricultural) increase, the signs of conflict between the various beneficiaries are becoming increasingly evident in most parts of the world. For example, in the western United States the conflict between urban-industrial users and the agricultural sector is becoming increasingly fierce. In the Philippines, the domestic and hydropower generation demands for metropolitan Manila are creating water shortages for irrigation in central Luzon. In India, increasing water demands for the city of Hyderabad are having an adverse impact on irrigation of neighboring areas. Similarly, in Indonesia, municipal water requirements for the Jakarta region and Surabaya are now in direct conflict with the existing irrigation arrangements for the surrounding areas. In South Africa, agricultural and domestic water requirements upstream are reducing flows of the rivers passing through the Kruger National Park, and as a result many adverse environmental impacts can already be observed in this world-famous protected area.

While conflicts between various water uses are increasing, analysts so far have generally not considered the flow rates necessary to preserve the ecosystems of the areas around the rivers. At present, adequate knowledge is not available on how to estimate what percentage of flow is necessary for the ecosystem needs of any particular river. As the societal concern for environmental conservation continues to increase, it is highly likely that ecosystem needs would become a legitimate form of water use during the early part of the twenty-first century. If so, the conflicts between all the various water uses can only intensify further.

Water Crisis

As all different modes of water use have continued to increase, many countries, especially those located in arid and semiarid regions, have started to face crises, although the magnitude, intensity, and extent of the crisis could vary from one country to another, or even within the same country, and also over time. Not surprisingly, the responses of individual countries, or even states or provinces within a large country, to reduce the impacts of that crisis could vary as well.

There are many, often interrelated, factors that could make the water crisis more pervasive in different parts of the world in the coming years. Only five major factors are discussed here.

First, it is an unfortunate fact that the amount of freshwater that is available to any country on a long-term basis is nearly constant for all practical purposes. Because of technical and economic considerations, only a certain percentage of the total water available can be used at any specific time. Even though technological advances have steadily increased the percentage of water available to any country that could be utilized economically, the fact still remains that a very high percentage of water would continue to remain undeveloped because of adverse economic conditions and environmental constraints. There is no question that extensive recycling and reuse can continue to increase the total volume of water that could be made usable in all countries over a given period of time. However, total recycling or reuse is neither technically possible nor economically feasible. Moreover, only limited knowledge is currently available as to how many times water can be reused without serious deterioration of its quality and attendant ecosystem degradation.

It should also be noted that the total volume of freshwater that can be used in any country generally cannot be significantly increased by artificial transportation over long distances. Low unit price of water means that unlike oil, it is generally not economic to transport water from one country to another. Furthermore, in contrast to the export of all other natural resources, even discussion of water export from a water-surplus to a water-deficient country generally generates strong public emotions. Even for two neighboring countries such as Canada and the United States that have historically had good mutual diplomatic and economic relations, the issue of export of water from Canada, a highly water-surplus country, to the United States has always become so emotional and politically charged that the successive Canadian governments have consistently considered even the technical analysis and discussions of possible water export unacceptable. The situation is no different in other parts of the world.

For economic reasons, water availability cannot be increased by desalination to any significant extent. Currently desalination can be a feasible option only under certain specific conditions and in very limited locations.

Second, water is an essential requirement for all human activities, ranging from drinking to agricultural production, and industrial development to all forms of large-scale energy generation. Accordingly, as the total global population increases, so do the aggregated human activities, which, in turn, would increase water requirements. This contributes to two contradictory trends which further complicate the water-management process. On one hand, a country's water requirements steadily increase with higher levels of human activities; on the other hand, per capita water available declines steadily since the total amount of freshwater available is limited. This is shown for a few select countries in Table 1.1.

Table 1.1 shows that for a country such as Nigeria, whose population is expected to increase significantly from about 108 million in 1994 to some 339 million by the year 2050, its per capita water availability is likely to decline from 2870 m^3/yr to only 910 m^3/yr by 2050. While this is likely to be the general global trend in the future, there would be some exceptions in a few countries such as Japan, whose total population is likely to decline modestly during this period.

While water planners have generally attributed higher water requirements in the future to increasing population numbers, another associated issue has been mostly ignored. This is the established fact that as the standard of living increases, lifestyles of people change, thus increasing per capita water requirement. For example, for England and Wales, if the present trends continue, the total water requirement by the year 2020 is expected to increase by more than 20 percent, even though increase in population is likely to be very modest. Lifestyle changes, primarily in terms of significant increases in dishwashers and washing machines, are primarily expected to account for this increase.

Similarly, per capita water use in Japan has exactly doubled between the 26-year period from 1965 to 1991, from 169 to 338 L per

TABLE 1.1 Population and Per Capita Water Availability for Selected Countries*

	Population				Annual renewable freshwater available (km^3)	Per capita freshwater availability (1000 m^3)		
	Millions			Growth rate % per annum				
	1994	2025	2050	1985–1994		1994	2025	2050
Argentina	34.2	46.1	53.1	1.4	994	29.06	21.56	18.71
Bangladesh	117.8	196.1	238.5	2.0	2357	20.00	12.02	9.88
Brazil	150.1	230.3	264.3	1.8	6950	46.30	30.18	26.30
Canada	29.1	38.3	39.9	1.3	2901	99.69	75.74	72.70
China	1190.9	1526.1	1606.0	1.4	2800	2.35	1.83	1.74
Egypt	57.6	97.3	117.4	2.0	59	1.02	0.60	0.50
India	913.6	1392.1	1639.1	2.0	2085	2.28	1.50	1.27
Indonesia	189.9	275.6	318.8	1.6	2530	13.32	9.17	7.94
Japan	124.8	121.6	110.0	0.4	547	4.38	4.50	4.97
Mexico	91.9	136.6	161.4	2.2	357	3.88	2.61	2.21
Nigeria	107.9	238.4	338.5	2.9	308	2.87	1.29	0.91
Turkey	60.8	90.9	106.3	2.1	203	3.34	2.23	1.91
United Kingdom	58.1	61.5	61.6	0.3	120	2.07	1.95	1.95
United States	260.6	331.2	349.0	1.0	2478	9.51	7.48	7.10

*The 1994 population estimates and population growth rates are from the *World Bank Atlas* (1996); population projections (medium variant) for 2025 and 2050 are from the United Nations (1994).

capita per day (Fig. 1.3). This increase in per capita water requirements is an important consideration for estimating future water needs by the planners of developing countries, whose water demands are accelerating at a very rapid and alarming rate. This aspect has thus far received very limited attention from the countries concerned as well as the appropriate international organizations.

The steady increase in per capita water requirement, especially in developing countries, can no longer be ignored. In a country such as India, around 10 percent of its current population can be considered to have an adequate standard of living compared to the rest of the world. Accordingly, the water requirement of this segment of the population is significantly higher than that of the greater Indian population. While 10 percent of India's population may not seem much, in absolute numbers it is more than 90 million, which is equivalent to more than 1.5 times the total population of the United Kingdom or 73 percent of the population of Japan. Furthermore, because of high population growth and rapid economic development, the situation in India can only become even more serious and complex in terms of water requirement in the future. Since the total number of affluent people in India is expected to rise rapidly, by the year 2025 they could be equivalent to that of 2.35 times the Japanese population, and their water requirement would increasingly become closer to the Japanese requirement. Hence, for future efficient water management, the increase in per capita water demand in developing countries due to improved lifestyles can no longer be ignored.

Third, throughout the world, for the most part all easily exploitable sources of water have already been, or are currently being, developed. This means that the costs of developing new water sources in the future are likely to be significantly higher in real terms than what have been observed in the past. For example, the average cost of providing storage for each cubic meter per second of river flow in Japan has increased nearly fourfold during the past 10 years. Approximately 20 percent of this additional cost can be attributed to new social and environmental requirements which were not required earlier. The major part of the additional cost, around 80 percent, is due to the fact that the new projects are inherently more complex technoeconomically, and thus significantly more expensive to construct. Similarly, World Bank (1992) analyses of domestic water supply projects from various developing countries indicate that the cost of development of each cubic meter of water for the next generation of projects is often two to three times higher than that of the present generation. Figure 1.4 shows the current and projected costs in 1988 constant dollars for supplying each cubic meter of water to many major urban centers of the developing world.

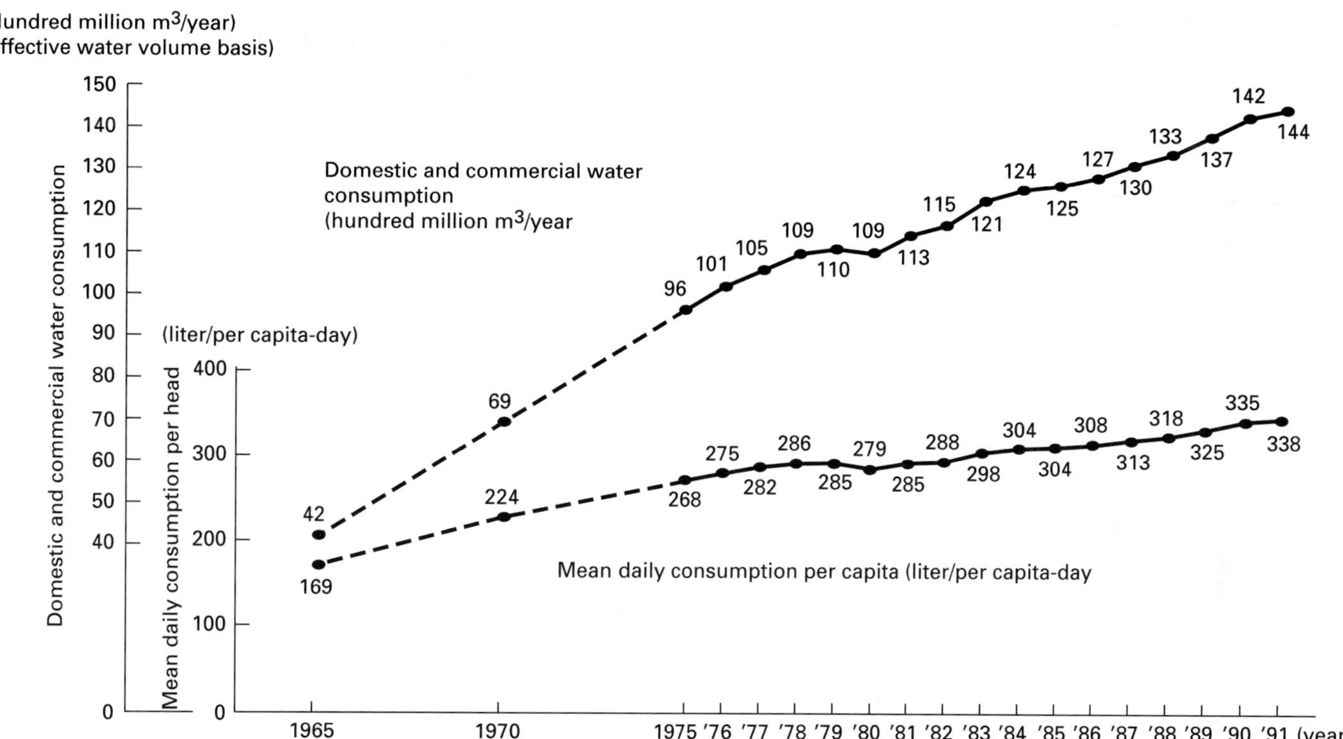

Figure 1.3 Annual domestic and commercial and daily per capita water consumption in Japan, 1965–1991. (*Source: National Land Agency.*)

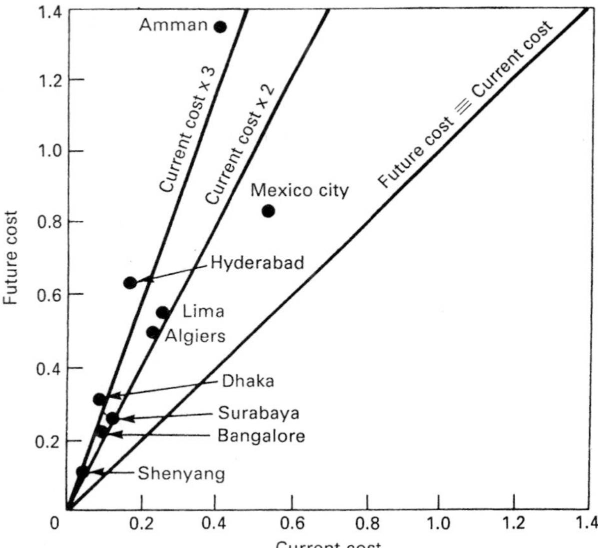

Figure 1.4 Current and projected future water costs per cubic meter (U.S.$, 1988).

The significant increase in the construction costs of future water-development projects has not yet been realistically considered by planners in nearly all parts of the world. This is likely to be an especially important factor for nearly all the developing countries, which are already facing high levels of external and internal debts. Accordingly, the amount of new investment funds that may be available to them in the future from different sources for the construction of new water projects would be somewhat restricted. Furthermore, the intersector competition as well as intrasector competition for the available funds are expected to be more intense in the future. All these economic issues are likely to have an adverse impact on the implementation and management of the next and later generations of water projects. A realistic assessment has to be that most new water projects will experience major delays in implementation, certainly much more than what their planners generally anticipate at present. Since water demands are expected to increase steadily in the future, there is likely to be considerable delay in constructing new projects, and the magnitude of the total water crisis is likely to be further exacerbated, at least in the near to medium-distant future.

Fourth, as human activities have increased, so have the effluent discharges to the environment, which have contaminated many currently used sources of surface and groundwater. The degree of contamination may vary from place to place, but the problem is serious in most parts of the world. Among the many contaminants are untreated or partially treated sewage, and all types of industrial effluents and agricultural chemicals. Since comprehensive global water-quality monitoring programs simply do not exist at present, a clear picture of global water contamination, and the extent to which water quality has been impaired for different purposes, simply is not available at present. However, on the basis of limited and anecdotal information available, it can be said that the problem is often serious near densely populated areas, especially for comparatively closed systems such as groundwater and lakes, and for stretches of rivers near such urban areas.

It should be noted that once water is contaminated, its decontamination may not be an easy or economic process. For example, cost-effective technologies currently simply do not exist for removing many pollutants such as nitrates or contaminated lake sediments.

Groundwater is extensively used as an important source of domestic water supply, ranging from 73 percent for the former Federal Republic of Germany, to 70 percent in the Netherlands and 30 percent in Great Britain. In the United States, groundwater is the primary source of potable water for over 90 percent of the rural population and 50 percent of the total population. Increased contamination could mean that some of the groundwater sources may no longer be appropriate for all types of municipal uses. For example, in the state of Nebraska, alone, concentrations of nitrates in domestic water supply in 38 towns are at present so high that babies under the age of 6 months have to be given bottled water for health reasons (Biswas, 1993). Thus, increasing contamination would further accelerate the difficulty of ensuring that adequate quantity and quality of water are available for municipal uses in the future.

Fifth, during the past two decades, the various environmental and social impacts of water-development projects have gradually become increasingly important issues. While the importance of environmentally sound water development can no longer be questioned, and must be considered as an integral component of any long-term sustainable development strategy, the fact remains that environmental analyses must be based on proven and scientific analyses and reliable data. However, it is not uncommon to find that objective and scientific analyses in recent years have been replaced by unsubstantiated and untested hypotheses, short-term political expediency, and emotional arguments. All these factors have not helped in the development of an objective and integrated framework within which assessment of the environmental and social impacts of the proposed water projects could be reliably carried out. This aspect is discussed further later in this chapter.

Of all types of major infrastructure development projects, oppositions to large water-development projects have been generally more intensive and more widespread globally compared to other types of projects. In certain Western countries, where major water-development projects have generally already been constructed, and additional projects are not considered essential, opposition to the construction of new water projects in other parts of the world is often widespread. Major international funding organizations such as the World Bank are currently under considerable political pressure from a few of their major donors not to support large-scale water-development projects in different parts of the world.

The very noticeable decline in the donors' interest in providing investment funds for major water-development projects in developing countries and the high capital requirements for new projects, mean that there may be significant delay in the implementation timetable of future projects. Unfortunately, these delays are yet to be recognized by the various national planning agencies, which means that additional new sources of water would not be available in the future as expected at present. This complacency, in all likelihood, would make the future water crises even more serious than what they are expected to be at present.

These five reasons and others associated with them mean that it would be an extremely difficult task to alleviate the water crisis of the future in many parts of the world in any significant manner within any reasonable timeframe. The technological problems may prove to be comparatively easy to solve; the really complex issues are likely to be in the environmental, social, political, and institutional arenas. Herein would probably lie the most difficult water-related challenge of the twenty-first century. The future welfare of millions of people all over the world would depend on how the water-management profession meets this challenge of providing an adequate quantity of water of the right quality in an environmentally and socially acceptable way. It would not be an easy task.

The Environment and Water Management

Interest in the environmental impacts of economic development has become an increasingly important consideration since the late 1960s, first in the United States and later in the rest of the world. Public interest in the various environmental issues peaked around 1972/73, and then started to decline steadily for much of the next 15 years. Overall interest in the various environmental issues was rekindled again in the late 1980s. It became a key political issue in the early 1990s, and probably peaked with the UN Conference on Environment and Development, which was convened in Rio de Janeiro, Brazil, in June 1992. The current anecdotal information available indicates that the interest has declined somewhat since its peak in 1992. While it is likely that the public, and thus political, interest in the environment will ebb and flow with time, it is equally likely that a certain minimum threshold of interest will be maintained on a permanent basis.

Interest in the environmental aspects of water development has also fluctuated with time during the past three decades. Not surprisingly, this interest has paralleled the general overall interest on the environmental issues. While forecasting future developments is always a hazardous task, one can say with almost total confidence that the days when water-development projects could be planned, designed, constructed, and managed without explicit consideration of environmental and social factors are now history in nearly all parts of the world. In some countries, a certain amount of lip service is still being given to environmental protection. However, current indications are that formal and real considerations of environmental factors in such countries are highly likely to become a reality in the near future.

This, however, does not mean that the environmental impacts of water projects were completely neglected in the past. A closer analysis would indicate that many specific environmental and social impacts, including development of salinity and waterlogging due to inappropriate irrigation management practices and resettlement of people due to inundation caused by the newly constructed reservoirs, have always been considered in the past. However, comprehensive and integrated environmental- and social-impact studies were not carried out in the past.

It should be noted that the environmental issues associated with water-development projects were treated no differently in the past than were all other types of infrastructural development activities. As interest in environmental issues accelerated, so did the concern with the environmental impacts of large development projects. No comprehensive environmental- and social-impact studies of major projects of any type were, in fact, carried out anywhere in the world during the pre-1965 era. Water projects were thus no exception. The current techniques for carrying out environmental-impact assessment were developed only after 1965, and the term *environmental-impact assessment* (EIA) first gained widespread use in the 1970s.

Sustainable Development

Contrary to popular belief, the concept of sustainable development is not new. The general philosophy behind the concept was expounded for centuries, if not millennia, earlier. The term *sustainable develop-*

ment or *sustainability* first became popular in the early 1980s, even though conceptually there is very little difference between this and other concepts such as ecodevelopment, which was prevalent in the early 1970s.

Sustainability has unquestionably been a popular concept in recent years, but unfortunately it means different things to different people. The situation is somewhat similar to the popular support witnessed for the conservation movement in the United States during the earlier part of the present century, when President Theodore Roosevelt exasperatedly exclaimed: "Everyone is for conservation, no matter what it means!" The situation is somewhat analogous at present for sustainable development.

It should be noted that the term *sustainability* has been used technically for several decades for harvesting reproducible natural resources, such as maximum sustainable yield of fisheries. This concept was broadened in the late 1970s by a small group of environmental scientists meeting under the aegis of the UN Environment Programme in Nairobi, Kenya. The broadened concept was expected to be a "new" idea for assessing and managing human impacts on the environment and natural resources.

The term became popular following the publication in 1987 of the report of the World Commission on Environment and Development (popularly known as the Brundtland Commission), entitled *Our Common Future*. The Commission defined sustainable development in a somewhat amorphous and ambiguous way as the "development that meets the need of the present without compromising the ability of the future generations to meet their own needs." Not surprisingly, with such a vague, simplistic, and internally inconsistent and static definition, the Commission was totally unable to specify what was to be sustained. The report is replete with references to sustainability, but was incapable of recommending how the concept could be operationalized. Sustainability was expected to be achieved in an unspecified and undetermined way.

Once the concept became popular, dozens of new definitions were offered. Currently one can readily identify more than 100 definitions of sustainable development without much difficulty. Even an organization such as the United Nations does not have a uniform, consistent definition for use by all its component organs.

In spite of the present rhetoric, it has to be admitted that operationally it has not yet been possible to define a development process which could be planned and implemented in such a way from the very beginning that it would become inherently sustainable; however, this may be defined. The best that can be achieved at present is to identify certain aspects of development which may contribute to unsustainability, and then take appropriate remedial steps to reduce, or even in some instances eliminate, these undesirable side effects. Unfortunately, it is simply not possible to devise a holistic process at present which could make a project intrinsically sustainable right from its very beginning.

For example, if sustainable water development is considered, it has been known for more than a century that irrigation without appropriate drainage would result in waterlogging and salinity, which would, in turn, progressively reduce agricultural yields over a period of time. Since the main objective of introducing irrigation is to increase agricultural yields, clearly any system that does not fulfill this purpose over the long term cannot be considered to be sustainable. Similarly, if farmers' extensive use of fertilizers increases the nitrate content of groundwater so that its potability is impaired, then the practice cannot be considered sustainable. However, ensuring that the system has proper drainage and that fertilizer use is efficient are not sufficient conditions to make an irrigation project automatically sustainable. There are numerous other factors which, singly or in combination, could contribute to its unsustainability. Some of these factors could be generic, and others may be project-specific; thus it may be difficult to identify all of them a priori. Accordingly, it is a very complex, and generally impossible, task to ensure that all the factors which could contribute to the unsustainability of a water-development project are identified during the planning and design phases and that appropriate remedial measures are taken to eliminate all their undesirable impacts.

While many issues need to be considered simultaneously within the context of sustainable water development, the following three factors are worth considering from a policy viewpoint.

1. *Short- versus long-term considerations.* One fundamental assumption behind the concept of sustainable development is that it would be viable over the long term. However, what constitutes "long term" has been neither clarified nor featured much in current discussions. The time factor, either inadvertently or because of its complexity, has basically been left fuzzy: Does sustainability cover 50 years, or 100, 500, or 1000 years or even more? Some have referred vaguely to several generations.

There are some fundamental dichotomies on the time framework for sustainability for practical application. For example, it is illogical to expect that the life period of a small check dam would be similar to that of a large multipurpose dam. Accordingly, however, the timeframe for sustainability is defined, there should be considerable flexibility in terms of the type of projects being considered.

Similarly, certain significant practical difficulties may have to be resolved if long time horizons are to be always considered in the real world. For example, if irrigated agriculture is to be considered, the normal planning horizon of farmers anywhere in the world extends to the next cropping season, or at most to the next two to three seasons. The overriding concern is how to maximize income within the timeframe of the next harvest. Thus, the farmers' mind-sets are inherently based on maximizing the profits over a continual series of single seasons. Clearly these successive short-term approaches could have long-term individual and societal costs (in terms of soil erosion, salinity development, etc.), but short-term individual benefits have almost always won over the long-term societal costs.

Accordingly, if the goal of society is long-term sustainable irrigation development, a main consideration has to be how to reconcile the short-term expectations of the main users (farmers) with the long-term needs of the society. Both conceptually and practically, it has not been easy to make these diverging needs converge.

2. *Externalities.* Externalities occur when private costs or benefits do not equal social costs or benefits. People operate primarily on the basis of their own private costs and benefits. If they perceive opportunities which could reduce their costs and/or increase potential benefits, they often take actions which could be beneficial to them but are unlikely to serve the common good. Common examples include use of excessive irrigation water by farmers in the headreaches of canals, which means that tailenders have an insufficient and/or unreliable water supply. This, in turn, could decrease the agricultural yields and thus incomes of the tailenders substantially. Similarly, wastes from municipalities could be discharged into canals and rivers, which could impair existing water uses downstream.

Such costs could be internalized, at least conceptually, through taxes, subsidies, and regulations. But in reality, even in developed countries, it has not been easy to internalize the externalities for four important reasons: (*a*) Methodologically, calculations of the precise value of externalities have been a very difficult task; (*b*) frequently there are politically powerful individuals and organizations who vociferously defend their own considerable private advantages against a large number of unorganized and disadvantaged individuals who may be experiencing additional costs somewhat indirectly; (*c*) externalities could develop steadily over time, and thus there could be a time gap before those affected realize the real costs they may have to pay, directly or indirectly; and (*d*) regulations to control such externalities in nearly all developing countries have proved to be somewhat ineffective and expensive.

3. *Risks and uncertainties.* A major issue confronting sustainable water-resource development is the risks and uncertainties inherently associated with such complex systems. For example, with the increasing population base of some Asian developing countries, there is no question that resources such as land and water have to be used intensively in order to maximize agricultural production. The fundamental question, for which there is no real clear-cut answer at the present state of knowledge, is: Up to what level can an agricultural production system be intensified without sacrificing sustainability? There are other difficult questions as well. For example, what early warnings could indicate the beginning of a transition process from sustainable to unsustainable? Which parameters should be monitored to indicate that such a transition is about to occur or, indeed, occurring? Clearly, existing knowledge bases and databases are inadequate even to identify the parameters that could indicate the passage from one stage to the other. Thus, currently it is not possible to accurately detect, much less predict, the transition of such sustainable system to an unsustainable one, or vice versa. In addition, water-resource systems are notoriously variable by nature. Their normal fluctuations could be so great that statistically significant data could be very expensive or even impossible to obtain in order to state categorically that such variations are due to natural causes or are signs of unsustainability. If, on already complex issues, additional factors such as potential climatic changes are superimposed, the degree of uncertainty in terms of detecting or predicting the transition process increases greatly (Abu-Zeid and Biswas, 1992). One is then confronted with the difficult issue of even identifying the direction of any change, let alone the magnitude of that change.

These types of fundamental issues need to be resolved successfully, before the concepts of sustainable water-resource development can be holistically conceived and then implemented. Unfortunately, while much lip service is given to sustainable water development at present, most of the published works on this subject are either somewhat general or a continuation of earlier "business as usual" undertakings that have only been given the latest trendy label of "sustainable development." If sustainable water development is to become a reality, national and international organizations will have to address many real and complex questions, which they have not done so far in any measurable and meaningful fashion. If not, and unless the current rhetoric can be translated effectively into operational reality, sustainable development will remain a trendy catchphrase for another few years, and then gradually fade away like the earlier concept of ecodevelopment.

Environmental Assessment

Whatever terminology one uses, or whichever concept is popular at any specific period of time, the fact remains that water-resource-development projects must be planned, designed, and operated in an environmentally sound manner. It is no longer adequate to consider only technical feasibility and economic efficiency of a project; it is equally essential to consider its environmental desirability and social acceptability. Simultaneous consideration of all major technical, economic, environmental, and social factors is a very difficult task, both conceptually and methodologically, but it is a task that needs to be accomplished. Current practices can be considered to be a useful initial step. However, these practices need to be objectively reviewed in terms of their application and then improved very substantially before water-resource projects can be managed in an environmentally optimal way. It is now becoming increasingly apparent that while some progress has been made during the past two decades on how to plan, design, construct, and operate water projects in an environmentally friendly way, much still remains to be done.

An essential ingredient for environmentally sound water management has to be proper environmental assessment. Experiences gained from different parts of the world during the past two decades indicate that reliable environmental assessment is the heart of any environmental management process. Good and comprehensive assessments during project planning, construction, and operation are preconditions for sound environmental management. Furthermore, assessments by themselves are not enough; they need to be properly integrated with the management and decision-making processes so that there are regular feedback between the results of the assessment and the management decisions. An objective analysis of the present situation indicates that there are four fundamental problems which need to be resolved before proper environmental management of water projects, from the planning stage all the way to the operational phase, can be carried out successfully with any degree of confidence:

1. Limited framework for environmental assessment
2. Absence of integrative approach
3. Lack of scientific knowledge
4. Absence of monitoring and evaluation during construction and operational phases

Although, for the most part, both water and environmental professionals often believe that it is possible to manage water projects optimally, until and unless these four problems are properly resolved, environmentally sound water management will remain suboptimal at best. Unfortunately, this fact has still not been properly recognized by both the professions. Only when the real problems are recognized, and deliberate attempts are made to solve them by the professions concerned, can some progress in this area be noted. The progress, however, is likely to be painfully slow.

Limited framework for environmental assessment

The assessment of environmental impacts became a formal requirement in the United States with the passage of that National Environmental Protection Act (NEPA) in 1970. Soon thereafter, environmental-impact assessment (EIA) became necessary in Canada and certain European countries. Many Asian countries followed suit. In fact, EIA became mandatory in certain Asian countries such as the Philippines and Thailand before some major western European countries such as the former Federal Republic of Germany, which adopted the process only in 1977. In general, Asian countries have made more progress in instituting EIA compared to their Latin American and African counterparts.

A careful and objective global analysis of the existing practices for carrying out EIA indicate that the general techniques that are being used at present have undergone very minor changes, both conceptually and methodologically, compared to what was practiced in the early 1970s, when EIA was first introduced. The major notable change during the past 25 years is the fact that computers available have become progressively more and more powerful; as a result, more parameters and data can now be analyzed cost-effectively than ever before. Also, the use of computers to carry out EIA has become widespread throughout the world.

In retrospect, the concepts used were acceptable and even laudable in the early 1970s, since they constituted a significant improvement over the then-prevailing practices. Unfortunately, since these techniques have been used continuously for more than 25 years, both the water-related and the environmental professionals have now automatically assumed that the EIA analyses and processes as practiced at present are reliable, usable, and contribute to significant environmental improvement. International organizations, ranging from the UN Environment Programme to the World Bank, now follow the same techniques and processes without having done any serious analyses as to what extent they actually preserve and/or improve environmental issues on a long-term basis. Nor have there been serious attempts to identify the real shortcomings of the techniques in terms of application, and then take steps to improve them. Interestingly, more than 100 EIA guidelines are available at present, but they vary only in very minor ways in terms of general approaches and concepts.

A careful and objective analysis would clearly indicate that significant changes and modifications are necessary if the techniques for environmental assessment, as used almost universally at present, could be considered appropriate for use during the late 1990s. Yet, unfortunately, these inappropriate techniques are still being used, with no serious questions asked regarding their effectiveness in terms of total environmental benefits. It appears that without any extensive review or discussion of the overall validity and desirability of the techniques, these are currently being accepted, at least implicitly, to be the only available alternatives. Since generally very limited, if any, reservations have been made on the efficacy of the current practices used for environmental assessment, there does not appear to be any long-term solution in sight.

There are at least three fundamental problems with the current environmental assessment techniques.

1. At the policy level, the linkages between environmental assessment and the social and economic aspects of water development are not clear. The linkages are generally fuzzy, even at the conceptual level. Normally, limited attempts, if any, are made to link environmental impacts at the policy level with economic and social issues. Even in the few instances in which such attempts have been made, the linkages were basically descriptive at best. Quantitative linkages are seldom made, and even theoretical methods to establish such interlinkages with any degree of reliability are conspicuous by their absence.

2. Although some progress has been made on the application of EIA at the project level, commensurate progress at policy and program levels has simply been missing. At the present state of knowledge, it is simply not possible to carry out environmental assessments of water policies and programs in any detailed and reliable fashion. Because of the generalized nature of the assessments, the results can at best be operationally used in a very limited way. In other words, usable EIA techniques for water policies and programs are still in their early stages. Much progress needs to be made before water policies and programs can be analyzed to make them environmentally sound before implementation. This area, unfortunately, has received very little research attention thus far.

3. During the past 25 years thousands of EIAs have been conducted throughout the world for various types of infrastructural development projects, including numerous ones for the water area. Irrespective of where these EIAs have been carried out, in Albania or the United States, the overall processes and the underlying philoso-

phies are somewhat similar—only the depth and rigorousness of the analyses may vary from one place to another.

A review of the EIAs carried out during the past 25 years would indicate that, uniformly and universally, the analysts have always concentrated on what is *not* environmentally sound water development rather than what is. The emphasis exclusively has been on certain aspects of development which may not be sustained. By trying to analyze and define sustainable water development by *only* those factors that could contribute to unsustainability, clearly the entire focus of the analyses has been on just one part of the equation. The other part, which could possibly be as important as the nonsustainable negative factors, has been completely ignored. The whole focus of sustainable water development, as it is construed at present, concentrates on what it is not, and then tries to ameliorate the negative impacts. A holistic approach to the issue would be to first consider what contributes to sustainable water development. This approach is simply missing at present. Instead, the overriding emphasis has been on how to identify and ameliorate the negative impacts. Clearly this approach to environmental assessment is highly skewed, and is unlikely to yield optimal environmental, economic, and social benefits.

Furthermore, it is worth noting than even though all major water-development projects would have many impacts, some of which would be positive and others negative, the word *impact* in the context of EIA has developed almost exclusively negative connotations. While any large project, irrespective of its nature, will unquestionably have both positive (otherwise why should a decision be taken to construct them?) and negative impacts, all current analyses of the environmental and social impacts invariably consider only the negative ones and their potential amelioration. Maximization of positive environmental impacts should be an equally important consideration, but it is completely ignored at present.

To a significant extent this overwhelming and almost total emphasis on the negative aspects of development projects can be explained by reviewing the historical conditions within which the EIA techniques were first developed and used. Before 1970, all project planning and analyses consisted primarily of technical and economic considerations; environmental and social analyses were mostly neglected. To the extent these were considered, they were limited to a very few select issues only, e.g., waterlogging and salinity developments and resettlement of people from the reservoir areas. Even these issues, although invariably analyzed, were seldom properly considered, especially in terms of implementation. Because of this general neglect of environmental and social factors, and numerous visible and adverse impacts of certain development projects on the environment and the society as a whole, a movement gradually developed for environmental protection in a few Western countries in the 1960s. Within a very short period, environmental protection became an increasingly important item in the political agendas of a few countries like the United States and Canada. Numerous environmental pressure groups and other nongovernmental organizations focused on the negative environmental impacts of various development projects to show, in many cases justifiably, that development projects were contributing to serious unwanted environmental and social side effects. Water projects were not an exception to this general overall trend.

Not surprisingly, this general attitude to and perception of environmental issues was reflected during the discussions of the UN Conference on the Human Environment held in Stockholm, Sweden, in 1972. This was the first of the several megaconferences that were convened by this world body during the 1970s at a very high governmental decision-making level. A retrospective analysis of the Stockholm Action Plan, as approved by all members of the United Nations, clearly indicates an overall negative approach to environmental management: (1) stop all pollution resulting from any development activity, (2) stop exhausting nonrenewable resources, and (3) stop using renewable resources faster than their regeneration. The plan focused on what should not be done, rather than what should be done and how these objectives could be achieved. The result was not surprising since the Stockholm Action Plan reflected the prevalent attitude of that time: that development activities have primarily adverse environmental impacts. Positive environmental impacts were basically ignored.

Accordingly, EIA techniques which were developed during this period reflected the concerns of that time, which were exclusively related to the identification and amelioration of negative impacts. The implicit assumption was that the large development projects can only have negative environmental impacts, and the positive impacts, if any, were so minor that they could be safely ignored. Because of this incorrect and somewhat unfortunate beginning, the term *impact* in the context of EIA very quickly assumed negative connotations only. This emphasis on only negative impacts has remained generally unquestioned and unchallenged during the past 25 years.

One undesirable side effect of this one-sided concern has been that while much is known about the adverse environmental impacts of large water-development projects, very limited progress has been made on identifying and quantifying positive impacts. The few examples that are currently available are case-specific, and accordingly may not be adequate to draw generic conclusions on the prevalence and extent of the occurrence of such positive impacts. These impacts are likely to be extensive and substantial.

In addition, and specifically for the area of water-resource development, another factor may have had a perceptible and continuing impact on the general thinking on the environmental impacts of large dams. This was the publication of a series of articles in the popular media in the United States by the well-known journalist Claire Sterling, on the adverse environmental impacts of the Aswan High Dam in Egypt. Her well-written articles at the beginning of the environmental movement, when the philosophies and techniques for the environmental-assessment processes were being formulated, caught the imagination of the general public, including most scientists. Her articles were a good interpretation of the then-prevailing views of the Western environmentalists as well as some of the prominent Egyptian scientists, that the Aswan High Dam has contributed to numerous environmental problems only, and these overwhelmed the positive benefits. These opinions were, of course, hypotheses since virtually no serious monitoring and evaluation of the environmental parameters were carried out for the dam on a comprehensive basis until the late 1980s. Sterling's articles reinforced the general bias that had started to develop in that era that large infrastructure development projects generally contributed to only major adverse environmental impacts.

The articles suited the times of a "small is beautiful" era very well for at least four important reasons:

- The Aswan High Dam was a large infrastructure which had the misfortune of being completed in 1968, exactly when the new and emerging environmental movement had started to flex its muscles.
- The United States, which had initially indicated that it would finance the construction of the Aswan High Dam, declined to do so primarily because President Gamal Nasser of Egypt became one of the four major personalities of the time who launched the new *nonaligned movement*. Egypt was thus no longer allied to the Western group. With the withdrawal of the U.S. offer, the (then) Soviet Union promptly stepped in to finance the project and to provide the necessary technical assistance. Since it was the first-ever major structure that was built in any African country by the Soviet Union, the dam generated considerable publicity in the West.
- Prime Minister Nikita Khruschev requested and received an official invitation from Egypt to participate in the opening ceremony of the dam when he boasted that the Soviet Union would drown capitalism in Africa, and their assistance in the construction of the dam was the beginning of this process. These political factors made the

reputation of the dam a casualty of the Cold War, and further contributed to its adverse public image in the West.

- It was much easier to severely criticize a new major structure built with the Soviet help in a far-off country compared to one in the West. Since technically the dam was properly constructed, and its economic benefits were never in any doubt, it was possible to criticize it only on environmental grounds, even though no scientific evidence was available for these criticisms.

Sterling's high-profile articles on only the negative environmental impacts of the Aswan High Dam found a very receptive audience in the West, who were mostly convinced that such large development projects could only be environmental disasters. Very few, if any, people realized that the articles were based on conjectures and not on facts. Sterling's writings simply reinforced the then-prevailing biases and helped to make the Aswan High Dam a cause célèbre among the environmentalists as a shining example of a bad, large development project. These generic criticisms were soon extended to many other major new water projects that were being constructed, nearly all of which were in developing countries. Sadly, this attitude does not appear to have changed much in the past two decades.

Thus, the Aswan High Dam very quickly became a symbol of everything that could be wrong with a major water-development project. Unfortunately, this view is still widely held, and most international publications available on this subject still do not provide an objective discussion of all the real benefits and costs of this dam. Extensive data that have now been collected over the past two decades, primarily with the support of the Canadian International Development Agency, indicate that many myths now surround the dam, which are generally accepted as facts. This is especially true outside Egypt, since the Egyptians had no doubt regarding the true benefits and costs of the dam.

The myths surrounding the dam have been repeated so many times that these are now accepted as facts outside Egypt. In reality, the Aswan has been a remarkably successful dam without which Egypt would have been in dire economic straits. It has unquestionably contributed to some adverse environmental impacts. However, the real issue can no longer be whether the dam should have been built, since without it Egypt would have been facing a continuing catastrophe, but rather what steps should have been taken to maximize the positive environmental impacts and reduce the negative ones (Biswas, 1992, 1996).

In retrospect, however, these developments had one major beneficial impact. The engineering profession, which dominates the global water-development field, has recognized explicitly that there are other important issues in addition to the regular technoeconomic analyses which need to be seriously considered to maximize human welfare. Accordingly, environmental and social assessments, which were mostly neglected prior to 1970, increasingly became accepted as an established procedure.

Absence of integrative approach

Environmental-assessment methodologies have continued to consider only certain selective aspects of water-development projects: an integrative approach has been basically missing. Many instances of this narrow and restrictive approach can be cited, but only one example is discussed here: health impacts of irrigation projects.

An objective and comprehensive analysis of the current practices and existing literature in this field will clearly indicate that almost exclusive emphasis has been on only *one* issue: vectorborne diseases such as malaria and schistosomiasis. Irrespective of the accuracy of the often quoted evidence for increases in waterborne diseases due to the construction of irrigation projects, an issue discussed later in this chapter, it is becoming increasingly evident that the present generally accepted theory is not only simplistic but also somewhat incorrect for the following reasons.

Viewed in any fashion, irrigation has to be considered an important component of the rural development process. With the expansion of an irrigated area due to a new project, agricultural activities and production increase as well. With higher per capita food availability and diversification of crop production in the area, significant improvement occurs in the food and nutritional status of the local people. New employment opportunities are generated by the intensification of agricultural and associated economic activities, which, in turn, improves the financial conditions of people, including landless laborers. Evidence from different parts of the world indicates that the nutritional status of rural people is often further improved through increased livestock holdings and the development of inland fisheries in the newly created reservoirs.

Furthermore, if irrigation is properly planned as an integral part of rural development, health of the rural population enhances significantly because of improvements in education, health-care and transportation facilities, and lifestyles of women. These overall interrelationships are shown diagrammatically in Fig. 1.5. All current environmental assessments of the health impacts of irrigation projects totally ignore these positive impacts, thus resulting in an incomplete and overall incorrect assessment; nor are these impacts monitored later.

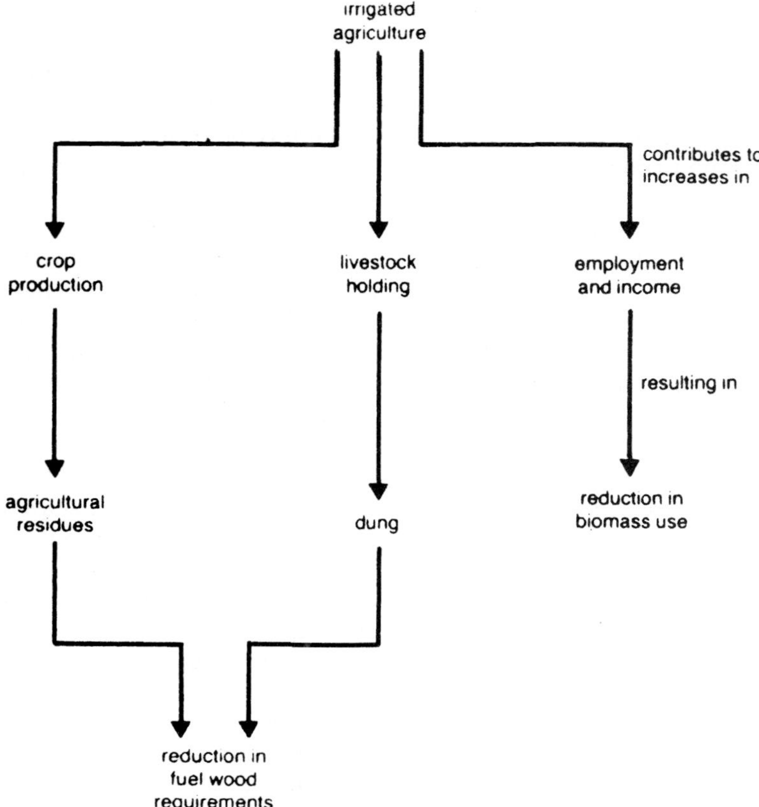

Figure 1.5 Impact of irrigated agriculture on biomass use.

At the present state of knowledge, methodologically it is possible to carry out postproject reliable and integrative environmental monitoring and assessment. For example, the evaluation carried out for a major international development agency for the Bhima Command Area Development Project in India is a good example of how such integrated analyses can be carried out (Biswas, 1987). Such analyses, however, are exceptions rather than rule.

Only two impacts of the Bhima project are discussed here as examples of the absence of an integrated approach which clearly are contributing to suboptimal results and incorrect conclusions.

First is women's education. Before irrigation was introduced in this economically disadvantaged region, landless laborers had to move constantly from one place to another searching for employment opportunities. Their daughters invariably moved with them during this nomadic lifestyle, and thus could never attend schools. With the introduction of irrigation, employment opportunities in rural areas have increased significantly. Consequently, such laborers could stay in one village and find employment within daily commuting distances. Because of this stability and increased economic opportunity due to irrigation development, the laborers began to send their daughters to schools without any formal encouragement from the government. Environmental assessments have not even cursorily considered women's education as a possible important benefit of introduction of irrigation.

The second environmental impact noted was the changing patterns of biomass fuel utilization in the project area as a direct result of irrigation. Large percentages of people purchasing fuelwood for cooking, or the total amount of fuelwood purchased per family, or both, in irrigated areas soon became significantly less than in nonirrigated areas for three reasons:

- Higher cropping intensities as well as yields significantly increased the availability of agricultural residues, which were then used for cooking.
- Increased livestock holdings in the irrigated areas produced more dung than ever before. Dry dung cakes became an important fuel for cooking.
- Increased employment opportunities and incomes encouraged people to move away from biomass fuels to more convenient commercial forms of energy.

The reasons for the reduction in the use of biomass fuels are shown diagrammatically in Fig. 1.6. These developments directly contributed to the following three major environmental and social benefits:

- Pressure on the forests in and around the region were reduced since fuelwood demands were noticeably reduced.
- Women and children spent less time collecting fuelwood.
- Money saved by many families in not buying as much fuelwood as before could be used for other productive purposes.

Women's education and biomass fuel utilization are just two examples of how an integrated approach could contribute to an objective and more accurate environmental assessment. Unfortunately, such an integrative approach is now for the most part missing.

Lack of adequate knowledge

In many areas of environmental assessment adequate scientific knowledge simply does not exist. There are also many areas in which

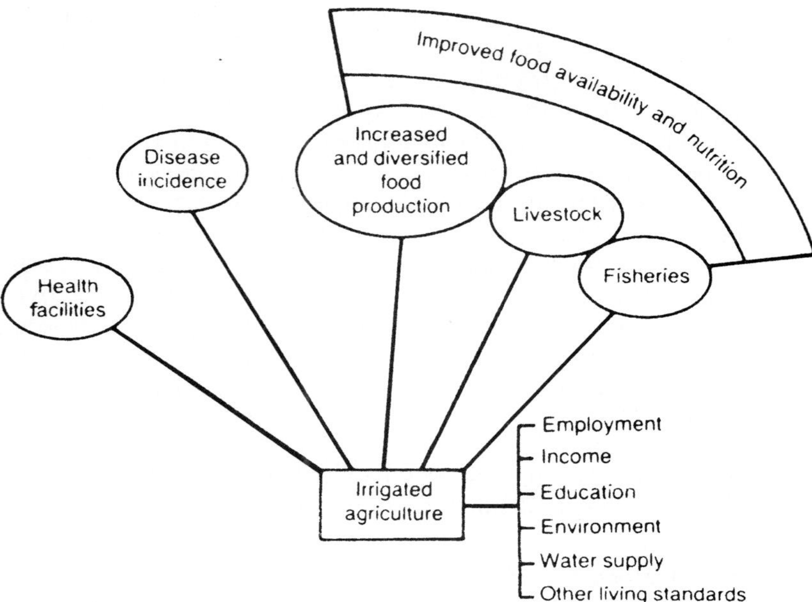

Figure 1.6 Interrelationships between irrigated agriculture and health.

current knowledge can at best be misleading and at worst totally erroneous. Regrettably, the right questions are seldom even being asked. For example, the two most widespread and important vector-borne diseases are probably malaria and schistosomiasis, but it is not possible at present to predict to what extent a water-development project per se may increase or decrease their incidence; nor is any attempt being made to answer such questions. The problem is further complicated by the site-specificity of the answers, and numerous interlinked associated factors.

An exhaustive study by the Indian Malaria Research Centre has indicated that the resurgence of malaria in that country occurred independently of irrigation expansion during the green revolution (Sharma, 1987). There is, however, no doubt that irrigation and agricultural practices, rice cultivation, and migration of agricultural workers have important bearings on the mosquito-vector fauna and malaria-transmission processes. The linkages are not clear, and there is no scientific evidence that a one-to-one relationship exists between irrigation development and malaria incidence.

Figure 1.7 shows a district-by-district average annual parasite rate (API) between 1982 and 1984 on a rice area map of India. Rice cannot be grown without extensive rain-fed or perennial irrigation. The API registers the number of malaria cases per thousand population in one

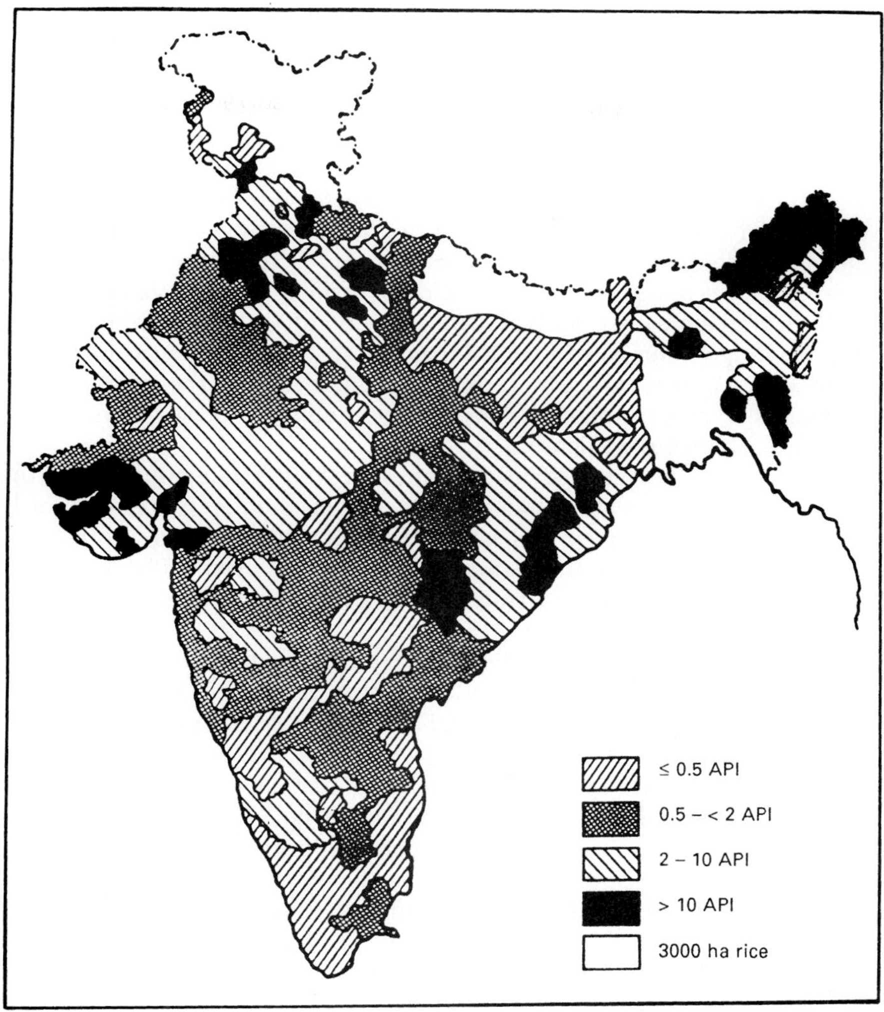

Figure 1.7 The relationship between area under rice cultivation and average API, 1982–1984. (*Source: Sharma, 1987.*)

year. The map indicates that for large parts of the country, with high areas of rice cultivation, malaria incidence rates are negligible (API < 0.5), or extremely low (API < 2). Equally there are some rice-growing areas in which the incidence of malaria is moderate (API 2 to < 10) or high (API > 10). This analysis indicates there is no specific direct interrelationship between irrigated areas and malaria-transmission rates. There appears to be many other important factors which govern disease transmission.

There are other issues which further complexify the linkages between irrigation and malaria. A study of two villages in the Kano plains of Kenya, one a newly established village within the Ahero irrigation scheme and the other an older village nearby in a nonirrigated area with traditional mixed agriculture, showed remarkable differences in terms of mosquito species present. In the new village, 65 percent of mosquito bites were from *Anopheles gambiae* (the principal vectors of malaria in tropical Africa), 28 percent were of *Mansonia* species (vectors of lymphatic filariasis and Rift Valley fever), and 5 percent were of *Culex quinquefasciatus* variety (another vector of lymphatic filariasis). In contrast, 99 percent of the mosquitoes in the older village belonged to the *Mansonia* species and less than 1 percent were *A. gambiae*. Thus, irrigation may change the transmission patterns of mosquitoborne diseases by changing species composition through habitat modification.

There are other stratification issues as well. The evaluation of the Bhima project indicated that malaria attacks women more frequently than men. How widespread is this type of stratification, in India or elsewhere, is simply unknown at present. Such specific questions have not been considered thus far, and hence no definitive answers are available.

If schistosomiasis is considered, there is no doubt that the construction of an irrigation system in a tropical or semitropical country, with extended reservoir shorelines and canal and drainage systems, creates favorable habitats for snails, which are the vectors of this disease. However, extensive studies in Egypt and several Asian countries indicate that much of the infection occurs not during the irrigation process but during domestic interactions with canal waters in the absence of water-supply and sanitation facilities. Improving the availability of domestic water supply and providing sanitation facilities, health education, and health care have reduced total schistosomiasis incidence in Egypt significantly. Currently, it is estimated that within a decade the disease will no longer be considered endemic for the first time since the pharaonic era.

A major difficulty in terms of how best to consider the interrelationships between irrigation and vectorborne diseases arises from a plethora of misleading and unreliable data, and absence of an adequate number of scientifically rigorous studies. This area is thus replete with poor and conflicting information, extensive repetition of data that have seldom been critically examined, and the elaboration of personal biases. To some extent, major international organizations have contributed, albeit not deliberately, to this sad situation. Scientists have automatically assumed that data and information published by such organizations are accurate. Unfortunately, many times they are not. For example, the current estimate of the World Health Organization (WHO) is that globally 200 million are infected with schistosomiasis. Incredibly, this number has remained constant at least for the last 26 years! The UN Environment Programme (UNEP) has incorrectly stated that schistosomiasis has been completely eradicated in China. The Food and Agricultural Organization (FAO) has stated that water development significantly increases onchocerciasis, whereas all available data indicate otherwise. Unquestionably, a major problem in this area has been caused by the uncritical acceptance and repetition of published information that is unreliable. As these dubious data are quoted and requoted by different scientists and institutions without any qualifications, they become "facts." Accordingly, the existing knowledge base needs to be substantially improved.

Monitoring and evaluation

While EIA has become a necessary preproject activity to obtain the necessary governmental approval for implementation of water-development projects, there is very little regular monitoring and evaluation of the important environmental parameters during project construction or after they become operational. This is the normal situation throughout the world. There are no institutional arrangements as to who should monitor and evaluate such projects, how these projects should be monitored, to whom such information should be provided, how the information should be used, and who should pay for the entire process. Regrettably, the entire emphasis at present is on preparation of EIAs and their clearance as a discrete activity; monitoring and evaluation of environmental parameters during operational phase receives only lip service at best.

This practice has contributed to major problems in terms of efficient environmental management of large water projects. This is because in many ways EIA is still an art and not a science. There are many interacting physical, technical, economic, environmental, and social factors, and it is impossible to predict in advance the net results of all the complex interactions. In addition, many impacts are site-specific, and thus may not be easy to predict in advance. Also, most impacts depend, at least to a certain extent, on the postproject management practices. For example, if the management of drainage systems is poor, salinity and waterlogging would become problems. However, when such problems could surface depends on the management practices; it may take 5, 15, 25, 50, or even more years. Accordingly, regular monitoring and evaluation at appropriate intervals is essential to identify when problems are developing so that the necessary remedial actions can be taken to rectify them.

In addition, because of the complexities involved, at the present state of knowledge it is simply impossible to predict *all* the environmental impacts of a major development project, the time when individual problems may surface, the magnitude of each impact, and the spatial distributions of the impacts over the project area. Accordingly, unless the project managers receive regular timely information on the problems during their early phases, appropriate remedial actions cannot be taken. By the time some of these problems become clearly visible, much of the possible damage may have already occurred. Normally it is significantly more expensive to rectify problems when they have become serious than to take remedial action during the earlier phases.

However, one of the most fundamental problems that has been created by this near-total absence of monitoring and evaluation of water-development projects is the validity of the hypotheses that are being used to make forecasts during the EIA process. Irrespective of the current rhetoric, the number of large dams anywhere in the world whose environmental and social impacts have been scientifically and objectively evaluated can be counted on the fingers of one's hands, with a few fingers left over. In the absence of monitored results, the hypotheses on which EIA forecasts are based cannot be validated, and thus the biases and errors are being continually perpetuated globally. The impact evaluation of the Aswan High Dam is a good example, which clearly indicates that the perceived wisdom at present on issues such as increases in schistosomiasis or rates of bed and bank erosion are clearly erroneous and require extensive revision. Without extensive monitoring of actual environmental and social impacts of large dams from different parts of the world, generic conclusions, which could be used to substantially improve the existing techniques for forecasting environmental impacts, cannot be drawn.

Thus, monitoring and evaluation of the environmental impacts of large water projects is necessary not only to improve their management practices and thus enhance their overall contributions to improve the socioeconomic conditions of the region but also to improve the existing EIA forecasting techniques significantly.

Clearly, proper environmental assessment of water projects is an important requirement to ensure that the economic and social benefits accrue as had been planned and expected earlier. However, the current almost-exclusive emphasis on carrying out environmental assessment only during the preconstruction phase is an important requirement but is not a sufficient condition to ensure that the environmental benefits of the project are maximized and the costs are minimized. Regular monitoring and evaluation at required intervals during the operational phase of a project and feedback of these results to the decision-making levels are absolutely essential to ensure long-term environmentally sound management of water projects.

References

Abu-Zeid, M. A., and Biswas, A. K., 1992, "Some major implications of climatic fluctuations in water management," in *Climatic Fluctuations and Water Management,* M. A. Abu-Zeid and A. K. Biswas, eds., Butterworth-Heinemann, Oxford, pp. 227–238.

Biswas, A. K., 1970, *History of Hydrology,* North-Holland, Amsterdam.

Biswas, A. K., 1987, "Monitoring and evaluation of irrigated agriculture: a case study of Bhima Project," *Food Policy,* **12**(1), 479–491.

Biswas, A. K., 1992, "The Aswan High Dam revisited," *Ecodecision,* Sept., pp. 67–69.

Biswas, A. K., 1993, "Water for agricultural development: opportunities and constraints," *Internatl. J. Water Resources Development* **9**(1), 3–12.

Biswas, A. K., 1994, "Considerations for sustainable irrigation development in Asia," *Internatl. J. Water Resources Development,* **10**(4), 445–455.

Biswas, A. K., 1996, "Water development and environment: a global perspective," Key Note Lecture, Annual Congress of Canadian Water Resources Association, Quebec City.

Falkenmark, M., and Biswas, A. K., 1995, "Further momentum to water issues: comprehensive water problem assessment," *Ambio* **22**(8), 556–560.

Sharma, V. P., 1987, "The Green Revolution in India and ecological succession of malaria vectors," 7th Annual Meeting, WHO/FAO/UNEP Panel of Experts on Environmental Management of Vector Control, Sept. 7–11, FAO, Rome.

World Bank, 1992, World Development Report, Oxford University Press, New York.

Chapter

2

Land Use, Erosion, and Water Resources

David Pimentel, C. Harvey, P. Resosudarmo, K. Sinclair, D. Kurz, M. McNair, S. Crist, L. Shpritz, L. Fitton, R. Saffouri, and R. Blair

College of Agriculture and Life Sciences
Cornell University
Ithaca, New York

Introduction

In terrestrial ecosystems, freshwater resources are essential to humans and all other species. About three-quarters of all species are associated with the terrestrial environment, and the remaining species are present in the oceans and other aquatic ecosystems (Pimentel et al., 1992). All living things require water and consist of 60 to 95 percent water.

Because 99 percent of the human food supply comes from land and associated freshwater, the ongoing loss of productive land by soil erosion from rainfall and wind threatens our food system and society (Pimentel et al., in press). To adequately feed the present world population of nearly 6 billion a diverse diet, we need about 0.5 ha of arable land per capita (Lal, 1989), yet only 0.27 ha per capita is available. In 40 years, only 0.14 ha per capita will be available (Pimentel et al., 1994). In many regions, this loss of land is a major cause of food shortages and malnutrition (WRI, 1992; Pimentel, 1993a). Over 1 billion humans (about 20 percent of the population) are now malnourished (World Bank, 1993; Speth, 1994). With the world population increasing at the rate of a quarter million per day and land degradation by erosion, escalating food shortages and malnutrition will only intensify and international instability result (Giampietro and Pimentel, 1993; Gore, 1994).

The world's human population is projected to reach about 15 billion by the year 2050 and reach a disastrous 40 billion by 2100, if the current rate of increase continues. At that time the earth would be adding 2 million people per day to its total population. Many leading scientists and public organizations are concerned about the rapid growth in population numbers and the deterioration of natural resources and the environment caused by human numbers and activities (CEQ, 1980; Keyfitz, 1984; Demeny, 1986; Hardin, 1986; Ehrlich, 1990; Holdren, 1992). As populations increase, basic resources are depleted, and this leads to environmental degradation while freedom of individual choice and quality of life decline (Durning, 1989; Durham, 1992). Worldwide at present, from 1.2 billion (Durning, 1989) to 2 billion people (V. Abernethy, Vanderbilt University, PC, 1992) are living in poverty, malnourished, diseased, and experiencing short lifespans. In the United States 36 million now are living in poverty, and this number grows daily (USBC, 1993).

The objectives of this assessment are to examine the ways in which erosion and water loss reduce crop productivity and also to assess the environmental and economic impacts. In addition, various worldwide agricultural techniques and practices that reduce erosion and conserve water are examined.

Water Use in Agriculture and Forestry

More than 13 times as much water is pumped for agriculture, forestry, and industry than is used for personal needs. In the United States, about 4900 L/d per person of water is used for all purposes (Pimentel et al., 1994). In China, only about 1200 L/d per person is used for all purposes, or only about one-quarter the use in the United States (Wen Dazong, Institute of Applied Ecology, Shenyang, China, PC, 1992). Worldwide the use of water is about 1800 L/d per person (WRI, 1991).

Agricultural production "consumes" more freshwater than any other human activity (Falkenmark, 1989). Worldwide about 87 percent of the freshwater is consumed (made nonrecoverable) by agriculture (Postel, 1992), while in the United States this figure is about 85 percent (NAS, 1989). Just during the growing season, for example, corn will transpire 4.2 million L of water (Leyton, 1983). This water and the water evaporating from the field is nonrecoverable water.

Little water is pumped for forest production. The only water that is pumped for trees exists in cities and suburbs. Thus, most of the freshwater that is pumped and consumed is for agriculture.

Although water is a renewable resource in one sense, it is also a finite resource. Rain falls each year, but the amount of rain that falls per location on earth is limited. This means that supplies of water per person decline as the population grows. Water availability per capita worldwide today is 33 percent lower than it was in 1970 because of the addition of 1.8 billion people (Postel, 1992). Veltrop (1991) projects that when the world population increases by 20 percent, the demand for water will double, thus further intensifying current world water shortages.

Several countries have already passed the level of water sustainability to provide the basic needs of their societies. In general, when the daily availability of water drops to 2740 L per person per day, then nations are considered water-short and there is a scarcity of water for food production, industry, and protection of the environment (Falkenmark, 1989). Worldwide with only about 1800 L per person per day available, a serious shortage of water already exists to supply the basic needs of the people (Pimentel et al., 1994).

Egypt, for example, receives practically no rainfall and depends almost totally (97 percent) on water flowing in from upstream neighbors (Gleick, 1993). Other nations in Africa also suffer severe water shortages, and these same nations have the highest rate of population growth (PRB, 1993). Piel (1994) reported that Malawi's population is doubling every 13 years.

Of the 14 countries in the Middle East, 9 already face significant shortages of water (Postel, 1992). Populations in 6 of these 9 nations are projected to double in about 25 years, further intensifying the water problem. In addition, almost all the rivers in the Middle East are shared by several nations. This means that tensions over water rights are a major political force throughout the region. Violence could explode in this region over water in this decade (Postel, 1992).

Erosion on Croplands and Pasturelands

Worldwide erosion rates

Of the world's agricultural land, about one-third is devoted to crops and the remaining two-thirds is devoted to pastures for livestock grazing (USDA, 1989; WRI, 1992). About 80 percent of world agricultural land suffers moderate to severe erosion, and 10 percent suffers slight to moderate erosion (Speth, 1994). Croplands and pastures are susceptible to erosion, but croplands are more vulnerable because the soil is repeatedly tilled and left without a protective cover of vegetation. However, natural erosion occurs throughout the world, especially on steep slopes, barren lands, and stream beds (Carson, 1985; Troeh and Thompson, 1993).

Although erosion rates in pasturelands are considerably lower than on croplands, erosion may exceed 100 t ha^{-1} yr^{-1} in severely overgrazed pastures (Lal, 1993). It is estimated that more than half of the world's pasturelands are overgrazed and are subject to erosive degradation (Worldwatch Institute, 1988).

Even though the problem is widespread on cropland, erosion rates are highest in Asia, Africa, and South America (ranging from 1 to 570 t ha^{-1} yr^{-1}), and lowest in the United States and Europe (ranging from 1 to 47 t ha^{-1} yr^{-1}) (Table 2.1). Although the sustainable soil-loss tolerance differs for soil types, a soil-loss rate of about 1 t ha^{-1} yr^{-1} is approximately sustainable (Hudson, 1981; Lal, 1984). A comparison of erosion rates by country is often misleading, because average erosion rates obscure the high degree of heterogeneity within each region. In Africa, for example, an average of 20 to 40 t/ha are lost each year, yet in many East African locations as much as 225 t ha^{-1} yr^{-1} may be lost annually (Lal and Stewart, 1990). Similarly, the United States currently has an average of combined wind- and water-erosion rate of about 17 t ha^{-1} yr^{-1}, but in some states, such as Iowa, erosion rates average 30 t ha^{-1} yr^{-1} (USDA, 1989).

Even these relatively low rates greatly exceed the sustainable rate of soil formation of approximately 1 t ha^{-1} yr^{-1} (the rate of conversion of parent material into soil in the A, E, and B horizons) (Troeh and Thompson, 1993). The T (tolerable soil loss) values range from 2 to 11

TABLE 2.1 Average Soil Erosion from Cropland (t ha^{-1} yr^{-1}) Worldwide

Country	Rate	Country	Rate
Argentina, Paraguay, Brazil[a]	19	Kenya[a]	5–47
Australia[b]	1–50	Madagascar[a]	25–250
Belgium[a]	10–25	Nepal[a]	40
China[c]	30	Niger[a]	35–70
England[d]	1–44	Papua New Guinea[a]	6–320
Ethiopia[e]	42	Peru (Andes)[f]	20–70
India[g]	30	Poland[h]	1–4
Italy[f]	0.2–47	Thailand (North)[i]	128
Ivory Coast[a]	60–570	United States[i]	17
Japan[j]	1–28	Zimbabwe[a]	50
Indonesia, Java[f]	60–120		

[a]Barrow (1991). [b]Edwards (1993). [c]Wen (1993). [d]Arden-Clarke and Evans (1993). [e]Hurni (1993). [f]Lal (1993). [g]Khoshoo and Tejwani (1993). [h]Ryszkowski (1993). [i]Napier et al. (1991). [j]USDA (1989).

t ha^{-1} yr^{-1}, but as Troeh and Thompson indicate, these T values are much higher than the actual rate of soil formation and conversion of parent material into soil. Throughout the world erosion rates are lowest in areas of undisturbed forests, ranging from 0.004 to 0.5 t ha^{-1} yr^{-1} (Bennett, 1939; Roose, 1988). However, once forest lands are converted to conventional agriculture, erosion rates soar to as much as 750 t ha^{-1} yr^{-1} because of tilling, vegetation removal, and overgrazing (Roose, 1988). As the world population continues to grow, the rates of deforestation and soil erosion are expected to increase (Kendall and Pimentel, 1994).

U.S. erosion rates

In the past 200 years of farming, the United States has abandoned an estimated 100 million ha (~30 percent) of farmland because of erosion, salinization, and waterlogging (Bennett, 1939; Pimentel et al., 1976; USDA, 1989). The soil-degradation problem associated with wind appears to be worsening, but water erosion appears to be declining (USDA, 1971, 1989; Lee, 1990). For instance, croplands lose an average of 17 t ha^{-1} yr^{-1} of combined water and wind erosion and pastures lose 6 t ha^{-1} yr^{-1} (USDA, 1989).

About 90 percent of U.S. cropland is losing soil above the sustainable rate of 1 t ha^{-1} yr^{-1} (Hudson, 1982; Lal, 1984), and about 54 percent of U.S. pastureland (including federal lands) is overgrazed and subject to accelerated erosion (Hood and Morgan, 1972; Byington, 1986). Soil-erosion losses, compounded by degradation caused by salinization and waterlogging, cause the abandonment of nearly 1 million ha of U.S. cropland per year (Larson and Ventullo, 1983; USDA, 1989; Soileau et al., 1990; Troeh et al., 1991). Some of the abandoned, degraded cropland may be used for either pasture or forest.

The tragic costs of soil erosion are well illustrated by the loss of rich U.S. soils. Iowa, which has some of the best soils in the world, has lost one-half of its topsoil after only 150 years of farming (Risser, 1981; Klee, 1991). Iowa continues to lose topsoil at an alarming rate of about 30 t ha^{-1} yr^{-1}, which is about 30 times faster than the rate of soil formation (USDA, 1989). The rich Palouse soils of the northwestern United States have similarly lost about 40 percent of their topsoil in the past century (Agricultural Research, 1979).

The work of the Soil Conservation Service of the USDA has decreased erosion rates in the United States only slightly during the past 50 years, because of increased erosion caused by commodity programs, crop specialization, widespread adoption of monocultures, abandonment of rotations, and sometimes the use of large, heavy farm machinery (OTA, 1982; NAS, 1989; Lee, 1990; Pimentel, 1990, 1993b). These changes have been accompanied by the creation of fewer and larger farms with increased mechanization (USDA, 1992).

During the past 50 years, the average farm size has more than doubled from 90 to 190 ha (USDA, 1967, 1992). To create larger farms and fields, the grass strips, shelterbelts, and hedgerows that once protected the soil have been removed, exposing the land to wind and water erosion (Lal, 1976, 1984; Quansah, 1981; El-Swaify et al., 1982; Hudson, 1982; Elwell, 1985). Because the larger farms typically specialize in one or two crops, such as corn and soybeans, the soil lacks vegetative cover for an appreciable part of each year and erosion rates are further accelerated. Crop specialization has also led to the use of heavier machines that sometimes inflict greater damage to the entire soil ecosystem (Buttel, 1982; OTA, 1982).

Although modern farming practices are contributing to the soil-erosion problem, the failure of farmers and governments to recognize and address this problem is equally important for the current soil crisis. Erosion is ignored by some farmers because it is difficult to measure visually in one storm or even in one season (Troeh et al., 1991; Stocking, 1994). Generally the damage goes unnoticed when a storm erodes 15 t/ha and removes slightly more than 1 mm of soil by sheet erosion. Governments also ignore erosion because of its insidious nature; that is, there are no major crises because the soil is slowly and quietly lost year after year.

Erosion and Water-Loss Processes

Erosive processes are set in motion by the energy transmitted from either rainfall, wind, or a combination of these forces. Although the effects of erosion are not easily observed on a daily basis, water and wind are both capable of quickly damaging the soil. Raindrops hit exposed soil with an explosive effect, launching soil particles into the air. For example, the energy in the raindrops falling on the state of Mississippi alone is the equivalent per year of nearly 1 billion tons of trinitrotoluene (TNT) (NAS, 1993). On sloping land more than half of the soil contained in the splashes is carried downhill. In most areas, raindrop splash and sheet erosion are the dominant forms of erosion (Allison, 1973; Foster et al., 1985). When rainfall is intense and rapid runoff occurs, gullies form ranging from 1 to 100 m deep and large volumes of water and soil are swept away (Lal and Stewart, 1990).

When wind speeds reach ≥25 mph (mi/h), the wind detaches soil particles from unprotected soil. Airborne soils can be transported thousands of miles. For instance, soil particles from eroded lands in Africa are transported as far as Brazil and Florida (Simons, 1992), and Chinese soil has been detected in Hawaii (Parrington et al., 1983).

Factors Influencing Erosion and Water Loss

Slope

Erosion increases dramatically on steep cropland because the increased angle facilitates water flow and soil movement (Fig. 2.1). It is not surprising, therefore, that mountainous regions, such as the Himalayas in Southeast Asia and the Andes in South America, suffer some of the world's highest erosion rates. The Philippines, where over 58 percent of the land has slopes greater than 11 percent, and Jamaica, where 52 percent of the land has slopes greater than 20 percent, similarly exhibit high soil losses (Lal and Stewart, 1990).

Areas with steep slopes historically have been cultivated only when arable land was scarce. However, steep slopes are routinely being converted from forests to agricultural use because of increasing population pressure and land degradation (Lal and Stewart, 1990). Once under conventional cultivation, these steep slopes suffer high erosion rates: in Nigeria, cassava fields on steep (~12 percent) slopes lost 221 t/ha per year, compared to an annual loss of 3 t/ha of soil on flat (<1 percent) land (Aina et al., 1977). Although steep slopes accelerate erosion rates, even land that is relatively flat (2 percent slope) suffers erosion (Troeh et al., 1991).

Vegetative cover

Vegetative cover reduces erosion in many ways and determines how vulnerable land is to erosion. Living- and dead-plant biomass reduce soil erosion by intercepting and dissipating raindrop and wind energy. Above-ground foliage slows the velocity of water running over the soil

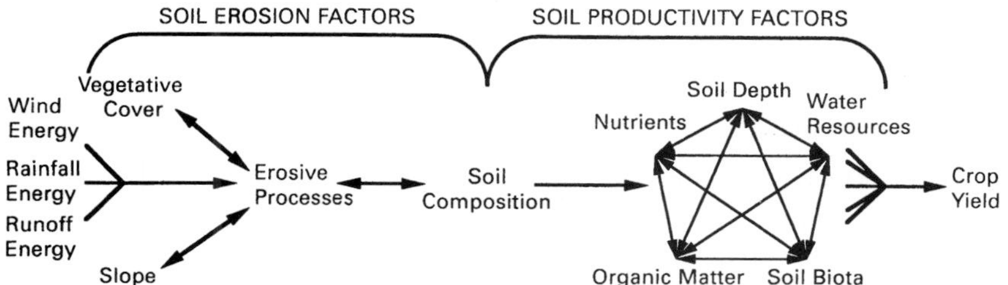

Figure 2.1 A simple model of the interrelationships between soil erosion and productivity. The energy from rainfall and wind initiates erosive processes such as water runoff and wind suspension. Factors such as slope, vegetative cover, and soil composition affect these erosive processes, and are affected by them.

and decreases the volume of water (and soil) lost in surface runoff (Langdale et al., 1992). In addition, plant roots physically bind soil particles, thus stabilizing the soil and increasing its resistance to erosion (Gray and Leiser, 1989). Plant roots enhance water conservation by creating pores in the soil surface that enable water to enter easily into the soil matrix. The uptake of water by plant roots also depletes the soil water content, and thereby further increasing infiltration rates.

Over the past 20 years, more than 200 million ha of tree cover worldwide has been removed, exposing the land to rain and wind (Brown, 1990). Grass cover has also been reduced. When vegetative cover is reduced, soil-erosion rates increase dramatically. In Missouri, for example, barren land lost soil 123 times faster than land which was covered with sod (<0.1 t ha^{-1} yr^{-1} erosion) (USFS, 1936). Similarly, in Oklahoma areas with ryegrass or wheat cover lost 2.5 to 4.8 times less water than land without cover (Sharpley and Smith, 1991). It is estimated that 40 million ha of valuable U.S. cropland is so highly erodible (USDA, 1989) that the only way to halt erosion on this land would be to convert it to a use that allows for permanent vegetative cover (WRI, 1992).

Loss of vegetative cover is extremely widespread in many third-world countries, where people collect leaves, roots, wood, and crop residues to provide household fuel, thus leaving the soil barren and most susceptible to erosion (Pimentel et al., 1986). About 60 percent of crop residues in China and 90 percent in Bangladesh are removed and burned for fuel each year (Wen, 1993). In areas where fuel is extremely scarce, even the roots of grasses and shrubs are collected (L. McLaughlin, the Badi Foundation, Macau, PC, 1991). Without the protection of crop residues and roots, soil-erosion rates can increase 10- to 100-fold (Fig. 2.2; Tables 2.2 and 2.3). As erosion reduces crop productivity, the degree of erosion control declines. Crop and residue cover are key to preventing erosion. Therefore, once crop productivity begins to decline because of soil degradation, the intensification of soil erosion becomes self-perpetuating.

Soil composition

Both the texture and the structure of soil influence its susceptibility to erosion. Soils with medium to fine texture, low organic-matter content, and weak structural development have low infiltration rates; thus, surface runoff increases (Foster et al., 1985).

Furthermore, both wind and water erosion alter the soil composition and make it more vulnerable to future erosion. Water erosion, for example, increases future erosion by exposing a less-granular and

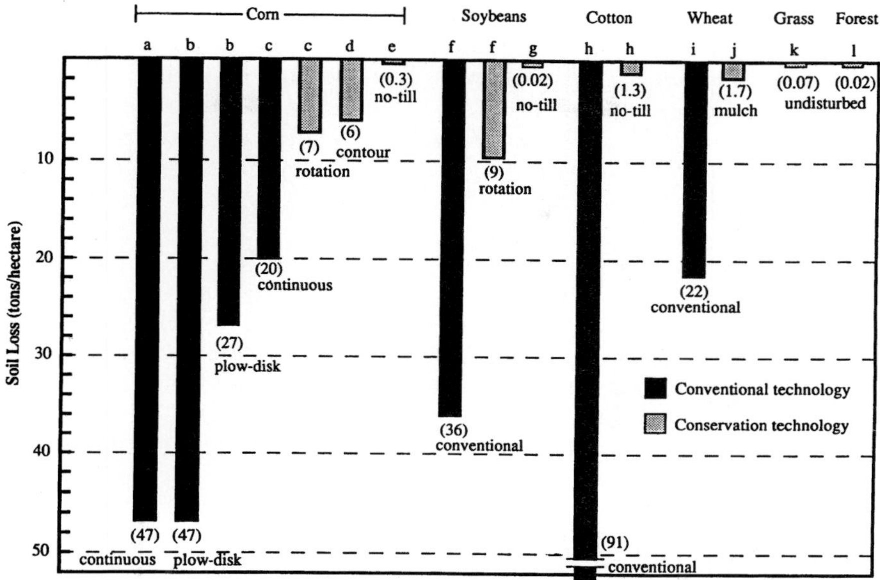

Figure 2.2 Annual soil loss (tons per hectare) by crop and technology in the United States: (*a*) Whitaker and Heinemann (1973) (Mo.); (*b*) Moldenhauer and Amemiya (1967) (Ind., Ohio); (*c*) Faeth et al. (1991) (Pa.); (*d*) Van Doren et al. (1950) (Ill.); (*e*) McGregor and Mutchler (1992) (Miss.); (*f*) McGregor et al. (1992) (Miss.); (*g*) Langdale et al. (1992) (Ga.); (*h*) Mutchler et al. (1985) (Miss.); (*i*) Johnson and Papendick (1968) (Wash.); (*j*) McGregor et al. (1990) (Miss.); (*k*) Bennett (1939) (Kans.); (*l*) Borman et al. (1974) (N.H.).

less-permeable subsoil, disintegrating aggregates through raindrop impact to create a compact surface crust, and plugging holes and pores with soil grains (Troeh et al., 1991). The abrasive energy of wind breaks down larger, more stable soil aggregates into smaller particles which then become easily detached and blown away.

Erosion, Water, and Reduced Productivity

Because of the erosion-associated loss of productivity and other factors, the per capita food supply has been reduced over the past 10 years and continues to fall (Kendall and Pimentel, 1994). The Food and Agricultural Organization (FAO) reports that per capita production of grains, which make up 80 percent of the world food supply, has been declining since 1984 (FAO, 1992).

Crop yields on severely eroded soil are much lower than those on protected soils because erosion reduces soil fertility and water availability. Corn yields on severely eroded soils are reduced by 12 to 21 percent in Kentucky, 0 to 24 percent in Illinois, 25 to 65 percent in the

TABLE 2.2 Water-Runoff Rates for Conservation Corn Plantings versus Conventional Plantings of Corn

Treatment	Water runoff (cm depth)	Conserved water, cm	Increased yield, t/ha*	Reference
Corn stover mulch vs. no stover residue	0.06 1.30	1.24	0.34	Ketcheson and Onderdonk (1973)
Rye cover mulch vs. residue burned	3.9 17.4	13.5	3.4	Klausner et al. (1974)
Manure mulch vs. no manure	9.0 13.1	4.1	1.1	Musgrave and Neal (1937)
Corn-oats-hay-hay vs. conventional continuous	0.58 3.08	2.50	0.6	Ketcheson (1977)
No-till in sod vs. conventional	3.7 10.7	7.0	1.8	Spomer et al. (1976)
Level terraced vs. contour planted	0.94 8.14	7.2	1.8	Schuman et al. (1973)
Dense planting vs. bare soil	2.49 3.32	0.97	0.2	Mohammad and Gumbs (1982)
Reduced till vs. conventional	2.1 3.6	1.5	0.4	McIsaac and Mitchell (1992)

*Increased yield based on Troeh et al. (1991).

TABLE 2.3 Estimated Annual Economic and Energy Costs per Hectare of Soil and Water Loss from Conventional Corn Assuming a Water and Wind Erosion Rate of 17 t ha^{-1} yr^{-1} over 20 Years

Factors	Annual quantities lost	Cost of replacement, $	Energetic costs (10^3 kcal)	Yield loss after 20 years of erosion, %
Water runoff	75 mm[a]	30[b]	700[c]	7[a]
Nitrogen	50 kg[d]		500[e]	
Phosphorus	2 kg[d]	100[d]	3[e]	8[f]
Potassium	410 kg[d]		260[e]	
Soil depth	1.4 mm[a]	16	—	7[h]
Organic matter	2.0 t[a]	—	—	4[i]
Water-holding capacity	0.1 mm[a]	—	—	2[j]
Soil biota	—	—	—	1[k]
Total on site		146	1460	20[l]
Total off site		50[m]	100	
Grand total		196	1560	

[a]See Table 2.4.
[b]The cost of replacing this much water using groundwater irrigation based on 1992 dollars (Hinz, 1985). The value is reduced by 40 percent because it is assumed that water erosion accounts for 60 percent of U.S. erosion (NAS, 1993). However, if rainfall were abundant, then this replacement cost would not be necessary.
[c]Energy required to pump groundwater from a depth of 30 m (Batty and Keller, 1980).
[d]Total nutrient loss based on Troeh et al. (1991).
[e]Energy required to replace the fertilizers lost (Pimentel, 1980).
[f]Based on the total loss of 340 t/ha of soil over 20 years and the mineralization and availability of the nutrients in this soil.
[g]Estimated.
[h]Based on reduced productivity of about 6 percent per loss of 2.5 cm of soil (Lyles, 1975).
[i]Organic matter content of the soil was assumed to decline from 4 to 3 percent over this period, resulting in a 4 percent decline in productivity.
[j]After the loss of 17 t/ha of soil per year, the water-holding capacity was assumed to decline by 1.9 mm and productivity declined by 2 percent; with severe erosion over time, plant-available water may decline by 50 to 75 percent (Schertz et al., 1989; Troeh and Thompson, 1993).
[k]Reductions in soil biota were assumed to reduce infiltration of water and reduce organic-matter recycling.
[l]Percentages do not add up because the impacts of the various factors are interdependent and some overlap exists (e.g., organic matter is interrelated with water resources, nutrients, soil biota, and soil depth). This loss would occur if lost nutrients and water were not replaced.
[m]See Table 2.7.

southern Piedmont, and 21 percent in Michigan (Frye et al., 1982; Olson and Nizeyimana, 1988; Mokma and Sietz, 1992). In parts of the Philippines, erosion has been responsible for corn productivity declines as high as 80 percent over the past 15 years (Dregne, 1992).

The majority (about 60 percent) of soil erosion on U.S. croplands is due to rainfall, but in the arid states, wind erosion is more important (WRI, 1992). For example, Texas cropland is affected mostly by wind erosion (average annual soil loss of 33 t/ha), while Tennessee cropland suffers primarily from water erosion (average annual soil loss of 32 t/ha) (Crosson, 1985).

Erosion by water and wind adversely affects soil quality and productivity by reducing infiltration, water-holding capacity, nutrients, organic matter, soil biota, and soil depth (OTA, 1982; El-Swaify et al., 1985; Troeh et al., 1991). Each of these factors influences soil productivity individually and interacts with the other factors, thus making the assessment of soil-erosion impacts on productivity extremely difficult (Fig. 2.1).

Water resources

All crops require enormous quantities of water for their growth and the production of fruit (NSESPRPC, 1981; Follett and Stewart, 1985; Falkenmark, 1989). For example, during a single growing season a hectare of corn (7000 kg/ha yield) transpires about 4 million L of water [based on Leyton (1983)], and an additional 2 million L/ha concurrently evaporate from the soil (Waldren, 1983; Donahue et al., 1990).

Because crops require such large quantities of water, it is vital that water soak into the soil instead of running off. When erosion occurs, soils absorb 10 to 300 mm/ha less water annually, or between 7 and 44 percent of total rainfall (Van Doren et al., 1950; Wendt and Burwell, 1985; Kramer, 1986; Wendt et al., 1986; Edwards et al., 1990; Hauser and Jones, 1991; Murphee and McGregor, 1991). This loss of water can severely reduce crop productivity; even a runoff rate of 20 to 30 percent of total rainfall results in significant water shortages for crops (Elwell, 1985). No-till culture increases the rate of water runoff but prevents erosion and soil loss.

When erosion increases the amount of water runoff, less water enters the soil matrix and is available for the crop. In the tropics, Lal (1976) reported that erosion may reduce infiltration by up to 93 percent and thus dramatically increases water runoff.

As erosion decreases the depth of soil and organic-matter content, its ability to retain water significantly declines (Buntley and Bell, 1976). Frye et al. (1982) reported that an uneroded soil 60 cm deep could hold 1500 m^3/ha of water whereas eroded soil 20 cm deep had a water-holding capacity of only 500 m^3/ha.

Water runoff and soil loss can be reduced by using cropping methods, such as intercropping, that increase soil organic matter and ground cover (Reid, 1985). For example, when silage corn is interplanted with red clover, water runoff can be reduced by 45 to 87 percent and soil loss can be reduced by 46 to 78 percent, compared with silage corn grown without clover (Wall et al., 1991).

Soil nutrients

In addition to causing water deficiencies, soil erosion causes shortages of basic plant nutrients, such as nitrogen, phosphorus, potassium, and calcium, which are essential for crop production (Fig. 2.1). A fertile agricultural topsoil typically contains 1 to 6 kg of nitrogen, 1 to 3 kg of phosphorus, and 2 to 30 kg of potassium per ton, whereas an eroded soil may have nitrogen levels of only 0.1 to 0.5 kg per ton (Alexander, 1977; Lal, 1980; Van Dijk, 1980; Foth and Ellis, 1988; Troeh et al., 1991). When erosion occurs, these nutrients are lost because the wind and water selectively remove the fine particles, leaving behind large particles and stones. Soil lost through erosion typically contains about three times more nutrients (range 1.6 to 10) than the soil left behind (Martin, 1941; Bhatt, 1977; Lal, 1980; Young, 1989).

When nutrient reserves are depleted by erosion, plant growth is stunted and crop yields decline (Table 2.3). Areas that suffer severe erosion may produce 15 to 30 percent lower corn yields than uneroded areas (Follett and Stewart, 1985; Olson and Nizeyimana, 1988). Under the current average annual soil-erosion rates (17 t/ha), the loss of nitrogen, phosphorus, and potassium could result in a long-term drop in corn yield of 800 kg/ha, or roughly 12 percent of the average corn yield (Tisdale et al., 1985).

Soil organic matter

Organic matter is a necessary component of soil because it facilitates the formation of soil aggregates, increases soil porosity, and thereby improves soil structure, water infiltration, and ultimately overall productivity (Greenland and Hayes, 1981; Tisdall and Oades, 1982; Chaney and Swift, 1984; Langdale et al., 1992). In addition, organic matter increases water infiltration, facilitates cation exchange, enhances root growth, and stimulates the proliferation of important soil biota (Allison, 1973) (Fig. 2.1). Also, it provides an essential source of both macro- and micronutrients for plant growth (Allison, 1973; Volk and Loeppert, 1982). Approximately 95 percent of the nitrogen, 25 to 50 percent of the phosphorus, and over 70 percent of the sulfur in soil is contained in soil organic matter (Allison, 1973).

Fertile topsoils typically contain about 100 t of organic matter (or 4 percent of total soil weight) per hectare (Follett et al., 1987; Young, 1990). Because the majority of organic matter is concentrated near the soil surface in the form of decaying leaves and stems, erosion of topsoil results in a rapid decrease in soil organic-matter levels. Several studies have demonstrated that the soil removed by either wind or water erosion is 1.3 to 5 times richer in organic matter than the soil left behind (Barrows and Kilmer, 1963; Allison, 1973). Thus, the loss of 17 t/ha of soil by rainfall removes nearly 2 t/ha of organic matter (Young, 1990). Since the inception of farming in the United States, an estimated one-half of soil organic matter has been lost from most soils (Curry-Lindahl, 1972). Some organic matter, of course, disappears because of mineralization.

Once the organic matter layer is depleted, soil productivity and crop yields decline because of the degraded soil structure and depletion of nutrients. The reduction of soil organic matter from 4.3 to 1.7 percent, for example, lowered the yield potential for corn by 25 percent (Lucas et al., 1977).

Many farmers apply inorganic fertilizers to their soils to replace the nutrients lost when organic matter is swept away; however, the continuous use of commercial fertilizers alone cannot sustain high crop yields everywhere. In tropical Africa, a soil that had lost over 60 percent of its organic matter produced corn yields at only 20 percent of its initial level despite the continuous application of 100 kg of nitrogen, 20 kg of phosphorus, and 50 kg of potassium per hectare each season (Agboola, 1990).

Soil biota

Although soil biota are often ignored in assessing the impact of erosion, they are a critical component of the soil and constitute a large portion of the soil biomass. Indeed, a large portion of the soil is living. One square meter may support populations of about 200,000 arthropods and enchytraeids (worms of genus *Enchytraeus*), and billions of microbes (Wood, 1989; Lee and Foster, 1991). A hectare of good-quality soil contains an average of 1000 kg of earthworms, 1000 kg of arthropods, 150 kg of protozoa, 150 kg of algae, 1700 kg of bacteria, and 2700 kg of fungi (Pimentel et al., 1980).

Soil biota have a profound effect on crop productivity because they recycle the basic nutrients required for plants in the ecosystem (Van Rhee, 1965; Pimentel et al., 1980). Earthworms and other soil biota also enhance productivity by increasing water infiltration. Through tunneling and burrowing activities, soil biota create thousands of pores and channels within the soil matrix which allow water to percolate efficiently into the soil. By increasing infiltration rates by as much as 21 times, earthworms simultaneously decrease surface-water runoff and reduce soil-erosion loss (Hopp, 1946; Teotia et al., 1950; Ehlers, 1975; Kladivko et al., 1986; Lee and Foster, 1991).

Soil biota further contribute to soil productivity by mixing the soil profile, enhancing aggregate stability, and preventing soil crusting. Earthworms commonly bring between 10 and 250 t ha^{-1} yr^{-1} of soil from underground to the soil surface (Edwards, 1981; Lee, 1985), while insects bring one-tenth of this amount (Hole, 1981; Zacharias and Grube, 1984; Lockaby and Adams, 1985). The churning and mixing of soil redistributes nutrients, aerates the soil, and increases infiltration rates, thus making the soil overall more favorable for plant growth.

The erosion that accompanies conventional agriculture may decrease the diversity and abundance of soil organisms (Atlavinyte, 1964, 1965). On the other hand, agricultural practices that maintain the soil organic-matter content at optimum levels favor the proliferation of soil biota (Reid, 1985). Thus, the simple practice of straw mulching may increase biota threefold (Teotia et al., 1950), while the application of organic matter or manure may increase earthworm and microorganism biomass as much as fivefold (Ricou, 1979).

Soil depth

The formation of soil is a very slow process; it takes between 200 and 1000 years to form 2.5 cm (1 in) of topsoil under cropland conditions, and even longer under pasture and forest conditions (Hudson, 1981; OTA, 1982; Lal, 1984; Elwell, 1985). In the United States, where an average of 2.5 cm of soil is lost every 16.5 years, soil is lost about 17 times faster than the rate at which it is formed (Troeh and Thompson, 1993). The average U.S. topsoil depth in 1776 is estimated to have been about 23 cm. Today, after about 200 years, or more likely the past 50 years of farming, the average depth has declined to two-thirds (~15 cm) of the original soil depth (Chapman et al., 1950). This loss of one-third of U.S. topsoil represents approximately the loss of 1500 years of U.S. land heritage (based on the rate of soil formation) (Kirchner et al., 1992).

The reduction in soil depth can depress crop yields by reducing the amount of water that soil can hold. Thin soils are unable to retain as much water as thick soils, and therefore exhibit lower crop yields (Beacher et al., 1985). The effect of soil depth on crop yield is particularly pronounced during periods of drought. Reid (1985) reported that although shallow and deep soils had similar corn yields during years of abundant rain the shallow soils had much lower corn yields (3.4 vs. 6.7 t/ha) than the uneroded, deep soils in drought years. By comparison, in tropical Nigeria the removal of 10 to 20 cm of topsoil reduced corn yields by 65 to 90 percent (Lal, 1985; Magrath, 1989).

At present average world topsoil depth is only 15 to 23 cm (Worldwatch Institute, 1988; Livingstone, 1992), and soil is being depleted at 2 to 3 mm per year (Pimentel, 1993a). The continued loss of soil from world agricultural lands can be expected to have serious long-term consequences on the ability of agriculture to provide food for the rapidly growing world population (Worldwatch Institute, 1988; Livingstone, 1992).

Model of Erosion and Water-Loss Effects on Crop Productivity

To assess how and to what extent erosion decreases crop productivity, it is necessary to consider the multiple factors that influence erosion rates, as well as the soil components that affect productivity.

We have developed empirical models that incorporate the numerous factors affecting both erosion rates and soil productivity. The slope of the land, soil composition, and extent of vegetative cover influence the rate of erosion, while the soil depth, presence of soil biota, organic matter, water-holding capacity, and nutrient levels influence the soil's productive capacity. These factors form a complex and interdependent system. Changes in one factor subsequently affect all or many others. The models demonstrate how soil erosion causes the loss of soil nutrients, depth, biota, organic matter, and water resources, and how these losses translate into reduced crop productivity. The models are based on the following set of assumptions: about 700 mm of rainfall, soil depth of 15 cm, slope of 5 percent, loamy soil, 4 percent organic matter, and soil erosion rate of 17 t ha^{-1} yr^{-1}. They provide a perspective on the interdependence of the various factors associated with the ecological effects of erosion.

Empirical evidence reveals that when soil erosion by water and wind occurs at a rate of 17 t ha^{-1} yr^{-1}, an average of 75 mm of water, 2 t of organic matter, and 15 kg of available nitrogen are lost from each hectare each year (Table 2.4). In addition, soil depth is reduced by 1.4 mm, the water-holding capacity is decreased by less than 0.1 mm, and soil biota populations are diminished. When combined, these losses translate into an 8 percent reduction in crop productivity over the short term (1 year). Note that the loss of water and nutrients account for nearly 90 percent of the loss in productivity (Table 2.4). If fertilizers were applied to the land, the estimated loss would be significantly reduced (Frye et al., 1982; Olson and Nizeyimana, 1988; Schertz et al., 1989; Mokma and Sietz, 1992).

Evaluated over the long term of 20 years, empirical evidence again confirms that water and nutrient loss continue to have the greatest effect on crop productivity, accounting for 50 to 75 percent of the reduced productivity (Young, 1989) (Table 2.3). A reduction in soil depth of 2.8 cm results in reduced productivity of about 7 percent. Soil depth is particularly critical, because it takes hundreds of years to replace a single centimeter of lost topsoil. The other factors, including soil biota, water-holding capacity, and soil depth, become signifi-

TABLE 2.4 Several Factors Contributing to Reduced Corn Yield Due to Soil Erosion of 17 t ha^{-1} yr^{-1} (10 t ha^{-1} yr^{-1} by Water and 7 t ha^{-1} yr^{-1} by Wind)

Factors	Quantities lost	Yield loss, %
Water runoff	75 mm[a]	7[a]
Nitrogen[b]	15 kg	
Phosphorus[b]	0.6 kg	2.4[c]
Potassium[b]	123 kg	
Soil depth	1.4 mm[d]	0.3[d]
Organic matter	2 t[e]	0.2[e]
Water-holding capacity	0.1 mm[f]	0.1[f]
Soil biota	—	0.1[g]
Total		8[h]

[a]Based on water-erosion rate of about 10 t ha^{-1} yr^{-1} on 5 percent sloping land under conventional tillage, water loss would be nearly 100 mm (Van Doren et al., 1950; Wendt and Burwell, 1985; Kramer, 1986; Wendt et al., 1986; Schertz et al., 1989; Murphee and McGregor, 1991). A conservative loss of 75 mm was assumed, and on the basis of this water loss, the estimated yield reduction was 7 percent (Shalhavet et al., 1979; Hanks, 1983; Feddes, 1984).

[b]Total nutrients lost are based on Troeh et al. (1991), but reduced on the basis of the nutrients that would not be immediately available because of a shortage of time for mineralization (Troeh and Thompson, 1993).

[c]The loss of N, P, and K nutrients was estimated to reduce yield by 2.4 percent (Kidd and Pimentel, 1992).

[d]Based on a bulk density of 1.25 g/cm and reduced yield of 6 percent per 2.5 cm of soil (Lyles, 1975).

[e]Based on a 4 percent organic-matter content of the soil and an enrichment factor of 3; the yield loss is minimal initially, but is significant in the long term.

[f]Water-holding capacity of the soil was calculated to be reduced by 0.1 mm, based on the loss of 17 t ha^{-1} yr^{-1} (Lyles, 1975).

[g]Reductions in soil biota were assumed to reduce infiltration of water and reduce organic-matter recycling, but have a minimal impact on yield for a single year.

[h]This estimated loss occurs after the annual loss of 17 t/ha. Percentages do not add up because the impacts of the various factors are interdependent, and overlap exists (e.g., organic matter is interrelated with water resources, nutrients, soil biota, and soil depth).

cant in the long term. If fertilizers were applied to the land, then the estimated loss would be reduced from one-quarter to one-third (Frye et al., 1982; Olson and Nizeyimana, 1988; Schertz et al., 1989; Mokma and Sietz, 1992).

On a yearly basis the effects of soil erosion often can be temporarily offset by the extensive use of fertilizers, irrigation, plant breeding, and other inputs. However, the long-term cumulative loss of soil organic matter, biota, soil depth, and water-holding capacity in some cases cannot be replaced by those interventions.

Erosion and Water-Loss Costs

Energy costs

Assuming an average annual erosion rate of 17 t/ha for combined wind and water erosion, we estimate that the on-site and off-site impacts of soil erosion and associated rapid water runoff require an additional annual expenditure of 1.6 million kcal/ha of fossil energy per hectare (Table 2.3). This suggests that approximately 10 percent of all the energy used in U.S. agriculture today is wasted just to offset losses of nutrients, water, and crop productivity caused by erosion. Although developed countries currently are using energy-based fertilizers, pesticides, and irrigation to mask the damage of soil erosion and to maintain their high crop productivity, this heavy dependence on fossil fuels is risky because fossil-energy supply is finite and prices are unstable. Developing nations, when using green-revolution technologies, are similarly relying intensively on the use of fossil-energy-based fertilizers, pesticides, and irrigation (Kendall and Pimentel, 1994). Recent studies suggest that the United States has only about 15 years of oil reserves remaining (Gibbons and Blair, 1991) and that the world's resources may be depleted within 30 to 50 years (Matare, 1989). Therefore, it is essential that we control erosion and conserve fossil energy resources.

On-site costs

The use of inappropriate agricultural practices and subsequent soil and water loss are responsible for significant economic and environmental on-site costs. The major on-site costs of erosion by both water and wind are those expended for replacing the nutrients and water losses due to erosion (Tables 2.2 to 2.6). When erosion by water and wind occurs at a rate of 17 t ha^{-1} yr^{-1}, approximately 100 mm/ha of water and 462 kg of nutrients are lost (Table 2.3). In the United States, if water had to be replaced, it would cost about \$30 ha^{-1} yr^{-1} to replace lost water by pumping groundwater for irrigation, and would require the expenditure of about 70 L of diesel fuel per hectare (assuming that water were available). An additional \$100 would be needed to replace the lost nutrients with fertilizers (Table 2.3). If the costs of lost productivity and off-site costs are summed, erosion costs the United States approximately \$196/ha (Table 2.3). In other parts of the world where irrigation is not possible and/or fertilizers too costly, the price of erosion is paid in reduced food production.

We estimate that the 4 billion t of soil and 130 billion t of water annually lost from 160 million ha of U.S. cropland translate into an onsite economic loss of more than \$27 billion each year [\$20 billion for nutrients (Troeh et al., 1991) and \$7 billion for lost water and soil

TABLE 2.5 Costs and Benefits of Soil and Water Conservation Technologies in Corn Production Compared with Conventional, Continuous Corn

Conservation technology	Total cost, $/ha[a]	Conser-vation cost, $/ha	Erosion rate, t/ha	Yield, kg/ha	Ratio of corn yield	Net benefits per ha, $
Conventional continuous	523	0	20	7400[b]	1.00	0
Contour planting	532	9[c]	2–6[d]	7900[e]	1.07	36[f]
Strip crop-contour	550	27[c]	2–6[d]	7900[e]	1.07	18[f]
Grass strips	569	46[g]	3[h]	7900[e]	1.07	−1[f]
Windbreaks	529	6[c]	5[e]	7600[i]	1.03	12[f]
No-till	493	−30[j]	<1[k]	8000[l]	1.08	85[f]
Ridge planting	493	−30[j]	1[m]	8000[n]	1.08	85[f]
Cover crop	562	39[c]	10[e]	7600[e]	1.03	−11[f]
Terraces	561	38[c]	1[e]	8000[e]	1.08	17[f]
Rotation[o]	563	40[p]	7[p]	8000[e]	1.08	15[f]

[a]The total cost per hectare includes the conservation cost in U.S. dollars per hectare as listed in the next column.

[b]USDA (1991).

[c]Barbarika (1987).

[d]Van Doren et al. (1950) and McIsaac and Mitchell (1992).

[e]Estimates based on information from Reid (1985) and Troeh et al. (1991).

[f]The net benefits were calculated by subtracting the annual cost of conservation from the returns from higher corn yield. Corn was priced at 9.2¢/kg ($2.30/bu) (USDA, 1991).

[g]The grass strips are estimated to occupy 23 percent of the crop area (Dillaha et al., 1989).

[h]Sediment loss is estimated to be reduced by 85 percent (Dillaha et al., 1989).

[i]Fifteen percent higher yield is reported for grain (Brandle et al., 1992); we assumed a 3 percent increase in yield for corn.

[j]Ridge planting and no-till cost about $30 less than conventional, continuous corn production (McIsaac et al., 1987); thus, the reason for the negative costs.

[k]McGregor and Mutchler (1992).

[l]Wagger and Cassel (1993).

[m]Pimentel (1993a).

[n]See Table 2.7.

[o]Faeth et al. (1991).

[p]Corn-soybean-wheat-fodder in Pennsylvania (Faeth et al., 1991).

depth (Table 2.3)]. The most significant component of this cost is the loss of valuable soil nutrients.

The costs of erosion are also high in other regions of the world. In India, for example, 6.6 billion t of soil (Lal, 1993) is lost each year, containing 5.4 million tons of fertilizer, worth $245 million (Chaudhary and Das, 1990). It is estimated that up to half of the amount of fertilizers applied each year in India is lost to erosive forces (M. S. Swaminathan, Swaminathan Institute, India, PC, 1993). In Costa Rica, erosion from farmland and pastureland annually removes nutrients worth 17 percent of the value of the annual crops and 14 percent of the value of livestock products (Repetto, 1992).

In addition to the loss of nutrients and water, erosion causes a loss of soil biodiversity that is impossible to value in terms of dollars

TABLE 2.6 Advantages of Conservation Ridge Planting–Rotation System Compared with Continuous Conventional–Tillage System in Terms of Economics, Energetics, and Soil Conservation

	Conventional			Ridge planting and rotations		
	Quantity	10^3 kcal	Cost, $	Quantity	10^3 kcal	Cost, $
Labor, h	10	7	50	12	9	60
Machinery, kg	55	1,485	91	45	1,215	75
Fuel, L	115	1,255	38	70	764	23
Nitrogen, kg	152	3,192	81	(27 t)*	559	17
Phosphorus, kg	75	473	53	34	214	17
Potassium, kg	96	240	26	15	38	4
Limestone, kg	426	134	64	426	134	64
Corn seeds, kg	21	520	45	21	520	45
Cover-crop seeds, kg	—	—	—	10	120	10
Insecticides, kg	1.5	150	15	0	0	0
Herbicides, kg	2	200	20	0	0	0
Electricity (10^3 kcal)	100	100	8	100	100	8
Transport, kg	322	89	32	140	39	14
Total		7,845	523		3,712	$337
Yield, kg	7,400	26,160		8,000	28,281	
Output/input ratio						
Energy ratio		3.33			7.62	

*A total of 27 t of cattle manure was applied to provide 152 kg of nitrogen plus quantities of P and K.

SOURCE: Pimentel (1993b).

(Pimentel et al., 1992), although it is of extreme ecological value. In addition to the economic losses of nutrients and water, erosion causes significant ecological damage. The siltation of aquatic systems destroys stream and lake ecosystems, whereas the removal of soil may affect plant composition and deplete soil biodiversity.

Off-site costs

Erosion not only damages the immediate area but also causes extensive damage to the surrounding environment. These off-site costs include roadway, sewer, and basement siltation; drainage disruption; undermining of foundations and pavements; gullying of roads; earth-dam failures; eutrophication of waterways; siltation of harbors and channels; loss of reservoir storage; loss of wildlife habitat and disruption of stream ecology; flooding; famine; damage to public health; and increased water-treatment costs (Gray and Leiser, 1989).

The largest proportion of off-site effects are caused by soil particles entering the water systems (USDA, 1990). Of the billions of metric tons of soil lost from U.S. cropland per year, approximately 60 percent is deposited in streams (USDA, 1989). Sediments harm aquatic plants and other organisms by contaminating the water with soil particles and altering habitat quality (Clark, 1987). In addition, some pesticides and fertilizers contained in the agricultural sediments may poison fish and cause eutrophication (Clark et al., 1985).

Siltation is a major problem in reservoirs because it reduces water storage, reduces electricity production, and shortens the lifetime or increases the maintenance costs of dams. Approximately 880 million t of agricultural soils is deposited into American reservoirs and aquatic systems each year, reducing their flood-control benefits, clogging waterways, and increasing operating costs of water-treatment facilities (WRI, 1992). To maintain navigable waterways, the United States annually spends over $520 million to dredge soil sediments from waterways (Clark, 1985).

Heavy sedimentation may also lead to river and lake flooding (Myers, 1993). For example, a significant amount of the flooding that occurred in the midwestern United States during summer 1993 was due to increased sediment deposition in the Mississippi River, Missouri River, and other rivers located in central United States. The combined damage of the flood to crops and homes was assessed by the government to be $20 billion (Allen, 1994).

The loss of aesthetic value of lakes and rivers due to turbidity also represents a substantial economic cost. Properties on turbid lakes and rivers are frequently valued at 15 to 20 percent less than properties located on clear freshwater (Young and Magleby, 1985). Also, many recreational activities such as boating, swimming, and fishing are prevented or reduced when turbidity is high.

Wind erosion similarly produces significant off-site damage and costs. It is estimated that it causes household property damage (due to the sandblasting of automobiles, buildings, and landscapes by wind-blown soil particles) and maintenance costs totaling over $4 billion/yr in the United States (Huszar and Piper, 1985; SCS, 1993; J. Werner, Soil Conservation Service, USDA, PC, 1994). The accumulation of soil in public and private buildings, roads, and railways similarly results in removal costs of over $4 billion/yr (Huszar and Piper, 1985; SCS, 1993).

A less-obvious erosion expense is the cost of health problems caused by increased soil particles in the air. Windborne dust irritates the respiratory tract and eyes, aggravates allergies and asthma, and carries with it toxic chemicals such as heavy metals and pesticides. The costs of these wind- and dust-related health problems in the United States exceed $5 billion/yr (Table 2.7).

In New Mexico, where wind erosion is a major problem, total off-site costs, including health and property damage, are estimated to be $465 million annually (Huszar and Piper, 1985). If we assume similar erosion costs in the western United States, the total off-site costs from wind erosion alone could be about $9.6 billion annually in the United States (Table 2.7).

TABLE 2.7 Damages by Wind and Water Erosion and Cost of Erosion Prevention Each Year

Type of damage	Cost, million $	Total, million $
Wind erosion[a]		
Exterior paint	18.5	
Landscaping	2,894.0	
Automobiles	134.6	
Interior and laundry	986.0	
Health	5,371.0[b]	
Recreation	223.2	
Road maintenance	1.2	
Cost to business	3.5	
Cost to irrigation and conservation districts	0.1	
Total wind erosion costs		9,632.5
Water erosion[c]		
In-stream damage		
Biological impacts	No estimate	
Recreational	2,440.0	
Water-storage facilities	841.8	
Navigation	683.2	
Other in-stream uses	1,098.0	
Subtotal in-stream	5,063.0	
Off-stream effects		
Flood damage	939.4	
Water-conveyance facilities	244.0	
Water-treatment facilities	122.0	
Other off-stream uses	976.0	
Subtotal off-stream	2,318.0	
Total water-erosion costs		7,381.0
Total costs of wind- and water-erosion damage		17,013.5[d]
Cost of erosion prevention[e]		8,400
Total costs (on and off site)[f]		44,399.0
Benefit/cost		5.24

[a]Huszar and Piper (1985), USBC (1992), SCS (1993); J. Werner, SCS, Washington, D.C., PC, 1994.

[b]Health estimates are based partly on Lave and Seskin (1977).

[c]Clark (1985), USBC (1992), SCS (1993); J. Werner, SCS, USDA, Washington, D.C., PC, 1994.

[d]Agriculture accounts for approximately two-thirds of the off-site effects.

[e]See text.

[f]The total on-site costs are calculated to be $27 billion (see Table 2.3 and text).

Combined on-site and off-site effects

In summary, the cost of all off-site environmental impacts of U.S. soil erosion are estimated to be about $17 billion per year (1992 dollars) (Table 2.7). An additional loss of $27 billion is due to reduced soil productivity. If off-site and on-site costs are combined, the total cost of erosion from agriculture in the United States is $44 billion per year (Table 2.7), or about $100/ha of cropland and pastureland. This erosion cost increases production costs by about 25 percent per year.

Worldwide, over 75 billion t of soil is lost each year (Myers, 1993); approximately two-thirds of this is lost from agricultural land. If we assume a cost of $3/t for nutrients (Troeh et al., 1991), $2/t for water loss (Table 2.3), and $3/t for off-site impacts (Table 2.7), this massive soil loss costs the world approximately $400 billion per year—more than $70 per person per year.

We estimate that it would take an investment of $6.4 billion per year ($40/ha for conservation) to reduce U.S. erosion rates from approximately 17 t ha^{-1} yr^{-1} to a sustainable rate of approximately 1 t ha^{-1} yr^{-1} on most cropland (Table 2.5). To reduce erosion on pastureland, an additional $2.0 billion per year ($5/ha for conservation) would be needed (Hart et al., 1988; USDA, 1992; Taylor et al., 1993; Fowler et al., 1994) (Table 2.7). The total investment for U.S. erosion control would therefore be $8.4 billion per year. Given that erosion causes $44 billion in damages each year, it seems that an $8.4 billion investment is a small price to pay; for every $1 invested, $5.24 would be saved (Table 2.7). This small investment would reduce U.S. agricultural soil loss by about 4 billion t and protect our current and future food supply.

Currently the United States spends $1.7 billion per year in the Conservation Reserve Program to remove highly erodible land from production, and this saves about 584 million tons of soil each year (USDA, 1991). Therefore, in this system $2.91 is invested to save one metric ton of soil, whereas in our proposed conservation system we assume a cost of $2.10 per metric ton of soil saved.

Erosion Control and Water Conservation

Techniques

Reliable soil-conservation technologies include ridge planting, no-till cultivation, crop rotations, strip cropping, grass strips, mulches, living mulches, agroforestry, terracing, contour planting, cover crops, and windbreaks (Carter, 1994). Although the specific processes vary, almost all conservation methods reduce erosion rates by maintaining a protective vegetative cover of the soil accompanied by reducing the frequency of plowing. Ridge planting, for example, reduces the frequency of tillage and leaves vegetative cover on the soil surface year round, whereas crop rotations ensure that some part of the land is continually covered with vegetation. Each conservation method may be used separately or in combination with other erosion-control techniques. To determine the most advantageous combination of appropriate conservation technologies, the soil type, specific crop or pasture, slope, and climate (rainfall and wind intensity), and socioeconomics of the people living in a particular site must be considered.

The implementation of appropriate soil- and water-conservation practices reduced erosion rates 2- to 1000-fold and water loss 1.3- to 21.7-fold (Fig. 2.2; Table 2.4). Conservation technologies also significantly reduce nutrient loss. For example, when corn-residue cover was increased by 10, 30, and 50 percent, the amount of nitrogen lost in surface runoff was reduced by 68, 90, and 99 percent, respectively (Palis et al., 1990).

By substantially decreasing soil and nutrient loss, conservation technologies preserve the soil's fertility and enable the land to sustain higher crop yields. In many cases, the use of conservation technologies may increase yields (Faeth, 1993). The use of contour planting, for example, has increased cotton yields by 25 percent (Texas), corn yields by 12.5 percent (Missouri), soybeans by 13 percent (Illinois), and wheat by 17 percent (Illinois) (Smith, 1946; Burnett and Fisher, 1954; Sauer and Case, 1954). On U.S. land with a 7 percent slope, yields from cotton grown in rotation increased by 30 percent, while erosion was reduced by nearly one-half (Hendrickson et al., 1963). In many areas where winds are strong, the establishment of tree and shrub shelterbelts help reduce wind energy by 87 percent and, in turn, decrease erosion by as much as 50 percent (Troeh et al., 1991).

Conservation technologies implemented in the tropics and elsewhere in the world are capable of decreasing erosion rates and improve crop yields (Borlaug et al., 1993). In Nigeria, yields from corn grown by no-till methods under favorable soil and climatic conditions are 61 percent greater than from corn grown with conventional tillage under similar field conditions (Wijewardene and Waidyanatha, 1984). In Mexico, the use of mulching has reduced erosion rates by 99 percent, with corn yields increasing by 30 percent (Maass et al., 1988), while contour cultivation in India increased corn yields by 46 percent (Magrath, 1989).

Benefits of conservation techniques

To demonstrate that agriculture can be made environmentally sound and sustainable without increasing operational costs, we have outlined a model conservation system in Table 2.6. In this system, corn and soybean are rotated with a rye cover crop, and crops are sown using ridge planting. A corn-soybean rotation system can be more profitable than raising either crop alone (Helmers et al., 1986; Lal et al., 1994). The rotation of crops reduced insect pests, diseases, and weed problems, and thereby improves the overall crop yield (Pearson, 1967; Mora and Moreno, 1984; NAS, 1968, 1989; Mulvaney and Paul, 1984; Paoletti et al., 1989; Pimentel et al., 1991; Ilnicki and Enache, 1992). The use of the rye cover crop and ridge planting reduces soil erosion and water runoff, and thereby conserves nutrients (Follett and Stewart, 1985; ASAE, 1985; Pimentel, 1993b). Erosion rates are reduced from about 20 t ha^{-1} yr^{-1} to a sustainable rate of 1 t ha^{-1} yr^{-1}, and crop yields are increased by about 8 percent because of improved insect pest control and soil conservation. In addition to decreasing soil loss and improving crop yields, this model system reduces the total production costs: a conventional-till system costs $523/ha, whereas our model system costs only $337/ha. This model clearly illustrates that we can both conserve soil resources and improve crop yields through the implementation of ecologically sound agricultural practices.

Conclusion

Soil erosion and water loss are critical environmental problems facing the world and threaten the sustainability of food production. Over 90 percent of all arable land is already eroded to some extent, and consequently soil productivity is decreasing at an alarming rate. Nearly one-third of the world's arable land area has been lost by erosion during the last 40 years. This major loss of productive agricultural land continues at a rate of more than 10 million ha per year. In the United States alone, over 1 million ha is lost each year because of erosion and soil degradation. Water loss associated with erosion is the single most important factor influencing crop productivity in the United States and worldwide.

With the daily addition of a quarter-million people, the world population's food demand is increasing at a time when per capita food productivity is beginning to decline. Although agricultural productivity can be bolstered partly by the use of fossil-based fertilizers, pesticides, and irrigation, the finite reserves of fossil energy preclude this alternative as a permanent solution to soil erosion.

The worldwide costs of soil erosion and water loss are enormous, totaling approximately $400 billion per year; U.S. losses amount to about $44 billion per year from on-site and off-site erosion effects. These costs are associated with losses of water resources, soil nutrients, organic matter, and soil biota, impacts on public health, lost food production, degradation of aquatic and terrestrial ecosystems, and damage to public and private property. These loss figures do not include the *permanent* damage to soil productivity.

With an estimated investment of $8.4 billion per year, erosion could be prevented on most U.S. cropland and pastureland and could save about $5.24 per dollar invested. Yet, instead of investing in erosion prevention, the U.S. government is proposing to reduce its investment. The 1994 budget of the U.S. Soil Conservation Service was set at $916.6 million. Thus, funding is proposed to be cut to only $720.6 million in the 1995 fiscal year.

Conservation technologies not only prevent soil erosion and water loss but also often have proved to increase crop yields. Indeed, soil- and water-conservation methods are cost-effective, with a relatively small dollar investment yielding significant benefits. The technologies for controlling erosion and rapid water runoff are well known and have been proved effective, yet these practices are not widely implemented because the public, government, and farmers either lack an awareness of the erosion problem or lack the commitment needed to deal with the issue.

When economic costs of soil and water loss and degradation and off-site effects are conservatively estimated into the cost/benefit analyses of agriculture, it makes sound economic sense to invest in programs that halt the widespread erosion and water loss. Human survival and wealth depend on adequate supplies of food, land, water, energy, and biodiversity. Infertile, poor-quality land and insufficient water will not sustain human food production. We should heed the warning that a nation that destroys its soils and water resources, destroys itself (Roosevelt, 1937).

Acknowledgments

We thank the following people for reading an earlier draft of this chapter, for their many helpful suggestions, and in some cases, for providing additional information: I. P. Abrol, Indian Council of Agricultural Research, New Delhi; Wen Dazhong, Chinese Academy of Sciences, Shenyang; R. H. Dowdy, USDA/ARS, St. Paul, Minn.; H. E. Dregne, Texas Tech University, Lubbock; W. Edwards, USDA/ARS, Coshocton, Ohio; M. T. El-Ashry, World Bank, Washington, D.C.; H. A. Elwell, Agricultural consultant, Harare, Zimbabwe; R. F. Follett, USDA/ARS, Fort Collins, Colo.; D. W. Fyrear, USDA/ARS, Big Spring, Tex.; G. E. Hallsworth, Glen Osmond, Australia; H. Hurni, University of Berne, Switzerland; T. N. Khoshoo, Tata Energy Research Institute, New Delhi, India; R. Lal, Ohio State University, Columbus; G. W. Langdale, USDA/ARS, Watkinsville, Ga.; W. B. Magrath, World Bank,

Washington, D.C.; K. C. McGregor, USDA/ARS, Oxford, Miss.; G. F. McIsaac, University of Illinois, Urbana; F. Mumm von Mallinckrodt, UNDP, New York; T. L. Napier, Ohio State University, Columbus; K. R. Olson, University of Illinois, Urbana; W. Parham, China Tropical Lands Project, Vienna, Va.; D. Southgate, Ohio State University, Columbus; B. A. Stewart, West Texas A&M University, Canyon; M. Stocking, University of East Anglia, Norwich, U.K.; M. S. Swaminathan, Swaminathan Research Foundation, Madras, India; D. L. Tanaka, USDA/ARS, Sidney, Mont.; G. B. Thapa, Asian Institute of Technology, Bangkok, Thailand; F. Troeh, Iowa State University, Ames; P. W. Unger, USDA/ARS, Bushland, Tex.; A. Young, Norwich, U.K.; and at Cornell University, Ithaca, N.Y.: R. Bryant, S. Bukkens, Dan Dalthorp, M. Giampietro, M. Pimentel, T. Scott, and N. Uphoff.

References

Agboola, A. A., 1990, Organic matter and soil fertility management in the humid tropics of Africa. In *Organic Matter Management and Tillage in Humid and Subhumid Africa,* E. Pushparajah, M. Latham, and C. R. Elliot, eds., pp. 231–243. Antananarivo, Madagascar: ISBRAM Proceedings of the Third Regional Workshop of the Africaland Programme.

Agricultural Research, 1979, Learning from the past. *Agricultural Research* **28**(4): 5–10.

Aina, P. O., R. Lal, and G. S. Taylor, 1977, Soil and crop management in relation to soil erosion in the rainforest of western Nigeria. In *Soil Erosion: Prediction and Control.* Ankeny, Iowa: Soil Conservation Society of America, publ. 21.

Alexander, M., 1977, *Introduction to Soil Microbiology.* New York: Wiley.

Allen, W., 1994, The flood of 1993 may rank as the worst weather disaster in U.S. history, *St. Louis (Mo.) Post Dispatch,* p. 01B.

Allison, F. E., 1973, *Soil Organic Matter and Its Role in Crop Production.* New York: Elsevier.

Arden-Clarke, C., and R. Evans, 1993, Soil erosion and conservation in the United Kingdom. In *World Soil Erosion and Conservation,* D. Pimentel, ed., pp. 193–216. Cambridge, U.K.: Cambridge University Press.

ASAE, 1985, *Erosion and Soil Productivity.* St. Joseph, Mich.: American Society of Agricultural Engineering.

Atlavinyte, O., 1964, Distribution of earthworms (Lumbricidae) and larval insects in the eroded soil under cultivated crops. *Pedobiologia* **4:** 245–250.

Atlavinyte, O., 1965, The effect of erosion on the population of earthworms (Lumbricidae) in soils under different crops. *Pedobiologia* **5:** 178–188.

Barbarika, A., 1987. Costs of soil conservation practices. In *Optimum Erosion Control at Least Cost,* pp. 187–195. St. Joseph, Mich.: American Society of Agricultural Engineers.

Barrow, C. J., 1991, *Land Degradation.* Cambridge, U.K.: Cambridge University Press.

Barrows, H. L., and V. J. Kilmer, 1963, Plant nutrient losses from soils by water erosion. *Adv. Agron.* **15:** 303–315.

Batty, J. C., and J. Keller, 1980, Energy requirements for irrigation. In *Handbook of Energy Utilization in Agriculture,* D. Pimentel, ed., pp. 35–44. Boca Raton, Fla.: CRC Press.

Beacher, H. H., U. Schwetmann, and H. Stumer, 1985, Crop yield reduction due to reduced plant available water caused by water erosion. In *Soil Erosion and Conservation,* S. A. El-Swaify, W. C. Moldenhauer, and A. Lo, eds., pp. 365–373. Ankeny, Iowa: Soil Conservation Society of America.

Bennett, H. H., 1939, *Soil Conservation.* New York: McGraw-Hill.

Bhatt, P. N., 1977, Losses of plant nutrients through erosion process—a review. *Soil Conserv. Digest* **5:** 37–45.

Borlaug, N. E., R. Lal, D. Pimentel, H. Popenoe, and N. D. Vietmeyer, 1993, *Vetiver Grass: A Thin Line against Erosion.* Washington, D.C.: National Academy of Sciences.

Bormann, F. H., G. E. Likens, T. C. Siccama, R. S. Pierce, and J. S. Eaton, 1974, Nutrient recovery of stable conditions following deforestation at Hubbard Brook. *Ecol. Monogr.* **44:** 255–277.

Brandle, J. R., B. B. Johnson, and T. Akeson, 1992, Field windbreaks—are they economical? *J. Product. Agric.* **5:** 373–378.

Brown, L. R., 1990, Assessing the planet's condition. *EPA J.* **16:** 2–6.

Buntley, G. J., and F. F. Bell, 1976, Yield estimates for the major crops grown on the soils of west Tennessee. *Tenn. Agric. Exp. Stn. Bull.* No. 561.

Burnett, E., and C. E. Fisher, 1954, The effect of conservation practices on runoff, available soil moisture and cotton yield. *Proc. Soil Sci. Soc. Am.* **18:** 216–218.

Buttel, F. H., 1982, Environmental quality in agriculture: some observations on political-economic constraints on sustainable resource management. *Cornell Rural Sociol. Bull. Ser. Bull.* 128.

Byington, E. K., 1986, *Grazing Land Management and Water Quality.* Harpers Ferry, W. Va.: American Society of Agronomy and Crop Science Society of America.

Carson, B., 1985, *Erosion and Sedimentation Processes in the Nepalese Himalaya.* Katmandu, Nepal: International Centre for Integrated Mountain Development.

Carter, M. R., ed., 1994, *Conservation Tillage in Temperate Agroecosystems.* Boca Raton, Fla.: Lewis Publishers.

CEQ, 1980, *The Global 2000 Report to the President of the U.S. Entering the 21st Century.* New York: Pergamon Press.

Chaney, K., and R. S. Swift, 1984, The influence of organic matter on aggregate stability in some British soils. *J. Soil Sci.* **35:** 223–230.

Chapman, P. W., F. W. Fitch, and C. L. Veatch, 1950, *Conserving Soil Resources: A Guide to Better Living.* Atlanta, Ga.: Smith.

Chaudhary, H. P., and S. K. Das, 1990, Nutrient status in relation to intensity of erosion in ravines of Yamuna. *J. Indian Soc. Soil Sci.* **38:** 126–129.

Clark, E. H., 1985, The off-site costs of soil erosion. *J. Soil Water Conserv.* **40:** 19–22.

Clark, E. H., 1987, Soil erosion: offsite environmental effects. In *Soil Loss: Processes, Policies, and Prospects,* J. M. Harlin and G. M. Bernardi, eds., pp. 59–89. New York: Westview.

Clark, E. H., J. A. Haverkamp, and W. Chapman, 1985, *Eroding Soils: The Off-Farm Impacts.* Washington, D.C.: The Conservation Foundation.

Crosson, P., 1985, National cost of erosion on productivity. In *Erosion and Soil Productivity.* pp. 254–265. St. Joseph, Mich.: American Society of Agricultural Engineers.

Curry-Lindahl, K., 1972, *Conservation for Survival.* New York: Morrow.

Demeny, P. G., 1986, *Population and the Invisible Hand.* New York: Paper No. 123, Center for Policy Studies, Population Council.

Dillaha, T. A., R. B. Reneau, S. Mostaghimi, and D. Lee, 1989, Vegetative filter strips for agricultural nonpoint source pollution control. *Am. Soc. Agric. Eng.* **32:** 513–519.

Donahue, R. H., R. H. Follett, and R. N. Tulloch, 1990, *Our Soils and Their Management.* Danville, Ill.: Interstate Publishers.

Dregne, H. E., 1992, Erosion and soil productivity in Asia. *J. Soil Water Conser.* **47:** 8–13.

Durham, D. F., 1992, Cultural carrying capacity: I = PACT. *Focus* **2:** 5–8.

Durning, A. B., 1989, *Poverty and the Environment: Reversing the Downward Spiral.* Washington, D.C.: Worldwatch Institute.

Edwards, C. A., 1981, Earthworms, soil fertility and plant growth. In *Workshop on the Role of Earthworms in the Stabilization of Organic Residues,* pp. 61–85. Lansing, Mich.: Beach Leaf Press.

Edwards, K., 1993, Soil erosion and conservation in Australia. In *World Soil Erosion and Conservation,* D. Pimentel, ed., pp. 147–170. Cambridge, U.K.: Cambridge University Press.

Edwards, W. M., M. J. Shipitalo, L. B. Owens, and L. D. Norton, 1990, Effect of *Lumbricus terrestis* L. burrows on hydrology of continuous no-till corn fields. *Geoderma* **46:** 73–84.

Ehlers, W., 1975, Observations on earthworm channels and infiltration on tilled and untilled loess soil. *Soil Sci.* **119:** 242–249.

Ehrlich, P., 1990, *The Population Explosion.* New York: Simon & Schuster.

El-Swaify, S. A., E. W. Dangler, and C. L. Armstrong, 1982, Soil erosion by water in the tropics. College of Tropical Agriculture and Human Resources, University of Hawaii, Resource Extension Series 024.

El-Swaify, S. A., W. C. Moldenhauer, and A. Lo, 1985, *Soil Erosion and Conservation.* Ankeny, Iowa: Soil Conservation Society of America.

Elwell, H. A., 1985, An assessment of soil erosion in Zimbabwe. *Zimbabwe Sci. News* **19:** 27–31.

Faeth, P., 1993, *Agricultural Policy and Sustainability: Case Studies from India, Chile, the Philippines and the United States.* Washington, D.C.: World Resources Institute.

Faeth, P., R. Repetto, K. Kroll, Q. Dai, and G. Helmers, 1991, *Paying the Farm Bill. U.S. Agricultural Policy and the Transition to Sustainable Agriculture.* Washington, D.C.: World Resources Institute.

Falkenmark, M., 1989, Water scarcity and food production. In *Food and Natural Resources,* D. Pimentel and C. W. Hall, eds., pp. 164–191. San Diego: Academic Press.

FAO, 1992, *FAO Quart. Bull. Statistics* **5**(2): 1–121.

Feddes, R. A., 1984, Crop water use and dry matter production: state of the art. In *Crop Water Requirements Conference International,* pp. 221–234. Versailles, France, Sept. 11–14, 1984: UNESCO under auspices of FAO and WMO.

Follett, R. F., and B. A. Stewart, eds., 1985, *Soil Erosion and Crop Productivity.* Madison, Wisc.: American Society of Agronomy and Crop Science Society of America.

Follett, R. F., S. C. Gupta, and P. G. Hunt, 1987, Conservation practices: relation to the management of plant nutrients for crop production. In *Soil Fertility and Organic Matter as Critical Components of Production Systems.* Madison, Wisc.: Soil Science Society of America and American Society of Agronomy.

Foster, G. R., R. A. Young, M. J. M. Ronkens, and C. A. Onstad, 1985, Processes of soil erosion by water. In *Soil Erosion and Crop Productivity,* F. R. Follett and B. A. Stewart, eds., pp. 137–162. Madison, Wisc.: American Society of Agronomy and Crop Science Society of America.

Foth, B. D., and B. G. Ellis, 1988, *Soil Fertility.* New York: Wiley.

Fowler, J. M., A. Torell, and G. Gallacher, 1994, Competitive pricing for the McGregor range: implications for federal grazing fees. *J. Range Manag.* **47:** 155–158.

Frye, W. W., S. A. Ebelhar, L. W. Murdock, and R. L. Bevins, 1982, Soil erosion effects on properties and productivity of two Kentucky soils. *Soil Sci. Soc. Am. J.* **46:** 1051–1055.

Giampietro, M., and D. Pimentel, 1993, The tightening conflict: population, energy use, and the ecology of agriculture. *The NPG Forum,* October. Teaneck, N.J.: Negative Population Growth, Inc.

Gibbons, J. H., and P. D. Blair, 1991, U.S. Energy transition: on getting from here to there. *Physics Today* **44:** 22–30.

Gleick, P. H., 1993, *Water in Crisis.* New York: Oxford University Press.

Gore, A., 1994, A proposal to reinvent policy on the environment. *The Earth Times* **15**(June): 1, 31.

Gray, D. M., and A. T. Leiser, 1989, *Biotechnical Slope Protection and Erosion Control.* Malabar, Fla.: Kreiger.

Greenland, D. J., and M. H. Hayes, 1981, *The Chemistry of Soil Processes.* New York: Wiley.

Hanks, R. J., 1983, Yield and water-use relationships: an overview. In *Limitations to Efficient Water Use in Crop Production,* H. M. Taylor, W. R. Jordan, and T. R. Sinclair, eds., pp. 393–411. Madison, Wisc.: American Society of Agronomy, Crop Science Society of America, and Soil Science Society of America.

Hardin, G., 1986, Cultural carrying capacity: a biological approach to human problems. *BioScience* **36:** 599–606.

Hart, R. H., M. J. Samuel, P. S. Test, and M. A. Smith, 1988, Cattle, vegetation, and economic responses to grazing systems and grazing pressure. *J. Range Manag.* **41:** 282–286.

Hauser, V. L., and O. R. Jones, 1991, Runoff curve numbers for the Southern High Plains. *Trans. ASAE* **34:** 142–148.

Helmers, G. A., M. R. Langemeir, and J. Atwood, 1986, An economic analysis of alternative cropping systems for east-central Nebraska. *Am. J. Alternative Agri.* **4:** 253–258.

Hendrickson, B. H., A. P. Barnett, J. R. Carreler, and W. E. Adams, 1963, *Runoff and Erosion Control Studies on Cecil Soil in the Southern Piedmont.* Washington, D.C.: USDA, Technical Bulletin no. 1281.

Hinz, W., 1985, Estimating irrigation. *Arizona Farmer-Stockman* **64:** 16.

Holdren, C., 1992, Population alarm. *Science* **255:** 1358.

Hole, F. D., 1981, Effects of animals on soil. *Geoderma,* **25:** 75–112.

Hood, L., and J. K. Morgan, 1972, Whose home on the range? *Sierra Club Bull.* **57:** 4–11.

Hopp, H., 1946, Earthworms fight erosion, too. *Soil Conserv.* **11:** 252–254.

Hudson, N. W., 1981, *Soil Conservation,* 2nd ed. Ithaca, N.Y.: Cornell University Press.

Hudson, N. W., 1982, Soil conservation, research and training requirements in developing tropical countries. In *Soil Erosion and Conservation in the Tropics,* pp. 121–133. Madison, Wisc.: American Society of Agronomy, Special Publication no. 43.

Hurni, H., 1993, Land degradation, famine, and land resource scenarios in Ethiopia. In *Soil Erosion and Conservation,* D. Pimentel, ed., pp. 27–62. Cambridge, U.K.: Cambridge University Press.

Huszar, P. C., and S. L. Piper, 1985, Off-site costs of wind erosion in New Mexico. In *Off-Site Costs of Soil Erosion: The Proceedings of a Symposium,* pp. 143–166. Washington, D.C.: The Conservation Foundation.

Ilnicki, R. D., and A. J. Enache, 1992, Subterranean clover living mulch: an alternative method of weed control, *Agric. Ecosyst. Environ.* **40:** 249–264.

Johnson, L. C., and R. I. Papendick, 1968, A brief history of soil erosion research in the United States and in the Palouse, and a look at the future. *Northwest Sci.* **42:** 53–61.

Kendall, H. W., and D. Pimentel, 1994, Constraints on the expansion of the global food supply. *Ambio* **23**(3): 198–205.

Ketcheson, J., 1977, Conservation tillage in eastern Canada. *J. Soil Water Conserv.* **32:** 57–60.

Ketcheson, J. W., and J. J. Onderdonk, 1973, Effect of corn stover on phosphorus in runoff from nontilled soil. *Agron. J.* **65:** 69–71.

Keyfitz, N., 1984, Impact of trends in resources, environment and development on demographic prospects. In *Population, Resources, Environment and Development,* pp. 97–124. New York: United Nations.

Khoshoo, T. N., and K. G. Tejwani, 1993, Soil erosion and conservation in India. In *Soil Erosion and Conservation,* D. Pimentel, ed., pp. 109–146. Cambridge, U.K.: Cambridge University Press.

Kidd, C. V., and D. Pimentel, eds., 1992, *Integrated Resource Management: Agroforestry for Development.* San Diego: Academic Press.

Kirchner, H. B., D. L. Wallace, D. Gore, and P. E. McNall, 1992, *Our National Resources and Their Conservation.* Danville, Ill.: Interstate Publishers.

Kladivko, E. J., A. D. MacKay, and J. M. Bradford, 1986, Earthworms as a factor in the reduction of soil crusting. *Soil Sci. Soc. Am. J.* **50:** 191–196.

Klausner, S. D., P. J. Zwerman, and D. F. Ellis, 1974, Surface runoff losses of soluble nitrogen and phosphorus under two systems of soil management. *J. Environ. Qual.* **3:** 42–46.

Kramer, L. A., 1986, Runoff and soil loss by cropstage from conventional and conservation tilled corn. *Trans. Am. Soc. Agric. Eng.* **29:** 774–779.

Lal, R., 1976, *Soil Erosion Problems on an Alisol in Western Nigeria and Their Control.* Lagos, Nigeria: IITA.

Lal, R., 1980, Losses of plant nutrients in runoff and eroded soil. In *Nitrogen Cycling in West African Ecosystems,* T. Rosswall, ed., pp. 31–38. Uppsala, Sweden: Reklan and Katalogtryck.

Lal, R., 1984, Productivity assessment of tropical soils and the effects of erosion. In *Quantification of the Effect of Erosion on Soil Productivity in an International Context.* F. R. Rijsberman and M. G. Wolman, eds., pp. 70–94. Delft, Netherlands: Delft Hydraulics Laboratory.

Lal, R., 1985, Soil erosion and its relation to productivity in tropical soils. In *Soil Erosion and Conservation,* S. A. El-Swaify, W. C. Moldenhauer, and A. Lo, eds., pp. 237–247. Ankeny, Iowa: Soil Conservation Society of America.

Lal, R., 1989, Land degradation and its impact on food and other resources. In *Food and Natural Resources,* D. Pimentel, ed., pp. 85–140. San Diego: Academic Press.

Lal, R., 1993, Soil erosion and conservation in West Africa. In *World Soil Erosion and Conservation,* D. Pimentel, ed., pp. 7–26. Cambridge, U.K.: Cambridge University Press.

Lal, R., and B. A. Stewart, 1990, *Soil Degradation.* New York: Springer-Verlag.

Lal, R., T. J. Logan, M. J. Shipitalo, D. J. Eckert, and W. A. Dick, 1994, Conservation tillage in the corn belt of the United States. In *Conservation Tillage in Temperate Agroecosystems,* M. R. Carter, ed. Boca Raton, Fla.: Lewis Publishers.

Langdale, G. W., L. T. West, R. R. Bruce, W. P. Miller, and A. W. Thomas, 1992, Restoration of eroded soil with conservation tillage. *Soil Technol.* **5:** 81–90.

Larson, R. J., and R. M. Ventullo, 1983, Biodegradation potential of groundwater bacteria. In *Proceedings of the Third National Symposium on Aquifer Restoration and Groundwater Monitoring,* May 25–27, D. M. Nielsen, ed. Worthington, Ohio: National Water Well Association.

Lave, L., and E. Seskin, 1977, *Air Pollution and Human Health.* Baltimore: Johns Hopkins Press.

Lee, E., and R. C. Foster, 1991, Soil fauna and soil structure. *Austral. J. Soil Res.* **29:** 745–776.

Lee, K. E., 1985, *Earthworms: Their Ecology and Relationships with Soils and Land Use.* Orlando, Fla.: Academic Press.

Lee, L. K., 1990, The dynamics of declining soil erosion rates. *J. Soil Water Conserv.* **45:** 622–624.

Leyton, L., 1983, Crop water use: principles and some considerations for agroforestry. In *Plant Research and Agroforestry,* P. A. Huxley, ed., pp. 379–400. Nairobi, Kenya: International Council for Research in Agroforestry.

Livingstone, D., 1992, Topsoil production: the missing environmental link. *J. Sustainable Agric.* **6**(1): 20–21.

Lockaby, B. G., and J. C. Adams, 1985, Pedoturbation of a forest soil by fire ants. *J. Soil Sci. Soc. Am.* **49:** 220–223.

Lucas, R. E., J. B. Holtman, and L. J. Connor, 1977, Soil carbon dynamics and cropping practices. In *Agriculture and Energy,* W. Lockeretz, ed., pp. 333–351. New York: Academic Press.

Lyles, L., 1975, Possible effects of wind erosion on soil productivity. *J. Soil Water Conserv.* **30:** 279–283.

Maass, J. M., C. F. Jordan, and J. F. Sarukhan, 1988, Soil erosion and nutrient losses in seasonal tropical agroecosystems under various management techniques, *J. Appl. Ecol* **25:** 595–607.

Magrath, W. B., 1989, *Economic Analysis of Soil Conservation Technologies,* Environment Department Division Working Paper no. 1989-4. Washington, D.C.: World Bank.

Martin, J. P., 1941, The organic matter in Collinton sandy loam and in the eroded material. *Soil Sci.* **52:** 435–443.

Matare, H. F., 1989, *Energy: Fact and Future.* Boca Raton, Fla.: CRC Press.

McGregor, K. C., and C. K. Mutchler, 1992, Soil loss from conservation tillage for sorghum. *Trans. ASAE* **35:** 1841–1845.

McGregor, K. C., C. K. Mutchler, and M. J. M. Romkens, 1990, Effects of tillage with different crop residues on runoff and soil loss. *Trans. ASAE* **33:** 1551–1556.

McGregor, K. C., C. K. Mutchler, and R. F. Cullum, 1992, Soil erosion effects on soybean yields. *Trans. ASAE* **35:** 1521–1525.

McIsaac, G. F., and J. K. Mitchell, 1992, Temporal variation in runoff and soil loss from simulated rainfall on corn and soybeans. *Trans. ASAE* **35:** 465–472.

McIsaac, G. F., J. K. Mitchell, J. C. Siemens, M. C. Hirschi, and M. J. Mainz, 1987, Yields, conservation and costs of tillage systems in Illinois. In *Optimum Erosion Control at Least Cost,* pp. 244–255. St. Joseph, Mich.: American Society of Agricultural Engineers.

Mohammad, A., and F. A. Gumbs, 1982, The effect of plant spacing on water runoff, soil erosion and yield of maize (Zea mays L.) on a steep slope of an Ultisol in Trinidad. *J. Agric. Eng. Res.* (London: Academic Press) **27**(6): 481–488.

Mokma, D. L., and M. A. Sietz, 1992, Effects of soil erosion on corn yields on Marlette soils in South-Central Michigan. *J. Soil Water Conserv.* **47:** 325–327.

Moldenhauer, W. C., and M. Amemiya, 1967. Save tomorrow's soils: control erosion from row cropping today. *Iowa Farm Sci.* **21:** 3–6, 21.

Mora, L. E., and R. A. Moreno, 1984, Cropping pattern and soil management influence on plant diseases: I. Diplodia macrospora leaf spot of maize. *Turrialbo* **341:** 35–40.

Mulvaney, D. L., and L. Paul, 1984, Rotating crops and tillage. Both sometimes better than just one. *Crop Soils* **367:** 8–19.

Murphee, C. E., and K. C. McGregor, 1991, Runoff and sediment yield from a flatland watershed in soybeans. *Trans. ASAE* **34:** 407–411.

Musgrave, G. W., and O. R. Neal, 1937, Rainfall and relative losses in various forms. *Trans. Am. Geophys. Union* **18:** 349–355.

Mutchler, C. K., L. L. McDowell, and J. D. Greer, 1985. Soil loss from cotton with conservation tillage. *Trans. ASAE* **28:** 160–163.

Myers, N., 1993, *Gaia: An Atlas of Planet Management.* Garden City, N.Y.: Anchor Press/Doubleday.

Napier, T. L., A. S. Napier, and M. A. Tucker, 1991, The social, economic and institutional factors affecting the adoption of soil conservation practices: the Asian experience. *Soil Tillage Res.* **20:** 365–382.

NAS, 1968, *Weed Control.* Washington, D.C.: National Academy of Sciences.

NAS, 1989, *Alternative Agriculture.* Washington, D.C.: National Academy of Sciences.

NAS, 1993, *Soil and Water Quality: An Agenda for Agriculture.* Washington, D.C.: National Academy of Sciences.

NSESPRPC (National Soil Erosion–Soil Production Research Planning Committee), 1981, Soil erosion effects on soil productivity: a research perspective. *J. Soil Water Conserv.* **32:** 82–90.

Olson, K. R., and E. Nizeyimana, 1988, Effects of soil erosion on corn yields of seven Illinois soils. *J. Product. Agric.* **1:** 13–19.

OTA, 1982, *Impacts of Technology on U.S. Cropland and Rangeland Productivity.* Washington, D.C.: U.S. Congress Office of Technology Assessment.

Palis, R. G., G. Okwach, C. W. Rose, and P. G. Saffigna, 1990, Soil erosion processes and nutrient loss. II. The effect of surface contact cover and erosion processes on enrichment ratio and nitrogen loss in eroded sediment. *Austral. J. Soil Res.* **28:** 623–639.

Paoletti, M. G., B. R. Stinner, and G. G. Lorenzoni, eds., 1989, Agricultural ecology and environment. *Agric. Ecosyst. Environ.* **27**(1–4).

Parrington, J. R., W. H. Zoller, and N. K. Aras, 1983, Asian dust: seasonal transport to the Hawaiian Islands. *Science* **246:** 195–197.

Pearson, L. C., 1967, *Principles of Agronomy.* New York: Reinhold.

Piel, G., 1994, Population: assessing the impact of development. *The Earth Times*, June 15, pp. 28–29.

Pimentel, D., ed., 1980, *Handbook of Energy Utilization in Agriculture.* Boca Raton, Fla.: CRC Press.

Pimentel, D., 1990, Environmental and social implications of waste in U.S. agriculture and food sectors. *J. Agric. Ethics* **3:** 5–20.

Pimentel, D., ed., 1993a, *World Soil Erosion and Conservation.* Cambridge, U.K.: Cambridge University Press.

Pimentel, D., 1993b, Environmental and economic benefits of sustainable agriculture. In *Socio-Economic and Policy Issues for Sustainable Farming Systems,* M. A. Paoletti, T. Napier, O. Ferro, B. Stinner, and D. Stinner, eds., pp. 5–20. Padova, Italy: Cooperativa Amicizia S.r.l.

Pimentel, D., E. C. Terhune, R. Dyson-Hudson, S. Rochereau, R. Samis, E. Smith, D. Denman, D. Reifschneider, and M. Shepard, 1976, Land degradation: effects on food and energy resources. *Science* **194:** 149–155.

Pimentel, D., E. Garnick, A. Berkowitz, S. Jacobson, S. Napolitano, P. Black, S. Valdes-Cogliano, B. Vinzant, E. Hudes, and S. Littman, 1980, Environmental quality and natural biota. *BioScience* **30:** 750–755.

Pimentel, D., D. Wen, S. Eigenbrode, H. Lang, D. Emerson, and M. Karasik, 1986, Deforestation: interdependency of fuelwood and agriculture. *Oikos* **46:** 404–412.

Pimentel, D., L. McLaughlin, A. Zepp, B. Lakitan, T. Kraus, P. Kleinman, F. Vancini, W. J. Roach, E. Graap, W. S. Keeton, and G. Selig, 1991, Environmental and economic impacts of reducing U.S. agricultural pesticide use. In *Handbook of Pest Management in Agriculture,* D. Pimentel, ed., pp. 679–718 (Vol. I). Boca Raton, Fla.: CRC Press.

Pimentel, D., U. Stachow, D. A. Takacs, H. W. Brubaker, A. R. Dumas, J. J. Meaney, J. O'Neil, D. E. Onsi, and D. B. Corzilius, 1992, Conserving biological diversity in agricultural/forestry systems. *BioScience* **42:** 354–362.

Pimentel, D., R. Harman, M. Pacenza, J. Pecarsky, and M. Pimentel, 1994, Natural resources and an optimum human population. *Popul. Environ.* **15**(5): 347–369.

Pimentel, D., C. Harvey, P. Resosudarmo, K. Sinclair, D. Kurz, M. McNair, S. Crist, L. Sphpritz, L. Fitton, R. Saffouri, and R. Blair, 1995, Environmental and economic costs of soil erosion and conservation benefits. *Science* **267:** 1117–1123.

Postel, S., 1992, *Last Oasis: Facing Water Scarcity.* New York: Norton.

PRB, 1993, *World Population Data Sheet.* Washington, D.C.: Population Reference Bureau.

Quansah, C., 1981, The effect of soil type, slope, rain intensity and their interactions on splash detachment and transport. *J. Soil Sci.* **32:** 215–224.

Reid, W. S., 1985, Regional effects of soil erosion on crop productivity—northeast. In *Soil Erosion and Crop Productivity,* R. F. Follett and B. A. Stewart, eds., pp. 235–250. Madison, Wisc.: American Society of Agronomy.

Repetto, R., 1992, Accounting for environmental assets. *Sci. Am.* **266:** 94–100.

Ricou, G. A. E., 1979, Consumers in meadows and pastures. In *Grassland Ecosystems of the World: Analysis of Grasslands and Their Uses,* R. T. Coupland, ed., pp. 147–153. Cambridge, U.K.: Cambridge University Press.

Roose, E., 1988, Soil and water conservation lessons from steep-slope farming in French speaking countries of Africa. In *Conservation Farming on Steep Lands,* pp. 130–131. Ankeny, Iowa: Soil and Water Conservation Society.

Roosevelt, F. D., 1937, *Letter from President to Governors, February 26, 1937.* Washington, D.C.: The White House.

Ryszkowski, L., 1993, Soil erosion and conservation in Poland. In *World Soil Erosion and Conservation,* D. Pimentel, ed., pp. 217–232. Cambridge, U.K.: Cambridge University Press.

Sauer, E. L., and H. C. M. Case, 1954, *Soil conservation pays off. Results of ten years of conservation farming in Illinois.* Urbana: University of Illinois Agriculture Station Bulletin 575.

Schertz, D. L., W. C. Moldenhauer, S. J. Livingston, G. A. Weesies, and E. A. Hinz, 1989, Effect of past soil erosion on crop productivity in Indiana. *J. Soil Water Conserv.* **44:** 604–608.

Schuman, G. C., R. G. Spomer, and R. F. Piest, 1973, Phosphorus losses from four agricultural watersheds on Missouri valley loess. *Soil Sci. Soc. Am. Proc.* **37:** 424–427.

SCS, 1993, *Wind Erosion Report* (Nov. 1992–May 1993). Washington, D.C.: Soil Conservation Service, USDA.

Shalhavet, J., A. Mantell, H. Bielorai, and D. Shimshi, 1979, *Irrigation of Field and Orchard Crops under Semi-Arid Conditions* (International Irrigation Information Center, Volcani Center, Bet Dagan, Israel). Elmsford, N.Y.: Pergamon Press.

Sharpley, A. N., and S. J. Smith, 1991, Effects of cover crops on surface water quality. In *Cover Crops for Clean Water,* W. L. Hargrove, ed., pp. 41–49. Ankeny, Iowa: Soil and Water Conservation Society.

Simons, M., 1992, Winds toss Africa's soil, feeding lands far away. *New York Times,* Oct. 29, pp. A1, A16.

Smith, D. D., 1946, The effect of contour planting on crop yield and erosion losses in Missouri. *J. Am. Soc. Agron.* **38:** 810–819.

Soileau, J. M., B. F. Hajek, and J. T. Touchton, 1990, Soil erosion and deposition evidence in a small watershed using Cesium-137. *Soil Sci. Soc. Am. J.* **54:** 1712–1719.

Speth, J. G., 1994, *Towards an Effective and Operational International Convention on Desertification.* New York: United Nations, International Convention on Desertification, International Negotiating Committee.

Spomer, R. G., R. F. Piest, and H. G. Heinemann, 1976, Soil and water conservation with western Iowa tillage systems. *Trans. Am. Soc. Agric. Eng.* **19:** 108–112.

Stocking, M., 1994, Soil erosion and land degradation. In *Environmental Science for Environmental Managers,* T. O'Riordan, ed. New York: Longman's.

Taylor, C. A., Jr., N. E. Garza, Jr., and T. O. Brooks, 1993, Grazing systems on the Edwards Plateau of Texas: Are they worth the trouble? I. Soil and vegetation response. *Rangelands* **15**(2): 53–57.

Teotia, J. P., F. L. Duky, and T. M. McCalla, 1950, *Effect of Stubble Mulch on Number and Activity of Earthworms.* Lincoln: Nebraska Agricultural Experiment Station Research Bulletin 165.

Tisdale, S. L., W. L. Nelson, and J. D. Beaton, 1985, *Soil Fertility and Fertilizers,* 4th ed. New York: Macmillan.

Tisdall, J. M., and J. M. Oades, 1982, Organic matter and water-stable aggregates in soils. *J. Soil Sci.* **33:** 141–163.

Troeh, F. R., J. A. Hobbs, and R. L. Donahue, 1991, *Soil and Water Conservation.* Englewood Cliffs, N.J.: Prentice-Hall.

Troeh, F. R., and L. M. Thompson, 1993, *Soils and Soil Fertility.* 5th ed. New York: Oxford University Press.

USBC, 1992, *Statistical Abstract of the United States 1990.* Washington, D.C.: U.S. Bureau of the Census, U.S. Government Printing Office.

USDA, 1967, *Agricultural Statistics.* Washington, D.C.: U.S. Department of Agriculture, Government Printing Office.

USDA, 1971, *Agriculture and the Environment.* Washington, D.C.: USDA, Economic Research Service.

USDA, 1989, *The Second RCA Appraisal. Soil, Water, and Related Resources on Nonfederal Land in the United States. Analysis of Conditions and Trends.* Washington, D.C.: U.S. Department of Agriculture.

USDA, 1990, *Fact Book of Agriculture.* Washington, D.C.: USDA, Office of Public Affairs.

USDA, 1991, *Agricultural Statistics.* Washington, D.C.: U.S. Department of Agriculture, Government Printing Office.

USDA, 1992, *Agricultural Statistics.* Washington, D.C.: USDA.

USFS (U.S. Forest Service), 1936, *The Major Range Problems and Their Solution.* Washington, D.C.: U.S. Government Printing Office.

Van Dijk, H., 1980, Survey of Dutch soil organic matter research with regard to humification and degradation rates in arable land. In *Soil Degradation* (Proceedings of Land Use Seminar on Soil Degradation), D. Boels, D. B. Davies, and A. E. Johnston, eds. Rotterdam: Balkema.

Van Doren, C. A., R. S. Stauffer, and E. H. Kidder, 1950, Effect of contour farming on soil loss and runoff. *Soil Sci. Soc. Am. Proc.* **15:** 413–417.

Van Rhee, J. A., 1965, Earthworm activity and plant growth in artificial cultures. *Plants Soil* **22:** 43–48.

Veltrop, J. A., 1991, There is no substitute for water. *Water Internatl.* **16:** 57.

Volk, B. G., and R. H. Loeppert, 1982, Soil organic matter. In *Handbook of Soils and Climate in Agriculture.* V. J. Kilmer, ed., pp. 211–268. Boca Raton, Fla.: CRC Press.

Wagger, M. G., and D. K. Cassel, 1993, Corn yield and water-use efficiency as affected by tillage and irrigation. *Soil Sci. Soc. Am. J.* **57:** 229–234.

Waldren, R. P., 1983, Corn. In *Crop-Water Relations,* I. D. Teare and M. M. Peet, eds., pp. 187–212. New York: Wiley.

Wall, G. J., E. A. Pringle, and R. W. Sheard, 1991, Intercropping red clover with silage corn for soil erosion control. *Can. J. Soil Sci.* **71:** 137–145.

Wen, Dazhong, 1993, Soil erosion and conservation in China. In *Soil Erosion and Conservation,* D. Pimentel, ed., pp. 63–86. New York: Cambridge University Press.

Wendt, R. C., E. E. Alberts, and A. T. Helmers, 1986, Variability of runoff and soil loss from fallow experimental plots. *Soil Sci. Soc. Am. J.* **50:** 730–736.

Wendt, R. C., and R. E. Burwell, 1985, Runoff and soil losses from conventional, reduced, and no-till corn. *J. Soil Water Conserv.* **40:** 450–454.

Whitaker, F. D., and H. G. Heinemann, 1973, Chemical weed control affect runoff, erosion and corn yields. *J. Soil Water Conserv.* **28:** 174–175.
Wijewardene, R., and P. Waidyanatha, 1984, *Systems, Techniques and Tools. Conservation Farming for Small Farmers in the Humid Tropics.* Colombo, Sri Lanka: Department of Agriculture, Sri-Lanka and the Commonwealth Consultative Group on Agric. for the Asia-Pacific Region.
Wood, M., 1989, *Soil Biology.* New York: Blackie, Chapman & Hall.
World Bank, 1993, *World Can Cut Hunger Rate in Half.* Washington, D.C.: World Bank Press Release, Nov. 29.
Worldwatch Institute, 1988, *State of the World.* Washington, D.C.: Worldwatch Institute.
WRI (World Resources Institute), 1991, *World Resources 1990–91.* New York: Oxford University Press.
WRI, 1992, *World Resources.* New York: Oxford University Press.
Young, A., 1989, *Agroforestry for Soil Conservation.* Wallingford, U.K.: C.A.B. International.
Young, A., 1990, Agroforestry, environment and sustainability. *Outlook Agric.* (Oxon: C.A.B. International) **19:** 155–160.
Young, C. E., and R. S. Magleby, 1985, Economic benefits of three rural clean water projects. In *The Off-Site Costs of Soil Erosion: The Proceedings of a Symposium,* pp. 134–142. Washington, D.C.: The Conservation Foundation.
Zacharias, T. P., and A. H. Grube, 1984, An economic evaluation of weed control methods used in combination with crop rotations: a stochastic dominance approach. *North Cent. J. Agric. Econ.* (North Dakota State University) **6:** 113–120.

Chapter

3

Sedimentation

Everett V. Richardson
Senior Associate, Ayres Associates, and
Professor Emeritus, Civil Engineering
Colorado State University
Fort Collins, Colorado

Introduction

Sediments, with their movement and deposition by water, can be either a benefit or detriment to the environment and utilization of water resources. The deposition of sediments by rivers forms the floodplains and deltas used by people for agricultural production and habitat. However, in contrast to popular belief, the fertility of these sediments is small, as shown by El-Tobgy (1976) for the Nile River. Negative effects such as the loss of storage in reservoirs; deposition on floodplains, in houses, and in cities during floods; and erosion of land are well known. *Sedimentation* is the engineering determination of sediment yield from the land surface, sediment transport by water, erosion and erosion control, and deposition of sediments. *Sediment deposition* is concerned mostly with reservoir sedimentation; however, deposition in lakes, rivers, tidal estuaries, harbors, and bays is also of concern, but beyond the scope of this chapter. *Sedimentation engineering* requires a knowledge of sediment properties, fluvial geomorphology, bed forms in alluvial sand channels, hydraulics of open-channel flow, and river mechanics.

Fluvial Geomorphology

Fluvial geomorphology is the study of the classification, description, planform, erosion history, nature, origin and development of present river forms and the underlying relationship to their present structure. Knowledge of the fluvial geomorphology of a stream is important in understanding sediment transport, response of the river to change in climate, tectonic activity, human activity, and changes in sediment and water discharge. Rivers are dynamic; they are continually modifying their shape, their position, and other morphological characteristics with long- and short-term changes in water or sediment discharge, climate, tectonic, and human activities. For example, through changes in water and sediment discharge, the construction of a dam can cause aggradation and degradation upstream and downstream.

Stream planform has been broadly classified into *meandering, straight, braided,* or *antibranching* (Leopold and Wolman, 1957; Schumm, 1972, 1977; Richardson et al., 1990; Lagasse et al., 1991) (Fig. 3.1). Other more extensive categories have been made (Culbertson et al., 1967; Brice and Blodgett, 1978), but the stream planform classification appears to be the most useful. A stream may be meandering in one reach; whereas, in another, may have a braided, straight, or anabranch planform. Stream fluvial geomorphology and

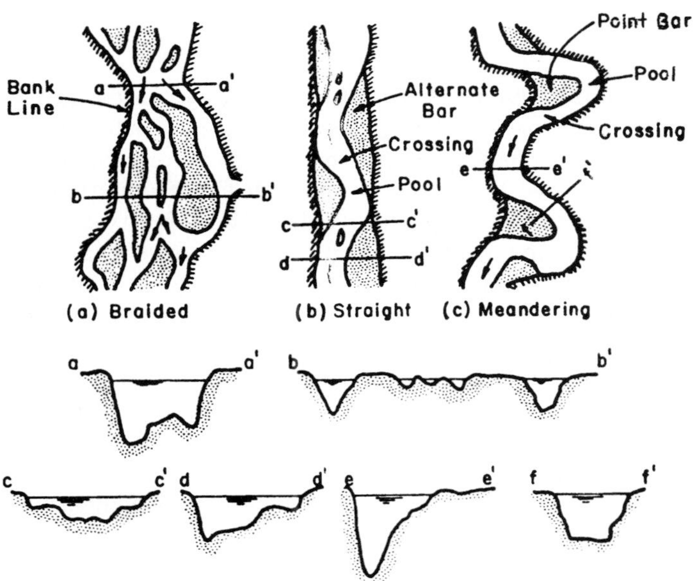

Figure 3.1 Stream planform (Richardson et al., 1990).

stability can affect stream size, flow habit, channel boundaries, bed material, valley or other settings, floodplain size, incision, bed and bank material, and tree cover. Lagasse et al. (1991) illustrate these factors in Fig. 3.2.

To determine whether the stream planform is meandering or braided, Lane (from a study of the relation between planform, discharge Q and slope S) found that when $SQ^{1/4} < 0.0017$, a sand-bed stream had a meander planform; when $SQ^{1/4} > 0.01$, a braided planform (Richardson et al., 1990). A stream with a value between 0.0017 and 0.01 could have either a meander or braided planform.

Leopold and Maddock (1953) determined regimen-type relations between the discharge and stream width ($W = aQ^b$), depth ($y = cQ^f$), velocity ($V = kQ^m$), bed-material discharge ($Q_s = pQ^j$), slope ($S = tQ^z$), and Manning's n ($n = rQ^y$). They determined values for the coefficients and exponents for many rivers in the United States. These values are also given by Richardson et al. (1990).

A qualitative relation to predict the response of a river system to change in water discharge Q, slope S, sediment discharge Q_s, and bed-material size D_{50} was given by Lane (1955), Schumm (1972), Richardson et al. (1990), and Simons and Sentürk (1992). The qualitative relationship for the response of a stream is

$$QS \sim Q_s D_{50} \tag{3.1}$$

This relation shows, for example, that if there is a change in discharge Q, there will be a corresponding change in one of the other three variables to bring the relationship into balance. For example, a decrease in Q and no change in sediment discharge or size of bed material would result in an increase in slope.

Sediment Properties

Sediments are classified by size into *clay, silt, sand, gravel, cobbles,* and *boulders* (Table 3.1). Along with size, the other important properties of sediment are fall velocity, density or specific weight, chemical attraction (cohesion) in clays, and shape. These properties are of the individual particles or the bulk properties of the mass, which are described by size frequency distributions, bulk specific weight, and porosity. Size distributions are determined by pipette (clays and silts), sieves (sand- to cobble-size material), and visual accumulation tube analysis (sand sizes), and pebble count (large gravel to boulders) (Brown, 1950; Vanoni, 1975; Richardson et al., 1990; Simons and Sentürk, 1992). The pebble-count method is the determination of the particle size distribution in the surface layer of a stream by the mea-

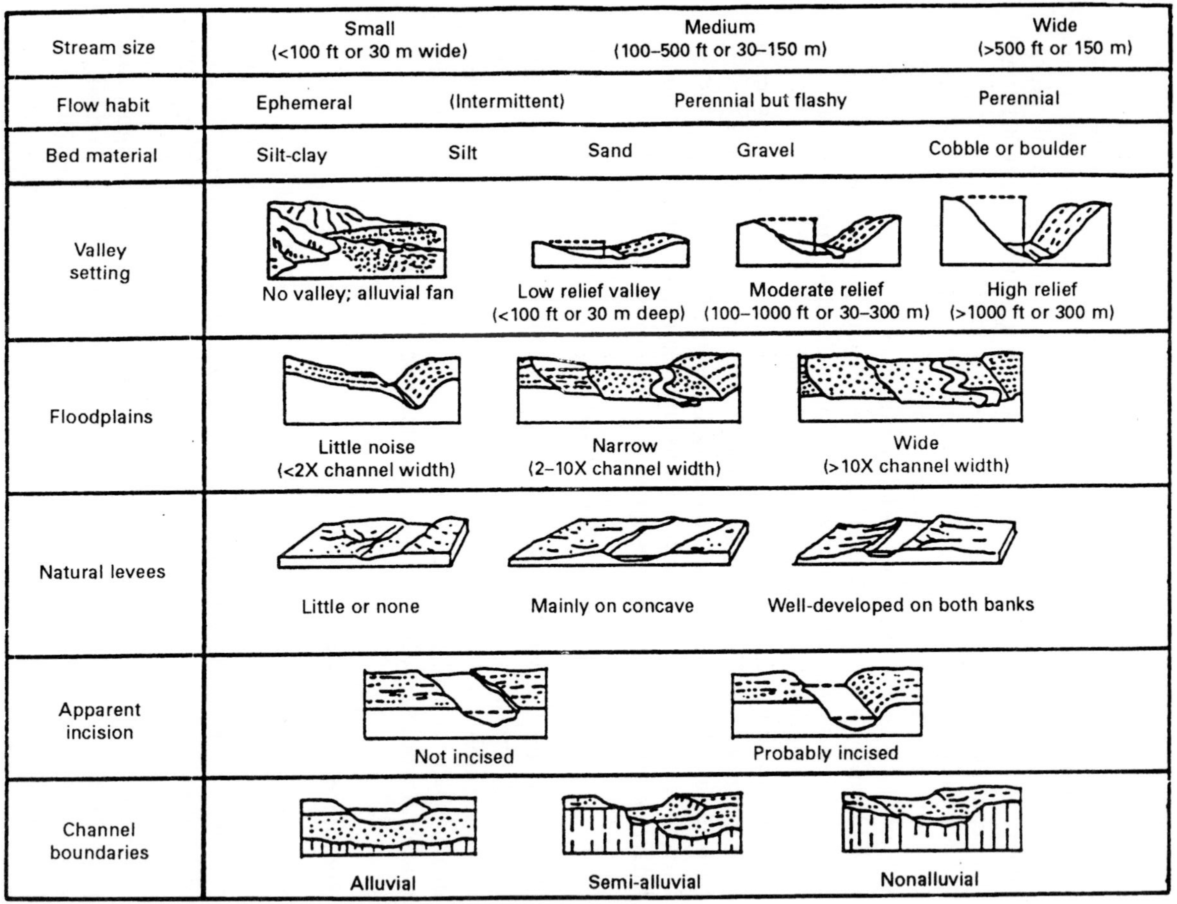

Stream size	Small (<100 ft or 30 m wide)	Medium (100–500 ft or 30–150 m)	Wide (>500 ft or 150 m)	
Flow habit	Ephemeral	(Intermittent)	Perennial but flashy	Perennial
Bed material	Silt-clay	Silt	Sand	Gravel	Cobble or boulder
Valley setting	No valley; alluvial fan	Low relief valley (<100 ft or 30 m deep)	Moderate relief (100–1000 ft or 30–300 m)	High relief (>1000 ft or 300 m)
Floodplains	Little noise (<2X channel width)	Narrow (2–10X channel width)	Wide (>10X channel width)	
Natural levees	Little or none	Mainly on concave	Well-developed on both banks	
Apparent incision	Not incised	Probably incised		
Channel boundaries	**Alluvial**	**Semi-alluvial**	**Nonalluvial**	

Figure 3.2 Classification of stream properties (Lagasse et al., 1991).

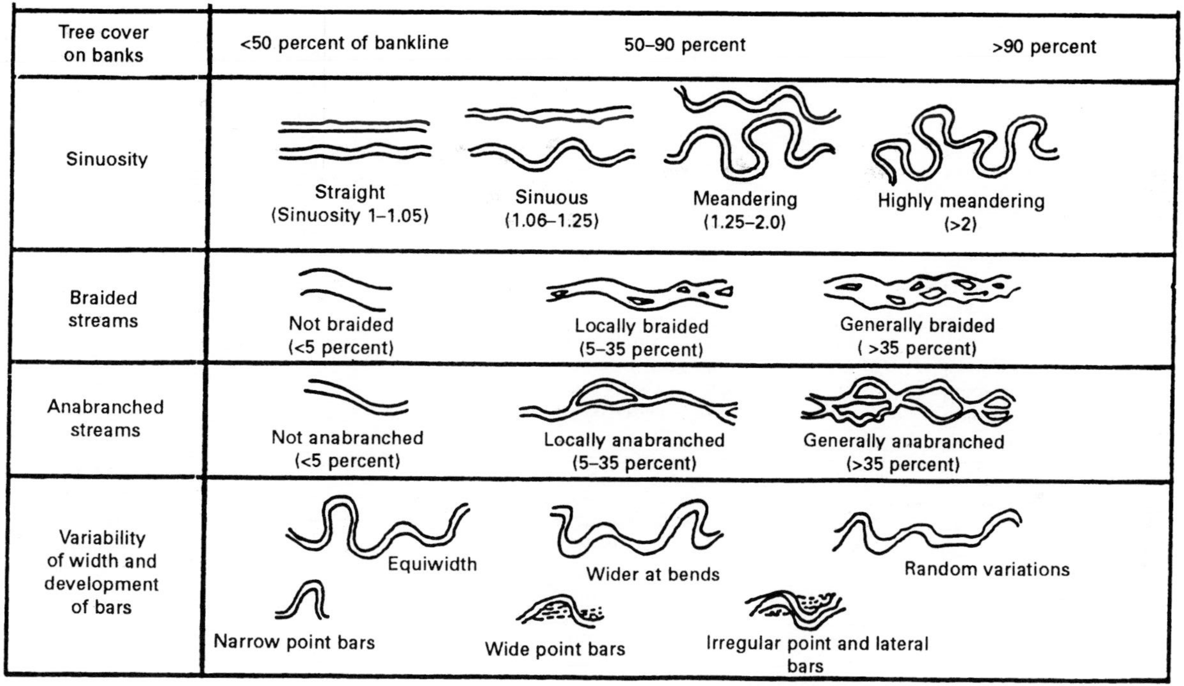

Tree cover on banks	<50 percent of bankline	50–90 percent		>90 percent
Sinuosity	Straight (Sinuosity 1–1.05)	Sinuous (1.06–1.25)	Meandering (1.25–2.0)	Highly meandering (>2)
Braided streams	Not braided (<5 percent)	Locally braided (5–35 percent)		Generally braided (>35 percent)
Anabranched streams	Not anabranched (<5 percent)	Locally anabranched (5–35 percent)		Generally anabranched (>35 percent)
Variability of width and development of bars	Equiwidth Narrow point bars	Wider at bends Wide point bars	Irregular point and lateral bars	Random variations

Figure 3.2 *(Continued)*

TABLE 3.1 Classifications of Sediments

Size				Approximate sieve mesh openings per inch		
Millimeters		Micrometers	Inches	Tyler	U.S. Standard	Class
4000–2000			160–80			Very large boulders
2000–1000			80–40			Large boulders
1000–500			40–20			Medium boulders
500–250			20–10			Small boulders
250–130			10–5			Large cobbles
130–64			5–2.5			Small cobbles
64–32			2.5–1.3			Very coarse gravel
32–16			1.3–0.6			Coarse gravel
16–8			0.6–0.3	2.5		Medium gravel
8–4			0.3–0.16	5	5	Fine gravel
4–2			0.16–0.08	9	10	Very fine gravel
2–1	2.00–1.00	2000–1000		16	18	Very coarse sand
1–1/2	1.00–0.50	1000–500		32	35	Coarse sand
1/2–1/4	0.50–0.25	500–250		60	60	Medium sand
1/4–1/8	0.25–0.125	250–125		115	120	Fine sand
1/8–1/16	0.125–0.062	125–62		250	230	Very fine sand
1/16–1/32	0.062–0.031	62–31				Coarse silt
1/32–1/64	0.031–0.016	31–16				Medium silt
1/64–1/128	0.016–0.008	16–8				Fine silt
1/128–1/256	0.008–0.004	8–4				Very fine silt
1/256–1/512	0.004–0.0020	4–2				Coarse clay
1/512–1/1024	0.0020–0.0010	2–1				Medium clay
1/1024–1/2048	0.0010–0.0005	1–0.5				Fine clay
1/2048–1/4096	0.0005–0.0002	0.5–0.24				Very fine clay

surement (weight or length) of 100 or more particles (Wolman, 1954; Ritter and Helley, 1968; Richardson et al., 1990). Size distributions are usually expressed as percent finer than a given size in the distribution. For example, D_i is the size of the particle in the distributions in which i percent of the particles is finer and G is the gradation coefficient. One measure of the gradation coefficient is given as

$$G = 0.5\left(\frac{D_{50}}{D_{16}} + \frac{D_{84}}{D_{50}}\right) \tag{3.2}$$

where D_i is the sediment diameter of the particle of which i percent of the sample is finer.

Fall velocity of sediment particles is a prime indicator of the interaction between fluid and sediments. The *fall velocity* of a particle is defined as the velocity of that particle falling alone in quiescent, distilled water of infinite extent. In most cases, the particle is not falling alone, and the water is not distilled or quiescent. However, measurement techniques are available for determining the fall velocity of an individual particle in groups of particles in a finite field in fluid other than distilled water (Vanoni, 1975). A particle falling at terminal velocity in a fluid is under the action of a driving force due to its buoyant weight and a resisting force due to the fluid drag. *Fluid drag* is the result of either the tangential shear stress on the surface of the particle, a pressure difference on the particle, or a combination of the two forces. The fall velocity of a particle is given by the following equation (Vanoni, 1975):

$$\omega = \frac{4\,Dg(S_s - 1)}{3\,C_d} \tag{3.3}$$

where ω = terminal fall velocity of the particle
C_d = coefficient of drag
g = acceleration due to gravity
D = diameter of the particle
S_s = specific weight of the particle

The coefficient of drag is a function of the particle Reynolds number ($\omega D/\nu$) and the shape factor of the particle, where ν is the kinematic viscosity and the shape factor (SF) is given by Albertson (1953) as

$$\text{SF} = \frac{c}{(ab)^{0.5}} \tag{3.4}$$

where a, b, and c are the lengths of the longest, intermediate, and shortest mutually perpendicular axes of a particle, respectively.

For values of the particle Reynolds number R less than approximately 0.1, $C_D = 24/R$, and the fall velocity is given by Stokes' law:

$$\omega = \frac{gD^2(S_s - 1)}{18\nu} \tag{3.5}$$

Clays and silt-size particles have fall velocities governed by Stokes' law. Sands and particles up to approximately 5 mm have fall velocities that are a function of particle Reynolds number and shape factor. Larger particles in clear water have fall velocities independent of the Reynolds number. The fall velocity of quartz sands is given in Fig. 3.3.

Specific weight is weight per unit volume. With granular materials such as soils, sediment deposits, or water-sediment mixtures, the specific weight is the weight of solids per unit volume of the material, including its voids. The measurement of the specific weight of sediment deposits is determined simply by measuring the dry weight of a known volume of the undisturbed material.

Porosity of granular materials is the ratio of the volume of void space to the total volume of an undisturbed sample. To determine porosity, the volume of the sample must be obtained in an undisturbed condition. Next, the volume of solids is determined either by liquid displacement or indirectly from the weight of the sample and

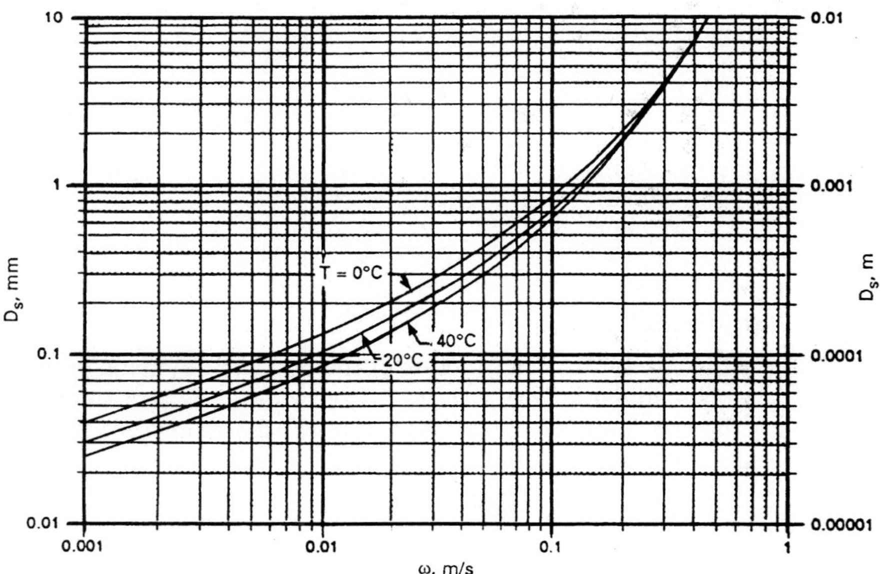

Figure 3.3 Fall velocity of sand-size particles (after Rouse, 1937).

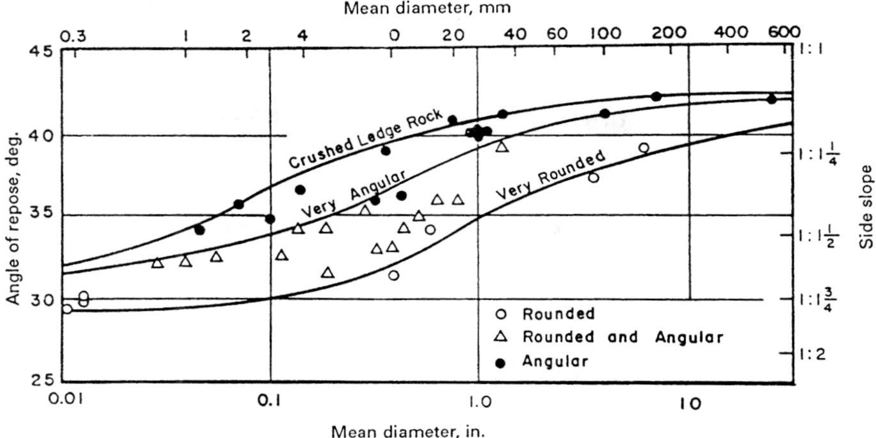

Figure 3.4 Angle of repose of noncohesive materials (Richardson et al., 1990).

the specific gravity of material. The void volume is then obtained by subtracting the volume of solids from the total volume.

Cohesion is the force by which particles of clay are bound together. This force is the result of ionic attraction among individual particles, and a function of the type of mineral, particle spacing, salt concentration in the fluid, ionic valence, hydration, and swelling properties of the constituent minerals.

Clays are alumina-silicate crystals composed of two basic building sheets: the tetrahedral silicate sheet and octahedral hydrous aluminum oxide sheet. Various types of clay result from different configurations of these sheets. The two main types of clay are kaolinite and montmorillonite. Kaolinite crystals are large (70 to 100 layers thick), held together by strong hydrogen bonds, and are not readily dispersed in water. Montmorillonite crystals are small (3 layers thick) held together by weak bonds between adjacent oxygen layers and are readily dispersed in water into extremely small particles.

Angle of repose is the maximum slope angle on which noncohesive material will reside without moving. It is a measure of the intergranular friction of the material. The angle of repose for dumped granular material is given in Fig. 3.4.

Bed Forms

Bed forms are formed on sand and gravel streams (0.062 to 64 mm of bed material) by the interaction of the fluid forces and sediment particles. Bed forms that commonly form are ripples, ripples on dunes, dunes, plane bed, antidunes, and chutes and pools (Fig. 3.5) (Simons

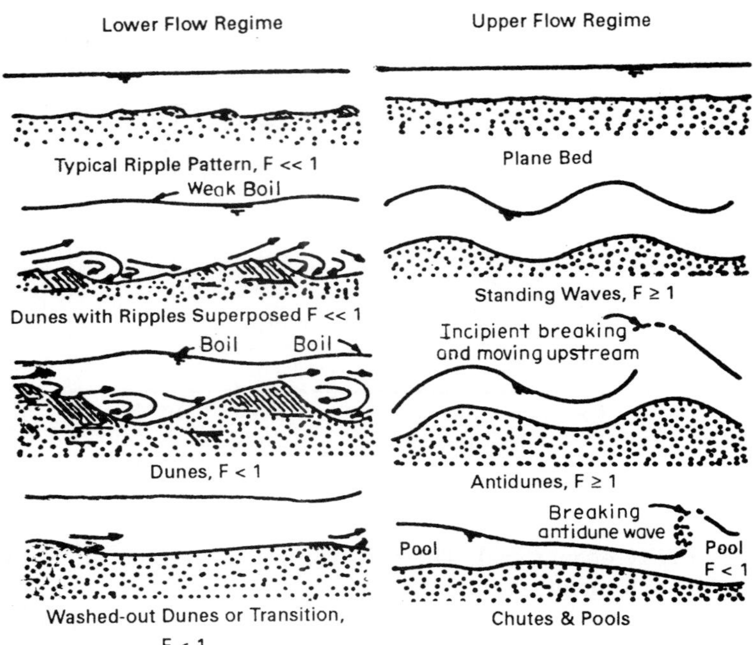

Figure 3.5 Forms of bed roughness in sand and fine gravel channels (Simons and Richardson, 1961, 1966).

and Richardson 1961, 1966; Vanoni, 1975; Richardson et al., 1990; Simons and Sentürk, 1992). These bed forms are listed in their order of occurrence with increasing values of stream power ($V\gamma y_0 S$) for bed materials having D_{50}<0.6 mm. For bed materials coarser than 0.6 mm, dunes form instead of ripples after beginning of motion at small values of stream power. Bed forms in alluvial streams depend on discharge, slope, depth of flow, size of bed material, and the viscosity of the fluid.

Simons and Richardson (1961, 1966) divided bed forms into lower and upper flow regimes separated by a transition zone on the basis of similarities in the shape of the bed configuration, mode and magnitude of sediment transport, resistance to flow, process of energy dissipation, and phase relation between the bed and water surfaces. The two regimes and their associated bed configurations are (see also Fig. 3.5):

Lower flow regime. Ripples, dunes with ripples superposed, and dunes.

Transitional flow regime. Bed roughness ranges from dunes to plane bed or antidunes.

Upper flow regime. Plane bed; antidunes with standing waves, antidunes with breaking waves, and chutes and pools.

In the lower flow regime, resistance to flow is large and sediment transport small. Water-surface undulations are out of phase with the bed surface, and there is a relatively large separation zone downstream from the crest of each ripple or dune. The most common mode of bed-material transport is for the individual grains to move up the back of the ripple or dune and avalanche down its face (contact-bed material discharge). After coming to rest on the downstream face of the ripple or dune, the particles remain there until exposed by the downstream movement of the dunes; they repeat this cycle of moving up the back of the dune, avalanching, and storage. Thus, most movement of the bed-material particles is in steps. The velocity of the downstream movement of the ripples or dunes depends on their height and the velocity of the grains moving up their backs.

Bed configuration in the transition zone may range from that typical of the lower flow regime to that of the upper flow regime depending mainly on antecedent conditions. If the antecedent bed configuration is dunes, the depth or slope can be increased to values more consistent with those of the upper flow regime without changing the bed form; or, conversely, if the antecedent bed is plane, depth and slope can be decreased to values more consistent with those of the lower flow regime without changing the bed form. Often in the transition from the lower to the upper flow regime, the dunes decrease in amplitude and increase in length before the bed becomes plane (washed-out dunes). Resistance to flow and sediment transport also have the same variability as the bed configuration in the transition.

In the upper flow regime, resistance to flow is small and sediment transport large. The usual bed forms are plane bed or antidunes. The water surface is in phase with the bed surface, and fluid does not separate from the boundary, except when an antidune breaks. Resistance to flow is the result of grain roughness with the grains moving, wave formation and subsidence, and energy dissipation when the antidunes break. The mode of sediment transport is for the individual grains to roll almost continuously downstream in sheets one or two grain diameters thick; however, when antidunes break, much bed material is briefly suspended, movement then stops temporarily, and there is some storage of the particles in the bed. Resistance to flow and range in concentration of bed material transport is given in Table 3.2.

The bed forms can change from those typical of the lower flow regime to those typical of the upper flow regime with a change in depth (discharge), slope of the energy-grade line, and viscosity of the water (temperature or concentration of silts and clays). With an increase in depth caused by an increase in discharge, a dune-bed form

TABLE 3.2 Resistance to Flow and Sediment Concentration for the Different Bed Forms

Alluvial sand-bed channels	Manning's n	Concentration, ppm
Lower flow regime		
Plane bed	0.014	0
Ripples $d \leq 0.7$ mm	0.018–0.028	10–200
Dunes	0.020–0.040	200–3,000
Transition		
Washed-out dunes	0.014–0.025	1,000–4,000
Upper flow		
Plane bed	0.010–0.013	2,000–4,000
Antidunes		
Standing waves Fr = 1	0.010–0.015	2,000–15,000
Breaking Fr $\geq$ 1	0.012–0.020	5,000–50,000
Chute and pools	0.018–0.035	5,000–50,000

SOURCE: Richardson et al. (1990).

can change to plane bed or antidune. A decrease in depth can reverse the change. With these changes, it is possible to experience a large increase in discharge with little or no change in depth. Most streams with sand beds will flow in the upper flow regime when the discharge and depth are large. The change in flow regimen with depth is illustrated in Fig. 3.6. The discontinuous rating curve (Fig. 3.6) was found to occur in many streams by Nordine (1964) and Dawdy (1961).

Slope of the energy-grade line is an important factor in determining the bed configuration which will exist for a given discharge and bed

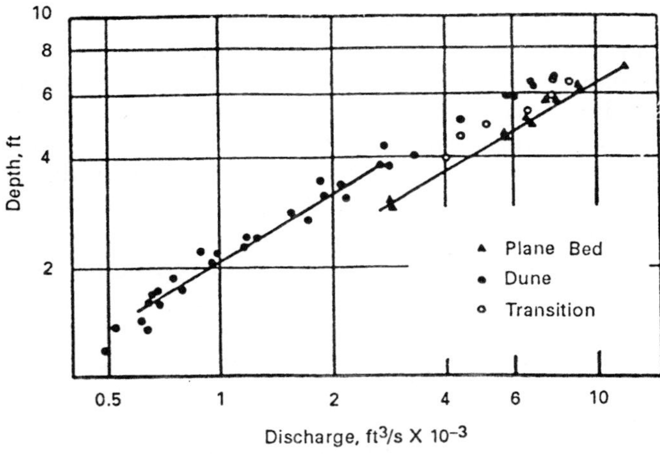

Figure 3.6 Correlation between depth and discharge for Elkhorn River near Waterloo, Nebraska (from Beckman and Furness, 1962) (Richardson et al., 1990).

material. In flume studies, with their limited range in depth and discharge, change in slope is the principal means of changing the bed forms. However, in stream and rivers, where slope is relatively constant, a change in bed forms occurs with a change in depth (discharge) and/or change in viscosity. Sand-bed streams with steep slopes may always flow in the upper flow regime.

A change in viscosity resulting from a change in temperature or the presence of large concentrations of fine sediment (clay) can change the bed form. A decrease in temperature can change a dune bed to washed-out dunes, plane beds, or antidunes. The Missouri River as it flows along the border between Iowa and Nebraska has a temperature between 21 and 27°C in summer and 1.7 and 15°C in fall. In summer, the bed form is dunes with a Manning n of 0.020; in fall, the bed form is washed-out dunes (transition) and the Manning n is 0.015 (U.S. Corps of Engineers, 1969).

An increase in the concentration of fine sediments can increase the apparent viscosity and density of the fluid-sediment mixture. This increase can change the bed form, resistance to flow, and concentration of bed-material transport. Flow without significant quantities of silts and clays in suspension will be limited to the concentration of bed material they can transport. The upper limit of fine sands that can be transported by these relatively clear-water flows is around 50,000 ppm. With large concentrations of silts and clays, concentrations of sand in suspension on the order of 200,000 ppm by weight have been observed (Beverage and Culbertson, 1964; Richardson and Hanly, 1965).

Apparent viscosity and density effects on bed forms, resistance to flow, and bed-material transport are described by Simons et al. (1963), Vanoni (1975), Richardson et al. (1990), and Simons and Sentürk (1992). The magnitude of the effect of fine sediment on the apparent kinematic viscosity of the mixture and the fall velocity of sands is large and depends on the chemical makeup of the clays (Figs. 3.7 and 3.8).

In addition to increasing the viscosity, fine sediment suspended in water increases the mass density of the mixture r and, consequently, the specific weight g. The specific weight of a sediment-water mixture is computed from the relation

$$\gamma_m = \frac{\gamma_w \gamma_s}{\gamma_s - C_s(\gamma_s - \gamma_w)} \tag{3.6}$$

where γ_m = specific weight of the water-sediment mixture
γ_w = specific weight of the water
γ_s = specific weight of the sediment

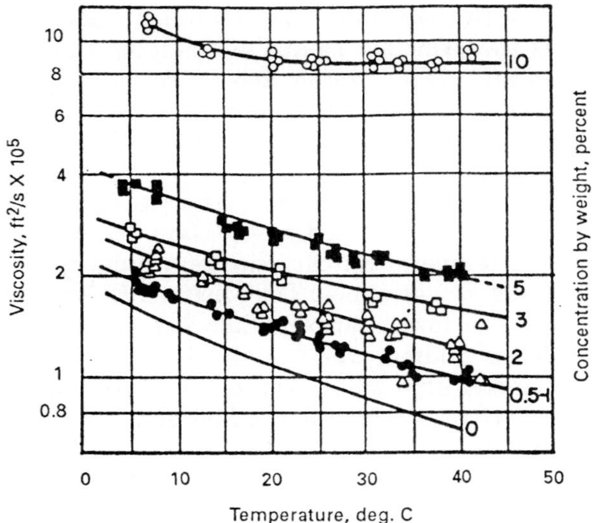

Figure 3.7 Apparent kinematic viscosity of water-bentonite dispersions (Simons et al., 1963).

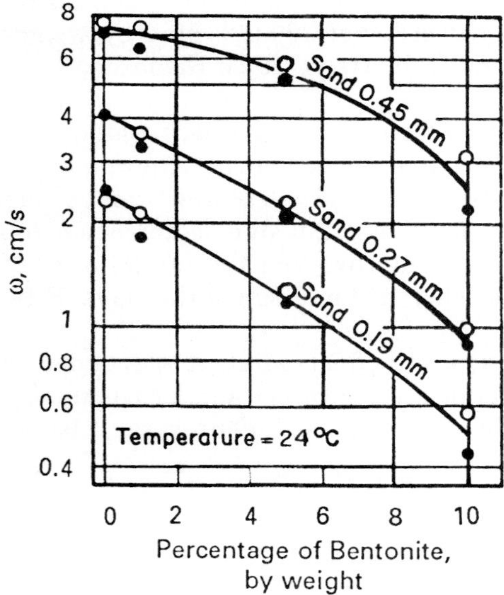

Figure 3.8 Variation of fall velocity of several sand mixtures with percent bentonite (Simons et al., 1963).

C_s = concentration by weight (in fraction form) of the suspended sediment

Sediment Transport

Beginning of motion of sediment particles, which is of great importance in (1) channel stability analysis, (2) scour of the foundations of bridges, (3) bed-load transport equations, and (4) design of riprap, can be related to either the shear stress on the sediment particles or the fluid velocity in the vicinity of the particles. When the grains are at incipient motion, these values are called the *critical shear stress* τ_c and critical velocity V_c. The choice of critical shear stress or velocity depends on (1) which is easier to determine in the field, (2) the precision with which the critical value is known for the particle size, and (3) the type of problem. In sediment-transport analysis, most equations use critical shear. In stable-channel design, either critical shear or critical velocity is used; whereas in bridge scour and design of riprap, critical velocity is commonly used. In addition to the forces on the particle resulting from the flowing water, waves and seepage into or out of the bed or banks affect the beginning of motion conditions.

Shields [cited in Brown (1950), Vanoni (1975), Richardson et al. (1990), and Simons and Sentürk (1992)] experimentally determined a relation between what is now called the *Shields parameter* and a particle-size, shear-velocity Reynolds number (Fig. 3.9). The Shields parameter is the ratio of the critical shear stress to the submerged weight of the particle $[\tau_c/(S_s-1)\gamma D]$, and the Reynolds number is VD/γ, where S_s is the specific gravity of water, $\gamma\_$ is the unit weight of water, D is the diameter of the particle, V is the shear velocity $[(\tau_c/\rho)]^{0.5}$, ρ is the density of water, and $\nu\_$ is the kinematic viscosity. The average shear stress on a boundary is given by γRS, where R is the hydraulic radius given by the area of the flow A divided by the

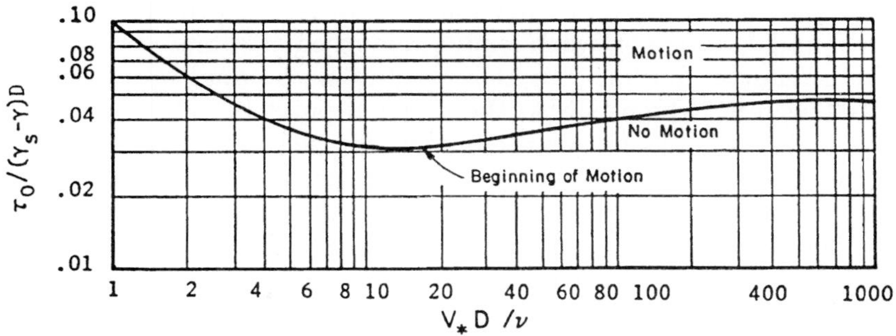

Figure 3.9 Shields relation as adapted by Gessler (1965).

wetted perimeter P and S is the slope of the energy-grade line. Also, the shear stress can be determined from the velocity of flow. Relations for determining the shear stress from the mean velocity or the velocity distribution are given in the publications cited above.

Although Fig. 3.9 shows a single-value function between the Shields parameter K_s and the Reynolds number, researchers have found other values for K_s. For sand-size material (0.062 to 2.0 mm), a value of 0.047 is used; for particle-size material (2.0 to 40 mm), a value of 0.03 is often used; and for particle-size material (coarser than 40 mm), 0.02 is used.

By setting the shear stress of the flow (τ_0) on the bed of a stream equal to the critical shear stress for the beginning of motion of a particle, the relation for the critical velocity V_c for the beginning of motion of a sediment particle results in

$$V_c = \frac{K_s^{1/2}(S_s - 1)^{1/2} D^{1/2} y^{1/6}}{n} \tag{3.7}$$

where the symbols are as defined previously.

Using $S_s = 1.65$, $n = 0.041D^{1/6}$, and $K_s = 0.047$, and all quantitative values in meters, the equation becomes

$$V_c = 6.79 y^{1/6} D^{1/3} \tag{3.8}$$

The sources of sediment transported by a stream are (1) the bed material that makes up the streambed and (2) the finer sediments that come from the banks and the watershed. This latter is termed *wash load* by Einstein (1950). Geologically, both materials come from the watershed; however, for the engineer, the distinction is important because bed material is transported at the capacity of the stream and is functionally related to measurable hydraulic variables. The wash load depends on availability, is not functionally related to measurable hydraulic variables, and must be measured.

Wash load is not found in appreciable quantities in the bed material. In sand-bed streams, wash load is composed of silts and clays and the bed-material discharge is composed of sizes equal to or greater than 0.062 mm, which is also the division point between sand and silt. In fast-flowing mountain streams with a bed of cobbles or coarser material, the wash load could be of sand or gravel. A reasonable criterion is to choose a sediment size finer than the smallest 1 or 2 percent of the bed material as the dividing size between wash load and bed-material sediment discharge.

The modes of sediment transport are (1) by the sediment particles rolling or sliding in contact with the bed (bed load or contact load) or (2) by suspension by the turbulence of the stream (suspended sediment discharge). Even as there is no sharp demarcation between bed-material discharge and wash load, there is no sharp line between contact load and suspended-sediment load. A particle may move part of the time in contact with the bed and at other times be suspended by the flow. The distinction is important because the two modes of transport follow different laws. The equations for estimating the total bed-material discharge of a stream are based on these laws.

The total sediment discharge of a stream is classified by (1) source of the sediments (bed-material discharge and wash-load discharge, (2) mode of transport (contact sediment discharge or suspended-sediment discharge), and (3) measurement (measured sediment discharge and unmeasured sediment discharge). The latter distinction occurs because suspended-sediment samplers can measure only to within a small distance to the bed and thus do not measure the contact sediment discharge and a small portion of the suspended-sediment discharge. An advantage of the modified Einstein method of determining total sediment discharge developed by Colby and Hembry [cited in Vanoni (1975) and Simons and Sentürk (1992)] is that it combines measured suspended-sediment discharge with a calculated unmeasured sediment discharge. Generally, the amount of bed material moving in contact with the bed of a large sand-bed river is 5 to 10 percent of the bed material moving in suspension, and the measured suspended-sediment discharge is 90 to 95 percent of the total sediment discharge. However, in shallow sand-bed streams with little or no wash load, the measured suspended-sediment load may be as small as 50 percent of the total load.

The magnitude of the suspended sediment discharge can be very large. Suspended-sediment concentrations as large as 600,000 ppm or 60% by weight have been observed (Beverage and Culbertson, 1964; Richardson and Hanley, 1965). Concentrations of this magnitude are largely fine sediments. By increasing fluid properties (viscosity and density), the fine material in the flow increases the capacity of the flow to transport bed material.

Many sediment transport equations are given in the literature. Brown (1950), Vanoni (1975), Richardson (1990), and Simons and Sentürk (1992) give many equations with references to others. The Meyer-Peter–Muller formula given in this literature is used to determine the contact bed-material discharge in coarse-bed streams (gravel sizes and larger). The modified Einstein procedure for determining total bed material and total sediment discharge is described in detail by Vanoni (1975) and Simons and Sentürk (1992). A useful graphic method of determining the total bed-material discharge in sand-bed streams developed by Colby (1964) and is given in the following paragraph.

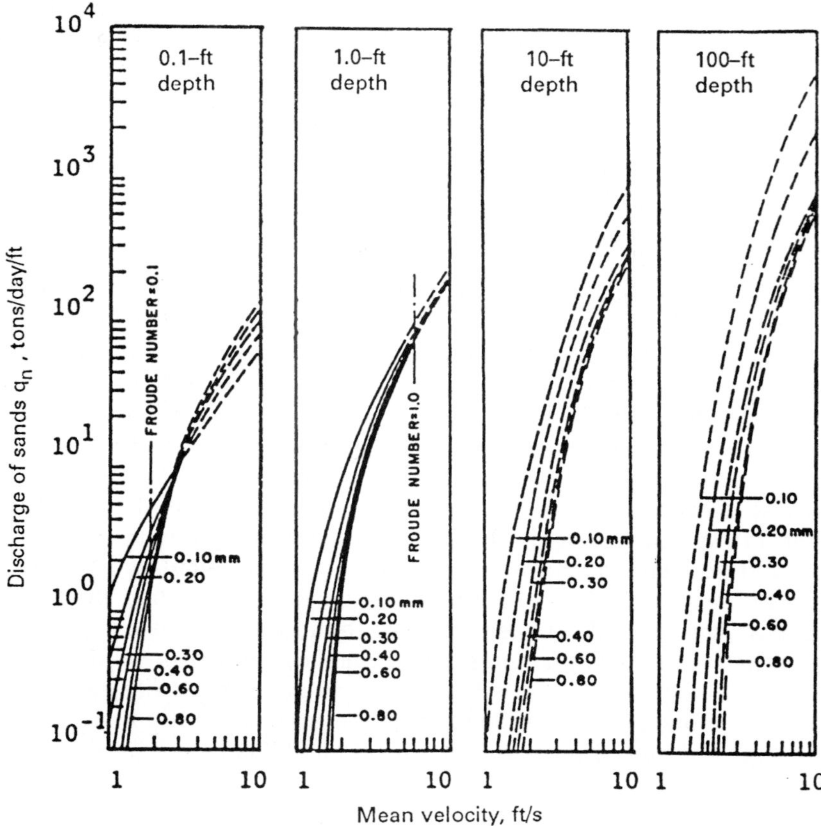

Figure 3.10 Colby's (1964) relation of discharge of sands to mean velocity, median size of bed material, depth of flow, and water temperature of 60°F.

Colby's method of estimating total bed material discharge uses four graphic relations developed from an extensive study of all pertinent variables (Figs. 3.10 and 3.11). In arriving at his curves, Colby was guided by the Einstein (1950) bed-load function and a large amount of total bed-material discharge from streams and flumes. However, it should be understood that all curves for 100 ft depth, most curves of 10 ft depth, and part of the curves of 1.0 and 0.1 ft depth are not based on data but are extrapolated from limited data and theory.

In applying Figs. 3.10 and 3.11 to compute the total bed-material discharge, the proposed procedure is as follows:

1. The required data are the mean velocity V, the mean depth y_0, the median size of bed material D_{50}, the water temperature T, and the fine sediment concentration C_f.

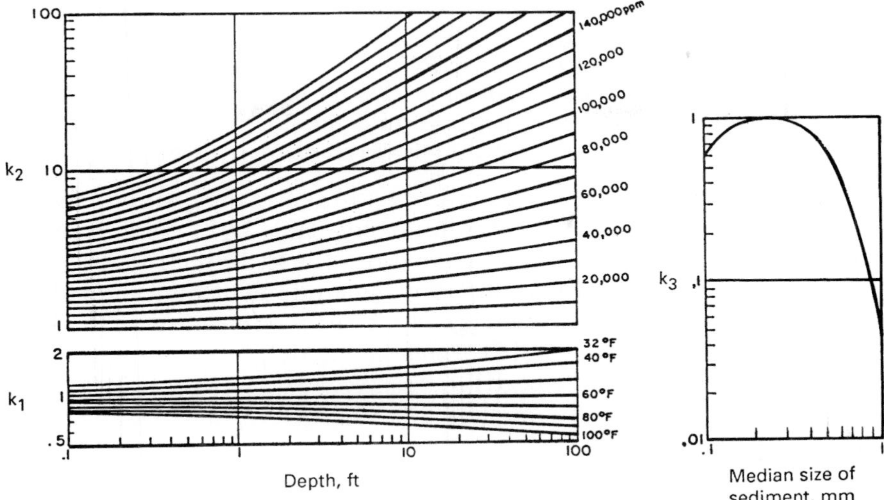

Figure 3.11 Colby's (1964) correction curves for temperature, fine sediment, and median size of bed material.

2. The uncorrected sediment discharge Q_n for the given V, y_0, and D_{50}, can be found from Fig. 3.10 by first reading q_n knowing V and D_{50} for two depths that bracket the desired depth and then interpolating on a logarithmic graph of depth versus q_n, to obtain the bed material discharge per unit width.
3. The two correction factors k_1 and k_2 shown in Fig. 3.11 are then used to account for the effect of water temperature and fine suspended sediment on the bed-material discharge. If the bed-material size falls outside the 0.20- to 0.30-mm range, the factor k_3 from Fig. 3.11 is applied to correct for the effect of sediment size.
4. The unit bed sediment discharge q_T corrected for the effect of water temperature, presence of fine suspended sediment, and sediment size is given by Eq. (3.9).

$$q_T = [1 + (k_1k_2 - 1)k_3]q_n \tag{3.9}$$

The total bed-material discharge is given by

$$Q_T = Wq_T \tag{3.10}$$

where W is the width of the stream.

In spite of many inaccuracies in the available data and uncertainties in the graphs, Colby (1964) found that

> about 75 percent of the sand discharges that were used to define the relationships were less than twice or more than half of the discharges that were computed from the graphs of average relationship. The agreement of computed and observed discharges of sands for sediment stations whose records were not used to define the graphs seemed to be about as good as that for stations whose records were used.

Sediment discharge is measured in natural streams using suspended-sediment samplers developed by the U.S. Interagency Sedimentation Committee (Richardson, 1994). Vanoni (1975) and Simons and Sentürk (1992) describe the equipment and methods of measuring the sediment discharge of a stream or river. Also, the U.S. Geological Survey issues publications giving detailed instructions for measuring sediment transport.

Sediment Yield

Erosion of land results from an overland and stream process. Overland processes are sheetwash, rilling, and gullying, which determine the sediment yield from a land surface. Stream processes are the bed and bank erosion of streams and rivers, as described in the section on sediment transport.

The U.S. Department of Agriculture developed a *universal soil-loss equation* (USLE) by regression analysis for the prediction of the sediment yield from overland flow. Vanoni (1975) and Simons and Sentürk (1992) describe the equation in detail. The USLE equation is

$$A = RKLSCP \tag{3.11}$$

where A = average annual soil loss, tons/acre
R = rainfall factor, in/h
K = soil erodibility factor, tons/acre per unit of rainfall factor R
S,L = topography factors with S for slope (land gradient) and L for length of slope
C = cropping management factor
P = erosion-control-practices factor

Vanoni (1975) and Simons and Sentürk (1992) give graphs for determining the factors and references to the basic documents.

Reservoir Sedimentation

Reservoir sedimentation is the deposit or trapping of the sediment discharge of a stream or river in the backwater upstream from a reservoir. The coarsest fraction of the sediment discharge is deposited in the backwater of the reservoir and at the upper end of the reservoir in a delta formation. The finer fraction of the sediment discharge is deposited in the reservoir and may even deposit at the face of the dam or flush through the reservoir. Fine sediments can move along the bottom of a full reservoir as density currents. The operation of the reservoir has a significant influence on the location and nature of the sediment deposits. If the reservoir is emptied or significantly lowered every year (or every few years), then the coarse sediments can be moved further down into the reservoir. If the dam that forms the reservoir has significant undersluices and can pass a significant amount of the mean annual flow (≥90 percent), the deposition of the sediment behind the dam will be small and the reservoir will have a low trap efficiency (1 to 20 percent). However, if the dam does not have undersluices that can pass a significant amount of the mean annual flow (the reservoir fills with the first part of the annual high flow and then passes excess water over its spillway, or the reservoir has significant volume for year-round storage), it will trap almost all the sediment discharge and will have a large trap efficiency (95 to 100 percent). The rate of depletion of reservoir storage from the storage of the sediment discharge of a stream depends on (1) the storage capacity of the reservoir in relation to inflow, (2) sediment inflow (quantity and size), (3) reservoir and dam characteristics, and (4) the specific weight (density) of the sediment deposits.

Reservoirs with large storage-capacity C ratios to mean annual inflow I (C/I ratios) have large trap efficiencies; however, if the C/I ratio is large (>100 percent), the reservoir will have a lower rate of storage loss (longer useful life) than will reservoirs with a low C/I ratio. For example, Lake Mead on the Colorado River and Lake Nassar on the Nile River have storage volumes that are approximately double their mean annual inflow ($C/I \sim 200$ percent), and trap efficiencies close to 100 percent. Their rate of sediment depletion of storage is so low that their useful life expectancy is measured in centuries. The reservoir behind Tarbella Dam on the Indus River has a storage volume that is less than 0.2 of the mean annual flow ($C/I \sim$ 13 percent) and a trap efficiency of about 100 percent. Its rate of storage depletion is so great that its useful life as a storage reservoir is less than 100 years. All three reservoirs have approximately the same mean annual sediment inflow. The reservoir behind Old Aswan Dam on the Nile downstream of Lake Nassar, with a C/I ratio of less than 3 percent, has an infinite useful life insofar as loss of storage from sedimentation is concerned. It has stored water for over 50 years before Lake Nassar was created by the High Aswan Dam with little or no loss of storage. Its trap efficiency is close to zero because it has undersluices across the total dam width, and most of the annual flood passes through the reservoir with only the tail of the annual flood stored.

Sediment inflow into a reservoir can be estimated by (1) using the measured annual total sediment discharge of the stream, (2) computation by the flow-duration–sediment-rating curve method, and (3) estimating the sediment yield from the watershed.

Use of measured annual total sediment discharge of a stream is limited to those sites that have a historical record based on frequent sampling to establish a reliable estimate of the sediment inflow into the reservoir. Normally, the record of sediment inflow is less than the streamflow record and is only suspended-sediment discharge. However, with studies the suspended-sediment record can be adjusted to obtain the total sediment discharge using the methods previously described.

The flow-duration–sediment-rating curve method extends the available total sediment discharge record to the historical streamflow record and is the most desirable. In this method, a sediment-rating curve is made relating the daily sediment discharge (normally expressed in tons per day) to the daily discharge. There will be a large scatter in the data, some of which may be seasonal, but with study, a reliable set of curves can be developed. A flow-duration curve is also prepared for the entire streamflow record. From these two sets of relations, the average annual sediment discharge is determined (Vanoni, 1975; Simons and Sentürk, 1992). The sediment-rating curve may be developed using measured suspended-sediment discharge, measured total sediment discharge (suspended-sediment plus unmeasured sediment discharge), or a measured suspended-sediment discharge corrected for the unmeasured sediment discharge.

The sediment yield from a watershed is estimated using the USLE given earlier [Eq. (3.11)]. This value is then used with the appropriate trap efficiency to determine sedimentation rates in a reservoir. This method is often used for small reservoirs.

Trap efficiency is the ratio of the sediment stored in a reservoir (trapped) to the quantity of the sediment inflow into the reservoir expressed as a percentage. It depends on (1) the ratio of reservoir volume to mean annual flow volume, (2) characteristics and operation of the dam, and (3) characteristics of the sediment inflow (size). Vanoni (1975) and Simons and Sentürk (1992) present curves that give trap efficiency as a function of the ratio reservoir storage capacity to mean annual inflow (C/I), showing that trap efficiencies increase with an increase in the C/I ratio. In using these curves, engineering judgment must be used to determine trap efficiency for a particular reservoir.

Reservoirs that are formed from large embankments (soil, concrete, or rock), with large storage volumes, over-year storage, and outlets that normally discharge less flow than the incoming floods have large trap efficiencies (close to 100 percent). These reservoirs may have some sediment removed by density currents or large velocities when the reservoir is low, but, considering the other approximations and uncertainties in the determination of the sediment inflow, these amounts are ignored. Small reservoirs with large velocities or reservoirs with undersluices that can pass large inflows will have low trap efficiencies. The Old Aswan Dam, described previously, is an example of the latter. Operation of the dam to have a low pool with high velocities part of each year will decrease trap efficiency.

The *specific weight* of the sediment (weight per unit volume) is used to determine the volume that the sediment inflow into a reservoir will occupy. Total sediment discharge is normally determined by weight. Simons and Sentürk (1992) give an equation suggested by Lane and Koelzer in 1943 and 1953. The equation is

$$\gamma_m = \gamma_1 X_1 + (\gamma_2 + B_2 \log T)X_2 + (\gamma_3 + B_3 \log T)X_3 \qquad (3.12)$$

where γ_m = average specific weight of sediment in the reservoir at the end of any given period (T)
γ_1 = initial specific weight of sand and coarser sediments
γ_2 = initial specific weight of silt
γ_3 = initial specific weight of clay
B_2 = constant rate of compaction of silt
B_3 = constant rate of compaction of clay
T = time, years
X_1 = percent of each class in the deposit

Values of γ_1 and B_1 are given in Table 3.3.

Distribution of sediments in reservoirs depends on (1) the size distribution of the sediment inflow, (2) size of the reservoir, and (3) the operation of the reservoir. Vanoni (1975) and Simons and Sentürk (1992) present methods to determine the location of the sediments in the reservoir.

TABLE 3.3 Values of γ_1 and B_1 (in lb/ft^3)

	Sand		Silt		Clay	
Reservoir operation	γ_1	B_1	γ_2	B_2	γ_3	B_3
Sediment always submerged or nearly submerged	93	0	65	5.7	30	16.0
Normal reservoir drawdown	93	0	74	2.7	46	10.7
Considerable reservoir drawdown	93	0	79	1.0	60	6.0
Reservoir emptied	93	0	82	0.0	78	0.0

Glossary

anabranch A stream whose flow is divided at normal or lower discharges by large relatively permanent islands. The channels are more permanent and more widely and distinctly separated than those of a braided stream.

bed form A relief feature on the bed of a stream, such as dunes, plane bed, or antidunes. Also called **bed configuration.**

bed material Material found on the bed of a stream. May be transported in contact with the bed or in suspension.

bed-material discharge The part of the total sediment discharge of a stream that is composed of grain sizes found in the bed and is equal to the transport capability of the flow.

braided stream A stream whose flow is divided at normal and low flow into several channels by bars, sandbars, or islands. The bars and islands change with time, sometimes with each runoff event. A braided stream has the aspect of a single large channel with several subordinate channels at normal and low flows.

contact-sediment discharge Sediment that is transported in a stream by rolling, sliding, or skipping along in contact with the bed. Also called **bed load** or **contact load.**

critical shear stress The minimum amount of shear (force) exerted by the flow on a particle or group of particles required to initiate particle motion.

discharge Time rate of the movement of a quantity of water or sediment passing a given cross section of a stream or river.

fine-sediment discharge (wash-load discharge) That part of the total sediment discharge of a stream that is not found in appreciable quantities in the streambed. Normally the fine-sediment discharge in a sand-bed stream is composed of particles finer than sand (0.062 mm) in coarse gravel, cobble- or boulder-bed stream silts, clays, and sand could be fine-sediment discharge or wash load.

flow-duration curve A graph indicating the percentage of time a given discharge is exceeded.

fluvial Related to stream or rivers.

geomorphology The branch of physiography and geology that deals with the general configuration (form) of the earth's surface and the changes that take place as the result of the forces of nature.

hydraulic radius The cross-sectional area of a stream divided by its wetted perimeter. Equals the depth of flow when the width is larger than 10 times depth.

meandering stream A stream with sinuous S-shaped flow pattern.

median diameter The particle diameter of which 50 percent of the particles are coarser and 50 percent are finer (D_{50}).

sediment yield The total sediment outflow from a unit of land (field, watershed, or drainage area) at a point of reference and per unit of time.

shear stress, tractive force The force or drag on the channel boundaries caused by the flowing water. For uniform flow, shear stress is equal to the unit weight of water times the hydraulic radius times the slope. Usually expressed as force per unit area.

shear velocity The square root of the shear stress divided by the mass density of water; has the units of velocity.

slope Fall per unit length of the channel bottom, water surface, or energy-grade line.

suspended sediment Sediment particles that are suspended in the flow by the turbulence of the stream.

suspended-sediment discharge The quantity of suspended sediment passing through a stream cross section per unit of time.

total bed-material discharge The sum of the suspended sediment bed-material discharge and the contact sediment (bed-load) discharge.

total sediment discharge The sum of the suspended sediment discharge and the contact sediment (bed-load) discharge, the sum of the bed-material discharge and the wash-load discharge or the sum of the measured sediment discharge and the unmeasured sediment discharge.

unmeasured sediment discharge The sediment discharge that is not measured by suspended-sediment samplers. It consists of the suspended-sediment discharge in the unsampled zone and the contact-sediment discharge. Suspended-sediment samplers normally do not measure to the bed of a stream.

References

Albertson, M. L., 1953, "Effect of Shape on the Fall Velocity of Gravel Particles," *Proc. 5th Hydraulics Conference,* University of Iowa, Iowa City.

Beckman, E. W., and Furness, L. W., 1962, *Flow Characteristics of Elkhorn River near Waterloo, Nebraska,* U.S. Geological Survey Water-Supply Paper 1498B, Washington, D.C.

Beverage, J. P., and Culbertson, J. K. 1964, "Hyperconcentrations of Suspended Sediment," *J. Hydraulics Div. ASCE* (New York) **90**(HY 6).

Brice, J. C., and Blodgett, J. C., 1978, *Countermeasures for Hydraulic Problems at Bridges,* Final Report, Federal Highway Administration, FHWA-RD-78-162, vols. 1 and 2.

Brown, C. B., 1950, "Sediment Transportation," in *Engineering Hydraulics,* H. Rouse, ed., pp. 769–858, Wiley, New York.

Colby, B. R., 1964, *Discharge of Sands and Mean-Velocity in Sand-Bed Streams,* U.S. Geological Survey, Professional Paper 462-A, Washington, D.C.

Culbertson, D. M., Young, L. E., and Brice, J. C., 1967, *Scour and Fill in Alluvial Channels,* U.S. Geological Survey, Open-File Report, Washington, D.C.

Dawdy, D. R., 1961, *Depth-Discharge Relations of Alluvial Streams—Discontinuous Rating Curves,* U.S. Geological Survey Water-Supply Paper 1948C, Washington, D.C.

Einstein, H. A., 1950, *The Bed-Load Function for Sediment Transportation in Open Channel Flows,* U.S. Department of Agriculture Technical Bulletin 1026, Washington, D.C.

El-Tobgy, H. A., 1976, *Contemporary Egyptian Agriculture,* Ford Foundation Report, 2d ed., Cairo, Arab Republic of Egypt.

Gessler, J., 1965, *The Beginning of Bedload Movement of Mixtures Investigated as Natural Armoring in Channels,* W. M. Keck Laboratory of Hydraulics and Water Resources, California Institute of Technology, Pasadena; also Report 69, Laboratory of Hydraulic Research and Soil Mechanics, Swiss Federal Institute of Technology, Zurich, Switzerland.

Lagasse, P. F., Schall, J. D., Johnson, F., Richardson, E. V., Richardson, J. R., and Change, F., 1991, "Stream Stability at Highway Structures," HEC-20, publ. no. FHWA-IP-90-014, McLean, Va.

Lane, E. W., 1955, "The Importance of Fluvial Morphology in Hydraulic Engineering," *Proc. ASCE,* no. 81, Paper 745, pp. 1–17, New York.

Leopold, L. G., and Maddock, T., Jr., 1953, *The Hydraulic Geometry of Stream Channels and Some Physiographic Implications,* U.S. Geological Survey Professional Paper 252.

Leopold, L. B., and Wolman, M. G., 1957, *River Channel Patterns: Braided, Meandering and Straight,* U.S. Geological Survey Professional Paper 282B, Washington, D.C.

Nordine, C. F., Jr., 1964, *Aspects of Flow Resistance and Sediment Transport: Rio Grande Near Bernalillo, New Mexico,* U.S. Geological Survey Water Supply Paper 1498H, Washington, D.C.

Richardson, E. V., 1994, "Sediment Measurement Instrumentation—a Personal Perspective," invited paper in *Fundamentals and Advancements in Hydraulic Measurements and Experimentation,* C. A. Pugh, ed., Proceedings of 1994 Symposium, ASCE, pp. 94–103, New York.

Richardson, E. V., and Davis, S. R., 1995, *Evaluating Scour at Bridges,* 3d ed., HEC-18, publ. no. FHWA-IP-90-017. McLean, Va.

Richardson, E. V., and Hanly, T. F., 1965, "Discussion of Hyperconcentration of Suspended Sediments," *J. Hydraulics Div. ASCE* (New York) **91**(HY 6).

Richardson, E. V., Simons, D. B., and Julien, P. Y., 1990, *Highways in the River Environment,* revision of 1975 publication by Richardson, E. V., Simons, D. B., Karaki, S., Mahmood, K., and Stevens, M. A., publ. no. FHWA-HI-90-016. FHWA, Washington, D.C.

Ritter, J. R., and Helley, E. J., 1968, *An Optical Method for Determining Particle Sizes of Coarse Sediment,* U.S. Geological Survey, Open-File Report.

Rouse, H., 1937, *Nomogram for the Settling Velocity of Spheres,* Report of the Commission on Sedimentation, National Research Council, Washington, D.C.

Schumm, S. A., ed., 1972, *River Morphology. Benchmark Papers in Geology,* Dowden, Huchinson & Ross, p. 429, Stroudsburg, Pa.

Schumm, S. A., 1977, *The Fluvial System,* Wiley, New York.

Simons, D. B., and Richardson, E. V., 1961, "Forms of Bed Roughness in Alluvial Channels," *J. Hydraulics Div. ASCE* (New York) **87**(HY 3), 87–105.

Simons, D. B., Richardson, E. V., and Haushild, W. L., 1963, *Some Effects of Fine Sediments on Flow Phenomena,* USGS Water Supply Paper 1498G, Washington, D.C.

Simons, D. B., and Richardson, E. V., 1966, *Resistance to Flow in Alluvial Channels,* U.S. Geological Survey Professional Paper 422J, Washington, D.C.

Simons, D. B., and Sentürk, F., 1992, *Sediment Transport Technology,* Water Resources Pub., p. 897, Littleton, Colo.

U.S. Corps of Engineers, 1969, *Missouri River Channel Regime Studies,* MRD Sedimentation Series No. 13.B, Omaha, Nebr.

Vanoni, V. A., ed., 1975, *Sedimentation Engineering,* ASCE Manual no. 54, p. 745, New York.

Wolman, M. G., 1954, "A Method of Sampling Coarse River Bed Material," *Trans. Am. Geophys. Union* (Washington, D.C.) **35**(6), 951–956.

Chapter

4

Waterlogging and Salinity

Safwat Abdel-Dayem
Drainage Research Institute
National Water Research Center
Cairo, Egypt

Introduction

Global food production has generally increased during this century because cropped area has expanded and productivity per unit area has increased. However, the growth rate has been decreasing for some time now. About 20 percent of the total population in developing countries are suffering from chronic malnutrition. In sub-Saharan Africa the proportion may increase to 37 percent and in some countries, even higher.

Irrigation and drainage contribute significantly to the increase of crop production. About 17 percent of the cropped lands at present are irrigated; however, they produce about 33 percent of the world food crops. Drainage enhances the productivity of another 10 percent of cropped lands (ICID, 1995).

FAO (1990) reported that out of the 235 million ha currently irrigated, 20 to 30 million ha is severely affected by salinity. An additional 60 to 80 million ha is affected to some extent. The percentage of area affected by salinity of the total irrigated lands in countries with high investments in irrigation is given in the World Bank reports as follows: 10 percent in Mexico, 11 percent in India, 21 percent in Pakistan, 23 percent in China, and 28 percent in the United States

(Umali, 1993). In spite of the decline in the global rate of growth of irrigated lands to 2 million ha per year, 2 to 3 million ha of the irrigated lands goes out of production every year because of salinization.

The extent of areas affected with salinity or in the process of salinization will continue to increase as long as irrigation is practiced unless preventive and corrective measures are considered. Control of water and soil moisture in the field are preconditions for successful crop production. Irrigation development, if done properly, will help contain or perhaps alleviate the threats of hunger, poverty, unemployment, and environmental degradation. Land drainage is essential for sustainability of irrigated agriculture. It is a means for protecting the irrigated lands against waterlogging and salinity. Many countries are now implementing effective drainage projects featuring their will to maintain high productivity of the agricultural lands.

Occurrence of Waterlogging and Salinity

Waterlogging occurs when the water table rises and eventually approaches the soil surface rendering the root zone unsuitable for crop growth. Excessive wetting and waterlogging are generally accompanied by oxygen (O_2) deficiency (Wesseling, 1974; Hillel, 1980; Smedema and Rycroft, 1983) because most pores are filled with water. This deficiency leads to increased resistance to transport of water and nutrients through the plant roots, reduction in root respiration, and formation of toxic compounds in soils and plants. Except for aquatic species, all plants will suffocate from excessive watering and waterlogging. Inadequate aeration in the root zone occurs when the effective air content falls below 5 to 10% by volume.

Plant sensitivity to wet conditions depends on temperature. Crops suffer more from waterlogging under warm weather because oxygen consumption is higher in this case than under cold weather. Moreover, evaporation rate in a warm climate is higher, and hence the hazard of salinity is likely to be greater than in a cool climate. Waterlogging also has an adverse effect on soil biological life and on its structure. This impairs mineralization and nitrification by microbes, resulting in nitrogen deficiency to crops. Farmers try to partially offset this effect by increasing the nitrogen-fertilizer application.

In humid and tropical regions excessive wetting and waterlogging results from intensive rainfall for prolonged periods. In many cold and moderate climates snowmelt water is another major source of excess water. Excess rainfall occurs also in semiarid climates as a result of intense, heavy storms in the rainy season. In arid regions overirrigation and canal seepage are the primary causes of excessive wetting and waterlogging. Waterlogging may also occur as a result of seepage from high-laying lands into neighboring valleys. It is common to see wet areas at the foot of a hill, particularly when poor irrigation is practiced at the higher elevations.

The aim of drainage in humid areas is not only to evacuate the surface runoff after a rain storm but also to lower the moisture content of the upper soil layers to allow air to penetrate more easily to the roots. This will provide the necessary O_2 to the root zone and facilitate transport of carbon dioxide (CO_2) produced by roots, microorganisms, and chemical reactions. Well-drained soils have higher temperatures, which helps crop growth, especially in spring. They also permit easy access to the fields for early planting dates during rainy seasons.

The term *soil salinity* is commonly used to refer to the soluble-salt content of the soil. When soil salinity increases to harmful levels, plants are subject to reduced osmotic potential of the soil solution and to toxicity of specific ions such as boron, chloride, and sodium. The original source of soluble salts are the primary minerals in soils and exposed rocks of the earth's crust (Fireman, 1957). Chemical decomposition and physical weathering release these minerals, which are usually transported by wind or water. Water from all natural surfaces or underground sources, even when of undisputed quality, contains dissolved salts.

In humid regions salinity is of little concern because soluble salts are carried downward by rain into groundwater and ultimately are transported to the oceans by streams. In arid and semiarid regions irrigation is the main cause of salinity problems. The application of irrigation water means an input of salts. The salt applied to the soil with irrigation water remains in the soil unless it is flushed out in drainage water or removed by harvested crops. The quantity of salts removed by crops is usually very small and makes insignificant contribution to salt removal (Schwab et al., 1981). The content of totally dissolved solids in irrigation water or the soil solution is usually expressed either as a concentration or as a corresponding electrical conductivity.

Upward movement of water from a water table may present a serious salinity hazard, especially when the groundwater is saline. The rate of upward movement increases when the water table becomes shallower. When water evaporates from the soil surface, salts accumulate at or near the surface. This process is defined as *secondary salinization*. Salts can be removed and prevented from accumulating only when the water table remains deep enough to permit leaching of salts without resalinization. Continuous irrigation is a main cause of water-table rise. The height of a water-table rise depends on the net deep percolation and the storage capacity of the soil expressed by its drainable porosity. In newly reclaimed lands where the water table is originally deep before the start of irrigation, the rate of water-table rise may be faster than many people would expect.

TABLE 4.1 Rise of Water Table Due to Irrigation

Irrigation project	Water table	
	Original depth, m	Rise, m/yr
Nubariya, Egypt	15–20	2.0–3.0
Beni Amir, Morocco	15–30	1.5–3.0
Murray-Darling, Australia	30–40	0.5–1.5
Amibara, Ethiopia	10–15	1.0
State Farm 29, Xinjang, China	5–10	0.3–0.5
Salt Valley, United States	15–30	0.3–0.5
SCAPP I, Pakistan	40–50	0.4
Bhatinda, India	15	0.6
SCARP VI, Pakistan	10–15	0.2–0.4
Khaipur, Pakistan	4–10	0.1–0.3

SOURCE: FAO (1990).

Table 4.1 shows the increase in groundwater levels resulting from the introduction of irrigation without adequate drainage in a number of projects in different countries (FAO, 1990). When drainage is implemented before introducing irrigation in a newly reclaimed area, it lowers the water table to the design depth in the required time (Fig. 4.1).

Calcium and magnesium are the principal cations found in the soil solution and on the exchange complex of normal soils in arid regions

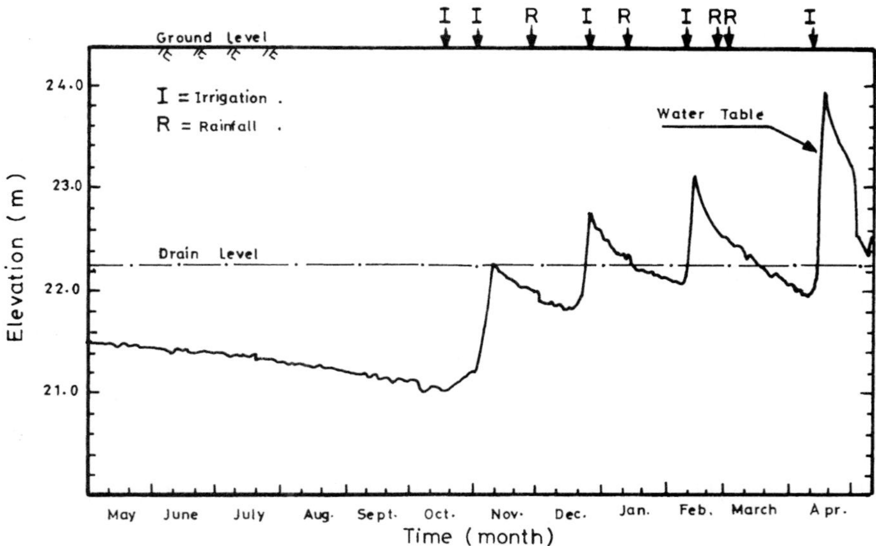

Figure 4.1 Groundwater rise due to irrigation in newly reclaimed lands in the Tigris Valley, Iraq.

(Richards, 1969). When excess soluble salts accumulate in these soils, sodium may be the predominant cation in the soil solution owing to the precipitation of calcium and magnesium compounds. Under such conditions a part of the original exchangeable calcium and magnesium is replaced by sodium. In general, half or more of the soluble cations must be sodium before significant amounts are adsorbed by the exchange complex.

The term *sodicity* refers to the presence of sodium ions on the exchange complex. When sodium ions predominate, the soil aggregates are unstable and are likely to disperse. The dispersement of soil particles causes the closure of the interaggregate pores, thus reducing the soil hydraulic conductivity. Sodicity is associated with an alkaline soil reaction, and therefore it is also referred to as *alkalinity.* When wet, the surface of sodic soils slakes and puddles. On drying, a hard crust is formed hampering seed emergence and crop growth. The sodic soils are further deteriorated by tillage practices such as ploughing and harrowing. They may form a compacted top layer with poor water-transmitting properties.

Characteristics of Salt-Affected Soils

The characterization of saline and alkali soils was first introduced by the Salinity Lab at Riverside, Calif. (Richards, 1969). Classification of salt-affected soils on the basis of these characteristics remains valid to date.

Saline soils are defined as those soils for which the saturation extract is more than 4 dS/m and exchangeable-sodium percentage is less than 15. The pH of these soils is normally less than 8.5. The dominant anions are chlorides and sulfates. Bicarbonates are present in small quantities. Sodium, as a rule, constitutes less than 50% of the soluble cations.

Saline soils are often covered by white crusts of salts on the surface. The presence of excess salts and the absence of significant amount of exchangeable sodium causes flocculation of saline soils. Thus, their hydraulic conductivity is equal to or higher than that of similar nonsaline soils. The reclamation of these soils can be achieved through providing adequate drainage and removing the soluble salts by leaching.

Saline-sodic soils are formed as a result of the combined processes of salinization and alkalinization. The conductivity of the saturated extract of these soils is greater than 4 dS/m, and the exchangeable-sodium percentage is greater than 15. The appearance and properties of these soils are generally similar to those of saline soils as far as excess salts are present. Under such conditions the pH of saline-sodic soils are seldom higher than 8.5 and the particles remain flocculated. Often, saline-sodic soils have a pH value near to neutral.

If the concentration of the soluble salts in saline-sodic soils is lowered by leaching, the exchangeable sodium may change to sodium carbonate. The continuation of this process may turn the soil highly sodic with a pH above 8.5. At this stage the particles disperse and the soil loses its structure and hence becomes unfavorable for the entry and movement of water. However, saline-sodic soils sometimes contain gypsum. When such soils are leached, calcium dissolves and the exchangeable sodium is replaced by calcium and sodium is removed with the excess soluble salts.

Soils for which the exchangeable-sodium percentage is greater than 15 and the conductivity of the saturation extract is less than 4 dS/m are described as non-saline-sodic soils. Their pH usually range between 8.5 and 10. These soils are sometimes referred to as "black alkali" soils. This is because dispersed and organic matter present in the soil solution of highly alkaline soils may be deposited on the soil surface, causing a black color. The soil often contains calcium carbonate, which has low solubility. Thus it is not useful for reclamation purposes unless soil pH is lowered.

Non-saline-sodic soils frequently occur in semiarid and arid areas in small irregular spots which are often described as "slick spots." Alkali conditions develop as a result of irrigation. The highly dispersed sodium-saturated clay at the soil surface may be transported downward through the soil and accumulate at lower levels. Such soils may develop a dense low permeable layer where the clay particles accumulate at a few inches of the soil surface below relatively coarse textured soil layer. The dense clay layer usually has a columnar or prismatic structure.

Water and Salt Balance in Irrigated Land

Water enters the root zone as a result of infiltration of irrigation or rainwater, lateral seepage from surrounding areas, and upward movement from a water table. On the other hand, water leaves the root zone by deep percolation, and evapotranspiration. The difference between the water entering and leaving the root zone in a given time will determine whether there is an increase or decrease of soil moisture stored in the root zone (Fig. 4.2). If a rather long period is considered so that the change in soil moisture may be neglected and assuming no lateral seepage, the water balance of the root zone could be expressed as follows:

$$I + R + U = \mathrm{ET} + P \qquad (4.1)$$

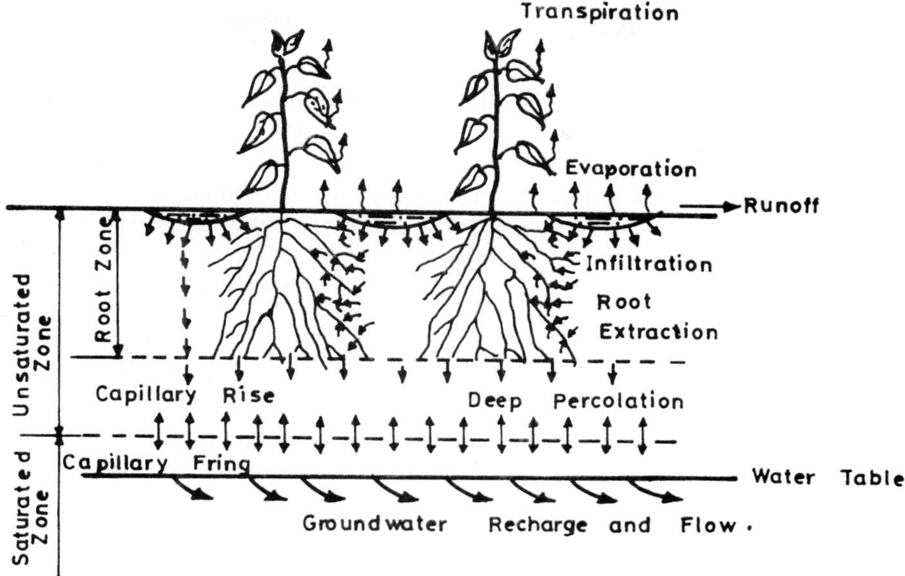

Figure 4.2 Water balance of the root zone.

where I = irrigation
R = rainfall
U = upward flux
ET = evapotranspiration
P = deep percolation

All quantities are expressed in millimeters. The time period should be the same for each component. In Eq. (4.1), I is defined as the effective quantity that really infiltrates into the soil.

The salt balance during the same period depends on the soluble salts in each component of Eq. (4.1). The change of salt content in the root zone is determined by the difference between salt inputs and outputs. If salts are highly soluble and do not precipitate and if the amounts of salts supplied by rainfall and fertilizers and exported by crops are negligible, the salt balance of the root zone then reads:

$$I_{C_i} + U_{C_g} = P_{C_p} + \Delta Z \tag{4.2}$$

where C is salt concentration in milliequivalents per liter; subscripts i, g, and p are suffixes denoting irrigation water, groundwater, and deep-percolation water, respectively; and ΔZ is the change in soil salt content.

If there is no seepage of foreign water to the area under consideration, the salt concentration of the percolating water in a field under irrigation for a long time can be considered equal to the salt concentration of the upward flux (Ritzema, 1994). When it is assumed that the irrigation water is thoroughly mixed with the soil water in the root zone, then the salt concentration of the percolating water will equal the salt concentration of the soil water at field capacity C_{fc}.

The net deep percolation D is the difference between the deep percolation and the upward flux during the time interval in which the water and salt balances are calculated and hence

$$D = P - U \tag{4.3}$$

If the salinity of the root zone is in a state of equilibrium (i.e., $\Delta Z = 0$) and Eqs. (4.1), (4.2), and (4.4) are combined and C_g and C_p are set equal to C_{fc}, the net deep percolation D can be expressed as

$$D = (\mathrm{ET} + R)\,\frac{C_i}{C_{\mathrm{fc}} - C_i} \tag{4.4}$$

The net deep percolation given by Eq. (4.4) represents the quantity of water that should be drained below the root zone in order to maintain its soil salinity unchanged when irrigated with an irrigation quantity I of salt concentration C_i. It is commonly known as the leaching requirement that should be applied as part of the irrigation water for salinity control in irrigated lands. The leaching requirement increases when the salinity of irrigation water is high and when the crop is sensitive to salinity. The drainage quantity D according to Eqs. (4.1) and (4.3) can be expressed as follows:

$$D = I + R + \mathrm{ET} \tag{4.5}$$

Thus, the irrigation quantity that is sufficient to maintain salt equilibrium in the root zone for certain quality of irrigation water (C_i) and crop tolerance level to salinity (C_{fc}) is given by

$$I = (\mathrm{ET} - R)\,\frac{C_{\mathrm{fc}}}{C_{\mathrm{fc}} - C_i} \tag{4.6}$$

The ratio of the net deep percolation D [Eq. (4.4)] and the irrigation quantity I [Eq. (4.6)] is known as the *leaching fraction* (LF), which can be expressed as

$$\mathrm{LF} = \frac{D}{I} = \frac{C_i}{C_{\mathrm{fc}}} \tag{4.7}$$

The value of C_{fc} is usually selected according to the tolerance level of the crop. The leaching requirement should be determined so that the salinity of the soil water cannot exceed the threshold value.

The leaching requirement represents, therefore, a necessary water use for a sustainable irrigated agriculture. It should be considered during the planning of water needs of any irrigation project. Maintaining a favorable salt balance in the soil requires proper and efficient irrigation management. Since drainage water carries salts to downstream users in a canal command or river basin, there is always a general tendency to reduce the leaching fraction to a minimum. The lower the design leaching fraction, the higher will be the requirements for water-management control.

Excessive leaching may be detrimental in that plant nutrients, especially nitrates, may be removed from the soil. It may also carry residues of pesticides. This may lead to pollution of groundwater and downstream river water. Overuse of water also add to the drainage problem by raising the water table. Water application should be sufficient to maintain a favorable salt balance in the soil, but without excessive leaching of plant nutrients and without materially adding to the drainage problem.

Prediction of Soil Salinity Changes

Salts accumulate in the root zone when the salt input is more than the salt output. Salts are brought in by irrigation water and groundwater moving upward and removed by water drained below the root zone [Eq. (4.2)]. Under practical conditions, the salts stored in the root zone at the end of the period under consideration are not the same as at the beginning, particularly for short-term changes. When irrigation or rainfall occurs, the salts are leached to lower layers. During periods of evapotranspiration, the concentration of salts rises in the root zone as soil water is removed by crop and by evaporation. Seasonal variations in climate, crops water applications, and water quality have significant influence on salt storage in the root zone.

If the net result at the end of the season is such that the salt storage is increased, inadequate water management is practiced in that area. On the contrary, proper and efficient water management during the crop growing season will avoid salt accumulation in the root zone. The change in salt quantity ΔZ [Eq. (4.2)] will have a positive value in this case. Downward movement of water is assumed to take place at a water content near field capacity. Thus, it can be assumed that the salt quantity ΔZ is dissolved in amount of soil water at field capacity W_{fc}. The incremental increase of salt concentration of the soil water at field capacity can be determined as:

$$C_{fc} = \frac{\Delta Z}{W_{fc}} \tag{4.8}$$

where C_{fc} is expressed in milliequivalents per liter (meq/L). The electrical conductivity is roughly proportional to the salt concentration and it can be determined in decisiemens per meter (dS/m) by dividing the salt concentration in meq/L by a value of about 12 (van Hoorn and van Alphen, 1994). Then Eq. (4.8) can be rewritten as

$$\mathrm{EC}_{fc} = \frac{\Delta Z}{12\ W_{fc}} \tag{4.9}$$

It is more common to express the soil salinity by the electrical conductivity of extract of the saturated paste. For most soils, with the exception of sand and loamy sand, the water content at saturation is about twice that at field capacity. The change in soil salinity expressed as change in the electrical conductivity of the saturated paste is given by

$$\mathrm{EC}_{e} = \frac{\Delta Z}{24\ W_{e}} \tag{4.10}$$

where EC_e is the electrical conductivity of the extract of the saturated paste and W_e is the water quantity of the root zone. The soil salinity in the root zone at the end of the period considered is then given by

$$\mathrm{EC}_{e1} = \mathrm{EC}_{e0} + \Delta\mathrm{EC}_{e} \tag{4.11}$$

where EC_{e0} and EC_{e1} are the initial and final soil salinities, respectively.

The salt-balance approach described above is too simplistic and approximate. It does not account for the effect and reciprocal effects of soil-water movement and solute transport (Hillel, 1980). Soil water carries its solute load in its convective stream, leaving some of it behind to the extent that the component salts are adsorbed, taken up by plants, or precipitated whenever their concentration exceeds their solubility. Highly soluble salts such as NaCl, $MgSO_4$, and $CaCl_2$ precipitate from soil solution only at salt concentrations that are far too high to impede plant growth. The slightly soluble salts precipitate at concentrations too low to cause damage to plants. The most slightly soluble salts are carbonates of magnesium and calcium (MgCO and $CaCO_3$) and gypsum ($CaSO_4 \cdot 2H_2O$). The presence of these salts in irrigation water may have a considerable influence on the salt balance in the soil.

A more elaborate deterministic approach for solute transport in the unsaturated zone is based on an adjective-dispersive-reactive three-dimensional equation, which reads:

$$\frac{\partial \theta C}{\partial t} = \Delta(\theta D_h \cdot \Delta c - q) + R(C) + \Gamma_c \tag{4.12}$$

where θ = soil-water content
C = volume-average solute concentration (mg/L)
D_h = second-rank hydrodynamic dispersion tensor (cm^2/t)
$R(C)$ = general solute reaction term (mg t^{-1} L^{-1})
Γ_c = solute sources (mg t^{-1} L^{-1})
t = time

For a one-dimensional flow system involving movement of nonreactive solute in a vertical direction, Eq. (4.12) reads

$$\theta \frac{\partial C}{\partial t} = \frac{\partial}{z}\left(\theta\, D_{hz} \frac{\partial C}{\partial z}\right) - q_z + \frac{\partial C}{\partial z} + \Gamma_c \tag{4.13}$$

where D_{hz} is a hydrodynamic dispersion coefficient in the vertical direction (cm^2/t) and q_z is the velocity (cm/t). Equation (4.6) is considered the basic differential equation for solute transport in the unsaturated zone. The solution of Eq. (4.13) gives the salt concentration at any depth in the soil profile.

Kandil (1992) used a mass-balance approach to solve Eq. (4.13) numerically for total dissolved salt concentrations in the soil profile. The solution was coupled with the field-scale water-management simulation model DRAINMOD (Skaggs, 1975) to estimate daily water-table depth, fluxes, and water-content profiles. The DRAINMOD-S (Kandil et al., 1992) model can predict the water-table levels induced by a certain water management on a specific soil type under prevailing climatic conditions. The corresponding changes in salt concentration of the different layers in the shallow soil profiles (Fig. 4.3) and drain effluent can also be predicted for the same period. DRAINMOD-S and similar simulation models are useful tools for design and evaluation of water-management systems. They are especially useful in the planning of irrigation and drainage systems on sustainable basis. They are helpful in assessing the future trends of water-table levels and soil salinity.

Waterlogging, Soil Salinity, and Crop Yield

Few plants can survive in a waterlogged or salt-affected soil. Plant species differ widely in their tolerance to a high water table or a high

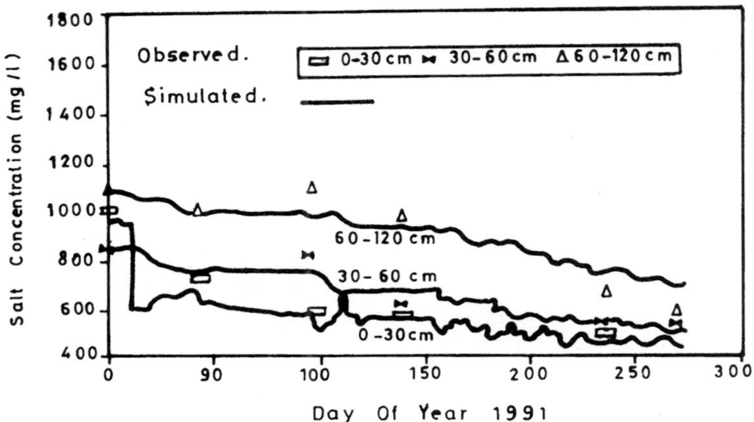

Figure 4.3 Prediction of long-term changes of salinity in the soil profile.

salt concentration. In a real soil water–plant system conditions vary continuously. Water management during the growing season determines the depth of the water table and its change with time and effects on salt accumulation in the soil profile.

The groundwater-table depth determines the state of moisture conditions during the entire growing season and therefore influences water supply, aeration conditions, and heat properties in soil. Thus, it integrates the effect of the prevailing physical conditions on plant growth. The effect of groundwater-table depth on crop yield can be easily determined by comparison with determinations of other soil properties such as aeration or thermal conductivity.

Crop yield decreases as depth of groundwater table decreases at shallow depths because the crops suffer from a lack of aeration. However, this effect depends on the soil physical properties such as water-holding capacity, pore size, and hydraulic conductivity. It also depends on plant species. Shallow-rooted crops demand a higher water table than do deep-rooted crops. The relation between yield and water-table depth depends largely on climatic conditions. In a dry year the optimum crop yield occurs at a shallower water-table depth. It is therefore difficult to transfer the results from one location to another.

An average water-table depth during the growing season cannot be used to assess the effect on yield because the effect of a temporarily high water table may not be recognized. A distinction should be made between a constant water table situated at some shallow depth below the soil surface at all times and a fluctuating water table that is allowed to rise into the root zone. Under fluctuating water-table conditions the concept of SEW_{30} (sum of excess water table rises above 30 cm depth) is used as a measure for excess moisture in the root zone during the growing season (Wesseling, 1974). SEW_{30} is expressed mathematically as

$$SEW_{30} = \sum_{i=1}^{n} (30 - x_i) \tag{4.14}$$

where x_i are daily water-table depths below surface during certain periods of the growing season. This computation considers only x_i values smaller than 30 (Wesseling, 1974). Large SEW values indicate poor drainage conditions.

The relationship between soil salinity and crop yield is determined in experiments under controlled environments. The results are published in many irrigation and drainage literature (FAO, 1985). Tolerance to soil salinity of some selected crops are given in Table 4.2. The published crop-tolerance values remain approximate values that can be used as guidelines for planning water-management systems or predicting future effects of water management on soils and crops. When the soil salinity exceeds a threshold value of a crop tolerance to salinity (Fig. 4.4), plant growth is suppressed. For any increase in soil

TABLE 4.2 Crop Tolerance to Soil Salinity (Saturated Extract dS/m)

	Expected yield decrease			
Crop	0%	10%	25%	50%
Barley	8.0	10.0	13.0	18.0
Cotton	7.7	9.6	13.0	17.0
Sugar beet	7.0	8.7	11.0	15.0
Wheat	6.0	7.4	9.5	13.0
Rice	3.0	3.8	5.1	7.2
Corn	1.7	2.5	3.8	5.9
Beans	1.0	1.5	2.3	3.6
Tomato	2.5	3.5	5.0	7.6
Potato	1.7	2.5	3.8	5.9
Pepper	1.5	2.2	3.3	5.1
Onion	1.2	1.8	2.8	4.3
Clover, berseem	1.5	3.2	5.9	10.3
Alfalfa	2.0	3.4	4.5	8.8
Fig	2.7	3.8	5.5	8.4
Orange	1.7	2.3	3.2	4.8
Apple	1.7	2.3	3.3	4.8
Peach	1.7	2.2	2.9	4.1
Apricot	1.6	2.0	2.6	3.7
Grape	1.5	2.5	4.1	6.7

SOURCE: FAO (1985).

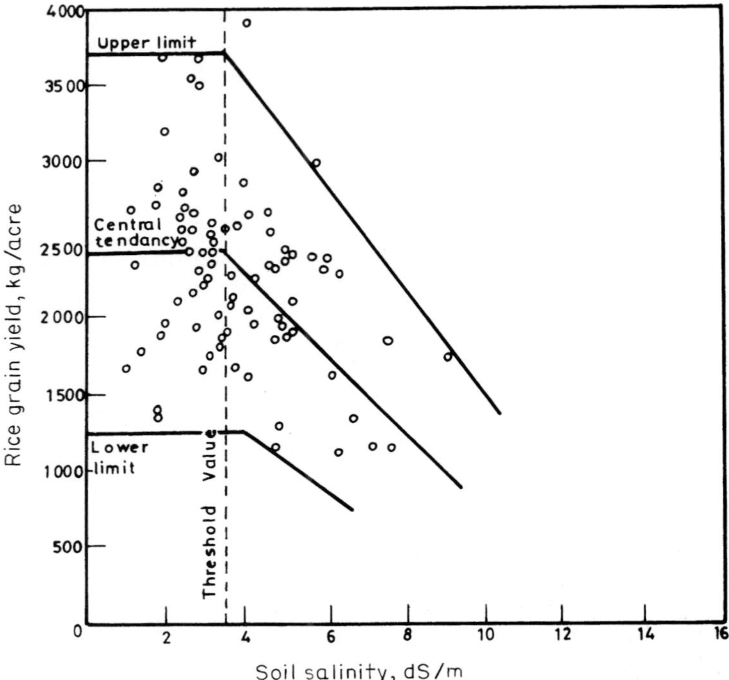

Figure 4.4 Relationship between crop yield and soil salinity from field observation.

salinity beyond the threshold value, the yield linearly decreases (Maas and Hoffman, 1976). Under real field conditions, yield is not influenced by salinity alone but also with many agronomic, climatic, and soil factors. This may explain the wide range of variations in the results given in Fig. 4.4.

It was noticed that a relatively shallow water table has insignificant effect on yield when a net downward flux is maintained during the crop season (Abdel-Dayem and Ritzema, 1990). A monitoring program in the Nile Delta showed that crop yields were not affected by a water table which fluctuates between the soil surface and a depth of 120 cm in the presence of an effective subsurface drainage system. Deeper water tables did not result in any increase in crop yield. It was even reported that a water table with an average seasonal depth may increase the deep percolation and reduce the soil water available to the crops leading to a reduced irrigation efficiency (Oosterbaan and Abu Sinna, 1990). The most important factor in such shallow watertable management is to maintain a low salinity level during the crop-growing season.

Oosterbaan (1994) reported that in semiarid regions with distinct wet and dry seasons, the drainage design may be restricted to the wet season. Heavy rains during the wet season would be sufficient to leach any salts that would accumulate in the soil profile during the dry season. However, personal communications of the author indicates that there are signs from India (RAJAD Project) that secondary salinization appeared in some project areas that were provided with drains to control the water table during the wet period. The results of the ongoing research in these areas are expected to develop more relevant design criteria for waterlogging and salinity control in such areas. Similarly, design criteria for drainage in the tropics are still needed. The International Program on Technology Research in Irrigation and Drainage (IPTRID) of the World Bank is initiating a research program to study the drainage requirements of the tropics in East Asia.

Effects of salinity may vary with the growing stage of the plant. For many crops the germination and early seedling stage is the most sensitive. Salinity of saturated extract in excess of 4 dS/m in the area of germinating seeds of sugar beets, rice, wheat, barley, and several vegetables may delay or inhibit their germination and early growth. However, the fact that salt accumulates more at seed depths than at lower depths, rather than increased sensitivity of plants to salts at the germination stage, might be the main cause of these problems. Special attention must be given to reduce soil salinity at seed planting depths to safe levels before planting. It is common practice in arid regions to apply a presowing irrigation to leach the salts accumulated during the previous fallow period, especially in the presence of a shallow water table. Seedling stages may be more salt-sensitive. Adequate leaching and drainage are therefore of special importance for improved crop production.

Reclamation of Salt-Affected Soils

Restoration of the productivity of waterlogged and saline soils that have developed in irrigated lands requires extensive investment of time and money. A careful assessment of the problem is required. It is necessary to start with detailed investigations that determine the causes of soil salinization which may be a result of the presence of a shallow water table, poor-quality irrigation water, presence of marine sediments, or poor water management. Topographic survey maps of adequate scale and aerial photographs are needed. Soil survey should be carried out in the project area to define the soil physical and chemical characteristics, and dominant ions. The state of deterioration of the soils, the extent of salinization, and the condition of the present irrigation and drainage infrastructure determine the costs and time of reclamation.

Poor drainage is a common feature of salt-affected soils. The first measure for reclamation is to install a drainage system to control the water table. The design of the drainage system should meet the soil and hydrologic conditions in the area during and after the reclamation phase. Drainage theories and design procedures to determine the drain depth and spacing have been developed and are well known (Smedema and Rycroft, 1983; Ochs and Bishay, 1992; Ritzema, 1994). Recently, more elaborate approaches have been developed for drainage-system design and evaluating their long-term effects. The new approaches aim toward satisfying the agriculture and environmental objectives at the same time (Skaggs, 1992). They remove the least amount of water necessary to satisfy the drainage functions that, if exceeded, it may cause drought stresses and increase pollutant loads in the receiving water streams. This requires a better understanding of drainage and water-table management suitable for the soil, crop, hydrological, and climatological conditions. Moreover, they consider the interactions of fertility, cultural, and water-management practices and their impact on water quality and productivity. These approaches depend on modeling water-management systems for crop production under different climatic conditions (Skaggs, 1975; Feddes et al., 1988; Kandil, 1992).

Reclamation of salt-affected soils requires the application of enough irrigation water to leach the salts from the soil. Leaching of saline soils is traditionally carried out by ponding water on the field for long periods of time. Intermittent leaching is also practiced, especially for clay soils. Saline water can be effectively used in fir stages of leaching, especially for saline-sodic soils. The presence of excessive amounts of exchangeable sodium requires application of chemical amendments containing calcium or products that enhance the solubility of calcium carbonate if present in the soil. Field leaching tests are usually carried out on the different soil types in the project to determine the leaching requirements and leaching efficiency coefficient (Dieleman, 1963). The depth of water required and the time needed for reclamation will vary from one place to another. They depend on soil texture and structure, soil pore geometry, cracking phenomena, initial salt content and the chemical composition of the soluble salts in the soil, salt content and chemical composition of the water used for leaching, leaching technique, and the scope for planting a salt-tolerant crop during reclamation.

Economic feasibility of the reclamation project depends largely on the prevailing conditions, which may differ and vary from project to project and from one country to another. The financial and economic appraisal is very important for decision making. It may be economically advisable to consider not reclaiming the area. The costs of reclamation in an area already under cultivation cover the construction of field and main drainage systems, adaptation or improvement of irrigation systems, land leveling, desalinization and improvement of soils, and other adjustment measures for developing the project area. If chemical amendments are required, additional costs will be required. The returns and benefits include higher gross yields in agriculture, increasing revenue in the ancillary industries, new employment opportunities, and enhanced purchasing power. In addition, the higher standards of living resulting from developing the area will improve hygienic conditions. There are other valid justifications for the high costs involved in land reclamation of salt-affected soils, such as realizing food security at the national level. On the other hand, reclamation of salt-affected soils would be difficult, if not impossible, when there is a shortage of water for leaching.

The drainage water resulting from leaching salt-affected soils contains high salt load. It has to be evacuated and disposed outside the project area. When this water is disposed to a river or a canal whose water is used downstream, it may have significant effects. The downstream users may seriously suffer. If the water is used for drinking purposes, there should be criteria specifying the upper limits of allowable salt concentrations and restrictions on certain elements which may have atoxic effects. If water is used in the downstream for irrigation, similar criteria will be required and moreover, the downstream user will be entitled to additional water quantities as the leaching requirements will increase with increase of the salinity of the irrigation water. When the drainage water is disposed into an inland lake without an outlet which acts as evaporation pond, the highly saline drainage water may disturb its salt balance and cause damage to the aquatic life in the lake. Disposal of the saline drainage effluent without considerable environmental side effects can be prohibitively expensive.

References

Abdel-Dayem, S., and Ritzema, H. P., 1990, Verification of drainage design criteria in the Nile Delta, Egypt. *Irrigation and Drainage Systems,* **4**(2), 117–131.

Dieleman, P. J., 1963, *Reclamation of Salt Affected Soils in Iraq.* ILRI publ. no. 11, International Institute for Land Reclamation and Improvement, Wageningen.

FAO, 1985, *Water Quality for Agriculture* (rev. 1). FAO Irrigation and Drainage Paper no. 29, Food and Agriculture Organization of the United Nations, Rome.

FAO, 1990, *Water and Sustainable Agricultural Development: An International Action Program. A Strategy for the Implementation of the Mar del Plata Action Plan for the 1990s,* Food and Agriculture Organization of the United Nations, Rome.

Feddes, R. A., Kabat, P., van Bavel, P. J. T., Bronswiijk, J. B., and Halbetsma, 1988, Modeling soil water dynamics in the unsaturated zone-state of the art. *J. Hydrol.* **100,** 69–111.

Fireman, M., 1957, Salinity and alkali problems in relation to high water tables in soils. In *Drainage of Agricultural Lands,* J. N. Luthin, ed., *Agronomy,* vol. 7, American Society of Agronomy, Madison, Wisc., pp. 505–513.

Hillel, D., 1980, *Application of Soil Physics,* Academic Press, New York.

ICID, 1995, *WATSAVE Activities, Some Lessons and Experiences,* International Commission on Irrigation and Drainage, New Delhi, India.

Kandil, H. M., 1992, DRAINMOD-S: A water management model for irrigated arid lands, Ph.D. thesis, Department of Biological and Agricultural Engineering, North Carolina State University, Raleigh.

Kandil, H. M., Skaggs, R. W., Abdel-Dayem, S., Aiad, Y., and Gilliam, J. W., 1992, *DRAINMOD-S: Water Management Model for Irrigated Lands, 1—Theories and Tests,* Paper no. 922566, ASAE winter meeting, Nashville, Tenn.

Mass, E. V., and Hoffman, G. J., 1976, Crop salt tolerance—current assessment. *Journal of Irrigation and Drainage Division* (ASCE) **103,** 115–134.

Ocks, W. J., and Bishay, G. B., 1992, *Drainage Guidelines,* World Bank Technical Paper no. 195, The World Bank, Washington, D.C.

Oosterbaan, R. J., and Abu Sinna, M., 1990, *Using SALTMOD to Predict Drainage and Salinity in the Nile Delta,* ILRI Annual Report 1989, Wagningen, pp. 63–75.

Oosterbaan, R. J., 1994, Agricultural drainage criteria. In *Drainage Practices and Applications,* H. P. Ritzema, ed., ILRI publ. no. 16, 2d ed., Wageningen, The Netherlands, pp. 635–687.

Ritzema, H. P., 1994, *Subsurface Flow to Drains. Drainage Principles and Applications,* ILRI publ. no. 16, 2d ed., International Institute for Land Reclamation and Improvement, Wageningen, The Netherlands, pp. 263–303.

Richards, L. A., 1969, *Diagnosis and Improvement of Saline and Alkali Soils,* Agriculture Handbook no. 60, United States Salinity Laboratory, U.S. Department of Agriculture, Washington, D.C.

Schwab, G. O., Frevert, R. K., Edminster, T. W., and Barnes, K. K., 1981, *Soil and Water Conservation Engineering,* Wiley, New York.

Skaggs, R. W., 1975, *A Water Management Model for Shallow Water Table Soils,* University of North Carolina Water Resources Research Institute, Technical Report no. 134, Chapel Hill.

Skaggs, R. W., 1992, Drainage and water management modeling technology. In *Drainage and Water Table Control, Proceedings of the Sixth International Drainage Symposium,* Nashville, Tenn., pp. 1–11.

Smedema, L. K., and Rycroft, D. W., 1983, *Land Drainage: Planning and Design of Agricultural Drainage Systems,* Batsford Academic and Educational Ltd, London.

Umali, D. L., 1993, *Irrigation-Induced Salinity: A Growing Problem for Development.* World Bank Paper no. 215, The International Bank for Reconstruction and Development/the World Bank, Washington, D.C.

van Hoorn, J. W., and van Alphen, J. G., 1994, Salinity control. In *Drainage Principles and Applications,* H. P. Resume, ed., ILRI Publ. no. 16, 2d ed., Wagningen, pp. 533–600.

Wesseling, J., 1974, Crop growth and wet soils. In *Drainage for Agriculture,* Jan van Schilfgaarde, ed., *Agronomy* no. 17, American Society of Agronomy, Madison, Wisc.

Chapter

5

Groundwater and the Environment

Ashim Das Gupta
Professor, Water Resources Engineering
School of Civil Engineering
Asian Institute of Technology
Bangkok, Thailand

Introduction

Groundwater has always been considered to be a readily available source of water for domestic, agricultural, and industrial use. In many parts of the world, groundwater extracted for a variety of purposes has made a major contribution to the improvement of the social and economic circumstances of humans. However, with increase in demand, this resource is being overexploited in many areas, resulting in permanent depletion of the aquifer system and associated environmental consequences such as land subsidence and water-quality deterioration. Moreover, with changes in land use and a vast increase in quantities and types of industrial, agricultural, and domestic effluent entering the hydrologic cycle, a gradual decline in water quality due to surface and subsurface pollution is observed. On the other hand, in some areas, water table builds up in aquifers underlying large irrigation areas as aquifers receive more water through recharge than they lose. As a consequence, productivity of irrigated lands decreases as a result of waterlogging and soil salinity.

Emergence of these problems, which have rightly captured the attention of all those concerned, has made it clear that planning and policymaking for the utilization of groundwater have not been rational and comprehensive enough. Since groundwater utilization will continue in the future, there is a need to stress the interaction between groundwater and the environment, to identify some critical gaps in our past efforts toward groundwater management if we want to avoid or significantly reduce the adverse consequences, while continuing to bear the fruits of the resource use on a sustainable basis over a long period of time.

The study of groundwater, in particular, where environmental implications are important is a multidisciplinary activity encompassing scientific, engineering, and economic considerations based on proper conceptualization of the functioning of the physical system followed by a detail quantitative analysis. With an introduction on the characteristics of groundwater flow, groundwater use, and the role of groundwater in the water-resource system, an elaboration on environmental issues and environmental assessment is provided. Groundwater management has historically been based on the safe or sustainable yield of the aquifer system. However, with the complexity of the groundwater system and the widespread use of the resource, a detail hydrogeological data collection, analysis, and evaluation of groundwater resources should be performed in an integrated manner (with feedback between activities) with proper respect to the objectives to be achieved for a particular system. The section on groundwater management therefore emphasizes general principles, data requirement, role of modeling and monitoring in management, and management strategy.

Flow-system characterization

Groundwater is an integral part of the hydrologic cycle. Water enters a groundwater system by infiltration of surface water through the ground surface to the saturated zone of the aquifer. Unless the top of the saturated zone is at or very close to the ground surface, the infiltration water migrates through the surface soil, called *unsaturated zone,* down to the saturated zone. Once in the saturated zone, the water moves laterally as well as vertically in response to hydraulic forces acting on it until it leaves through the ground surface in the form of springs or is removed through abstraction points such as boreholes or is released directly into surface water bodies such as rivers, lakes, and oceans.

A groundwater flow system, therefore, consists of aquifers which are water-yielding strata and of confining beds which are geologic units restricting flow of water. Water enters the flow system in recharge areas and moves through the aquifers and confining units, according to their hydraulic properties and the hydraulic gradients, to discharge areas (Heath, 1983). Climate, topography, and geology determine the nature and location of areas of recharge and discharge. Aquifers in humid environments are replenished from infiltration of rainfall through downward percolation of soil water past the plant root zone; seepage from streams, rivers, and lakes also contributes to groundwater recharge. However, for arid environments, percolation from the beds of streams, rivers, and alluvial fans constitutes the source of groundwater recharge.

Toth (1963) observed that, in many cases, groundwater flow systems can be qualitatively subdivided into paths of local, intermediate, and regional flows. He defined a local flow path as a relatively shallow path extending from a recharge area to an adjacent lower discharge area, sometimes hundreds of meters or less in length. Intermediate flow paths include at least one local flow system between their respective points of recharge and discharge and are somewhat deeper and longer than local flow paths, perhaps on the order of one or few kilometers. Regional flow paths are the longest and deepest paths. They begin at the major groundwater divide and traverse the entire region to the major drain.

Hydrogeologic framework. Groundwater occurs in many different geologic formations. Nearly all rocks in the upper part of the earth's crust, whatever their type, origin, or age, possess openings called *pores* or *voids*. In unconsolidated granular materials, voids are spaces between the grain, which may become reduced by compaction and cementation. In consolidated rocks, the only voids may be the fractures or fissures, which are generally restricted but may be enlarged by solution. The volume of water contained underground depends on the percentage of these openings or pores or voids in a given volume of the formation, which is termed the *porosity* of the underground formation.

The geometric arrangement of aquifers and confining beds in an area is determined by the stratigraphy and geologic structure. *Stratigraphy* is the branch of geology that is concerned with the composition, sequence, and correlation of stratified formations. *Geologic structure* is concerned with folding and fracturing of rocks due to movements of the earth's crust. Folding of the rocks can cause jointing or cleavage that increases permeability, or it can close up primary openings and decrease permeability. Faults can disrupt the continuity of aquifers or confining units and thereby restrict groundwater flow. Alternatively, fault zones can be permeable and enhance groundwater flow.

Groundwater occurrence and availability is governed largely by the state of compaction of the formation and cementation which controls the pore volume and by its transmissive property. With these considerations, the geologic formations encountered in nature can be broadly divided into three categories: the unconsolidated, the semiconsolidated, and the consolidated formation. Unconsolidated formations are usually of alluvial, glacial, lacustrine, marine, or aeoline deposition. In many areas aquifers of alluvial origin are important sources of water supply. These deposits are composed of particles of gravel, sand, silt, or clay size, laid down by physical processes in river channel or on floodplains. Sandstone and conglomerate are the consolidated equivalent of sand and gravel, sometimes in semiconsolidated form. About 25 percent of the sedimentary rock of the world is sandstone. In many countries, sandstone strata form regional aquifers that have vast quantities of potable water. Sandstone bodies of major hydrologic significance owe their origin to various depositional environments, including floodplain, marine shoreline, and deltaic, aeoline, and turbidity-current environments. Consolidated formations are hard rock formations composed mainly of crystalline rocks, granites, gneisses, schists, and basaltic formations. The rocks in these formations have no primary pore spaces. However, appreciable quantities of groundwater are available in many places in cracks and fissures of these formations.

Physical properties of the system. Groundwater movement is dependent on the hydraulic properties of the formation and the hydraulic energy gradients across the flow domain. Groundwater flow rates are generally very slow and are often measured in meters per year. Thus, residence times for a groundwater system may range from months to many hundreds of years. For certain types of media, the rate of groundwater flow may be significantly higher. In fractured media in which the dominant flow is through the fracture network or in karstic formations containing enlarged conduits, groundwater may travel hundreds or thousands of meters in a few days. The directions of flow are dependent largely on the geometry and hydraulic characteristics of the geologic formations constituting the aquifer system and the topography of the land surface overlying the formation coupled to the available sources of infiltration or recharge water. The physical characteristics of aquifer which govern the rate of flow in aquifer as well as the water-release capability of aquifer matrix are hydraulic conductivity and storage coefficient. Because of the heterogeneous nature of the formation, hydraulic conductivity and storage-coefficient values are functions of space. In addition, hydraulic conductivity is a function of direction of measurement, and, in general, for sedimentary deposits, horizontal hydraulic conductivity is greater than vertical hydraulic conductivity. Table 5.1 provides the range of values for physical characteristics of different formations. The vertical hydraulic conductivities listed in section II of Table 5.1 indicate that the vertical

TABLE 5.1 Representative Hydraulic Characteristics

I. Material	Horizontal hydraulic conductivity, m/d
Gravel	40–1200
Basalt	4×10^{-8}–800
Limestone	8×10^{-4}–800
Sand	4–120
Sand, dune	4–20
Peat	0.16–4
Sandstone	4×10^{-3}–2
Loess	8×10^{-5}–0.8
Clay	8×10^{-6}–8×10^{-2}
Till	2×10^{-5}–4×10^{-2}
Shale	4×10^{-7}–4×10^{-3}
Coal	4×10^{-2}–40
II. Material	**Vertical hydraulic conductivity, m/d**
Sand, gravel, and clay	4×10^{-3}–4×10^{-2}
Clay, sand, and gravel	4×10^{-4}–3×10^{-3}
Clay	2×10^{-5}–4×10^{-5}
Shale	4×10^{-9}–4×10^{-5}
III. Material	**Specific yield**
Peat	0.30–0.50
Sand	0.20–0.40
Gravel	0.10–0.35
Loess	0.15–0.35
Silt	0.01–0.30
Sandstone	0.10–0.40
Sandy clay	0.03–0.20
Clay	0.01–0.20
Igneous, weathered	0.20–0.30
Siltstone	0.01–0.35
Limestone	0.01–0.25
IV. Material	**Confined storage coefficient (m = aquifer thickness)**
Clay	$1 \times 10^{-4} \times m$
Sand and gravel	$1 \times 10^{-5} \times m$
Rock, fissured	$1 \times 10^{-6} \times m$

SOURCE: Adapted from Walton (1987).

hydraulic conductivity of a clay containing sand and gravel will have a value of about 1 mm/d. Therefore, with a typical vertical hydraulic gradient in a heavily exploited aquifer of 1:1, the downward vertical flow can be of the order 1 mm/d. Another important fact implicit in section IV of the table is that the confined storage coefficient of an alluvial aquifer 100 m thick containing clays and sands will be between 0.001 and 0.01. This is far higher than the confined storage coefficient of consolidated aquifers (~0.0001).

TABLE 5.2 Data Pertinent to Prediction of Groundwater Flow

Physical framework
Hydrogeologic map showing areal extent and boundaries of aquifer
Topographic map showing surface-water bodies
Water table, bedrock configuration, and saturated thickness map
Hydraulic conductivity map showing aquifer and boundaries
Hydraulic conductivity and specific storage map of confining bed
Map showing variation in storage coefficient of aquifer
Relation of stream and aquifer (hydraulic connection)
Stresses on system
Type and extent of recharge areas (irrigated areas, recharge basins recharge wells, impoundments, spills, tank leaks, etc.)
Surface-water diversions
Groundwater pumpage (distributed in time and space)
Streamflow (distribution in time and space)
Precipitation and evapotranspiration
Observable responses
Water levels as function of time and position
Other factors
Economic information about water supply
Legal and administrative rules
Environmental factors
Planned changes in water and land use

SOURCE: Adapted from Bouwer et al. (1988).

The groundwater flow distribution within an aquifer may vary substantially over time in response to changes in the natural recharge/discharge distribution and through development of the aquifer system as a source of water supply. The data required to assess the flow processes are given in Table 5.2. In general, these data requirements include a geometric description of water-bearing formations, storage and transmissive properties, and source and sink information, such as wells. For any particular area, it is rare to have all information available. A field data-collection program has to be initiated to gather as much information as possible.

Transport processes. Shallow aquifers are usually the most important sources of groundwater for water-supply purposes, but the upper portions of these aquifers are also most susceptible to pollution. The entry of contaminant to shallow aquifers occurs directly through wells, by downward percolation through the zone of aeration, by induced recharge from surface water bodies, by interaquifer flow, and by upconing of deeper saline water due to overpumping of wells. The extent of groundwater contamination due to waste percolation from the land surface depends strongly on the rate and volume of recharge water. In a semiarid or arid environment, contaminants may be retained above the water table in a nearly permanent fashion. On the other hand, in humid areas, contaminants may be rapidly carried downward from the land surface to the water table. In a homogeneous porous medium, percolating water will pass vertically through the unsaturated zone. However, in a heterogeneous, stratified medium, percolating water may become perched above layers of low permeability. In this situation, lateral movement for substantial distances can occur above the water table.

As with flow-system characterization, contamination characterization begins with understanding the controlling processes and the data required to define these processes. Nonreactive (conservative) dissolved contaminants in saturated porous media are controlled by the following factors:

1. *Advection (or convection).* This mechanism causes contaminants to be transferred by the bulk motion of the groundwater.
2. *Mechanical (or kinematic) dispersion.* This process involves mechanical mixing caused by pore-scale heterogeneity.
3. *Molecular diffusion.* This mechanism causes the contaminant molecules or ions to move opposite the direction of concentration gradient. This movement is caused by the random kinetic motion of the ions or molecules.

The combined effect of mechanical dispersion and molecular diffusion is known as *hydrodynamic dispersion. Dispersion* is the overall process that causes the zone of contaminated groundwater to occupy a greater volume than if the contaminant distributions were influenced only by advection. If a slug of contaminant enters the groundwater system, advection causes the slug to move in the direction of groundwater flow. Hydrodynamic dispersion causes the volume of the contaminated zone to increase and the maximum concentration in the slug to decrease.

For reactive contaminants, there are additional processes which affect transport. They include adsorption, desorption, and chemical or biological reactions.

1. *Adsorption or desorption.* These processes involve mass transfer of contaminants. *Adsorption* is the transfer of contaminants from the groundwater to the soil; *desorption* is transfer of contaminants from the soil to the groundwater.
2. *Chemical reactions.* These processes involve mass transfer of contaminants caused by various chemical reactions (e.g., precipitation and dissolution, oxidation and reduction). For some contaminants, degradation is also an important process that may need to be characterized.

TABLE 5.3 Data Pertinent to Prediction of Pollutants in Groundwater (in Addition to Those Listed in Table 5.2)

Physical framework
Estimates of the parameters that comprise hydrodynamic dispersion
Effective porosity distribution
Information on natural (background) concentration distribution (water quality) in aquifer
Estimates of fluid density variations and relationship of density to concentration
Stresses on system: sources and strengths of pollutants
Chemical-biological framework
Mineralogy of media matrix
Organic content of media matrix
Groundwater temperature
Solute properties
Major-ion chemistry
Minor-ion chemistry
Eh-pH environment
Observable responses
Areal and temporal distribution of water quality in aquifer
Streamflow quality (distribution in time and space)

SOURCE: Adapted from Brouwer et al. (1988).

The processes of adsorption-desorption and chemical reactions play important roles in controlling the migration rate as well as concentration distributions. These processes tend to retard the rate of contaminant migration and act as mechanisms for concentration attenuation. Because of their effects, the plume of a reactive contaminant expands more slowly and the concentration changes more slowly than those of an equivalent nonreactive contaminant.

The data requirements for contamination characterization are listed in Table 5.3. These requirements are in addition to the requirements in Table 5.2. The flow system must first be understood in order to characterize advective transport. These data requirements provide a broad view of the factors affecting contaminant transport from a site.

Groundwater use

It is customary to think of groundwater as being more important in arid or semiarid areas and surface water as more important in humid areas. However, inventories of groundwater and surface-water use reveal the worldwide importance of groundwater. The reasons for this include its convenient availability close to where water is required, its excellent natural quality (which is generally adequate for potable supplies with little or no treatment) and the relatively low capital cost of development. Development in stages, to keep pace with rising demand, is usually more easily achieved for groundwater than surface water (Novotny and Olem, 1994).

In the United States, groundwater is important in all climatic regions; it accounts for about 50 percent of livestock and irrigation water use and just under 40 percent of public water supplies. In rural areas of the United States, 96 percent of domestic water is supplied from groundwater (Todd, 1980). Some very large cities are totally dependent on groundwater. In Latin America, many of the continent's large cities—Mexico City, Lima, Buenos Aires, and Santiago—obtain a significant proportion of their municipal water supply from groundwater (Foster et al., 1987). In Europe, also, groundwater has always played a major part in water supplies. The proportion of groundwater in drinking-water supplies in some European countries in 1988 was (UNEP, 1989):

Denmark	98%
Portugal	94%
German Fed. Rep. (former)	89%
Italy	88%
Switzerland	75%
Belgium	67%
Netherlands	67%
Luxembourg	66%
Sweden	49%
United Kingdom	35%
Spain	20%
Norway	15%

Many of the major cities of Europe are, therefore, dependent on groundwater. In Africa and Asia, most of the largest cities utilize surface water, but many millions of people in the rural areas are dependent on groundwater. For many millions more, particularly in sub-Saharan Africa, who do not as yet have any form of improved supply, untreated groundwater supplies from protected wells with handpumps are likely to be their solution for many years to come (Novotny and Olem, 1994).

A survey conducted by the Economic and Social Commission for Asia and the Pacific (ESCAP) in 1990 revealed that a large number of countries in the Asia and Pacific region depend on groundwater for drinking, municipal, industrial, and agricultural water supply. The percentages of the total water supply contributed by groundwater for various sectors of water use in selected countries of the ESCAP region are given in Table 5.4. The contribution of groundwater to the drink-

TABLE 5.4 Percentage of Total Water Supply Contributed by Groundwater in Each Sector of Water Use in Selected Countries of the ESCAP Region

Country or area	Drinking-water supply	Municipal water supply	Industrial water supply	Agricultural water supply	Others
Australia		20 (for municipal, industrial)		13	
Bangladesh	2 (for drinking, municipal, industrial)			20	
China	18 percent of total				
India	80 (rural)	50	50–80	45	—
Japan	14	—	6	4	1
Republic of Korea	35 (rural)	8 (urban)	—	8	—
Malaysia	25 (for drinking, municipal)		3	2	1
Maldives	80	100	50	—	—
Mongolia	82	100	99	60	—
Myanmar	30	—	—	1	—
Nepal	60	80	80	20	—
Pakistan	—	—	—	44	—
Philippines	60	80	50	—	—
Samoa	70 (rural)	3	—	—	—
Thailand	50	10	20	15	5

SOURCE: Information provided by countries in the returns to the questionnaire circulated by ESCAP, 1990. [Adapted from Das Gupta (1991).]

ing-water supply is quite significant to the range of 80 percent in the Maldives and 82 percent in Mongolia. Also, municipal water supply is derived mainly from groundwater resources in the Maldives and Mongolia, and in Nepal, 80 percent of the municipal water supply comes from underground aquifers. Groundwater is also used to a significant extent for industrial water supply in these countries. In India, 50 to 80 percent of industrial water requirements are met from groundwater. Groundwater contributes significantly to the water-supply requirements for agriculture in several countries, including India, the Maldives, Mongolia, and Pakistan. Agricultural water use in the Maldives is mainly from groundwater, whereas in India and Pakistan, respectively, 45 and 44 percent of water used in agriculture is from groundwater resources. Approximately 20 to 25 percent of Bangladesh's water supply comes from groundwater, of which approximately 8 percent is used for drinking, municipal, and industrial water supplies (this amounts to 2 percent of the total water supply required) and about 90 percent for agricultural water supply (about 20 percent of the total agricultural water requirement). Groundwater is the primary potable-water source for the coastal regions of

Bangladesh. Groundwater in northern China, which is an arid or semiarid region, is the main water resource for municipal, industrial, and agricultural use (Das Gupta, 1991; Mawla, 1991; Yangqun, 1991).

Interaction with surface water

When a stream and an aquifer are in direct hydraulic contact, an interchange of surface water and groundwater occurs. If the water table stands higher than the river stage, groundwater may enter the stream as baseflow. Normally groundwater discharge can make up most of the streamflow during dry months. On the other hand, a substantial percentage of total recharge to groundwater in some arid and semiarid areas can be from the leakage of water through the streambed. Many perennial streams recharge the subsurface formation in some portion upstream of their reach, while groundwater discharge appears in the streams farther downstream. Gaining and losing reaches can be influenced by human activities. Similarly, any factor which can alter the river stage or the water table could change the rate of the baseflow. For example, it has been commonly observed in irrigation areas bordering a river, that part of the irrigation water which percolates below the root zone discharges into the river as irrigation return flow. When artificial recharge activities are conducted near a stream, part of the recharge water may find its way into the stream. As observed in Iran, ganat flow rates have been increased as a result of artificial recharge activities conducted nearby (Zomorodi, 1990).

Freeze and Cherry (1979) pointed out that the streamflow hydrographs reflect two very different types of contributions from the watershed. The peaks, which are delivered to the stream by overland flow and subsurface stormflow, and sometimes by groundwater flow, are the result of a fast response to short-term changes in the subsurface flow systems in hillslopes adjacent to channels. The baseflow, which is delivered to the stream by deeper groundwater flow, is the result of a slow response to long-term changes in the regional groundwater flow systems. In the upper reaches of a watershed, subsurface contributions to streamflow aid in the buildup of the flood wave in a stream. In the lower reaches, a different type of groundwater-streamflow interaction, known as *bank storage,* often moderates the flood wave.

Role of groundwater in water-resource systems

Groundwater plays an ever increasing role in water-resource systems. The aquifers may serve as a source of water supply, a medium for storage of water or a means of upgrading water quality. Schwarz (1989) elaborated on functions of aquifers and outlined various options for management of aquifers as components of water-resource systems. As a source of water supply, groundwater may be used as the single source to meet a specific demand or may be used in conjunction with surface water, as a regular complementary source to meet water quantity and quality requirements. It may also serve as the only source of water in certain areas during dry seasons or dry spells. As a storage facility, aquifers are used to store excess surface water by artificial recharge for the purpose of supplying water in low-flow seasons and dry spells. As a means of water-quality upgrading, aquifers are used to take advantage of various processes occurring naturally in groundwater: filtration, aeration, mixing with indigenous water, and providing retention time for biodegradation, decomposition of organisms, and decay of pathogenic microorganisms.

The coordinated operation of a groundwater basin involves the use of a basin for transmitting and storing varying portions of the area's water supply by coordinating the aquifer functions with surface facilities such as reservoirs and pipelines to meet the water requirements of an area. The integration of groundwater-development plans into the total water-resource system should be considered while formulating the water master plan of an area. The first step is to rationally combine the surface water and groundwater facilities to allow their coordinated use.

Environmental Issues

The present use of groundwater is quite extensive. In many areas, particularly in arid regions and in islands, groundwater is the only source of freshwater supply. Groundwater users often overlook groundwater problems until they are well advanced. These problems sometimes take the form of continual lowering of potentiometric levels or water table, which could result into severe environmental consequences such as land subsidence, water-quality degradation, and saltwater encroachment. On the other hand, rising groundwater levels may adversely create the drainage problems and affect the agricultural production. The scope and severity of groundwater problems and environmental issues vary from place to place depending on the hydrogeologic, geotechnical, and geochemical environment. These problems and environmental issues can be grouped into three main categories: issues of overcharging, issues of overexploitation, and issues of groundwater contamination.

Overcharging

In alluvial plains, occurrence and distribution of surface water and groundwater are interdependent. Water usage with one of the resources may have impact on the other. For example, irrigation only with surface water will result in gradual buildup of shallow groundwater table in alluvial plains. Water-table rise can be substantial moving into the root zone of the plants, causing waterlogging and salinity problems. A continuous recurrence of this problem in many areas resulted into significant reduction in agricultural production. The classic example is from Punjab, where the introduction of irrigation at the beginning of this century has caused extensive waterlogging and salinization. Before the introduction of a canal irrigation system, the water table underneath most of the flat stretches of land was very deep. The intensive application of irrigation water, coupled with inadequate subsurface drainage, resulted in a gradual rise of the water table, as illustrated in Fig. 5.1. In some areas, the water-table rise was 24.4 m or even more with an average rate of rise of 0.46 m/yr.

A series of *salinity control and reclamation projects* (SCARPs) have been initiated by the Water and Power Development Authority (WAPDA) of Pakistan, most of which involved installation and operation of tube wells with the dual purpose of producing additional water for irrigation and lowering the water table. In areas where there was a suitable groundwater aquifer but the water was too saline to be used for irrigation, it was envisaged to install tube wells for drainage only, disposing of the pumped water into the rivers, wasteland, or sea.

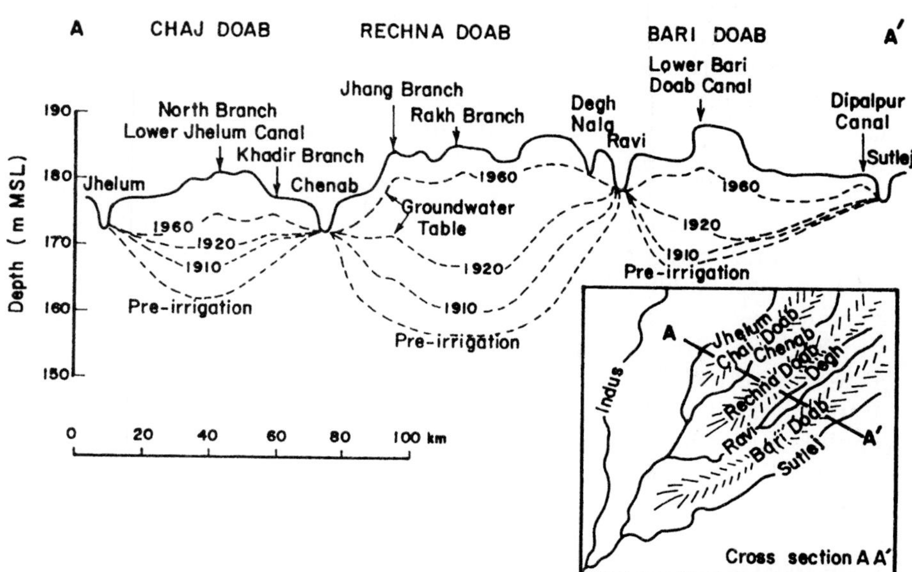

Figure 5.1 Groundwater profiles in Punjab, northeastern Pakistan. MSL = mean sea level. [*Adapted from Oosterbaan (1988).*]

In areas where the aquifer was not suitable for installation of wells, tile drains were proposed. Priority was, however, assigned to areas having usable groundwater quality since the pumped water could be used in conjunction with river water for irrigation, with a resultant increase in agricultural production.

In a recent report at a national workshop critically examining the drainage problem in the Indus basin, Ahmad (1995) indicated that the national experts have called for a comprehensive strategy for sustained management of waterlogged and saline soils in the Indus plain. These problems require varying degrees of attention, management, and utilization. He pointed out that out of 41.2 million acres in the command of the Indus basin irrigation system, 14.8 million acres are provided with drainage facilities. Although these efforts have somewhat alleviated the problem, about 19.2 million acres still require drainage. Out of these, about 5.9 million acres have water tables less than 5 feet (1.5 m) below ground surface and thus require immediate attention.

Overexploitation

Overexploitation refers to a state of pumping of groundwater when the total abstraction exceeds the long-term recharge of the groundwater system, resulting in progressive depletion of the resource and continual decline of potentiometric level or water table. This leads to increased cost of pumping even with a reduction in well yield. Other associated adverse consequences would be land subsidence, deterioration of groundwater quality, and seawater encroachment in coastal areas. The reduction in pore water pressure due to groundwater pumping results in increase in intergranular pressure (effective stress) on the aquifer matrix from the overburden formations in unconsolidated deposits. *Land subsidence* is the accumulated effect of compression of different strata at different depth below the ground surface due to this increase in intergranular pressure. Changes in groundwater quality due to groundwater overdraft result from geochemical evolution and induced anthropogenic pollution. In coastal aquifers, freshwater and saltwater are in dynamic equilibrium. Under natural conditions of flow, freshwater discharges into the sea. Pumping groundwater in coastal aquifers will disturb this dynamic equilibrium, and overpumping a well will cause the encroachment of seawater into aquifers contaminating the freshwater resource in the production areas.

Land subsidence

Subsidence of the land surface is an important environmental consequence of groundwater overdraft. As mentioned above, it is caused by the consolidation of the soil deposits due to the lowering of the groundwater level or the potentiometric pressure. For confined aquifers, the increase in effective stress is due to the reductions in upward hydraulic pressure against the bottom of the upper confining layers caused by the drop in the potentiometric head. In a multilayer system, this drop in potentiometric head induces release of water from the confining layers, whereby confining layers become compressed. For unconfined aquifers, the increase in intergranular pressure is due to the loss of buoyancy of solid particles in the zone dewatered by the falling water table. In addition, initiation or acceleration of lateral flow of groundwater can cause lateral compression of the aquifer and, hence, lateral movement of the land surface, due to an increase in the seepage force or frictional drag exerted by the flowing water on the solid particles. Magnitude of subsidence of ≥30 cm over confined aquifer systems have been reported in many parts of the world. Poland (1969) reported that the deposits which are compacting in the areas of major subsidence as a result of groundwater extraction are unconsolidated to semiconsolidated clastic deposits of late Tertiary Age. Most of them are alluvial and lacustrine deposits. All areas are characterized by confined aquifer systems containing permeable aquifers of sand and/or gravel of low compressibility, interbedded with clayey aquitards and aquicludes of low permeability, high compressibility, and varying thickness. The depth of groundwater withdrawal ranges from 200 to 900 m. Clays rich in montmorillonite are more porous and more compressible than are clays consisting mainly of illite or kaolinite. The earliest definite observation of subsidence due to groundwater withdrawal was made in the Santa Clara Valley in California (Rappleye, 1933). Table 5.5 gives a brief description of some selected examples of subsidence areas. Mexico City and the San Joaquin Valley of California are classic examples of regional subsidence due to groundwater withdrawal. Rivera et al. (1991) indicated that in Mexico City, by 1980, the total pumping rate exceeded 21 m^3/s with more than 600 wells, and by that time a maximum of more than 6 m of land subsidence was observed at some location, constituting one of the most remarkable cases in the world because of its magnitude and extension. Johnson (1981), in his foreword to papers on land subsidence published in the *Journal of the Irrigation and Drainage Division of the American Society of Civil Engineers,* commented, "Most areas of known subsidence are along coasts of water bodies where it becomes quite obvious when the ocean or lake starts moving further up on the shore. In some such areas, the usual heavy population and intensive industrial development are protected from inundation by many feet of water only by construction of an extensive system of dikes, flood walls, locks and pumping stations costing as much as millions of dollars."

TABLE 5.5 Subsoil Conditions and Instrumentation Employed in Some Major

Location, country	Area of subsidence, km²	Depth range of compaction, m	Time of principal occurrence	Maximum subsidence,† m
Mexico City, Mexico	25 (1973)	10–50	1948–1960	9 (1973)
Osaka, Japan	120 (1960)	10–500	1948–1965	3–4 (1965)
Tokyo, Japan	230 (1974)	10–500	1938–1975	4.6 (1975)
Taipei Basin, Taiwan	100	30–200	1961–1975	1.8
South-Central Arizona, USA	925 (1967)	100–300	1948–1967	3.2 (1975)
Houston-Galveston Texas, USA	6,475 (1976)	50–600	1943–1973	2.3 (1973)
San Joaquin Valley, California, USA	13,500 (1976)	90–900	1935–1966	8.8 (1976)
Venice, Switzerland	10 (1971)	100–300	1930–1973	0.11 (1973)
Bangkok, Thailand	285 (1980)	10–200	1950–1980	0.80 (1980)

*Year in the bracket indicates the period in which the area of subsidence was measured. The present size of the area may be larger or smaller than that indicated.

†As of year indicated.

Land-Subsidence Areas

Subsoil conditions	Instruments for observation			References
	Absolute subsidence	Pore-water pressure	Compression between soil layers	
Sediments of alluvial and lacustrine origin, but some places are made up largely of artificial fills	Leveling	Observation well + piezometer	None	Poland (1969), Figueroa Vega (1976)
Alluvial layer 20–30 m underlain by diluvial layer several hundred meters thick	Leveling	Observation well	Compression type	Murayama (1969), Yamamoto (1976)
Alluvial layer 20–60 m underlain by diluvial layer several hundred meters thick	Leveling	Observation well	Compression type	Ishii et al. (1976)
Alternating layers of clay, silt, sand, and gravel to a depth of 280 m	Leveling	None	None	Wu and Hwang (1969), Wu (1976)
Alluvial plain, permeable alluvial deposits, about 700 m deep	Leveling	Observation well	Extension type	Schumann and Poland (1969), Winika and Wold (1976)
Thick strata of unconsolidated lenticular deposits of sand and clay	Leveling	Pumping well	None	Gabrysch (1969, 1976)
Unconsolidated alluvial lacustrine and marine deposits, about 700 m deep	Leveling	Observation well	Extension type	Poland (1969), Poland et al. (1975), Lofgren (1969, 1976)
Unconsolidated fluvial deposits of sand, clay, and peat	Leveling	Observation well	None	Gambolati and Freeze (1973), Gambolati et al. (1974)
Unconsolidated fluvial and deltaic sediments several hundred meters thick	Leveling	Observation well + piezometer	Compression type	AIT (1981)

Subsidence depends on many factors, but of particular importance are the compression characteristics of the subsoil. If the clay is soft and highly compressible, settlement will be large and rapid. On the other hand, if the clay is hard and has previously been consolidated, the rate of subsidence will be slow and may escape notice because a large area has settled uniformly. The general pattern of subsidence is a characteristic bowl shape, with the greatest subsidence at the center of the well field and with the area of subsidence considerably greater than the well field. The difference in settlement within a region can be attributed to natural variation in the stratigraphy of the soil and the difference in the evolution and intensity of pumping of groundwater. The time-consolidation characteristics of subsidence are complex because the effective stress increase is gradual according to the increased pumping that occurs over the years. Even if changes in potentiometric levels have ceased, there will still be some subsidence caused by the consolidation of clays as a result of the long-term effect of the increased loading (Dawson, 1963). In areas of substantial subsidence, many associated problems could occur, including sinking of benchmarks; collapse of well casing; reversal of gradients for drainage and sewerage systems; damage to roads, railways, storm sewers, or other underground pipelines; and cracks in buildings. Also, for coastal cities having a low ground-surface elevation, land subsidence may lead to a serious threat of seawater flooding during every high-tide season. Surface subsidence due to groundwater overdraft is an essentially irreversible phenomenon. It can be controlled and further subsidence can be checked by halting declines in groundwater levels. However, rebound of the ground surface normally is insignificant, even if groundwater levels are restored to the presubsidence height.

Observation. Subsidence has been monitored in various parts of the world by measuring one or more of the following quantities: (1) absolute subsidence, (2) compression of soil layers, and (3) pore-water pressure. Types of measuring instrumentation used for some case histories are given in Table 5.5. The *absolute subsidence* refers to the variation in ground elevation against a fixed reference level, such as the mean sea level (MSL), and leveling with respect to a surficial fixed point has been the most common monitoring technique. Since the total subsidence of a multilayer deposit is the summation of the compression of the component layers, it is essential to measure the compression of each individual layer. Soil-layer compression results from effective stress increase due to reduction in pore-water pressure. Measurement of pore-water pressure is therefore needed for the evaluation of subsidence. The soil profile in the subsidence areas is usually composed of alternate layers of sand or gravel (constituting the aquifers) and clay (constituting the aquitards). As the permeability of these two types of layers is significantly different, the instrumentation of measuring pore-water pressure is selected accordingly.

As an example of recent trends, large-scale groundwater utilization in the Bangkok area resulted in adverse economic-environmental problems such as continual decline of potentiometric levels, land subsidence, and groundwater quality deterioration by saltwater encroachment. Many associated and potential problems such as flooding, loss of property and human life, severe deterioration of infrastructure facilities, groundwater pollution, and health hazards have been attributed to the effects of excessive groundwater withdrawal and land subsidence (Nutalaya et al., 1989; Ramnarong and Buapeng, 1991). The groundwater pumpage is from a multiaquifer system underlying the Bangkok metropolitan area. Prior to 1954, potentiometric levels of all aquifers were few meters below ground surface and by 1982, levels in aquifers being pumped went down to 50 m below ground surface. Results of the comprehensive study during 1987 and 1981 indicated that the land-subsidence rates were 10 cm/yr in the eastern suburbs and 5 to 10 cm/yr in central Bangkok. After the remedial measures for controlling groundwater pumpage were introduced in 1983, the total pumpage dropped sharply from 1983 to 1987, resulting in a recovery of potentiometric levels and reduced rates of subsidence. However, there has been an increasing trend in total pumpage since 1988. These trends can be seen from the observed subsidence at benchmark CI 10-1 and changes in potentiometric levels in three aquifers at nearby wells as shown in Fig. 5.2. From 1987 to early 1992, maximum land subsidence of 85.3 cm was observed at this benchmark and as much as 1.7 m of subsidence occurred in the eastern suburb of Bangkok between 1940 and 1992 (Ramnarong and Buapeng, 1992).

Groundwater contamination

The quality of groundwater in aquifers can be affected by natural and human activities, and the extent to which the quality is affected by either natural processes or human activities varies with the hydrogeologic and climatic settings. In aquifers unaffected by human activity, the quality of groundwater results from geochemical reactions between the water and rock matrix as the water moves along flow paths from areas of recharge to areas of discharge. Contaminants may reach groundwater from a variety of sources (see Table 5.6). Some wastes are, by intentional design, discharged to the subsurface; examples include septic systems, spray irrigation, and land disposal of sludge. Other wastes may reach groundwater unintentionally. Wastes may, and often do, migrate to groundwater from impound-

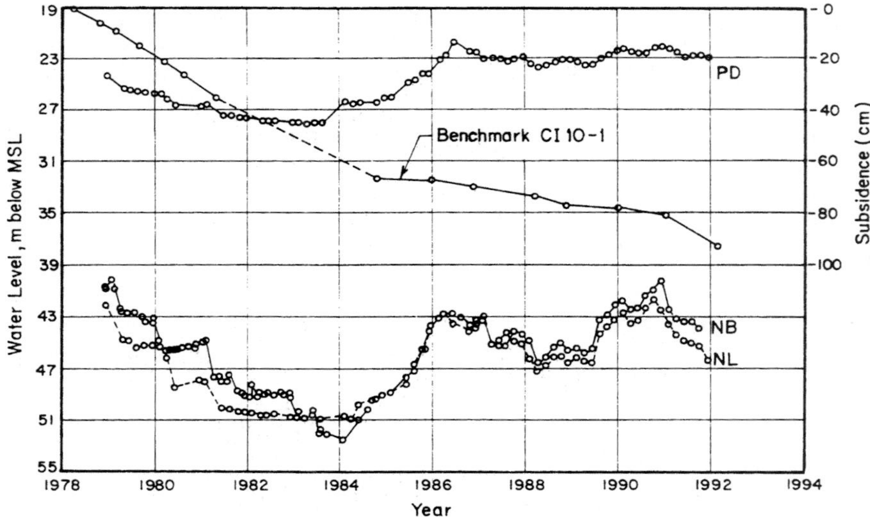

Figure 5.2 Observed subsidence at benchmark CI 10-1 and changes in potentiometric level in three aquifers (PD—Prapradaeng aquifer; NL—Nakhon Luang aquifer; NB—Nonthaburi aquifer) in Bangkok area. [*Adapted from Ramnarong and Buapeng* (*1992*).]

TABLE 5.6 Sources of Groundwater Contamination

Wastes		Nonwastes
Sources designed to discharge waste to the land and groundwater Spray irrigation Septic systems, cesspools, etc. Infiltration or percolation basin Waste-disposal wells Brine-injection wells	Sources that may discharge waste to the land and groundwater unintentionally Surface impoundments Landfills Animal feedlots Acid mine drainage Mine spoil piles and tailings	Sources that may discharge a contaminant (not a waste) to the land and groundwater Buried-product storage tank and pipelines Accidental spills Highway deicing salt stockpiles Ore stockpiles Application of highway salts Product storage ponds Agricultural activities

SOURCE: Adapted from USEPA (1977).

ments, landfills, animal feedlots, leaky sewer lines, and other sources. Unprotected wastes from a variety of sources may enter groundwater systems as indicated in Fig. 5.3.

The severity and extent of groundwater contamination is determined by the hydrogeologic setting, the nature of the contaminant, and the effectiveness of regulatory action. The groundwater setting

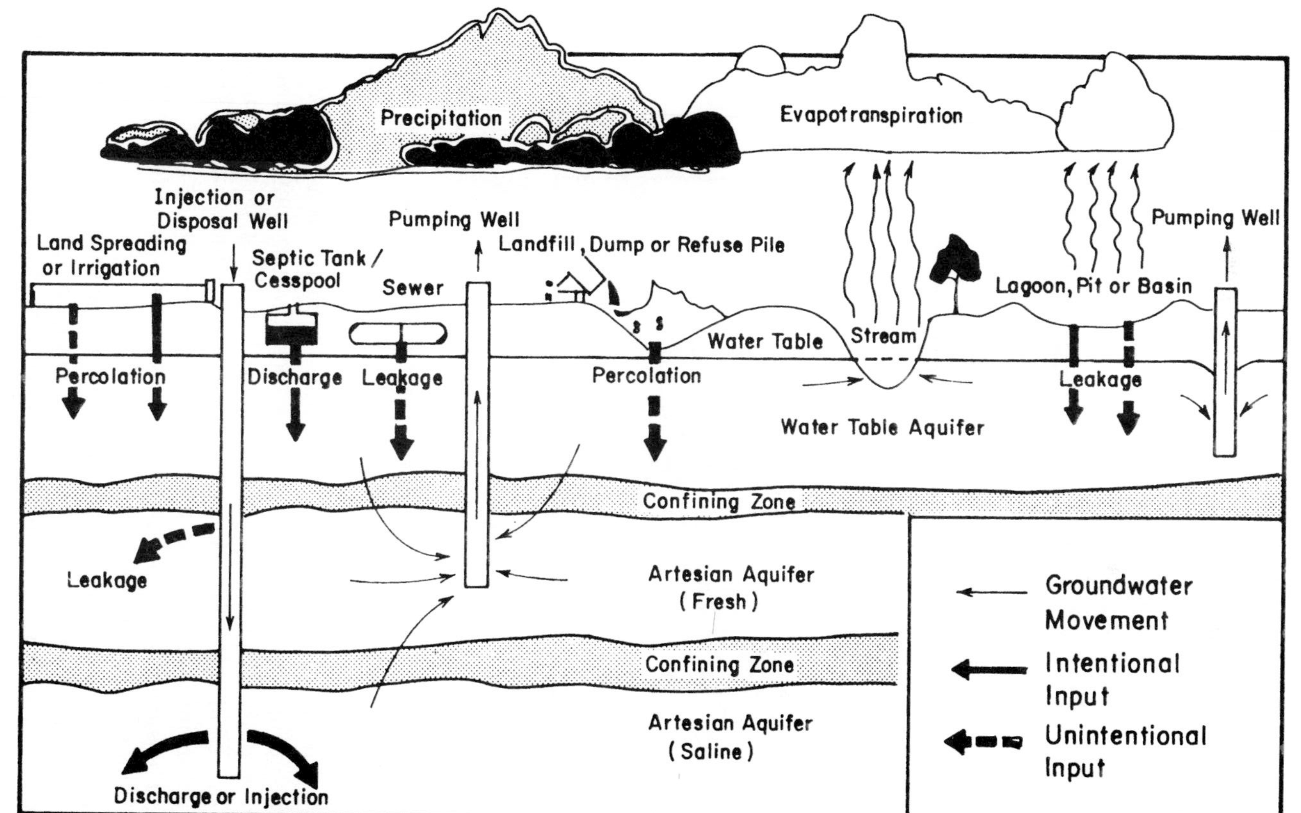

Figure 5.3 How waste disposal may contaminate groundwater. [*Adapted from Roscoe Moss Company (1990).*]

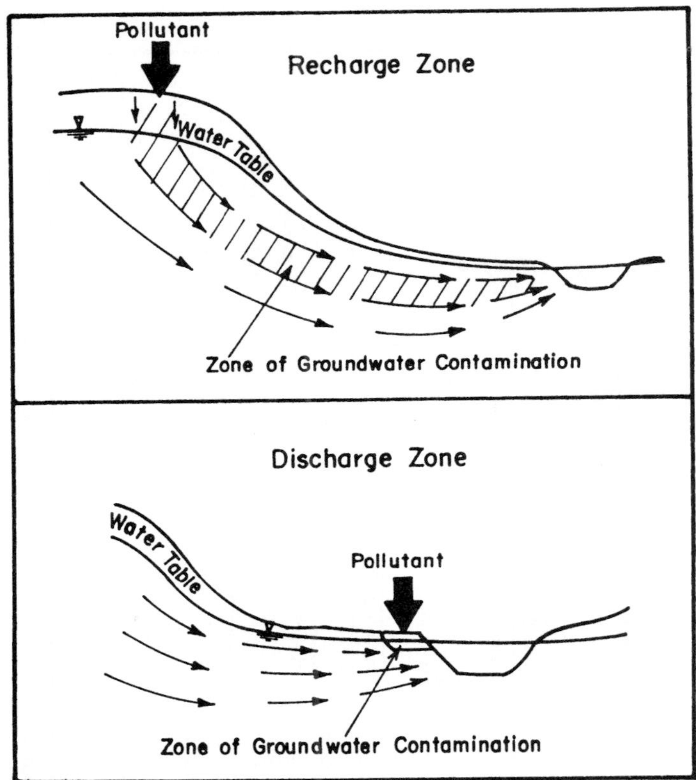

Figure 5.4 Extent of groundwater contamination from pollutant entering recharge and discharge zone. [*Adapted from Lindorff, (1979)*.]

determines the potential extent of contamination. As shown in Fig. 5.4, contaminants introduced into the surface on an upland recharge site potentially may move a great distance and may affect a major portion of an aquifer. This is especially true in areas underlain by coarse-grained sediments or fracture rocks, where contaminants may move rapidly through the subsurface with little or no attenuation. If the source of contamination is in or near a discharge zone (Fig. 5.4), the potential extent of groundwater contamination is much more limited. The severity of a pollution incident also is dependent on the volume and nature of the contaminant.

One of the most common methods of waste disposal is by sanitary landfills. Solid wastes are spread on land, compressed, and covered with soil. When precipitation takes place over such an area, the percolating water dissolves many organic and inorganic salts which may be transported to nearby aquifers. Similarly, gases such as methane, carbon dioxide, and hydrogen sulfide produced on site may be transported by gaseous diffusion through unsaturated media to adjacent terrains.

Sewage is another source of groundwater contamination. Both raw and treated sewage has been used for irrigation. In arid regions the water content of sewage is an important resource in itself in addition to supplying valuable nutrients such as phosphorus, nitrogen, and heavy metals. Leachates from such lands introduce bacteria, viruses, and chemical compounds in local groundwater. Similarly, seepage from septic tanks, cesspools, and lagoons can be hazardous.

Continual deep cultivation coupled with the extensive use of fertilizers, insecticides, and pesticides has made agriculture a major source of groundwater pollution. Also, storage and disposal of wastes from poultry farms have introduced large quantities of nitrates, potassium, and phosphates in aquifers. Contamination of aquifers by petroleum products can occur when these leak from storage tanks or underground pipelines or by accidental spills.

Lastly, large withdrawals from aquifers hydraulically connected to the sea can lead to a landward hydraulic gradient. This induces saline seawater to displace freshwater in coastal aquifers. This phenomenon is called *saltwater intrusion*. Almost every coastal area suffers from this contamination problem.

From the hydrodynamic point of view, these sources can be classified according to their geometry:

1. Point source (nonpenetrating injection wells, leakage from tanks, etc.)
2. Line source (saltwater intrusion, seepage from canals carrying wastewater, etc.)
3. Plane or diffuse source (agricultural waste)

Broadly, considering the characteristics of constituents, pollutions can be classified into four groups.

1. Physical (heat, suspended solids, etc.)
2. Chemical (inorganic chemical contaminants—brines, acid, nitrates, sulfates, chlorine, etc.; organic chemical contaminants—gasoline, hydrocarbons, phenol, etc.)
3. Biological (derived mainly from sewage and decomposable by microorganisms such as bacteria, virus, and protozoa)
4. Nuclear (uranium, radioactive elements derived from nuclear power plants, chemical reprocessing plants, etc.)

In the case of deep aquifers, pollution by chemicals, radioactive materials, and heavy metals is important, whereas for shallow wells, biological pollution may also be of significance. As the pollutants move through the subsurface formation, they are subjected to a number of attenuation effects. With increasing cases of groundwater-quality degradation in many areas, it is necessary to understand the mechanism by which pollutants can enter the groundwater system and to develop proper management policies.

Observations. Some approximations of the extent of groundwater contamination in the United States were made on the basis of estimates of the number and average spatial extent of different types of contamination. For example, Lehr (1982) estimated the volume of contaminated groundwater resulting from waste-disposal sites of all types and compared this volume to an estimate of the volume of available groundwater. He estimated that less than 1 percent of the groundwater resources had been contaminated by these sources. Lehr acknowledged that any of the large number of assumptions that were required to develop this estimate could be readily challenged. His key point was that a relatively small percentage of the groundwater resource had been contaminated by point sources and that use of appropriate preventive measures in the future would keep the overall percentage low. Subsequent articles, for example, those by Pettyjohn and Hounslow (1983) and Lehr (1984), elaborated on different sources of pollution, stressing the recognition of the existing problem, development of alternate technologies for remediation, and establishment of a long-term strategy to contain and reduce existing pollution by the end of this (twentieth) century through detection, correction, and prevention methodologies.

In the 1980s, issues related to pollution of groundwater became apparent in many European countries. An increasing number of wells were found to be contaminated by chemical wastes from point sources, and also a widespread non-point-source pollution from fertilizers was threatening the water resources in extensive areas (Dyhr-Nielsen, 1989; Lagas et al., 1989; Meier and Mull, 1989).

The nature of the groundwater-quality problem in Asia and the Pacific region varies from country to country. Chemical contamination in deep wells, mainly from fertilizers, industrial wastes, and dumped solid wastes as well as biological contamination, related particularly to human waste, is common to several countries of the region. Also, saltwater contamination problems exist in coastal areas as well as in inland areas in some countries. The extent of groundwater-quality problems, derived from Das Gupta (1991), is summarized in Table 5.7.

Many cases of groundwater pollution in India were reported by Verma (1989). Important causes of pollution were identified as haphazard urban development without adequate attention to sewage and waste disposal, rapid industrialization without proper treatment of

TABLE 5.7 Extent of Groundwater-Quality Problem in Asia and the Pacific Region

Country	Nature of contaminant	Area or region affected
Bangladesh	Physical, chemical	Southern belt: salinity and iron Regional: high iron content
China	Chemical	Southern and northeastern China: high iron content Coastal zone and low-lying areas of plain: salty water
India	Physical, chemical, biological	Coastal areas, semiarid western and northwestern parts: salinity Semiarid regions: fluoride Humid and forested: iron Urban and rural areas: nitrates
Malaysia	Chemical	Coastal aquifers: salinity Alluvial aquifers: iron
Maldives	Chemical, biological	Densely populated areas
Mongolia	Chemical	Southern part of the country
Myanmar (formerly Burma)	Chemical	Central part of the country
Nepal	Chemical, biological	Terai: dug well contamination—human Kathmandu valley: dissolved gases, iron and manganese contamination
Pakistan	Physical, chemical	Scattered areas in different provinces
Philippines	Physical, chemical, biological	Coastal areas, areas adjacent to tidal effects
Thailand	Physical, chemcial	Northeast region, coastal plain: high salt concentration Southern peninsula: effect of mining

SOURCE: Adapted from Das Gupta (1991).

waste products, unsanitary dumping of refuse and other solid wastes, and excessive use of fertilizers and pesticides for agricultural soils. Somasundaram et al. (1993) reported on groundwater pollution of the Madras urban aquifer in India by a range of species, including nitrates, heavy metals, and microorganisms. Leachate migration from open dump sites has been identified as a potential threat of groundwater pollution in metropolitan Manila (Pinlac, 1992). Also, seawater intrusion is becoming significant in coastal areas of metropolitan Manila, Cebu, Bulacan, and Pampanga, and in Capiz Province of Panay Island. Groundwater contamination is also observed in several major cities and rural areas in Indonesia, caused primarily by saltwater intrusion and domestic, industrial, and agricultural wastes

(Soetrisno and Hehanussa, 1992). Ramnarong (1991) reported the incidence of groundwater pollution by toxic substances such as lead in certain parts of Thailand. Also, arsenic in groundwater was found to be the cause of groundwater pollution in several villages in southern Thailand as these villages are situated near tin mines. In addition, incidence of nitrate contamination of groundwater from fertilizers and saltwater intrusion in coastal areas were reported. Khan (1991) provided a summary on groundwater quality of countries under the Global Environmental Monitoring System (GEMS/Water) program in Asia and the Pacific during the period 1979–1987, and the water-quality status indices assigned by the author, as given in Table 5.8, indicate the severity of pollution problems in Asia and the Pacific region.

Even though a number of incidents of groundwater-quality degradation in different countries of the Asia and the Pacific region have been reported, a comprehensive management program for groundwater is yet to be initiated in most of those countries. One reason could be that the groundwater plays a supplemental role in the overall water-use sector, even though there are specific subsectors and areas in which groundwater plays a significant role. The other reason would be the absence of proper institutional and legislative structure to authorize a broad effective management of groundwater. Surface water is usually publicly owned, while groundwater is privately owned in most cases. With this private-ownership concept, the idea of groundwater management through "government control" becomes untenable as it encroaches on private rights. However, as more demands are made on the groundwater basin, there is an ever-increasing logic in support of the desirability of its management.

TABLE 5.8 Status of Water Quality in Asia and the Pacific

Pollution type	Indian subcontinent	Southeast Asia	Pacific Islands	China	Japan, Australia, and New Zealand
Pathogenic agents	1–3	1–2	2–3	1–3	0–1
Organic matter	1–3	0–2	0–1	1–3	0–1
Nitrate	0–1	0–1	1–2	0–2	0–1
Fluoride	0–1	0	0	0–2	0
Heavy metal	0–1	0–2	0–1	0–2	0–2
Pesticides	0–1	0–1	0–1	0–1	0–1
Eutrophication	0–1	0–3	0	0–2	0–1
Salinization	0–1	0–1	0–3	0–2	0–2
Acidification	0	0–1	0	0–1	0–1
Sediment load	0–2	0–2	0–1	0–3	0–1
Iron	0–1	NA	NA	0–2	NA

Key: 0—very little pollution; 1—some pollution; 2—major pollution; 3—severe pollution; NA—data not available or not obtained.

SOURCE: Adapted from Khan (1991).

Specifically, every groundwater aquifer is subject to the possibility of depletion for some period of time and to some extent. This common characteristic strongly recommends the practice of intelligent and planned authoritative management with the view of obtaining the maximum benefit from the underground water source. With the realization of the general public of the ensuing long-term impact through the support of technical information and demonstration of consequences of not managing the resource properly, a management program could be formulated and funding could be secured for its implementation.

Saltwater intrusion

Saline water is the most common type of pollution in fresh groundwater. Intrusion of saline water occurs where saline water displaces or mixes with freshwater in an aquifer. Saline water in aquifers may be derived from any of several sources:

1. Encroachment of seawater in coastal areas
2. Seawater that has entered aquifers during past geologic eras
3. Salt in salt domes, thin beds, or disseminated in geological formations
4. Water concentrated by evaporation in tidal lagoons, playas, or other enclosed areas
5. Return flows to streams from irrigated lands
6. Human saline wastes

The mechanisms responsible for saline-water intrusion fall into three categories. One involves the reduction or reversal of groundwater gradients, which permits denser saline water to displace freshwater. This situation commonly occurs in coastal aquifers in hydraulic continuity with the sea when pumping of wells disturbs the natural hydrodynamic balance. A second method stems from the destruction of natural barriers that separate freshwater and saline water. An example would be the construction of a coastal drainage canal that enables tidal water to advance inland and to percolate into a freshwater aquifer. The third mechanism occurs where there is subsurface disposal of saline water such as into disposal wells, landfills, or other waste repositories.

The freshwater requirement of many coastal areas, particularly for oceanic islands, is met partially or fully by the available groundwater resource in coastal aquifers. This resource is vulnerable to seawater intrusion, which must be taken into consideration in formulating plan for development. Countries having seawater intrusion problem includes Australia, Belgium, France, Greece, Germany, Indonesia, Iran, Israel, Lebanon, Libya, the Netherlands, Philippines, Spain, Thailand, the United Kingdom, and the United States. The list of vulnerable islands is much longer and includes Bermuda, the Canary Islands, Christmas Island (Kiritmati), the Cocos Islands, Guam, the Hawaiian islands, Mauritius, Seychelles; and South Pacific islands, including the Cook Islands, Ellice Islands, Fiji Islands, Gilbert Islands, Solomon Islands, and Tarrawa (Falkland, 1991; Ghassemi et al., 1993).

Freshwater and saltwater are miscible fluids. In the intruded region, a brackish transition zone of finite thickness separates the two fluids. This zone develops from dispersion by flow of the freshwater plus unsteady displacements of the interface by external influences such as tides, recharge, and pumping of wells. In general, the greatest thickness of transition zones are found in highly permeable coastal aquifers subject to heavy pumping. Observed thicknesses vary from less than 1 m to more than 100 m. An important consequence of the transition zone and its seaward flow is the transport of saline water to the sea. This water originated from the underlying saline water; hence, from continuity considerations, there must exist a small landward flow in the saline-water region (Todd, 1980). Figure 5.5 illustrates the flow patterns in the three subsurface zones. Within the transition zone the salinity of the groundwater increases progressively with depth from that of the freshwater to that of the saline water.

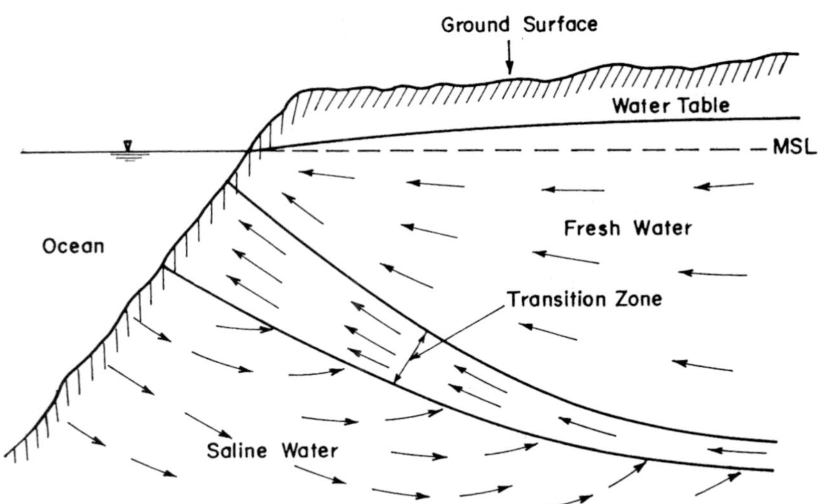

Figure 5.5 Vertical cross section showing flow pattern of fresh and saline water in an unconfined coastal aquifer. [*Adapted from Todd (1974).*]

The intruding saltwater renders large zones of the aquifers unsuitable for all human needs. When an aquifer contains an underlying layer of saline water and is pumped by a well penetrating only the upper freshwater portion of the aquifer, a local rise of the interface below the well occurs. This phenomenon is known as *upconing* of the interface between freshwater and saltwater.

Problems of saline-water intrusion are acute not only in coastal aquifers but also in many inland areas. The basic mechanisms causing mixing of saline water with freshwater in inland aquifers are change of pressure heads, where the two fluids are hydraulically connected, and destruction of natural barriers separating the two fluids. It is common to find an aquifer underlaid by saline water and separated from it by fairly impermeable stratum. Destruction of this barrier (either during drilling or by leakage through abandoned wells) or the development of a steep pressure gradient toward the freshwater (due to overpumping) can cause the saline-water intrusion. Many geologic formations were deposited in marine environments. The water-bearing strata in those formations may contain entrapped saline-water bodies, called *connate water,* adjoining the freshwater bodies. Under natural condition, there is equilibrium without much mixing between the freshwater and saline water. The pumping in the freshwater bodies will disturb the dynamic equilibrium between the two liquids, and the saline water may eventually encroach into freshwater bodies, resulting in gradual degradation of freshwater source. This phenomenon contributes significantly to the increase in salt content of freshwater aquifers underlying metropolitan Bangkok. As evident from field study and investigations considering the flow and transport phenomenon, the major sources of groundwater-quality degradation are the seawater near the coastline and the connate water in different locations entrapped under marine conditions in pore spaces of sediments at, or subsequent to, the time of their deposition. The areal distribution of chloride concentration in one of the aquifers underlying metropolitan Bangkok is shown in Fig. 5.6.

The encroachment of seawater can be controlled by properly selecting the location of shallow wells and infiltration galleries where drawdown can be maintained within a reasonable amount not inducing significant encroachment or upconing. On the other hand, by maintaining groundwater levels well above sea level in the vicinity of coasts, seawater encroachment can be prevented. However, this will impose restrictions on groundwater utilization, and, if necessary, the pumping pattern may need to be modified; even the natural replenishment to the basin may need to be augmented by artificial recharge. Todd (1974) elaborated on the methods of controlling seawater intrusion by (1) reducing groundwater extractions, (2) creating physical

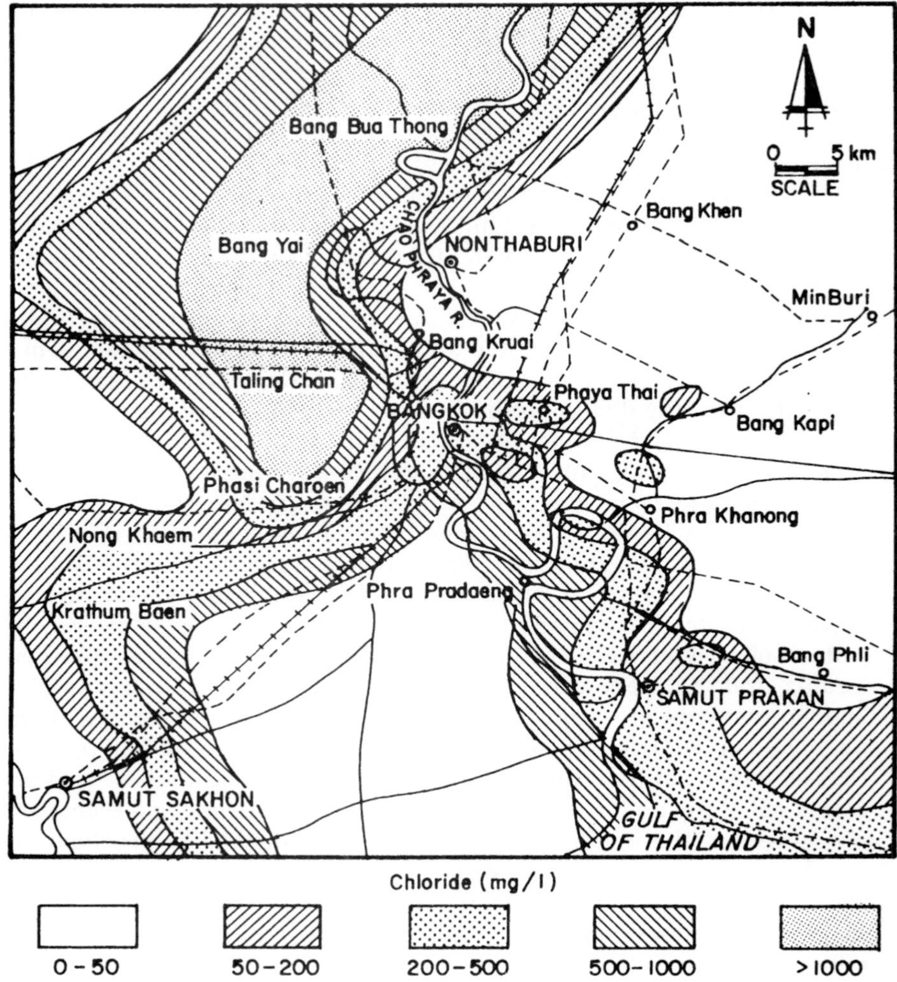

Figure 5.6 Areal distribution of chloride concentration (in milligrams per liter) in Nakhon Luang aquifer in Bangkok and surrounding areas. [*Adapted from Ramnarong and Buapeng (1984).*]

barriers of impermeable membrane of some type placed between the seawater source and the pumping well, (3) installing a pumping trough for intercepting seawater moving landward toward the supply wells, (4) constructing a hydraulic ridge to provide necessary gradient to push back the seawater front, and (5) providing a combination barrier of pumping trough and hydraulic ridge. An evaluation of the physical system is needed, considering different factors of control measures with due regard to environmental and socioeconomic aspects for determining the most feasible method of protecting the groundwater resource from seawater intrusion.

Environmental Assessment

Groundwater development progresses in stages with the introduction of new wells as demand arises. In many countries, legal and administrative measures were related mainly to protecting the existing wells so that new developments would not affect the operation of the existing ones. Permission to drill a well was given on a case-by-case basis with no comprehensive regional planning efforts involved. In some cases, development plans were prepared on the basis of simplified hydrogeologic analyses as aquifers are extensive and abundant resources were available. However, as development progressed, emergence of severe adverse consequences in different environments due to continual pumping in an unplanned manner indicated the need for initiation of a comprehensive evaluation planning at the earliest stage so that necessary control measures could be instituted to minimize the impact. Groundwater development by installing more wells as one aspect of development process and the assessment of the functioning of the groundwater system as another aspect of evaluation of the state of the system should be carried out concurrently. However, this second limb of evaluation of the state of the system lags way behind the development process, which is not desirable as groundwater problems remain unnoticed until some inference can be drawn from production well water in case groundwater is contaminated or from indirect effects on infrastructure in case of land subsidence.

Development planning of groundwater resources involves allocation of groundwater to competing water demands and uses at required quality. Additional consideration should be given to the possible emergence of any environmental consequences such as land subsidence and water quality deterioration as development progresses. The environmental assessment program is a continuous, methodological, and technical assessment process with observations, measurements, and field and laboratory analysis of selected components and variables of a groundwater system with the following objectives:

1. To collect, process, and analyze groundwater-level data, and to correlate the external stresses on the system with groundwater levels and properties of the system in order to assess the spatiotemporal change of the groundwater system's response and to project the future dynamic behavior of the system.
2. To collect, process, and analyze data on environmental aspects such as land subsidence and water quality for assessing present

conditions and to integrate the spatiotemporal changes with the changes in the dynamic behavior of the system due to external stresses, and then to assess the anticipated changes in environmental consequences due to potential future stresses on the groundwater system and hence impacts in time and space.

3. To provide feedback to the observational network for modification and improvement of the data-collection process, and to provide information to authorities and policymakers on proper policy planning and strategy for groundwater use with a view to protect the resource for sustainable use.

Methodologies

An understanding of the underlying physical, chemical, and hydraulic transport processes occurring within the groundwater system is necessary for proper environmental assessment. One can come across various methods and approaches for assessment in the literature. An appropriate selection of method or a combination of methods is based on site condition, expected level of development, availability of time, and funding for investigation. The long-term objective is to assess the water availability and potential for safe development with due regard to environmental concern. Geologic, hydrogeologic, and geochemical investigations, coupled with estimates of recharge, provide the basis for the evaluation of groundwater potential. Estimates of recharge can be made using a variety of techniques, but, again, these estimates may not reflect the condition in the aquifer when intensive exploitation occurs. Initial estimates are therefore to be updated as development progresses through model study conforming to the dynamic behavior of the groundwater system. Various methods and investigations are reviewed under six phases of activity.

1. *Hydrogeologic investigation.* The purpose is to assess

- The geohydrologic setting with delineation of different strata and their extent through surface mapping and surface and subsurface geophysical investigations
- The hydrologic parameters of different strata through laboratory testing on undisturbed soil samples collected and through field pumping tests

This phase of activity is to be carried out at all stages of groundwater development with a view to define a conceptual model of geohydrologic setting and to define the hydrogeologic conditions in the best possible way.

2. *Hydrometeorological method.* The purpose is to assess the natural replenishment of groundwater from rainfall. Models for balanc-

ing the components of hydrologic cycles, like any watershed model, are used to estimate the deep-percolation component resulting from the soil-moisture-accounting component. This estimate is made on long-term basis with approximation of balancing various hydrologic components.

3. *Observation of groundwater levels and water level hydrograph analysis.* With observation of water levels at different locations, maps of the groundwater table and potentiometric levels are prepared to infer the resulting flow pattern, to locate the recharge areas and drainage bases of aquifers. Such maps may further reveal the existence of impermeable boundaries, and variations in hydraulic characteristics. Comparison of water-level maps prepared on a seasonal basis may provide some estimate of natural replenishment from rainfall. The water-level hydrographs prepared from long-term records of groundwater levels along with groundwater-withdrawal data can be used to estimate safe yield.

4. *Geotechnical investigation.* Any proposed investigation of land subsidence due to groundwater depletion needs to assess the magnitude and rate of subsidence due to present and future patterns of groundwater pumping. The geotechnical investigation in this respect consists of

- Measurement of soil compressions at different depths to correlate with potentiometric pressure drop
- Determination of compression characteristics of soil samples collected from different depths

5. *Groundwater balance.* When no closed cycle of water levels may be drawn, a more general method of groundwater balance has to be applied. The balance area should preferably be a closed area with impermeable boundaries. Sometimes a streamline may also be considered as an impermeable boundary. The balance time should be short enough to include small variations of flow rates but long enough to ensure that errors in water-level measurements are small in comparison to time variations in water levels. The groundwater balance includes the following components:

- Aquifer inflows with known geometry and recorded gradients but with transmissivities often unknown
- Aquifer outflows
- Change of water in storage with known geometry and recorded changes of water levels but with storage coefficients often unknown
- Leakage from overlying and underlying aquifers with recorded gradients but with leakage coefficients often unknown

- Withdrawal of water by wells and springs often recorded or estimated by irrigated areas or electrical power consumption
- Return flow from irrigation
- Natural replenishment from rainfall with unknown subsurface and surface parameters.

Many unknown parameters are handled by a set of balance equations which are solved to find the value of the unknowns. The balance method depends on the availability of a long record of water levels, pumping, rainfall, and other parameters.

6. *Model study.* A *model* expresses the conceptual representation of the system in casual relationships among various components of the system and between the system and its environment. Models provide a quantitative framework for synthesizing hydrogeologic conditions with spatial and temporal trends of the system's response. Although engineering decisions at the local level can be based on field observations, however, inference on the regional scale can be done only by integrating all local-level observations through a mathematical model. A model is first formulated according to the available information about

- The extent of aquifers
- The nature of its boundaries
- The hydraulic properties of the aquifer system and their variation
- The natural replenishment and its distribution in time and space

A steady-state undisturbed flow condition is first simulated on the model to verify and calibrate it under its preexploitation state of flow. Sometimes even in this early stage some of the model parameters such as transmissivity, replenishment, and boundary conditions may be verified. In the second stage of model calibration the recorded history of aquifer utilization is simulated. In this stage, aquifer parameters and their distribution and spatiotemporal distribution of pumpage are tested and adjusted if necessary to get the model response to agree with the historical record of the field response. A further stage of model studies includes simulations of tentative future exploitation plans. Aquifer response, i.e., the drawdown of water levels or potentiometric levels in response to a given plan, is estimated from the model study. When drawdown exceeds tolerable levels, the tentative plan has to be revised. The revision may be a decrease of overall exploitation, or sometimes redistribution of exploitation may suffice to solve these drawdown constraints. In such a way, a future groundwater exploitation plan is tested and verified in a trial-and-error procedure. Several operating scenarios in the future could be defined, and response of the aquifer system can then be simulated. A proper management strategy could then be defined on the basis of response evaluation considering socioeconomic and environmental aspects.

Selection of methods. The evaluation methodologies outlined are complementary and usually successive. The groundwater potential for development is initially estimated by a hydrometeorological method. After some years of data collection, hydrograph analysis and groundwater balance are used to revise the estimate. Finally the estimate is revised through model study considering all probable environmental implications.

Mathematical models

Mathematical models are essential to analyze complex problems in groundwater environment. The range of alternative feasible decisions for operation is determined by analyzing the response of the groundwater system through model study satisfying the constraints that are imposed on the dynamics of the system and from environmental considerations. Both simulation and optimization have been used in the analysis of groundwater-system dynamics. To evaluate alternative management decisions, a simulation model is recomputed for each alternative considered. The results of alternative strategies are compared in terms of criteria used, and, accordingly, an appropriate management decision is made. On the other hand, optimization models use a mathematically based search procedure to find the policy that maximizes the criteria without actually simulating the model for every possible alternative.

One basic requirement of any scientific method is to demonstrate in an objective way that a model provides an acceptable description of reality. This implies that the model should be able to predict observations with some degree of accuracy. In environmental applications this fairly straightforward concept is complicated by a number of factors which are familiar to modelers. First, natural systems are often very heterogeneous and difficult to characterize in a simple way. Second, the field data needed to formulate and test theories about such systems are difficult to obtain. As a result, most environmental modeling efforts are hampered by data limitations. This applies to both the input data used to generate model predictions and the validation data needed to check model performance.

Most groundwater models in use today are deterministic mathematical models. Deterministic models are based on the conservation of mass, momentum, and energy relating fluxes of mass, momentum, and energy to measurable state variables such as hydraulic head and solute concentration. Deterministic groundwater models generally require the solution of partial-differential equations. There are two basic ways to solve these equations. The analytical methods embody classic mathematical approaches leading to the exact solution to a flow or other environmental problems in terms of controlling parameters. While numerical methods yield approximate solutions to the governing equation(s), they require discretization of space and time. Because of the functional form of the solution and the interrelationships between parameters, analytical models provide a great deal of physical insight into how the processes control flow, transport, and other environmental conditions. But analytical models require that the parameters and boundaries be highly idealized. Another useful way in which the analytical solutions are used is to provide a check on the accuracy of numerical models, which can be subject to a variety of different errors. A brief introduction on flow, transport, and consolidation models is provided in the following sections.

Flow model. The fundamental equation of flow in the saturated porous media was empirically derived by Henry Darcy in 1856. The generalized Darcy equation

$$q = -\frac{k}{\mu}(\nabla p - \rho g \nabla z) = -K \nabla h \tag{5.1}$$

where q = Darcy velocity
k = permeability tensor
μ = dynamic viscosity of the fluid
∇ = differential operator
p = fluid pressure
ρ = fluid density
g = acceleration due to gravity
z = vertical distance above a datum (positive upward)
K = hydraulic conductivity
h = hyraulic head

Equation (5.1) describes a linear relationship between groundwater flow and the hydraulic gradient, and this condition holds for flow situations encountered in a wide range of hydrogeologic environments. Combining the Darcy equation of flow [Eq. (5.1)] with the continuity equation, the groundwater-flow equation for a three-dimensional flow situation is derived as

$$\frac{\partial}{\partial x}\left(K_x \frac{\partial h}{\partial x}\right) + \frac{\partial}{\partial y}\left(K_y \frac{\partial h}{\partial y}\right) + \frac{\partial}{\partial z}\left(K_z \frac{\partial h}{\partial z}\right) = S_s \frac{\partial h}{\partial t} + \hat{Q} \tag{5.2}$$

where K_x, K_y, K_z = directional hydraulic conductivity components coinciding with the cartesian-coordinate directions
S_s = specific storage
$\hat{Q}$ = specified source or sink expressed as volumetric flow rate per unit volume of aquifer (positive for fluid withdrawn)

Various groundwater-flow-model structures with varying degrees of sophistication have been proposed, such as (1) single-horizontal-layer model (Pinder and Bredehoeft, 1968), (2) vertical-section, axisymmetric model (Javandel and Witherspoon, 1969), (3) quasi-three-dimensional (or coupled-layers) model (Bredehoeft and Pinder, 1970; Chorley and Frind, 1978; Premchitt and Das Gupta, 1981), and (4) three-dimensional model (Freeze, 1971; Gupta and Tanji, 1976; Frind and Verge, 1978; McDonald and Harbaugh, 1984). The first two model categories were for generalized and simplified situations where aquifers are considered to be separated from each other by aquitards. In some practical situations, this assumption may be oversimplified and may lead to significant errors in the model simulation of real groundwater basins. On the other hand, a full three-dimensional model is applicable for situations with great variability in the distribution of water-bearing strata where the concepts of aquifer and aquitard are not applicable. However, for most of the field situations, observations on water levels or potentiometric levels are carried out with respect to different aquifers (in case of multilayer system) and also field measurements are not adequate for defining three-directional subsurface properties. A quasi-three-dimensional model structure has therefore been adopted for multiaquifer systems in many practical applications with areal horizontal flow in aquifers, and the linkage between aquifers is being simulated by vertical leakage fluxes. The flow equation [Eq. (5.2)] for quasi-three-dimensional flow conditions in a multiaquifer system is expressed as

$$\frac{\partial}{\partial x}\left(T_x \frac{\partial h}{\partial x}\right)_i + \frac{\partial}{\partial y}\left(T_y \frac{\partial h}{\partial y}\right)_i = S\left(\frac{\partial h}{\partial t}\right)_i + Q_i + W_i \qquad (5.3)$$

where subscript i = index assigned to the aquifer under consideration
T_x, T_y = directional transmissivity of the aquifer
S = storage coefficient of the aquifer
Q = source/sink term expressed as volumetric flow rate per unit aquifer area

The term W_i corresponds to the rate of groundwater flow to the upper and lower aquifers in the domain which includes (1) the direct leakage between aquifers and (2) the water yielded from storage in clay layers. The general expression for the term W_i is of the form (Premchitt, 1981)

$$W_i = \sum \frac{k}{l} \int_o^t M(t - \tau)\frac{\partial h'}{\partial \tau}\, d\tau \tag{5.4}$$

where k, l = hydraulic conductivity and thickness of intervening aquitard, respectively

M = flow-rate function involving a time-rate factor c_v/H^2, where c_v is the coefficient of consolidation and H represents half of the thickness of the aquitard

h' = a function of potentiometric levels in the aquifer under consideration and the two adjacent aquifers

t = time since the start of the simulation

In simplified cases where the effect of time-delayed yield from the aquitard is neglected, the term W_i is given by

$$W_i = \frac{k_{i-1}}{l_{i-1}}(h_i - h_{i-1}) + \frac{k_i}{l_i}(h_i - h_{i+1}) \tag{5.5}$$

where the indices i, $i - 1$, and $i + 1$ refer to the specific aquifer under consideration and the adjacent aquifers and aquitards and the term k/l is the *seepage factor* for the confining layer.

Solute-transport model. Solute transport models deal primarily with groundwater quality. They are used to simulate and predict the movement and concentration of various contaminants in aquifers, such as leachates from landfills and irrigated areas, and saltwater intrusion in coastal areas. To accomplish this, the models incorporate mathematical approximations of the transport, by means of fluid flow and/or mixing of one or more chemical constituents in the groundwater. Transport models that describe the movement of contaminants without reactions are called *conservative transport models,* while those taking into account reactions are termed as *nonconservative transport models.* Solute-transport models are much more complex than flow models in that they consider quality in conjunction with quantity. In principle, a solute-transport model contains a flow submodel which computes the fluid-flow velocity and then utilizes this velocity in a quality submodel which transports the contaminant in the flow field. Under certain circumstances, such as low concentrations of contaminants, flow and quality submodels can operate independently. In other cases, dealing with high contaminant concentrations or saltwater effects, flow and quality submodels are to be solved in a coupled manner.

The different mathematical models recently applied to groundwater transport problems have been summarized in a number of references. The model survey of van der Heijde et al. (1985) reviewed a total of 84 numerical mass-transport models. Only five of the surveyed models had fully coupled flow and quality submodels which accounted for the effects of varying density on the flow field. Of 35 management models surveyed, four dealt with the quantity and quality of groundwater and two models addressed both the quantity and quality of a conjunctive groundwater–surface water system. Javandel et al. (1984) compiled and demonstrated some of the best and most usable mathematical models for predicting the extent of subsurface contamination in a format useful to field-response personnel. The methods presented ranged from simple analytical and semianalytical solutions to complex numerical codes. A comprehensive review of solute transport modeling was presented by Naymik (1987). Subsequently a summary of selected subsurface hydrological and hydrochemical models was provided by Mangold and Tsang (1991).

The various mass-transport processes that affect the transport of contaminants in the subsurface environment include advection, molecular diffusion, mechanical dispersion, biochemical transformations, and interphase mass transfer. An elaboration of these processes has been provided. Considering the mathematical basis for expressing these processes and taking the mass balance of chemical species over an elementary aquifer volume, the solute-transport equation is expressed as

$$\frac{\partial}{\partial x}\left(D_x \frac{\partial C}{\partial x}\right) + \frac{\partial}{\partial y}\left(D_y \frac{\partial C}{\partial y}\right) + \frac{\partial}{\partial z}\left(D_z \frac{\partial C}{\partial z}\right) - \frac{\partial}{\partial x}(V_x C) - \frac{\partial}{\partial y}(V_y C) - \frac{\partial}{\partial z}(V_z C) = \frac{\partial C}{\partial t} + \frac{\rho_b}{\phi}\frac{\partial S}{\partial t} + Ck_l + S\frac{k_s}{\phi}\rho_b \quad (5.6)$$

where C = concentration in the liquid phase as mass of contaminant per unit volume of solution

S = concentration in solid phase as mass of contaminant per unit mass of dry soil

V_x, V_y, V_z = directional seepage velocity components

D_x, D_y, D_z = directional hydrodynamic dispersion coefficients

ρ_b = bulk density of soil

k_l, k_s = first-order decay rate in the liquid phase and soil phase, respectively

ϕ = effective porosity

Equation (5.6) has two unknowns, S and C; therefore, it is necessary to specify the relationship between S and C in order to solve the equation. Assuming linear adsorption isotherm of the form $S = K_d C$, Eq. (5.6) is expressed in the form

$$\frac{\partial}{\partial x}\left(D_x \frac{\partial c}{\partial x}\right) + \frac{\partial}{\partial y}\left(D_y \frac{\partial C}{\partial y}\right) + \frac{\partial}{\partial z}\left(D_z \frac{\partial C}{\partial z}\right) - \frac{\partial}{\partial x}(V_x C) - \frac{\partial}{\partial y}(V_y C) - \frac{\partial}{\partial z}(V_z C) = R\frac{\partial C}{\partial t} + kC \quad (5.7)$$

where R is the retardation factor $= 1 + (\rho_b K_d)/\phi$, where K_d is the distribution coefficient; and k is the overall first-order decay rate $= k_l + (k_s \rho_b K_d)/\phi$.

In the conventional formulation of Eq. (5.7), assuming that the diffusion is negligible compared to the large-scale mixing processes of hydrodynamic dispersion, Scheidegger (1961) related the groundwater velocity to the dispersion tensor as

$$D_{ij} = \alpha_{ijmn} \frac{V_m V_n}{|V|} \quad (5.8)$$

where
α_{ijmn} = aquifer dispersivity tensor (fourth-order)
V_m, V_n = components of seepage velocity in m and n directions
$|V|$ = magnitude of resultant seepage velocity

The dispersion tensor for two-dimensional flow is defined as

$$D = \begin{bmatrix} D_{xx} & D_{xy} \\ D_{yx} & D_{yy} \end{bmatrix} \quad (5.9)$$

with

$$\begin{aligned} D_{xx} &= \frac{1}{V^2}(d_L V_x^2 + d_T V_y^2) \\ D_{yy} &= \frac{1}{V^2}(d_T V_x^2 + d_L V_y^2) \quad (5.10) \\ D_{xy} = D_{yx} &= \frac{1}{V^2}(d_L - d_T)(V_x V_y) \end{aligned}$$

In Eq. (5.10), d_L and d_T are, respectively, the longitudinal and the transverse dispersion coefficients. Both Scheidegger (1961) and Bear (1979) showed that the dispersivity of an isotropic medium can be defined by two constants: the longitudinal dispersivity of the medium

α_L and the transverse dispersivity of the medium α_T. The dispersion coefficients are dependent on average velocity magnitude in the flow system and are defined as

$$d_L = \alpha_L V \qquad d_T = \alpha_T V \tag{5.11}$$

The dispersivity parameter is scale-dependent, and a proper characterization of this parameter is needed for assessing the movement and spreading of the contaminant plume at field scale. Gelhar et al. (1992), in a critical review article, provided an outline of the theoretical description of dispersive mixing in porous media, as well as a tabular summary of existing data on values of field-scale dispersivity and related site information reported in the literature.

For contamination problems, aquifer analysis is focused on one of two general types of situations: (1) assessment of already contaminated sites and (2) planning to minimize contamination hazards from future activities. Both types of situations require the capability to predict the behavior of chemical contaminants in flowing groundwater. Figure 5.7 illustrates in a general manner the role of models in providing input to the analysis of groundwater contamination prob-

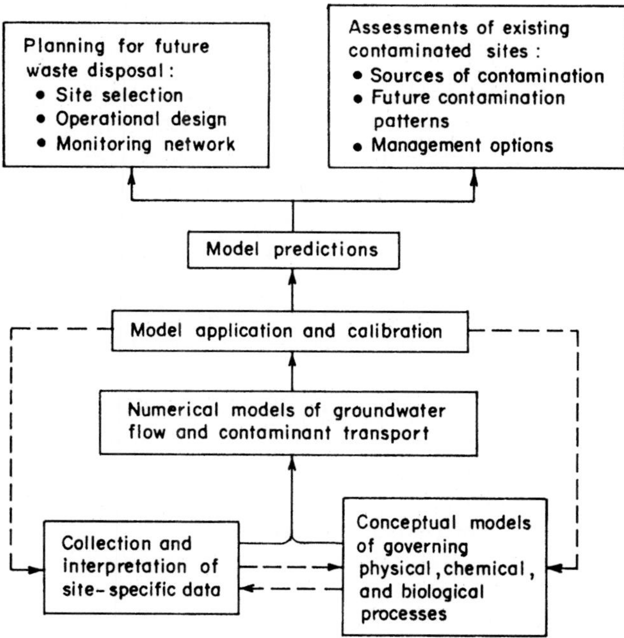

Figure 5.7 The use of simulation models in evaluation groundwater contamination problems. [*Adapted from Konikow (1981)*.]

lems. There is a major difference between evaluating existing contaminated sites and evaluating new or expected sites. For the former, if the contaminant source can be reasonably well defined, the history of contamination itself can, in effect, serve as a surrogate long-term tracer test that provides critical information on velocity and dispersion on a regional scale. However, when a contamination problem is recognized and investigated, the locations, timing, and strengths of the contaminating sources are for the most part unknown, because the release to the groundwater system occurred in the past when there was no monitoring. In such cases it is often desirable to use a model to determine the characteristics of the source on the basis of the present distribution of contaminants; thus, the requirement is to run the model backward in time to assess where the contaminants came from. Although this is theoretically possible, in practice there is usually so much uncertainty in the definition of the properties and boundaries of the groundwater system that an unknown source cannot be uniquely identified. At new or expected sites, historical data are seldom available to provide a basis for model calibration and to serve as a control on the accuracy of predictions. As indicated in Fig. 5.7, there should be allowances for feedback from the stage of interpreting model output to both the data collection-analysis phase and the conceptualization and mathematical definition of the relevant governing processes.

Land-subsidence model. The basic theories that describe the two phenomena involved in land subsidence due to groundwater withdrawal are those of Theis (1935) and Terzaghi (1925). Later Biot (1941) presented concise mathematical expressions that fully described the groundwater flow and stress-strain relationships in a soil in a three-dimensional domain based on the theory of elasticity. Sandhu (1979) reviewed the development of land-subsidence models and noted certain differences of opinion in setting up constitutive relationships. Helm (1982) compared major conceptual models of subsidence (double-porosity, half-space, viscoelastic, and clastic aquitard drainage) and developed an alternative approach. In most situations, a one-dimensional deformation model is coupled with a two- or three-dimensional hydrologic model (defined by a groundwater-flow model) to calculate consolidation at selected points in the system. The hydrologic model provides the variations of hydraulic head in time and space in response to the water withdrawal, and these values are then used to calculate the time-dependent consolidation at any point in the system.

Full mathematical simulation of regional subsidence has been carried out in many areas, for example, for the case of Venice by Gambolati and Freeze (1973), and Gambolati et al. (1974); for the case of Mexico City by Rivera et al. (1991); and for the case of metropolitan Bangkok by AIT (1981). A generalized model for the simulation of land subsidence in a multiaquifer system is provided in this section. The model couples a quasi-three-dimensional hydrologic model; expressed by Eq. (5.3) with one-dimensional consolidation model. The hydrologic model links together a set of horizontal aquifer models by means of the leakage terms through the intervening aquitards. The fundamental assumption of this approach is that the flow is horizontal in the aquifers and vertical in the aquitards. This assumption is satisfactory as long as the hydraulic conductivity contrasts are greater than two orders of magnitude (Neuman and Witherspoon, 1969). The hydraulic-head depletion in aquitards considering storage effect is estimated by a one-dimensional consolidation model using the calculated hydraulic heads in aquifers as the boundary conditions for the consolidation model. With known hydraulic-head decline in aquifers and aquitards, along with known compression characteristics, the compaction of several layers can be estimated, and the cumulative compaction is the amount of subsidence at a particular point in the system. The relationship between the consolidation model and the hydrological model is schematically shown in Fig. 5.8.

The consolidation model is used to find (1) the potentiometric-head distribution throughout the clay layer and (2) the magnitude of the compression of each slice Δz of the clay layer (due to the drop in potentiometric head) and to sum these to represent the layer compression. The governing equation for the potentiometric head in the clay layer is

$$\frac{\partial}{\partial z}\left(k_z \frac{\partial h_c}{\partial z}\right) = S_{sv} \frac{\partial h_c}{\partial t} \tag{5.12}$$

which is equivalent to

$$\frac{\partial}{\partial z} c_v \frac{\partial h_c}{\partial z} = \frac{\partial h_c}{\partial t} \tag{5.13}$$

where h = potentiometric head in the clay layer
S_{sv} = specific storage of clay layer
$c_v = k_z / S_{sv}$ = coefficient of consolidation, where kz is the vertical permeability of aquitard

The potentiometric head in the aquifers derived from the hydrologic model are used as the boundary heads for the consolidation model for any individual clay layer at any location. The compression of a layer

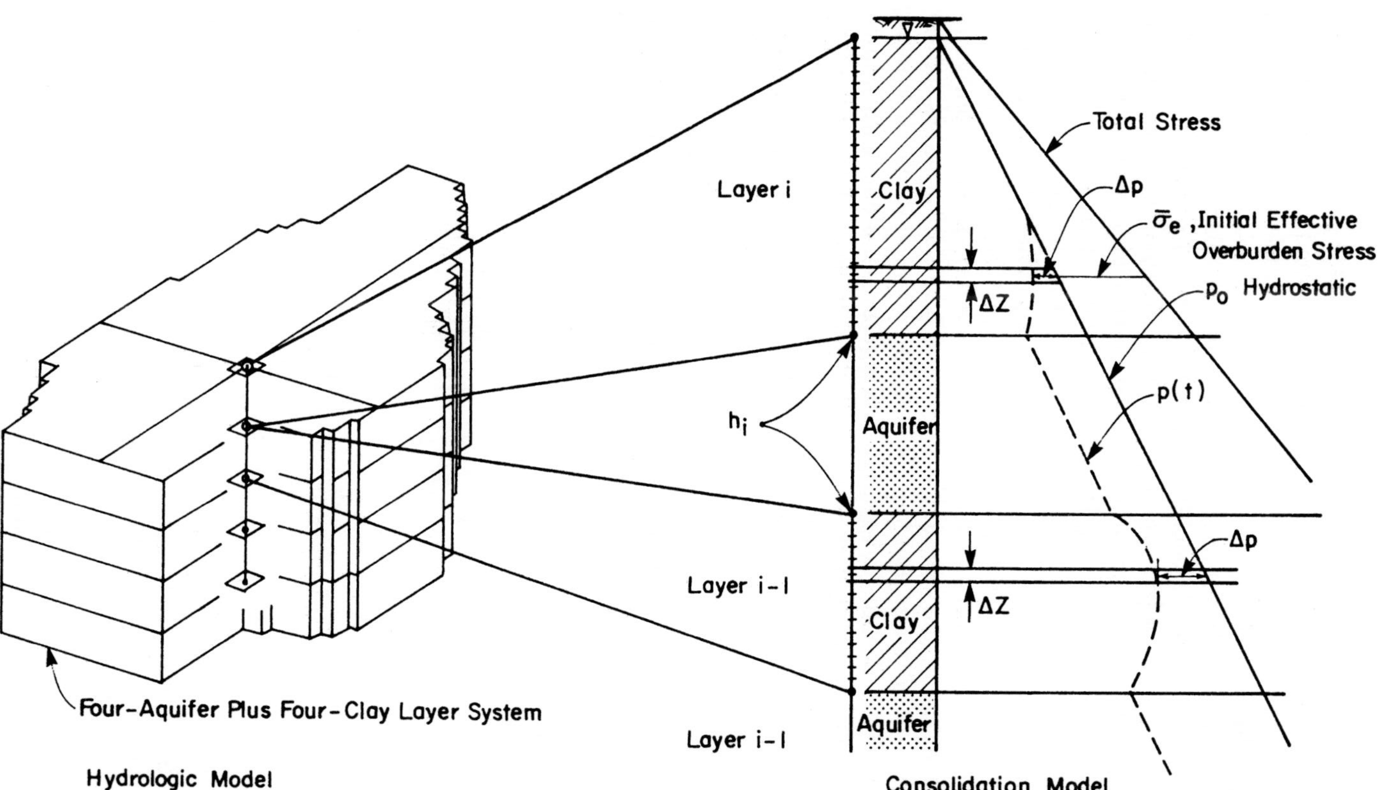

Figure 5.8 Relationship between consolidation model and hydrologic model.

is calculated using the logarithmic theory by Terzaghi (Terzaghi and Peck, 1948), which is based on the fact that when void ratio e is plotted against log σ_e (where σ_e is effective stress), a sigmoid curve is generally obtained with flat portions for the very low and very high values of σ_e, but with an essentially straight portion for the midrange σ_e values. The slope of this linear section is called the *compression index* C_c. Considering a clay element of thickness Δz, the vertical strain $\Delta\varepsilon_z$ due to a potentiometric-head drop of $-\Delta h_c$ can be found from the e/log σ_e relationship as

$$\Delta\varepsilon_z = \frac{C_c}{1 + e_0} \log\left(1 + \frac{\Delta\sigma_e}{\bar{\sigma}_e}\right) \tag{5.14}$$

or alternatively

$$\Delta\varepsilon_z = CR \log\left(1 - \frac{\Delta h_c \gamma_w}{\bar{\sigma}_e}\right) \tag{5.15}$$

where $\Delta\sigma_e$ = effective stress increment which in subsidence study can be taken as $-\Delta p = -\Delta h_c\gamma_w$, where γ_w is the specific weight of water; $\bar{\sigma}_e$ = initial effective overburden stress; e_0 = initial void ratio. The magnitude of vertical strain for each slice of thickness, Δz, can also be calculated using the specific storage parameter S_{sv} for clay in the compression range

$$\Delta\varepsilon_z = -S_{sv}\,\Delta h_c \tag{5.16}$$

In situations where potentiometric head is rising instead of declining, the swelling (or recompression) parameters have to be used in place of C_c, CR, and S_{sv}, because of the characteristic two modules of compression of the soil. This is also the case where the effective stress at a moment is less than the maximum effective stress that the soil has experienced in the past. The change of parameter is also applied to cv, where the coefficient of consolidation should be changed to the coefficient of swelling c_s. The total compression of a clay layer $\Delta\rho_c$ is then the summation of compression of slices considered:

$$\Delta\rho_c = \sum_l \Delta\varepsilon_z\,\Delta z \tag{5.17}$$

The compression of sand layers occurs immediately when a pore-pressure drop takes place as a result of groundwater pumping. The compression can be found directly from the pressure drop, the layer compressibility, and the layer thickness. The compression of an aquifer $\Delta\rho_a$ due to the change in head Δh is found from

$$\Delta\rho_a = -S\,\Delta h \tag{5.18}$$

where S, the storage coefficient of the aquifer, can be employed to represent the compressibility of the aquifer. Finally, the subsidence at a point on the land surface is the summation of compression of all clay and sand layers through the depth range beneath that point

$$\rho_t = \Sigma(\Delta\rho_c + \Delta\rho_a) \tag{5.19}$$

where ρ_t is the total subsidence at a given point.

Modeling process

The modeling study is an integral part of the investigation process related to groundwater development and management. Models are being extensively used for synthesizing many factors involved in analyzing complex groundwater problems. The steps in the modeling process are indicated in a flow diagram in Fig. 5.9. The purpose of a model study should be clearly defined as it will indicate the extent of information needed on the physical characteristics of the system, the processes involved, and interaction of the system with the external factors. Initially, through collection of field data and analysis, a conceptual model is defined, and this provides a hypothesis for how a system or process operates. The conceptual model is then translated into a mathematical model defined by a set of governing equations and boundary conditions that contain mathematical variables, parameters, and constants. This stage of model development is accomplished in steps improving on the model concept based on initially available data through field data collection and observation of system response. Once a mathematical model is formulated, the next step is to obtain a solution using one of two general approaches: the analytical approach and the numerical approach. Most of the groundwater models applied in practice are based on the numerical approach.

When a numerical algorithm is implemented in a computer code to solve one or more partial-differential equations, the resulting computer code can be considered a generic model. When the parameters, boundary conditions, and grid dimensions of the generic model are specified to represent a particular geographic area, the resulting computer program is a site-specific model (Konikow and Bredehoeft, 1992). The modeling process consists of using the computer code to solve a site-specific problem. The model is then calibrated and verified with field data. During calibration a set of values for aquifer parameters and stresses is found so that the model approximately reproduces observed field conditions. There are two basic approaches to model calibration. One approach is a trial-and-error process in which certain model parameter values are carefully adjusted until the model output

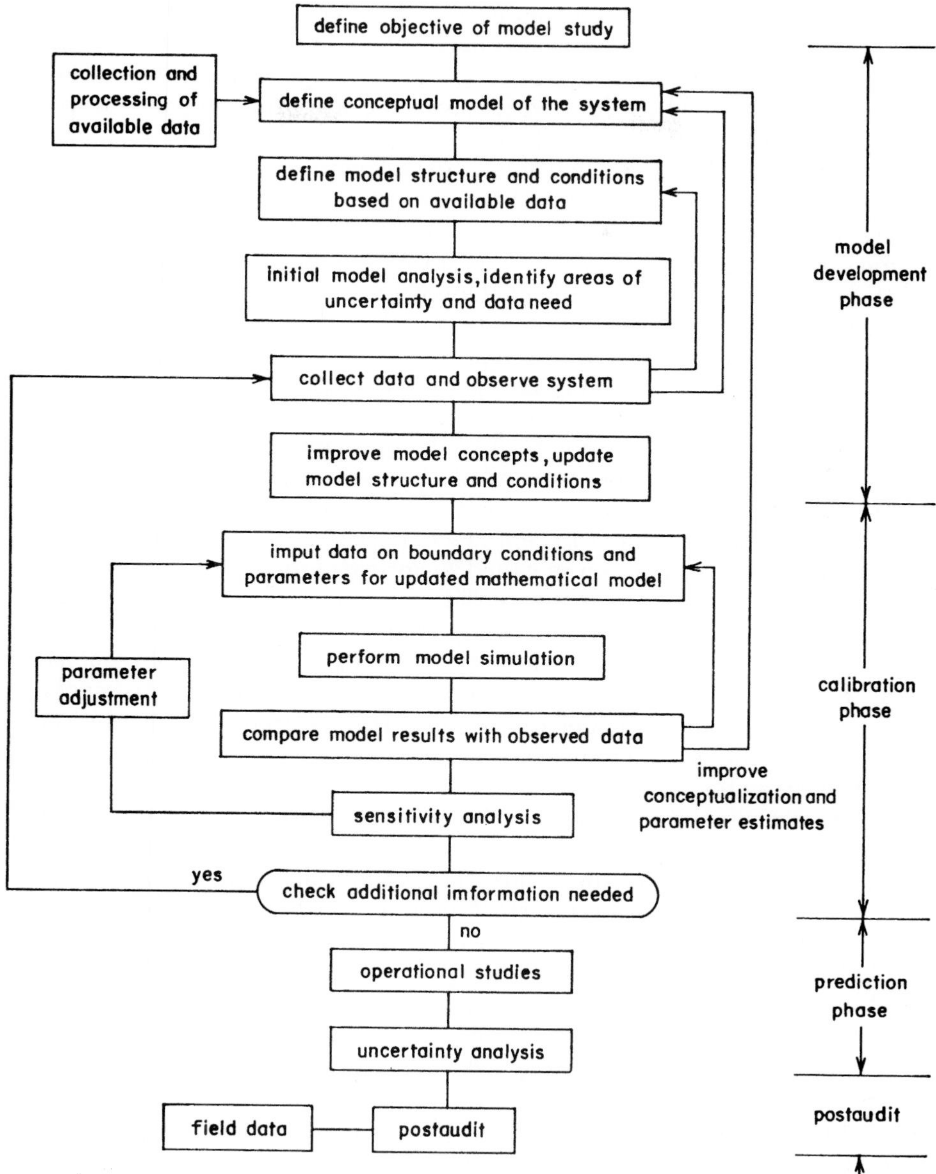

Figure 5.9 Steps in different phases of modeling process.

reproduces some observed conditions as closely as possible. The outcome of this type of calibration is highly dependent on the user's experience and can vary substantially from user to user. The other approach to model calibration involves the use of a computer-search algorithm that computes model parameter values by minimizing some measures of the difference between model solution and available field observations. This procedure is known as *parameter identification* (Yeh, 1986). The model is calibrated when it reproduces historical data within some subjectively acceptable level—there are no rules other than one's own judgment. One does not obtain a unique set of parameters. A poor calibration will indicate (1) an error in the conceptual model, (2) an error in the numerical solution, or (3) a poor set of parameter values. One may not be able to distinguish among several sources of error. However, further improvement on conceptualization and parameter estimate can be introduced as inferred from sensitivity runs, and model calibration can be better accomplished. With the set of calibrated parameter values and stresses, the response of the system is further simulated to reproduce a second set of field data. This step of model verification is called *historical data validation* and the purpose is to establish greater confidence in model use.

Once an adequate match between historical data and model output is achieved, the model is then commonly used to predict the response of the system when subjected to anticipated conditions or stresses in the future. In recent model studies, uncertainty analysis is performed to place confidence bounds on predictions arising out of the uncertainty in parameter estimates.

The model calibration and verification demonstrate that the model can reproduce past behavior. It is necessary to check to what extent the model can predict the future. The validation of the model in this respect is called a *predictive validation* or a *postaudit*. In a postaudit, new field data are collected several years after the completion of the modeling study to determine whether the predicted events come true. If the model's prediction was accurate, the model may be considered valid for that particular site for the conditions simulated. The postaudit should occur long enough after the prediction was made to ensure that there has been adequate time for significant changes to occur. Postaudits have not been considered a routine part of the modeling process, and, in fact, they may not be necessary if the purpose of the model is to analyze current steady-state behavior or to make short-term predictions. However, when the model is used to make predictions on the order of tens of years, a postaudit is necessary to ensure that the model produces meaningful results.

Bredehoeft and Konikow (1993) reported that several postaudits have been performed to evaluate the accuracy of predictions made using supposedly validated models. Compared to the number of model studies, the number of postaudits is small. There are numerous problems in examining past predictions; often the stress placed on the system was quite different from that used in the model analysis, or errors may arise from mistakes in the conceptual model, model structure, and parameter estimates.

Planning model

The objective of the groundwater planning model is to manage conjunctively the water supply and quality resources of a groundwater basin so as to minimize or maximize the given planning objectives while satisfying the hydrologic, economic, and environmental constraints of the problem. The management of groundwater quality is a multiobjective planning problem, a problem characterized by conflicting objectives, constraints, and policies.

Assuming a deterministic groundwater system, a single chemical constituent, and a fixed planning horizon of discrete time period T^*, the objective function of the optimization model can be expressed as the weighted sum of the individual objectives (Willis and Yeh, 1987).

$$\max Z = \sum_{k=1}^{T^*} \sum_{l} \lambda_l F_l (h^k, C^k, Q_w^k, Q_r^k) \tag{5.20}$$

where F_l is the lth objective of the planning problem and it is assumed that a particular objective is a function of the state variables (the head and mass concentration in planning period k) and the pumping Q_w^k and injection Q_r^k policies. The λ_l represents the weights or preferences associated with the planning objectives.

The management model is constrained by

1. The hydraulic and water-quality response equations deduced from the flow equation and the solute-transport equation, respectively. These response equations may be expressed as

$$f^k(h^k, h^{k-1}, Q_w^k, Q_r^k) = 0 \tag{5.21}$$

$$g^k(C^k, C^{k-1}, Q_w^k, Q_r^k, \hat{C}^{k-1}) = 0 \tag{5.22}$$

where h^{k-1}, C^{k-1} = head and concentration, respectively, at beginning of kth planning period

h^k, C^k = head and concentration, respectively, at end of planning period

$\hat{C}^{k-1}$ = waste-injection concentration at beginning of kth planning period, a decision variable of the problem

2. The water target requirements within any planning period

$$\Sigma_{w \varepsilon \pi} Q_w^k \geq D^k, \forall k \tag{5.23}$$

where D^k is the assumed water target in period k and the index π defines the location of all pumping sites in the system.

3. The waste-disposal constraint

$$\Sigma_{r \varepsilon \psi} Q_r^k \geq WL^k, \forall k \tag{5.24}$$

where WL^k is the magnitude of waste load in period k and index ψ defines the location of injection sites in the system.

4. The maximum feasible pumping and injection rates Q_w^* and Q_r^*, respectively, and limiting hydraulic head level h^*

$$Q_w^k \leq Q_w^* \qquad \omega \varepsilon \psi, \forall k \tag{5.25}$$

$$Q_r^k \leq Q_r^* \qquad r \varepsilon \psi, \forall k \tag{5.26}$$

$$h_i^k \leq h_i^* \qquad i \varepsilon \Delta, \forall k \tag{5.27}$$

where Δ is an index set defining the control locations in the aquifer.

5. The groundwater-quality constraint introduced at each pumping well to prevent degradation of the aquifer's water supply

$$C_w^k \leq C_w^* \qquad w \varepsilon \pi, \forall k \tag{5.28}$$

6. The injection constraints to prevent possible clogging problems

$$\hat{C}_r^k \leq \hat{C}_r^* \qquad r \varepsilon \psi, \forall k \tag{5.29}$$

7. The nonnegative restrictions of the decision variables

$$C^k, Q_w^k, Q_r^k, h^k, \hat{C}^k \qquad \geq 0 \tag{5.30}$$

Equations (5.20) through (5.30) constitute the mathematical model for the optimal conjunctive management of groundwater-quality and -quantity resources.

Computationally, the model is extremely difficult to solve for several reasons. First, the constraint set of the model is nonconvex. The pumping and injection decisions affect the velocity field and, consequently, the convective and dispersive mass transport occurring in the aquifer. In the solute-transport equations, these terms are multiplied by the mass concentrations. Because the constraints and the response equations are equalities, the nonconvex feasible region is restricted. The size of the constraint set may also be inordinately large because of the temporal discretization of the response equations. The discretization affects the optimality of the planning policies. Finally, if the objectives are nonlinear or nonconvex, convergence properties of solution algorithms are difficult to prove. Consequently, local solutions to the planning problem can be expected using nonlinear programming algorithms. However, there also exists the possibility of obtaining only stationary solutions to the optimization model. In this event, simulation may be a more viable solution approach (Willis and Yeh, 1987).

Framework of assessment approach

Each groundwater system is unique in its physical setting. As such, it is not possible to describe a simple outline of an assessment plan which will inevitably lead to the best groundwater program. It is obvious that an assessment plan in the context of a regional system will require more effort and will consider more factors than in the case of local-level development. However, initial strategies adopted at local-level development need to be upgraded and modified as development progresses through evaluation of field data and the system response to external stresses. In the context of regional groundwater resources, one needs to appraise the hydrogeologic state of the groundwater system with its resource availability and evaluate the system's response to external stresses and other environmental conditions. In a chronological manner, one has to meet the following basic requirements:

1. Define the groundwater system by mapping the stratigraphic information to identify aquifers, confining layers and their extent; interconnect different aquifers; and establish the hydraulic nature of boundaries.
2. Define the flow system by identifying all components of inflow to the groundwater system, the flow pattern within the system, and all components of outflow from the system; also identify the mechanisms controlling each flow component.
3. Determine aquifer properties, estimate inflow and outflow components, and prepare groundwater balance.
4. Define the baseline values for groundwater quality parameters and identify probable sources of contamination.

5. Determine the susceptibility of the groundwater system to overexploitation and related environmental consequences.
6. Develop in stages mathematical models for assessment, and verify these models with observed responses of the groundwater system.
7. Perform operational studies with mathematical models subject to well-defined criteria and constraints to develop a resource-utilization plan.

It has to be realized that substantial amounts of data and information are required to accomplish these requirements. Organizations overseeing the groundwater development in a particular area should possess the capability and expertise to institute a cost-effective data collection-evaluation scheme as groundwater development takes place. In this regard, a close cooperation and coordination are essential among the various government departments dealing with groundwater. Also, the team responsible for resources assessment and evaluation of development strategies taking into consideration the environmental implications should be both multidisciplinary and interdisciplinary. The protocol of data collection, processing, and evaluation for environmental assessment involves the following steps:

1. Collect and review existing data on geology, hydrology, and meteorology of the region; review all literature and information available pertaining to the groundwater condition of the region.
2. Define the hydrogeologic flow system with available observation on groundwater flow gradient, potentiometric surface maps, and available water-quality data; make preliminary conceptualization of the system based on simple model of hydrologic water balance.
3. Carry out additional field investigations on geologic and hydrogeologic aspects for delineating the aquifer system, defining hydrologic characteristics and stresses on the system in order to provide an updated conceptualization of the system and an updated estimate of available groundwater resource using groundwater balance.
4. Plan field monitoring on the groundwater-level water quality to establish baseline information and to observe spatiotemporal trends. Also conduct field and laboratory investigations on the extent of compressibility of different formations, in particular clay layers within the system.
5. Correlate observed water-level trends on potentiometric-level change with stresses on the system and the spatial variability of hydrologic characteristics, water-quality degradation, and other environmental consequences such as land subsidence.

6. Formulate mathematical models conforming to the conceptual model for assessment of the dynamic response of the system (initial level of mathematical model development).
7. Evaluate the model performance by comparing model response with the field-observed response of the system. As data collection on system's response progresses, with feedback of additional information, conceptualization is further improved and mathematical models are updated through stages of improvement and modification.
8. Calibrate the mathematical models using past records of the system's response and stresses. Evaluate effectiveness of models to reproduce environmental consequences.
9. With calibrated mathematical models, evaluate alternative schemes of development and predict various effects of future groundwater withdrawals.
10. Identify the acceptable operating scheme subject to criteria and constraints imposed on the dynamics of the system.
11. Evaluate the adequacy of the monitoring system to provide necessary information, and continue monitoring the response and stresses on the system to check how the state of the system conforms with the assessment.
12. Periodic revision of assessment as groundwater utilization continues and more data become available.

These chronological steps imply that the groundwater-resource development should be staged, allowing progressive data collection and resource evaluation. A conceptual framework for such a program is illustrated in Fig. 5.10. With additional data collected, conceptualization of the system can be improved, any environmental consequences of development could be incorporated in the modeling process, and assessment could be upgraded.

Biswas (1992) outlined the process through which environmental-impact assessment (EIA) of groundwater-development projects can be carried out in arid and semiarid developing countries. EIA can be considered as a planning tool which assists planners in anticipating potential future impacts of alternative groundwater-development activities, both beneficial and adverse, with a view to selecting the optimal alternative which maximizes beneficial effects and mitigates adverse impacts on the environment. It is necessary that EIA study be implemented at the initiation of the development project and the environmental changes be monitored once the project is operational so that necessary adjustment can be incorporated as the development progresses. This will help achieve beneficial effects and mitigate

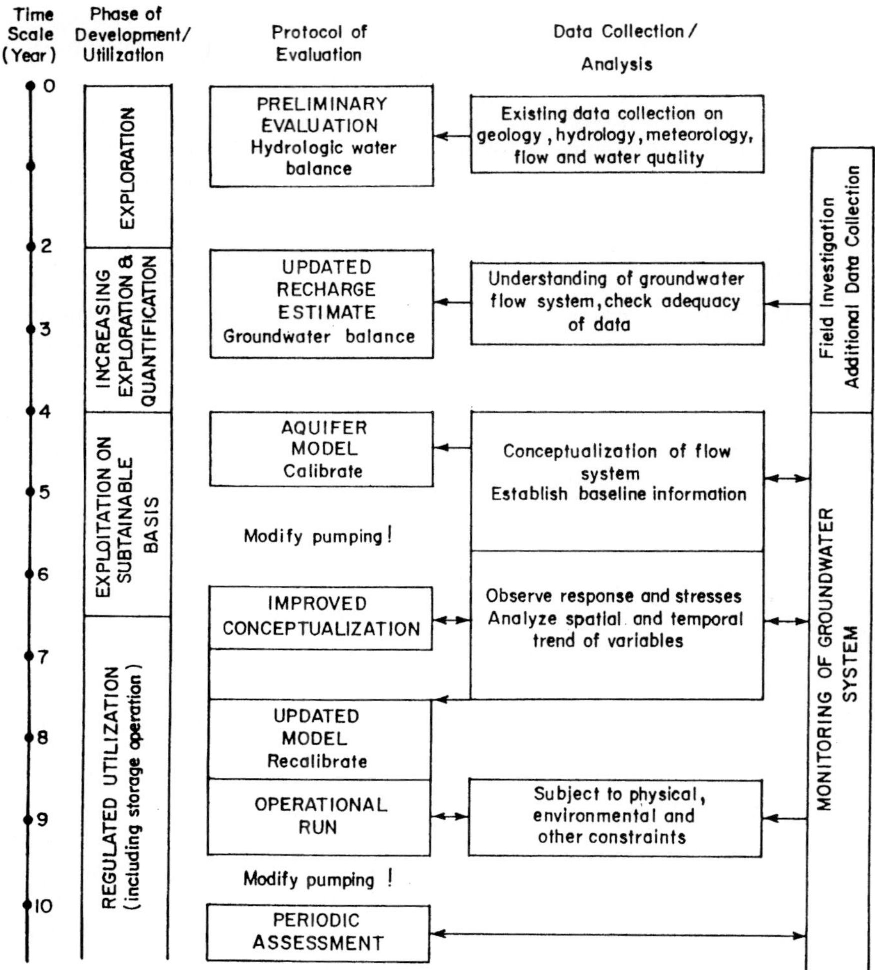

Figure 5.10 A conceptual framework of staged development, data collection, assessment, and evaluation. *Note:* Time scale and protocol of evaluation provided as a guide.

adverse consequences. The objectives of applying EIA to groundwater management could be to (Biswas, 1992)

1. Identify adverse environmental problems that may be expected to occur
2. Incorporate into the development action appropriate mitigation measures for the anticipated adverse problems

3. Identify the environmental benefits and disbenefits of the project, as well as its social and environmental acceptability to the community
4. Identify critical environmental problems which require further study and/or monitoring
5. Examine and select the optimal alternative from the various relevant options available
6. Involve the public in the decision-making process related to groundwater management
7. Assist all the parties involved in the specific development project to understand their individual roles, responsibilities, and overall relationships with one another

It is understandable that a mathematical model study integrating with data collection on system response would provide the best framework for environmental assessment. However, the cost of hydrogeologic, hydrochemical, and geotechnical (if necessary) investigation is very high, and the human and financial resources available for such work, particularly in developing nations, are generally limited. Moreover, this level of investigation and assessment is usually done or can be done only when an aquifer system has been stressed, and by that stage, some negative impact has already been felt. This leads to the fact that only those assessment procedures which utilize basic data that can be collected with limited funding and limited human resources are likely to be applied. In view of this situation, empirical assessment approaches would be the most feasible and general evaluation procedures for a preliminary and rapid environmental assessment of groundwater development. In recent years, researchers dealing mainly with analysis and evaluation of field problems have come up with approaches for determining the susceptibility of aquifers to continual decline of water level, to water-quality degradation, and to land subsidence.

Adams and MacDonald (1995) developed a simple technique for determining the susceptibility of aquifers to water-level decline, saline intrusion, and land subsidence. The technique is based on assigning numeric values to contributing factors and then summing to give an overall grade of susceptibility. Only relative values are used in the final designation (high, medium, low), because of wide parameter variability at individual locations. As a large number of variables contribute to these effects, authors decided to establish three scales separately for water-level decline, saline intrusion, and land subsidence instead of having only one measure of susceptibility. For water-level decline, four parameters are used: hydraulic diffusivity (ratio of transmissivity to storage coefficient), heterogeneity (both lateral and vertical changes), annual recharge, and the volume of aquifer. Since recharge has probably the greatest effect on an aquifer's susceptibility, it has been given double the weight of the other parameters, which have been given equal weight. For saline intrusion, three parameters have been included: hydraulic gradient, hydraulic conductivity, and the presence or effectiveness of any hydraulic barrier. Each parameter has been given the same weight. The susceptibility of an unconsolidated aquifer to subsidence is estimated on the basis of the three parameters: the stratification of the aquifer, the combined thickness of the saturated aquifer (including the confining layer), and the aquitard compressibility. Each parameter has again been given the same significance. The authors also pointed out that once the susceptibility of an aquifer to three different effects of overexploitation has been determined, a more coherent chronology of events can be proposed for the specific case under consideration.

On the basis of field and laboratory investigation, Prokopovich (1991) proposed an approach of theoretically determining the susceptibility of an aquifer sediment to subsidence and to estimate the ultimate amount of such subsidence. First, the total weight of overburden should be estimated using field densities of typical overburden sediments obtained at different depth intervals and at the maximum thickness of the overburden. Then a consolidometer test is conducted on typical samples of sediments of the aquifer system for two loadings. The first loading should be equal to the total weight of overburden minus the relief of the loading due to the existing potentiometric head. The second consolidometer test is then carried out on the already-precompacted samples using the total weight of overburden without subtraction. Additional compaction during this second test would indicate (1) the susceptibility of sediments to subsidence following a decline in potentiometric head and (2) the ultimate amount of possible subsidence. In the case of unconfined aquifers, a decline in water level increases grain-to-grain stress within the aquifer. A similar approach can be used for unconfined aquifers, but the initial loading in the consolidometer should represent the total weight of overburden plus the weight of drained aquifer sediments, i.e., the wet density of the aquifer at the point of their specific retention. This approach requires undisturbed soil samples of the overburden and of aquifer sediments obtained by drilling at different locations and carrying out a series of consolidometer tests.

Several empirical assessment methodologies have been developed for evaluating the groundwater-pollution potential from various point or diffuse sources of contamination at the land surface. The risk of groundwater pollution can be conceived of as the interaction between (Foster and Salas, 1991):

1. The contaminant load that will be or might be applied to the subsurface environment as a result of human activity
2. The aquifer pollution vulnerability, which depends on the natural characteristics of the strata separating the groundwater from the land surface

The aquifer vulnerability is a function of

1. The hydraulic conductivity of the saturated zone, in a hydraulic sense, to the penetration of pollutants
2. The attenuation capacity of the strata overlying the saturated zone, as a result of physical retention of and chemical reaction with contaminants

Foster and Hirata (1991) elaborated on a groundwater-pollution risk assessment methodology developed by the Pan American Center for Sanitary Engineering and Environmental Sciences for the Latin American–Caribbean region. On the basis of data most likely to be available or most easily collected, this methodology provides pollution vulnerability indexing considering

1. The depth to the groundwater table in the case of phreatic aquifers and the depth to the top of the aquifer in the case of confined groundwater systems
2. The lithological character and degree of consolidation of the strata separating the saturated zone of the aquifer from the land surface

Canter and Knox (1985) described two empirical assessment methodologies with application in specific field situations. The first one is the surface impoundment-assessment method based on work by LeGrand (1964). The index is based on four factors: the unsaturated zone, the availability of groundwater (saturated zone), groundwater quality, and the hazard potential of the waste material. The waste–soil-site interaction matrix is the other method developed for assessing industrial solid or liquid waste disposal on land (Phillips et al., 1977). This method involves summation of the products of various waste–soil-site considerations with 10 factors related to the waste and 7 factors associated with the site of potential waste application.

Empirical assessment methodologies should be utilized for relative evaluations and not absolute considerations of environmental consequences. Considerable professional judgment is needed in the interpretation of results. However, they do represent approaches which can be used, on the basis of minimal data input, to provide a structured procedure for preliminary evaluation. Such an assessment is the first step in the systematic evaluation of susceptibility of groundwater systems to adverse consequences of development and in many instances will require follow-up field investigation, monitoring, and modeling study, if necessary, before realistic control measures can be specified.

Groundwater Management

Groundwater management involves the planning, implementation, and operation necessary to provide safe and reliable groundwater supplies. Objectives typically focus on aquifer yield, recharge, water quality, and other environmental factors and on legal, socioeconomic, and political factors. In the past, management strategies have focused on the development of groundwater resources, and projects of various types and scales have been developed and managed in response to the growing demand for water by communities and industries. In spite of bringing many benefits, several of these projects have also yielded serious problems. In some areas all exploitable groundwater resources have been developed and sometimes even overdeveloped. As mentioned earlier in this chapter, overexploitation of groundwater resources has resulted in continual decline of potentiometric levels in aquifers in many areas associated with land subsidence and groundwater-quality deterioration by saltwater encroachment. As the world is becoming more concerned with sustainable and environmentally sound development, the issue of groundwater quality and other environmental consequences due to groundwater overdraft is gaining increased importance and must be integrated with quantity issues in the development, control, and protection of groundwater resources, alone or in conjunction with surface-water resources.

Groundwater utilization has medium- and long-term impacts which affect the future availability and access to the resource. It is very likely that in the near future, in many areas, the emphasis will be one of demand—rather than supply—management. *Demand management* focuses on the water-use side of water-resource management. To reduce water pollution, an attempt should be made to reduce water requirements by water conservation, water pricing, wastewater reuse, and land-use planning. To address this multidimensional requirement, institutional organizations and legal procedures will be particularly important. Comprehensive planning, impact assessment, and containment of adverse consequences are the central themes of groundwater management.

An understanding of the functions of a groundwater system with quantitative and qualitative evaluation of input to and output from the system forms the base for groundwater management. The storage changes in aquifers depend on recharges of aquifers and discharges from aquifers. A very large amount of water may be stored in aquifers. Of this amount, however, often only a small part can be withdrawn without adversely affecting specific conditions in a specific groundwater system. For example, if certain springs in a system are required to continue flowing, this demand will certainly introduce a limitation on extraction. Similarly, environmental considerations such as water-quality deterioration may require a certain level of groundwater head to be maintained which will introduce constraint on withdrawal. In this regard, an understanding of the storage behavior of confined and unconfined aquifers is very necessary. In situations where overexploitation is expected, the only feasible way to sustain the resources is through an attempt to increase the input, e.g., through artificial recharge. Such an attempt should be carefully evaluated. If no increase in recharge can be achieved, the resulting groundwater mining due to overexploitation would be of limited duration. It should be envisaged how long such developments can be continued and what has to be done after the storage has been depleted or the water quality has deteriorated too far.

If water quality is becoming a potential future problem, some strategies are possible to postpone such adverse development. Again, artificial recharge would be of help in specific situations subject to the condition that groundwater movements are extremely slow. It is not correct to imply that groundwater development always has only negative aspects. Under careful management, such as controlled recharge and abstraction, aquifers have proved to be very useful storage reservoirs. The extensive volume available may be used for the temporary storage of excess water available at times, which would otherwise have been wasted. This stored water may then be used effectively in periods of surface-water scarcity without having any adverse consequence on the groundwater system.

Human activities associated with changes in society, and often combined with inadequate land management, adversely affect the recharge, storage, yield, and quality of water. The consequences for groundwater resources of inappropriate planning and use of land are manifold, depending on natural, environmental, and socioeconomic conditions. The land-water system is most seriously affected by spillage of chemicals, leakage from waste-disposal sites, pollution from faulty septic-tank systems, and various agricultural activities. Land-use planning in relation to groundwater-protection management should be seen as a dynamic process. The difficult task in accommodating the various demands for the use of available space, while at the same time preserving the land and protecting the groundwater, can be achieved only through an integrated effort covering local, regional, and national planning.

Key issues and problems

Skilled and well-organized planning is essential for optimal groundwater management. The nature and extent of groundwater planning and management depend largely on the goals or objectives, problem situations, and local conditions. The objectives usually fall under one of the following: (1) to limit withdrawals to the safe yield so as to extend the life of the aquifer, (2) to protect the groundwater basin from some hazards such as intrusion of seawater or pollution from other undesirable substances, (3) to provide water supply to maximize benefits, at minimum cost, and (4) to avoid land subsidence. Too often, however, objectives are not clearly stated. Even if some objectives are stated, criteria and their specific values for evaluating any alternative plan are far from being precise. Apart from natural factors such as safe yield, geohydrologic regime, land use, and geographic location, as well as technical factors such as capacity of wells and well fields, groundwater planning must address socioeconomic factors and administrative factors which impose constraints on development and implementation processes. The socioeconomic factors include water-use patterns, water-demand characteristics, price elasticity, and available funds. The administrative factors include existing laws and regulations, the institutional organization, and decision-making procedures.

Until recently, groundwater management was normally governed by economic-efficiency considerations in the use of water and capital. However, as indicated, there are a number of social and environmental considerations which must be accounted for in the groundwater planning process. The involvement of various agencies and individuals at different levels (e.g., national, sector, project, user) and the existence of various physical as well as nonphysical impacts imply that groundwater planning is often characterized by multiple, often conflicting, objectives and criteria. The relative importance or preference of these objectives and criteria is therefore needed to arrive at a satisfactory tradeoff. It is extremely important to strengthen the institutions involved in the specific sector, supported by appropriate legal measures. But it is also necessary to have an effective mechanism whereby a horizontal coordination (interagency) as well as a vertical communication (various levels) takes place properly. Such a mechanism, although difficult to establish, is of critical importance for the implementation of any management plan.

Planning and decision making are continuous, dynamic processes which must take into account changing conditions and circumstances. All inputs to the water-resource system are subject to variations such as change in demand (due to population growth, economic development, land use), change in quantity and quality of the resources (due to natural development and human activities), the technological options, and the social and political preferences regarding objectives and criteria. The planning process, therefore, is characterized in terms of a set of activities that take place over time and that interact through the transmission and feedback of information. At all stages of this process, questions arise which, if properly posed and adequately answered, will lead to plans that in some sense are better than if questions had not been posed or adequately answered. It should be appreciated that objectives and concerns themselves change over time, and planning must adjust with changing conditions. Here again, a need arises to balance the short-term and the long-term objectives. The short-term socioeconomic gains may have to be traded for the long-term sustainability, with its varied dimensions. This, again, is not an easy task because of the complex interactions involved and the difficulty of specifying various noncommensurate criteria.

In basin planning studies, surface waters, because of their abundance and ready replenishment, are considered the primary source of water supplies. However, because of intermittence, scarcity, and uneven distribution of surface waters, groundwater plays a very important supplementary role. The main difficulty experienced in groundwater management stems from the fact that groundwater is often considered an inexhaustible resource and by and large there is a tendency to overestimate the possibilities offered by groundwater. Some of the management techniques for optimum use of available water resources which have not received due attention in planning are the conjunctive use of surface water and groundwater, the artificial recharge, the indirect recharge, the interbasin water transfer, and the intraregional transfer of water.

Generally, water is considered as a free good; no charge is levied to the consumers. In some cases it is expected that water should be provided at low cost in the amounts and quality desired. Management has been resource-utilization-oriented. However, in order to sustain the availability of resources, the emphasis should gradually change to water-conservation and water-demand management. The mechanism of introducing demand management could be through an economic avenue such as charging for water utilization directly related to the production cost, as well as through legal means of licensing for water abstraction and effluent charging which could be highly effective in reducing industrial water use. Various recent works in this field (Kindler, 1991; Leusink, 1992) have indicated that the application of demand-oriented measures will play a crucial role in future groundwater-resource planning. However, its implementation is not readily feasible because of the administrative effort that would be required to bring about a change in this.

Water users play a dual role in water-resource planning: on one hand they are the ultimate beneficiaries, but on the other hand they are the ultimate managers, whose behavior plays a dominant role. Planning that does not pay sufficient attention to this issue has little chance of being implemented successfully. The desired behavior of water users can be stimulated through proper economic and legal incentives, but it can never be completely enforced by government agencies and will always depend on cooperation. Public participation and community-based management are required for virtually every water-resource-development project. It is desirable, for strengthening the awareness of the user, to involve the beneficiaries at an early stage of planning and implementation and to take social issues into consideration.

Water-resource planning, by nature, takes place under conditions of risk and uncertainty. Uncertainties in groundwater systems planning arise for several reasons, such as inherent variability as well as lack of knowledge of the natural system, processes, and parameters; data collection and monitoring; data estimation and analysis; modeling procedures and conditions; specification and change of objectives and/or constraints; and socioeconomic factors and other operational factors. Traditionally, these uncertainties have been ignored in groundwater planning. It is clear, however, that these uncertainties, when ignored, may adversely affect planning and management decisions and, when built into the planning process, may result in better alternative solutions. Since it is extremely complicated to consider explicitly all the factors causing different types of uncertainties, a pragmatic approach is to incorporate the predominant uncertainties characterizing the specific problem at hand. A review of various approaches suggested in the literature for considering these uncertainties is provided by Das Gupta and Onta (1994).

Issues of groundwater-quality management

Most contaminants are detected some time after entering the subsurface. Weeks, months, or years may pass before a problem is noticed. The contaminant may travel a great distance and affect a large portion of an aquifer before pollution is recognized. Even if the source is identified and removed, and no further contaminants enter the groundwater flow system, contaminants may continue to adversely affect groundwater for a long time. With no remedial action, tens, hundreds, or thousands of years may be necessary to flush the contaminants out of the groundwater flow system. Existing groundwater-contamination problems will become more troublesome in the future because of the long-term factors involved in decay of the pollutants, the slow movement of the affected groundwater body, exhaustion of the soil's ability to reduce the concentrations of the pollutants or to remove specific pollutants, and the ever-increasing volume and complexity of contaminating fluids. Six factors, all local or regional in nature, determine the extent of groundwater pollution. Four of them—soil, geology, climate, and hydrology—exist in nature and are beyond human capacity to control, although they may be modified. The other two factors—land use and growth patterns—are subject to human control through planning and management. Because there are so many complexities involved, the concept of groundwater protection always advocates prevention over remedial measures or rehabilitation of contaminated water. This calls for impact evaluation as development progresses, with adaptation of the protection strategy, policy, and management. Various methodologies, including groundwater models, as reviewed in the previous sections, are used for impact assessment. Vrba and van Waegeningh (1991) elaborated on groundwater-protection strategy, policy, and management, discussed in the following sections.

Protection strategy. The principal objective of a groundwater-protection strategy is to preserve the natural quality of groundwater, particularly for drinking purposes, for the benefit of the present and future generations. This requires the building of an organizational structure with the necessary powers and resources for the creation, coordination, and implementation of a comprehensive groundwater protection strategy and policy. In turn, legislation is needed to regulate the management and control of groundwater protection and quality-conservation programs. It is necessary to have a government institution of groundwater protection, within the ministry or agency responsible for the protection of the whole environment. Vrba and van Waegeningh (1991) described the main goals of groundwater-protection strategy as follows:

- To define the value of groundwater resources, taking into account local, regional, and national interests and needs
- To define the extent, degree, and criteria of groundwater protection and assign their implementation to all levels of relevant government authorities (national, regional, local)
- To establish the legal and institutional bases, regulatory statutes, and standards for groundwater protection
- To establish a system of inspection and an effluent-control system, including fines on the "polluter pays" principle
- To provide methodological and technical guidelines for activities related to groundwater protection and rehabilitation

- To support and coordinate research programs for the development of methods of groundwater protection and technologies for detecting and eliminating pollution in the underground
- To educate the general public about groundwater-protection programs

With respect to the objectives of groundwater-protection strategies, they also emphasized the following activities related to technical aspects:

- Identification and listing of past, existing, and potential pollution sources, and evaluation of their nature and extent
- Operation of efficient groundwater-quality-monitoring programs and implementation of the data obtained
- Provision of technical and scientific assistance in solving the problems involved in groundwater protection and pollution
- Availability of experienced and qualified personnel

At present, the fund necessary for groundwater-protection strategies comes mainly from governmental financial support. In situations where water charges are introduced and fines for failure to observe the laws and regulations for groundwater protection are collected, these provide additional financial support. However, the latter aspect is very difficult to introduce. Currently many countries lack water-charging systems. Regarding contamination, there is some difficulty in identifying polluters. Moreover, relevant laws and regulations are not always implemented. In any case, with time, a charging system is expected to come into existence; fines and charges will be introduced when one fails to abide by the rules and regulations for groundwater protection. This might be the driving means for introducing the protection strategy when the management and control system is under development.

Protection policy. The need for a protection policy has been realized only recently, as development of the resources, which were taken for granted as being available at sufficient quantity and good quality in many places, has resulted in undesirable consequences such as the continual decline of the water table and water-quality degradation. Considering the variability of hydrogeology, protection of all groundwater resources at the same level would be unrealistic in terms of management and control. The criteria for defining a groundwater protection policy depend on (Vrba, 1991) (1) the value of groundwater and its vulnerability, (2) the volume of utilizable groundwater resources, and (3) the current and expected demand for water resources in a given region.

TABLE 5.9 Classes of Groundwater in the United States

Class	Basic criteria	Level of groundwater protection	Remarks
I. Special groundwater	Highly vulnerable, irreplaceable, or ecologically vital	Extremely high	Significant water resources value
II. Current and potential sources of drinking water and waters having other beneficial uses	All other groundwater use or available for drinking or other purposes	High to moderate prevention of contamination, based on technological remedies rather than through restrictions	Majority of usable groundwater in USA
III. Groundwater not considered potential sources of drinking water and of limited beneficial use	Heavily saline or heavily polluted	Usually low migration to class I or II groundwater or discharge to surface water must be precluded	Limited beneficial use

SOURCE: USEPA (1984).

Classification of groundwater resources according to these criteria is always complicated and questionable and requires good knowledge of the hydrogeologic, economic, and social aspects of the concerned area. A good example of comprehensive groundwater protection policies can be found in the United States (Table 5.9), in which three classes of groundwater are defined (USEPA, 1984). A follow-up guidance document (USEPA, 1986) describes the procedures and information needs for classifying groundwater into hierarchical categories on the basis of their value to society, use, and vulnerability to contamination. This, in turn, provides a factor in deciding the level of protection or remediation that the resource is to be provided. Groundwater-protection policy should always be interrelated with the protection of the remaining components of the hydrologic cycle, soil protection, land-use planning, and abstraction of other natural resources, with a view to integrated management and improved allocation. For any region, protection measures should be in conformity with the environmental, social, and economic values of different entities in that region. It is rather a difficult task to come up with a suitable strategy defining priorities and preferences which strike a balance between environmental protection and economic well-being.

Protection management. Groundwater-protection management should go along with the groundwater development, as a part of a water-management plan. Reconnaissance followed by investigations and pilot studies should be carried out to assess the available resources and to evaluate the economically usable resources. As has been pointed out, the abstraction increases in stages and should be consonant with the evaluation process for sustainable development. Groundwater development is a dynamic process; as such, the protection criteria should always be examined and determined as development progresses. Vrba and van Waegeningh (1991) identified two categories of groundwater protection management as (1) general protection of groundwater resources and (2) comprehensive protection of groundwater around public water supplies. An elaboration, given by the authors, on these two categories is provided in the following sections.

General protection of groundwater. This is based on the assumption that all effectively accessible groundwater resources are or will be tapped for drinking or other purposes, and therefore their protection is desirable. Under water-management plans, national, regional, or local governments and water authorities should therefore bear the responsibility for, and financially support the protection of, unused groundwater resources.

Implementation of general protection of groundwater resources calls for the following activities:

1. Investigation of the groundwater system and determination of its vulnerability
2. Identification, listing, and assessment of the existing and potential pollution sources
3. Monitoring of the groundwater system

Comprehensive groundwater protection. This is concerned mostly with public water supplies. The groundwater resources utilized for public drinking-water supplies are protected by protection zones, referred to as *wellhead protection areas* in the United States, usually comprising two or three levels of protection. The main purpose of groundwater protection zone delineation is to protect drinking-water supply wells or wellfields from pollution and provide the population with water which meets the standards for drinking water. In several countries, wellhead protection is an obligatory part of groundwater protection policy and strategy programs, and is based on the relevant legislation.

Comprehensive protection of public water supply wells requires

1. Cooperation between national, regional, or local water authorities; waterworks companies; and land users (particularly farmers)
2. Establishment of a general concept of groundwater and land management in protection zones taking into account rural community needs

3. Implementation of technical, institutional, legislative, and control measures and regulations in protection zones

The techniques and methods employed for delineating groundwater protection zones of water supply wells depend primarily on aquifer permeability and hydraulic conductivity (porous, fissured, or karstic), complexity and vulnerability, properties and thickness of the unsaturated zone, quantities, and properties and degradability of potential pollutants and their distance from the wells or wellfield.

First-level groundwater-protection zones protect the well and its immediate environment from mechanical damage and direct pollution. Their extent is usually small—several tens of square meters at the maximum. They exclude all activities.

Second- and third-level groundwater-protection zones are extensive (several hundreds of square meters to several square kilometers) and include the discharge areas, the cones of depression (zone of influence) around pumping wells, the recharge and contribution areas, and other vulnerable areas of a given water-supply system.

In several European countries, the second-level zones cover areas having a delay or residence time of 50 to 60 days. It should be emphasized that the delay time has been determined so as to protect water-supply wells from the risk of groundwater microbial contamination, but it may become inadequate for viruses and certain chemical pollutants.

In general, the level of restrictions and prohibitions in groundwater-protection zones decreases with the distance from a well or wellfield. Second- and third-level protection zones cover significant areas of ground, frequently including arable land. For this reason, the overprotection of water-supply wells is not desirable because restrictive measures of exclusion of land from farming activities lead to economic losses. On the other hand, the underproduction of wells and wellfields may cause groundwater pollution requiring long-term and costly remedial action. Overall, a delicate cost/benefit balance is involved when determining the extent of protection areas.

Conjunctive use of surface water and groundwater

Conjunctive use refers to the coordinated and planned utilization of surface-water and groundwater resources in an integrated manner, appreciating interdependence as well as different characteristics of the two sources of water. Conjunctive or integrated water use has long been recognized as an effective strategy for the optimum development and management of water resources. The role of conjunctive use is particularly significant today, more than ever before, when there is a growing need to satisfy the ever-increasing water demand within the limited resources available while considering various social and environmental impacts of water utilization and protecting the resources for sustainability. Most conjunctive-use schemes throughout the world have developed for one of two reasons: (1) the demand on either surface water or groundwater resource was too large or (2) the quality of extracted groundwater was sufficiently poor to require mixing of groundwater with surface water. To some extent, conjunctive use has been forced on countries where surface irrigation sources have been fully utilized and the way forward must involve supplementing surface sources by groundwater pumping. This is the common feature of many projects on the Indo-Gangetic plain. On the other hand, conjunctive use is being considered to alleviate environmental consequences of overexploitation of groundwater, such as land subsidence in large cities. A classification of conjunctive use for managing surface water and groundwater is presented in Fig. 5.11. However, efforts are still to be delegated to demonstrate in practice an orderly plan of utilizing the two resources considering timing, requirements, and constraints; and to achieve a benefit more than what could be achieved from utilizing either of the resources.

Both surface water and groundwater are part of the hydrologic cycle and are interconnected. It is interesting to note that water-resource-utilization projects, from either surface water or ground-

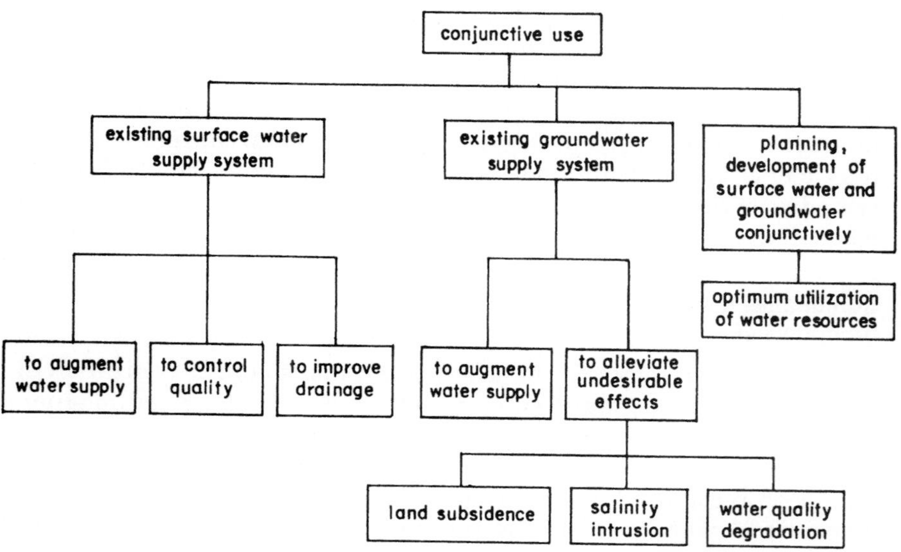

Figure 5.11 Classification of conjunctive use.

water resources, tend to increase this interconnection. Impounding water in a reservoir or barrages on rivers definitely causes some additional seepage, while lowering the water table by well fields situated near surface-water bodies or rivers will induce recharge. Besides these technical interdependencies, another important factor to consider is the time lag between the occurrence of maximum flows in surface systems and the maximum groundwater levels. While the seasonal surface runoff is already receding, the groundwater level might continue to rise. When viewing the watershed concept, with the surface watershed and the underlying groundwater watershed consisting of two parallel reservoirs, one might consider one watershed to have quick-response characteristics (surface runoff) and the other, a very slow response (groundwater movement). Because of this difference between their respective regimes, the conjunctive use of surface water and groundwater offers a unique chance to improve the performance (output, reliability, versatility) of a water-resource system. The possible applications of this concept of integrated use of surface water and groundwater are indicated in the following sections.

Development of storage. Although surface reservoirs have been widely used to regulate variable streamflows, in many regions proper storage sites are not available or are no longer as plentiful as they once were and construction costs of such developments have increased considerably. Moreover, the anticipated environmental consequences are likely to make the new storage schemes less attractive. Underground aquifers can play an important role in such cases. Comparison of the capabilities and characteristics of an underground system with those of a surface system shows that

1. The storage capacity of a groundwater basin is analogous to the storage capacity of a surface reservoir, with the loss of water to evaporation that is characteristic of surface storage. There may be internal losses, depending on the geologic characteristics of the basin. The life of a surface reservoir depends on the sediment load of inflowing streams, whereas the life of an underground reservoir may be affected by the water quality.
2. The rate of deep percolation and subsurface inflow to a groundwater basin corresponds to the rate of inflow into a surface reservoir.
3. A subsurface delivery system has a starting point (natural and human-made recharge areas) and a terminal point (wells), just as does a surface system. The potentiometric pressure or the groundwater table may be linked to the hydraulic gradient line elevations in surface distribution system.
4. Capabilities can be calculated for a groundwater system with the use of an equation that describes the flow characteristics of the basin.

With proper operation of surface-water and groundwater systems, additional storage of water in subsurface environment can be developed for the excess surface water which would have otherwise been lost.

Drought mitigation. Groundwater in most places is treated primarily as a water-supply source rather than as a potential long-term carryover storage source which could be utilized for drought mitigation purposes. By relying primarily on surface water, when available, a potentially cheaper source of water is utilized, while falling back on the groundwater to overlap periods of water shortages. Because of the considerable carryover storage of the groundwater body, groundwater can be used over an extended period of consecutive drought years. If the joint management of both resources ensures periods of replenishment of the aquifer during years of abundant surface flow, the amount drawn from groundwater on a short-term basis might exceed the natural recharge rate; i.e., groundwater mining occurs. Within the scope of conjunctive use it is important to facilitate artificial recharge of the aquifer to ensure greater efficiency of storage during the generally short periods of abundant surface flows. Groundwater aquifers can be separated into upper reserves to be used for storage capacity in nondrought years and lower reserves to be used in drought years only. These capacities should be replenished when excess surface flow is available.

Prevention of waterlogging and salinity. Conjunctive use of groundwater with surface water for irrigation is imperative from the consideration of waterlogging, salinity, and subsurface drainage in many areas. For many alluvial plans, the water table exists at shallow depths, and with irrigation exclusively from surface water, these aquifers are recharged unintentionally with further rise of the water table, creating waterlogging problems in the absence of adequate subsurface drainage. In the long run, such an irrigation practice may even be counterproductive. Groundwater utilization lowers the water table, helps subsurface drainage, and is thus an antidote for this impending adversity. Thus, it is necessary to resort to groundwater utilization to maintain the subsurface hydrologic equilibrium for optimum agricultural production and making irrigation sustainable on a long-term basis.

Other possible applications. The reliance on two types of source to meet the water demand of a region increases the number of development options considerably. It could be possible to plan a staged development considering both sources. This helps reduce funds needed at a given time and allows an implementation schedule to follow the development trend of demand closely. Since the incorporation of the conjunctive-use concept helps avoid oversize project elements which have unused excess capacity, it also contributes to the overall profitability of development. Moreover, with the consideration of storage in subsurface environment, the liver-storage volume of the surface storage components can be reduced, or more space can be used for flood control. This also contributes to conservation of water within the region since flood can be withheld, instead of being discharged downstream. Furthermore, because groundwater can be used as an alternative resource, seepage losses assume a different importance within the water resources system and in some cases can be regarded as artificial recharge.

A special field of application of the conjunctive-use concept is concerned with the reduction of pumping costs. Having different charges for electricity for daytime and nighttime consumption, pumping could concentrate on the hours of low-priced energy to fill the surface storage facilities, from where water demand can be met via gravity distribution. Thus surface storage units are used to back the groundwater utilization. Since pumping requires energy input, it is particularly beneficial if the surface-water-resource-development component of the project also includes hydropower generation. The refill of reservoirs during nighttime might be used again to generate electricity the next day; thus conjunctive use can, within certain limits, also be interpreted as a pumped storage scheme. Another application of the conjunctive use is to improve water quality. Treated sewage might be recharged into groundwater. Through the passage in porous media, various processes tend to purify the water, thus recycling the water even for multiple usage.

Implementation issues. The conjunctive use and management of surface water and groundwater is the most appropriate and prospective strategy for optimum utilization of water resources in many areas. Because of certain hydrologic, topographical, technological, and socioeconomic factors, both sources have certain limitations as well as potentialities. The development and operation of a conjunctive-use system will have to deal with one of the following situations:

1. *Systems which are planned and designed as conjunctive-use systems.* In these situations, there will be ample scope to exercise management options right from the inception to the operation phase.
2. *Systems which are planned and designed as either surface-water or groundwater systems, usually the former, and where the complementary system of conjunctive use is in the process of development.* Here the management options are constrained by the inflexibilities of the developed system.

3. *Systems in which both surface water and groundwater have developed in the course of time with only an ad hoc, rather than a planned and designed, conjunctive-use system.* In these cases, the management options are limited mostly to the operation of the system.

Although there has been increasing realization of the suitability of the conjunctive-use system for water development, particularly in the irrigation sector, very few projects have been implemented so far on the basis of this concept. Lack of understanding of the functioning of and interaction between water-resource subsystems, lack of data and information, lack of proper institutional structure, and conflict of interest among different water-use sectors could be cited as some of the constraints. Coe (1990) elaborated on physical and operational constraints, financial and institutional constraints, and legal aspects in implementing conjunctive-use programs. A substantial amount of literature is available on conjunctive-use study using different modeling approaches, particularly considering the engineering aspects (Onta et al., 1994). All these studies presume the existence of a central management body, responsible for all water-related problems, and have the proper authority to enforce actions defined by optimal development strategy. In reality, the lack of such a management unit can be regarded as the chief hurdle to implement conjunctive use of different types of water resources. There is hardly a project area where at least a few wells or surface-water-diversion facilities would not exist. Thus the basis is generally given to start with, and the need for a dedicated agency is to be stressed to develop a comprehensive utilization strategy. A conjunctive-use model should identify and incorporate important environmental, socioeconomic, and institutional factors, in addition to the physical ones, characterizing the physical problem.

It would be difficult to define the scope and organization of a water-resource-management body, since the legal and historical background of the countries varies. However, a few suggestions can be made emphasizing the need of a basic assessment of resources and potential allocations. In order to ensure a balanced development for a region and the country as a whole, both the regional and the central government should be represented. Since water supplies for communities are always a priority of water-resource management, the municipalities involved should be represented as well. Major water users (private, collective, and state-owned) are involved since they are expected to access better and bigger shares from water. Environmental aspects should be included in all considerations either by the direct involvement of the corresponding governmental groups or by environmental groups. The coordination of the objectives of such a heterogeneous group is a formidable task. In order to ensure evenhandedness and independence of the water management agency, the setup of an autonomous organization with technical departments and a decision-making body, representing the interest groups concerned, is suggested.

Role of models in groundwater management

Groundwater modeling has emerged as a powerful tool for groundwater planning and management. Modeling can serve multiple purposes: providing a framework for organizing hydrologic data, attaining a consistent qualitative understanding of the hydrologic system, quantifying the properties and behavior of that system, and allowing quantitative prediction of the response of that system to externally applied stresses. Evaluation of hydrogeologic data and analysis of monitoring data on water level and concentration can provide an understanding of the flow and transport processes in the physical system. However, simulation models provide a more appropriate and rigorous method for integrating all the available data together and for evaluation of the response of the physical system when subjected to changes in conditions and external stresses. However, it is to be cautiously noted that modeling is never a straightforward exercise and without limitations in practice. What hydrologists, engineers, and managers may be wary of is that all too often misstatement or overstatement of the modeling application occurs. Just as it is with many disciplines, so it is with modeling—a little knowledge can be a dangerous thing. Mathematical models are approximations which simplify to gain a better understanding of complex, natural phenomena. As such, certain assumptions are made in the development of the model which must be adhered to in its application.

The ultimate use of models should be to assist in making management, planning, and policy decisions by providing useful information. It should be recognized that modeling is only part of an overall groundwater investigation that integrates various aspects of data collection and analysis. The accuracy of a model prediction depends on the correctness of the conceptualization, the accuracy of the hydrogeologic information (e.g., initial and boundary conditions), the reliability of the estimated parameters, and the accuracy of the historical response of, and stresses on, the system. Because of the dynamics of the planning process, it may happen that the answers derived from the models will suggest that the original problem was not well conceived and needs to be reformulated; hence the role of models is iterative. They are used to provide information to aid in decision making. With equal value, they provide information that is fed back to aid in redefining the problem. A further essential point concerning the role played by models in the planning process is that the information they provide must be quantitative and in sufficient detail and accuracy to provide guidance to the decision maker.

There are two basic techniques of modeling for groundwater planning and management: simulation and optimization. Simulation models approximate the behavior of a complex groundwater system, representing all its characteristics by mathematical relationships, and provide the operating policy of the system when subjected to a given management alternative. For real-world problems, various numerical models using finite-difference, finite-element, methods or characteristics or boundary-element methods have been used (van der Heijde et al., 1985; Anderson et al., 1993). Numerous standard textbooks exist on this subject (e.g., Bear and Verruijt, 1990; Anderson and Woessner, 1992). A simulation model is used repeatedly to predict the response of the system for each management alternative, whereas an optimization model describes how best to utilize groundwater resources under given objectives and constraints. Multiobjective optimization approaches have also been proposed to consider multiple objectives and criteria and their tradeoff. However, optimization is normally feasible only for well-defined problems under some simplifying assumptions. In response to the need to systematically relate the hydraulic behavior of the groundwater system to the optimal design/operating policy under the prevalent management objectives and constraints, there has been a tendency to combine simulation and optimization (e.g., Gorelick, 1983; Willis and Yeh, 1987). A review of groundwater-management models is provided by Das Gupta and Onta (1994). Whatever modeling technique is adopted, the objective is to provide useful information and insights for decision making within the limitations of each technique. Thus, models are regarded as essential tools to supplement, but not replace, the experience and judgment of planners and managers.

Knowledge gap and research needs. With the extensive development (and sometimes overdevelopment) of groundwater resources, attention is being diverted to a quantitative analysis integrating both quantity and quality aspects and other environmental consequences depending on the practical situation. The current state is such that the function of any mathematical model study is to provide an absolute rather than a relative evaluation. The present and future theoretical and experimental investigations at both field and laboratory levels should address the following main issues: (1) how to better understand individual processes and reactions, (2) how to translate these processes into equivalent mathematical form to facilitate site-specific simulation capability, (3) how to validate the model and ascertain the reliability of prediction considering uncertainties of the physical systems, and (4) how to integrate interdisciplinary technologies needed to solve groundwater-contamination problems.

There is a lack of understanding of many important physical processes and system dynamics in groundwater environments. Processes in the unsaturated zone—those associated with the flow of immiscible contaminants, coastal aquifer saltwater intrusion, and complex chemical interactions and phase changes—have received little attention. Consequently, engineers and scientists have limited ability to predict the effect of remediation on the fate and transport of a contaminant in an aquifer, the effect of changes in pesticide and fertilizer applications for crop yields, and the effect of salinity buildup on crop losses and human health. The initial need for research lies in understanding the physical and chemical processes taking place in different subsurface environments and in developing mathematical representations of these processes. With proper understanding of a system's dynamics, management models can then be formulated and applied.

Most groundwater models consider systems to be deterministic and represent only the average behavior. Along with recognition of the need for proper management of groundwater resources to ensure sustainable development, there is an increasing concern about the reliability of making long-term predictions, taking into consideration the uncertainties of system components. The only types of model error that have received some attention are those associated with model parameters, especially those related to physical flow processes. There have been very few studies exploring uncertainty associated with chemical and biological process parameters and how they influence groundwater-quality management. Further research is needed in this area. All parameters must be viewed in terms of their spatiotemporal variability and the uncertainty this generates in managing groundwater systems. In most cases, the values of aquifer parameters and variables are based on laboratory tests, and there is no well-defined relationship between these values and the actual field values. Moreover, the spatiotemporal discretization (also the representative values of the parameters or variables) used in the simulation and optimization models significantly influences the results and optimal decisions. Research on this aspect of scale representation toward understanding field-scale models would be highly beneficial.

There are also serious problems associated with numerical computation of the best management alternative. Simulation, especially under a stochastic environment, can be infeasible for evaluation of a large number of management decisions. Optimization methods are in some cases limited either because they require a special mathematical structure or the current optimization algorithms are not computationally efficient enough to analyze highly complex groundwater-management problems. There is also a need to expand our understanding of the criteria by which alternative management programs should be judged and the types of constraints that need to be imposed. The subject of groundwater management, within the broader context of environmental management, is interdisciplinary in nature, and it would be beneficial to pool the knowledge of experts in related disciplines. There are many issues (e.g., ecological, institutional, social, legal, political) which cannot be easily quantified and considered in the modeling framework. Yet, these multidimensional issues must be addressed simultaneously for any major environmental policy formulation and resource development. Research that integrates issues of water quantity and quality with these different management factors would be valuable.

There is a huge gap between the state of the art and the state of practice in the use of management models for water management, in general, and groundwater management, in particular. Very few real-world applications of management models have been noted (Rogers and Fiering, 1986; Willis and Yeh, 1987; Biswas, 1991). Some reasons for this state of affairs are the lack of knowledge about the problem, lack of financial resources and technical expertise, complexity and limitations of optimization models leading to their lack of credibility especially in implementing the model results, simplifying assumptions, lack of good monitoring and information on the system, and various other institutional factors. In an overview paper on modeling applications, Loucks (1992) stated that communication between the system analysts and planners or managers is the most important factor to be improved to close this gap. Lack of understanding of the decision-making process was also considered by Biswas (1991) to be one of the most critical issues in the application of models for environmental management, especially in developing countries.

Some new developments in the field of water-resource-systems modeling—motivated by the developments in computer technology that are aimed at facilitating interaction and communication—are interactive approaches, computer graphics, geographic information systems, and expert systems. These are also active areas of current research. Applied research considering these new developments and emphasizing real-world applications is recommended. Field testing of simulation-optimization methodologies has also been limited, and results are not always presented in the scientific literature. Feedback from field investigations is essential.

Data requirement and role of monitoring

Data on both the hydrogeologic aspects and dynamic aspects of the physical system are needed for the management of a groundwater system. These data are collected from a network of observation wells called *monitoring wells* which extends over a specified geographic area. The functions of the groundwater-monitoring network are threefold. First, it needs to establish the basic structure, geometry, and hydrologic characteristics of the flow field. Second, it needs to establish the dynamic reaction of the system under natural and stressed conditions. It is only after these two steps have been determined and a decision has been made as to which critical quantities to monitor, that the third step, the final network density and configuration, can be defined. Because of the inherent uncertainty and heterogeneity of natural hydrogeologic systems, the true monitoring network cannot be specified before gaining some basic knowledge about system configuration and dynamics. At the initial stage, the information goal of the monitoring program is to facilitate understanding the hydrologic processes. As development progresses, the data-collection system must meet the needs of users and help them make right decisions on management issues. The most important technical data and information needs for groundwater planning and management are listed in Tables 5.2 and 5.3.

Monitoring methodology. The basic purpose of a methodology for monitoring groundwater systems is to provide a framework for the planning and development of a monitoring program. The methodology should serve two roles simultaneously: (1) assist a designated local monitoring agency in designing and implementing a monitoring program and (2) guide governmental agencies at the regional and national levels in establishing realistic monitoring priorities, not only in terms of what should be monitored and when but also in terms of timing and funding. One can expect an infinite number of combination of water use, probable consequences of unplanned development, hydrogeologic situations, and monitoring program. Therefore, persons involved in a monitoring program will be required to exercise professional judgment in order to interpret and apply the methodology to the specific local situation they encounter.

The methodology needs to be phased to accommodate the fact that the monitoring programs are dynamic. As a monitoring program develops, it must be revised to meet the needs of changing objectives and new information. The primary objectives of the first phase are to

1. Divide up a region on the basis of administrative, physiographic, and priority considerations
2. Identify and make a preliminary ranking of priority concerns in each area which needs immediate attention
3. Examine those concerns showing the greatest potential of adverse socioeconomic and environmental consequences and determine what data and information already exist, what data and informa-

tion are needed, and what it will cost to obtain the needed information

The second phase would deal with the design and implementation of new monitoring programs to provide additional data and substantiate some of the observations made in the first phase. Finally, the third phase would concentrate on the implementation of any control or abatement procedures, monitoring their effectiveness, and revising the monitoring activities as needed. Tinlin (1981) provided a detailed elaboration of monitoring methodology under these three phases of activity.

Monitoring strategy. Groundwater data collection and monitoring can be viewed as a time-based evolutionary process. The process begins with the mere collection of data, and with time, it focuses on the information product of the system. In fact, the data-collection program along with data analysis and evaluation should be viewed as a system for management decision. The personnel operating the monitoring effort should have a clear understanding of the information expectation of the data-analysis procedure (say, model studies) so that the data collection and observation network can be modified or upgraded accordingly as time progresses. The collection of data should be undertaken with a good understanding of what, and how much, is needed, and a high degree of precision and accuracy should be consistently maintained. The observations of groundwater levels are usually carried out at points distributed over a large area, and measurements are taken at much shorter intervals, while the monitoring network for environmental observations such as water quality and land subsidence would be according to the specific problem situation, and the sampling could be at different intervals. However, the quality of data is impaired if there are gaps in the records. Records, even those of high quality, may become useless if they are not continuous or observed frequently enough. As more decision makers become aware that their decisions are sensitive to the quality of data that they use, the issue of quality assurance is becoming more important.

The design of a groundwater data-acquisition system and the scale of operations should conform to the physical framework of a geohydrologic setup, taking into account the prevalent meteorologic and hydrogeologic conditions. The dynamic nature of the groundwater regime should also be kept in view. The input and external stresses on an aquifer vary widely in time and thereby modify the geometry, boundaries, and storage of an aquifer. The conceptualization of an aquifer system at the initial stage of data collection would be an approximate one. The groundwater data-acquisition program should be planned in such a manner that measurements and information on a wider base are available to enable an improved conceptualization and analysis of the system as development progresses.

The monitoring strategy addresses all phases of activities to acquire information about the functioning of the physical system in order to facilitate the decision-making process for management purposes. This basically includes the following:

1. Delineation of the monitoring area
2. Investigation of the groundwater system to be monitored
3. Site selection for monitoring stations
4. Identification of critical concerns and inventory of potential and actual contaminating sources
5. Design of the monitoring network
6. Establishment of a data-acquisition system, specifically the methods, frequency, and extent of measurements, observations, and analyses in the field and in the laboratory and methods of water sampling
7. Selection of field and laboratory instruments and computer hardware

The software part includes (1) database management, especially methods of data handling (tranmission, processing, storage and retrieval) and analysis (statistical and visual method charts, maps, and models, e.g., conceptual, stochastic, transport forecasting) and (2) methods of data transmission into information and mechanisms of information transmission to the user.

Management strategy

Developing a national groundwater management strategy is an immensely complex task in both technical and institutional terms. There must be sufficient flexibility to adapt to circumstances and to design the ultimate regulatory framework as one learns about the system and its interaction with the environment. Circumstances should shape the legislative framework, and not the other way around. The principal components of national strategy are

1. To provide for regional, state, and local management of groundwater resources with necessary central backup, periodic assessment, overview, and support in terms of funding and staffing
2. To preserve the present quality of groundwater available for drinking-water supplies and other uses requiring high-quality water

3. To coordinate regional or state programs and facilitate interstate dialog and other arrangements to cope with interstate groundwater problems.

The fragmentation of authority in and among various agencies is a common and often perplexing issue across a country. Conflicting goals of water agencies and users should be reconciled through interaction in the development of a groundwater-management plan by all interested parties or their elected or appointed representatives. The key to a strategy's success is a cooperative relationship among all levels of governments. This will not be an easy task. Cooperation requires incentives, and one needs to ask what incentives there are for other central government agencies, multiple state agencies, district and local government to cooperate. A closely related issue is the matter of authority itself—how much uniformity or variety is needed in the matter of jurisdiction over groundwater management—in other words, what the roles of different water agencies are as far as the management is concerned.

From the technical viewpoint, one has to deal with interrelationships: between quality and quantity of groundwater; groundwater vis-à-vis surface water. A pertinent issue here is whether there is enough water of the right quality for the use citizens want to make of it. Groundwater and surface water are closely related in the hydrologic cycle and must be considered together in a comprehensive water-resource-development and management program.

Legislation. Water legislation in general and groundwater legislation in particular provide the means for implementing and enforcing any desired water policy effectively; but legislation by itself does not constitute a panacea for solving problems. Any legislation, to be effective, must be the result of water policy decisions which should precede its enactment and should be based on the political, technical, economic, social, legal, and institutional factors prevailing in any one country. What is more, water legislation is strongly influenced by the legal systems of the countries and must take into consideration the sociological, religious, and philosophical character of the people of any particular country or region. The purpose of water legislation is to ensure, on the basis of water availability, the optimum use of that resource and its conservation, in order to satisfy present and future water demands for every type of utilization. This may be achieved by bringing under unified, coordinated, or centralized administrative control the existing and future uses of water. A basic "water act" should include a statement of national policy on water-pollution control so that there is a clearcut legal basis on which to promulgate subsequent detailed regulations.

In most countries, at the initial stage, existing environmental laws are used to protect groundwater. Laws intended especially to protect groundwater usually start only when great concern is expressed, particularly by the public, for the protection of groundwater, in particular its quality, as being degraded by pollution. In many cases, considering the incidence of threats, legislative regulatory programs are designed around a series of permits aimed at specific categories of human activities. This form of legislative control may not work effectively. First, it is not possible to come up with a comprehensive list of activities to be regulated; second, local variability in groundwater vulnerability is extreme, thus complicating general consideration of groundwater consequences of various activities. Some aquifers are highly vulnerable to exploitation and contaminants released at the land surface, while some are virtually immune to local surface conditions. As such, without having a thorough evaluation, if legislation is directly based on a few incidences of threats, we will have a system which does not respond effectively to local variability. The consequences of such an approach would be to either set such a high level of protection that progress and growth would be excessively inhibited or to set such minimal standards that significant deterioration could occur while we believe that protection exists.

The preceding discussion clearly indicates that groundwater laws and legislation should be of rolling type requiring modification every few years in the light of operational experience and as a result of improved understanding, increased knowledge, and availability of data. Virtually every human activity has the potential to affect groundwater quality to some extent. The risk of contamination is controlled by the vulnerability of the aquifer to contamination and the type, amount, and location of contaminant discharges to the aquifer that might result from the land-use activity. Of particular concern are discharges and land-use activities in major aquifer recharge areas or near water-supply wells. Thus, another approach to the control of contamination sources is to prohibit or restrict certain land-use activities within designated critical areas. Such controls typically are applied at the local level to provide different levels of protection to aquifers on the basis of their classification.

Another factor considered in the groundwater-protection act is the water-quality standard. There are numerical standards for water quality for different uses. For groundwater, geohydrologic conditions determine both the quality of natural groundwater and the attenuation of contaminants: natural or synthetic (human-made). These conditions and their effects vary tremendously, such that some groundwater is naturally contaminated beyond any possible use while other groundwater is nearly pristine. Some contaminants in groundwater move rapidly with little attenuation, while under other conditions the same contaminants are rapidly removed as water passes through the ground. As such, determination of acceptable groundwater quality should be exercised on a case-by-case basis, using natural-quality drinking water and stream standard as an index for guidance. One cannot, for the sake of economy and effectiveness, regulate an entire aquifer by requiring groundwater quality to meet the uniform conditions of quality. Aldwell et al. (1991) dealt with the history, development, and content of groundwater-protection legislation and regulations as they relate to land-use planning.

Political support. Political support of the groundwater management and protection efforts is a must in developing and implementing a successful management program. There are two primary reasons why strong political support is essential: (1) developing and implementing groundwater management and protection programs is expensive, and (2) effective programs require cooperation of several governmental and nongovernmental entities. Publicizing groundwater problems and efforts to protect groundwater is very important, and political authorities can play a significant role in creating a general public awareness. For most people, groundwater is "out of sight, out of mind," so there can be only a low level of awareness among many decision-makers and the general public.

Growth management and land-use control are two basic factors to be considered in developing groundwater protection programs. Wherever there are strong pressures for industrial, commercial, or residential development, the authority concerned with groundwater management must seek support of the political constituencies for growth management in its effort in developing and implementing groundwater-protection programs.

Political support for groundwater-management/protection planning can be developed in different ways. In some situations, political support develops from the general public being affected. The people's voice, through organizational meetings and campaigns, provides the thrust to develop groundwater-protection programs. In other situations, the government or nongovernment authorities take the initiative, which helps in organizing public support to the adaptation of groundwater-protection programs. A combination of top-down and bottom-up activities is needed to accord political support for groundwater management and protection planning.

Training needs. One crucial issue in groundwater-development planning and formulation of management strategies considering the quantity as well as the quality aspect is the scarcity of personnel adequately educated and trained at all levels in hydrogeology and different branches of geosciences and engineering. The critical disciplines include surface hydrology, study of flow phenomena below the land surface, study of transport of substances by moving water, and analytical skills in the broad field of decision making. The technical cadres must span a broad spectrum of trained professionals, from surface hydrologists, hydrogeologists, hydrogeochemists, and water-resources engineers, to regional managers of water-resource systems and to technicians working in the field. In this way, one can approach the optimal utilization of water resources and maintain adequate water quality with a reasonable chance of success.

Academic training of groundwater professionals needs some reorientation. Groundwater training has historically focused more on that part of the soil-aquifer system that lies below, rather than above, the water table. Investigations of groundwater pollution also involve the vadose zone, particularly for contaminants originating from land surface with diffuse sources. In certain situations, pollution sources and the vadose zone should receive more consideration than the aquifer itself. Soil chemistry, water chemistry, and geochemistry are all important aspects of groundwater-pollution evaluation. A new approach for the academic training of hydrogeologists who specialize in groundwater quality includes specific training on pollution sources, soil chemistry, and geochemistry, including courses in sanitary engineering, mining, agriculture, and other fields. Water and pollutant movement through the vadose zone must receive greater attention in academic education.

The public lack of knowledge of groundwater is extensive. Presently there is little concern for problems that seem decades away. To a large extent, the solid earth is being used as the disposal medium sometimes with forethought, but often without much thought at all. Sooner or later, contaminants will appear in groundwater, and it may take decades before the general public is aware that contamination exists. There is certainly a need for developing an understanding in groundwater environment and how the transport process takes place. Methods of providing public information certainly depend on the level of education, culture, and traditions of the people; the staffing and financial resources available; and the message itself. This could be through talks, displays, and news media using television, radio, and newspapers. Another long-term proposition of developing public awareness is to introduce this at an early age in primary-school curricula, especially within a general introduction to protection of the environment.

An integrated approach. The interdisciplinary nature of the groundwater-management problem requires integrated approaches to its solution. Korfiatis (1991), on the basis of experience in the United

States, analyzed the multidisciplinary nature of the groundwater-pollution problem and proposed an integrated approach to address the problem entities and their interrelationship. To provide a background on this aspect, salient features of this approach taken from this reference are incorporated in this section. Figure 5.12 identifies eight entities which have a major role in the water sector, particularly groundwater management and quality maintenance. These entities often have conflicting interests and are driven by different motivations. Their common interests and needs are to be identified, and groundwater-management policies must be formulated considering their major concerns and responsibilities.

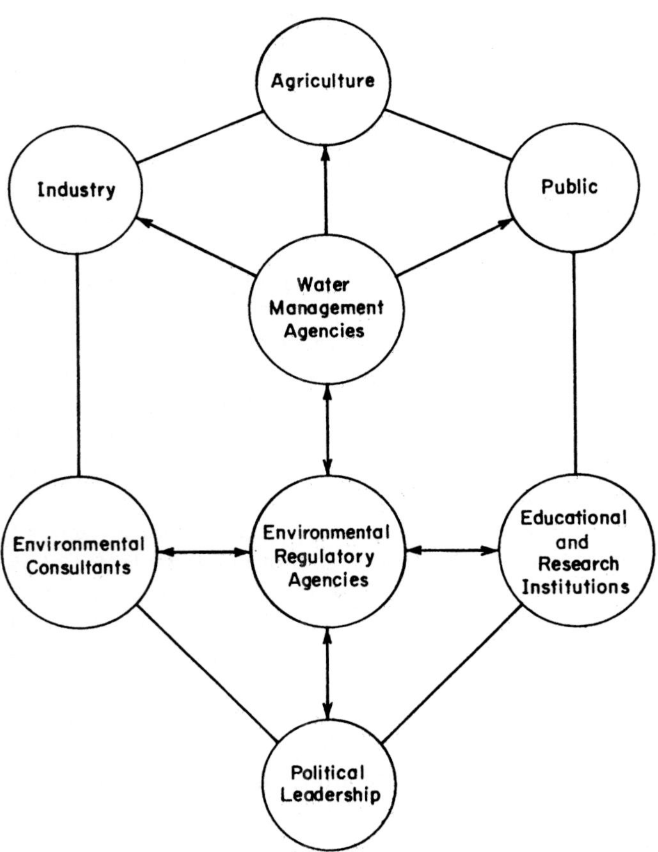

Figure 5.12 Components of an integrated approach to groundwater management.

Environmental regulatory agencies are vested with the authority to formulate, promulgate, and enforce regulations and develop policies and management strategies to ensure sustainable use of groundwater and for the protection and restoration of groundwater quality. Water-management agencies deal with all water-related problems responsible for formulating and implementing the integrated water-development policies with due consideration to socioeconomic and environmental aspects. The public plays a very vital role in all aspects of groundwater protection and management. The active organized participation of the public can have a positive impact on waste reduction, rational use of water, allocation of public funding for groundwater protection, policymaking, and legislation development. Irrational and irresponsible public involvement, however, may be detrimental. In this regard, understanding of and education on the groundwater system and its environment is very essential. Education and research institutions contribute significantly to solutions of groundwater development and management problems by providing training and technological support, developing new technology, and facilitating technology transfer. Industries that produce, store, or handle products that can potentially become soil or groundwater pollutants should be aware of the high cost of groundwater pollution. The effectiveness of industry as a contributor in the development of integrated approaches for soil- and groundwater-quality protection can be enhanced by (1) taking steps to reduce the waste-stream quantity and strength and (2) developing long-range plans for environmentally sound waste disposal.

Agricultural practices, although considered to be a fairly harmless activity, are becoming one of the most serious entities contributing nitrates and other soluble salts linked to fertilizers to the groundwater environment and thereby contaminating the resource. Serious consideration is to be given to limiting or regulating the use of undesirable chemicals, particularly in areas designated as the vulnerable areas as far as groundwater quality is concerned. Political leadership is quite important when one talks about environmental protection, especially groundwater protection. Their support would be very effective in addressing this long-term problem within the context of realistic expectation, technical soundness, and optimum benefits. Introduction of competent environmental consultants will play a vital role in the development of expertise within the concerned organizations and transfer of technical know-how for tackling the problem. These individuals are expected to be well versed with the latest developments and state-of-the-art technology. They should interact effectively with educational and research institutions on human-resource development. Also they should interact and provide advice to government water management and regulatory agencies.

References

Adams, B., and MacDonald, A. (1995), *Overexploitation of Aquifers—Final Report,* Technical Report WC/95/3, British Geological Survey, United Kingdom.

Ahmad, R. (1995), National workshop critically examines the drainage problems in the Indus Plain, *DRIP* (Drainage and Reclamation Institute of Pakistan) *Newletter* **16**(3).

Aldwell, C. R., Alfoldi, L., and van Waegeningh, H. G. (1991), Legislation and regulation, in *Integrated Land Use Planning and Groundwater Protection in Rural Areas,* International Hydrological Programme, UNESCO, Paris, pp. 55–63.

Anderson, M. P., and Woessner, W. W. (1992), The role of the post-audit in model validation, *Adv. Water Resources* **15**(3), 167–173.

Anderson, M. P., Ward, D. S., Lappala, E. G., and Prickett, T. A. (1993), Computer models for subsurface water, in *Handbook of Hydrology,* D. R. Maidment, Editor-in-Chief, McGraw-Hill, New York, pp. 22.1–22.34.

Asian Institute of Technology (AIT) (1981), *Investigation of Land Subsidence Caused by Deep Well Pumping in the Bangkok Area,* Comprehensive Report 1978–1981, submitted to National Environment Board, Bangkok, Thailand.

Bear, J. (1979), *Hydraulics of Groundwater,* McGraw-Hill, New York.

Bear, J., and Verruijt, A. (1990), Modeling Groundwater Flow and Pollution, Reidel, Dordrecht, Netherlands.

Biot, M. A. (1941), General theory of three dimensional consolidation, *J. Appl. Phys.* **12,** 155–164.

Biswas, A. K. (1991), Environmental modeling for developing countries: problems and prospects, in *Environmental Modeling for Developing Countries,* A. K. Biswas, T. N. Khoshoo, and A. Khosla, eds., Ty Cooly, London.

Biswas, A. K. (1992), Environmental impact assessment for groundwater management, *Water Resources Devel.* **8**(2), 113–117.

Bouwer, E., Mercer, J. W., Kavanaugh, M., and Digiano, F. (1988), Coping with groundwater contamination, *J. Water Pollution Control Federation,* **60**(8), 1414–1428.

Bredehoeft, J. D., and Konikow, L. F. (1993), Groundwater models—validate or invalidate, *Groundwater,* **31**(2), 178–179.

Bredehoeft, J. D., and Pinder, G. F. (1970), Digital analysis of areal flow in multiaquifer groundwater systems: a quasi-three-dimensional model, *Water Resources Res.* **6**(3), 883–888.

Canter, L. W., and Knox, R. C. (1985), *Septic Tank System Effects on Groundwater Quality,* Lewis Publishers, Mich.

Chorley, D. W., and Frind, E. O. (1978), An iterative quasi-three-dimensional finite element model for heterogeneous multiaquifer systems, *Water Resources Res.* **14**(5), 943–952.

Coe, J. J. (1990), Conjunctive use-advantages, constraints and examples, *J. Irrigation Drainage Eng. ASCE,* **166**(3), 427–443.

Darcy, H. (1856), *Les Fontaines Publiques de la ville de la ville de Dijon,* Victor Dalmont, Paris.

Das Gupta, A. (1991), Study on groundwater quality and monitoring in Asia and the Pacific, in *Groundwater Quality and Monitoring in Asia and the Pacific,* Water Resources Series no. 70, Economic and Social Commission for Asia and the Pacific, Bangkok, pp. 13–99.

Das Gupta, A., and Onta, P. R. (1994), Groundwater management models for Asian developing countries, *Water Resources Devel.* **10**(4), 457–474.

Dawson, R. F. (1963), Land subsidence problem, *J. Surv. Map. Div. ASCE,* **89,** 1–12.

Dyhr-Nielsen, M. (1989), Application of groundwater modeling in water resources management in Denmark, in *Groundwater Management: Quantity and Quality,* IAHS publ. no. 188, IAHS Press, Wallingford, United Kingdom, pp. 559–570.

Falkland, A. ed. (1991), *Hydrology and Water Resources of Small Islands: A Practical Guide,* UNESCO, Paris.

Figueroa Vega, G. E. (1976), Subsidence of the City of Mexico, a historical review, in *Proceedings 2nd International Symposium on Land Subsidence,* Anaheim, Calif., pp. 35–38.

Foster, S. S. D., and Hirata, R. (1991), *Groundwater Pollution Risk Assessment,* Pan American Health Organization, WHO, Pan American Center for Sanitary Engineering and Environmental Sciences, Lima, Peru.

Foster, S. S. D., and Salas, H. J. (1991), Identification and ranking of surface and groundwater pollution risks, in *Criteria for and Approaches to Water Quality Management in Developing Countries,* Natural Resources Water Series no. 26, United Nations, New York.

Foster, S. S. D., Ventura, M., and Hirata, R. (1987), *Groundwater Pollution: An Executive Overview of the Latin America-Caribbean Situation in Relation to Potable Water Supply,* Pan American Center for Sanitary Engineering and Environmental Services, Lima.

Freeze, R. A. (1971), Three-dimensional, transient, saturated-unsaturated flow in a groundwater basin, *Water Resources Res.* **7**(2), 347–366.

Freeze, R. A., and Cherry, A. (1979), *Groundwater,* Prentice-Hall, Englewood Cliffs, N.J.

Frind, E. O., and Verge, M. J. (1978), Three-dimensional modeling of groundwater flow systems, *Water Resources Res.* **14**(5), 844–856.

Gabrysch, R. K. (1969), Land surface subsidence in the Houston-Galveston Region, Texas, *Proceedings, Tokyo Symposium on Land Subsidence,* Tokyo, vol. 1, pp. 43–54.

Gabrysch, R. K. (1976), Land surface subsidence in the Houston-Galveston Region, Texas, *Proceedings 2nd International Symposium on Land Subsidence,* Anaheim, Calif., pp. 16–24.

Gambolati, G., and Freeze, R. A. (1973), Mathematical simulation of the subsidence of Venice, 1 theory, *Water Resources Res.* **9**(3), 721–733.

Gambolati, G., Gatto, P., and Freeze, R. A. (1974), Mathematical simulation of the subsidence of Venice, 2 results, *Water Resources Res.* **10**(3), 563–577.

Gelhar, L. W., Walty, C., and Rehfeldt, K. R. (1992), A critical review of data on field-scale dispersion in aquifers, *Water Resources Res.* **28**(7), 1955–1974.

Ghassemi, F., Chen, T. H., Jakeman, A. J., and Jacobson, G. (1993), Two- and three-dimensional simulation of seawater intrusion: performance of the "SUTRA" and "HST3D" models, *AGSO J. Austral. Geol. Geophys.* **14**(2/3), 219–226.

Gorelick, S. M. (1983), Review of distributed parameter groundwater management modeling methods, *Water Resources Res.* **19**(2), 305–319.

Gupta, S. K., and Tanji, K. K. (1976), A three-dimensional Galerkin finite element solution of flow through multiaquifers in Sutter Basin, California, *Water Resources Res.* **12**(2), 155–162.

Heath, R. C. (1983), *Basic Ground-water Hydrology,* U.S. Geological Survey Water Supply Paper 2220.

Helm, D. C. (1982), Conceptional aspects of subsidence due to fluid withdrawal, in *Recent Trends in Hydrology,* T. N. Narasimhan and R. A. Freeze, eds., Special Paper no. 189, Geological Society of America.

Ishii, M., Kuramochi, F., and Endo, T. (1976), Recent tendencies of the land subsidence in Tokyo, *Proceedings 2nd International Symposium on Land Subsidence,* Anaheim, Calif., pp. 25–34.

Javandel, I., and Witherspoon, P. A. (1969), A method of analyzing transient fluid flow in multilayered aquifers, *Water Resources Res.* **5**(4), 856–869.

Javandel, I., Doughty, C., and Tsang, C.-F. (1984), *Groundwater Transport: Handbook of Mathematical Models,* Water Resources Monograph Series no. 10, American Geophysical Union, Washington, D.C.

Johnon, A. I. (1981), Foreword to five papers on land subsidence, *J. Irrigation Drainage Eng. ASCE,* **107**(IR2), 113–114.

Khan, K. R. (1991), Groundwater quality issues in Asia and the Pacific, in *Groundwater Quality and Monitoring in Asia and the Pacific,* Water Resources Series no. 70, Economic and Social Commission for Asia and the Pacific, Bangkok, pp. 100–107.

Kindler, J. (1991), Water pollution implications of urban population growth in developing countries, in *Criteria for and Approaches to Water Quality Management in Developing Countries,* Natural Resources Water Series no. 26, United Nations, New York.

Konikow, L. F. (1981), Role of numerical simulation in analysis of groundwater quality problems, in *Proceedings, International Symposium on Quality of Groundwater,* Noordwijkerhout, Netherlands.

Konikow, L. F., and Bredehoeft, J. D. (1992), Groundwater models cannot be validated, *Adv. Water Resources* **15**(20), 75–83.

Korfiatis, P. (1991), Integrated approaches to soil and groundwater pollution abatement and remediation, in *Integrated Approaches to Water Pollution Problems,* J. Bau, J. P. Lobo, Ferreira, J. D. Henriques, and J. de Oliveira Raposo, eds., Elsevier Applied Science, London, chap. 17, pp. 171–178.

Lagas, P., Verdam, B., and Loch, J. P. G. (1989), Threat to groundwater quality by pesticides in the Netherlands, *Groundwater Management: Quantity and Quality,* IAHS publ. no. 188, IAHS Press, Wallingford, United Kingdom, pp. 171–180.

LeGrand, H. E. (1964), Systems for evaluation of contamination potential of some waste disposal sites, *J. Am. Water Works Assoc.* **56**(8), 959–974.

Lehr, J. H. (1982), How much groundwater have we really polluted? *Groundwater Monitoring Rev.* **2**(1), 4–5.

Lehr, J. H. (1984), The state of the nation's groundwater: 1984, *Groundwater Monitoring Rev.* **4**(1), 4.

Leusink, A. (1992), The planning process for groundwater resources management, *Water Resources Devel.* **8**(2), 98–102.

Lindorff, D. E. (1979), Groundwater pollution: a status report, *Groundwater,* **17**(1), 9–17.

Lofgren, B. E. (1969), Field measurement of aquifer system compaction, San Joaquin Valley, Calif., in *Proceedings, Tokyo Symposium on Land Subsidence,* Tokyo, vol. 1, pp. 272–284.

Lofgren, B. E. (1976), Hydrogeologic effects of subsidence, San Joaquin Valley, Calif., in *Proceedings 2nd International Symposium on Land Subsidence,* Anaheim, Calif., pp. 113–123.

Loucks, D. P. (1992), Water resources systems models: their role in planning, *J. Water Resources Plan. Manag. ASCE,* **118**(3), 214–223.

Mangold, D. C., and Tsang, C.-F. (1991), A summary of subsurface hydrological and hydrochemical models, *Water Resources J.* **ST/ESCAP/SER.C/168,** 36–62.

Mawla, G. (1991), Bangladesh: groundwater quality and monitoring, in *Groundwater Quality and Monitoring in Asia and the Pacific,* Water Resources Series no. 70, Economic and Social Commission for Asia and the Pacific, Bangkok, pp. 150–155.

McDonald, M. G., and Harbaugh, A. W. (1984), *A Modular Three Dimensional Finite Difference Groundwater Flow Model,* USDI, U.S. Geological Survey, National Center, Reston, Va.

Meier, W., and Mull, R. (1989), Nitrate in groundwater, in *Groundwater Management: Quantity and Quality,* IAHS publ. no. 188, IAHS Press, Wallingford, United Kingdom, pp. 181–190.

Murayama, S. (1969), Land subsidence in Osaka, in *Proceedings, Tokyo Symposium on Land Subsidence,* Tokyo, vol. 1, pp. 105–130.

Naymik, T. G. (1987), Mathematical modeling of solute transport in the subsurface, *Crit. Rev. Environ. Control* **17**(3), 229–251.

Neuman, S. P., and Witherspoon, P. A. (1969), Theory of flow in a confined two-aquifer system, *Water Resources Res.* **5**(4), 803–816.

Novotny, V., and Olem, H. (1994), *Water Quality: Prevention, Identification and Management of Diffuse Pollution,* Van Nostrand-Reinhold.

Nutalaya, P., Yong, R. N., Chumnankit, T., and Buapeng, S. (1989), Land subsidence in Bangkok during 1978–1988, in *Workshop, Bangkok Land Subsidence—What's Next,* Bangkok.

Onta, P. R., Das Gupta, A., and Paudyal, G. N. (1994), Conjunctive water use planning and management models: a state-of-the-art review, in *Proceedings, Ninth Congress of APD-IAHR,* Singapore, vol. I, pp. 40–47.

Oosterbaan, R. J. (1988), Effectiveness and environmental impacts of irrigation projects: a review, in *Annual Report,* International Institute for Land Reclamation and Improvement, Netherlands, pp. 18–34.

Pettyjohn, W. A., and Hounslow, A. W. (1983), Organic compounds and groundwater pollution, *Groundwater Monitoring Rev.* **3**(4), 41–47.

Phillips, C. R., Nathwani, J. S., and Mooji, H. (1977), Development of a soil-waste interaction matrix for assessing land disposal of industrial wastes, *Water Res.* **11**(10), 859–868.

Pinder, G. F., and Bredehoeft, J. D. (1968), Application of the digital computer for aquifer evaluation, *Water Resources Res.* **4**(5), 1069–1093.

Pinlac, E. M. (1992), Groundwater contamination assessment in the Philippines, in *Technical Workshop, Groundwater Contamination in Sub-Humid and Humid Tropical Asia,* UNESCO/ROSTSEA Thailand National Committee for IHP, Department of Mineral Resources, Bangkok.

Poland, J. F. (1969), Land subsidence and aquifer system compaction, Santa Clara Valley, Calif., in *Proceedings, Tokyo Symposium on Land Subsidence,* Tokyo, vol. 1, pp. 285–294.

Poland, J. F., Lofgren, B. F., Ireland, R. L., and Pugh, R. G. (1975), *Land Subsidence in the San Joaquin Valley, California, as of 1972,* U.S. Geological Survey Professional Paper 437-H, U.S. Geological Survey.

Premchitt, J. (1981), A technique in using integrodifferential equations for model simulation of multiaquifer systems, *Water Resources Res.* **17**(1), 162–168.

Premchitt, J., and Das Gupta, A. (1981), Simulation of a complex groundwater system and an application, *Water Resources Res.* **17**(3), 673–685.

Prokopovich, N. P. (1991), Detection of aquifers susceptibility to land subsidence, in *Land Subsidence,* IAHS publ. no. 200, IAHS Press, Wallingford, United Kingdom, pp. 27–33.

Ramnarong, V. (1991), Thailand: groundwater quality monitoring and management, in *Groundwater Quality and Monitoring in Asia and the Pacific,* Water Resources Series no. 70, Economic and Social Commission for Asia and the Pacific, Bangkok, pp. 218–227.

Ramnarong, V., and Buapeng, S. (1984), Groundwater quality problems in Thailand, in *Proceedings, International Symposium on Hydrochemical Balance of Fresh Water Systems,* Sweden.

Ramnarong, V., and Buapeng, S. (1991), Mitigation of groundwater crisis and subsidence in Bangkok, *J. Thai Geosci.* **2,** 125–137.

Ramnarong, V., and Buapeng, S. (1992), Groundwater resources of Bangkok and its vicinity: impact and management, in *Proceedings, National Conference on Geologic Resources of Thailand: Potential for Future Development,* suppl. vol., Department of Mineral Resources, Bangkok, pp. 172–184.

Rappleye, H. S. (1933), Recent areal subsidence found in releveling, *Eng. News Record,* **110,** 845.

Rivera, A., Ledoux, E., and de Marsily, G. (1991), Nonlinear modeling of groundwater flow and total subsidence of the Mexico City aquifer-aquitard system, in *Proceedings, 4th International Symposium on Land Subsidence,* IAHS publ. no. 200, IAHS Press, Wallingford, United Kingdom, pp. 45–58.

Rogers, P. P., and Fiering, M. B. (1986), Use of systems analysis in water management, *Water Resources Res.* **22**(9), 1465–1585.

Roscoe Mass Company (1990), *Handbook of Groundwater Development,* Wiley-Interscience, New York.

Sandhu, R. S. (1979), Modeling land subsidence, in *Evaluation and Prediction of Subsidence,* S. K. Saxena, ed., American Society of Civil Engineers, pp. 565–579.

Scheidegger, A. E. (1961), General theory of dispersion in porous media, *J. Geophy. Res.* **66**(10), 3273–3278.

Schumann, H. H., and Poland, J. F. (1969), Land subsidence, earth fissures and groundwater withdrawal in South-Central Arizona, U.S.A., in *Proceedings, Tokyo Symposium on Land Subsidence,* Tokyo, vol. 1, pp. 295–302.

Schwarz, J. (1989), Aquifers as components of water resources systems, in *Groundwater Management: Quantity and Quality,* IAHR Publication no. 188, IAHS Press, Wallingford, United Kingdom, pp. 433–441.

Soetrisno, S., and Hehanussa, P. (1992), Groundwater contamination in Indonesia, in *Technical Workshop, Groundwater Contamination in Sub-Humid and Humid*

Tropical Asia, UNESCO/ROSTSEA Thailand National Committee for IHP, Department of Mineral Resources, Bangkok.
Somasundaram, M. V., Ravindarn, G., and Tellam, J. H. (1993), Groundwater pollution of the Madras urban aquifer, *Groundwater,* **31**(1), 4–11.
Terzaghi, K. (1925), *Erdbaumechanik auf Bodenphysikalischer Grundlage,* Fraz Deuticke, Vienna.
Terzaghi, K., and Peck, R. B. (1948), *Soil Mechanics in Engineering Practice,* Wiley, New York.
Theis, C. V. (1935), The relation between the lowering of the piezometric surface and the rate and duration of discharge of a well using groundwater storage, *Trans. Am. Geophys. Union* **16,** 519–524.
Tinlin, R. M. (1981), Methodology for monitoring groundwater quality degradation, *Groundwater Monitoring Rev.* **1**(2), 27–32.
Todd, D. K. (1974), Salt water intrusion and its control. *J. Am. Water Works Assoc.* **66,** 180–187.
Todd, D. K. (1980), *Groundwater Hydrology,* Wiley, New York.
Toth, J. (1963), A theoretical analysis of groundwater flow in small drainage basins, *J. Geophys. Res.* **68**(16), 4795–4812.
United Nations Environment Programme (UNEP) (1989), *Environmental Data Report,* 2d ed., Blackwell Reference, Oxford.
United States Environmental Protection Agency (USEPA) (1977), *The Report to Congress: Waste Disposal Practices and Their Effects on Groundwater: Executive Summary,* USEPA, Las Vegas, Nev.
United States Environmental Protection Agency (USEPA) (1984), *Groundwater Protection Strategy,* Office of Groundwater Protection, USEPA, Washington, D.C.
United States Environmental Protection Agency (USEPA) (1986), *Guidelines for Groundwater Classification under the EPA Groundwater Protection Strategy,* Office of Groundwater Protection, USEPA, Washington, D.C.
van der Heijde, P. K. M., Bachmat, Y., Bredehoeft, J. D., Andrews, B., Holtz, D., and Sebastain, S. (1985), *Groundwater Management: The Use of Numerical Models,* Water Resources Monograph 5, 2d ed., American Geophysical Union, Washington, D.C.
Verma, R. D. (1989), Safeguards for groundwater pollution in India, in *Proceedings, International Workshop on Methodologies for Development and Management of Groundwater Resources in Developing Countries,* National Geophysical Research Institute, Hyderabad, India, pp. 841–851.
Vrba, J. (1991), Hydrogeological aspects of groundwater protection and pollution, in *Criteria for and Approaches to Water Quality Management in Developing Countries,* Natural Resources Water Series no. 26, United Nations, New York.
Vrba, J., and van Waegeningh, H. G. (1991), Groundwater protection, *Integrated Land Use Planning and Groundwater Protection in Rural Areas,* International Hydrologic Programme, UNESCO, Paris.
Walton, W. C. (1987), *Groundwater Pumping Tests, Design and Analysis,* Lewis Publishers, Mich.
Willis, R., and Yeh, W. W.-G. (1987), *Groundwater Systems Planning and Management,* Prentice-Hall, Englewood Cliffs, N.J.
Winika, C. C., and Wold, P. D. (1976), Land subsidence in Central Arizona, *Proceedings, 2nd International Symposium on Land Subsidence,* Anaheim, Calif., pp. 95–103.
Wu, C. M. (1976), Groundwater depletion and land subsidence in Taipei Basin, in *Proceedings, 2nd International Symposium on Land Subsidence,* Anaheim, Calif., pp. 389–398.
Wu, C. M., and Hwang, J. M. (1969), Land subsidence problems in Taipei Basin, in *Proceedings, Tokyo Symposium on Land Subsidence,* Tokyo, vol. 1, pp. 21–34.
Yamamoto, S. (1976), Recent trend of land subsidence in Japan, *Proceedings, 2nd International Symposium on Land Subsidence,* Anaheim, Calif., pp. 25–34.
Yangqun, W. (1991), China; groundwater quality and monitoring, in *Groundwater Quality and Monitoring in Asia and the Pacific,* Water Resources Series no. 70, Economic and Social Commission for Asia and the Pacific, Bangkok, pp. 156–157.

Yeh, W. W.-G. (1986), Review of parameter identification procedures in groundwater hydrology: the inverse problem, *Water Resources Res.* **22**(2), 95–108.

Zomorodi, K. (1990), Optimization of streamflow regulation through artificial recharge, *Memoires of the 22nd Congress of IAH, xxii,* Lausanne, Switzerland, pp. 1026–1033.

Chapter

6

Water-Quality Monitoring

Deborah V. Chapman
Environment Consultant
Kinsale, County Cork, Ireland

Introduction

The need to define the quality of water has developed with the increasing demand for water which is suitable for specific uses and conforms to a desired quality. The most fundamental need is for water suitable for drinking, personal hygiene, and food preparation and that poses no risk to human health. Increasing development and its associated industrialization have, in addition, introduced the need for water quality with specific physical, chemical, or biological characteristics. At the same time, however, water bodies offer a convenient option for the disposal of domestic, agricultural, and industrial effluents and wastewaters, all of which can significantly affect the natural physical, chemical, and biological characteristics of receiving waters. In many world regions water resources serving as a waste-disposal facility for one activity are the source of water for another activity, necessitating even more thorough and complex monitoring and assessment activities.

Accurate assessment of water quality, whether in relation to the requirements of intended water uses or in order to determine the impacts of an activity on the water resource (such as waste disposal or abstraction), depends on the results generated by specific monitoring activities which define the physical, chemical, and/or biological condition of the resource. Consequently, the quality of the data generated during a monitoring program is crucial to the resultant assessment and the eventual effectiveness of any recommendations for management action.

Although many monitoring programs traditionally have been designed to gather data from a single water body or aquatic system, monitoring and assessing water quality can no longer be confined to local or national boundaries. Large rivers may pass through several nations, major aquifers may traverse national boundaries, and two or more countries may share the shores of a large lake or reservoir. In addition, some water-quality issues have become transnational (e.g., acidification of lakes and rivers due to atmospheric deposition popularly known as "acid rain") or even global (such as deposition of lead and accumulation in freshwater ecosystems) in nature. Consequently, there is an increasing need to design monitoring and assessment programs with the purpose of providing information for environmental management and protection at the international and global scale.

The need to protect water resources by purpose-orientated assessments harmonized for natural basins or catchments was highlighted by the International Conference on Water and the Environment, held in Dublin in January 1992. The recommendations from this conference were endorsed by the United Nations Conference on Environment and Development (UNCED), popularly referred to as the Earth Summit, held in Rio de Janeiro in June 1992. Representatives of 178 governments attended this conference highlighting the global commitment to integrating development with environmental protection. An important outcome of the conference was a program outline, known as Agenda 21, for future action relating to environment and development. In its section on freshwater, Agenda 21 presents objectives which include all countries establishing the institutional arrangements required to ensure the efficient collection, processing, storage, retrieval, and dissemination to users of information about the quality and quantity of available water resources, at the level of catchments and groundwater aquifers, in an integrated manner.

Since all uses of water generate, directly or indirectly, wastewater which differs in quality for the original source and which is returned to the hydrologic cycle, the quality of the world's water resources is gradually changing. These changes can be assessed only by the standardized monitoring of water quality in relation to major water bodies (river catchments, aquifers, etc.) and over long time periods (decades rather than years). The need for proper integration of water-resource and water-quality management has been stressed many times in recent years (Biswas et al., 1993) and was included in the objectives of Agenda 21.

Monitoring programs can range from those using the simplest techniques for a few sites (e.g., the use of a Secchi disk in a lake or reservoir), and analyzing the results with a pocket calculator and hand-drawn graph, to those using advanced laboratory facilities and techniques capable of measuring low concentrations of any variable in large numbers of samples, and processing the results in a powerful computer. Monitoring the quality of freshwater can be carried out in a variety of ways: by making quantitative measurements of physical, chemical, and biological characteristics and by qualitative descriptions of some features such as odor, transparency, and vegetative changes. All monitoring activities can generate enormous amounts of information. The time, effort, and financial resources spent in collecting this information can be immense, and it is essential that the most effective use be made of the resources required to collect the information and of the information thus collected. This can be achieved only by carefully defining the objectives of a monitoring program and including the anticipated outcome, in terms of the assessments achieved, in these objectives. Monitoring without clear purpose and without a final assessment of the generated data is a very wasteful activity. A properly conducted program with fully assessed data should be able to provide useful information for environmental management and should be able to make recommendations for the improvement of the monitoring program itself.

Definition of Water-Quality Monitoring

The term *monitoring* is often used very loosely to encompass all aspects of the collection and evaluation of information relating to water quality. For the purposes of this discussion, a distinction is made between monitoring and the evaluation of the resultant data.

Monitoring is defined as being the actual collection of information at set locations and at regular intervals in order to provide the data which may be used to define current conditions, establish trends, etc. (Meybeck and Helmer, in press).

The complete process of monitoring, data evaluation, and reporting the results of the monitoring, can be defined as an assessment. The regular collection of water-quality information is an important component of monitoring and distinguishes monitoring activities from surveys, which may be conducted only once or for a specific time period (e.g., a single season) and for a specific purpose (such as during research projects or to obtain preliminary information prior to the detailed design of a monitoring program).

Monitoring in relation to use

Traditionally, water-quality monitoring has been carried out to assess the suitability of a water resource for a particular use. Suitability is assessed in relation to acceptable concentrations of selected water-quality variables which are defined by guidelines, standards, or maximum allowable concentrations (MACs). Perhaps the best-known set of variables defined for a specific use of water are the World Health Organization (WHO) guidelines for drinking-water quality (WHO, 1993). The guideline values are set in relation to an extensive review of the possible health effects of different concentrations of natural variables and of contaminants most usually encountered in water (WHO, 1984). Many national agencies responsible for providing drinking water have based their own monitoring criteria on WHO guidelines and have set maximum allowable concentrations MACs of specific variables. Provided the concentrations of the measured variables are below the MAC values, the water is considered safe for human use.

Table 6.1 includes examples of some of the more common uses for which guidelines and standards have been set in different countries, specifically, drinking-water supplies, livestock watering, irrigation, and protection of fisheries and aquatic life. In addition to nationally set standards, each specific industrial use has its own set of water-quality criteria which depend on the nature of the process for which the water is used.

Although acceptable water quality may need to be defined principally in relation to the use with the highest quality requirements, and defined by any associated quality standards, the quantity of the resource available may have an important role in defining the acceptable quality. In general, the uses with the highest demands on water quality (Table 6.1) are often those which demand relatively less in terms of quantity, such as drinking water. Domestic use accounts for only 8 percent of global freshwater withdrawals, compared with 69 percent for agricultural use and 23 percent for industrial use (World Resources Institute, 1992). Where total water resources are scarce, it is often necessary to accept a quality which does not meet the guidelines or standards in every respect, and, although this is not recommended for drinking-water supplies, it may be a feasible option for agricultural or certain industrial uses.

Monitoring solely in relation to guidelines or standards simplifies the design of the monitoring program because the objectives of the program are already defined (see below) and assessment is relatively straightforward. The main task facing the program designers is the selection of number and location of monitoring sites, but for some water uses this aspect is also well defined (e.g., sampling for drink-

TABLE 6.1 Examples of Maximum Allowable Concentrations of Selected Variables Specified in National and International Guidelines

Variable	Drinking water		Fisheries and aquatic life		Livestock watering, Canada	Irrigation, Canada
	WHO	Canada	Russia	Canada		
Total dissolved solids, mg/L	1,000	500			3,000	500–3,500
Turbidity (NTU)	5	5				
pH	<8.0[a]	6.5–8.5		6.5–9.0		
Dissolved oxygen, mg/L			4.0[b]–6.0	5.0–9.5		
Nitrate as N, mg/L		10				
Nitrate, mg/L	50		40		100[c]	
BOD, mg/L			3			
Sodium, mg/L	200		120			
Chloride, mg/L	250	250	300			100–700
Sulfate, mg/L	250	500	100		1,000	
Fluoride, mg/L	1.5	1.5	0.75		2.0[d]	1.0
Aluminum, mg/L	0.2			0.005–0.1[e]	5.0	5.0
Boron, mg/L	0.3	5.0			5.0	0.5–6.0
Cadmium, mg/L	0.003	0.005	0.005	0.0002–0.0018[f]	0.02	0.01
Chromium, mg/L	0.05[g]	0.05	0.02–0.005	0.02–0.002	1.0	0.1
Copper, mg/L	2[g]	1.0	0.001	0.002–0.004[f]	0.5–1.0[h]	0.2[i]/1.0[j]
Lead, mg/L	0.01	0.05	0.1	0.001–0.007[f]	0.1	0.2
Mercury, mg/L	0.001	0.001	0.00001	0.0001	0.003	
Zinc, mg/L	3	5.0	0.01	0.03	50	1.0[k]/5.0[l]
DDT, μg/L	2	30.0		1 ng/L		
Lindane, μg/L	2	4.0				
Fecal coliforms, n 100 mL^{-1}	0	0				100

[a]For effective disinfection with chlorine.
[b]Lower level acceptable under ice cover.
[c]Nitrate plus nitrite.
[d]1.0 if animal feed contains fluoride.
[e]Depending on pH.
[f]Depending on hardness.
[g]Provisional.
[h]Depending on the animal.
[i]Sensitive crops.
[j]Tolerant crops.
[k]Soil pH <6.5.
[l]Soil pH >6.5.

SOURCES: Environment Canada (1987), WHO (1993), and Chapman and Kimstach (in press).

ing-water quality at points of supply for domestic use such as standpipes, treatment works outlets).

Impact monitoring

The need to monitor the impact of human activities on water quality is constantly growing with the pressure of increasing human populations. Human activities in one part of a watershed may impair water quality and restrict the use of the resource by others elsewhere in the same watershed. An immediate reason for impact monitoring is to determine the changes inflicted on the water body by the disposal of wastes and wastewaters, agricultural runoff, water withdrawals, etc. In addition, such monitoring is required to assess the extent of any effects, recommend remedial measures, regulate the sources of impacts, and assess the success of regulatory and remedial measures. The latter activity is often required in relation to legislated standards and may use prescribed monitoring programs and methods.

For an efficient and informative monitoring program designed to determine the effects of a specific impact, some knowledge of the anticipated effects of that impact is essential, such as the substances being added to the water body and their likely concentrations, as well as the anticipated or known physical, chemical, and biological (e.g., toxicological) characteristics or effects in aquatic ecosystems. If this information is not already available in the published literature, specific studies such as laboratory-based bioassays or preliminary surveys in the receiving water body may be necessary.

Impact monitoring can take many forms, from traditional chemical or biological analysis of water samples to advanced physiological biotests or complex dynamic, in situ monitoring systems combining continuous chemical analysis with sensitive biotests using fish or other aquatic organisms (see the example of the Rhine below).

Impact monitoring is particularly important when a water resource is also used for drinking-water supplies or sensitive uses, as in many large rivers which cross national boundaries. Those responsible for setting discharge standards for wastewaters have an obligation to ensure that the water downstream is fit for later use and that the wastewater-treatment operations (if present) are adequate in relation to the natural self-purification mechanisms of the water body. This kind of impact monitoring requires regular or continuous analysis of the receiving water body and frequent evaluation of the results.

Trend monitoring

Significant and long-term changes in water quality can be assessed only by trend monitoring which is carried out at regular intervals over long time spans. Many water-quality issues have been empha-

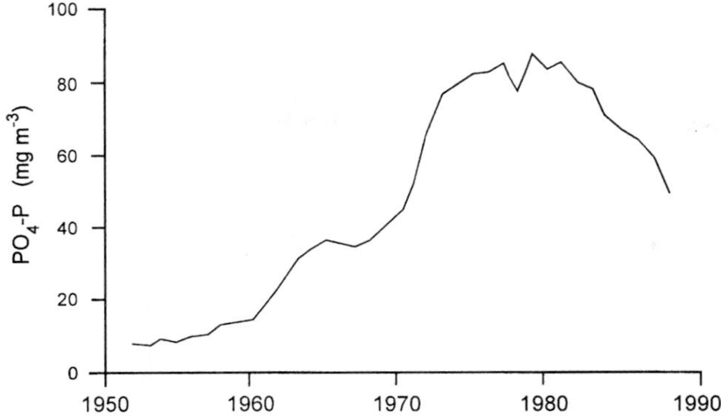

Figure 6.1 Development and control of eutrophication in Lake Constance, Switzerland, as indicated by long-term trends in phosphorus. (*After Thomas et al., in press.*)

sized by trend monitoring, such as the development of eutrophication in Lake Constance (Fig. 6.1). Such long-term monitoring activities are now an essential component of water-quality assessment programs because of the global deterioration in water quality caused by human activities, even in areas where human populations are very low.

Although the detection of trends in specific water-quality variables can constitute the prime objective, many monitoring programs with other objectives can also be used to detect trends, provided the data are collected at suitable, regular intervals over long enough time periods. Normally, trend assessments are made using data collected over decades, but, occasionally, a clear trend in water quality can be observed over a period of about 10 years as, for example, in the case of the increases in nitrate concentrations in public supply wells in Bermuda between 1973 and 1982. These increases in nitrate, reaching concentrations in excess of the WHO guideline value, were caused by unsewered sanitation contaminating the groundwater (Foster et al., 1987).

Flux monitoring

Increasing awareness of the transport of contaminants has led to a recent growth in flux monitoring in order to estimate the flux, or quantity in mass per unit time, of nutrients and contaminants discharged by rivers or groundwaters to lakes and oceans and across international boundaries. Flux monitoring requires a knowledge of the water discharge and the concentration of the compound of interest and closely links hydrologic and water-quality monitoring activities. The importance of the transport of contaminants has been recognized by the inclusion of river-flux monitoring stations in the Global

Environment Monitoring System (GEMS) global water-quality monitoring and assessment program known as GEMS/Water (WHO, 1991) (see later).

Multipurpose monitoring

The agencies responsible for the design and implementation of monitoring programs are frequently faced with having to monitor for uses and for impacts within the same water bodies. The financial and technical resources required to carry out individual monitoring programs designed for specific water uses and impacts could be prohibitive, even for a wealthy nation. Thus, in practice, many regular and long-term monitoring programs serve several purposes. Typically, measurements will be made which can be used to assess the health of the ecosystem, determine the presence and effects of contaminants, and assess the quality of the water against any standards set for specific uses of the water. In the past, however, there was a tendency for regional and national agencies to install such multipurpose monitoring programs without considering the final use of the information, and often without carrying out the necessary preliminary investigations which are necessary to aid optimum program design (Meybeck et al., in press b) (see below). Such programs generated large quantities of data which were of little value for specific assessment purposes, much of which has probably never been interpreted or used. Although the need for multipurpose monitoring is acknowledged in certain circumstances, the general use of multipurpose monitoring is no longer recommended.

Defining Objectives

The time and effort devoted to the setting of objectives can be repaid by the more focused and cost-effective monitoring program which results from them. In the past, monitoring was frequently carried out without a clear impression of the purpose of the exercise and of how the data obtained might be interpreted. The setting of clear objectives before designing a monitoring program is essential to the overall success of the program and has been emphasized for all types of monitoring programs, in all types of water bodies (Chapman, in press). The objectives must state the reason for the monitoring program and the anticipated information to be gained from it. The anticipated outputs must be clearly described in order to aid those involved in the design and implementation of the program to choose the appropriate methods and media to sample (physical, chemical, or biological), suitable sampling locations, number of samples, and data-analysis techniques. Inappropriate choices resulting from poorly defined objectives can lead to wasted effort and resources. In some situations the objectives cannot be clearly stated without some preliminary investigations into all factors and influences on a water body, such as:

- Climatological, geologic, and hydrologic influences
- Water uses and withdrawals
- Sources of contaminants (point and diffuse sources)

Objectives should always be realistic in relation to the resources available to carry out the optimum program design. Every water body can be described in terms of its physical, chemical, and biological characteristics, and there are many methods available to do this. Much monitoring in the past has been based on traditional physical and chemical analyses of the water in situ or of samples collected and transported to the laboratory. However, in recent decades, other approaches have been refined which are often more suited to describing the effects of particular water-quality issues, such as the determination of contaminant transport using suspended particles or the determination of the health of an ecosystem by examining the biological communities present. Unfortunately, there is often a tendency to be tempted by new, and often sophisticated, methods without critically examining whether these methods will actually fulfill the objectives of the monitoring program. Within the constraints of the resources available, the most suitable methods should be selected and, if necessary, be tested during a preliminary program. Embarking on a new method will almost always incur extra expense in terms of equipment and the training of personnel.

Single-objective monitoring addresses one issue only and can often be confined to a limited set of variables from well-defined locations and time intervals. An example of a single objective is to check whether water being extracted from a well was fit for domestic supply as set by national guidelines. The design of such a monitoring program is fairly straightforward because the sample site is defined (the point of abstraction) and the variables (and probably the frequency of samples) to be measured are also specified by the legislated standards. If the data are to be submitted to a national or international database, the methods to be used for analysis may also be set by recognized "standard methods" (such as issued by the International Standards Organization) and subjected to regular quality checks.

An example of single-objective monitoring in relation to an environmental impact is the International Co-operative Programme on Assessment and Monitoring of Acidification of Rivers and Lakes which became fully operative in 1987. The objective was to establish, on regional bases, the degree and geographic extent of acidification of surface waters. Participating countries in the Northern Hemisphere followed methods approved by the program task force and were subject to intercalibration exercises. Chemical measurements consisted of some general variables (such as discharge, temperature, and oxygen concentration) pH, acidity, alkalinity, salts, nutrients, and aluminum. Data were submitted to a central database for storage and analysis (NIVA, 1992).

As mentioned above, a single agency may be charged with the responsibility for several types of monitoring on a single water body, such as impact and trend monitoring, as well as checking water quality against standards for specific uses. In reality, resources are often limited, and the design of a monitoring program frequently has to be a compromise between its objectives and the resources available to undertake it. In such situations it may be necessary to have a multi-objective monitoring program providing data for several assessments.

When setting objectives it is necessary to consider the ideal monitoring program in relation to the socioeconomic and technical and scientific development of the country. Although it might be ideal to measure trace metals in water, the laboratory and technical facilities to carry it out may be beyond the resources of the responsible agency. In such circumstances it may be possible to adopt a different approach based on less-expensive techniques, such as some biological approaches or the study of particulate matter.

An example of an advanced monitoring program with multiple objectives can be illustrated by the National Water-Quality Assessment (NAWQA) Program, which was initiated in the United States in the mid-1980s (Hirsch et al., 1988). The program had specific, but wide-ranging, goals:

- To provide a nationally consistent description of current water-quality conditions for a large part of the water resources of the United States
- To define long-term trends (or lack of trends) in water quality
- To identify, describe, and explain, as far as possible, the major factors which affect observed water-quality conditions and trends

The program required a complex design involving many people and separate agencies. Therefore, it was initiated as a pilot phase which could be used to assess its effectiveness and to make recommendations for its further implementation or modification. River basins and aquifers were chosen in study areas covering areas of thousands, to tens of thousands, of square miles. Use was made of local agencies already collecting relevant information, and existing data were also used to supplement field studies. The expected products of the program are clearly listed and include:

- Identification of key substances for possible regulation
- Designing of local monitoring programs (with respect to variables to be analyzed, sampling locations, frequency, and timing)
- Determination of relative effects of point sources and non–point sources

In addition, the requirement for certain statistical descriptions of water-quality conditions and changes have been identified for certain constituents among the outputs of the program. The details relating to the design of this multiobjective program are described in Hirsch et al. (1988).

Monitoring is a dynamic process which should be kept constantly under review. Many monitoring programs may go on for years or decades. Regular review of the program helps to ensure that the program design evolves to meet the changing uses and impacts on the water body over time. Land-use practices may change, industries may introduce new chemicals into their processes and effluents, and more-stringent standards may be required for a given water use. A program originally conceived with simple objectives may need to be expanded into a multiobjective program or vice versa.

Components of a Monitoring Program

The design of a monitoring program is largely responsible for whether the program generates valid and meaningful information which is useful for management purposes. As has already been discussed, the first and extremely important step is the setting of realistic objectives which state the anticipated outputs (quantitative or qualitative, type of information needed for management, etc.). However, the final design of a monitoring program depends not only on the objectives but also on the financial and technical resources of the responsible agency. With the help of the information obtained during the preliminary surveys, the sampling media, methods of sample analysis, location of sample sites, frequency of sampling, and data-handling techniques all have to be selected. The steps involved in a complete monitoring and assessment program are summarized in Table 6.2, and the links between the steps are illustrated in Fig. 6.2. A detailed guide to the methods available for, and the practical implementation of, freshwater monitoring programs is available in Bartram and Ballance (in press).

Preliminary surveys

Before the specific components of the program can be planned, it may be necessary to conduct a preliminary survey to help determine the

TABLE 6.2 The Principal Activities Associated with Each Component of a Monitoring and Assessment Program

Program component	Principal activities
Setting objectives	Consider water uses, and legislation or guidelines Consider economic and technological constraints
Preliminary surveys	Survey literature or databases for existing physical, chemical, biological, or hydrological data, information on methods, etc. Test field and laboratory methods if necessary Carry out special survey to select sites and/or methods for long-term use and evaluate results Evaluate technical and financial resources required
Monitoring program design	Select monitoring media (water, sediments, biota), variables (physical, chemical, biological), field sites, sampling frequency, specific methods, and equipment Produce final program design and guidelines for technical personnel
Implementation of monitoring program	Field operations: collection of samples and in situ measurements Laboratory operations: sample bottle preparations and pretreatments, analysis of samples Hydrologic measurements: collection of information on discharge, levels, etc.
Quality control	Checking of field and laboratory techniques, e.g., with sample blanks Checking in-house analytical techniques with sample blanks and spiked samples Participation in interlaboratory quality-assurance exercises Regular checking for suspect data in databases
Data manipulation	Data storage: transferring results from field and laboratory operations into database Analysis of data: application of statistical methods, e.g., correlations, trend analysis Assimilation and presentation of data: tables of results, data summaries, graphs
Assessment	Interpretation of results: establishment of causes and effects, degree of compliance with guidelines, determination of trends, etc. Evaluation of achievement of objectives
Recommendations for management	Recommendation for action, legislation, etc. Recommendations for program continuation, modification, etc.

most suitable media for the monitoring in conjunction with the variables to be measured or to obtain information about the water body (e.g., hydrologic characteristics) which may affect the intended methods. These surveys might consist of, for example, a short-duration program based on the full-scale design or a series of tests of different methods in the situations in which they are likely to be used. Such surveys help determine the technical and financial resources required

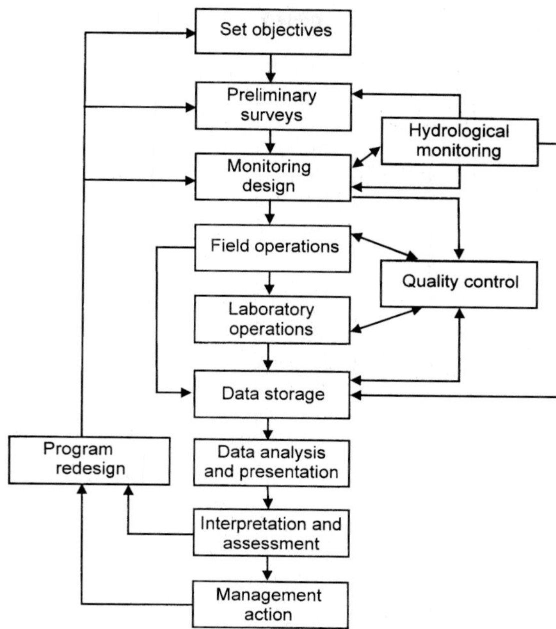

Figure 6.2 Links between the main components of a water-quality monitoring and assessment program.

for the full program, as well as its feasibility, and provide essential information to the monitoring program designers.

In some circumstances existing data from other monitoring programs or surveys may be available which give an indication of the natural variability of the water quality, sources of impacts, past trends, hydrologic and chemical variability, etc. Examples of the types of preliminary surveys which provide relevant information for the design of monitoring programs related to the assessment of particulate-matter quality in rivers, lakes, and reservoirs are given in Table 6.3. Many of the surveys listed in Table 6.3 are also relevant to most monitoring programs; for example, an inventory of sources of wastes and wastewaters, together with their chemical composition, aids the selection of chemical variables and the possible choice of biological approaches and the most suitable sites for sample collection. An initial survey of existing boreholes, wells, and springs together with any existing hydrogeologic data (see Table 6.4) is particularly important to the design of groundwater monitoring programs, because using existing and suitable sampling points can help limit the cost of collection of groundwater samples considerably. Preliminary surveys for groundwater monitoring design will always require an initial, short,

TABLE 6.3 Preliminary Surveys Pertinent to Particulate-Matter Quality Assessments

Water bodies	Type of survey	Information obtained
Rivers	Water discharge Q	River regime Extreme discharge statistics
	Suspended sediment (TSS*)	TSS variability Relationship TSS = $f(Q)$ Annual sediment discharge
	Inventory of major pollutant sources	Location of pollutant sources Types of pollutants Estimated quantities discharged
Lakes and reservoirs	Bathymetric survey	Volume Hypsometric curve Deepest points
	Temperature and O_2 profiles	Thermal structure Turnover period Intensity of vertical mixing
	Chlorophyll and transparency	Periods of algal production Resuspension of sediments
	Sedimentologic survey (grain-size)	Area of deposition Occurrence of fine deposits
	Inventory of major pollutant sources	As for rivers

*Total suspended solids
SOURCE: Thomas and Meybeck (in press).

and intensive sampling program in order to identify the features of the natural groundwater quality, act as a supplement to a pollutant source inventory, and identify or confirm seasonal, lateral, or vertical variations in groundwater quality (Chilton, in press).

Sampling media

Three media are used mainly for water-quality assessment: the water itself, suspended and deposited particulate matter, and living organisms. The choice of medium is governed not only by financial and technical restraints but also by its suitability to characterize a particular water-quality issue. Monitoring water quality in relation to use governed by standards is most often carried out by chemical analysis of water samples. Monitoring to determine the effects of a wastewater discharge may combine all three media, each for a different purpose. Chemical analysis of water and suspended particulate-matter samples may be used to determine dispersion, transport, and deposition, whereas analysis of the biological communities may indicate ecological effects. The analysis of contaminant accumulation in the tissues of organisms may also indicate possible human health risks.

Whereas the analysis of a water sample gives an indication of concentration for an instant in time, analysis of biological communities

TABLE 6.4 Hydrogeologic Data Requirements for Groundwater-Monitoring System Design

Data	Possible source material
Aquifer locations, depths, and aerial extent	Geologic and hydrogeologic maps and reports
Physical properties of aquifers, especially transmissivity	Test pumping, geologic and hydrogeologic maps
Areal distribution of groundwater levels	Borehole archives and observation wells
Areal distribution of depth to groundwater	Borehole archives and observation wells
Areas and magnitudes of natural groundwater recharge	Climatic and hydrologic data, soils, and land use
Areas and magnitudes of artificial recharge	Irrigation records, municipal recharge schemes
Areas and magnitudes of natural groundwater discharge	Streamflow and water-level data
Locations and magnitudes of groundwater abstractions	Borehole archives, municipality and irrigation department records, well owners
Directions and velocities of groundwater flow	Geologic and hydrogeologic maps, water level, and transmissivity data

SOURCE: Chilton (in press).

and tissues, or of deposited sediments, indicates the accumulated effects of fluctuating concentrations over much longer time spans (e.g., weeks to years depending on the organisms). The analysis of sediment cores taken from lakes and rivers can indicate the historical emissions of certain metals and compounds in the watershed (Alderton, 1985) as well as the distribution of contaminants from point sources of emissions (Meybeck et al., 1989; Thomas and Meybeck, in press).

The characteristics of the principal media used in water quality monitoring are summarized in Table 6.5. Discussions of the principles and examples of uses of particulate matter and biological methods are given in Thomas and Meybeck (in press) and Friedrich et al. (in press), respectively, as well as many other specialized guidebooks. A detailed treatment of the chemical analysis of water samples is available in Bartram and Ballance (in press).

Selection of water-quality variables

In order to assess whether human use has any actual or potential impact on a water resource, or whether the resource is suitable for a specific use (prior to or without treatment), it is first necessary to

TABLE 6.5 Principal Characteristics of Media Used for Water-Quality Assessments

Characteristics	Water	Particulate matter: Suspended	Particulate matter: Deposited	Living organisms: Tissue analyses	Living organisms: Biotests	Living organisms: Ecological surveys[a]	Living organisms: Physiological determinations[b]
Type of analysis or observation	← Physical → ← Chemical[c] →			← Biological →			
Applicability to water bodies	Rivers, lakes, groundwater	Mostly rivers	Lakes, rivers	Rivers, lakes	Rivers, lakes	Rivers, lakes	Rivers, lakes
Intercomparability[d]	← Global →			Depends on species occurrence	Global	← Local to regional →	
Specificity to given pollutant	← Specific →				← Integrative →		
Quantification	← Complete quantification of concentrations and loads →		Concentrations only	Quantitative	Semi-quantitative	← Relative →	
Sensitivity to low levels of pollution	Low	← High →			Variable	Medium	Variable
Sample contamination risk	High[e]	Medium	← Low →		Medium	← Low →	
Temporal span of information obtained	Instantaneous	Short	Long to very long (continuous record)	Medium (1 month) to long (>1 year)	Instantaneous to continuous[f]	← Medium to long →	

Levels of field operators	Untrained to highly trained[g]	Trained	Untrained to trained	Trained	←——— Medium to highly trained ———→		
Permissible sample storage duration[g]	Low	High	High	High	Very low	High	NA[h]
Minimum duration of determination	Instantaneous (in situ determination) to days	Days	Days to weeks	Days	Days to months	Weeks to months	Days to weeks

[a]Including algal biomass estimates.
[b]Histologic, enzymatic, etc.
[c]Including BOD determination.
[d]Most biological determinations depend on the natural occurrence of given species and are, therefore, specific to a given geographic region. Chemical and physical descriptors are globally representative.
[e]For dissolved micropollutants.
[f]For example, organisms continuously exposed to water.
[g]Depending on water-quality descriptors.
[h]Not applicable.
SOURCE: Meybeck et al. (in press b).

determine the background or "natural" water quality, i.e., unaffected by human activities. Background quality reflects the physical and chemical conditions created by the natural climatological, geologic, and biological influences and can be determined by a set of standard variables which can be modified by a knowledge of local geologic conditions (Table 6.1). This basic set of variables is then included in the monitoring program together with variables chosen in relation to the specific use or anticipated impact. In practice, even water bodies remote from human settlement are not totally unaffected by human activities because of the long-range transport of pollutants that occurs in the atmosphere (Meybeck et al., 1989; Meybeck and Helmer, in press). For most purposes, background water quality can be determined by sampling prior to abstraction or wastewater disposal begins, but in water bodies already affected by human activity it may be necessary to sample some distance away (e.g., upstream) or even in an adjacent watershed.

Where monitoring is conducted in relation to guidelines or standards, the minimum set of variables to be measured is defined in the guidelines or standards, but in other circumstances the selection of variables may have to be guided by the findings of a preliminary survey of the natural physical and chemical characteristics of the water body and the anticipated contaminants from point or diffuse sources. A full understanding of water-quality characteristics and sources, behavior of quality variables, and so on can help one make the most appropriate selections. A description of the origins, sources, behavior, and transformations of the principal water-quality variables and some guidance for selection in relation to monitoring with respect to water uses and impacts is given in Chapman and Kimstach (in press). A detailed description of the study and interpretation of physical and chemical variables in freshwater is available in many specialized texts (e.g., Environment Canada, 1987; APHA, 1989; Hem, 1989).

Monitoring in relation to drinking-water supply has two main aspects: the monitoring of the supply which may be used for treatment and/or disinfection prior to distribution and the monitoring of the actual water that is delivered to the consumer (i.e., distributed by pipes or extracted from a well). The important variables are those relevant to human health, such as pathogens (particularly fecal coliforms) and toxic substances, such as metals and pesticides, but those relevant to aesthetic quality may also be included. Monitoring prior to treatment helps determine the required level of treatment and can identify substances or contaminants which may interfere with the treatment process. Monitoring after treatment checks the efficiency of the treatment process and whether the water is suitable for distribution to consumers and usually includes the variables defined by the drinking-water standards and possibly variables associated with the treatment process (e.g., residual chlorine). Many groundwater sources require little or no treatment prior to consumption, and once the initial suitability of the water has been determined, these sources need to be checked only occasionally for pathogens or contaminants, such as pesticides or organic solvents. Recreational and nonconsumptive use of water bodies requires monitoring mainly for pathogens, toxic substances, or variables associated with the aesthetic quality of the water, such as suspended solids, color, and odor.

Agricultural activities demand large quantities of water, mainly for irrigation. The quality of the water with respect to pathogen content is important where it is sprayed directly onto crops and where it is used to water livestock. Many of the variables included for monitoring the quality of water for livestock watering are the same as those for drinking-water supplies, although livestock can often withstand higher concentrations of suspended solids and salinity. The salt content is important for major irrigation activities because it can affect soil quality and fertility. Some countries have prepared guidelines for irrigation water quality, such as Canada (see Table 6.1).

The quality requirements of water for industrial use are often stringent, and depend on the nature of the industrial process. Although some guidelines have been proposed for major industrial processes (Chapman and Kimstach, in press) each monitoring program needs to be based on the specific industrial needs.

Management of an aquatic environment with respect to its suitability to sustain aquatic life and fisheries depends on the provision of conditions which allow growth and reproduction, principally the availability of oxygen, adequate nutrients or food supply, and the absence of toxic chemicals. Guidance for the appropriate selection of variables for monitoring waters of importance as fisheries can be obtained from the European Inland Fisheries Advisory Commission (EIFAC) of the Food and Agriculture Organization of the United Nations (FAO).

Variable selection for monitoring in relation to a known or anticipated impact is based on knowledge of, or information relating to, the pollution source, the nature of the discharge and its possible effects, transformations, and behavior within the water body. One of the most common forms of impact monitoring is related to disposal of municipal wastewater, which usually contains domestic sewage, but may also contain urban runoff and collected wastewaters from industries. The selected variables, therefore, usually consist of those indicative of organic matter [biochemical oxygen demand (BOD), chemical oxygen demand (COD), suspended matter, conductivity, chloride, ammonia, and nitrogen compounds] and fecal matter (pathogens), together with those selected according to the nature of any other wastewater inputs. For example, collected urban runoff can contain high concentrations of lead and oil products arising from the use of automobiles.

Agricultural activities give rise to diffuse inputs of nutrients (from fertilizers), pesticides, and inorganic matter (associated with runoff during land clearance) as well as occasional point sources of wastewaters high in organic matter arising from intensive livestock rearing practices. An inventory of the agricultural activity in the area surrounding the water body would help to define the optimum set of variables for impact monitoring, although in most situations nutrients, suspended solids, and oxygen or BOD would always be included.

The history of poor planning and control of land disposal sites for solid municipal and hazardous wastes has led to problems of groundwater contamination in some situations (Chilton, in press). Variable selection depends on the nature of the material disposed to the landfill site and could include pathogens for domestic waste disposal and metals and organic or radioactive chemicals for industrial waste disposal.

The atmospheric transport of contaminants, together with the acidification of susceptible waters by acid deposition (acid rain), is constantly increasing the number of variables that need to be included in monitoring programs assessing the effects of atmospheric sources of pollution. Assessment programs principally for water bodies at risk of acidification should include alkalinity, pH, sulfate, and nitrate, whereas those at risk from industrial emissions to the atmosphere need to consider local and regional industrial activity, for example, by means of an inventory of atmospheric emissions. Widespread atmospheric transport has been proved for lead, cadmium, arsenic, certain pesticides, and other organic compounds.

Monitoring to assess the effects of industrial effluents requires a knowledge of the nature of the effluent and its chemical composition in order to aid the selection of the appropriate variables in addition to those variables which indicate the quality of the water prior to its disturbance by the effluent (e.g., the background quality variables). However, in some circumstances the variables may be defined by discharge regulations imposed on some industrial processes which produce particularly, or potentially, harmful effluents. In these situations the final concentrations of contaminants, or other relevant variables, after disposal and mixing in the water body are compared with the defined maximum allowable concentrations set in the regulations or consents. In the case of a major accidental release of industrial effluents it may be necessary to conduct a full survey using an extensive list of organic, inorganic, and general variables in order to determine the nature and impact of the accidental release, prior to choosing the most appropriate variables for monitoring the recovery of the water body.

Ultimately, the choice of variables will also have to take into consideration the resources available to carry out the necessary sampling, and to perform the analytical techniques for the selected variables at the level of accuracy required to fulfill the objectives. In practice, a compromise often needs to be made between the ideal selection of variables and the resources available to measure them.

Site selection, sampling frequency, and number of samples

Natural water quality can vary spatially, as well as temporally, within all water bodies. Groundwater can show very little natural variation over time (years to decades), whereas some rivers and shallow lakes can show variability in biological and chemical characteristics within hours. Lakes are particularly subject to seasonal influences, resulting in warming and cooling of surface waters, mixing of the water column, growth of plants and algae, and the associated changes in chemical quality variations (Wetzel, 1975). Slow-flowing, lowland rivers and canals can also show seasonal variations in algal populations, as in the River Meuse (Descy, 1992). High seasonal rainfall or snowmelt conditions can increase river flows and the associated fluxes of particulate matter, nutrients, etc. All these temporal variations need to be determined and taken into consideration when planning the frequency and timing of a sample collection. The impact of wastewater discharges can be worse during warm-water, low-flow periods, and more intensive sampling may be required during these times. Flux determinations may require additional sampling during high-rainfall seasons.

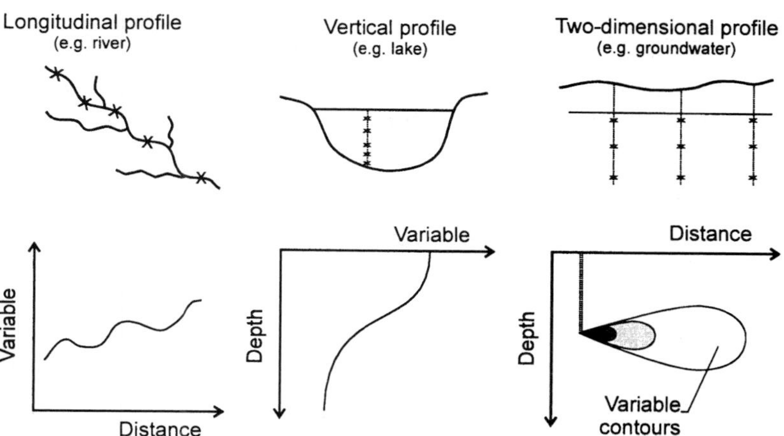

Figure 6.3 Different types of spatial variability in water bodies. (*After Meybeck and Helmer, in press.*)

Spatial variability can occur within horizontal and vertical dimensions as illustrated by Fig. 6.3. Turbulent, well-mixed rivers exhibit variations in water quality principally along their length and very little variation vertically. However, major effluent discharges or river confluences can lead to variations across the width, as well as along the length, of the river. Lakes and reservoirs subject to stratification can show significant vertical variations in water quality during stratified periods but little or no variability when the water column is fully mixed. Other than lakes and reservoirs which are very regular in shape and bathymetry, most lakes show some horizontal variations in biological characteristics and the associated chemical variables. Many lakes and reservoirs can be considered as a series of connected water basins, each of which may have slightly different water-quality characteristics. This is particularly true for very sinuous reservoirs formed by damming natural river valleys. At a minimum, a sampling point would be required in each main basin of the lake, and preliminary surveys would be required to establish the degree of variability between basins before selecting the optimum number of, and locations for, sampling points. Groundwater can show horizontal and vertical variations in water quality, and the nature of groundwater flow systems have to be established by hydrogeologists before the number and location of sample sites can be chosen (Chilton, in press).

The degree and type of natural variability in a water body can greatly influence the number of samples required to adequately characterize the water quality for a specific sampling interval. A single sample, for example, is usually completely inadequate to describe conditions in the vertical dimension of a water body which thermally stratifies. The variability in the results obtained during a preliminary survey taking many samples and the desired precision of the estimated value can be obtained using statistical methods (Demayo and Steel, in press). Such approaches can help determine the number of samples required in relation to the objectives of the program and, thereby, reduce unnecessary use of financial and technical resources.

Choice of methods for data collection

Many variables can be measured by more than one method, ranging from simple, inexpensive approaches (some of which can be carried out in the field) which give relatively low precision determinations to highly sophisticated and expensive laboratory-based, analytical methods offering a high degree of precision. The choice of method depends on the necessary degree of accuracy to meet the monitoring-program objectives, together with financial and technical restraints. Using expensive and technically demanding methods simply because the facilities are there, or because the methods seemed attractive to try, are wasteful of resources and may not enhance the success of the monitoring program if the objectives and the final assessment do not warrant their use. A program is more likely to fail its objectives if overly complex methods are attempted unnecessarily or, conversely, if the approaches used to collect data are too simple and of inadequate precision. The appropriate choices can be aided by some experimentation and preliminary surveys. Descriptions of some methods for field and laboratory determination of water quality, which can aid method selection, are given in Bartram and Ballance (in press).

A long-term monitoring program which relies on technical resources requires adequate financial and technical backup to cover any personnel or mechanical difficulties which may arise. This is particularly true where personnel are specifically trained in techniques or where only a single piece of analytical equipment is available. When it becomes necessary to replace, temporarily or permanently, personnel or methods (including a piece of analytical equipment), it is advisable to allow a period of overlap where both old and new equipment (or personnel) collect data for the same sites and water-quality variables. This allows the two data sets to be compared during the overlap period and reduces the risk of a sudden anomaly appearing in the data.

Monitoring programs which submit data to a central database to which other laboratories, agencies, countries, etc. also submit data are often subject to quality assurance (see below). Even where no interlaboratory scheme exists, quality assurance is a very important aspect of the data-collection process and should be implemented at all stages (field, laboratory, data storage).

Integration of hydrologic and water-quality data

The importance of understanding the hydrologic characteristics of water bodies when planning and implementing water-quality monitoring programs has been stressed several times above. Routine hydrologic monitoring is sometimes the responsibility of an agency separate from that which carries out water-quality monitoring. If the information from existing hydrologic monitoring is not available or is not sufficiently detailed, it may be necessary to include hydrologic measurements specifically in the water-quality monitoring program. However, unnecessary duplication of effort can sometimes be avoided if it is possible to take water-quality samples from points at which hydrologic measurements are made and if arrangements can be made for access to the data collected. Table 6.6 summarizes the hydrologic

TABLE 6.6 Hydrologic Information Required for Water-Quality Assessment

Level*	Rivers	Lakes/reservoirs	Groundwaters
		Basic Information	
A	Watershed map	Thermal regime	Major aquifer type
B	River seasonal regime	Bathymetric map	Aquifer map
C	Flow-duration statistic	Water balance and current patterns	Hydrodynamic characteristics
		Hydrologic Monitoring	
A	River level at sampling	Lake level at sampling	Piezometric level
B	River discharge at sampling	Lake level between sampling	Piezometric level between sampling
C	Continuous river discharge	Tributary discharge and lake water budget	Full knowledge of groundwater hydrodynamics

*Levels A, B, and C are increasing orders of assessment program complexity.
SOURCE: Meybeck et al. (in press b).

data that are required for monitoring and assessment programs of increasing complexity for the major water-body types.

River flows are influenced by the surrounding land catchment area which the river drains. Fluctuations in runoff, caused primarily by variations in rainfall but also by changes in land use, result in fluctuations in the discharge regime of the river. The concentration of suspended material carried by rivers generally increases with increased river discharge, although this simplified relationship is usually complicated by various other factors (Thomas and Meybeck, in press). Water quality in rivers is, therefore, also linked to their discharge and the land-based activities in the surrounding catchment which affect discharge, such as deforestation and urbanization. The extreme variability in discharge which can be observed in rivers can complicate monitoring activities and lead to misleading results if sampling is carried out at inappropriate times (Meybeck et al., in press a).

The concept of residence time (the period it takes for the water in a water body to be completely exchanged) is very important, for example, in the recovery of a water body after an accidental pollution incident, and defines the time period for which it may be necessary to continue monitoring for possible effects within the water body. Residence time is particularly relevant in lakes, reservoirs, and groundwaters. Residence times in most lakes and reservoirs can vary from weeks to years and in groundwater, from weeks to thousands of years.

Data handling and interpretation

Once the fieldwork has been conducted and the samples analyzed, it is necessary to store and interpret the data obtained. The importance of this aspect should not be overlooked because the final assessment of the water quality, which determines the suitability of the water for use or supports recommendations for management of the water body, are all made using the data collected by the monitoring program. Only if the results obtained by monitoring programs are regularly interpreted and reported is it possible to ascertain whether a program is meeting its objectives in the most efficient way. If the accuracy of the generated results is not maintained, the interpretation of the data may be impaired or could even lead to misleading conclusions. In addition, data which are difficult to retrieve or cumbersome to manipulate are less likely to be consulted when further information is required.

Data can be stored in handbooks, in filing systems, or, more usually in recent times, on desktop or mainframe computers. Extensive monitoring programs generating large quantities of data benefit from the use of a computer to store and manipulate the data which could otherwise become unmanageable. Nevertheless, a program limited to measuring a few variables can be managed satisfactorily using manual filing systems, a pocket calculator, and hand-drawn graphic representations of results. The use of a computer database, especially where large quantities of data for many variables at many locations need to be stored, makes retrieval of information for any particular monitoring site, date, or variable much faster. Although commercially available computer databases may form an economic and useful starting point, it is often preferable for major monitoring programs (such as national water-quality programs) to build a database specifically suited to the data stored and the anticipated uses of those data.

Care must be taken when transferring the results of field or laboratory measurements to the data-storage system to ensure that no errors due to transcription are introduced. To some extent this can be avoided with certain analytical or automatic monitoring techniques which can transfer results directly to a portable electronic storage mechanism (such as computer diskette) which eliminates the need to reenter each value onto a computer database by hand.

Data interpretation involves such activities as checking values for compliance with guidelines, comparing results between sampling stations, analyzing trends, and determining cause-effect relationships between water quality and other activities (such as land use and pollutant sources). An important aspect of data interpretation is presenting the data in a form which can be easily understood by those not closely involved in the monitoring program (such as environmental managers and the general public).

The plotting of a graph of changes in concentration over time is a simple example of data interpretation and representation of the results of a monitoring program (see Fig. 6.1). However, many assessments require statistical treatment of the results to determine, for example, trends, significant variations between sites, and correlations between variables. Many statistical methods have preconditions, such as the number of samples required. Nevertheless, it is common for inexperienced workers to attempt to use statistical methods on inappropriate data sets because they did not consider the statistical treatment of the samples before planning the monitoring program and select the sampling frequency, sample density, etc., accordingly. As a general rule, it is preferable to consult a statistician at the program-design phase rather than only at the data-analysis phase (when it may be already too late to obtain more samples if needed). Consideration of the statistical approaches to be used can also indicate the number of samples required to give a particular level of precision to the results.

Management recommendations and program evaluation

The prime objective of a monitoring program is to provide information which directly or indirectly aids the management of water resources, whether this is by establishing trends, checking compliance with guidelines, or determining the effects of contaminant transfer. Consequently, much effort is focused on this aspect when the analysis and interpretation of the monitoring results is undertaken. However, it is equally important to determine whether the program design is adequate to meet its objectives, and to use the results obtained to make any necessary recommendations for modifications to the program, such as increased or decreased frequency of sampling, inclusion of more or less variables and other relevant modifications. This reevaluation of the program design is an important aspect of the overall monitoring and assessment process and helps to ensure the long-term cost-efficiency of the activity. Reevaluation of the early-warning monitoring program for the Rhine River was carried out after the major chemical-spill accident at the Sandoz chemical plant in Basel, in 1986. The early-warning system consisted of eight stations designed to protect drinking-water intakes from accidental pollution by continuous monitoring, including the use of biological methods such as the observation of the behavior of fish contained in tanks through which the river water is diverted as a continuous flow. A wider range of sensitive "in-line" bioassays with organisms from all trophic levels is being evaluated for eventual incorporation into the early-warning monitoring program (Schmitz et al., 1994).

Monitoring programs spanning several years may need to be reevaluated with respect to the methods in use and new approaches or techniques which may have been designed or introduced during the routine running of the program. For example, in the past decade there has been an increasing awareness of the suitability of some biological approaches and of the use of particulate matter in contaminant monitoring. In addition, it may be necessary to review programs which have been in place for several decades and to make them more specific in their objectives. Together with advances in monitoring techniques, there has been an increased awareness of some environmental issues, such as acidification and groundwater contamination, which require specific approaches and/or the monitoring of specific variables. There is now much specialized literature available (see other chapters in this volume) which explains these issues, summarizes the findings to date, and indicates which water bodies may be at risk. A review of the appropriate literature, combined with the interpreted results of a monitoring program, can give an indication of any modifications required to the program design.

Quality Assurance and Control

Even the most carefully designed monitoring programs can lead to inadequate results if sufficient care is not taken to control the quality of data generated. Errors can be introduced at almost any stage of the monitoring, sample analysis, and data-handling stages (see example for chemical monitoring in Table 6.7). *Internal quality control* includes all operational techniques used by laboratory staff for continuous assessment of the quality of individual analytical procedures, whereas *external quality control* (sometimes known as *quality assurance*) is a method of establishing the accuracy of analytical techniques by comparing one laboratory's results with another analyzing the same sample (Bartram and Ballance, in press). The importance of quality control should not be underestimated in any monitoring program because it ensures that the data generated, particularly the data obtained by the chemical analysis of water-quality variables, are valid and can be used for accurate and meaningful assessments of water quality. Therefore, a significant proportion of the resources available for the monitoring program (e.g., 10 to 20 percent of the total financial, technical, and personnel resources available) should be allocated for this purpose (Meybeck et al., in press b).

Care and strict adherence to recommended procedures when sampling are very important to ensure that the samples collected are truly representative. This can mostly be achieved through the rigorous training of those responsible for sample collection. For example, simple precautions such as standing downstream and collecting the sample in an upstream direction are very important. Similarly, following the recommended method of sampler operation, or depth for

TABLE 6.7 Some Possible Sources of Error in the Water-Quality-Assessment Process with Special Reference to Chemical Methods

Assessment step	Operation	Possible source of error	Appropriate actions
Monitoring design	Site selection	Station not representative (e.g., poor mixing in rivers)	Preliminary surveys
	Frequency determination	Sample not representative (e.g., unexpected cycles or variations between samples)	
Field operations	Sampling	Sample contamination (micropollutant monitoring)	Decontamination of sampling equipment, containers, preservatives
	Filtration	Contamination or loss	Running field blanks
	Field measurement	Uncalibrated operations (pH, conductivity, temperature)	Field calibrations Replicate sampling
		Inadequate understanding of hydrologic regime	Hydrologic survey
Sample shipments to laboratory	Sample conservation and identification	Error in chemical conservation Lack of cooling Error in biological conservation Error in and/or loss of label Break of container	Field spiking Appropriate field pretreatment Field operator training
Laboratory	Preconcentration	Contamination or loss	Decontamination of laboratory equipment and facilities
	Analysis	Contamination Lack of sensitivity Lack of calibration Error in data report	Quality control of laboratory air, equipment, and distilled water Quality-assurance tests (analysis of control sample; analysis of standards) Check internal consistency of data (with adjacent sample, ionic balance, etc.)

Computer facility	Data entry and retrieval	Error in data handling	Checks by data-interpretation team
Interpretation	Data interpretation	Lack of basic knowledge Ignorance of appropriate statistical methods Omission in data report	Appropriate training of scientists
Publication	Data publication	Lack of communication and dissemination of results to authorities, the public, scientists, etc.	Setting of goals and training to meet the need of decision makers

SOURCE: Meybeck et al. (in press b).

collection of the samples, is essential for meaningful samples. Nevertheless, taking replicate samples is also recommended as a method of checking that samples are representative.

The handling and storage of the samples prior to transport and analysis are also particularly important in ensuring that they give meaningful results. Sample containers should be prepared according to the requirements of the later analysis techniques (e.g., acid-washed containers for samples to be used for later metal-concentration analysis) and should be clearly labeled in advance to ensure that they are used for the correct samples. Field analysis on samples should be carried out in the correct sequence, and such samples cannot normally be used for any later analyses. For example, conductivity should be carried out before pH measurements on the same sample because the pH electrode may release concentrated electrolytes which would affect a conductivity measurement. It is also recommended that one blank sample (usually distilled water) be included for every 10 water samples in all sample handling and treatments (filtration, storage, and preservation) to help identify any sources of contamination.

For checking the quality of the laboratory analytical techniques internally, validity checks should be made with every batch of samples or at frequent and regular intervals. These consist of calibration checks, the inclusion of blank samples, and the use of samples spiked with a particular substance or compound for analysis. External quality control, or quality assurance, is particularly important where more than one laboratory and/or agency are submitting measurements for the same variables to a central database for analysis and interpretation. Normally, once or twice a year, the same control samples (for which the concentrations of the variables are known only to the organizers of the quality-assurance exercise) are distributed to all participating laboratories from a central point. The results of the analyses of the samples are returned to the distribution point for comparison. When problems are found, they are referred back to the laboratories for investigation and correction.

Even the final stages of storage and treatment of the data obtained are open to sources of error. Mistakes can occur when data are recorded by hand from instruments, or transferred by hand from notebooks to record sheets or from record sheets to computer databases. Transcribed data should always be checked back against the original copy immediately after transfer, and regular and careful checks should be carried out on databases. Database checks should be performed by individuals with a knowledge of water quality who would be capable of recognizing possible errors. Some instrumental techniques allow direct transfer of the results to a computer, reducing the risk of transcription errors, and some computer data-entry systems can be set up to recognize and "flag" data which do not fall within an expected range. Even the final analysis of the data held in a database could be subject to mishandling. Statistical analysis of data requires that particular conditions be met by the data set and proper procedures be followed (Demayo and Steel, in press). The easy availability of a wide range of statistical and graphic techniques on personal computers has allowed inexperienced or untrained individuals to use inappropriate data-interpretation and data-presentation techniques.

Loss of data is still a relatively common and serious problem. Unless samples can be stored for long periods of time after aliquots have been removed for analysis, the lost data are often irretrievable, although careful storage of all notebooks, data sheets, and record books can sometimes facilitate the regeneration of data from original analytical results. However, the increasing use of computers to manipulate all stages of data handling, storage, and interpretation increases the risk of total data loss or data corruption if proper precautions are not taken. At least two backup copies of data held on computer systems should always be stored on a removable medium, such as diskettes, tapes, and magnetic or optical disks. One copy should be stored in a different location to ensure that a copy is preserved in the event of fire or theft. All backups should be updated regularly so that if it is necessary to reconstruct the database, only a minimum of effort would be required to include data added since the last backup was made. Accidental corruption of the main database should be avoided by careful control over access, such as by the use of confidential passwords.

Progress in Water-Quality Monitoring in Developing Countries

As has been stressed several times above, monitoring programs should be designed with realistic objectives and in relation to the availability of resources to meet those objectives. For many developing countries with principally agriculturally based economies, the main objective of water-quality monitoring at the local or regional scale is to determine whether water resources present a risk to human health. Monitoring is, therefore, focused on fecal-coliform detection in samples from wells, water holes, and distribution systems. At the national level there is usually a need to ensure that there are sufficient water resources of adequate quality to supply the main population centers.

The ability to provide wastewater and excreta disposal facilities often cannot keep pace with the rapid growth occurring in some urban centers (in both developed and developing countries), leading to contamination of important water resources required for human use, particularly for drinking water. Contamination of groundwater supplies by unsewered sanitation (septic tanks and pit latrines) has proved to be the cause of numerous disease outbreaks, as well as the cause of increased nitrate concentrations, in localities in South America, Africa, and India (Chilton, in press). The problem is most severe where water for consumption is drawn from relatively shallow wells and used without treatment, such as in small and medium-size towns or densely populated periurban and rural areas. Fortunately, many of the methods and approaches available to determine contamination of water by human effluents are relatively simple and cheap, making them feasible for most agencies to adopt with suitable training.

Preliminary surveys are particularly important where resources for monitoring are limited, such as in developing countries, because they help identify the key issues to be addressed, the causes of the water-quality problems, and the sources and nature of contaminants. A strategy for developing a microbiological monitoring program for drinking-water supply wells in rural areas has been described by Lloyd and Helmer (1991) and tested in several developing countries. Preliminary, and fairly simple, sanitary inspections combined with the collection of water samples helps identify the sources and severity of contamination of wells. This information can then be used to identify wells needing regular monitoring and to plan and monitor the effects of remedial protection measures for the groundwater sources.

Rapidly industrializing countries are finding themselves faced not only with the pressures on water resources arising from high population densities and municipal wastewaters but also with widespread deterioration of water resources as a result of poorly controlled and monitored industrial effluents and emissions. For many agencies in such nations the financial and technical resources required to effectively monitor industrial impacts on their water resources are beyond their means. The careful setting of realistic objectives and the appropriate selection of monitoring media, methods, and variables are particularly important for these agencies. Participation in regional, national, or even global monitoring programs can sometimes be helpful in defining objectives, selecting sites and methods, and providing access to training, quality control, and data-interpretation procedures (see below).

Water-quality indices or grades

Some developing countries are already using, testing, or considering adaptations and combinations of simple and inexpensive methods such as those already in use in many developed countries for making regional or nationwide assessments of the general quality of surface-water resources. Examples of the approaches being considered include the water-quality assessment of the rivers in Nordrhine-Westfalen, Germany (Meybeck et al., in press a) and the rivers and canals of England and Wales (National Rivers Authority, 1994). The former is a biological classification, whereas the latter is based on a few simple chemical measurements but is being expanded to include a biological component. These approaches allocate quality grades to the monitored waters, ranging from highly polluted (or degraded) to unpolluted, or from good to bad, and are sufficiently sensitive to show improvements in water quality over time achieved through the introduction of, for example, wastewater treatment.

The highly industrialized state of São Paulo, Brazil, has also been regularly assessing the quality of its rivers for some years using a water-quality index based on simple chemical and physical measurements, together with fecal-coliform counts (CETESB, 1991; Chapman, 1992). The results of measurements of pH, BOD, total nitrogen, total phosphate, temperature, turbidity, total residue, dissolved oxygen, and fecal coliforms are used to calculate an index between 0 and 100 (CETESB, 1991). The higher the index, the better the quality of the water. The index values are divided into five categories of water quality which can be illustrated on maps by indicating rivers with specific colors associated with each category of water quality. This popular form of presentation of water-quality assessment has the advantage that no expert knowledge is required to understand it, making it particularly suitable for management personnel and the general public alike. Influences of the highly populated and industrialized centers are clearly indicated by the color banding.

Biological approaches

There is a growing interest worldwide in monitoring methods which integrate the effects of all influences on the aquatic environment. Many of the methods being developed and tested are less dependent on the advanced instrumental methods used for chemical monitoring programs and may help the developing and rapidly industrializing countries make the best use of their limited monitoring resources. Techniques and approaches that indicate deterioration in general water quality, rather than give an indication of precise concentrations of numerous chemical variables, are particularly suitable for developing countries wishing to monitor the spatial and/or temporal changes in water quality of their surface waters.

Approaches being considered and developed include the study of aquatic community structure (e.g., diversity and abundance; similarity), bioassays adapted and developed for local species, simple physiological measurements with selected communities (e.g., oxygen production by phytoplankton, growth rates of bacteria), and simple morphological studies (e.g., gill filament damage in fish) (Friedrich et al., in press). Such methods are mainly qualitative but, when applied rigorously at regular intervals over a suitable time span, can indicate trends or highlight sudden changes in water quality. For example, bioassays or toxicity tests performed regularly in situ in water bodies receiving continuous or regular discharges of industrial effluents, or carried out in the laboratory using samples of water taken from the water body, can give an indication of gradual or sudden deterioration or improvement. A sudden increase in toxicity in samples from selected sites can help identify the possible need for special chemical water sampling. Preliminary surveys using bioassay techniques may have the added benefit of reducing the number of samples required for expensive chemical analysis by eliminating sites which may be inappropriate or unnecessary.

Assessment systems based on bioindicators (organisms which reflect specific ecological conditions and/or environmental pollution) can be adapted to regional or local conditions by determining the statistical correlations between the organisms to be used and physicochemical data measured at the same sites. Systems for running waters are being developed and tested in Brazil (Rio de Janeiro and Curitíba/Paraná) using the genus level of taxonomic identification for macrozoobenthic organisms (Friedrich, 1989). Benthic organisms reflect the water quality for a long period because they are relatively long-lived and mostly fairly immobile. They can also be easily sampled and then preserved for later identification and counting.

An alternative approach for monitoring the impact of point sources of effluents is the use of artificial substrates which organisms can colonize and that can be placed in the water body at suitable sites (e.g., upstream and downstream of an effluent discharge). Artificial substrates have the advantage that they eliminate any differences in results which may be caused by natural differences in substrate (and the associated biological communities) between sites. The samplers can be periodically removed and/or checked and are particularly useful if there is a sudden or accidental increase or change in the nature of an effluent discharge.

Where limited analytical capability exists to measure contaminants such as metals (e.g., by means of atomic-absorption spectrophotometry), the necessity to measure the very low concentrations which may occur in water samples can sometimes be overcome by monitoring the concentrations in an alternative medium. Some biological tissues and suspended or deposited particulates accumulate metals over long time periods to concentrations higher than those of the water from which they are sampled, and their suitability as environmental indicators is now well established. The higher metal concentrations are easier to measure with some precision, although the preparation of the material may be quite complex or require strict adherence to defined methods to avoid contamination and misleading results. However, reference material is available for different sediments and biological tissues which can be used to determine the precision of the analytical techniques being used.

Water-Quality Monitoring on a Global Scale

Awareness of the widespread deterioration of water resources on a global scale in the 1970s, together with the transnational nature of many major water bodies, highlighted the need for a global assessment of the trends in freshwater quality which could be used as a basis for rational international action. In 1977, the United Nations Environment Programme (UNEP) and the World Health Organization (WHO), together with the UN Educational, Scientific and Cultural Organization (UNESCO) and the World Meteorological Organization (WMO), responded to this need by initiating the first global monitoring program for freshwater, known as GEMS/WATER (a component of the Global Environment Monitoring System). The program aimed to help generate the necessary data for a global assessment of freshwater quality by

- Collaborating with member states in establishing new monitoring systems and strengthening old ones
- Improving the validity and comparability of water-quality data within and between member states
- Assessing the incidence and long-term trends of water pollution by selected persistent and hazardous substances

This was achieved by establishing carefully selected monitoring stations throughout the world which would return data to a central database. Baseline stations were chosen where there was little or no influence from diffuse or point sources of pollutants. Impact stations were chosen to represent water bodies discharging to the oceans or where there was at least one major use of the water (stations were selected with respect to drinking-water supply, irrigation use, aquatic life, or multiple impacts), and trend stations were chosen to represent large areas of diverse human activity. Data collection started in 1978 and is continuing to the present day. Measurements were made for about 50 different water-quality variables, including dissolved oxygen, BOD, fecal coliforms, nitrates, and selected chemicals and trace contaminants. Analytical quality-control exercises were carried out within laboratories as well as in the form of interlaboratory comparisons. A review of the first phase of the GEMS/WATER program and its achievements is available in WHO (1991). National agencies were given assistance with methodology, training, and analytical quality control which, for many developing countries, proved to be of great benefit to their own national water-quality monitoring-program development.

The first attempt at a global assessment of freshwater quality using the results of the GEMS/WATER program (UNEP and WHO, 1988) highlighted the lack of data for many world regions, especially the developing countries. A more comprehensive assessment (Meybeck et al., 1989) used many other sources of data, including published data from special surveys and other monitoring networks, because of the lack of data in the global database for significant variables relevant to global water-quality issues (especially from developing countries). The results of the assessment, which highlighted some of the world's major water-pollution problems as well as some of the shortcomings in global freshwater-quality monitoring, were considered in the reviewing and redesigning of the global program (WHO, 1991).

The revised objectives of the GEMS/WATER program can be summarized as follows:

- To provide water-quality assessments of the world's freshwater relative to human health, aquatic ecosystem health, and other global environmental concerns which define the status of water quality, identify and quantify trends, define the causes of observed conditions and trends, identify types of water-quality problems in specific geographic areas, and provide information and assessments in a form useful for evaluating management alternatives and making management decisions
- To provide information on the transport and, where appropriate, the assessments of fluxes of toxic chemicals, nutrients, and pollutants from major river basins to the continent-ocean interfaces
- To strengthen national water-quality monitoring in developing countries, including the improvement of analytical capabilities and data-quality assurance

In accordance with these objectives, monitoring stations were redesignated or newly selected to provide information on baseline water quality (as before), long-term trends, and global river fluxes. Table 6.8 gives the minimum variables to be measured at each type of station. In response to the program review in 1990, the groundwater component of the program is being developed and expanded, and the

TABLE 6.8 Variables Used in the GEMS/WATER Program for Basic Monitoring

Measured variable	Streams: baseline and trend	Headwater lakes: baseline and trend	Ground-waters: trend only	Global river flux stations
Water discharge or level	x	x	x	x
Total suspended solids	x	—	—	x
Transparency	—	x	—	—
Temperature	x	x	x	x
pH	x	x	x	x
Electrical conductivity	x	x	x	x
Dissolved oxygen	x	x	x	x
Calcium	x	x	x	x
Magnesium	x	x	x	x
Sodium	x	x	x	x
Potassium	x	x	x	x
Chloride	x	x	x	x
Sulfate	x	x	x	x
Alkalinity	x	x	x	x
Nitrate	x	x	x	x
Nitrite	x	x	x	x
Ammonia	x	x	x	x
Total phosphorus (unfiltered)	x	x	—	x
Phosphorus, dissolved	x	x	—	x
Silica, reactive	x	x	—	x
Chlorophyll *a*	x	x	—	x
Fluoride	—	—	x	—
Fecal coliforms (trend stations only)	x	x	x	—

SOURCE: WHO (1991).

sampling and analytical methods will include particulate matter. Methods for using biota as indicators of integrated environmental effects are also being developed.

For many developed countries, participation in the GEMS/WATER program involves little additional effort because designated stations are often chosen from existing national networks using standardized and harmonized techniques. Other participants can benefit from the training provided in monitoring and assessment methods, the analytical quality-control exercises and the special data-analysis software designed for the program and adapted to aid in management of freshwater resources. Emphasis is being placed on encouraging developing countries to use techniques which are commensurate with their financial and technical means.

Conclusions and Recommendations

The success of a monitoring program can be judged by the fulfillment of its objectives and its ability to provide meaningful information which can be used to manage the water resource. To achieve these aims, the program must be supported by the practical and administrative framework and the legal instruments to back up any recommendations arising from the program. Five main reasons why a monitoring-and-assessment program may fail to provide the expected information for environmental managers have been identified by Meybeck et al. (in press a), on the basis of some suggestions by Wilkinson and Edworthy (1981):

- The objectives of the assessment were not defined properly.
- The monitoring program was installed with insufficient knowledge of the water body.
- There was inadequate planning of the sample collection, handling, storage, and analysis.
- The information obtained was poorly archived.
- The data were not properly or adequately interpreted and reported.

All these factors must be considered prior to embarking on a new monitoring program to ensure that the most efficient use of resources can be made. The techniques presently available, and those constantly being developed or refined, can ensure that an appropriate monitoring program should be feasible for virtually all agencies responsible for the quality and quantity of their water resources. Current understanding and progress in water-quality assessment have, however, highlighted the need for basin- or catchmentwide, as well as regional, approaches to water-quality monitoring. For such programs to provide useful data, harmonization and standardization of techniques are essential. As a result of the increasing interest and cooperation of several international agencies, assistance and training schemes are available which will help all nations achieve these goals.

References

Alderton, D. H. M., 1985, Sediments. In *Historical Monitoring*. MARC Report No. 31, Monitoring and Assessment Research Centre, University of London.

APHA, 1989, *Standard Methods for the Examination of Water and Wastewater,* 17th ed. American Public Health Association, Washington, D.C.

Bartram, J., and Ballance, R., eds., (in press), *Water Quality Monitoring: A Practical Guide to the Design and Implementation of Fresh Water Quality Studies and Monitoring Programmes.* Chapman and Hall, London.

Biswas, A. K., Jellali, M., and Stout, G., 1993, *Water for Sustainable Development in the 21st Century.* Water Resources Management Series 1. Oxford University Press, New Delhi.

CETESB, 1991, *Relatório de Qualidade das Águas Interiores do Estado de São Paulo—1990.* Companhia de Tecnologia de Saneamento Ambiental, São Paulo.

Chapman, D., 1992, Water quality assessment around the world: the role of chemistry with particular reference to Brazil and Malaysia. In Coomber, D. I., Langer, S. S., and Pratt, J. M., eds., *Chemistry and Developing Countries.* Commonwealth Science Council, London, and Royal Society of Chemistry, London, pp. 55–61.

Chapman, D., ed., in press, *Water Quality Assessments. A Guide to the Use of Biota, Sediments and Water in Environmental Monitoring,* 2d ed. Chapman and Hall, London.

Chapman, D., and Kimstach, V., in press, The selection of water quality variables. In Chapman, D., ed., *Water Quality Assessments. A Guide to the Use of Biota, Sediments and Water in Environmental Monitoring,* 2d ed. Chapman and Hall, London, pp. 59–126.

Chilton, J., in press, Groundwater. In Chapman, D., ed., *Water Quality Assessments. A Guide to the Use of Biota, Sediments and Water in Environmental Monitoring,* 2d ed. Chapman and Hall, London, pp. 413–510.

Demayo, A., and Steel, A., in press, Data handling and presentation. In Chapman, D., ed., *Water Quality Assessments. A Guide to the Use of Biota, Sediments and Water in Environmental Monitoring,* 2d ed. Chapman and Hall, London, pp. 511–612.

Descy, J.-P., 1992, Eutrophication in the River Meuse. In Sutcliffe, D. W., and Jones, J. G., eds., *Eutrophication: Research and Application to Water Supply.* Freshwater Biological Association, Ambleside, pp. 132–149.

Environment Canada, 1987, *Canadian Water Quality Guidelines.* Prepared by (with updates) the Task Force on Water Quality Guidelines of the Canadian Council of Resource Ministers, Environment Canada, Ottawa.

Foster, S. F., Ventura, M., and Hirata, R., 1987, *Groundwater Pollution: An Executive Overview of the Latin American-Caribbean Situation in Relation to Potable Water-Supply.* Pan American Center for Sanitary Engineering and Environmental Sciences (CEPIS), Lima, Peru.

Friedrich, G., 1989, Use of biomonitoring in water protection with special reference to Northrhine-Westfalia/FRG and Brazil. Paper presented at the Workshop International em Estratégias de Monitoramento Ambiental, Universidade Federal de Bahia, Salvador, Bahia, Dec. 4–9, 1989.

Friedrich, G., Chapman, D., and Beim, A., in press, The use of biological material. In Chapman, D., ed., *Water Quality Assessments. A Guide to the Use of Biota, Sediments and Water in Environmental Monitoring,* 2d ed. Chapman and Hall, London, pp. 175–242.

Hem, J. D., 1989, *Study and Interpretation of Chemical Characteristics of Natural Waters.* Water Supply Paper 2254, 3d ed. U.S. Geological Survey, Washington, D.C.

Hirsch, R. M., Alley, W. M., and Wilber, W. G., 1988, *Concepts for a National Water-Quality Assessment Program.* U.S. Geological Survey Circular 1021. U.S. Geological Survey, Denver, Colo.

Lloyd, B., and Helmer, R., 1991, *Surveillance of Drinking Water Quality in Rural Areas.* Longmans Scientific and Technical, Harlow, United Kingdom.

Meybeck, M., and Helmer, R., in press, An introduction to water quality. In Chapman, D., ed., *Water Quality Assessments. A Guide to the Use of Biota, Sediments and Water in Environmental Monitoring,* 2d ed. Chapman and Hall, London, pp. 1–22.

Meybeck, M., Chapman, D., and Helmer, R., 1989, *Global Freshwater Quality: A First Assessment.* Blackwell Reference, Oxford.

Meybeck, M., Friedrich, G., Thomas, R., and Chapman, D., in press a, Rivers. In Chapman, D., ed., *Water Quality Assessments. A Guide to the Use of Biota, Sediments and Water in Environmental Monitoring,* 2d ed. Chapman and Hall, London, pp. 243–318.

Meybeck, M., Kimstach, V., and Helmer, R., in press b, Strategies for water quality assessment. In Chapman, D., ed., *Water Quality Assessments. A Guide to the Use of Biota, Sediments and Water in Environmental Monitoring,* 2d ed. Chapman and Hall, London, pp. 23–57.

National Rivers Authority, 1994, *The Quality of Rivers and Canals in England and Wales (1990 to 1992). As Assessed by a New General Quality Assessment Scheme.* Water Quality Series no. 19. HMSO, London.

NIVA, 1992, *Evaluation of the International Co-operative Programme on Assessment and Monitoring of Acidification of Rivers and Lakes.* Norwegian Institute for Water Research, Oslo.

Schmitz, P., Krebs, F., and Irmer, U., 1994, Development, testing and implementation of automated biotests for the monitoring of the River Rhine, demonstrated by bacteria and algae tests. *Water Sci. Technol.* **29**(3), 215–221.

Thomas, R., and Meybeck, M., in press, The use of particulate material. In Chapman, D., ed., *Water Quality Assessments. A Guide to the Use of Biota, Sediments and Water in Environmental Monitoring,* 2d ed. Chapman and Hall, London, pp. 127–174.

Thomas, R., Meybeck, M., and Beim, A., in press, Lakes. In Chapman, D., ed., *Water Quality Assessments. A Guide to the Use of Biota, Sediments and Water in Environmental Monitoring,* 2d ed. Chapman and Hall, London, pp. 319–368.

UNEP and WHO, 1988, *Assessment of Freshwater Quality.* Report on the results of the WHO/UNEP programme on health-related environmental monitoring. United Nations Environment Programme, Nairobi, Kenya.

Wetzel, R. G., 1975, *Limnology.* Saunders, Philadelphia.

WHO, 1984, *Guidelines for Drinking-Water Quality,* vol. 2, *Health Criteria and Other Supporting Information.* World Health *Organization, Geneva.*

WHO, 1991, *GEMS / WATER 1990–2000, the Challenge Ahead.* WHO/PEP/91.2, World Health Organization, Geneva.

WHO, 1993, *Guidelines for Drinking-Water Quality,* vol. 1, *Recommendations,* 2d ed. World Health Organization, Geneva.

Wilkinson, W. B., and Edworthy, K. J., 1981, Groundwater quality systems—money wasted? In van Duijvenbooden, W., Glasbergen, P., and van Lelyveld, H., eds., *Quality of Groundwater.* Proceedings of an International Symposium, Noordwijkerhout. Studies in Environmental Science no. 17, Elsevier, Amsterdam, pp. 629–642.

World Resources Institute, 1992, *World Resources 1992–93.* Oxford University Press, New York, Oxford.

Chapter

7

Water-Quality Prediction and Management

Christian Jokiel
IWW, University of Technology
Aachen, Germany

Daniel P. Loucks
Cornell University
Ithaca, New York

Introduction

The quality of natural water bodies impacts those using or living within those water bodies. Discharges of pollutants can degrade the quality of the water, and hence adversely affect the water's beneficial uses as well as the health of its aquatic ecosystem. To manage the quality of natural water bodies that are subject to pollutant inputs, one must be able to predict the degradation in quality that results from those inputs. Many models have been developed to assist managers in making these predictions. These models differ in their detail, and hence in their data requirements. This chapter briefly reviews the basic principles on which many such predictive models are based. This brief review limits itself to the modeling of the impacts resulting from point-source and non-point-source pollutants on inland rivers and lakes, and focuses on simulation rather than analytical methods. Readers interested in more detail—and there is a considerable amount of it—may refer to any of the references listed at the end of this chapter. Literature on water-quality modeling and prediction is expanding relatively rapidly, as our knowledge of water-quality processes, and how to model or predict them, improves.

Natural water bodies, if of adequate quality, can provide many benefits. Natural water bodies can be the sources of water supplies for agricultural, individual, and industrial uses; provide recreational opportunities; and provide a habitat for fish and other aquatic plant and animal species of value to humans. The benefits derived from these uses, especially, are heavily dependent on access to a sufficient quantity and quality of water. Natural water bodies also provide a means of waste disposal and transport. The management challenge is to control this particular beneficial use so that the resulting water quality does not cause unacceptable losses to the other beneficial uses of the resource.

What is considered an acceptable compromise between the cost of pollutant reduction and pollutant discharge control and the quality of a natural water body subject to point- and non-point-source pollutant discharges differs from location to location. It is, indeed, a political decision. To make informed decisions, however, the tradeoffs between costs and quality must be identified and known to those involved in any negotiation among competing water users. Models can be used to identify these tradeoffs.

Before investments are made to improve or protect water quality, it is prudent to predict, to the best of one's ability, the impact—i.e., the water-quality improvement—associated with each alternative water-quality management decision. Admittedly, these predictions will not be perfect. Since future pollutant loads and water quantities are unknown, since any prediction model will always be a simplification of reality, and since the parameter values of any water-quality model will likely be uncertain, there will always be uncertainty associated with any model prediction. Water-quality prediction is not an exact science. Thus, to the extent possible, prediction uncertainty should also be estimated and presented, along with the quality prediction, to those responsible for making management decisions.

Important water-quality impacts of rivers and lakes include the effects of oxygen depletion, eutrophication, pathogenic organisms, acidification, salinization, toxic contamination, and turbidity and sedimentation. Oxygen depletion results from the oxidation of organic matter by aerobic aquatic organisms. Organic matter, from domestic and many industrial wastes, is food for these organisms. As they metabolize this "food," they require oxygen. This oxygen is obtained from the water. The amount of oxygen required to fully metabolize this organic matter is termed the *total biochemical oxygen demand* of this organic matter. If the water's dissolved oxygen concentration drops below that required by certain organisms living in the water, these organisms will die. This is sometimes evidenced by fish kills, and in the extreme, by the production of obnoxious gasses such as methane and hydrogen sulfide.

Eutrophication, or the production of blooms of algae, typically results from the presence of sufficient amounts of nitrogen, phosphorus, and other nutrients. Algae both consume and produce oxygen, depending on the absence or presence of light. Algae that die add to the biochemical oxygen demand that may already exist in the water.

Waterborne diseases remain the major cause of human sickness and death. It is because of the threat of contamination by pathogenic bacteria, protozoa, and viruses that water treatment is often implemented, even though in some cases it is not otherwise deemed necessary.

Airborne deposition of SO_2 and NO_x particulates, in either solid or liquid form, is a major cause of land and water acidification. This, in turn, can cause the mobilization of elements from soils that, where buffering is inadequate, can become toxic to fish and other aquatic life. Acidification of land can also be damaging to plant life, especially certain tree species. Entire forests have been destroyed (especially in eastern Europe) by excessive acidification.

In areas where irrigation is practiced, the return flows from soils containing salt can be high in various calcium, sodium, chloride, and sulfate ions that, in turn, can contribute to the salinity of receiving water bodies. Considerable individual as well as governmental costs are incurred in various places throughout the world to reduce salinity concentrations and the damages caused by high salinity levels in natural water bodies and in water-supply distribution systems.

Heavy metals, pesticides, herbicides, and many other complex agricultural and industrial chemicals can be highly toxic to living organisms. These toxic substances often attach themselves to sediments, further complicating their removal once they are discharged into natural water bodies. The prediction and management of sediment, both suspended and bed load, is difficult at best. The transport of bed-load sediments depend on high flow velocities, and the duration and extent of these (often storm-driven) events are not easily predictable. High suspended-matter concentrations can result in deposition in river reaches and lakes, reduction of light for photosynthesis, degradation of aquatic ecosystems, and additional costs for water consumers requiring clear water.

Models to predict these and other water-quality impacts began to be developed in the mid-1920s. Of primary interest then, and of continued interest today, was and is the fate of biochemical oxygen demand (BOD) and its impact on the dissolved-oxygen concentration, (DOC) in natural water bodies. The first of these models, developed by Streeter and Phelps (1925), has been extended by numerous modelers and is, indeed, the basis of many of todays more comprehensive multiconstituent models. [For a discussion of model development since 1925, readers may refer to any number of excellent books and papers on this subject, e.g., Jorgensen (1982, 1986), Orlob (1983), Reckhow and Chapra (1983a, b), Beck (1987), Thomann and Mueller (1987), and USEPA (1987), and Somlyody and Varis (1994).]

In addition to physically based models, statistical "black box" models have also been developed and used successfully to predict water quality. These models range from relatively simple regressions and their extensions (Box and Jenkins, 1970; Young, 1974, 1979) to linguistic and pictorial models (Cámara et al., 1987, 1990), neural networks (Beale and Jackson, 1990; Hall and Minns, 1993; Rumelhart et al., 1994), and various forms of decision analyses (Chapra and Reckhow, 1983; von Winterfeld and Edwards, 1986; Varis et al., 1990).

This discussion focuses on the more physically based models (and the relatively simple less data-demanding ones) without implying that they are any better or worse than other modeling approaches. It is often the skill of the individual who develops or selects and applies (calibrates and verifies) a model of a particular water body that determines the quality of the predictions needed for the particular management issue more than the type and complexity of the model itself.

This discussion is a review of some common mass-balance simulation methods for predicting the flows, volumes, and constituent concentrations in natural surface-water bodies. This involves a description of alternative transport (advection and dispersion) and reaction schemes. The methods discussed are relatively simple, and hence more appropriate for models used for management and policy analyses than for an improved scientific understanding of the hydrodynamic-driven water-quality processes that take place among various constituents and aquatic ecosystems.

Model Selection and Complexity

Hydraulic and accompanying water-quality models range from those that are relatively simple (such as zero or one-dimensional first-order reaction models) to those that are much more complex (three-dimensional, higher-order reaction models). Those who are applying these various model types must decide what degree of complexity is required to address the issues and questions being asked and to represent the water body of concern. Clearly, increased data, and hence increased costs, are associated with the more complex models available today for anyone's use.

Most model users will not be model developers. They will be obtaining the models and their accompanying computer simulation programs from any of a large number of organizations that develop and maintain such models and their programs and documentation. We suspect that readers of this chapter, whether future developers or users, will not be looking for the state of the art in model complexity, but will prefer an introduction to the subject, especially with respect to modeling approaches commonly used in simulation programs. We believe having a knowledge of some of the basic predictive (governing) equations and a familiarity with methods and assumptions used to convert water-quality constituent inputs to resulting water quality concentration outputs will facilitate not only the selection of the appropriate model type but also some understanding of its limitations.

We leave it to those who have more space than what we are allocated to discuss the numerical methods used to solve many of the water-quality predictive equations presented here (see, e.g., Abbott, 1979, Thomann and Mueller, 1987). We will also leave it to others to discuss the stochastic versions of many of the deterministic models presented here (e.g., Fedra, 1983; DiToro, 1984; Beck and van Straten, 1983; Beck, 1987; Somlyody and Wets, 1988). But we mention them here because they do exist and are important contributions to the theory and practice of water-quality modeling and prediction.

Classification of Water-Quality Models

Water-quality models can be classified according to three general criteria:

- Representation of mathematical model
- Representation of time
- Representation of biological, chemical, and physical processes

Representation of mathematical model

With regard to the mathematical approach, water-quality models can be divided into

- Empirical or statistical models
- Deterministic models
- Stochastic models

Empirical or statistical (black-box) models are based on experiences. Interrelations between various parameters are deduced statistically from observed data. A theoretical description of interrelations based on physical, biological, or chemical processes is not made. Rather, coefficients of various regression equations or neural networks are determined, and then these equations or networks are used to predict events within the range of events observed. How good a model is for such predictions can be estimated using any of various measures involving differences between observed and calculated values, as are discussed later.

In contrast to empirical models, deterministic physically based models are mathematical descriptions of natural processes. *Input data* are fixed quantities without accompanying explicit description of their statistical variation. The results of deterministic calculations are fixed values without accompanying information on their probability distributions.

Stochastic mathematical water-quality models are also based on the description of natural processes. However, variation of input and output data is taken into account by statistical analyses and syntheses of time series. In contrast to deterministic models, the results of stochastic models provide information on the probabilities of various predictions. The random variation of output data of stochastic models is due to the variation of the various hydraulic and biochemical processes and, in addition, to the random variation of many of the input variables. Both occur simultaneously, and hence are difficult to determine individually. Stochastic models usually require considerably more sample data to calibrate and verify than do deterministic models. It is for this reason, usually, that they are rarely used.

Representation of time

With regard to the representation of time, it is necessary to consider the model components hydraulics and quality separately.

Hydraulic models can be divided into stationary and nonstationary models, whereas quality models are divided into static and dynamic models. Stationary models assume the flow to be constant during the simulation time period. In a nonstationary model the discharge and thus the flow varies over time, but is typically assumed constant in each time step. In a static quality model both input and biological, chemical, and physical processes are regarded as constant, whereas in a dynamic model they vary with time.

For effective water-quality management the ability to simulate critical situations that may occur in a water body can be of special interest. Such critical situations could be caused by relatively long, hot, dry periods in the summer, or by relatively short-time heavy-rainfall events which result in high nonpoint loadings. For relatively long-term dry periods, a quasi-dynamic simulation may be sufficient, since the hydraulic characteristics needed for the simulation of the water quality are usually relatively constant. Short-term impulse loadings from intensive rainfall-runoff events, however, can best be represented using a fully dynamic model (Shanahan and Somlyody, 1995).

Many analytical models are static with respect to their inputs and processes. Many simulation models over multiple discrete time periods are quasi-dynamic. A few models have been developed that are fully dynamic. Clearly, their data requirements for model calibration and verification exceed those of the other model types. The more realistic (and hence complex) the model, the greater will be its data needs and hence costs of application.

Representation of biological, chemical, and physical processes

Water-quality models can be grouped according to their representation of biological, chemical, and physical processes:

- Kinetic models
- Biocoenotic models
- Ecological models

Constituent concentration transformation processes in kinetic models are represented by reaction-rate constants. These processes consist of constituent transformation (from one type of constituent to another), production (growth), and decay processes. Boundary conditions influencing these processes and interactions between pollutants should be taken into account.

Nutrient cycles (or portions of them) are simulated in kinetic models using a combination of reaction rates (T^{-1}) and transformation coefficients (M/M). For example, the oxidation of ammonium to nitrite and then to nitrate involves the metobolic processes of bacteria and is often simulated using reaction-rate constants whose values may depend on several other conditions, such as water temperature, pH (hydrogen-ion concentration), and dissolved oxygen. The reaction-rate constants for the conversion of ammonium to nitrate, and the transformation rate for the uptake of oxygen in the process, represent biological activities of the bacteria. Kinetic models are the most commonly used of all water-quality models and are described in more detail later.

Biocoenotic models focus on the organisms that consume or produce the various constituents rather than on the concentrations of constituents themselves. The organisms that make up the aquatic food chain can be subdivided into production (primary production), consumption (secondary production), and destruction (decomposition) organisms. The populations of these various organisms and their metabolic processes are modeled and, in turn, determine the concentrations of various constituents which are of interest to water-quality managers.

This biocoenotic approach is typically carried out for only selected groups of organisms, i.e., for only a portion of the biocoenosis. Not only are such models computationally intensive; rarely are there adequate field data required for model calibration. The modeling of short-term impulse loadings, such as from storm events, is also especially difficult using this modeling approach. For an example of a biocoenotic approach to water-quality prediction in the Rhine River, refer to Rinaldi et al. (1979).

Ecological models are typically extensions of the kinetic models. In addition to reaction rates and transformation coefficients for simulating various constituent concentrations, including nutrients, portions of the aquatic food chain such as phytoplankton and zooplankton that consume these nutrients are also included. Thus ecological models are particularly applicable for the simulation of eutrophication processes that can occur in nutrient-rich water bodies. Attempts have been made to extend these ecological models up the food chain to include fish, but with only limited success.

Hydraulic Considerations for Water-Quality Modeling

Water quality is determined largely by the flows and volumes characterizing natural water bodies. These flows and volumes need to be modeled because it is within those flows and volumes that all water-quality processes take place. Hydraulic characteristics of water bodies can considerably influence water quality, but rarely vice versa. The hydraulic features or parameters that influence water quality include the

- Flow or discharge
- Flow velocity
- Depth of flow
- Water-surface area

Knowledge of the flow in a water body is necessary for computing mass balances and for determining the degree of mixing.

Knowledge of the flow velocity permits the calculation of the advective part of pollutant-transport processes and determines the period of time a constituent particle is in the water and thus, the period of time in which biological, chemical, and physical processes can act on that particle. Moreover, flow velocity influences the degree of sedimentation and erosion. A high flow velocity, for instance, can exceed the critical bottom shear stress and result in a resuspension of sediment and a remobilization of fluid particles bound in the sediment. The depth of flow can affect the amount of light available in the water and thus on the biological activity of various organisms that require light.

All physical exchange processes between the air and water occur at the water surface. Hence the extent of the exchange of gas and temperature depend in part on the area of the water surface and also, but rarely considered explicitly, on its elevation as a result of changing atmospheric pressures and solar radiation with height.

Additional impacts of the hydraulic characteristics on the calculation of the water quality, such as scouring of benthos deposits on the bottom surfaces of water bodies or the extinction of various species of algae, are possible. The extent to which these impacts need be integrated into the simulation depends on the degree of detail desired or required for the particular water body and for the particular water-quality management problem being addressed.

Physical processes of pollutant transport

The transport of waterborne constituents in open-channel flow (Fig. 7.1) is influenced by three physical processes:

- Transport by current velocity (advection)
- Turbulent and molecular diffusion
- Dispersion.

The term *advection* describes the downstream pollutant transport at the mean flow velocity. Actual flow velocity varies temporally and

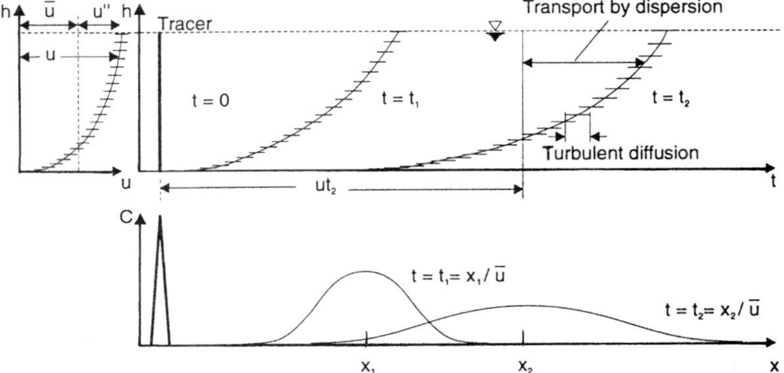

Figure 7.1 Pollutant transport in open-channel flow (Markowsky, 1980).

spatially and this results in mixing. Mixing processes are caused by turbulent and molecular diffusion and by dispersion. *Turbulent diffusion* describes pollutant mixing due to *temporal* deviations of the flow velocity from the local mean value, whereas *dispersion* describes mixing due to *spatial* velocity fluctuation. Compared to turbulent diffusion and dispersion, molecular diffusion, which describes the transport due to molecular forces, is usually negligible.

Normally, dispersion has the greatest impact on constituent mixing. The lower the depth of flow is in comparison to the width of the channel, the greater is the impact of dispersion. Therefore, the term *dispersion* is often used to describe all three types of mixing processes. For a more detailed description of these mixing processes see, for example, Orlob (1983) and Markowsky (1980).

Advection-dispersion equation

Since transport processes in open surface-water bodies are dominated by advection and dispersion, the advection-dispersion equation is used to model pollutant transport. In a one-dimensional (longitudinal) model the equation is

$$\frac{\partial C}{\partial t} = E_x\left\{\frac{\partial^2 C}{\partial x^2}\right\} - u\left\{\frac{\partial C}{\partial x}\right\} \pm S_C \tag{7.1}$$

This equation is the fundamental equation for the calculation of pollutant transport in one dimension. The dispersion coefficient E_x (L^2/T) describes the mixing processes of diffusion and dispersion. The parameter u is the mean flow velocity (L/T). The combined source and sink terms, all denoted by S_C (M/L^3T), include the loading as well as the characteristic biological, chemical, and physical reactions that affect the concentration of the constituent C, as applicable. Sink terms do not apply to conservative constituents that are not transformed. Various source and sink terms are discussed in more detail in the sections that follow.

The extent of advection and dispersion can be influenced by numerous external factors, including Coriolis forces, surface roughness, and wind-induced water-surface stresses. Detailed water-quality models take some of these processes into account, especially when they substantially affect the prediction of water quality in particular water bodies. Derivations of the advection-dispersion equation are given in Markowsky (1980) and Fisher et al. (1979).

Solution techniques used in water-quality simulation programs for the calculation of pollutant transport in open-flow water bodies include the approach of Taylor (1921) together with flow-routing procedures, as applicable. The approach of Taylor is the classic approach for the one-dimensional advection-dispersion equation. Assuming an impulse-like injection of a mass M_0 of conservative pollutant at time t and distance X equal to 0 within the whole cross-sectional area A and a constant channel geometry, the resulting concentration distribution at a point X over time t, or vice versa, is

$$C(X,t) = M_0 \frac{e^{-[(X-ut)^2/4E_x t]}}{A(4\pi E_x t)^{1/2}} \tag{7.2}$$

This equation gives a symmetrical Gaussian distribution, as shown in Fig. 7.2, which does not always correspond to the distributions observed in natural water bodies. In natural rivers dead zones result in a further reduction and retention of the pollutant cloud, which is not represented in Taylor's approach.

Becker and Sosnowski (1969) give an analytical solution for the one-dimensional advection-dispersion equation for any mass input function $M_0(t)$. The concentration distribution in a uniform river reach or channel results from the convolution

$$C(X,t) = \int_0^t \left(\frac{M_0(\tau)}{uA}\right) \frac{X \cdot \exp\left\{\dfrac{[u(t-\tau)-X]^2}{4\pi E_x(t-\tau)}\right\}}{(t-\tau)[4\pi E_x(t-\tau)]^{1/2}}\, d\tau \tag{7.3}$$

Unlike the approach of Taylor, the routing procedure can take any inflow distribution into account. However, it does not capture or describe well the influence of dead zones.

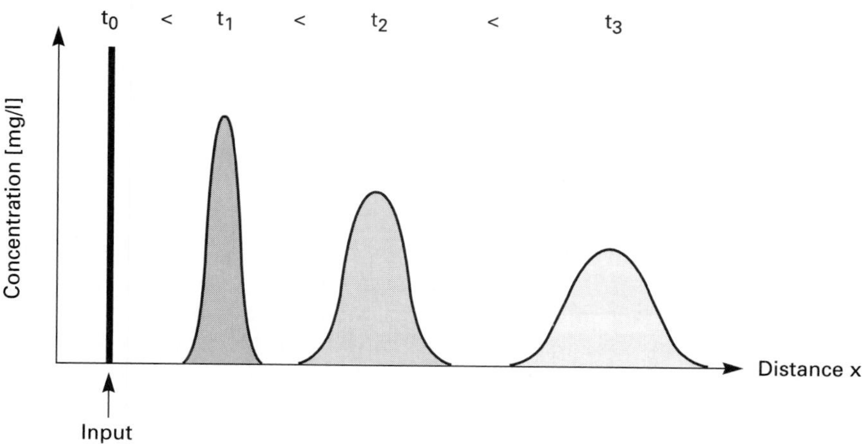

Figure 7.2 Gaussian distribution resulting from the approach due to Taylor.

Mass-balance modeling

Alternatively to the approaches based on Taylor, some simple mass-balance methods can be used for modeling flows, volumes, and the water-quality constituents within those waters. These modeling approaches represent simplifications of the hydraulic processes taking place in many water bodies. Hence, if the particular problems being addressed require more precision than is possible from the use of these simpler mass-balance models, then the additional costs must be incurred to develop the model and collect the necessary additional data to permit more realistic water-quality predictions. The advantages of these simpler methods, of course, is that if they are adequate for the particular issues, questions, and water body of concern; are less costly; and require less data for calibration and verification.

Among the simplest hydraulic descriptions in water-quality models are the plug-flow and continuously stirred tank reactor (CSTR) methods. Plugged-flow reactor hydraulics is often applied in river models, as it was in the earliest ones developed by Streeter and Phelps (1925) for describing the biochemical oxygen demand (BOD) and dissolved oxygen (DO) in the Ohio River in the United States. Assuming no inputs or outputs along a river reach, the change of constituent concentration (e.g., of BOD) in a "plug" of water as it travels along a river reach is assumed to be some function of the time that the plug travels along that reach. Plug flow may also occur in some lakes, especially if the lake is long and narrow or the flow is under ice.

The fully mixed CSTR method is commonly used for lakes, but can also apply to sections of river reaches as well, especially when there is dispersion present. The number of these CSTRs along any given river reach, or defining a lake, can determine the degree of dispersion that is being modeled in that river reach or lake, as is discussed in more detail later.

It is possible to combine, in one, two, or three dimensions (i.e., in the longitudinal, lateral, and/or vertical directions), both the plug flow and the CSTR methods (see, e.g., Chen and Orlob, 1972; Frisk, 1989). Before doing this, we consider in more detail the two types of modeling approaches.

CSTR modeling. Among the best known of all CSTR models are those of Vollenweider (1975) who used them as simple guides for estimating lake eutrophication potential. In such models an input of some constituent N, expressed in units of mass per unit time (M/T), is assumed to mix immediately throughout the lake. Hence the outflow concentration (M/L^3) of that constituent equals the average lake concentration. If the lake volume, V (L^3), is constant, the rate of change in concentration C over time t is

$$\frac{\partial C}{\partial t} = \frac{N}{V} - \frac{QC}{V} - kC^n \tag{7.4}$$

where Q is the outflow (L^3/T) and k is a nth-order reaction rate constant (T^{-1}). Assuming a first-order reaction rate ($n = 1$) and integrating yields

$$C_{t+1} = \frac{N_t}{Q_t + kV} - \left(\frac{N_t}{Q_t + kV} - C_t\right) e^{-\Delta t[(Q_t/V) + k} \tag{7.5}$$

where Q_t, N_t = average flow and constituent mass inputs in period t
C_t = concentration at the beginning of period t
Δt = duration of period t relative to the units used for Q and N and k

The average detention time T of the constant volume lake in each period t is the volume V divided by the inflow or outflow Q_t:

$$T_t = \frac{V}{Q_t} \tag{7.6}$$

The flushing coefficient is the inverse of the detention time T:

$$\rho_t = \frac{1}{T_t} \tag{7.7}$$

The average detention time T_{N_t} of the constituent N is based on the input N_t, rather than the output. For each period t

$$T_{N_t} = \frac{(C_t + C_{t+1})V}{2N_t} \tag{7.8}$$

If the concentration reaches a steady state, i.e., $C_t = C_{t+1}$, then

$$T_{N_t} = \frac{kN_tV}{Q_t + kV} = \frac{1}{\rho_t + k} \tag{7.9}$$

If the lake volume as well as the flow and mass inputs are changing, then the inflow Q_t^I need not equal the outflow Q_t^O in each period t. Denoting V_t as the initial volume in period t, the final volume equals

$$V_{t+1} = (Q_t^I - Q_t^O)\Delta t + V_t \tag{7.10}$$

The concentration of the constituent at the end of the period t is approximately

$$C_{t+1} = \frac{C_t\left[V_t - \dfrac{\Delta t(Q_t + kV_t)}{2}\right] + N_t\,\Delta t}{V_{t+1} + \dfrac{\Delta t(Q_t + kV_{t+1})}{2}} \tag{7.11}$$

where the expression $\Delta t(Q_t/V_t + k)$ must be no greater than 2. The smaller the time step Δt, the closer this finite-difference approximation will be to the integrated form of the mass-balance differential equation, assuming, of course, that the water body can be modeled as a fully mixed continuously stirred tank reactor.

Plug-flow reactor modeling. One-dimensional (usually longitudinal) plug-flow reactor modeling assumes no mixing in that dimension and complete mixing in the other dimensions. The change in concentration is dependent only on the time elapsed. The mass-balance differential equation for an nth-order reaction with no additional inputs of a single constituent concentration of C is

$$\frac{\partial C}{\partial t} = -kC^n \tag{7.12}$$

Assuming a constant first-order reaction ($n = 1$) and a single initial constituent concentration C_0 at some time $t = 0$, integrating the differential equation [Eq. (7.12)] permits the definition of the concentration some time Δt:

$$C_t = C_0 e^{-k\,\Delta t} \tag{7.13}$$

The travel or detention time $T_t = V/Q$ for water in the river or lake reach remains the same as in the case of the CSTR; the constituent detention time in a reach requiring a travel time of Δt can be derived by dividing the mass of pollutant in a river or lake section (requiring some integration) by the constituent-mass input $N_t = Q_t C_0$. The result is

$$T_{N_t} = \frac{\dfrac{Q_t C_0(1 - e^{-k\Delta t})}{k}}{Q_t C_0} = \frac{1 - e^{-k\Delta t}}{k} \tag{7.14}$$

for a nonzero, and non-time-varying in this case, reaction-rate constant k.

If nonpoint additions of constituent mass occur along the reach at the constant rate of N_{np} per unit length (M/LT) associated with a non-point-flow input of Q_{np} per unit length (L^3/LT) and the average

model a stream or river reach is viewed as having sections in which the water travels at the mean flow velocity from the upstream to downstream, and where advection dominates, and other sections in which the water flow is reduced, or even in the reverse direction, and where there may exist relatively stagnant pools having very little inflow and outflow, at least for short periods of time. In these sections of a stream, dispersion may dominate. Hence, each stream or river reach is a combination of plug-flow segments and completely mixed segments (Wallis et al., 1989a, 1989b). The completely mixed segments represent dispersive mixing in the "dead zones" and the plug-flow transport segments represent the advective transport in the reach. Analogous to the source and sink terms S is the first-order reaction-rate constant k that describes the processes that alter the concentration of the constituent C^O at the end of the dead-zone segment.

The ADZ model assumes that all the advective portions of the reach can be combined into one, and all the dead zones in the reach can be combined into a single aggregate one. Hence, each reach is a combination of an advective portion and a dead-zone portion—i.e., a plug-flow reactor and a continuously stirred tank, fully mixed, reactor. Constituent reactions can occur in each portion.

For each pair of advective and mixed segments that represent a reach or subreach, given an inflow concentration of C^I, the rate of change in the outflow concentration, C^O, is first determined by advection, and then by dispersion in the dead zone. For just the dead zone portion, having a volume of V_{ADZ}, an inflow of Q^I and an outflow of Q^O

$$\frac{\partial V_{\text{ADZ}} C^O}{\partial t} = Q^I C^I - Q^O C^O - k V_{\text{ADZ}} C^O \tag{7.24}$$

where C^I is the influent concentration. This C^I, in turn, results from the advection of an initial concentration C_0 in the reach. The concentration of a nonconservative constituent at the end of the advection portion of the reach equals the concentration C^I entering the aggregated dead zone. If the travel time in the advective portion of the reach is T, then

$$C^I = C_0 e^{-kT} \tag{7.25}$$

Combining Eqs. (7.24) and (7.25), the rate of change in the mass of a constituent at the end of a reach assumed to contain one plug-flow portion followed by an aggregated dead zone (completely mixed tank reactor) is

$$\frac{\partial V_{\text{ADZ}} C^O}{\partial t} = Q^I C_0 e^{-kT} - Q^O C^O - k V_{\text{ADZ}} C^O \tag{7.26}$$

The transport between successive dead zones represents the temporal delay of mixing. The degree of dispersion in the dead zone is determined by the dead-zone volume, V_{ADZ}.

It should be noted that the representation of a river reach by means of a transport reach and a reactor is not a geometrically correct representation, i.e., the reactor volume V_{ADZ} does not correspond to the volume of the river reach in question and the transport time T is not a function of its length and mean flow velocity. Thus, these parameters have to be determined by tracer experiments or—if no data for calibration are available—estimated. Since the representation is not geometrically correct, changes of flow, such as those caused by a variation of discharge, cannot be immediately taken into account. For each discrete range of flows, the model has to be calibrated anew. On the other hand, in contrast to the longitudinal dispersion coefficient which varies by a factor of 1000 for different types of open channels, the ADZ model detention and time-of-flow times are very stable. In addition, ADZ models may result in more accurate predictions of water quality for some water bodies (Young and da Costa, 1986; Wallis et al., 1989b; CRES, 1990; da Costa, 1992).

Figure 7.3 illustrates the ADZ modeling approach for a series of successive stream or river reaches.

Figure 7.4 compares data of Leibundgut's survey (Leibundgut, 1990) with the results of the ADZ model Ghaeizadeh (1992). As can be seen, the measured concentration distribution very closely corresponds with the results of the ADZ model. Wallis et al. (1989b) have shown that simulation results differ very little from the actual concentration distributions even with inaccurately estimated parameters.

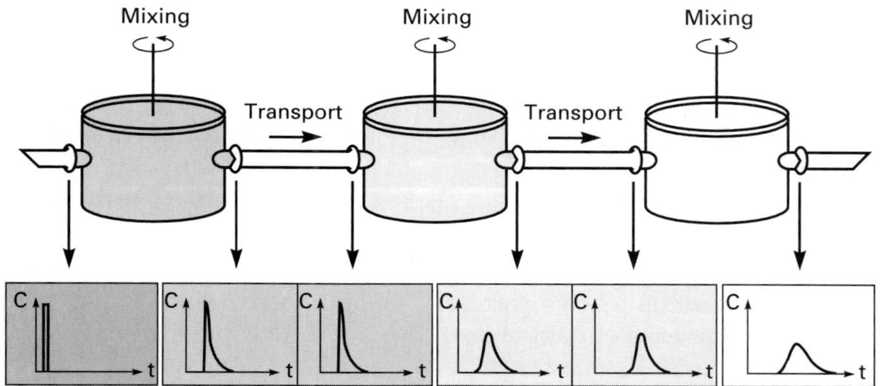

Figure 7.3 Diagram of the ADZ modeling approach.

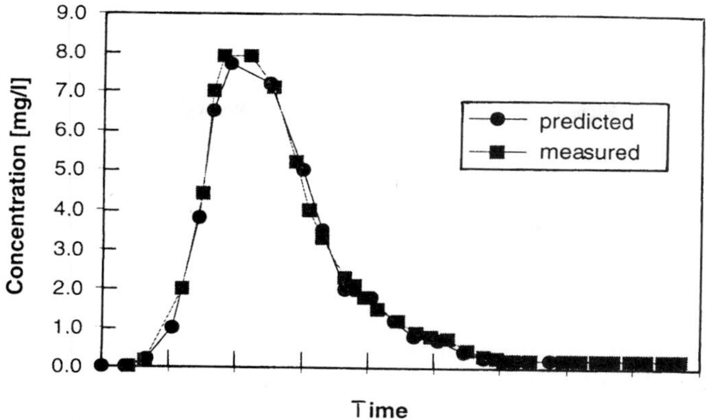

Figure 7.4 Concentration distribution of the Rhine River at Marckolsheim, km 241.

One-Dimensional Water-Quality Simulation throughout a Water-Resource System

Although not considered here, some one-dimensional lake water-quality models assume complete mixing in the longitudinal and lateral directions, and focus on the changing concentrations with depth. Here we are concerned with the longitudinal—as well as time-varying—constituent concentrations throughout a water-resource system.

Water-resource systems that are being modeled for simulation studies are typically represented as a node-link network. The nodes are the endpoints of each link. Links can represent river or stream reaches, diversion canals, or channels transporting wastewater to the system. The nodes can represent junctions of links, diversion sites, monitoring sites, lakes, reservoirs, wetlands, aquifers, gauge sites, demand sites, wastewater-treatment sites, and any other feature of the system being modeled. One-dimensional (longitudinal) water-quality simulation models typically focus on the prediction of constituent concentrations at the beginning and end of each link, and at each node, at the beginning and end of each simulation time step. They assume complete mixing in the lateral and vertical directions. To do this, the constituent concentrations have to be calculated throughout each link and node.

Plug-flow reactor simulation

Using the plug-flow method of water-quality simulation, a water body is subdivided into multiple segments. The simulation model simulates each successive segment in each simulation time step. It does this by taking into account the concentrations at the end of the last time step in the segment and in the flows entering the segment from adjacent connected segments. The simulation must also consider any additional inputs or outputs. The time of flow through each segment must not exceed the duration of the simulation time step; otherwise at each step of the simulation, i.e., at each particular segment and time step, multiple segments must be simulated. The complexity of keeping track of what is going on in this situation usually discourages one from trying to permit this in any simulation program.

Plug-flow reactor procedures are easier to simulate when the flows are constant in successive simulation time steps. If the flows within, and hence volumes of, water bodies are constant, the simulation model need focus on only the variation of constituent mass inputs. Given the duration of the simulation time steps, the number of volume segments in each water body is determined by the time of flow, or distance flow travels, during each simulation time step, compared to the total distance or time of flow. For storage nodes, the total time of flow is the detention time (volume per outflow). (If the computed number of segments is not an integer, then the duration of the simulation time steps can be reduced, automatically by the program, to reduce any errors caused by large fractional final segment volumes making up a water body.)

The concentration of a constituent at the end of each segment is computed using the following governing or predictive equation containing only advective and source and sink terms:

$$\frac{\partial C}{\partial t} = u\frac{\partial C}{\partial x} \pm S_c \tag{7.27}$$

The distance X is the total time of travel times the flow velocity within each segment. Assuming a conservative pollutant, Fig. 7.5 illustrates the process of segmentation using a plug-flow reactor.

While there is no dispersion in the plug-flow reactor model, changes in concentration in the longitudinal direction may result from additions, withdrawals, and various biological, chemical, and physical processes.

Completely mixed reactor simulation

In contrast to the plug-flow reactor model, where no dispersion is assumed, the CSTR, or completely mixed method, assumes that each segment of each water-body component is fully mixed, and hence its constituents are completely dispersed. Hence, there are no concentration gradients in each segment, but there can be concentration differences between segments. During the time constituents spend in any

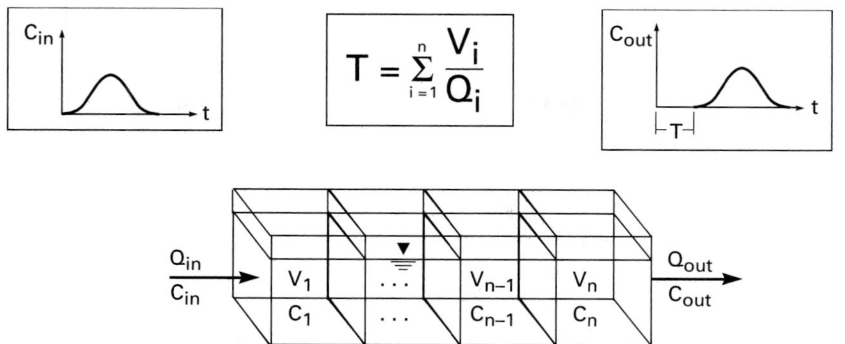

Figure 7.5 Diagram of the plug-flow-reactor modeling approach. $i, \ldots, n$ = index for time interval Δt; Q_i = flow; C_i = concentration in volume i; V_i = volume i; T = total detention time.

segment, their concentrations may change due to any of numerous processes previously mentioned. This modeling approach involves a sequence of segments as shown in Fig. 7.6.

As with the plug-flow segment, the detention time of the completely mixed segment depends on the segment volume and the discharge. Determining the number of segments in each water body, however, is a way of calibrating the extent of dispersion present in the water body. The more segments, the less dispersion present. The spacing of segments along a reach, or the relative volume in each segment, influences the skewness of the distribution of constituent concentrations at any time within the reach. For water bodies being represented by a sequence of completely mixed segments, the simulation time step must not exceed the time of flow through any single segment.

The model of the complete-mix reactor can be used to represent river or stream reaches, wetlands, reservoirs, and natural lakes. To include a possible remobilization of sediment, Troubounis (1985) extended this modeling approach by subdividing these completely mixed segments into vertical layers. With the help of such an extension, seasonal layers in dead-water zones can be taken into account.

Combined plug-flow–continuous-mix reactor simulation

Nothing requires, of course, that a simulation model be a completely plug-flow or a completely mixed reactor type. The two methods can be combined in various ways. To simulate open-channel flow between

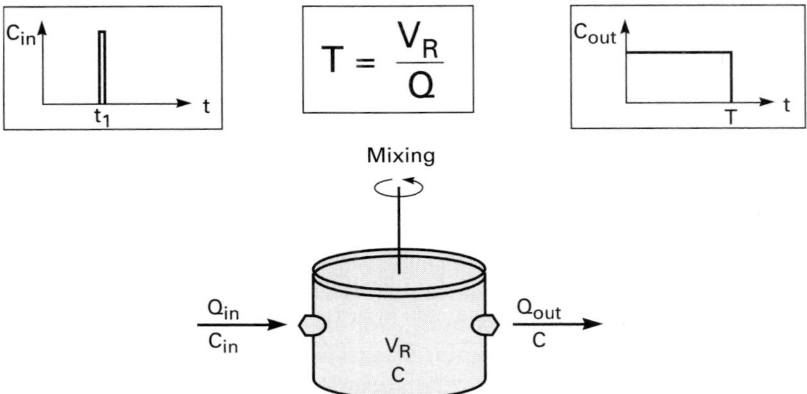

Figure 7.6 Diagram of the complete-mix reactor modeling approach. Q = flow; C = concentration in the reactor; V_R = reactor volume; T = total detention time.

water-storage bodies, for example, a sequence or cascade of complete-mix reactors can be used to represent reaches, or portions of reaches, where some dispersion is present, and a sequence of plug-flow reactors can be used where no dispersion is present. If both types of processes are prevalent throughout a reach or other water body, then a mix of the two approaches may be able to produce accurate predictions for the concentrations at the discharge end of the reach or other water body, if not throughout the water body.

Modeling Water-Quality Reactions: Some Extensions

Temperature

While temperature is not influenced by any of the biological or chemical transformation processes of other constituents, the rates at which these processes take place are considerably affected by temperature. In addition, with increasing temperature, the oxygen-saturation concentration of a water body decreases.

Temperature is subject to seasonal, daily, and periodical variation. The impact of these variations increases with increasing time of travel. The higher the time of travel, the longer the water body is subject to the heat-exchange processes with the surrounding atmosphere, with various inflows, and with the streambed or riverbed. Other heat sources such as friction, mechanical work, or heat of reaction are negligible in comparison.

TABLE 7.1 Heat Fluxes of the Temperature Budget

Heat flux	Range (kJ m^{-2} h^{-1})
Short-wave radiation of the atmosphere	+170 to +1220
Long-wave radiation of the atmosphere	+1000 to +1500
Long-wave radiation of the water body	−1000 to −1700
Evaporation	−960 to −3500
Convection	−160 to +170
Heat exchange at the water-soil interface	−300 to +300

The exchange processes with the environment consist of solar shortwave radiation (global radiation), atmospheric longwave radiation, reflected longwave radiation from the water body, evaporation, convection, and heat exchange with the water-soil interface. Each of these processes can be modeled (see, e.g., Orlob, 1983; Thomann and Mueller, 1987). In this case, a considerable amount of data will be required. It is thus often convenient to consider the total heat flux H as an exogenous input dependent on the existing water temperature.

Table 7.1 gives the magnitudes of the different heat fluxes. A positive sign indicates that heat is conducted into the water body; a negative sign, that heat is conducted out of the water body.

The rate of change in water temperature T_W (in kelvins) is dependent on the total heat flux H [in kilojoules per square meter per hour (kJ m^{-2} h^{-1})], the water density ρ (kilograms per square meter), the specific-heat capacity of the water body φ (in kilojoules per kilogram per kelvin), and the depth of flow h (in meters):

$$\frac{\partial T_W}{\partial t} = \frac{H}{h\varphi\rho} \tag{7.28}$$

Oxygen balance

Reaeration-rate constant. One of the more sensitive parameters in any water-quality model that includes the dissolved-oxygen constituent is the reaeration rate constant. The reaeration-rate coefficient accounts for the effects of flow velocity, turbulence, and channel geometry, all of which impact on the reaeration of water bodies from the atmosphere through the water surface. There are numerous approaches for determining the reaeration coefficient (Orlob, 1983). Four commonly used approaches based on flow velocity u and depth h are those proposed by O'Connor and Dobbins (1958)

$$k_r = 3.951\left(\frac{u^{0.5}}{h^{1.5}}\right) \tag{7.29}$$

by Churchill et al. (1962)

$$k_r = 5.013\left(\frac{u^{0.969}}{h^{1.673}}\right) \tag{7.30}$$

by Owens et al. (1964)

$$k_r = 5.364\left(\frac{u^{0.67}}{h^{1.85}}\right) \tag{7.31}$$

and by Langbein and Durum (1967)

$$k_r = 5.133\left(\frac{u}{h^{1.33}}\right) \tag{7.32}$$

Figure 7.7 compares these approaches for different velocity/depth (u/h) ratios.

Apart from the approaches mentioned above, other reäeration coefficients may be assigned to individual portions of water bodies. Thus high oxygen transfers from air into water due to rapids or waterfalls

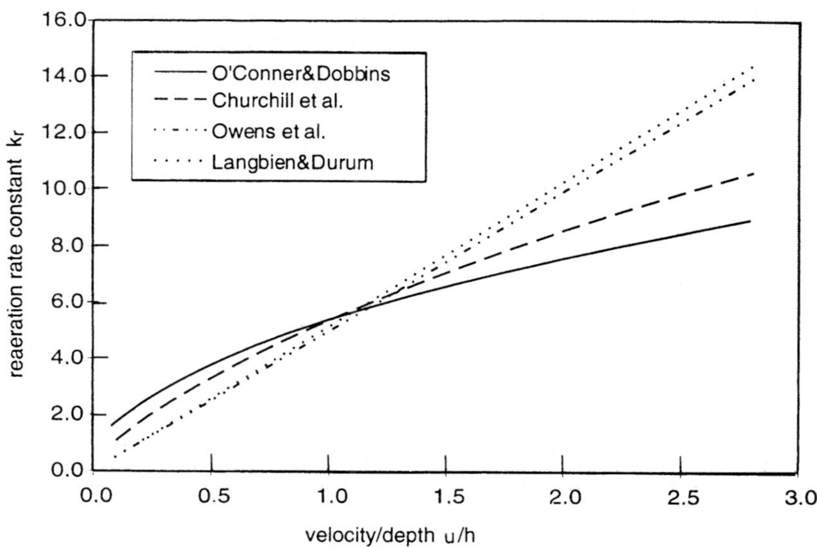

Figure 7.7 Comparison of approaches for the calculation of the reaeration coefficient k_r.

within a river or stream reach can be taken into account when appropriate. Aeration can also result from biological processes, such as those caused by autotrophic organisms (primary producers) who produce cell biomass by photosynthesis. Apart from biomass, oxygen is produced. Important primary producers are algae, which can be subdivided into phytoplankton and benthic algae. Phytoplankton floats in water, whereas benthic algae are fixed to the bottom.

Part of the oxygen produced by photosynthesis is consumed again by respiration. Respiration occurs both in the absence and in the presence of light and increases with increasing intensity of photosynthesis and the growth rate of algae. Approaches for the simulation of these processes and detailed explanations are given in EPA (1985), Brown and Barnwell (1987), or BLU (1984).

Multiple-constituent impacts. The rate of change of any constituent may also be determined by a combination of *n*th-order reactions. Consider, for example, the first-order reactions involved in the change in the dissolved-oxygen-deficit (DOD) concentration due to the oxidation of carbonaceous BOD (CBOD) and of ammonium-nitrogen (NH_4-N), due to reaeration from the atmosphere. Defining k_B and k_{NH_4} as the first-order reaction-rate constants for oxidation of CBOD and NH_4-N, and k_r as the reaeration-rate constant, the rate of change of dissolved-oxygen deficit equals

$$\frac{\partial \mathrm{DOD}}{\partial t} = k_B(\mathrm{CBOD}) + 4.57k_{\mathrm{NH_4}}(\mathrm{NH_4\text{-}N}) - k_r(\mathrm{DOD}) \qquad (7.33)$$

The transformation coefficient 4.57 in this equation equals the mass of oxygen required per unit mass of oxidized ammonium-nitrogen.

Reaction-rate functions

The reaction-rate constant k used in kinetic water-quality models may represent purely physical processes, such as sedimentation or volitization or reaeration, where flow velocities and depths, temperatures, atmospheric pressures, turbulence, etc. may influence its value. It may also represent various biochemical processes. The reaction-rate "constant" may actually be a function of the concentrations of one or more different constituents. One of the commonly used approaches for modeling ecological (living) constituents whose growth is dependent on the concentrations of various nutrients, as well as light and temperature, is to use the Michaelis-Menton formulation. In its simplest form, the reaction rate k governing the rate of growth of a constituent concentration is expressed as

$$k = K^{\max}\left[\text{operator} < \frac{C_i}{C_{2i} + C_i} > \right] \tag{7.34}$$

where $K^{\max}$ defines the maximum growth rate constant for constituent C (T^{-1}), C_{2i} is the concentration (M/L^3) that would result in half the maximum rate of growth, and C_i is the actual concentration of the ith constituent that influences the rate of change of concentration C. The term *operator* in Eq. (7.34) may represent the selection of the minimum among each of the terms denoted by the index i in the angle brackets if each independently influences the time rate of growth, or an expectation operation, or a product operation. QUAL2 uses the latter and assumes first-order reactions ($n = 1$). In this case, if the C_i concentrations are low, the growth rate is a first-order process. If the C_i concentrations are increased, the process converges to a zero-order process, reflecting the fact that there is a maximum limit to the rate of growth of the concentration C.

Nutrient cycles

Deterministic water-quality models that include portions of nutrient cycles and some ecological constituents such as phytoplankton and zooplankton involve more complex processes than discussed so far. In this section we review some of the approaches taken to model such processes.

Nitrogen. Nitrogen compounds of interest to water-quality managers include the inorganic forms of ammonia (NH_3), ammonium (NH_4^+), nitrite (NO_2^-), and nitrate (NO_3^-). Water-quality models usually model the nitrogen content of these compounds. These compounds are closely related. The nitrification, i.e., the oxidation of ammonium-nitrogen to nitrate-nitrogen, which involves oxygen uptake, is such a reaction. Moreover, ammonium is an important nutrient for algae and—in higher concentrations—may foster excessive algae growth or eutrophication.

The main ammonium source in open-channel flow, apart from the discharge of municipal wastewaters, is the runoff from agriculture. The main cause of oxygen uptake is due to the oxidation of ammonium to nitrate. This process occurs in two distinct steps in which two distinct genera of bacteria are involved. The first step, the oxidation of ammonium-nitrogen to nitrite-nitrogen, is carried out by bacteria of the genera *Nitrosomas*. Bacteria of the genera *Nitrobacter* carry out the oxidation of nitrite-nitrogen to nitrate-nitrogen.

Oxidation of NH_4^+ to NO_2^- is as follows:

$$NH_4^+ + 1.5O_2^- \xrightarrow{\text{nitrosomas}} NO_2^- + H_2O + 2H^+ \tag{7.35}$$

Oxidation of NO_2^- to NO_3^-:

$$NO_2^- + 0.5O_2 \xrightarrow{\text{nitrobacter}} NO_3^- \tag{7.36}$$

Since the duration of nitrite-nitrogen in water bodies is relatively short, these two processes are often combined into a single process of converting ammonium to nitrate:

$$NH_4^+ + 2O_2 \longrightarrow NO_3^- + H_2O + 2H^+ \tag{7.37}$$

This oxidation process draws dissolved oxygen from the water body. For each mole of ammonium-nitrogen, 4 mol of oxygen (O) is required. With the atomic weights (N, 14 g/mol; O, 16 g/mol), the oxygen uptake and exertion is $4(16)/14 = 4.57$ g of oxygen per g of nitrogen.

Both growth and reaction rates of the bacteria involved depend on the temperature. A temperature between 20 and 30°C is optimal. At temperatures lower than 5°C, nitrification rarely occurs.

If insufficient dissolved oxygen is available, *denitrification,* i.e., the bacterial reduction of nitrate to nitrogen, occurs, releasing oxygen, which is consumed again for the oxidation of organic matter to carbon dioxide:

$$4NH_3^- + 4H^+ + 5C_{org} \longrightarrow 5CO_2 + 2N_2 + 2H_2O \tag{7.38}$$

Thus, the increase of oxygen does not increase the amount of dissolved oxygen but decreases the biochemical oxygen demand. Denitrification is not explicitly modeled in many water-quality models.

The following two equations give the source and sink terms for the combined nitrification process. In almost all water-quality models the transformation rates of the nitrogen cycle are represented by first-order reactions, but these reaction-rate constants, k_{NH_4} and k_{NO_3}, may be functions of other constituent concentrations as discussed elsewhere in this section:

$$\frac{\partial NH_4\text{-N}}{\partial t} = k_{NH_4}(NH_4\text{-N}) \tag{7.39}$$

$$\frac{\partial NO_3\text{-N}}{\partial t} = k_{NH_4}(NH_4\text{-N}) - k_{NO_3}(NO_3\text{-N}) \tag{7.40}$$

For example, in this case the rate constant k_{NO_3} could represent the uptake of nitrate by algae, which, in turn, could be a function of the concentration of algae as well as the concentrations of other nutrients, the available light and temperature, as discussed elsewhere in this section.

Apart from the transformation processes described above, there are further sources and sinks within the nitrogen cycle. Ammonium results from both the decomposition of detritus and organic sediment and the transformation processes due to the growth of biological mass and respiration. Primary producers use ammonium and nitrate as nutrients. Approaches for the simulation of these transformations are given in Orlob (1983), Thomann and Mueller (1987), BLU (1984), and EPA (1985), and are similar to the approach used for modeling the phosphorus cycle, described below.

Phosphorus. For example, consider only two forms of phosphorus as modeled by QUAL2E, and as outlined in more detail in Somlyody and Varis (1994). These two forms are organic phosphorus in algae detritus and dissolved phosphorus in the water.

Let C represent, as before, phosphorus concentration. The next subscript letters, A, AD, and D, will represent algae, algae detritus, and dissolved. Thus C_A denotes the concentration of phosphorus in algal biomass, C_{AD} the concentration of phosphorus in algae detritus, and C_D the concentration of dissolved inorganic phosphorus. The processes to be modeled include uptake, growth, and death of algae as influenced by phosphorus, light, temperature, mineralization, sedimentation, and sorption exchange. The rates of change in the respective concentrations are

$$\frac{\partial C_A}{\partial t} = \left(G_A - D_A - \frac{v_A}{h}\right)C_A \tag{7.41}$$

$$\frac{\partial C_{AD}}{\partial t} = D_A C_A - \left(M_{AD} - v_{AD}\frac{1-f}{h}\right)C_{AD} \tag{7.42}$$

$$\frac{\partial C_D}{\partial t} = G_A C_A - M_{AD} C_{AD} + k_x(C_{D,\text{eq}} - C_D) \tag{7.43}$$

where G_A is the first-order algal growth reaction-rate constant (T^{-1}) and is the product of the maximum growth rate $G_{\max}$, the dissolved-phosphorus limitation factor G_D, the temperature-reduction factor G_T, and the light-reduction factor G_L. No other nutrients or conditions are considered limiting in this example. Hence, assuming the product form of Eq. (7.34), we obtain

$$G_A = G_{\max} G_D G_T G_L \tag{7.44}$$

The nutrient (phosphorus) limitation factor term G_D is a Michaelis-Menton expression of the dissolved-phosphorus concentration C_D:

$$G_D = \frac{C_D}{C_{2D} + C_D} \tag{7.45}$$

ranging from 0 to less than 1 and requiring a half-saturation (maximum growth) concentration constant C_{2D} (M/L^3), and

$$G_T = \begin{cases} 1 & \text{if the temperature is at some optimum level} \\ 0 & \text{if the temperature is above some maximum critical level} \end{cases}$$

As the average temperature deviates from the optimum level, the value of G_T decreases from 1 and approaches 0.

Finally, the light-attenutation factor G_L equals a function (ranging from 0 to 1) of the optimal light intensity of algae growth, the daily total global radiation, the length of the photoperiod, and the extinction coefficient characterizing the light penetration with depth.

This completes the description of the first parameter for algal growth G_A in the first of the three differential equations for phosphorus above. Now on to the others:

- D_A = a temperature-dependent first-order death rate (T^{-1}) for algae
- M_{AD} = a temperature-dependent first-order mineralization rate (T^{-1}) for algae detritus
- v_A and v_{AD} = settling velocities of algae and detritus, respectively (L/T)
- f = the dissolved fraction of detritus not subject to sedimentation
- h = depth of water body (L)
- $C_{D,\text{eq}}$ = the equilibrium concentration of phosphorus in suspended solids (M/L^3)
- k_X = the transport coefficient of sorption exchange (T^{-1})

This illustrates one approach, and one that requires a considerable number of parameter values. Somlyody and Koncsos (1991) describe an application of an alternative approach, and one that requires fewer parameters. In any event, the calibration of water-quality models, especially those containing multiple constituents, and some of them ecological species, is a challenging task for even the experienced.

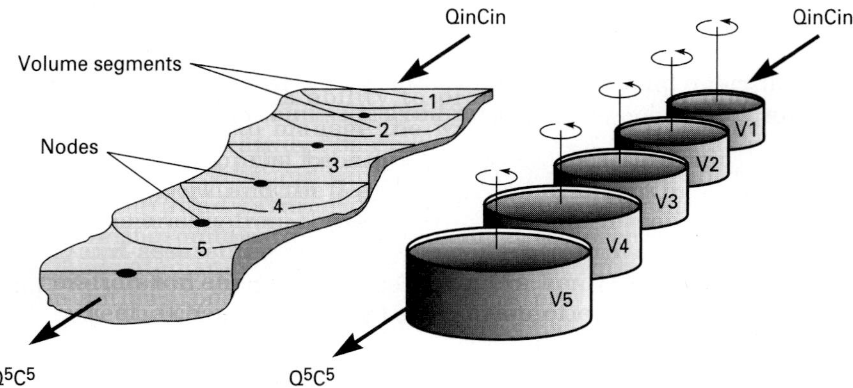

Figure 7.8 Continuous-flow stirred-tank reactor model.

Model Development and Solution Procedure: An Example

In this section a model for the prediction of constituent concentrations in open-channel flow along a river is developed. The main focus will be the derivation of the explicit equations for the approach chosen and the explanation of the crucial parameters.

For the calculation of constituent transport, the river system is defined as a series of links and connecting nodes. Each link, representing a river reach, is divided into a number of successively connected segments or elements (completely mixed reactors) (Fig. 7.8).

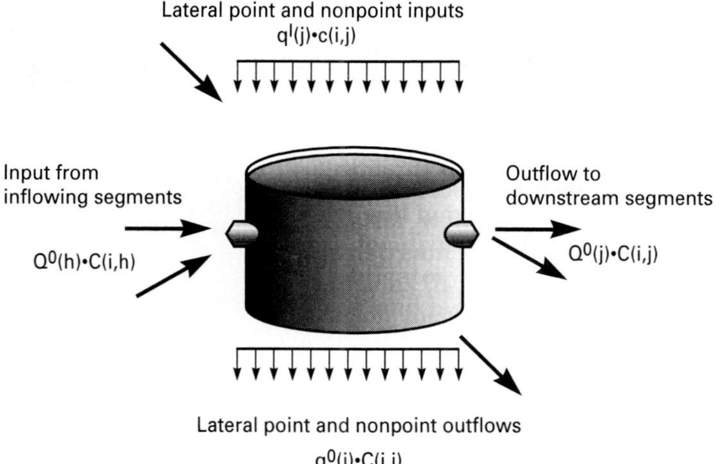

Figure 7.9 Mass balance of an individual reach or storage segment.

The change of mass $M(i,j)$ of constituent i within the jth segment's reactor volume $V(j)$ is the sum of the inflowing mass $M^I(i,j)$, the outflowing mass, $M^O(i,j)$, and the applicable reactions $S(i,j)$:

$$\frac{\partial[C(i,j)V(j)]}{\partial t} = M^I(i,j) - M^O(i,j) \pm S(i,j) \tag{7.46}$$

As can be seen in Fig. 7.9, the mass fluxes result from the multiplication of the flows with the respective concentrations. Both the inflowing mass and the outflowing mass consist of three components. The inflowing mass equals the sum of the total mass entering from the upstream inflow element(s), and the mass entering from lateral-point and nonpoint (or diffuse) sources. The outflowing mass flux consists of the total mass exiting the segment and entering the outflow element(s), and the lateral-point and nonpoint outflowing mass.

Thus, the mass inflows $M^I(i,j)$ and outflows, $M^O(i,j)$ result from the sum of products of flows and their respective concentrations. Let the index h represent all inflow segments connected to segment j and the index k represent all outflow segments connected to segment j. The total mass input of constituent i to segment j, $M^I(i,j)$, equals the input from all inflowing segments h, and any lateral or nonpoint inputs, $q^I(j)c^I(i,j)$:

$$M^I(i,j) = \sum Q^O(h)C(i,h) + q^I(j)c^I(i,j) \tag{7.47}$$

The mass output of constituent i from segment j is that portion entering the immediate downstream segments k, and any other discharge for local consumption or to account for any losses:

$$M^O(i,j) = Q^O(j)C(i,j) + q^O(j)C(i,j) = \sum_k Q^I(k)C(i,j) + q^O(j)C(i,j) \tag{7.48}$$

Note that the sum of inflows $\Sigma_k\ Q^I(k)$ to all outflowing segments k equals the total outflow $Q^O(j)$ to those segments from segment j. Flows from segments h or to segments k do not include lateral or local inflows $q^I(j)$ and outflows $q^O(j)$.

Since there is no concentration gradient over space within the reactor, the outflow concentration of the point and nonpoint outflows correspond with the concentration in the reactor. This concentration can change over time; hence denote $C(i,j,t)$ as the concentration of constituent i of segment j at the beginning of time period t. Similarly, denote $c^I(i,j,t)$ as the concentration of constituent i entering segment j during time period t. Assuming a constant flow in each time step, and a small duration Δt between two successive discrete simulation time steps t, the equations for the inflowing mass and outflowing mass per unit time (M/T) have the form

$$M^I(i,j,t) = \sum_h Q^O(h,t)[C(i,h,t)] + q^I(j)[c^I(i,j,t)] \tag{7.49}$$

$$\begin{aligned} M^O(i,j,t) &= Q^O(j,t)C(i,j,t) + q^O(j)C(i,j,t) \\ &= [Q^O(j,t) + q^O(j)]C(i,j,t) \end{aligned} \tag{7.50}$$

Note that only the initial segment concentrations are used to compute mass inflows and outflows to and from each segment. Complete mixing is assumed to occur at the end of each simulation time step, not during the time step. This assumption is critical in preventing some portion of a constituent introduced at the upstream end of a water-resource system from finding its way to the most downstream end of the system in a single time step regardless of the time of flow through the system.

The source/sink term $S(i,j)$ in Eq. (7.46) includes the time rate of change of mass of a constituent in a reactor due to biological, chemical, and physical processes. The change of mass of a constituent i in a period t is a function of its concentration $C(i,j,t)$, the reactor segment volume $[V(j,t) + V(j,t+1)]/2$, and the reaction-rate constant $k(i,j,t)$, for that constituent i in segment j in period t. The latter is a measure of the rate that the pollutant i is transformed. The changing concentrations of other constituents i^* could also affect the rate of change of constituent i, and vice versa, as discussed previously. Defining the stoichiometric transformation constant (mass of i consumed per unit change in the mass of i^*) as $T(i^*,i)$ and assuming a first-order reaction process:

$$\begin{aligned} S(i,j,t) = &-k(i,j,t)C(i,j,t)\frac{[V(j,t) + V(j,t+1)]}{2} \\ &-\sum_{i^*} k(i^*,j,t)T(i^*,i)C(i^*,j,t)\frac{[V(j,t) + V(j,t+1)]}{2} \end{aligned} \tag{7.51}$$

By substituting the differential term in Eq. (7.46) with its discrete approximation, we obtain

$$\frac{\partial C(i,j)V(j)}{\partial t} = \frac{[C(i,j,t+1)V(j,t+1) - C(i,j,t)V(j,t)]}{\Delta t} \tag{7.52}$$

and by substituting the right-hand side of Eq. (7.52) into Eq. (7.46) and using Eqs. (7.49) to (7.51), we can solve for the final concentration $C(i,j,t+1)$ of each constituent i in each successive segment j in each successive time step t:

$$C(i,j,t+1) = \frac{C(i,j,t)V(j,t)}{V(j,t+1)} + \frac{\Delta t}{V(j,t+1)} \times [M^I(i,j,t) - M^O(i,j,t) + S(i,j,t)] \qquad (7.53)$$

To illustrate how these equations work in a simulation over space and time, consider a simplified example in which a conservative constituent of concentration C_0 is introduced in the first time period into the first of a series of connected segments ($j = 1,2,3,\ldots, n$). Assume the flows Q and volumes V of each segment to be constant and that the duration of each simulation time step Δt is some fraction f of the detention time of each segment V/Q. Hence

$$\Delta t = f\left(\frac{V}{Q}\right) \qquad (7.54)$$

Equation (7.53) now becomes

$$\begin{aligned} C(i,j,t+1) &= C(i,j,t) + f[C(i,j-1,t) - C(i,j,t)] \\ &= (1-f)C(i,j,t) + fC(i,j-1,t) \end{aligned} \qquad \text{for} \quad j = 2,3,\ldots \quad (7.55)$$

For the first segment, $j = 1$, assuming no initial concentration at the beginning of period 1, and no additional concentration introduced after period 1, the concentration in any subsequent period t is

$$C(i,l,t+1) = C_0(1-f)^{t-1} \qquad (7.56)$$

Figure 7.10 illustrates the spatial distribution at the end of various time steps based on $C_0 = 1$ for various values of f.

If the detention time fraction $f = 1$, the constituent concentration would exist in only one segment at each time step, and dispersion would exist only within that single segment. The entire mass of the constituent would move from segment to segment in each time step. If $f = 0$, then no flow occurs and the concentration in each segment remains the same. Note that the spatial distributions over multiple segments are not Gaussian for values of f not equal to 1 or 0.5.

System parameters

Time-step interval Δt. The *detention time* in any segment is the time that a particle needs to flow through that segment. It equals the ratio

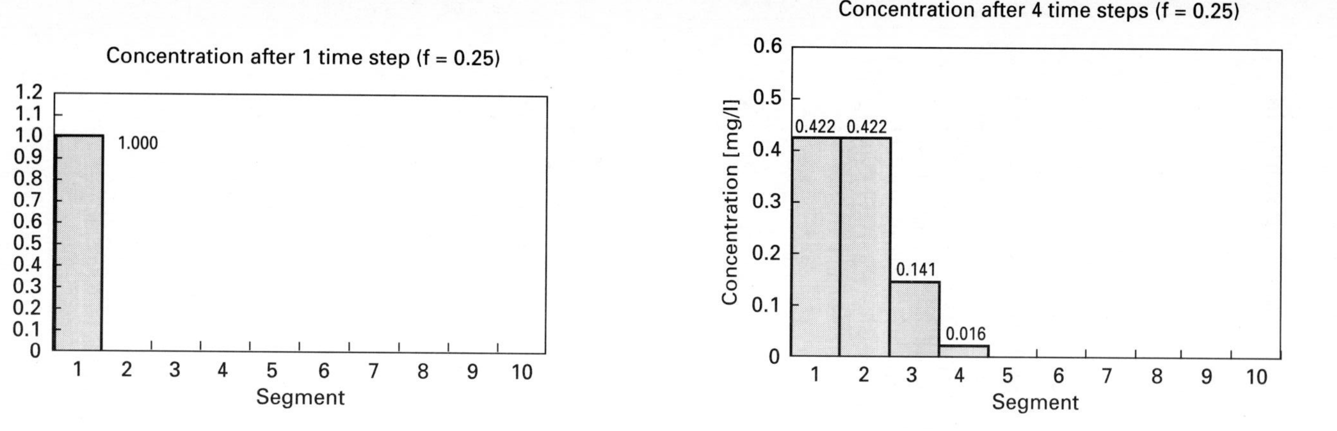

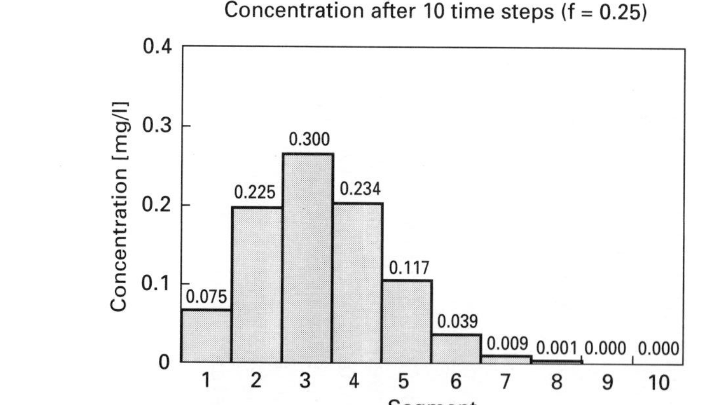

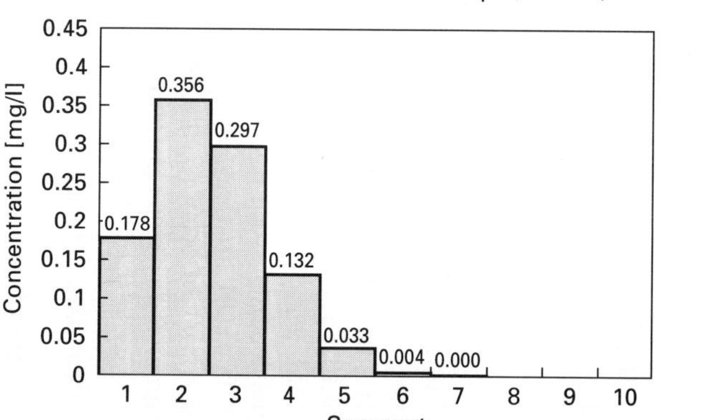

Figure 7.10 Spatial distributions of end-of-period constituent concentrations at various times and for various time-of-flow fractions of detention time.

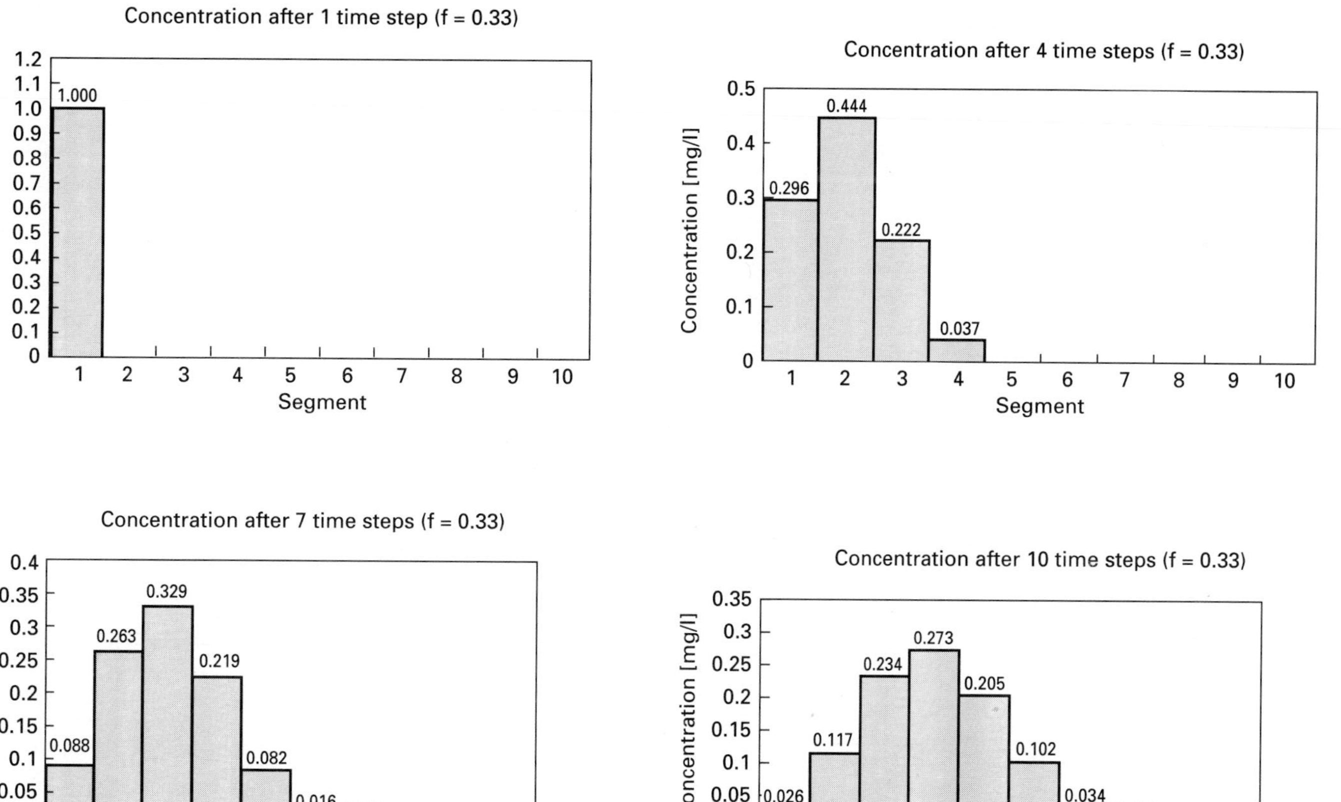

Figure 7.10 *(Continued)*

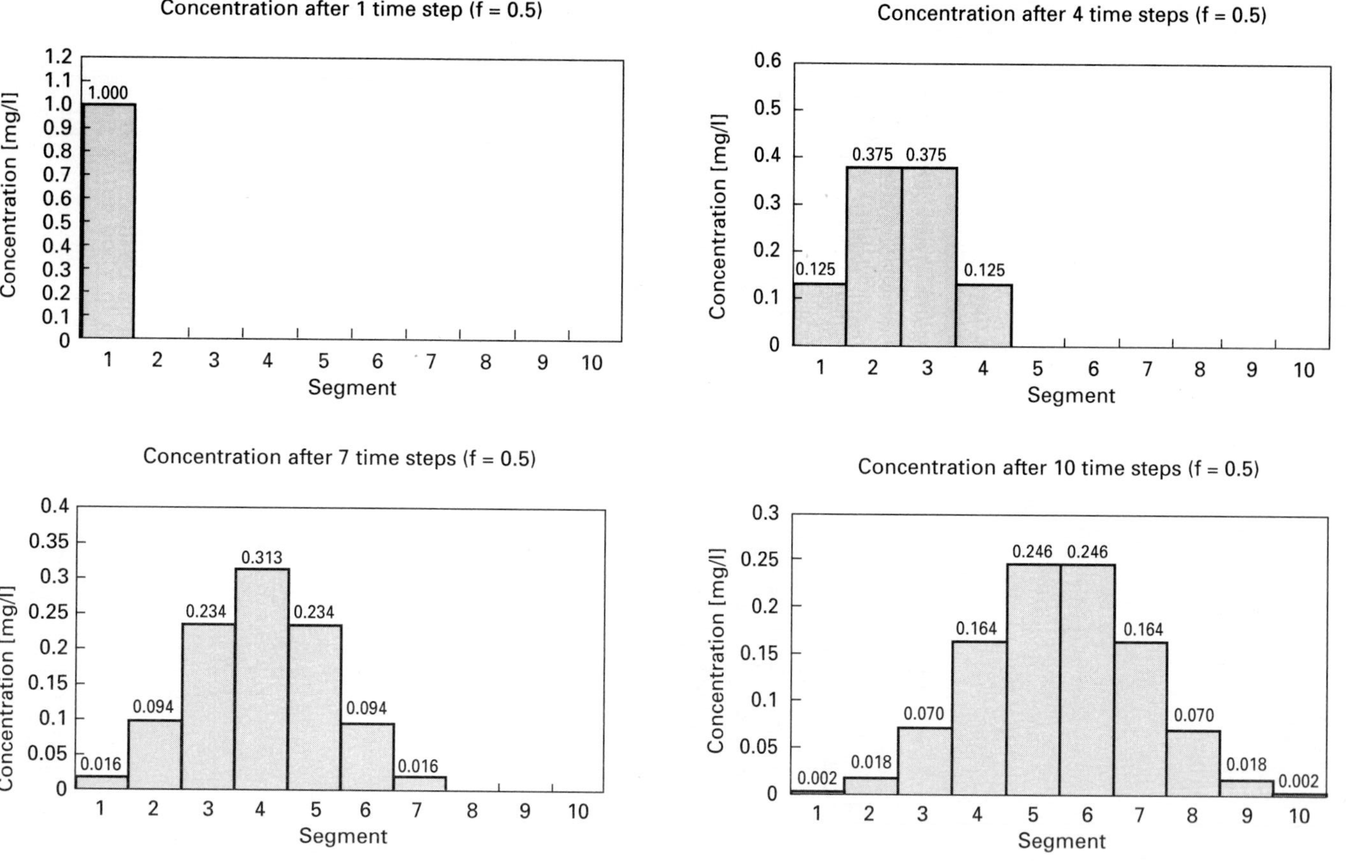

Figure 7.10 *(Continued)*

of the segment volume $V(j,t)$ to the discharge $Q^O(j,t)$. As discussed previously, the simulation time-step duration Δt cannot exceed the shortest detention time of any segment in the system. Otherwise, a particle might be transported completely through one segment and into another within one time step, and this would have to be taken into account in the simulation of each element.

Note that

$$\frac{V(j,t)}{Q^O(j,t)} = \frac{\Delta X(j)}{u(j,t)} \tag{7.57}$$

where $u(j,t)$ is the mean travel time velocity in segment j during period t and $\Delta X(j)$ is the distance along a river or stream reach represented by the segment j. This relationship may be helpful when deciding on the segmentation of a water body.

Consider the impact of the simulation time-step interval Δt on the time-based dispersion within a segment of a fixed volume and flow, i.e., a segment having a constant detention time. For this, the concentration distribution of a conservative pollutant in the form of a symmetric Gaussian distribution with an area of 1 is input into the segment. Figure 7.11 shows several resulting concentration/time

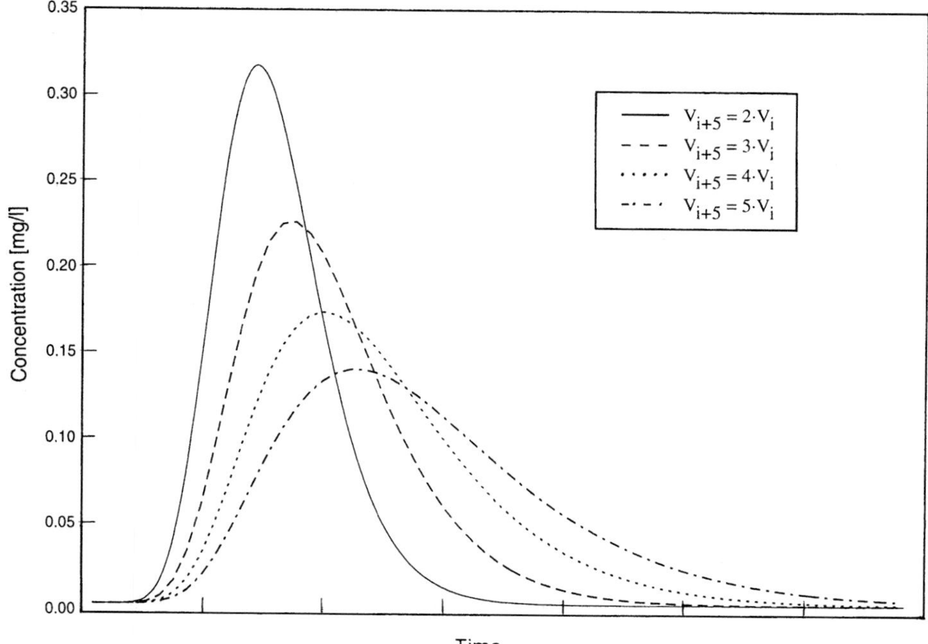

Figure 7.11 Concentration distribution of a Gaussian-distributed induction depending on the ratio $\Delta t / T$.

distributions depending on the ratio of time-step interval Δt to detention time $T = V/Q$.

With decreasing time-step intervals Δt, the time distribution of constituent concentration flattens and spreads in longitudinal direction. Analogous to the classic solution of the advection-dispersion equation, these results correspond with the application of a bigger dispersion coefficient E_X. The area under each distribution is constant for all ratios $\Delta t/T$; i.e., conservation of masses is preserved for this conservative constituent. In each case the concentration crest longitudinally moves with mean velocity.

Because of the regular segmentation, a symmetrical concentration time distribution occurs which does not always correspond with reality. As mentioned earlier, the distribution of a pollutant cloud in rivers is often skewed to the right rather than being symmetrical as shown in Fig. 7.11.

The skewness of a pollutant cloud can be simulated by the completely mixed reactor model with an apt segmentation.

Segmentation. Regular segmentation results in a symmetrical concentration time distribution. With a variable segmentation, the skewness of a constituent concentration in natural flows can be simulated. Figure 7.12 shows the effect of variable segment volumes on the concentration distribution. In this figure the volumes of every fifth segment have been increased relative to the volumes in the other segments.

As can be clearly seen, the skewness increases with the increasing nonhomogeneity of segmentation. Figure 7.4 illustrates the results of experiments on the Rhine River (Leibundgut, 1990) that confirm the existence of such skewed concentration distributions.

Calibrating and Assessing Model Applicability and Precision

The parameters of a water-quality model whose values need to be determined can often be considered as the unknown values of an optimization model that includes the water-quality model as constraints. The objective function of such an optimization model is some measure of the model prediction error that is to be minimized. These measures might be the sum of squared differences, the mean absolute difference, or the maximum absolute difference between the predicted (calculated) and observed concentration values.

In very general terms, consider any water-quality model that predicts the quality concentration C_X at some site X as some function f $(\mathbf{I},\mathbf{P},X)$, where $\mathbf{I}$ is the set (vector) of known (usually point-source)

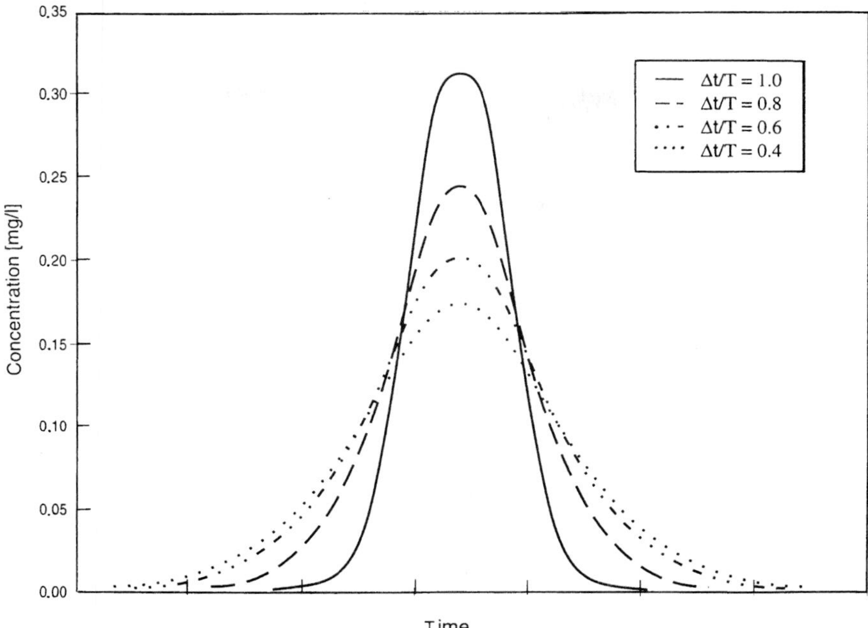

Figure 7.12 Concentration distribution of a Gauss-distributed constituent input depending on segmentation.

inputs, and **P** is the set (vector) of unknown parameters and the portion of the total concentration at point X resulting from the unknown (usually non-point-source) wasteload inputs:

$$C_X = f(\mathbf{I}, \mathbf{P}, X) \tag{7.58}$$

For example, consider the simplest BOD−DO predictive equations proposed by Streeter and Phelps (1925) applied to a single river reach of distance X with time of flow Δt determined by the flow and its velocity. Assume that there exist a number of measurements of BOD_0 and BOD_X and DO_0 and DO_X at the beginning and end of the reach and that the flows, and hence time of travel, of the flow Δt are known and identical for each of these measurements. Define $\mathrm{BOD}_{X,\mathrm{np}}$ and $\mathrm{DO}_{X,\mathrm{np}}$ to be the unknown incremental BOD and DO constituent concentrations resulting from nonpoint inputs to the reach and $\mathrm{DO}_{\mathrm{sat}}$ as the known saturation dissolved-oxygen concentration. (The term $\mathrm{DO}_{X,\mathrm{np}}$ could be negative, reflecting an additional dissolved-oxygen deficit due to nonpoint inputs.) In the predictive equations,

$$\mathrm{BOD}_X = \mathrm{BOD}_0[e^{-k_B \Delta t}] + [\mathrm{BOD}_{X,\mathrm{np}}] \tag{7.59}$$

$$\mathrm{DO}_X = \mathrm{DO}_{\mathrm{sat}} - \mathrm{BOD}_0\left[\frac{k_B}{k_r - k_B}(e^{-k_B\,\Delta t} - e^{kr\,\Delta t})\right]$$

$$+ (\mathrm{DO}_{\mathrm{sat}} - \mathrm{DO}_0)[e^{-k_r\,\Delta t}] + [\mathrm{DO}_{X,\mathrm{np}}] \tag{7.60}$$

the inputs **I** are the measured BOD_0 and BOD_X and DO_0 and DO_X values together with $\mathrm{DO}_{\mathrm{sat}}$, and the unknown parameters **P** are all the terms in square brackets. For each combination of BOD_0 and BOD_X and DO_0 and DO_X values, there are two equations. Given a sufficient number of measurements, and hence a sufficient number of equations, the constant parameter values, i.e., the values of the expressions within the square brackets, can be determined using optimization. Once the values of the expressions within the square brackets are determined, knowing the Δt values permits the determination of the actual deoxygenation-rate constant k_B and the oxygen reaeration-rate constant k_r in this illustration.

In this example, the predictive equations are linear, so linear programming (optimization) could be used to find the best values of these and the other parameters. A number of objective functions could be used. For example, the objective might be to minimize some function of the error E_i between calculated and observed concentrations for each set of measurements i, such as

$$\text{Minimize} \sum_i E_i^2 \tag{7.61}$$

$$\text{Minimize} \sum_i |E_i| \tag{7.62}$$

$$\text{Minimize maximum } \{|E_i|\} \tag{7.63}$$

The error terms E_i are associated with each equation containing a set of input values. Each equation, i.e., each set of input values, is denoted by the index i. Hence for some constituent C at site X, the predictive equation for each sample set i is

$$E_i + f(\mathbf{I},\mathbf{P},X) = C_{xi} \tag{7.64}$$

where C_{xi} is the measured value of the concentration of C at site X. The unknown error terms E_i for each equation i are unrestricted in sign.

If the function $f(\mathbf{I},\mathbf{P},X)$ is too nonlinear for easy solution by available mathematical programming (constrained optimization) methods, various search techniques, including genetic algorithms, may prove to be useful. In any event, the greater the number of measured sets of input values (flows and corresponding concentrations) for each point X, the more accurate the computed parameter values may be. The minimum number equals the number of unknown parameters and concentrations in the vector **P.**

Once computed, the parameter values should be checked for reasonableness. If some parameter values are not within their expected ranges, some other model structure may be required, or some of the model's parameters may have to be set, perhaps arbitrarily at first, before calculating the remaining parameter values. If these values are satisfactory, then some of them might be set to recalculate better values for the remaining ones, including those parameter values that were first set. Model calibration is as much of an art as a science, especially in practice where accurately sampled and measured data are often few.

Models developed for water-quality prediction and management can also be used to determine the sensitivity of their input and parameter values to the resulting predictions. Some input and parameter values will likely influence the predicted concentrations quite significantly, and if so, their values need to be as accurate as possible. For those parameters that do not significantly influence the predicted concentrations, or if the predicted concentrations are considerably below some critical level, then, for the purposes of water-quality management, there is little need to invest money in obtaining more accurate values.

Models are useful for both predictions and for determining the needed accuracy of their input and parameter data. They should be used for both purposes, and too often they are not. It is easy to spend money collecting data. But if limited funds are to be used to collect data, it is prudent to try to spend those funds efficiently and effectively. During the planning of any effective data-collection program, models can help identify how sensitive decisions are to various data, and hence help identify which data need to be collected and how accurate those data need be.

A brief description of the many water-quality models available today could occupy this entire chapter. Some short references to a few of these that span the range of available models and their data requirements conclude this review.

Water-Quality Modeling in Practice

In the 1920s Streeter and Phelps (1925) were the first to predict the distribution of dissolved oxygen in rivers using mathematical models. Their oxygen-balance model is based on two processes: the physical reaeration of water from the atmosphere, serving as oxygen source; and the oxidation of organic material, causing an oxygen sink. Both processes were regarded as first-order reactions. The authors assumed that the deoxygenation reaction rate was only a function of the water temperature. To the credit of Streeter and Phelps, their modeling approach forms the basis of many of the water-quality models used today.

Later, with the development in computer technology, more detailed oxygen-balance models (Thomann, 1963) and multiparameter models (Thomann et al., 1976; Beck, 1985) were developed. In addition to dissolved oxygen, these more recent models include a number of significant conservative and nonconservative constituents. Examples of such models include the water-quality models QUAL-I and WQRRSQ developed by the Water Resources Engineers, Inc. The first version of the series of QUAL models, available in the early 1970s, modeled temperature-dependent BOD-DO relationships. Reservoir models developed by Water Resources Engineers, Inc. and Massachusetts Institute of Technology were one-dimensional in the vertical direction, and included effects of temperature stratification. Later versions permitted the simulation of the impacts of nutrients and photosynthesis on the oxygen balance. The current version of the QUAL models, the 10-constituent QUAL2E model (Brown and Barnwell, 1987); and a quasi-dynamic model for streams, rivers, lakes, and reservoirs, WQRRSQ (Smith, 1986); and successive versions of a hydrodynamic water-quality model called the Water Quality Analysis Simulation Program (WASP) rank among the most widely used water-quality models today. This is in no small measure due to their continued maintenance, documentation, and distribution by agencies of the U.S. government (Ambrose et al., 1988).

For additional references on the considerable development of successively more comprehensive water-quality models for surface-water systems, especially in the 1970s and 1980s, that focused on both hydrodynamics as well as the ecological and chemical aspects of water quality, one can refer to Rinaldi et al. (1979), Scavia and Robertson (1979), Jorgensen (1982, 1986), Straskraba and Grauck (1985), Orlob (1983), and McCutcheon (1989), in addition to the references already cited. Among those considering uncertainty in model predictions, see, for example, Young (1979), Fedra et al. (1981), Beck and van Straten (1983), Beck (1987, 1991), and Hornberger and Spear (1981).

Troubounis (1985) has developed a stationary flow–dynamic loading model for the detailed simulation of the carbon, nitrogen, phosphorus, and oxygen balance of nonflowing or slowly flowing water. The model accounts for the exchange processes at the sediment–water column interface which have to be taken into account for slowly flowing or nonflowing water. The transport processes are based on the CSTR model. The reactor is vertically divided into two layers to permit examination of the active sediment layer in more detail.

A fully dynamic water-quality model was developed by da Costa (1992) using the ADZ method. It was applied to the river Ave in

Portugal, which had the necessary database for model calibration. His one-dimensional first-order kinetic model included conservative pollutants, the nitrogen cycle, biochemical oxygen demand, and dissolved oxygen. The ADZ model successfully modeled the asymmetrical concentration distributions observed by in situ tracer experiments on that river.

Perhaps representative of the most sophisticated water-quantity and water-quality modeling approaches are those similar to MIKE 11, a modeling system for rivers and channels developed by the Danish Hydraulics Institute (DHI). It includes several modules for the simulation of surface runoff, flow, sediment transport, and water quality in river or channel systems. It permits, but does not require, fully dynamic water-quality simulation. The influence of wind, surface flow over floodplains, and tidal phenomena can be simulated. Obviously for detailed simulations, detailed input data sets are required.

References

Abbott, M. B., 1979, *Computational Hydraulics, Elements of the Theory of Free Surface Flows,* Pitman, London.

Ahlgren, I., Frisk, T., and Kamp-Nielsen, L., 1988, Empirical and theoretical models of phosphorus loading, retention and concentration vs. lake trophic state. In Persson, G., and Jansson, M., eds., *Phosphorus in Fresh Water Ecosystems,* vol. 48, *Developments in Hydrobiology,* pp. 285–303.

Ambrose, R. B. Jr., Wool, T. A., Conolly, J. P., and Schanz, R. W., 1988, *WASP4, a Hydrodynamic and Water Quality Model—Model Theory, Users Manual and Programmer's Guide.* Report no. EPA/600/3-87/039, U.S. Environmental Protection Agency, Athens, Ga.

Ambrose, R. B., Jr., P.E., Barnwell, T. Jr., 1989, Environmental software at the U.S. Environmental Protection Agency's Center for Exposure Assessment Modeling. *Environ. Software* **4,** 76–93.

Andreu, J., Capilla, J., and Sanchis, E., 1991, AQUATOOL, a computer assisted support system for water resources research management including conjunctive use. In Loucks, D. P., and daCosta, J. R., eds., *Decision Support Systems: Water Resources Planning,* NATO ASI Series, vol. G26, Springer-Verlag, Berlin.

Aris, R., 1956, The longitudinal diffusion coefficient in a flow through a tube with stagnant pockets. *Proc. Roy. Soc. London, A,* **235,** 67–77.

Bayerisches Landesamt für Umweltschutz (BLU), 1984, *Die Gewässergütesimulation als Planungsinstrument, dargestellt am Beispiel der wasserwirtschaftlichen Rahmenuntersuchung Donau und Main,* R. Oldenbourg Verlag, Munich-Vienna, *Schriftenreihe,* vol. 65.

Beal, R., and Jackson, T. 1990, *Neural Computing: An Introduction,* Institute of Physics, Bristol, United Kingdom.

Beck, M. B., 1983, Uncertainty, system identification, and the prediction of water quality. In Beck, M. B., and van Straten, G., eds., *Uncertainty and Forecasting of Water Quality,* pp. 3–68, Springer-Verlag, Berlin.

Beck, M. B., 1983a, A procedure for modeling. In Orlob, G. T., ed., *Mathematical Modeling of Water Quality: Streams, Lakes, and Reservoirs,* Wiley, New York.

Beck, M. B., 1983b, Sensitivity analysis, calibration and verification. In Orlob, G. T., ed., *Mathematical Modeling of Water Quality: Streams, Lakes, and Reservoirs,* Wiley, New York.

Beck, M. B., 1985, *Water Quality Management: A Review of the Development and Application of Mathematical Water Quality Models,* Lecture Notes in Engineering, no. 11, Springer-Verlag, New York.

Beck, M. B., 1987, Water quality modeling: a review of the analysis of uncertainty. *Water Resources Res.* **23,** 1393–1442.

Beck, M.B., 1991, Forecasting environmental change. *J. Forecasting* **10**(1), 3–10.

Beck, M. B., and van Straten, G., 1983, *Uncertainty and Forecasting of Water Quality,* Springer-Verlag, Berlin.

Beck, M. B., and Young, P., 1975, A dynamic model for DO-BOD relationships in a non-tidal stream. *Water Resources Res.* **9,** 769–776.

Becker, M., and Sosnowski, P., 1969, Eine impulsantwort fuer flussabschnitte zur durchflussvorherbestimmung. *Wasserwirtschaft–Wassertechnik,* **19** (12), 410–418.

Beer, T., and Young, P., 1983, Longitudinal dispersion in natural streams. *ASCE J. Environ. Eng.* **109**(5), 1049–1067.

Box, G. E. P., and Jenkins, G. M., 1970, *Time Series Analysis, Forecasting and Control,* Holden-Day, San Francisco.

Brown, L. C., and Barnwell, T. O., Jr., 1987, *The Enhanced Stream Water Quality QUAL2E and QUAL2E-UNCAS: Documentation and User Manual,* Report no. EPA/600/3.87/007, U.S. Environmental Protection Agency, Athens, Ga.

Câmara, A., Pinheiro, M., Antunes, P., and Seixas, J., 1987, A new method for qualitative simulation of water resources systems. 1. Theory. *Water Resources Res.* **23,** 2015–2018.

Câmara, A., Cardoso da Silva, M., Carmona Rodrigues, A., Remédio, J. M., Castro, P. P., Soares de Oliveira, M. J., and Fernandes, T. F., 1991, Decision support system for estuarine water-quality management. *ASCE J. Water Resources Plan. Manag.* **116,** 417–432.

Câmara, A., Ferreira, F. C., Loucks, D. P., and Seixas, M. J., 1990, Multidimensional Simulation Applied to Water Resources Management. *Water Resources Res.* **26**(9), 1877–1886.

Chapra, S. C., 1991, Toxicant-loading concept for organic contaminants in lakes. *J. Environ. Eng. ASCE,* **117**(5), 656.

Chapra, S. C., and Reckhow, K. H., 1983, *Engineering Approaches for Lake Management,* vol. 2, *Mechanistic Modeling,* Butterworth Publishers, Boston.

Chapra, S. C., and Tarapchak, S. J., 1976, A chlorophyll a model and its relationship to phosphorus loading plots for lakes. *Water Resources Res.,* **12**(6), 1260–1264.

Chen, C. W., and Orlob, G. T., 1972, *Ecologic Simulation for Aquatic Environments,* Final Report for the Office of Water Resources Research, U.S. Department of the Interior, Walnut Creek, Calif., Water Resources Engineers, Inc.

Chen, C. W., and Orlob, G. T., 1975, Ecologic simulation for aquatic environments. In Patten B. C., ed., *Systems Analysis and Simulation in Ecology,* vol. 3, pp. 476–588, Academic Press, New York.

Chow, Ven Te, 1959, *Open-Channel Hydraulics,* McGraw-Hill Kogakusha Ltd., Tokyo.

Churchill, M. A., Helmore, H. L., and Buckingham, R. A., 1962, The prediction of stream reaeration rates. *J. Sanit. Eng. Div., ASCE,* **88**(SA4), 1.

CRES, 1990, *The ADZ Analysis Manual,* Center for Research on Environmental Systems, Lancaster, United Kingdom.

Cunningham, P. A., 1988, *Nonpoint Source Impacts on Aquatic Life—Literature Review,* prepared for Monitoring and Data Support Division, Office of Water Regulations and Standards, U.S. Environmental Protection Agency, Research Triangle Institute, Research Triangle Park, N.C.

daCosta, J. R., 1992, Dynamic modeling of transport and mixing in rivers using the ADZ theory, dissertation, Lancaster University, United Kingdom.

daCosta, J. R., Loucks, D. P., and Santos, M. A., 1989, *Methodologies for Water Resources Policy Analysis,* Laboratório Nacional de Engenharia Cicil, Lisboa, Portugal.

delFuria, L., Rizzoli, A., and Arditi, R., 1995, Lakemaker: a general object-oriented software tool for modeling the eutrophication process in lakes. *Environ. Software,* **10**(1), 43–64.

Dillon, P. J., 1974, A critical review of Vollenweider's nutrient budget model and other related models. *Water Resources Bull.,* **10**(5), 969–989.

DiToro, D. M., 1984, Probability of stream quality due to runoff. *J. Environ. Eng., ASCE,* **110**(3), 607.

Dobrinski, Krakau, Vogel, 1987, *Physik fâr Ingenieure,* B. G. Teubner, Stuttgart, 8. Auflage.

Doetsch, P., 1987, Ermittlung und exemplarische Anwendung eines Verfahrens zur nutzungsadäquaten Quantifizierung von Gewässergâte, dissertation, RWTH Aachen.

Driscoll, E. D., Mancini, J. L., and Mangarella, P. A., 1983, *Technical Guidance Manual for Performing Waste Load Allocations,* book II, *Streams and Rivers,* chap. 1, Biochemical oxygen demand/dissolved oxygen, Report no. EPA/440/4-84-020, Office of Water Regulations and Standards, U.S. Environmental Protection Agency, Washington, D.C.

Environmental Protection Agency (EPA), 1985, *Rates, Constants and Kinetics—Formulations in Surface Water Quality Modeling,* 2d ed., Environmental Research Laboratory, U.S. Environmental Protection Agency, Athens, Ga.

Fedra, K., 1983, A Monte-Carlo approach to estimation and prediction. In *Uncertainty and Forecasting of Water Quality,* Springer-Verlag.

Fedra, K., 1983, *Environmental Modeling under Uncertainty: Monte Carlo Simulation,* IIASA RR-83-28, Laxenburg, Austria.

Fedra, K., van Straten, G., and Beck, M. B., 1981, Uncertainty and arbitrariness in ecosystems modeling: a lake modeling example. *Ecol. Modeling,* **13,** 87–110.

Fisher, H. B., 1967, The mechanics of dispersion in natural streams. *Proc. ASCE, J. Hydraulics Div.,* **93,** 187–216.

Fisher, H. B., List, E. J., Koh, R. C. Y., Imberger, J., and Brooks, N. H., 1979, *Mixing in Inland and Coastal Waters,* Academic Press, New York.

Frisk, T., 1981, New modifications of phosphorus models. *Aqua Fennica,* **11,** 7–17.

Frisk, T., 1989, *Development of Mass Balance Models for Lakes,* publications of the Water and Environment Research Institute, National Board of Waters and the Environment, Helsinki, Finland.

Frisk, T., Niemi, J. S., and Kinnunen, K. A. I., 1981, Comparison of statistical phosphorus-retention models. *Ecol. Modeling,* **12**(1), 11–27.

Ghaeizadeh, H. R., 1992, Schadstoffausbreitung in fliessgewaessern, thesis, Institut fuer Wasserbau und Wasserwirtschaft, RWTH Aachen, Aachen, Germany.

Grobler, D. C., Rossouw, J. N., van Eeden, P., and Oliveira, M., 1987, Decision support system for selecting eutrophication control strategies. In Beck, M. B., ed., *System Analysis in Water Quality Management,* pp. 219–230, Pergamon Press, Oxford.

Guariso, G., and Werthner, H., 1989, *Environmental Decision Support Systems,* Ellis Horwood, Chichester (United Kingdom).

Hall, M. J., and Minns, A. W., 1993, Rainfall-runoff modeling as a problem in artificial intelligence: experience with neural network. *Proceedings of BHS 4th National Hydrology Symposium,* Cardiff, Wales.

Hornberger, G. M., and Spear, R. C., 1981, An approach to the preliminary analysis of environmental systems. *J. Environm. Mgmt.*, **12**, 7–18.

Janus, L. L., and Vollenweider, R. A., 1984, Phosphorus residence time in relation to trophic conditions in lakes. *Verh, Internatl. Verein. Limnol.,* **22,** 179–184.

Jørgensen, S. E., 1980, *Lake Management,* Pergamon Press, Oxford.

Jørgensen, S. E., 1982, Modeling the ecological processes, In Orlob, G. T., ed., *Mathematical Modeling of Water Quality,* pp. 116–149, Wiley/IIASA, Chichester.

Jørgensen, S. E., 1986, *Fundamentals of Ecological Modeling,* Elsevier, Amsterdam.

Jørgensen, S. E., and Gromiec, M. J., eds., 1989, *Mathematical Submodels in Water Quality Systems,* Elsevier, Amsterdam.

Langbein, D. B., and Durum, W. H., 1967, *The Aeration Capacity of Streams,* U.S. Geological Survey Circular 542, U.S. Government Printing Office, Washington, D.C.

Leibundgut, C., 1990, Markierversuch rhein 04/89—abschlussbericht, Institut fuer Physische Geo-graphie der Universitaet Freiburg, Auftrag der Bunesanstalt fuer Gewaesserkunde, Koblenz.

Loucks, D. P., and daCosta, J. R., eds., 1991, *Decision Support Systems: Water Resources Planning,* NATO ASI Series, vol. G26, Springer-Verlag, Berlin.

Luettich, R. A., Jr., Harleman, D. R. F., and Somlyody, L., 1990, Dynamic behavior of suspended sediment concentrations in a shallow lake perturbed by episodic wind events. *Limnol. Oceanogr.* **35,** 1050–1067.

Markowsky, M., 1980, *Strömungsmechanische Aspekte der Wasserqualität,* R. Oldenbourg Verlag, Munich-Vienna.

McCutcheon, S., 1989, *Water Quality Modeling,* vol. 1, *Transport and Surface Exchange in Rivers,* CRC Press, Boca Raton, Fla.

Nordin, C. F., and Troutman, B. M., 1980, Longitudinal dispersion in rivers: the persistence of skewness in observed data. *Water Resources Res.,* **16,** 123–128.

O'Connor, D. J., 1988, Models of sorptive toxic substances in freshwater systems. I: Basic equations. II: Lakes and reservoirs. III: Streams and rivers. *J. Environ. Eng. ASCE,* **114**(3), 507, 533, 552.

O'Connor, D. J., and Dobbins, W. E., 1958, Mechanism of reaeration in natural streams. *Trans. Am. Soc. Civ. Eng.,* **2934**.

O'Melia, Ch. R., 1972, An approach to modeling of lakes. *Schweiz. Z. Hydrol.,* **34**(1), 1–33.

Orlob, G. T., ed., 1983, *Mathematical Modeling of Water Quality: Streams, Lakes and Reservoirs,* Wiley, Chichester.

Owens, M., Edwards, R. W., and Gibbs, J. W., 1964, Some reaeration studies in streams. *Internatl. J. Air Water Pollution,* **8**.

Park, R. A., O'Neill, R. V., Bloomfield, J. A., Shogart, H. H., Jr., Boorth, R. S., Goldstein, R. A., Mankin, J. B., Komce, J. F., Scavia, D., Adams, M. S., Clesceri, L. S., Colon, E. M., Dettman, E. H., Hoopes, J. A., Huff, D. D., Katz, S., Kitchell, J. F., Kohberger, R. C., LaRow, R. J., McNaught, D. C., Peterson, J. L., Titus, J. E., Weiler, P. R., Wilkinson, J. W., and Zahorak, G. S., 1974, A generalized model for simulating lake ecosystems. *Simulation,* **23,** 33–50.

Peters, R. H., 1986, The role of prediction in limnology. *Limnol. Oceanogr.,* **31**(5), 1143–1159.

Postma, L., 1990, *DELWAQ, User's Manual, Version 3.0,* Delft Hydraulics, Delft, Netherlands.

Reckhow, K. H., and Chapra, S. C., 1983a, *Engineering Approaches for Lake Management,* vol. 1, *Data Analysis and Empirical Modeling,* Butterworth, Boston.

Reckhow, K., and Chapra, S., 1983b, Confirmation of water quality models. *Ecol. Modeling,* **20,** 113–133.

Rinaldi, S., Soncini-Sessa, R., Stehfest, H., and Tamura, H., 1979, *Modeling and Control of River Water Quality,* McGraw-Hill, New York.

Ruland, P., Arnold, U., and Rouvé, G., 1991, Pollutant transport with fine cohesive sediments modeling sediment erosion and transport processes in lakes and river basins. In Yu, S. L., and Shih, K. K., eds., *Proceedings of the International Conference on Computer Applications in Water Resources,* July 3–6, 1991, vol. 2, pp. 665–672, Tamkang University, Tamsui, Republic of China.

Rumelhart, D. E., Widrow, B., and Lehr, M. A., 1994, The basic ideas in neural networks. *Commun. ACM,* **37**(3) 87–92.

Ryding, S.-O., and Rast, W., 1989, *The Control of Eutrophication of Lakes and Reservoirs,* UNESCO, Paris, and Parthenon Publishing Group, Carnforth, Lancaster (United Kingdom) and Park Ridge, N.J.

Scavia, D., and Robertson, A., 1979, *Perspectives on Lake Ecosystem Modeling,* Ann Arbor Science Publ., Ann Arbor, Mich.

Schnoor, J. L., Sato, C., McKechnie, D., and Sahoo, D., 1987 (June), *Processes, Coefficients, and Models for Simulating Toxic Organics and Heavy Metals in Surface Waters,* Report EPA/600/3-87/015, U.S. Environmental Protection Agency, Athens, Ga.

Shanahan, P., and Harleman, D. R. F., 1984, Transport in water quality modeling. *ASCE J. Environ. Eng.* **110,** 42–57.

Shanahan, P., and Somlyody, L., 1995, *Modeling the Impacts of Diffuse Pollution on Receiving Water Quality,* WP-95-2, IIASA, Laxenburg, Austria.

Smith, D. J., 1986, *Water Quality for River-Reservoir Systems, Computer Program Description WQRRSQ,* U.S. Army Corps of Engineers, Water Resources Support Center, Davis, Calif.

Smith, V. H., 1985, Predictive models for the biomass of blue-green algae in lakes. *Water Resources Bull.,* **21**(3), 433–439.

Somlyody, L., and Koncsos, L. 1991, Influence of sediment resuspension on the light conditions and algal growth in Lake Balaton. *Ecol. Modelling,* **57,** 173–192.

Somlyody, L., and van Straaten, G., eds., 1986, *Modeling and Managing Shallow Lake Eutrophication,* Springer-Verlag, New York.

Somlyody, L., and Varis, O., 1994, *Modeling the Quality of Rivers and Lakes,* IIASA, Laxenburg, Austria.

Somlyody, L., and Wets, R. J.-B., 1988, Stochastic optimization models for lake eutrophication management. *Oper. Res.,* **36**(5).

Straškraba, M., and Gnauck, A. H., 1985, *Freshwater Ecosystems: Modeling and Simulation,* Elsevier, Amsterdam.

Streeter, H. W., and Phelps, E. B., 1925, *A Study of the Pollution of Natural Purification of the Ohio River,* U.S. Public Health Service, Bulletin 146, Washington, D.C. (reprinted 1958).

Taylor, G. I., 1921, Diffusion by continuous movements. *Proc. London Math. Soc. A,* **20,** 196–211.

Taylor, G. I., 1954, The dispersion of matter in turbulent flow through a pipe. *Proc. Roy. Soc. London,* **A223,** 446–468.

Thomann, R. V., 1963, Mathematical model for dissolved oxygen. *Proc. ASCE, J. Sanit. Eng. Div.,* **89**(SA5).

Thomann, R. V., 1982, Verification of water quality models. *J. Environ. Eng. Div. ASCE,* **108**(EES), 923–940.

Thomann, R. V., 1987, Systems analysis in water quality management—a 25 year retrospect, in Beck, M. B., ed., *Systems Analysis in Water Quality Management,* International Association of Water Pollution Research and Control.

Thomann, R. V., and Mueller, J. A., 1987, *Principles of Surface Water Quality Modeling and Control,* Harper & Row, New York.

Thomann, R. V., O'Connor, D. J., and DiToro, D. M., 1976, *Mathematical Modeling of Natural Systems,* Technical Report, Environmental Engineering and Science Program, Manhattan College, New York.

Troubounis, G., 1985, Strukturorientierte simulation des kohlenstoff-, stickstoff-, phosphor- und sauerstoffhaushaltes flacher gewaesser, dissertation, Institut fuer Siedlungswasserwirtschaft der Universitaet Karlsruhe, Karlsruhe, Germany.

USEPA, 1987 (July), *Selection Criteria for Mathematical Models Used in Exposure Assessments: Surface Water Models,* Report no. EPA/600/8-87/041, Office of Health and Environmental Assessment, U.S. Environmental Protection Agency, Washington, D.C.

Valentine, E. M., and Wood, I. R., 1977, Longitudinal dispersion with dead zones. *Proc. ASCE, J. Hydraulics Div.,* **103,** 975–990.

Varis, O., 1991, Computational modeling of the environment with applications on lake eutrophication, thesis for Doctor of Technology, Laboratory of Hydrology and Water Resources Management, Helsinki University of Technology, Helsinki, Finland.

Varis, O., Kettunen, J., and Sirviö, H., 1990, Bayesian influence diagram approach to complex environmental management including observational design. *Comput. Stat. Data Anal.,* **9**, 77–91.

Vollenweider, R. A., 1975, Input-output models with special references to the phosphorus loading concept in limnology. *Schweiz. Z. Hydrol.,* **37**(1), 53–84.

von Winterfeld, D., and Edwards, W., 1986, *Decision Analysis and Behavioral Research,* Cambridge University Press, Cambridge, United Kingdom.

Wallis, S. G., Guymer, I., and Bilgi, A., 1989a, A practical engineering approach to modeling longitudinal dispersion, *Hydraulic and Environmental Modeling of Coastal, Estuarine and River Waters,* vol. 1, *Coastal Waters, Hydrodynamics, Mathematical Models,* in Falconer, R. A., Goodwin, R. G. S., and Matthew, eds., University of Bradford, Computational Hydraulics and Environmental Modeling Research Group.

Wallis, S. G., Young, P., and Beven, K., 1989b, Experimental investigation of the aggregated dead zone model for longitudinal solute transport in stream channels. *Proc. Inst. Civ. Eng.,* **87**(pt. 2), 1–22.

WES, 1986a, *CE-QUAL-R1: A Numerical One-Dimensional Model of Water Quality: User's Manual,* Instruction Report E-82-1, Waterways Experiment Station, U.S. Army Corps of Engineers, Vicksburg, Miss.

WES, 1986b, *CE-QUAL-W2: A Numerical Two-Dimensional, Laterally-Averaged Model for Hydrodynamics and Water Quality: User's Manual,* Instruction Report E-86-5, Waterways Experiment Station, U.S. Army Corps of Engineers, Vicksburg, Miss.

Wright, J. R., and Wang, J. Y., 1991, Emerging technologies for water resources modeling: new tools and challenges for the classroom. In Yu, S. L., and Shih, K. K., eds., *Proceedings of the International Conference on Computer Applications in Water Resources,* July 3–6, 1991, vol. 2, pp. 495–502, Tamkang University, Tamsui, Republic of China.

Young, P.C., 1974, A recursive approach to time-series analysis. *Bull. Inst. Math. Appl.,* **10**, 209–224.

Young, P.C., 1979, Self-adaptive Kalman filter. *Electron. Lett.,* **15**, 358–360.

Young, P., and Costa, J. R., 1986, ADZ water quality model—a new versatile approach for transport and dispersion modeling for the Ave River. *Proceedings of the Workshop in Methodologies for Water Quality Assessment, Planning and Management,* Laboratório Nacional de Engenharia Civil, Lisbon, Portugal.

Chapter

8

Eutrophication

Mitsuru Sakamoto
Professor
School of Environmental Science
University of Shiga Prefecture
Hikone, Japan

Introduction

Because of the pressing need for freshwater for human society and development, sound management of freshwater resources is an inevitable task. Despite the importance of freshwater resources, the impacts of extending human activities on water bodies during this century have resulted in considerable deterioration of water quality and impairment of daily water uses in many freshwater bodies. Accelerated eutrophication, one example of the anthropogenic deterioration of water quality, has become a worldwide problem in a considerable number of lakes, reservoirs, rivers, and coastal waters in many countries (OECD, 1982; Ryding and Rast, 1989; Vollenweider, 1989).

Until the early 1960s, eutrophication was an unfamiliar phenomenon to the public. In the mid-1960s, people in highly populated and industrialized countries became aware of considerable changes in water quality such as changing color and increasing turbidity due to anthropogenic eutrophication. To understand the causes and processes of these environmental changes and to find effective remediation measures, intensive examinations and research have been conducted. The International Symposia on eutrophication at Madison, Wisconsin, in 1965 (National Academy of Sciences, 1969), at Uppsala in 1968

(OECD, 1970), and in Michigan (ASLO, 1972), as well as the presentation of an overview report on the scientific fundamentals of the eutrophication of lakes and flowing waters with particular references to nitrogen and phosphorus (Vollenweider, 1968), provided a valuable basis and stimulus for subsequent extensive research and enactment of effective countermeasures to solve eutrophication problems. The objective of the present chapter is to review the major developments in eutrophication studies since the 1970s and to illustrate both the important problems and urgent tasks for studies on eutrophication management.

The process of eutrophication involves changes in chemical and biotic components and their complex interactions within a lake under the influence of physical environments such as temperature and vertical mixing of water mass. The eutrophication process is derived mostly from impacts outside the lake, especially nutrient inputs. To reveal the causes and processes of eutrophication and provide reliable measures for management of lakes, a deep understanding of the limnological and ecological processes of eutrophication in lakes and their watersheds is necessary. In this chapter, therefore, the limnological and ecological processes of the major problems of eutrophication and rehabilitation are described in detail.

Specific Features of Eutrophication Studies

Most of the eutrophication research is characterized by a holistic approach (Rigler, 1973). Because of the diversity and variability of components and their interactions within a lake ecosystem, generally it is not easy to define cause-and-effect relationships between any changes within lakes. Thus, investigators who conduct research on eutrophication attempt to consider a lake ecosystem as a black box to obtain definitive information from both the responses of ecosystem components to given environmental impacts and the dynamics of these components. The concept of phosphorus loading and the input/output model by Vollenweider (1969) represent a holistic approach to eutrophication research.

For this feature of lake eutrophication research, a number of excellent studies have been conducted involving whole-lake fertilization experimentation in selected lakes (Schindler et al., 1971, 1973; Schindler, 1974, 1977; Schindler and Fee, 1974; Lundgren, 1978; Jansson, 1978; Hillbricht-Ilkowska and Zdanowski, 1983) and by mass-balance examination of limiting nutrients (Schindler and Nighswander, 1970; Møller Andersen, 1974; Jansson, 1978; Jensen et al., 1990) as well as by correlation analysis of eutrophication-related variables in many lakes and throughout many seasons (Sakamoto,

1966; Vollenweider, 1968, 1976; Dillon and Rigler, 1975; Oglesby, 1976; Pace, 1986). Worldwide and nationwide surveys on trophic state and its relationship to nutrient loading in many lakes have clarified the limnological aspects eutrophication in relation to nutrient loading and lake morphometry, and have helped to establish reliable guidelines for predicting trophic state (Clasen and Bernhardt, 1979; Nordfiorsk, 1980; Fricker, 1980; OECD, 1982; Lee and Jones, 1984; Salas and Martino, 1991). These studies identified phosphorus as the nutrient most capable of controlling eutrophication and also suggested a possible level of phosphorus reduction for eutrophication control. Actually, reduced phosphorus inputs into lakes by sewage diversion and other techniques were successful in recovering water quality in some lakes and reservoirs (Ahlgren, 1978; Edmondson, 1972a, 1972b; Edmondson and Lehman, 1981; Baricca, 1989, 1990). However, some trials for recovery of water quality by sewage diversion or phosphorus elimination from effluents and sewage (Faafeng and Nilssen, 1981; Varis, 1990; van Liere et al., 1990; Boers et al., 1993) have been less successful. The continued supply of nutrients from lake-bottom sediments and uncontrolled diffused sources was proposed to account for these less effective responses (Larsen et al., 1961; Cullen and Fiorsberg, 1988; Boers et al., 1993).

It must be noted that in these studies, changes in both the biotic community and water quality in relation to nutrient supply were analyzed under the assumption of a steady state for each variable during a given period of time. This is not the case in natural lakes, where biotic community and environments display pronounced short-term changes with mutual interactions in each lake (Ryding and Fiorsberg, 1980). Mathematical modeling partially enables us to make a simulation analysis of the short-term changes of ecosystem components in a lake (Los et al., 1984; Los and Brinkman, 1988; Jørgensen, 1989). However, there have often been significant discrepancies between the observed changes and the calculated values by these models, probably due to limited knowledge of both modeling and characteristics of the components of their interactions.

One characteristic change associated with eutrophication of lakes is the alteration in phytoplankton community structure, especially domination of eutrophic species of cyanobacteria (blue-green algae). As for the changes in algal populations in relation to changed nutrient supply, much information has been presented along with background information on the behaviors and competition among component algal species (Tilman, 1977; Sommer, 1981; Sommer et al., 1986; Reynolds, 1984). Higher concentration and/or continued internal loading of phosphorus even after reduced point-source loading could account for continued development of cyanobacterial blooms (Fiaafeng and

Nilssen, 1981; Varis, 1990; Klein and Chorus, 1991; Boers et al., 1993). For reliable management of lake environments, further intensive limnological and ecological studies on the response of ecosystems are needed with more attention to changes of component organisms and their interactions within an ecosystem.

Trophic State and Eutrophication

Definition

There seems to be no agreement on the definition of the trophic state among societies and individuals (Rast and Kerekes, 1981). Although this is due partly to individual differences in the understanding of natural-lake environments, the major reason seems to be the fact that the trophic-state concept has been qualitatively defined to denote the state of nourishment of lakes for plant growth. The terms *eutrophic* and *oligotrophic* were introduced into limnology from bog science by Naumann (1919) and employed to classify lakes according to the relationship between plant production and nutrients. In his research on Eifelmaare lakes, Thienemann (1913–1915) categorized shallow fertile lakes in the lowland along the Baltic Sea as Baltic type and deep infertile lakes on the mountain region as alpine type and later (1921, 1925) equated them to eutrophic and oligotrophic types after the nomenclature of Naumann (1919). In this study of the trophic type (Table 8.1), Thienemann (1925, 1926) focused on the close relationships between epilimnetic plant production and nutrients, hypolimnetic oxygen, benthic fauna (benthos), and lake depth, and presented the idea that nutrients could affect epilimnetic plant production and result in the control of both bypolimnetic oxygen content through decomposition or organic matter and bottom fauna with a species-specific oxygen requirement. This synthetic understanding of a biotic community–environment system in a lake as a whole contributed to the later applications of the concept of trophic dynamics (Lindemann, 1942).

In nature, a lake is always receiving a supply of silt and nutrients from the watershed, and becomes shallow and more fertile. Increasing plant production by nutrient enrichment and increasing sedimentation of undecomposed plant detrutues and silt on the lake bottom accelerate lake-basin shoaling. Thus, a lake evolves to a more eutrophic state. Natural eutrophication takes place at a very low rate and takes a long period of time, sometimes several hundred or thousand years. However, the recent increase in human activities on the watershed has necessitated the increased supply of plant nutrients into lakes and resulted in the enhanced growth of phytoplankton and macrophytes. These biological changes in lakes caused by nutrient enrichment have resulted in changes in other organisms and water

TABLE 8.1 Characteristics of Two Lake Types*

	Oligotrophic	Eutrophic
Distribution	Alpine and foothills	Baltic lowlands and Alps
Morphpometry	Deep, narrow littoral	Shallow, broad littoral
Hypolimnion and epilimnion	Large	Small
Water color	Blue-green	Brown-green, Green-yellow
Transparency	Great	Small, or very small
Water chemistry	Poor in N and P; humic material absent; Ca viable	Rich in N and P; humic material slight; Ca usually high
Suspended matter	Minimal	Rich, planktonic
Bottom mud	Nonsaprobic	Saprobic (gyttja)
O_2 in summer	60–70% minimum decrease with depth	0–40% minimum Decrease sharply in metalimnion
Plant production in littoral	Low	Rich
Plankton	Low in quantity, presence at all depths, large diurnal migration, seldom blooms	Large quantity, diurnal migration, often blooms
Benthos	*Tanytarsus* fauna; no *Chaoborus*	*Chironomus* fauna; *Chaoborus* usually present
Succession	To eutrophy	To pond or meadow

*Dystrophic type excluded.
SOURCE: Modified from Thienemann (1925).

quality, such as decreased hypolimnetic oxygen content, and have greatly affected the normal use of natural water bodies by human society. Table 8.2 shows representative impairment of water use due to accelerated eutrophication. It must be noted that eutrophication is not the problem; the resultant changes in aquatic environments pose several serious environmental problems. However, this does not mean that eutrophication is totally undesirable. In certain environments, eutrophication of oligotrophic waters could be desirable for increasing production of fishery resources; even whole-lake fertilization has been conducted to enhance fish production (Stockner, 1981; Johannessen et al., 1984; Stockner and Shortreed, 1985).

Criteria and indicators of trophic state

Quantitative criteria. As described in the previous section, the terms *eutrophic* and *oligotrophic* quantitatively denote the level of lake fertility, and there has been no clear delineation between two trophic types. For scientific necessity in limnological studies of lakes, several attempts have been made to delineate these trophic types. For example, Sakamoto (1966), using the lake-classification system of Yoshimura (1935), reported the ranges of total nitrogen (TN) and total

TABLE 8.2 Initial Biological Changes and Resulting Impairments of Water Uses Associated with Accelerated Eutrophication of Lakes and Reservoirs

Water-use impairment	Algal blooms: changes in species composition	Excess growth of macrophytes and littoral algae
Water-quality impairment		
Taste, odor, color, filtration, flocculation, sedimentation, and other difficulties in treatment for drinking water supply	Very frequent	At times
Hypolimnetic oxygen depletion: iron, manganese, CO_2, NH_4, H_2S, CH_4, etc., formation	Frequent	At times
Recreational impairment		
Unsightliness	Frequent	At times
Odor	Frequent	At times
Unpleasant to bathers	Frequent	Frequent
Increased health hazards	At times	At times
Boat travel and sailing impairment	At times	Frequent

SOURCE: Modified from Vollenweider (1989).

phosphorus (TP) of Japanese lake waters in spring (1.0 to 1.2 mg of N and 0.013 to 0.078 mg of P per liter in eutrophic lakes, 0.12 to 0.58 mg of N and 0.008 to 0.031 mg of P per liter in mesotrophic lakes, and 0.15 to 0.22 mg of N and 0.003 to 0.017 mg P per liter in oligotrophic lakes). Similar delineations were presented by USEPA (1975) and Carlson (1977) on North American lakes and by Forsberg and Ryding (1980) for Swedish lakes. There was a considerable discrepancy in the delineation values among researchers (Forsberg and Ryding, 1980). This arose from individual differences in opinion on the trophic state of the lakes studied because it was evaluated as a whole on the basis of information on several environmental variables and biotic commnunities in a lake. For this reason, some overlap of borderlines between trophic types is inevitable (Ryding and Rast, 1989).

OECD (1982) proposed two systems to quantitatively define trophic categories: one with fixed boundaries and the other with open-boundary values (Tables 8.3 and 8.4). The fixed-boundary system was based on the best judgment as to the transition between two categories based on OECD data. Considering the diversity of geographic, ecological, and sociological environments related to lake eutrophication, the validity of arbitrary or rigid categorization seems uncertain. Total information rather than limited information on a few parameters must be the basis for judgment of thetrophic state (OECD, 1982). The fixed-boundary system has an advantage in its easy application for environmental management with limited limnological background

TABLE 8.3 Classification of Trophic Types by Fixed Boundary Values

Trophic category	TP	Mean Chl	Maximum Chl	Mean Secchi	Minimum Secchi
Ultraoligotrophic	<4.0	<1.0	<2.5	>12.0	>6.0
Oligotrophic	<10.0	<2.5	<8.0	>6.0	>3.0
Mesotrophic	10–35	2.5–8	8–25	6–3	3–1.5
Eutrophic	35–100	8–25	25–75	3–1.5	1.5–0.7
Hypertrophic	>100	>25	>75	<1.5	<0.7

Explanation of terms: TP = mean annual in-lake total phosphorus concentration, μg/L; mean chlorophyll = mean annual chlorophyll *a* concentration in surface waters (μg/L); maximum chlorophyll = peak annual chlorophyll *a* concentration in surface waters (μg/L); mean Secchi = mean annual Secchi depth transparency, m; minimum Secchi = minimum annual Secchi depth transparency, m.

SOURCE: OECD (1982).

information. The open-boundary system has an advantage in flexible application. In the OECD system, boundary values based on statistical analysis of the OECD data pool are presented in association with mean values and ± 1 and ± 2 standard deviations. Such an open-boundary system developed with probabilistic aspects has a great advantage for quantitative categorization with less risk of overlap in the classification of trophic types.

Even if the OECD system is of wider applicability, the concepts of eutrophic and oligotrophic are unfamiliar to the decision maker and environmental technologist. To reduce uncertainty in communicating to the public on the trophic state of lakes with the traditional trophic classification concepts, Carlson (1977) developed a numerical *trophic-state index* (TSI) system with a scale of 0 to 100 from any data of chlorophyll *a* (Chl *a*) concentration, total TP concentration, and Secchi disk transparency. Because of the simple and objective evaluation method, Carlson's index has been widely employed in many studies to assess the status of lakes in relation to eutrophication (e.g., Cooke, 1981), although the index values are affected by other substances that attenuate underwater light (Lorenzen, 1980).

For delicate management of lake environments, other measures on the trophic state of lakes might be needed in some cases. Hillbricht-Ilkowska (1985) described two assessment systems of the lake trophic state which are employed in Poland and applicable to temperate lowland lakes dominating in middle-eastern Europe. Under the *lake-quality evaluation system,* lakes are classified into three categories and monitored for 12 parameters, including TP, Chl *a*, and Secchi disk transparency. Under the *lake ecosystem monitoring system,* lakes are divided into two groups according to the mixing regime of lake water, dimictic lakes, and polymictic lakes. Each group is divided into three categories for which water quality is monitored for TP, Chl *a*,

TABLE 8.4 Classification of Trophic State* in the OECD Eutrophication Program

Variable		Annual mean values†			
		Oligotrophic	Mesotrophic	Eutrophic	Hypertrophic
Total phosphorus, mg/m^3	$\bar{x}$	8.0	26.7	84.4	
	$\bar{x} \pm 1$ SD	4.85–13.3	14.5–49	48–189	
	$\bar{x} \pm 2$ SD	2.9–22.1	7.9–90.8	16.8–424	
	Range	3.0–17.7	10.9–95.6	16.2–386	750–1200
	n	21	19 (21)	71 (72)	2
Total nitrogen, mg/m^3	$\bar{x}$	661	753	1875	
	$\bar{x} \pm 1$ SD	371–1180	485–1170	861–4081	
	$\bar{x} \pm 2$ SD	208–2103	313–1816	395–8913	
	Range	307–1630	361–1387	393–6100	
	n	11	8	37 (38)	
Chlorophyll a, mg/m^3	$\bar{x}$	1.7	4.7	14.3	
	$\bar{x} \pm 1$ SD	0.8–3.4	3.0–7.4	6.7–31	
	$\bar{x} \pm 2$ SD	0.4–7.1	1.9–11.6	3.1–66	
	Range	0.3–4.5	3.0–11	2.7–78	100–150
	n	22	16 (17)	70 (72)	2
Chlorophyll a peak value, mg/m^3	$\bar{x}$	4.2	16.1	42.6	
	$\bar{x} \pm 1$ SD	2.6–7.6	8.9–29	16.9–107	
	$\bar{x} \pm 2$ SD	1.5–13	4.9–52.5	6.7–270	
	Range	1.3–10.6	4.9–49.5	9.5–275	
	n	16	12	46	
Secchi depth, m	$\bar{x}$	9.9	4.2	2.45	
	$\bar{x} \pm 1$ SD	5.9–16.5	2.4–7.4	1.5–4.0	
	$\bar{x} \pm 2$ SD	3.6–27.5	1.4–13	0.9–6.7	
	Range	5.4–28.3	1.5–8.1	0.8–7.0	0.4–0.5
	n	13	20	70 (72)	2

*Trophic state based on the opinion of the investigators of each lake. The geometric mean based on log 10 transformation was calculated after removing values or 2 SD where applicable in the first calculation.

†*Key:* $\bar{x}$ = geometric mean; SD = standard deviation; values in parentheses refer to the number of variables (n) employed in first calculation.

SOURCE: Rast and Kerekes (1981).

and Secchi disk transparency, and, if necessary, for abundance and composition of plants and animals. Although not simple, this monitoring system might be of benefit to provide actual knowledge on responses of the ecosystem and biota to trophic stress.

In most of the trophic-state classification system, static parameters such as TP concentration were employed for the criteria of assessment. Considering the dynamic characteristics of a lake ecosystem, it is logical to employ dynamic, metabolic parameters of a lake ecosystem as the criteria. In view of the trophy which involves the concept of organic-matter supply rate, Rodhe (1969) described the trophic state of lakes as a function of primary production during the vegetation period. In the OECD monitoring program on eutrophication, annual rates of primary production were measured in many monitored lakes, and a significant correlation was found between annual production rates and the annual average concentrations or annual loading rates of TP (OECD, 1982). While primary production rate actually provides valuable information on the dynamic status of the ecosystem, the routine measurement of this rate is not easy in terms of instrumentation and is time-consuming. For this reason, the dynamic measure of primary production may be represented by TP loading rates or TP level of lake water because phosphorus was identified as the dominant limiting factor of primary production.

In shallow eutrophic lakes, generally dense macrophyte vegetation is found together with filamentous algae and periphyton in littoral areas. Aquatic macrophytes like *Potamogeton* have roots which reach down into the bottom mud to absorb nutrients into the plant bodies. For settlement of rooted macrophytes, the development of muddy bottom is necessary at the shallow littoral region by sedimentation of silts and plant detritus. In a process of lake aging, a lake becomes more shallow by deposition of silt and plant detritus to provide a basis for the development of rooted aquatic plant vegetation. Actually, the more rich vegetation of submerged macrophytes is found on the more silted lakes and the more silted bottom sands (Pearsall, 1920). The abundance of rooted aquatic plants is generally decreased with increasing density of phytoplankton, especially in hypertrophic lakes (Carpenter, 1982; Best et al., 1984). Although further studies are needed on the reason for reduced abundance of macrophytes, this fact is of great importance for management of eutrophication because of the dominant occurrence of macrophytes in shallow eutrophic lakes. Macrophytes often account for more than 50 percent of plant biomass in lakes (Wetzel, 1975; Westlake, 1980; Howard-Williams and Allanson, 1981).

On the basis of overall surveys of Chinese lakes, Liu (1990) classified lakes into two types: a *responding* type, in which plant production increases with nutrient concentration and a *nonresponding* type, in which plant production remains steady irrespective of enrichment or reduction of nutrients. In China, most lakes are of the responding type. Responding-type lakes are further classified into phytoplankton responding type and macrophyte responding type. The phytoplankton responding-type lakes are distributed throughout China, and their environmental changes due to eutrophication have been under intensive studies. Macrophyte-responding-type lakes are shallow, and macrophyte occupies a dominant position in plant vegetation, and phytoplankton is at an inferior position in growth competition with phytoplankton. Among 24 lakes studied in the eutrophication survey, only two were of the macrophyte responding type and 73 percent were of the phytoplankton responding type (Liu, 1990). There is insufficient accumulation of information on the macrophyte-responding-type lakes.

Biological indicators. There is numerous information on the changes of aquatic communities in association with eutrophication. If the changes and occurrence of some species or some group of organisms are identified as specific through a given trophic state, we can use them as a biological indicator of its trophic state. Pearsall (1932) was the first to note a close relation between the water chemistry of lakes and the relative dominance of a distinctive phytoplankton group. On the basis of the fact that species composition of phytoplankton differs between unproductive and productive lakes, Thunmark (1945) and Nygaard (1949) proposed phytoplankton quotients, the relative ratio of species numbers of two algal groups, as indicators of the trophic state. However, subsequent studies have suggested the limited applicability of phytoplankton quotients because of their less significant correlation with the trophic state (Brook, 1965). Because of long-term preservation of silicious frustules in bottom sediments, diatoms have often been employed as indicators of previous lake environments. Following analysis of diatom fossil distribution in lake sediment cores along with a survey of both the distribution and changes of diatom species or groups in lake waters with increasing eutrophication, Stockner (1971, 1972) proposed the *ratio of Araphinidineae to Centrales* (the A/C index) by which lakes were categorized as either oligotrophic (A/C = 0 to 1.0), mesotrophic (A/C = 1.0 to 2.0), or eutrophic (A/C $>$ 2.0). This index has been successfully employed to assess the past trophic-state changes of some lakes (Brungum, 1979), while further examinations indicated the limited applicability of this index to waters other than low-alkalinity lakes (Brungum and Patterson, 1983). Recently, a weight-averaging calibration model on diatomic data has been developed and, in conjunction with application of the transfer function to the chemical data, provided new insight into historical analysis of previous nutrient levels of lakes

(Hall and Smol, 1992; Anderson et al., 1993). Another diatom-based model relating to the previous trophic state of lakes was developed for Florida lakes (Whitmore, 1989).

There are numerous reports on the changes of zooplankton community structure in association with the changing trophic state of lakes. Generally, with increasing lake productivity zooplankton exhibits increase of total biomass and replacement of species within Cladocera and Copepoda (Hall et al., 1970). In eutrophic waters, small cladocerans, cyclopoid copepods, rotifers, and ciliates predominate, while macrozooplankton becomes less abundant (Brooks, 1969; Bays and Crisman, 1982). Observed changes of zooplankton with increasing eutrophication seem to be caused mostly by changes in food relationships. Beaver and Crisman (1990) suggested that bacterivorous microzooplankton might be a sensitive indicator of trophic-state change because of their concurrent increase with bacteria which shows quick response to increasing phosphorus load. Many eutrophic lakes are characterized by a high density of rotifers. This seems to be related to increased organic-matter supply and associated changes in food chains due to increased eutrophication.

As Thienemann (1925) observed, some distinctive zoobenthos species can reflect the trophic state of lakes. This is caused mainly by a different response of each species to changed hypolimnetic dissolved-oxygen (DO) concentration with increasing eutrophication. Decreased DO initially causes growth retardation of zoobenthos, and a further decrease in DO results in their negative growth (Jónasson, 1984). Difference in the tolerance to decreased DO was found among species of zoobenthos (Jónasson, 1969). With increasing eutrophication due to sewage inflow in Lake Estrom, there was a marked decrease in the density of the midge *Chironomus anthracinus* and the mussel *Pisidium* species, while the tubificid *Potamothrix hammoneiensis* remained unaffected (Jónasson, 1984). Oligochaete and chironomid larva communities can exhibit a marked change of dominant species with increasing eutrophication (Lang, 1984; Lods-Crozet and Lachavanne, 1994). A significant negative correlation was found between the TP concentration of lake water and the relative abundance of oligotrophic species of oligochaete (Lang, 1990). Lang and Reymond (1992, 1993) found an increased relative abundance of oligotrophic species associated with decline of the TP level of lake water in two lakes.

Many factors are involved in the changes of distinctive species and community structure of aquatic organisms. However, the dominance of a species or group in a given lake may not always represent a trophic state of the lake in which its organism was dominant. Many studies on lake remediation suggested delayed responses of biotic communities to nutrient reduction (Willén, 1975; Millard and

Johnson, 1986). It is recommended that assessment of the lake trophic state using biotic indictors be supplemented with that by chemical and biochemical indices which can exhibit a sensitive response to changing nutrient load.

Factors involved in the control of eutrophication

Watershed and lake morphometry as integral factors of lake trophy control. Eutrophication caused by enrichment of water bodies results in enhanced plant growth and subsequent changes of other organisms and water quality through food chains and nutrient cycling. Many publications have indicated that to control eutrophication, we must focus on the key agent causing these changes, namely, the limiting nutrients to enhance the growth of algae and macrophytes.

In general, natural waters are characterized by a limited supply of nitrogen (N) and phosphorus (P) for plant growth. This situation is especially true in deep oligotrophic lakes located on drainage basins underlaid with less-weathered rocks as in the Archean plateau of northern Sweden (Ahl, 1972), the Precambrian Shield of Canada (Gray et al., 1990), and the Andes Mountains (Campos, 1984). On the other hand, lakes and rivers in the fluvial lowlands such as eastern European lowland lakes (Straškrabá and Straškrabová, 1969) are eutrophic primarily due to edaphic control by the substratum of drainage basins. In addition, fluvial lowland is generally characterized by a high percentage of arable land and higher population density, suggesting great anthropogenic influences on lowland lakes. A significant linear relationship between algal biomass and the percentage of arable area in the English Lake District (Gorham et al., 1974) suggests that environmental factors favoring agricultural production lead to high algal production in these lakes. There were a number of reports on increased nitrogen and phosphorus loads from point sources (rural and urban areas with higher population densities and several industries) and resulted in the accelerated eutrophication of lakes. The increased nutrient load from these sources on the watershed has been a major cause of accelerated eutrophication. Attempts to reduce nutrient from sewage flowing into lakes were often unsuccessful in recovering the lake condition due to continued nutrient loads from agricultural farmlands and other nonpoint sources on the watershed as well as internal loads from the bottom sediments on lakes, suggesting the necessity of overall environmental actions through the watershed and lake to reduce nutrient discharge.

After nutrients are supplied in lakes, several factors can affect nutrient availability by algal and primary productivity. Since

Thienemann's studies (1925), the morphometry of lakes, especially mean depth, has been considered as an important factor in controlling lake productivity. Rawson (1939, 1955) found a significant inverse relationship between mean depth and the biomass of plankton and fish in many Canadian lakes. A similar inverse correlation of lake mean depth to in-lake trophic parameters was found for total chlorophyll (approx. Chl *a* to *b*), TN, and TP concentrations (Sakamoto, 1966a, 1966b) (Fig. 8.1) and for primary production in ELA [Experimental Lake Area (Canada)] lakes (Fee, 1979) and Amazonian tropical lakes (Miller et al., 1984). This is due primarily to higher efficiency of nutrient availability to biological production in shallower lakes. Light might often be a factor involved in morphometric control of nutrient availability and productivity. In deeper lakes, if lake water is vertically mixed down to deeper than the critical depth for photosynthesis, the photosynthetic growth of phytoplankton population is light-limited (Talling, 1971). In some lakes, nonliving suspended matter (silts) and colored substances supplied from watershed as well as high algal density caused decreased light penetration in lake water and resulted in light-limited algal growth even in shallow lakes (Smith, 1986; Yin et al., 1994).

Geographic location represented by latitude and altitude is a rough determinant of climate and solar-energy supply, which greatly affect algal production (Brylinsky and Mann, 1973). Latitudinal and altitudinal differences are associated with the different seasonality of thermal stratification of lake waters and the seasonality of algal productivity and population dynamics (Rawson, 1939). The process of eutrophication might be affected by such geographic differences in environmental conditions. These complex relationships between many factors affecting the trophic nature and algal productivity through watershed-lake ecosystems are illustrated in Fig. 8.2.

Causative nutrients of accelerated eutrophication. As for the nutrients causing eutrophication, a considerable number of studies have been conducted. There are four approaches to the identification of causative nutrients through the studies.

The *first,* widely employed, method involved comparative examination of temporal changes of algae and nutrient concentrations in natural waters. This type of study has generally been conducted together with other approaches and helped to identify possible causative nutrients in many surveys.

The *second* method involved examination of chemical composition and physiological activities of algae. On the basis of data on the chemical composition of algae cultured in nutrient-limited media, Healey (1978) and Healy and Hendzel (1979, 1980) proposed the boundary

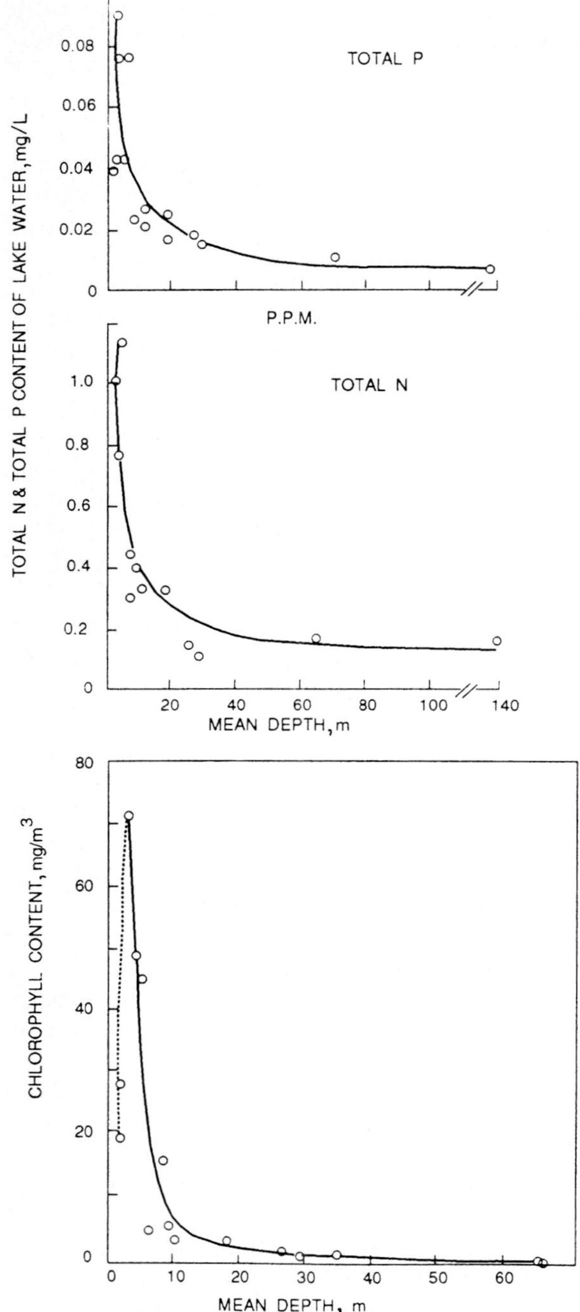

Figure 8.1 Total P, total N, and chlorophyll concentrations as a function of mean depth of Japanese lakes in spring (Sakamoto, 1966b).

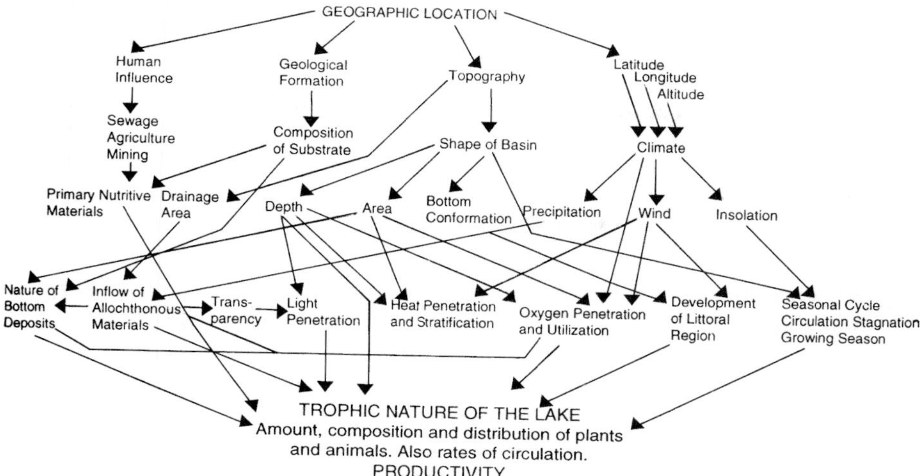

Figure 8.2 Interrelationships between factors affecting the trophic nature of lakes [Rawson (1939); cited from Ryding and Rast (1989)].

values of carbon, nitrogen, and phosphorus contents of algal cells to assess the nutrient status of algae in lakes. However, because of contamination by detritus and other nonliving materials in plankton samples, there is a possibility that carbon, nitrogen, and phosphorus data of crude natural plankton samples might give a false idea of the nutrient status of natural phytoplankton. To overcome this problem, several methods have been proposed to measure metabolic activities which are specific to living cells and display significant changes in the nutrient-deficient condition. One of them involves measurement of hot-water extractable cellular surplus phosphorus which can be accumulated in excess in plant cells when there is a rich supply of available phosphorus in the water (Fitzgerald and Nelson, 1966). This method has been successfully employed to assess the state of phosphorus deficiency of natural plankton in lakes (Serruya and Berman, 1975; Eriksson and Pettersson, 1984). Another promising method of identifying the nutrient status of living cells is to measure intracellular enzymatic activities which vary in accordance with conditions of nutrient supply. Measuring alkaline phosphatase activity of plankton samples is the representative method (Fitzgerald and Nelson, 1966; Berman, 1970). Lean and Pick (1981) proposed a P-deficiency index (relative rate of optimal photosynthesis to the maximum P uptake rate) as a promising physiological indicator of P deficiency. Comparative studies on nutrient deficiency of phytoplankton using alkaline phosphatase activity, surplus phosphorus, and P-deficiency index clearly demonstrated considerable temporal changes in P-deficiency of plankton associated with seasonal change of lake-water mixing (Pettersson, 1980, Pettersson et al., 1990; Istvánovics et al., 1992.)

The *third* method for assessing limiting nutrients in controlling algal growth and eutrophication is the nutrient enrichment experiment on natural lake water. The assay is generally conducted by short-term incubation of lake-water samples in small containers with enrichment of possible limiting nutrients. Published enrichment experiments indicated that phosphorus alone or together with nitrogen and/or often with some trace elements such as iron and molybdenum were limited for algal production in many lakes (Goldman, 1960, 1964, 1981; Kalff, 1971; Sakamoto, 1971; Schelske and Staermer, 1972; Powers et al., 1972; Lin and Schelske, 1981; Dodds and Priscu, 1990). In some lakes, nitrogen was demonstrated as a primary limiting nutrient (Goldman and Armstrong, 1969; Goldman, 1981, 1988; Wurtsbaugh et al., 1985). Seasonal change of limiting nutrients was often observed (Lehman and Sandgren, 1978). Nutrient enrichment bioassay is an operational test, and the test result often varied with different test conditions. Comparative studies of short-term and long-term bioassays in an oligotrophic lake indicated that long-term assays resulted in contemporaneous N and P deficiency which was not revealed by short-term bioassay (Dodds and Priscu, 1990). A short-term bioassay may not reflect the effect of nutrient enrichment on the entire community which took place over long incubation periods at long-term assay. Short-term incubation also may represent deficiency of only one element on sensitive populations, while long-term incubation may indicate simultaneous deficiencies (Dodds and Priscu, 1990).

One problem of enrichment bioassay is a partial confinement of more than one or two elements on total population of natural communities in small containers. The longer the samples are incubated, the more the unnatural state is created in containers as a result of unbalanced development of the community at near-natural state over several weeks and months (Hecky and Kilham, 1988). Thus enclosure and whole-lake enrichment experiments made a valuable contribution to reliable information on nutrients limiting algal growth and controlling eutrophication. In particular, whole-lake experiments in the Experimental Lake Area (ELA) in northwestern Ontario, Canada, definitely confirmed phosphorus as a dominant factor in accelerating algal growth and controlling eutrophication (Fig. 8.3) (Schindler et al., 1973; Schindler and Fee, 1974; Schindler, 1974, 1975; Levine and Schindler, 1989). In the late 1960s there was an acrimonious debate on the possibility that carbon (C) rather than P or N might be the critical limiting nutrient in freshwater lakes (ASLO, 1972). Whole-lake fertilization experiments in Lake 227 of the ELA demonstrated that C limitation was seldom observed because of the atmospheric

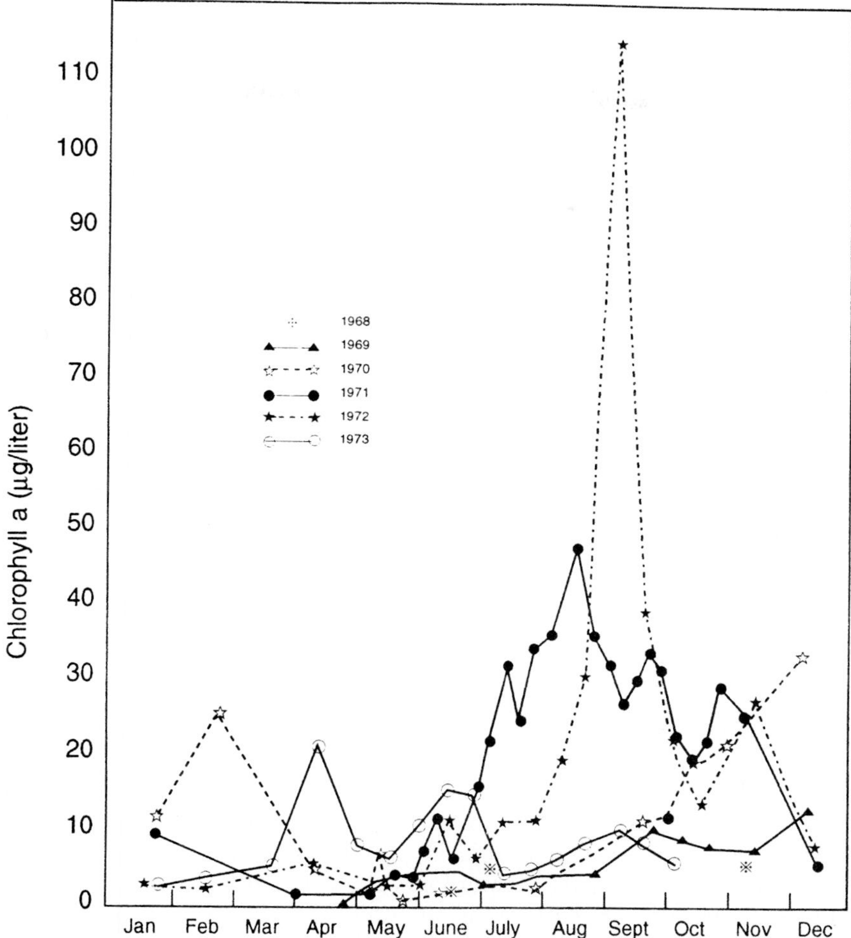

Figure 8.3 Chlorophyll *a* concentrations in Lake 304 of the Experimental Lake Area, Canada. In 1968, 1969, and 1970, the lake was fertilized. In 1971 and 1972, the lake received annual additions of 0.4 g of phosphorus, 5.2 g of nitrogen, and 5.5 g of carbon per square meter. In 1973, additions of nitrogen and carbon were continued at the same rate but phosphorus was not added (Schindler, 1974).

supply of carbon dioxide through the lake surface (Schindler, 1971, 1975). Another notable finding from whole-lake fertilization experiments was that with severe N limitation, nitrogen-fixation cyanobacteria (blue-green algae) were developed to relieve the N deficiency (Schindler, 1977), suggested the possible regulation mechanism of a natural ecosystem in a N-limited condition.

The *fourth* method to assess limiting nutrients involves analysis of the quantitative relationships between nutrient level (TP and TN levels) and algal biomass or algal biomass and/or productivity in natural

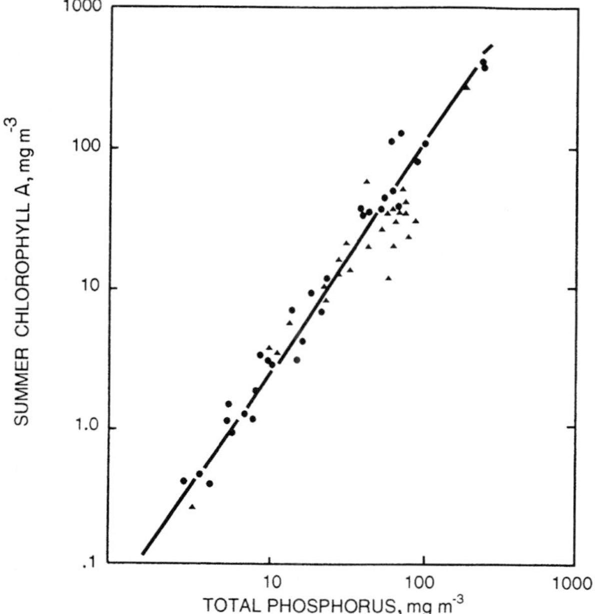

Figure 8.4 Summer-average chlorophyll concentration versus total phosphorus concentration at spring overturn. Circles represent data from Sakamoto (1966a); regression line represents Sakamoto's data (Dillon and Rigler, 1974).

waters (Sakamoto, 1966a; Dillon and Rigler, 1974; Forsberg and Ryding, 1980; OECD, 1982; Smith, 1982). Significant linear correlation between summer Chl *a* and TP in epilimnion or surface waters of temperate lakes in spring through a wide range of trophic states (Fig. 8.4) indicated that phosphorus is the dominant factor limiting phytoplankton growth. The statistically significant relationship between mean annual concentrations of Chl *a* and TP in fertilized and unfertilized ELA lakes (Schindler, 1975; Schindler et al., 1977) supported the critical role of phosphorus in eutrophication control.

The phosphorus limitation of algal growth in natural lakes is due to the shorter supply of phosphorus relative to its requirement for algal growth. There is also much evidence of a shortage of nitrogen supply for algal growth in freshwater lakes (Goldman and Armstrong, 1969; Lehman and Sandgren, 1976; Goldman, 1981, 1988; Wurtsbaugh et al., 1985). On the basis of correlation analysis of Chl to TP and TN concentrations in Japanese lakes, Sakamoto (1966a) concluded that when the TN/TP ratio (weight ratio) of lake waters was <10, Chl yield varied with the TN level but at the TN/TR >17 Chl, phosphorus was the critical nutrient. Identical conclusions were drawn from algal bioassay in Swedish lakes (Forsberg et al., 1978; Forsberg and

Ryding, 1980). Using data in 228 north-latitude lakes, Smith (1982) developed a multiple-regression model of chlorophyll involving TP and TN in terms of an equation. Data analysis using this model demonstrated a highly significant effect of TN on observed Chl values through 127 lakes except in lakes with the TN/TP > 35, suggesting a significant influence of TN on phytoplankton yield even in lakes where phosphorus limited algal yield only moderately. Published evidence indicated decreasing trend of Chl concentration with the increasing TN/TP of lake water through many lakes (Forsberg and Ryding, 1980). Forsberg and Ryding (1980) described that in lakes with Chl concentrations of <20 mg/m^3, the TN/TP ratios were ≥17 and phosphorus was the critical limiting nutrient, but where Chl was ≥70 the TN/TP was below 10 and nitrogen limited algal yield.

It is interesting to remark in connection with the significant relationship between the TN/TP ratios and Chl concentrations that the TN/TP ratios declined with increasing TP level of lake water toward more eutrophic lakes. On analysis of environmental monitoring data in Japanese lakes, Sakamoto (1985) found that the TN/TP ratios ranged from 3:1 to 500:1 in oligotrophic lakes with lower TP concentrations, but with increasing TP level the TN/TP values gradually fell to less than 10:1. An identical pattern of decreasing TN/TP ratios with increasing TP level was reported by Downing and McCauley (1992) (Fig. 8.5) from the examination of data on epilimnetic waters of

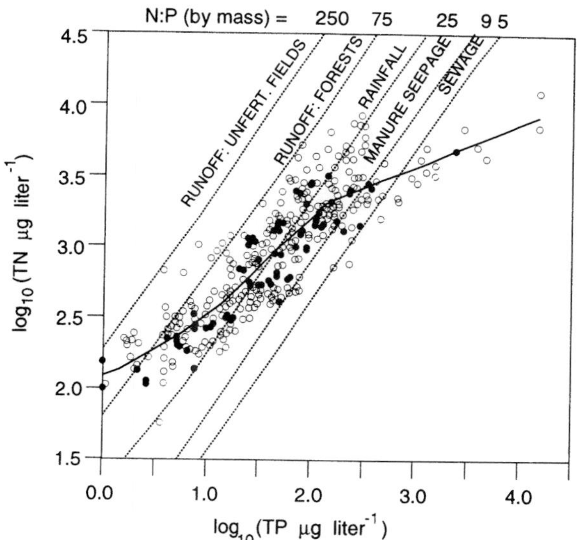

Figure 8.5 Relationships between mean summer TN and TP concentrations (in micrograms per liter) in epilimnetic waters of world lakes. Dashed lines show N/P ratios (Downing and McCauley, 1992).

world lakes. Bottom sediments of eutrophic lakes are usually anaerobic. In such anaerobic environments, nitrogen loss due to denitrification as well as phosphorus return from bottom sediment could be accelerated and result in the decrease in TN/TP ratios of the lake water toward more eutrophic lakes. In the mountain lakes in central Europe, the TN/TP ratios were otherwise considerably affected by TN level of the lake waters (Kopáček et al., 1995).

Causative factors of cyanobacterial bloom. A typical nuisance in eutrophied lakes is massive development of cyanobacterial (blue-green algal) bloom. Schindler (1977) showed that low N/P ratios in fertilized lakes favored the development of nitrogen-fixing cyanobacteria. By compiling published data from 15 natural lakes worldwide, Smith (1983) detected that cyanobacteria tended to dominate at the TN/TP <29 (weight). The N/P ratio alone was not sufficient to explain the dominance of cyanobacteria. On the basis of published data from 15 natural lakes worldwide, Smith (1986) found that at a fixed TN/TP ratio of lake waters the relative biomass of cyanobacteria increased as the availability of light decreased (as estimated from Secchi transparency) and at a fixed light level the relative biomass of cyanobacteria increased as the TN/TP ratio decreased. There were many reports on the high phosphorus requirement by cyanobacteria in pure culture. Lam (1979) indicated that the growth rate of *Anabena oscillarioides* increased proportionally with increasing phosphorus concentrations up to 100 μg/L, beyond which increasing phosphate concentration affected only the final yield. A similar dependency of algal grown on P level was observed in the relative distribution of cyanobacterial biomass as a function of TP level of lake water in Norwegian lakes (Faafeng and Hessen, 1993).

Temperature, light, and physical stability of water mass were also suggested as important factors responsible for the development of cyanobacterial bloom. Hammer (1964) observed considerable influence of water temperature on the time of appearance and the sequence of bloom formation of cyanobacteria in lakes of Saskatchewan, Canada. Most of the intensive bloom of cyanobacteria was observed at or about 20°C, although there was some variation among species. Light-saturated photosynthetic rate of *Microcystis aeruginosa* increased with increasing temperature, but declined considerably below 11°C, and no photosynthesis occurred below 4°C (Takamura et al., 1985). Many planktonic blue-green algae can regulate their buoyancy by formation of gas-vacuolate and photosynthetic products in their cells and colony so as to control their movement in an illuminated water column (Reynolds, 1984). Solar radiation affected the photosynthetic production and buoyancy of algae as well as the thermal stability of the water column, thus controlling algal bloom development. Representative bloom-formation species, *M. aeruginosa,* is easily concentrated at the surface of stratified water column under calm conditions. In studies of *Microcystis* bloom in a hypertrophic reservoir where weak thermal stratification of water column was frequently interrupted by wind-induced mixing, Köhler (1992) found that positive growth of *Microcystis* was limited to the period with thermal stratification and fell to near zero in mixing conditions, although *Microcystis* biomass remained at a high level. These features of *Microcystis* might be responsible for development of the considerable bloom of this species at nutrient-rich conditions in summer.

Eutrophication of World Lakes and Reservoirs

Overall pattern

There are a large number of lakes and reservoirs through different latitudes and altitudes. Because of the differences in climate and geologic characteristics with latitude and altitude, phytoplankton productivity could display considerable variation along with these axes. On the basis of data from a worldwide survey of primary production in lakes during the IBP (International Biological Programme), Brylinsky and Mann (1973) indicated that production was smaller in lakes of high latitude and greater in those of lower latitudes. In high latitudes, solar radiation can be more important as a limitation factor, but in lower latitudes the factors related to nutrient availability could be more critical in determining the level of primary production (Fig. 8.6). In a similar analysis on the effects of latitude, Kalff (1991) detected a significant effect of latitude on the annual phytoplankton productivity and total P of lakes. Schindler (1978) found a significant linear correlation between annual average Chl-*a* and annual average total P from the data on many lakes over 38°S to 75°N. These suggest that nutrients, especially phosphorus, can be a major factor limiting phytoplankton production and their yield, and hence the key factor controlling eutrophication in many lakes.

On the basis of this report, Vollenweider (1989) overviewed the general pattern of eutrophication of inland waters in the world. We discuss the characteristics of lakes worldwide in relation to the trophic state. Table 8.5 shows distribution of yearly phosphorus loading and mean annual chlorophyll concentrations from 335 case studies in OECD cooperative studies. A percentage of >50 (185 of 335 lakes) scattered along the main axis, indicating a close correlation between algal production and phosphorus supply through many lakes in the world. In this same group of lakes, 27 percent (91 lakes) lie below this

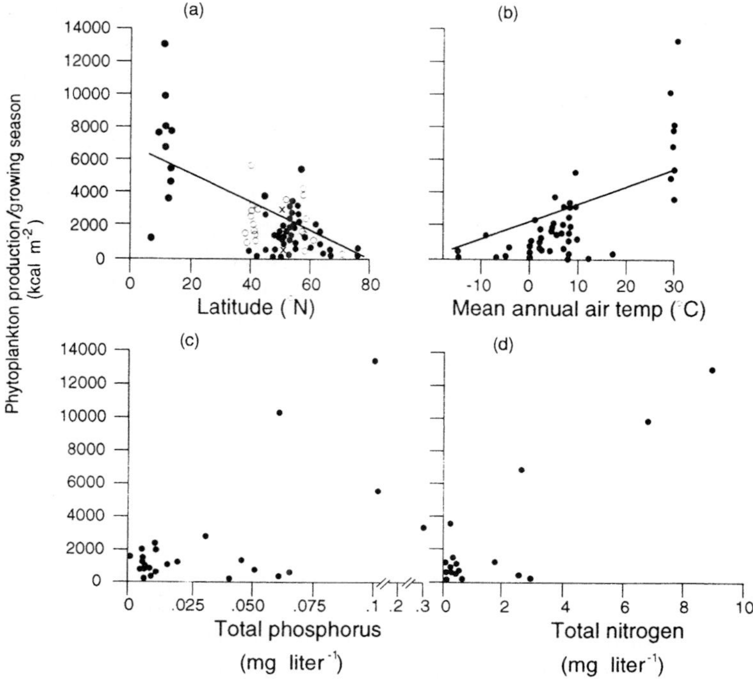

Figure 8.6 Relationships of phytoplankton production to (*a*) latitude (closed circles represent IBP sites; open circles, study sites outside IBP; x marks, Lake Washington); (*b*) mean annual air temperature; (*c*) total phosphorus; (*d*) total nitrogen (Brylinsky and Mann, 1973).

main axis, suggesting that factors other than phosphorus could control production or colimit algal growth. Deficiency of nitrogen might account for the chlorophyll level below the main axis. Light limitation due to turbid water and vertical mixing and grazing pressure might also be responsible for the low chlorophyll level below the main axis.

Table 8.6 illustrates the percentage distribution of oligotrophic, mesotrophic, and eutrophic lakes throughout the world and in different countries, based on data from the OECD Monitoring Survey and nationwide and continentwide surveys (Vollenweider, 1989). Of the lakes in 18 countries in the OECD Monitoring Survey, 65 percent were eutrophic and 18 percent were oligotrophic. Similarly, in the United States, 70 percent were eutrophic and only 7 percent were oligotrophic. Conversely, in Canada, 12 percent of the lakes were eutrophic and 73 percent oligotrophic. This shows that 750,000 Canadian lakes are still oligotrophic because they distribute on less fertile land with low population density as represented by Precambrian Shield lakes. On the other hand, most of OECD lakes in the United States are small

TABLE 8.5 Distribution of Annual Phosphorus Load and Mean Chlorophyll Concentration in 335 Lakes in OECD Eutrophication Studies*

Chlorophyll, mg/m^3	$[P]_\lambda = [P]_j/(1 + \sqrt{T_w})$, mg/m^3					Σ cases (chlorophyll)
	<4	4–10	10–35	35–100	>100	
>25	—	—	2.3 1 1.0	25.0 11 16.2	72.7 32 48.5	44 13.1
8–25	—	2.7 2 2.3	14.7 11 11.2	44.0 33 48.5	38.7 29 43.9	75 22.4
2.5–8	—	23.3 27 30.7	51.7 60 61.2	20.7 24 35.3	4.3 5 7.6	116 34.6
1–2.5	8.4 7 46.7	62.7 52 59.1	28.9 24 24.5	—	—	83 24.1
<1	47.1 8 53.3	41.2 7 8.0	11.8 2 2.0	—	—	17 5.1
Σ cases (phosphorus)	4.5 15	26.3 88	29.3 98	20.3 68	19.7 66	(100) 335 (100)

*Phosphorus load is represented by the flushing-corrected phosphorus in-flow concentration. $[P]_\lambda$ = annual average in-lake concentration of total P. $[P]_j$ = annual average inflow concentration of total P. Underlined numbers are the number of lakes in the boxed category. Nonunderlined numbers in high and low position indicate, respectively, the percentage that the number of lakes directly below represents of the total number of lakes giving mean chlorophyll concentration shown at the far right of the row, and the percentage that the number of lakes directly above represents of the total number of lakes with phosphorus load shown at the very bottom of the column.

SOURCE: Modified from Vollenweider (1989).

lakes surrounded by farms and highly populated land, and thus they are prevalently eutrophic. In Italy and Germany, the situation is comparable to that in the United States. Many of these lakes were originally oligotrophic. Changes in environments of watershed-oriented lakes toward more eutrophic or—in some cases to more oligotrophic—states by remedial reduction of nutrient input were observed. The data from Japan and other countries included data on reservoirs. In Spain, around one-third of 300 reservoirs are eutrophied. Similar eutrophication problems of reservoirs have been recorded in South and Central America (Vollenweider, 1989).

In the foregoing discussion on the distribution of lakes, we considered a number of lakes with different trophic states. In view of sustainable management of water resources, we also need to examine the effects of eutrophication on the distribution of water resources in vol-

TABLE 8.6 Eutrophication of Lakes and Reservoirs: Percent Oligotrophic, Mesotrophic, and Eutrophic (+ Hypertrophic) as Classified in Various Comparative Studies and Reports

Region and/or countries	Percent			Number of cases
	Oligotrophic	Mesotrophic	Eutrophic	
18 OECD countries (Western Europe, United States, Japan, Australia)	18	17	65	101
OECD and Canada	48	16	36	230
OECD and Canada + other countries	30	35	35	335
Canada	73	15	12	129
United States	7	23	70	493
Italy	29	28	43	65
Germany	8	38	54	72
Baltic Belt countries	15	38	31(+ 16)	130
Japan and other countries	25	39	36	36
China	44	32	24	34
11 PAHO countries (South and Central America)	24	20	56	25
South African reservoirs	31	41	28	32

SOURCE: Vollenweider (1989).

ume. As shown in Table 8.7, only 2 percent of the world lakes in terms of water volume and 12 percent in terms of water surface area were affected by eutrophication, although 30 percent were eutrophic. Sixty-eight percent of the world lake water volume was still oligotrophic.

As one of the tasks of global lake environmental management, Lake Biwa Research Institute (1984) and the International Lake Environment Committee (ILEC, 1987–1989) conducted worldwide surveys of the data on the present lake environmental conditions. Tables 8.8 and 8.9 show data on the present status of lakes and their watershed from these surveys. Important parameters related to eutrophication are tabulated along with information on the limnological characteristics of the lakes surveyed. In terms of Secchi disk transparency, Lake Bikal in eastern Siberia, Lake Tupo in New Zealand, and Lake Tahoe and Lake Huron in North America are still in good condition, showing the lowest values of TN, TP, and Chl con-

TABLE 8.7 Trophic Conditions of 180 Natural Lakes (OECD-Canada Study; Excluding Reservoirs)

	Oligotrophic	Number of samples	Mesotrophic	Number of samples	Eutrophic	Number of samples	Total
Area, km	144,470	71	101,201	43	32,600	66	278,331
Area, %	51.9	39.4	36.4	23.9	11.7	36.7	100
Volume, km	16,000	71	7,132	42	525	66	23,657
Volume, %	67.6	39.7	30.2	23.5	2.2	36.9	100
Average mean depth, m	22.8	—	35.9	—	24.1	—	—
Average water residence time, years	15.8	—	6.0	—	3.1	—	—

SOURCE: Vollenweider (1989).

TABLE 8.8 Limnological Characteristics of Selected Lakes in the World*

Continents and lakes	Altitude, m	Lake surface area, 10^3 km^2	Mean depth, m	Water residence time, years	Drainage basin, 10^3 km^2	Secchi disk reading, m	Total N, mg/L	Total P, µg/L	Chl *a*, mg/m^3
Asia and Oceania									
Lake Baykal	456	31.500	730	—	560.0	41	—	—	—
Dongting-hu	33.5	2.740	6.5	—	259.43	0.2	—	2	—
Tai-hu	3.1	2.425	2.1	0.65	—	—	—	—	—
Languna de Bay	1.8	0.900	3.6	—	3.820	0.4	—	—	—
Bung Boraped	23.8	0.1064	2.6	—	10.623	1.3	—	—	—
Lake Sonkuhla	0	0.987	1.5	0.3	7.0	0.6	1.35	110	—
Lake Kinneret	−209	0.170	23.5	4.8	2.730	2.8	—	—	4.2
Lake Burley Griffin	555	0.007	4.7	0.2	1.865	1.7	—	42	—
Lake Taupo	357	0.616	97.4	10.6	3.327	16	—	9	1.4
Lake Biwa	85	0.674	40.8	5.5	3.174	4.3	0.3	14	6.4
Lake Chuzennji	1269	0.0115	104.3	5.2	0.121	8.0	0.19	6	1.6
Lake Kasumigaura	0.2	0.230	3.9	0.55	1.949	0.9	0.96	61	47.8
Lake Sagami	167	0.0033	19.2	0.05	1.016	2.0	1.48	82	24.2
Lake Kawaguchi	832	0.0060	9.3	0.33	0.126	3.3	0.35	12	4.9
Lake Suwa	759	0.0133	4.9	0.13	0.531	1.0	2.36	120	98.1
Lake Kizaki	764	0.0014	17.9	0.5	0.024	3.0	0.41	15	2.8
Lake Shinji	0.3	0.0803	4.6	0.25	1.996	1.1	0.42	56	21.8
North and South America									
Lake Washington	0	0.0876	32.2	2.4	1.274	5.8	0.26	18	4.97
Lake Tahoe	1890	0.499	75.2	700	0.841	27.6	0.21	6	0.24
Lake Mendota	850	0.0394	12.2	6.0	0.522	3.7	1.8	120	13.9
Lake Superior	183	82.367	24.0	191	124.838	—	—	4	1.5
Lake Huron	176	59.570	58.8	22.6	128.464	13	—	5	1.3
Lake Michigan	176	58.016	82.7	99.1	117.845	—	—	—	1.3
Lake Erie	174	25.821	17.7	2.6	78.769	2.6	—	16	4.3
Lake Ontario	75	19.009	84.2	7.9	75.272	4.1	—	—	5.8
Lake Winnipeg	217	23.750	12.0	3.6	953.250	1.1	—	—	—

Lake Titicaca	3812	8.100	106.9	70	42.670	7.4	—	—	—
Lake Broa	695	0.0068	3.2	0.14	0.228	1.8	—	8	17
Europe									
Lake Paajarvi	103	0.0135	15.3	3.5	0.255	—	1.16	10	2.3
Lake Malaren	0.3	1.096	12.8	2.6	22.603	2.4	0.81	35	4.5
Lake Trummen	161	0.001	1.4	0.5	0.013	0.6	0.67	15	59.5
Lake Esrom	9	0.0173	12.6	8.5	0.0076	4.0	0.88	238	9.5
Windermere	39	0.0148	21.3	—	0.231	—	—	20	10.0
Tjeukemeer	−1.0	0.021	1.8	0.2	—	—	—	—	150
Bodensee	398	0.476	100	4.5	10.90	5.0	—	66	2.5
Lake Lunzer	604	0.0007	18.6	0.3	0.027	—	—	7	—
Lake Balaton	105	0.6	3.0	5.5	5.18	0.7	—	53	—
Lake Zurich	406	0.0651	50.7	1.4	1.74	5.1	—	—	2.8
Lake Leman	372	0.5842	152	11.8	7.975	6.0	—	30	5.2
Lake Maggiore	194	0.2125	176	4	6.387	6.1	0.80	16	5.4
Lake Skadar	5	0.3723	5.1	—	5.490	4	—	—	—
Africa									
Lake Victoria	1134	68.8	40	23	184	—	—	—	3.0
Lake Tanganyika	773	33.0	580	—	250	—	—	—	5.1
Lake Chilwa	622	0.60	—	—	7.5	0.07	—	—	100
Lake Sibaya	23	0.0775	12.7	—	0.465	—	—	—	3.0

*The annual mean values are given for water-quality variables.
SOURCE: Modified from Lake Biwa Research Institute (1984).

TABLE 8.9 Environmental Parameters of Drainage Basin and Trophic Conditions of Some Selected Lakes in the World

Continents and lakes	Population density per lake volume, 10^6 km^{-3}	Land usage of drainage basin (areal %)			N and P loading per lake volume, 10^4 t m^{-3} yr^{-1}		Trophic state, algal bloom
		Forest	Farm	Urban	N	P	
Asia and Oceania							
Lake Baykal	3.8	—	—	—	1.2	0.3	O Δ
Dongting-hu	670	25.9	28.8	13.2	—	—	— —
Dong-hu	3,300	—	35.3	27.6	800	141	● ▲
Languna de Bay	770	23.8	52.0	6.5	120	29	— ▲
Bung Boraped	3,300	16.3	68.3	—	—	—	— —
Lake Sohkuhla	710	25.6	68.1	6.0	—	—	● ▲
Lake Kinneret	64	3.7	38.5	1.8	62	4.7	◎ P
Lake Burley Griffin	9,000	37.0	59.8	3.2	1400	96	● ▲
Lake Taupo	0.42	54.6	1.4	0.1	1.1	0.2	O Δ
Lake Biwa	41	63.2	4.9	8.9	33	3.9	◎ ▲
Lake Chuzennji	1.9	99.3	0.3	0.3	19	1.7	O Δ
Lake Kasumigaura	240	28.6	50.9	12.7	500	54	● ▲
Lake Sagami	3,300	78.9	4.4	—	6900	680	● ▲
Lake Kawaguchi	3.2	72.2	8.5	3.3	6.1	0.8	◎ Δ
Lake Suwa	2,500	29.2	13.5	9.1	4300	620	● ▲
Lake Kizaki	60	86.6	4.6	2.7	200	11	◎ Δ
Lake Shinji	710	47.5	9.1	3.9	1100	84	● ▲
North and South America							
Lake Washington	530	30.0	10.0	60.1	—	—	◎ Δ
Lake Tahoe	0.33	84.0	1.2	7.7	—	—	O Δ
Lake Mendota	360	—	—	—	110	9.6	● —
Lake Superior	0.045	94.5	1.4	0.1	—	0.04	O Δ
Lake Huron	0.35	65.7	22.4	1.8	0.7	0.14	O Δ
Lake Michigan	2.8	49.8	23.4	3.5	0.5	0.13	O Δ
Lake Erie	30	14.6	59.1	9.2	53	0.38	◎ Δ
Lake Ontario	4.5	55.8	31.6	4.4	11	0.7	◎ Δ
Lake Winnipeg	14	40.0	50.0	—	27	2.4	◎ Δ
Lake Titicaca	1.0	0.1	5.9	0.02	—	—	O Δ
Lake Broa	46	—	—	—	—	—	● —
Europe							
Lake Paajarvi	0.73	56.5	17.6	1.7	36	2.0	◎ Δ
Lake Malaren	81	62.8	19.9	—	89	6.0	◎ Δ
Lake Trummen	33,000	—	—	—	—	—	● ▲
Lake Esrom	110	0	100	0	—	—	● Δ
Windermere	—	—	—	—	—	—	◎ —
Tjeukemeer	—	0	—	—	—	—	● —
Bodensee	1.4	—	—	—	—	—	O Δ
Lake Lunzer	23	64.4	6.7	0.7	—	3.1	— Δ
Lake Balaton	220	25.9	46.2	8.5	148	14	— ▲
Lake Zurich	150	2.6	6.6	3.7	—	5	O ▲
Lake Leman	8.3	—	—	—	—	1	◎ Δ
Lake Maggiore	18	20.4	—	—	29	1.5	◎ Δ
Lake Skadar	—	—	—	—	—	—	— Δ

TABLE 8.9 Environmental Parameters of Drainage Basin and Trophic Conditions of Some Selected Lakes in the World (*Continued*)

Continents and lakes	Population density per lake volume, 10^6 km^{-3}	Land usage of drainage basin (areal %)			N and P loading per lake volume, 10^4 t m^{-3} yr^{-1}		Trophic state, algal bloom
		Forest	Farm	Urban	N	P	
Africa							
Lake Victoria	0.18	—	—	—	—	—	O Δ
Lake Tanganyika	0.077	—	—	—	—	—	◎ Δ
Lake Chilwa	—	—	—	—	—	—	● Δ
Lake Sibaya	1.5	—	—	—	—	—	O Δ

Key: ● eutrophic; ◎ mesotrophic; O oligotrophic; ▲ with bloom of blue-green algae; Δ no bloom of blue-green algae; P, peridinium bloom.
SOURCE: Modified from Lake Biwa Research Institute (1984).

centrations. In general, high concentrations of TP and TN are found in small shallow lakes and low concentrations are in large deep lakes. Comparison of Tables 8.8 and 8.9 indicates that lakes with low Secchi transparency values and a great abundance of Chl are characterized by the occurrence of cyanobacterial blooms and high N and P loading. The clear correlation between P loading and human population density on the watershed (Fig. 8.7) indicates considerable influence of

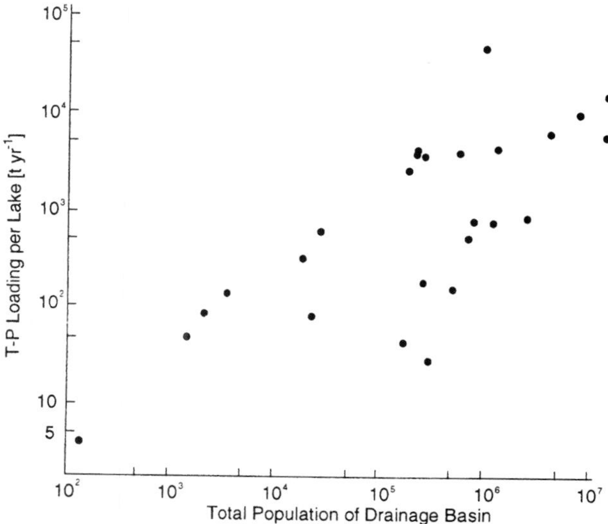

Figure 8.7 Relationship between specific loading rate of TP and human population of the drainage basin over many lakes in the world (Lake Biwa Research Institute, 1984).

human activities on eutrophication of lakes. The pattern of land use does not seem to have significant correlation to nutrient loading.

Case studies

Western Europe. A considerable number of studies have been conducted on the accelerated eutrophication of lakes and reservoirs in western Europe. These studies can be categorized into three groups: (1) monitoring surveys of trophic states in many lakes, (2) long-term surveys of environmental changes in selected lakes, and (3) follow-up studies on the remediation processes of lakes after nutrient reduction.

Monitoring surveys of eutrophication are represented by OECD eutrophication projects which were conducted in four regional projects to determine the present state of eutrophication of lakes and to present measures for prediction and remediation of lake environments. The OECD monitoring surveys in western Europe were conducted in the Nordic Project, Alpine Project, and Shallow Lakes/Reservoirs Project.

Of 10 lakes in Scandinavian countries involved in the OECD Nordic Project, 6 lakes were eutrophic, 1 mesotrophic, and 3 oligotrophic (NORDFORSK, 1980). Lakes Esrom, Mossø, and Tuusulamjärvi located on the watershed of moraninic clay or clay earth were naturally eutrophic, but the other three eutrophic lakes were significantly affected by municipal wastewater and nutrient loads from agricultural sources. In Lake Tuusulamjärvi, the non-point-source load has increased considerably mainly because of the increasing application of fertilizers since 1950. The relative contribution of non-point-source loads to the total loads was 22 to 53 percent for P and 35 to 70 percent for N during the mid-1970s. Two oligotrophic lakes (Pääjärvi and Vattern) were deep, and larger parts of their catchment areas were underlaid with granite or moraine, and the clay areas were very limited. Of the drainage areas, 10 to 30 percent were agricultural lands with a limited supply of nutrients from paper mills and residential areas.

The OECD Nordic Project report gave us a general idea of the trophic state of the lakes studied. Summer average values of TN, TP, and Chl concentrations in six eutrophic lakes (Esrom, Mossø, Päijänne I, Tuusulamjarvi, Boren, and Ekoln) ranged from 440 to 1770 mg N/m^3, from 17 to 180 mg P/m^3, and from 7 to 117 mg Chl/M^3, respectively, with the highest values in Lake Tuusulanjärvi and Lake Mossø, characterizing the eutrophic state. The corresponding values in oligotrophic lakes (Päijänne II, Päijänne IV, Päijänne V, Pääjärvi, and Vattern) were 410 to 970 mg N/m^3, 8 to 12 mg P/m^3, and 1 to 6 mg Chl/m^3 with the bottom values in Lake Vattern representing the oligotrophic state. Significant correlations were found between TP and chlorophyll concentrations or primary production and also between TP loading and the biological parameters in many lakes. In about half of the lakes, phosphorus was a limiting nutrient, and in the other half nitrogen alone or together with phosphorus was a primary limiting element.

Regional information on the trophic state and eutrophication of lakes in each area has been presented in national reports. In surveys of 347 Norwegian lakes (Faafeng and Hessen, 1993), the median values of TP, TN, and Chl concentrations were 6.5 mg P/m^3, 250 mg N/m^3, and 2.2 mg Chl/m^3, respectively, suggesting that most Norwegian lakes remain rather undisturbed by human activities. Of these lakes, 85 percent were P-limited and 6 percent were N-limited, while 9 percent were both N- and P-limited. The percentage abundance of cyanobacteria changed with the TN/TP ratio; it was low in lakes with TN/TP $>$ 80 but equally distributed in lakes with TN/TP $<$ 80. This percentage was also positively correlated with TP concentration and approached 80 percent or more in lakes with TP $>$ 50 mg P/m^3, suggesting an importance of TP concentration of lake water in the development of cyanobacteria.

Many lakes in agricultural and densely populated areas had been progressively eutrophied. Lake Mjøsa, the largest lake in Norway, had been eutrophied during the 1970s with gradual increase of algal biomass (Holtan, 1978, 1979). In 1975 and 1976 the cyanobacteria *Oscillatoria bornettii* f. *tenuis* predominated during summer. Organic-matter geosmin emitted by this alga gave an unpleasant taste and odor to potable water. A regional survey revealed that 64 percent of the P load was supplied from point sources located around the lake (36 percent from municipalities and 28 percent from industrial areas), 11 percent agricultural areas, and 22 percent forested and unproductive areas (Holtan, 1978, 1979). Under a new action plan, 42 new sewage-treatment plants were constructed, and 280 km of new sewer pipeline was installed. Ninety-one percent of the population in the densely built-up areas were connected to the treatment plants with P elimination over 90 percent efficiency by the end of 1980 (Holtan, 1981). TP concentration of the surface water decreased from 15 to 18 μg/L prior to 1976 to 11 μg/L in 1980. The primary production rate decreased to 110 mg C m^{-2} d^{-1}, 20 percent of the 1975 level, along with a decrease in Chl concentration. The phytoplankton community was dominated by large species of diatoms with a minor contribution of cyanobacteria to the total algal biomass.

In Swedish lakes, there was little increase of in-lake TP concentration in 11 of 25 lakes surveyed in the 1930s and 1975, but an obvious TP increase in the other lakes under sewage supply (Wallsten, 1978).

Of four large lakes (Mälaren, Hjälmaren, Vättern, and Vänern) representing 25 percent of the Swedish lake surface, Lake Mälaren and Lake Hjälmaren have been under increasing nutrient input as a result of increasing population density, expanding industries, and intensive farming in the areas around the lakes during the 1960s. This resulted in increased P and Chl concentrations and frequent development of heavy cyanobacteria bloom in summer (Ahl, 1975; Willén, 1975; Winderholm, 1978). Comparisons of the observed loading with the natural loading estimated from atmospheric precipitation loss from forests and soils indicated that artificial loading constituted 53 to 81 percent of the observed TN load and 57 to 89 percent of the observed TP load with increasing percentages in the following order: Lake Vättern < Lake Vänern < Lake Hjälmaren < Lake Mälaren (Ahl, 1975). Increasing TN and TP loading in seven bays of Lake Mälaren were associated with increases of Chl concentration, shares of cyanobacteria in total algal biomass, and shares of zooplankton biomass (Ahl, 1975).

Because of the importance of lake resources for human society, a large expansion of advanced wastewater-treatment plants with chemical precipitation for P removal started in 1968 in Sweden. By the early 1980s, over 90 percent of the densely populated area was connected to wastewater-treatment plants with chemical precipitation (Willén, 1975; Forsberg et al., 1978; NORDFORSK, 1980). In Lakes Boren and Ekoln, after the improved water treatment, P load from the wastewater-treatment plants was considerably reduced, and diffused P loading increased from 20 to 60 percent, resulting in decreased in-lake concentrations of Chl and TP and increased transparency (Forsberg et al., 1978). Reduced P was also associated with the changes of the benthic community with increasing biomass and diversity of chironomids in Lake Mälaren (Winderholm, 1978).

Lake restoration by reduced nutrient loading is well exemplified by the case studies in Lake Norrviken in central Sweden and Lake Trummen in southern Sweden. Lake Norrviken had received sewage and wastewater from domestic and industrial sources for many years and been heavily eutrophicated until 1969 when all the sewage effluents were diverted from the lake to the new sewage-treatment plants (Ahlgren, I., 1972, 1978; Ahlgren, G., 1978). After diversion, a pronounced decline of P and N (both total and inorganic) concentrations was observed in association with decreased phytoplankton biomass and primary production. *Oscillatoria agardhii* Gom., which dominated in the polluted lake, diminished after the diversion. The other cyanobacterial species, diatom and green algae, dominated with the increasing diversity of phytoplankton community.

Lake Trummen had been utilized as the recipient of sewage from towns and wastewater from factories and was severely polluted until 1958 (Björk et al., 1972). Diversion of the sewage from the lake in 1959 did not improve the water quality. Considering the role of bottom sediment affected with pollutants as an important internal nutrient source, the upper 1-m sediment layer was pumped up from the lake in 1970 and 1971. After dredging, the surface sediment was oxidized, and the lake concentrations of N and P considerably decreased (Bengtsson et al., 1975) (Fig. 8.8). Prior to the restoration the lake was characterized by vernal and autumnal blooms of *Melosira* spp. and *Synedra* spp. and heavy summer bloom of cyanobacteria

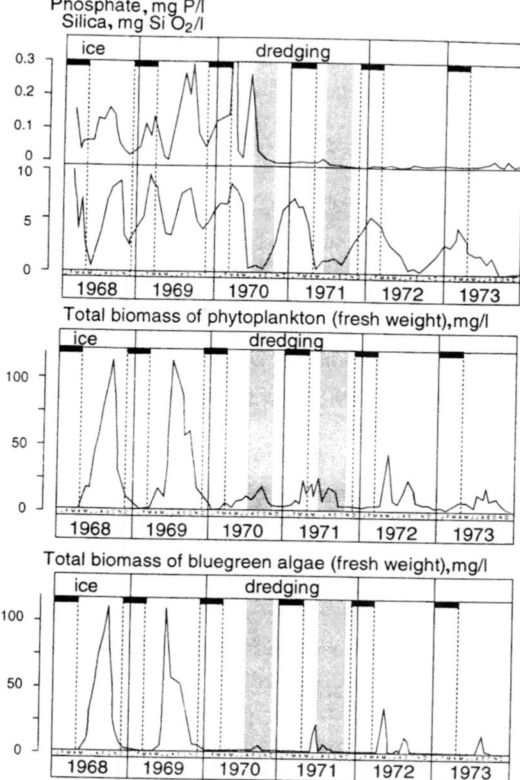

Figure 8.8 Concentrations of phosphate and silicate and total biomass of phytoplankton and cyanobacteria (blue-green algae) at 0.2 m depth in Lake Trummen, Sweden (modified from Bengtsson et al., 1975; Cronberg et al., 1975).

(Cronberg et al., 1975). After restoration, spring and autumn diatom bloom failed to appear, and summer phytoplankton was dominated by diatoms together with small miscellaneous forms, green algae, and cyanobacteria. The abundance of rotifers and cladocerans decreased, presumably due to a decrease in the production and biomass of phytoplankton. The abundance of tubificids increased (Andersson et al., 1975). The total fish catches declined to the lowest values as found in oligotrophic lakes, mainly as a result of the shortage of food supply.

Of the 28 lakes surveyed in the OECD Alpine Project (Fricker, 1980; Ambühl, 1981), 22 were eutrophic. TP concentration, P load, and annual primary production rate were 33 to 484 mg P/m^3, 1100 to 10,010 g P m^{-2} yr^{-1}, and 180 to 800 g C m^{-2} yr^{-1}, respectively, with the highest values in Lakes Lugano and Greifensee characterizing the eutrophic state. In the other six lakes, the corresponding values were 7.6 to 25.8 mg P/m^3, 60 to 4000 g P m^{-2} yr^{-1}, and 30 to 320 g C m^{-2} yr^{-1} with the lowest values in Lunzer Untersee and Feldsee, suggesting lakes in an oligomesotrophic state. Over the decades, TP concentrations in several Swiss lakes showed a remarkable increase, with a maximum rate of 20 percent increase per annum. Thus, even the large Lakes Geneva and Constance, which were previously oligotrophic, are now eutrophic.

Lake Zürich, a representative lake in which intensive eutrophication studies have been conducted, is the first lake in the world to undergo sewage treatment using chemical precipitation processes to eliminate P, which were introduced with much success (Thomas, 1969; Ambül, 1981). Lake sediment data suggested that the eutrophication of this lake commenced around the 1890s. Sewage discharge from the highly populated communities, including Zurich City, and tourists had caused progressive eutrophication and reached its peak state in the 1960s. Cyanobacteria *Oscillatoria rubescens* predominated until 1965 and reappeared in 1976. From 1900 onward, mass development of *Melosira islandica* spp. *helvetica, Stephanodiscus hantzschii,* and *Mougeotia* were observed (Schanz and Thomas, 1981). The phosphate concentration, which was less than 30 μg/L before 1950, increased to 90 μg/L in 1955–1967. Considerable polimnetic oxygen consumption during the stagnation period was observed not only in the deepest layers but also in the upper hypolimnion. To improve water quality, the sewage treatment plants were constructed within the drainage basin, and the majority of domestic and industrial effluents receive effective treatment today. The precipitation of P with ferric or aluminum salts achieved P reduction of 85 percent or more (Thomas, 1969). TP loading decreased from 5.5 g in 1960 to 1.9 g m^{-2} yr^{-1} in 1980.

According to a nationwide survey of Italian lakes (Gaggino and Provini, 1988), 85 percent of 131 Italian lakes had N/P ratios > 15:1

and phosphorus was limiting. Of these, 8.2 percent were ultraoligotrophic, 10.9 percent oligotrophic, 40.1 percent mesotrophic, 30.6 percent eutrophic, and 10.2 percent hypertrophic. Hypertrophic Lakes Lugano, Varese, and Grande di Avigliana were under heavy anthropogenic eutrophication. The trophic state of the largest lakes was in increasing order as follows: Garda, Bracciano, Bolsena, Iseo, Maggiore, Como, and Lugano. In 13 of 69 lakes, external loads of TP were at an acceptable level and in 28, objective oligotrophy may be achieved with suitable intervention. The remaining 28 lakes could become mesotrophic only by drastic P reduction. Comparison of natural TP levels calculated as a morphoedaphic index (Vighi and Chiaudani, 1985) with the observed TP level of lake waters suggested that 48 percent of the lakes surveyed were nearly in the natural trophic state. In the other 52, a considerable P reduction may be needed for return to their natural state. In Lakes Maggiore and Como, P reduction of 29 to 44 percent is needed before treated sewage can be discharged.

Eutrophication of Lake Maggiore was caused by a marked increase of population and industrialization in the drainage area for the past few decades. The lake is a renewed tourist resort, and more than 12 million tourists per year visit the lake area. Commercial fishing, boating, and agricultural water uses for rice fields and dairy pastures represent the major uses of the lake (Calderoni et al., 1993). Phosphate concentration of lake water increased from less than 10 μg/L in the early 1960s to peak values of 32 to 42 μg/L in the late 1970s (Mosello and Ruggiu, 1985). Algal productivity and biomass increased about three times, and cyanobacterial blooms developed. Because three large lakes (Lugano, Varese, and Orta), which are located within the watershed of Lake Maggiore, are important nutrient sources for Lake Maggiore, integrated restoration measures through these three lakes and their inflowing rivers must be registered.

The majority of lakes and reservoirs in Europe surveyed in the OECD Shallow Lakes/Reservoirs Project have a mean depth of less than 10 m and a water residence time shorter than 1 year. According to the final report of the project (Clasen and Bernhardt, 1979), 70 percent of the lakes were eutrophic. Their TP concentrations were greater than 30 mg P/m^3, with peak values up to 100 mg P/m^3. More than half of the semiartificial reservoirs were mesotrophic with TP concentration of $<$30 mg P/m^3. In pump-storage reservoirs, TP concentration was greater than 100 mg P/m^3, corresponding to a highly eutrophic state. The TP/TP ratios of lake water were smaller than 15:1 throughout the lakes and reservoirs surveyed, indicating P limitation of algal growth. Inland waters in Spain are characterized by a larger number of reservoirs. According to the report from the nationwide survey on the trophic state of Spanish reservoirs (Oritz and

Pena, 1984), Spanish reservoirs are characterized by (1) large fluctuations in mean depth and water residence time because of climate and dam operation, (2) withdrawal of hypolimnetic water, (3) long narrow shape causing significant nutrient retention, and (4) shorter hydraulic retention time, all of which should affect the trophic state of the reservoirs. Regression equations for summer Chl vs. TP concentration differed from those for OECD countries. Good correlations of mean summer Secchi disk transparency with the flushing corrected P loading or TP concentration were detected. Hypolimnetic oxygen depletion did not show correlation with TP loading. On analysis of published and unpublished data on many Spanish inland waters, Alvarez et al. (1992) concluded that 80 percent of lakes, 70 percent of reservoirs, and 60 percent of streams were eutrophic (average TP > 80 mg P/m^3 for stagnant water and >150 mg P/m^3 for running water). Correlation analysis of TP concentrations of the stream water and watershed features in the streams indicated that agricultural eutrophication affected eutrophication in the majority of inland water bodies in Spain.

Eastern Europe. Lake Balaton is a representative, large freshwater lake of eastern Europe. The lake has been under extensive study since the early 1960s (Herodek, 1984; Tóth and Padisak, 1984; Herodek et al., 1988; Hajos, 1992; Vörös and Göde, 1993). It was oligomesotrophic in 1961, but for the subsequent 20 years, algal biomass showed a twofold increase and primary production, a two- to eightfold increase with the predominance of colony-forming and filamentous cyanobacteria. One-third of the total nutrient loads flows into the western basin from the Zala River. The Sio Canal at the east end of the lake drains the water into the River Danube. Because of the long shape of the lake, the western basin was more eutrophic. In the eastern and western basins, average concentrations of TN were 850 to 1650 mg N/m^3, and those of Chl were 30 and 100 mg Chl/m^3, respectively. The corresponding values of the TN/TP ratios were 28 and 16, indicating P limitation and westward of trophic-state increase. One source of the increased nutrient load was growing tourism because of the favorable condition of the lake for bathing and holidaying throughout the year. At least half of the nutrient load was diffuse in origin, mostly agricultural origin.

In Russia, artificial (human-made) eutrophication was under extensive survey later than in western Europe because of the later spread of serious problems. Most reports of eutrophication in Russia are available in Russian, but some of them are provided in English. The state of eutrophication of inland waters was summarized by Ghilarov (1982). His report indicates that many lakes and reservoirs in western Russia currently show eutrophication symptoms due to increasing loading from agricultural land and municipality wastes.

In Lake Krasnoe of the Karelia Isthmus, primary production and phytoplankton biomass doubled in about 10 years with the appearance of typical eutrophic species. Because 80 percent of the watershed was covered with forest, the increasing agricultural activity was considered a significant cause of eutrophication. Ladozhskoye Ozero, the largest lake in northwest Russia, despite the great nutrient supply of anthropogenic origin, had remained oligotrophic because of the enormous lake volume. But phosphorus concentration, which was less than 10 mg/m^3 in the early 1960s, increased to 20 to 40 mg/m^3 in the late 1970s. In Upper Volga reservoirs, including Ivanjkovskoe Reservoir, the drinking-water source for Moscow, a clear trend of eutrophication was not observed, although there was an increased relative abundance of cyanobacteria. The agricultural land of the Upper Volga supplied a TN of 2.5 kg/ha and a TP of 0.13 kg/ha in summer, and 30 percent of the total supply came from municipal wastes, more than a half of which was detergent phosphorus. Lake Beloje in northwest Russia changed from oligotrophic to mesotrophic as a result of increasing navigation. Primary production increased from 4 to 9 mg C/m^3 to 5 to 30 mg C/m^3, and phosphate concentration doubled for around one decade.

In Dnieper reservoirs situated in the Steppe of fertile Chernozoem soils and warm climate, besides natural nutrient factors, the nutrient supply from runoff from agriculture lands and municipal sewage were important. Nutrient flux from agricultural lands exceeded that from the natural drainage. Enormous development of cyanobacterial bloom was observed in these reservoirs from spring at 14 to 16°C through late summer at 20 to 25°C with seasonal alternation of dominant species. Considerable cyanobacterial blooms accumulated on the surface to form scum and interfered with the use of water for domestic supply and recreational use. Some alpine lakes also suffered from increased eutrophication. In the largest Armenian Lake Seven, sewage discharge and lowering water level for hydroelectric power generation were the main factors of increased eutrophication. In large Asian lakes (Baikal, Issyk-Kul), eutrophication symptoms were not detected, although Lake Baikal has been greatly threatened by the pollution of toxic wastes from many industries (Galazy, 1989; Galazy and Tarasova, 1993).

North America. Studies on lake eutrophication in North America were conducted by nationwide surveys and long-term surveys of many lakes. In the OECD Monitoring Program in the United States, the data compiled from the national surveys were analyzed (Rast and

Lee, 1978; Lee and Jones, 1984). Phosphorus loading in 37 of 57 lakes was above the excessive loading level (0.1 g P m^{-2} yr^{-1}) described by Vollenweider (1976), indicating the predominance of eutrophic lakes. Only 14 percent were below the permissible loading, oligotrophic lakes. In the majority of studied lakes, mean summer Chl concentration ranged from 10 to 100 mg/L, Secchi transparency was less than 5 m, and hypolimnetic oxygen-depletion rates ranged from 0.2 to 2.0 g O_2 m^{-2} d^{-1}.

Eutrophication and its recovery in Lake Washington (Edmondson, 1969, 1972a, 1972b; Edmondson and Lehman, 1981) have often been referred to as a representative example of cultural eutrophication and successful lake remediation. Lake Washington could have been originally oligotrophic. With the increasing population of Seattle, the lake received increasing amounts of raw sewage supply until an early diversion system completed in the late 1930s. Starting in 1941, the lake had been again enriched with an increasing volume of effluent from the secondary sewage-treatment plants which were built along the outfalls entering the lake. There was an additional wastewater supply from septic tanks and other sources. The relative contribution of these sewage sources to the total P supply in the lake was 56 percent in 1957 and 75 percent in 1963. Considerable enrichment of the lake with sewage resulted in increased production and biomass of algae. To recover the lake environment, the works of sewage diversion from Lake Washington took place over 5 years and decreased effluent from treatment plants entering the lake to zero by 1968. The data on nutrient inputs to the lake (Edmondson and Lehman, 1981) indicated that the annual average TP loading after the diversion was around 40 percent of the loading in 1962–1966, although the corresponding value for TN was only 90 percent. After the diversion, fluvial source accounted for 86 percent of the total TP loading and 92 percent of the total TN. To such a dramatic decrease of TP loading, the lake clearly responded with gradual decreases in concentrations of TP, particulate P, phytoplankton, and the proportion of cyanobacteria in the lake water (Fig. 8.9). The lake was regarded as having recovered from eutrophication by 1975.

The environmental changes in Lake Tahoe, California for the past five decades (Goldman and Amenzaga, 1975, 1984; Goldman, 1981, 1988) provide a typical example of eutrophication in an oligotrophic alpine lake. Because of the relatively small forested watershed consisting of decomposed granite and volcanic formation, Lake Tahoe, with a mean depth of 313 m and a surface area of 500 km^2 at an altitude of 1898 m, had extremely low fertility and high transparency (Secchi transparency to 40 m). After about 1940, the human population had increased gradually and disturbed the watershed by construction of

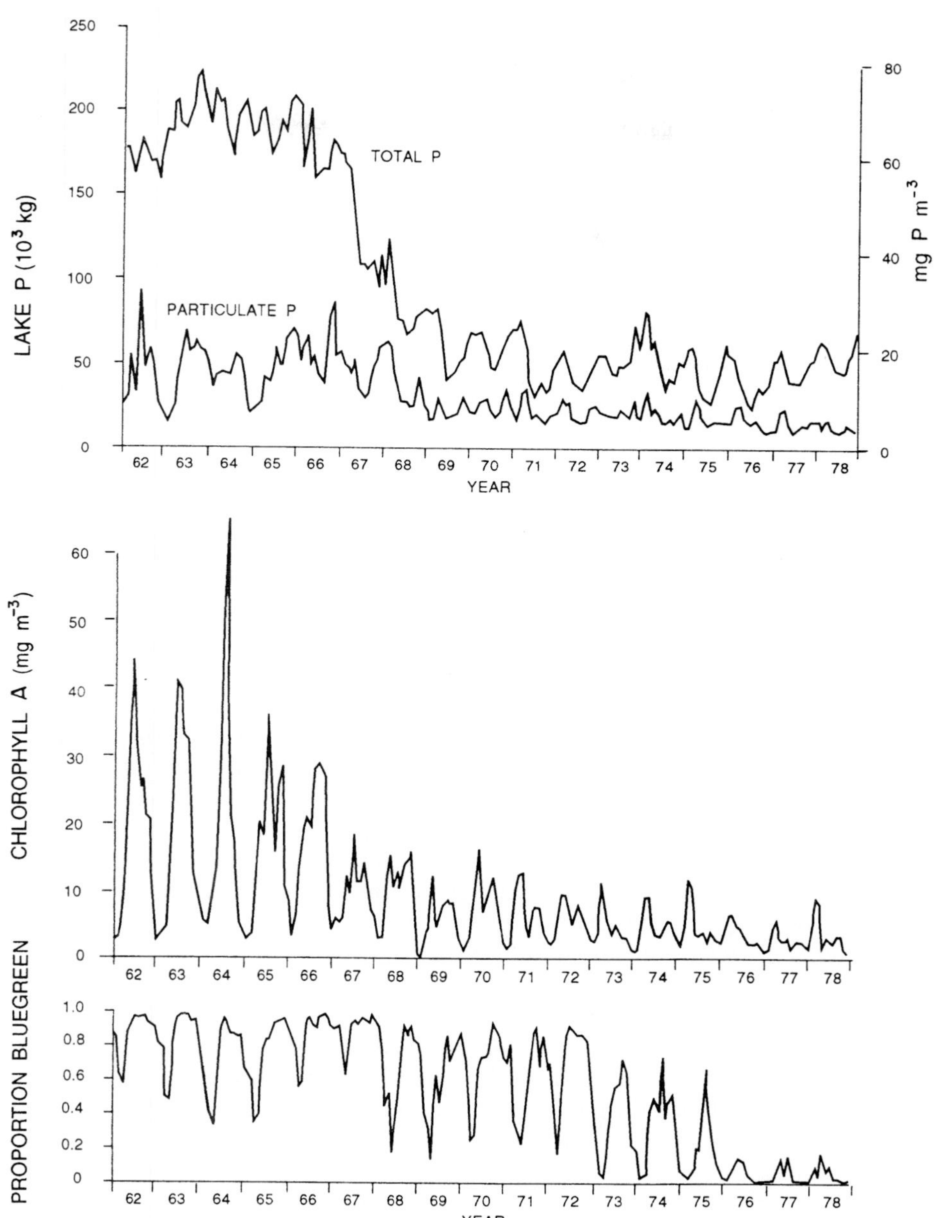

Figure 8.9 Changes in total mass of TP concentrations of particulate P and chlorophyll *a*, and proportion of cyanobacteria (blue-green algae) in total phytoplankton in the surface-water samples in Lake Washington (Edmondson and Lehman, 1981).

roads, buildings, and golf courses. Although the sewage was completely diverted from the lake, these land disturbances and fertilizer applications in lawns provided a continued nutrient supply into the lake. Starting with the visible increases of periphyton in the littoral zone, aquatic plant production displayed a considerable increase. More than a twofold increase of phytoplankton production (from 39 g C m^{-2} yr^{-1} in 1959 to ~115 g C m^{-2} yr^{-1} in 1985/86) was observed during 27 years associated with decreasing Secchi transparency. While Lake Tahoe was characterized by N limitation of primary production, total nitrate concentration in the lake water increased 1.3 times from 1973 to 1986, but a corresponding change of TP was observed. This indicates that Lake Tahoe has progressively eutrophied because of increased loading of N and has evolved from N to P limitation.

The Laurentian Great lakes, which account for about 20 percent of freshwater resources on the earth surface, represent a major water resource for the United States and Canada. Increasing population density and agricultural and industrial activities in the lake basin since early in this (twentieth) century had a marked impact on the lakes and resulted in considerable deterioration of water quality. Barica (1989, 1990) and Dobson (1981) described a history of water pollution in the Great Lakes as follows. The water pollution of the Great Lakes was started with epidemics due to discharge of human waste at the beginning of this century, followed by oil pollution in the 1940s and accelerated eutrophication in the 1960s due to increasing nutrient loading from industrial, residential, and agricultural sources. Lake Erie was sensitive to the increased nutrient loading because of lake morphometry and high P regeneration from sediments. Extensive algal blooms and hypolimnetic oxygen depletion occurred and often resulted in massive fish die-off. The Great Lakes Water Quality Agreement was signed between the United States and Canada in 1972, and an extensive P reduction program, including phosphate-based detergent ban and P removal at sewage-treatment systems, was carried out.

As a result, P loading from municipal sources dramatically declined so as to approach almost the scheduled levels by the early 1980s. P reduction resulted in considerable decrease of in-lake TP and Chl concentrations in Lake Erie and Lake Ontario (Fig. 8.10), while nitrate concentration increased with higher rates in densely populated and agriculturally productive basins of Lake Ontario and Lake Erie. A significant increase of nitrate concentration with TP decline resulted in an increasing TN/TP ratio from 20 to 60 (Lean et al., 1990), shifting from N limitation or N and P colimitation to P limitation of algal growth. Despite this decrease in P loading, there were few changes of primary production and only a slight increase in net hypolimnetic

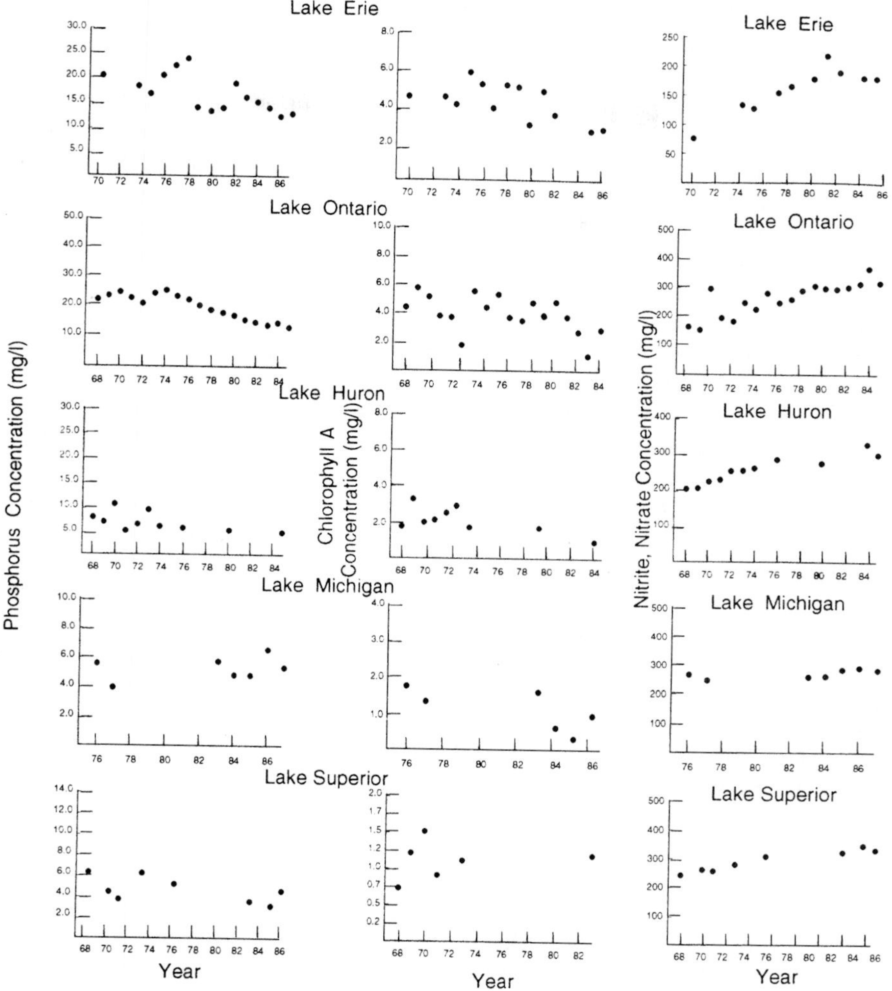

Figure 8.10 Long-term trend in total phosphorus, chlorophyll *a*, and nitrate plus nitrite nitrogen concentrations in the Great Lakes (Barica, 1989).

oxygen depletion rate, as well as little change of abundance and species composition of zooplankton (Lean et al., 1990). Diverse and delayed biological response to P reduction may be responsible for the unbalanced changes of biotic communities and environments.

In Canada, which has the largest number of freshwater lakes in the world, the lakes vary in area from tiny ponds to the largest lake, Lake Superior. The OECD Eutrophication Program in Canada, compiled by Janus and Vollenweider (1981), indicated that 110 lakes in the southern portion of Canada behaved in a manner similar to the pattern for the world OECD lakes with respect to the Chl-vs.-TP relationship. After the OECD surveys, the National Water Research Institute, Canada, conducted the data surveys on the current state of lakes in Canada at the request of the ILEC. On the basis of these surveys, Gray et al. (1990) assessed the environmental state of 42 lakes in Canada. The effects of human disturbance were lowest in lakes in Arctic regions and taiga plains because of the extremely low population densities, while eutrophication may be expected if there is high nutrient input even at Arctic latitudes (Schindler et al., 1974). In the boreal and taiga shield region making up the Precambrian Shield, human impact was little, because of minimal agricultural development due to limited soil development and harsh climate. But, a lake within a half-day drive from towns in the southeastern boreal area, with summer cottages and towns along its shorelines, and with inflowing rivers, caused eutrophication problems (Lake Muskoka).

In boreal plains and prairies with phosphorus-rich soils, lakes can be naturally eutrophic. Extensive agricultural development accelerated eutrophication in most lakes in these areas. In these nutrient-rich lakes, production of phytoplankton, zooplankton, and fish was very high, although fish cannot survive over a winter because of oxygen depletion under ice (Barica, 1987). In saline lakes located in the prairie, Chl was low irrespective of high TP concentration, due to inhibited algal growth by high salinity. In the cordillera regions of western Canada, measurable eutrophication occurred in the lakes with urban sewage-treatment plants and industries along the lakes and rivers in the valleys. Lake size and flushing rate significantly affected the extent of eutrophication. In Lake Kootenay, British Columbia, TP concentration was initially more than 100 μg/L by discharge from a fertilizer factory 200 km upstream, but followed by a remarkable P reduction by discharge control and nutrient reduction in a reservoir subsequently constructed upstream (Gray et al., 1990).

Temperate Asia. In association with considerable increases of population density and industrial activities in Japan after the 1950s, the water quality of lakes and rivers in the Japanese islands considerably deteriorated because of wastewater discharge from municipal and industrial point sources. Deterioration began with heavy-metal pollution of rivers, which had serious effects on human health, followed by organic-matter pollution due to wastewater discharge from pulp and food factories and subsequent accelerated eutrophication due to discharge of sewage and wastewater rich in nutrients. Enrichment effects were more pronounced in shallow lakes located in fluvial plains with large populations and active agricultural productions. Typical examples were Lake Teganuma, Lake Kasumigaura, and Lake Suwa, which had been highly eutrophied since the mid-1960s following the occurrence of dense cyanobacterial blooms every summer (Sakamoto, 1985; Environment Agency of Japan, 1993). Concentrations of TP and TN in lake water were 0.44 mg P/L and 4.3 mg N/L in Lake Teganuma, 0.066 to 0.072 mg P/L, and 0.84 to 1.3 mg N/L in Lake Kasumigaura, and 0.14 mg P/L and 1.6 mg N/L in Lake Suwa. The corresponding values in mesotrophic Lake Biwa were 0.008 mg P/L and 0.29 mg N/L [annual averages in 1990 (Environment Agency of Japan, 1993)]. Among the 148 lakes surveyed, cyanobacterial blooms developed with a high frequency in 28 percent of lakes with a mean depth of less than 5 m; freshwater red tides of *Peridinium* or *Uroglena* occurred in 30 percent of the lakes with a mean depth of >20 m, and drinking-water-supply problems resulted in 50 percent of the lakes with a mean depth of <20 m (Sakamoto, 1985). According to a nationwide survey, 33 percent of 480 lakes were eutrophic, 17 percent mesotrophic, 26 percent oligotrophic, 13 percent dystrophic, and 4.4 percent acidic (Environment Agency of Japan, 1987).

For protection of water quality from deterioration due to accelerated eutrophication, several legal measures to reduce P and N loading were formulated, and construction of sewage systems has been promoted. The percentage of the population connected to sewage-treatment plants was 50 percent in 1991. Programmed control measures of eutrophication were effective in reducing TP and Chl concentrations in some eutrophic lakes but not in some mesooligotrophic lakes.

A case history of eutrophication and the effects of control measures can be seen for Lake Biwa (Kira, 1984; Morioka, 1984; Yamaguchi, 1984; Shiga Prefectural Government, 1995). Lake Biwa is the largest lake in Japan and an important water resource for the Osaka and Kyoto regions, the second largest industrial and population centers in Japan. The lake was originally oligotrophic. Increased industrialization and increased population have adversely affected the water quality, and the lake has rapidly eutrophied. Transparency of >10 m in 1930 decreased to 2 to 5 m in recent years because of the increased density of phytoplankton. An unpleasant odor was often evident in the tap water. Starting in 1977, considerable blooms of a flagellate, *Uroglena americana* has developed early every summer. Hypolimnetic oxygen in the northern basin in summer decreased after around 1950, falling 33.3 to 50 percent of the original level observed in the late 1980s. To recover the lake, the Shiga Prefectural Government enacted "the Ordinance for Prevention of Eutrophication in Lake Biwa" in 1980 to reduce TP and TN loading from industries. The ordinance imposed an obligation to control effluents from agricultural and livestock farms. It was also associated with a ban on the use of synthetic detergent containing phosphate so as to assure 20 percent reduction of TP loading. These measures were effective in decreasing TP and TN loading from rivers into the southern basin, but minimal changes have been observed following TP and TN loading into the northern basin. A significant decrease in in-lake TP and TN concentrations has not been observed in either basin. Rather, a gradual increase of COD (chemical oxygen demand) level of the lake water has been recorded. Blooms of *Uroglena americana* have developed every year with no tendency for decline. Continued nutrient loading from nonpoint sources such as agricultural farms and individual septic tanks on the watershed of the northern basin might be responsible for the small decrease in TN and TP concentrations in the lake water. Further extensive countermeasures to reduce nutrient loading from nonpoint sources are required.

In China, there are around 2300 natural lakes with a total water storage capacity of 708 billion m^3, of which 29 percent is freshwater (Jin, 1994). In addition, there are 86,852 reservoirs with water-storage capacity of 413 billion m^3. In some of these lakes and reservoirs, eutrophication has become considerable. In particular, many lakes and reservoirs in urbanized areas are heavily polluted as a result of rapid urbanization and cannot be used as a source of drinking water. Fish production in many eutrophied lakes has been seriously affected by decreased dissolved oxygen and reduced food available (edible plankton and macrophytes) induced by eutrophication. According to the National Survey on Lake Eutrophication (1987–1990) for 25 lakes and reservoirs (Jin, 1990), annual averages of TN of the surveyed lakes ranged from 0.4 to 20.8 mg N/L with higher values in the spring. In the majority of lakes, the values were between 1.0 and 5.0 mg N/L, and in 20 percent of them, TN exceeded 5.0 mg N/L. Corresponding values of TP were 0.018 and 0.97 mg P/L, with higher values above 0.02 mg P/L in 92 percent of the lakes. Observed TN and TP concentrations were 5 to 12.5 times and 10 to 50 times greater than the critical TN and TP levels for eutrophication (0.2 mg N/L and 0.02 mg P/L). Eutrophic urban lakes are characterized by higher concentration of NH_4-N than NO_3-N, suggesting that ammonification due to organic-matter decomposition exceeded nitrification. Secchi disk transparency in 50 percent of the lakes surveyed was less an 0.6 m with lowest values of 0.2 to 0.4 m in urban lakes. The high turbidity of the Chinese lakes is attributable to phytoplankton in high density and clay minerals supplied from the drainage by soil erosion. Chlorophyll concentrations ranged from 30.2 to 240 mg/m^3 in urban lakes, and from 3.3 to 23.8 mg/m^3 in suburban lakes. In all the lakes studied, cyanobacteria predominated together with green algae, suggesting eutrophic and hypertrophic states in the Chinese lakes.

In the nationwide survey of 1982 (Jin, 1994), 5 percent of the lakes surveyed were eutrophic, 91.8 percent mesotrophic, and 3.2 percent oligotrophic. In the more recent survey (1987–1990), 55 percent were eutrophic, 44.5 percent mesotrophic, and 0.5 percent oligotrophic. This indicates that most Chinese lakes have undergone progressive eutrophication for less than one decade. There are a number of point and nonpoint nutrient sources in densely populated drainage basins. These basins have been intensively settled and modified for centuries, and have resulted in considerable discharge of sewage and nutrients to rivers and lakes. Traditional use of manure as fertilizer has also contributed to the increased nutrient supply in lakes. Nutrients in lake sediment are another major nutrient source for progressive eutrophication. Phosphorus contents were very high in the lakes studied, especially in urban lakes, presumably due to discharge of industrial wastes and municipal sewage into the lakes for prolonged periods. To clear the lake, Lake Liuhau was drained and cleaned by dredging bottom mud. However, within months, *Microcystis* started to grow rapidly and the water quality returned to the previous hypertrophic condition. This example illustrates how point and nonpint nutrient sources were associated with eutrophication.

Oceania. Limited information is available on the trophic state and eutrophication of freshwater lakes in the islands and Archipelago in the Pacific. Limnological features affecting the trophic state and current lake conditions in Australia and Indonesia are reviewed.

Although nutrient levels in Australian waters often exceed the level to cause eutrophication, eutrophication problems are less common. This is due to the characteristic environmental conditions of Australian lakes (Cullen and Smalls, 1981). Most commonly, Australian lakes are warm monomictic, with one complete mixing in winter, or polymictic. There are no dimictic lakes. The mean annual rainfall of 420 mm and a mean annual runoff of 45 mm are very low in comparison to other countries. Moreover, the distribution of rainfall is heterogeneous. Such climatic conditions make the majority of land surface vulnerable to some degree of aridity. Under this condition, the human demand for water has resulted in the construction of impoundment in all Australian states. Most Australian studies on eutrophication have been conducted on these impoundments, because of their economic importance.

Two environmental features affecting eutrophication are identified. The first is the great variability of water flow. Remarkable differences in TP export rates were recorded between drought years and normal years in Lake Burley Griffin. Flow rate of the water was more important than TP concentration in controlling TP export. The high erodability of Australian soils significantly affects TP export. If the drainage basin is disturbed, large quantities of soils can enter waterways and be transported in lakes so as to significantly influence lake eutrophication. Excessive growth of macrophytes is another factor affecting eutrophication in inland waters. After a national survey of eutrophication in Australian inland waters, aquatic macrophytes create a major environmental problem as well as cyanobacterial bloom, which impaired recreational uses and drinking-water supply in inland waters.

An example of lake eutrophication under such environmental influences has been well described on Lake Burley Griffin located in Canberra (Cullen, 1991). Lake Burley Griffin is a shallow elongate artificial lake characterized by highly variable water residence time, ranging from 1.9 days during flood periods to over 100 days during droughts. As a result of sewage inflow from the upstream town and flooded nutrient supply, the lake has been nutrient-rich with TP concentrations of 60 to 130 mg P/m^3 in the upper basin and 30 to 103 mg P/m^3 at the lower dam wall. The lake was dominated by nuisance growth of submerged macrophytes from the late 1960s through the 1970s, followed by the collapse of those macrophages during the 1980s. Phytoplankton density was low when macrophytes were prevalent, but increased after collapse of the macrophytes. Intensive surveys were conducted on the phosphorus budget and its contribution to the eutrophication of this lake. These studies revealed that removal of phosphorus from the sewage upstream resulted in decreases of effluent TP concentrations from 6 to 0.5 mg P/m^3 and of annual loading from 18 to 1.5 t/yr (metric tons per year).

The trophic state of freshwater lakes in Indonesia was intensively studied by the Sunda Expedition in 1928/29 (Ruttner, 1931; Thienemann, 1931) to provide evidence on limnological factors to control the trophic state in tropical lakes. No further limnological study was undertaken until 1950–1970, when studies were undertaken to support fisheries development (Nontji, 1994). By the late 1970s, many institutes, including BIOTROP and IPB, were involved in limnological studies. The Institute of Limnology, established in 1986, began active research and provided much information on the present state of Indonesian lakes and their environmental management (Giesen, 1994; Nontji, 1994).

Indonesia has over 500 lakes unevenly distributed throughout the archipelago. Tectovolcanic lakes are numerous and are located in volcanically active regions. Floodplain lakes cover more than 55 percent of the total lake area, and permanent lakes (including the largest, Lake Toba) account for 33 percent. Of 35 lakes surveyed, 12 were eutrophic, 3 mesotrophic, and 8 oligotrophic (Giesen, 1994).

Considerable human-induced changes of hydrology and water quality have been observed in the Indonesian lakes. Widespread deforestation or conversion of forests led to significant changes of lake hydrology and siltation in one-third and one-fourth of the lakes, respectively. In floodplain lakes, silt deposition resulted in fertile lake banks (Giesen, 1994). Another human-induced change is the pronounced infestation of exotic weeds. Water hyacinth (*Eichhornia crassipes*), introduced in 1984, and *Salvinia molesta,* first recorded in 1951, have spread widely in eutrophic waters and suppressed the growth of native aquatic plants. The problems of aquatic weeds are their wide distribution through Indonesia, nuisance occurrence, and negligible utilization value (Nontji, 1994). Heavy infestations of aquatic weeds reflect their rapid growth rates (increase rate in fresh weight of *Eichhornia* and *Salvinia* 3.8–27.9 percent per day). Excessive growth of aquatic weeds is harmful for fisheries, irrigation, water transportation, electricity generation, aesthetics, and human health. Aquatic weeds also enhance evapotranspiration and silting in lakes, reservoirs, and wetlands. Almost 60 percent of open waters of Kalimantan are densely covered with *Eichhornia* and *Salvinia* (Nontji, 1994). In Atiluhur Reservoir in western Java (Lake Maninjau in Sumatora and Lake Sentani in IrianJaya), blooms of cyanobacteria, *Microcystis,* cause a serious nuisance. Aquatic weed control using mechanical, chemical, and biological methods has been attempted with partial success. However, the rate of weed growth has overcome most of these efforts. Biological control using herbivorous fish and tawes has also been attempted to a limited extent (Nontji, 1994).

Africa. The ecosystems of lakes in tropical Africa are conditioned by high temperature of the water (~22 to 30°C) and seasonal variability in thermal stratification with relatively small vertical differences in water temperature, which is strongly influenced by wind-induced vertical mixing of the upper water mass. Phytoplankton seasonality in the deeper African lakes is related to the region-specific thermal and wind regime (Talling, 1966, 1969). Continued high temperature creates "endless summer" in which the biological control dominates the cycling of essential elements through the year (Kilham and Kilham, 1990). In Lake Victoria, Lake Malawai, and Lake Tanganyika, the concentrations of inorganic nitrogen and phosphorus of the lake waters are significantly lower than those of TN and TP, including organic forms, suggesting the importance of remineralization of organic matter in algal production (Hecky and Bugenyi, 1992). In Lake Tanganyika, vertical mixing contributes to the supply of more than 90 percent of phosphate in the eutrophic zone. The role of wind mixing in nitrogen supply is negligible because of a considerable dentrifiction loss in the oxic-anoxic interface. Biological nitrogen fixation dominates the nitrogen budget in the upper layer.

Nitrogen has been considered the most possible limiting nutrient for phytoplankton production in eastern African lakes (Talling and Talling, 1965; Viner, 1973), while limitation of algal growth by phosphorus has been demonstrated in hypertrophic Lake Naivasha, Lake Oloidien, and Lake Sonachi, Kenya (Kalff, 1983) and by phosphorus with nitrogen in Lake Chilwa (Moss, 1969).

As described by Kilham and Kilham (1990) and Hecky and Bugenyi (1992), there have been considerable changes in water quality in Lake Victoria since the 1960s, when exotic fish species were introduced. Silicon concentrations have not changed, and the Chl values of the inshore waters have increased two to four times during the past two decades. Explanations for these changes are based on the consequences of introducing exotic fish species and the edicts of intensive lakeside agriculture development and urban development (Bugenyi and Balirwa, 1989). Another explanation is the stimulated algal production due to the increased atmospheric supply of nitrate and sulfate. Rainfall comprised a greater part of the external nitrogen and phosphorus supply in Lake Victoria (Heckey et al., 1991). Atmospheric contamination from savanna burning or a high level of acid rain over tropical Africa might account for increased deposition of N and onto Lake Victoria (Kilham and Kilham, 1990; Hecky and Bugenyi, 1992).

In Lake Naivasha, rivers contributed about 78 and 71 percent of the external TN and TP loading, respectively. Because about 80 percent of the river TP was retained by a lakeside swamp, rainfall phosphorus falling on the lake surface was the major source of P to the lake (Kalff and Brumelis, 1993). According to Harper et al. (1993), the water level of Lake Naivasha declined between 1980 and 1987 to expose about 35 percent of the littoral soils, resulting in clearance of lakeside vegetation (*Cyperus papyrus* and other emergent plants). The long rain of 1988 caused a rapid water-level increase, resulting in an increase of conductivity and chlorophyll biomass (to 101 μg/L). Littoral vegetation during high water levels has an ability to filter silts and nutrients. Clearance of the vegetation allows inflow of silts and nutrients to the lake and results in high algal biomass and eutrophication of the lake.

Many reservoirs in South Africa have suffered from deterioration of water quality by discharge of secondary treated sewage effluent from high-density urbanized areas and industrial effluent (Toerien and Steyn, 1975; Toerien and Walmsley, 1978). In four reservoirs studied, N and P were the limiting nutrients for algal growth. Loading of dissolved N and P to these impoundments exceeded the permissible loading level recommended by Vollenweider. The relatively lower N/P ratio in loading could account for the development of the dense blooms of N-fixing cyanobacteria.

On the basis of limnological information from tropical Africa and other regions, Thornton (1987) summarized the general characteristics of eutrophication in lakes and reservoirs of tropical and subtropical regions. Productivity of tropical lakes and reservoirs at year-round higher temperature is higher with steady occurrence at phytoplankton blooms than that of temperate lakes. Thornton suggested higher boundary concentrations of 50 to 60 μg P/L and 20 to 100 μg N/L for mesotrophic and eutrophic lakes in tropical systems, as compared to those for the OECD temperate water bodies. Lakes and reservoirs in the tropics are also characterized by low nitrogen as a major limiting nutrient for algal production.

South America. The extensive range of latitude (12°N to 55°S) and the Andes Mountains across South America give the continent considerable topographic relief and a diverse climate. Thus, lakes and reservoirs in this continent are situated in varying climatic conditions ranging from temperate to tropical. In the tropical lakes of South America, the effects of high temperature on lake-water stratification and physiology and production of aquatic organisms are substantially identical to those in African lakes. In shallower lakes of flooded regions, the seasonal stratification pattern is complex with water mass mixing by surface cooling (Tundisi, 1984). In Lake Don Helvécio and Lake Carica, located in the Rio Doce Valley, mixing due to surface cooling is responsible for enrichment of the euphotic zone by nutrients in deeper layers (Tundisi and Matsumura-Tundisi, 1984). In Amazon Lake Jacaretinga under a strong influence of the river water inflow, diurnal change of lake water stratification was observed with the complete mixing of the lake water at night when the lake water level was low (Tundisi and Matsumura-Tundisi, 1984).

In Brazil, a great number of reservoirs have been constructed during the past three decades for hydroelectric power and water supply. Even in Sáo Paulo State there are more than 100 reservoirs. While most of them were constructed recently, in some reservoirs, eutrophication was considerable. Fifty-two reservoirs in Sáo Paulo State had Chl ranging from less than 1 to 50 mg/m^3 and primary productivity from 0.02 to 2 g C m^{-2} d^{-1} (Tundisi, 1981; Matsumura-Tundisi et al., 1981). Henry et al. (1985) reported that phosphorus was the limiting nutrient for algal growth in these reservoirs because of high concentration of nitrate by contamination with fertilizer and sewage, while in Labo Reservoir nitrogen was identified as the primary limiting nutrient (Henry and Tundisi, 1983). Recent studies of the water quality in five reservoirs illustrated that TP and TN concentrations ranged from 16 to 47 mg P/m^3 and from 770 to 1700 mg N/m^3, respectively, being at mesotrophic level according to the eutrophication criteria for tropical lakes (Tundisi et al., 1991).

Even within the tropical zone, the trophic state of lakes could differ with the location. In the survey of primary production of Andean and Amazonian tropical lakes of Ecuador (Miller et al., 1984), Chl concentrations and primary production were at very low levels in the low-altitude jungle lakes and the high-elevation mountain lakes. These lakes are persistently covered with cloud or fog with much rain. Cloudy weather could limit algal productivity of these lakes. In contrast, the lakes of inter-Andean plateau were productive, with an average 6.1 g C m^{-2} d^{-1}. Increased nutrient loading from the basins under considerable cultural impact seems to have contributed to the high productivity. There are a large number of lakes in the Andes Mountains. Surveys of water quality and biotic communities in these Araucanian lakes (Campos, 1984) indicated the orthograde distribution of dissolved oxygen with less hypolimnetic deficit, low phosphate concentration (1 to 16 μg/L), low Chl concentration (0.007 to 49.4 mg/L) and community structure of oligotrophic lakes. Thus, the Araucanian lakes are still in an oligotrophic state.

In temperate Argentina there are more than 400 lakes and reservoirs. The nationwide survey of 1984–1986 (Quiros, 1991) illustrated that Chl levels were below 1.0 mg/m^3 in the deep lakes of the Patagonean Andes Region, 1.0 to 80 mg/m^3 in shallow lakes of the Patagonian Plateau, 1.0 to 218 mg/m^3 in lakes and reservoirs of the central-western and northeastern arid regions, and as high as 400 mg/m^3 in ponds or shallow lakes on the Pampa Plain. In lakes and reservoirs with TN/TP ratios <22, P was the most important variable to explain Chl variation. Of all the lakes studied, 30 percent were limited by N by midsummer.

For the development of a simplified methodology to evaluate eutrophication in warm-water tropical lakes and reservoirs of Latin America and the Caribbean, CEPIS and PAHO/WHO have conducted a regional program since 1981 on about 40 warm-water tropical lakes and reservoirs of the regions. A *warm-water tropical lake* was defined as a water body with a minimum temperature of 10°C under normal conditions and a minimum annual mean of 15°C. The outline of the program and the survey results up to 1990 were presented by Salas and Martino (1991). The 29 lakes and reservoirs from which the data were employed for the model analysis described had an average depth of 1.0 to 26.4 m, TP concentration of 10 to 680 mg/m^3, and TP loading of 0.046 to 142.9 g m^{-2} yr^{-1}. Judging from the TN/TP ratios of the lake water, the majority of these warm-water tropical lakes and reservoirs in the CEPIS program were P-limiting.

The geometric-mean TP concentration for each trophic-state CEPIS lake was 119 mg/m^3 for eutrophic lakes, 39.6 mg/m^3 for mesotrophic lakes, and 21.3 mg/m^3 for oligotrophic lakes. These values were higher than those for OED temperate lakes (see Table 8.4). Fixed boundary values of TP to separate eutrophic/mesotrophic and mesotrophic/oligotrophic were, respectively, about 70 mg/m^3 and 30 mg/m^3, higher than those for the temperate lakes (see Table 8.3). This result is consistent with the report of Baker et al. (1981) on the boundary values for Florida lakes and Thornton's (1987) proposal for tropical and subtropical lakes. The model analysis used to predict the trophic response to TP loading illustrated that the overall loss rate of TP in lakes after inflow was double the magnitude of that for OECD temperate lakes. This might be due to the fact that most of the CEPIS program water bodies were reservoirs. Consequently, a higher sedimentation rate could be expected. The best-fit multiple-regression equation was derived for the in-lake TP concentration as a function of areal TP loading rate, detention time, and mean depth. Using this equation, the phosphorus concentrations of lakes and reservoirs in CEPIS regions as well as those in Africa could be predicted with less standard error.

Eutrophication Models

Model for environmental management

Along with the extension of knowledge on limnological and ecological processes, there has been considerable progress in the sciences of environmental management of lake ecosystems for the last 30 years. Eutrophication modeling is one of them and has been developed to understand reactions of the lake ecosystem to anthropogenic perturbation. Because of the great diversity and complex interactions of components in a lake ecosystem, it is difficult to predict the responses of a lake ecosystem to an environmental impact through integration of knowledge on respective issues in the lake ecosystem. Holistic analysis of the system by modeling provides an excellent tool to predict ecosystem reactions and ascertain unknown factors in the regulation of ecosystem dynamics (Jøorgensen and Mejer, 1976). The modeling approaches in eutrophication research on freshwater lakes are briefly reviewed here.

Summarizing the different approaches and objectives, the eutrophication models of lakes may be categorized into two groups: empirical models and conceptual models. Both approaches have provided new insight and excellent predictive tools for the decisions on environmental policies and actual management practices in many lakes. In many cases, correlation analysis of a few variables was employed to develop the models. Conceptual models are based on conceptual diagrams of the ecological and limnological processes in a given lake ecosystem. This approach may be further divided into two subgroups: mass-balance models and dynamic models. Mass-balance models are used to establish general rules in control of matter dynamics and are applicable to many lake systems. Generally, the models are simple so as to be easily understandable. Because of their failure to reproduce subtle change, the environmental analyses using these models are conducted on long-term bases such as an annual basis. For this feature, the empirical approach is often employed for parameterization of a mass-balance model. A representative example is the well-known eutrophication models of Vollenweider (1969, 1976). Dynamic models entail systems of differential equations with a number of variables. Each term reflects a respective important limnological and ecological process. It is also generally a function of a few subcomponental processes. Because of the limited knowledge of subcomponental processes in the natural lake ecosystem, the applicability of dynamic models has been restricted to a few water bodies from which the data were taken, although development of more general models is currently in progress (Jørgensen, 1986, 1989).

Biswas (1975) stressed the importance of developing the simplest model, which is not all-purpose and understandable for environmental managers. He also suggested developing the model in parallel with data collection and trying to update the models with more data with the best knowledge on the ecosystems modeled. Jørgensen (1989, 1990) also documented the importance of thorough knowledge of the ecosystems modeled and understanding both the purpose of the model and the data employed for model development.

A considerable number of models have been developed for prediction and management of eutrophication and its remediation for the last 20 years (Jørgensen, 1976, 1989; Reckhow, 1979; Jørgensen et al., 1981; Ryding and Rast, 1989). For details, see the cited references. Here, I review case studies on the modeling for two basic issues related to lake eutrophication: the control of algal growth by in-lake phosphorus concentration and the control of in-lake water quality by reduced phosphorus loading.

Phosphorus vs. Chl models

For environmental assessment of lake eutrophication, it is essential to predict responses of phytoplankton to increased or reduced loading of nutrients due to environmental changes and management practices. As a step to attain this goal, quantitative information on effects of the changing concentration of limiting nutrients on phytoplankton yield in natural lakes is of paramount importance. The first information on this subject was provided by Sakamoto (1966) on Japanese lakes. He noted that Chl yield was a linear logarithmic function of TP and TN concentrations in the surface lake waters in late spring, early summer, and early autumn. He also suggested that at the TN/TP <15 to 17 and TN/TP >9 to 10 Chl yield was nearly balanced with phosphorus and nitrogen but at the TN/TP <10 nitrogen was limiting, and at the TN/TP >17 Chl was limited by phosphorus. On the basis of Sakamoto's finding and other published data as well as their own data, Dillon and Rigler (1974) developed a general TP-vs.-Chl relationship to predict average summer Chl from spring TP. A considerable number of Chl-vs.-TP empirical relationships had been presented afterward. There was a great difference in Chl yield for a unit TP among the reports. Nicholls and Dillon (1978) documented that different slopes of the regression lines accounted for differences in sampling time and analytical techniques as well as variability of cellular Chl content and TP availability.

Because of the ecological importance of phosphorus as a key factor of eutrophication and phytoplankton production, a number of further studies have been conducted on this subject. In the OECD Monitoring Surveys, regional features of the regressions of TP to Chl and other variables were examined (Table 8.10). There was a significant difference in the exponent and intercept among regions. What is responsible for the variation of TP-vs.-Chl relationships? On the basis of empirical and theoretical analysis of the data of 228 temperate lakes, Smith (1982) developed the multiple-regression model of Chl as a function of TP and TN and suggested the importance of the TN/TP ratio for Chl yield. In the remediation process of 16 temperate lakes by P reduction, each lake responded differently to P reduction. Smith and Shapiro (1980) noted that the difference resulted from changes in the TN/TP ratio of lake water after P reduction. In oligomesotropic Canadian lakes, a regression model as a function of TP and TN/TP explained 92 percent of the Chl variance during the summer stratification period (Dillon et al., 1988). The response of annual average Chl during 9 years in P-limited lakes in central Ontario to the change of TP was highly variable. This could not be explained by the TN/TP ratio (Molot and Dillon, 1991). On the basis of an analysis of the published data from a number of lakes, Prairie et al. (1989) showed that the slope of the Chl-vs.-TP and Chl-vs.-TN regression equations varied concomitantly with the TN/TP ratios and was highest with the TN/TP ratios of 23 to 28, beyond which the slope declined. This indicated that the change of Chl with TP and TN can be described by a sigmoid curve. In many reports the Chl-vs.-TP relationship was described by a sigmoid curve rather than a linear model (Forsberg and Ryding, 1980; Canfield, 1983; McCauley et al., 1989; Prairie et

TABLE 8.10 OECD Relationships between Annual Phosphorus Load and Water-Quality Parameters

OECD project	Derived relationship	n	r	SE
	Annual Mean Total Phosphorus Concentration, μg/L			
Combined OECD study	$1.55X^{0.82}$	87	0.93	0.192
Shallow lakes and reservoirs	$1.02X^{0.88}$	24	0.95	0.185
Alpine lakes	$1.58X^{0.83}$	18	0.93	0.212
Nordic lakes	$1.12X^{0.92}$	14	0.86	0.252
U.S. study	$1.95X^{0.79}$	31	0.95	1.60
	Annual Mean Chlorophyll Concentration, μg/L			
Combined OECD study	$0.37X^{0.79}$	67	0.88	0.257
Shallow lakes and reservoirs	$0.54X^{0.72}$	22	0.87	0.238
Alpine lakes	$0.47X^{0.78}$	12	0.94	0.189
Nordic lakes	$0.13X^{1.03}$	13	0.82	0.329
U.S. study	$0.39X^{0.79}$	20	0.89	0.261
	Annual Maximum Chlorophyll Concentration, μg/L			
Combined OECD study	$0.74X^{0.89}$	45	0.89	0.284
Shallow lakes and reservoirs	$0.77X^{0.86}$	21	0.88	0.276
Alpine lakes	$0.83X^{0.92}$	11	0.96	0.191
Nordic lakes	$0.47X^{1.00}$	13	0.77	0.373
U.S. study	—	—	—	—
	Annual Mean Secchi Depth, m			
Combined OECD study	$14.7X^{-0.39}$	67	−0.69	0.237
Shallow lakes and reservoirs	$8.47X^{-0.26}$	26	−0.55	0.237
Alpine lakes	$15.3X^{-0.30}$	18	−0.74	0.171
Nordic lakes	—	—	—	—
U.S. study	$20.3X^{-0.52}$	22	−0.82	0.196
	Areal Hypolimnetic Oxygen-Depletion, g O_2 m^{-2} d^{-1}			
Combined OECD study	$\sim 0.1X^{0.55}$	—	—	—
Shallow lakes and reservoirs	—	—	—	—
Alpine lakes	—	—	—	—
Nordic lakes	$0.085X^{0.467}$	—	—	—
U.S. study	$0.115X^{0.67}$	—	—	—

Key: $X = \{[L(P)/q_s]/(1 + \sqrt{t_w})\}$ = flushing corrected average annual phosphorus inflow concentration (OECD, 1982)—see Eq. (8.6); n = number of data points; r = correlation coefficient; SE = standard error of estimate; dash (—) indicates data not available.
SOURCE: Modified from OECD (1982) and Ryding and Rast (1989).

al., 1989; Watson and McCauley, 1992; Mazumder, 1994a, 1994b). Watson and McCauley (1992) found that a large inedible phytoplankton exhibited a nonlinear response with increasing TP level while the small edible algae did not vary with TP level. There might be a change of the herbivore community toward higher TP level, which might account for the observed sigmoid relationship of Chl vs. TP.

Mazumder (1994a, 1994b), who studied the effect of herbivores on the Chl-TP relationship in both experimental systems and natural lakes, noted the generation, at a given TP, of greater Chl yields in systems with a prevalence of smaller algae and without large *Daphnia* but lower Chl yields in systems with large *Daphnia*. He also detected that a sigmoid pattern of the Chl-TP relationship was related to the transition from stratified to mixed systems along with trophic gradient.

Mass-balance models

Lake eutrophication is triggered by the increase of the concentration of the limiting nutrient(s) in a water body due to its (their) increased supply. For the management purpose of eutrophication, it is necessary to know the mass balance of the limiting nutrient(s) in a lake. According to Vollenweider (1969), in the lake as a mixed reactor in which there are input, output, and deposition of nutrients, the mass balance of P as total P in the lake can be described as

$$\frac{d[\mathrm{P}]}{dt} = \frac{J}{V} - \sigma[\mathrm{P}] - \rho[\mathrm{P}] \tag{8.1}$$

where $[\mathrm{P}]$ = in-lake P concentration
J = annual P loading rate
V = lake volume
σ = sedimentation rate (yr^{-1})
ρ = flushing rate (yr^{-1})

The flushing rate ρ is equal to the total hydraulic load Q (yr^{-1}) divided by lake volume V.

The steady-state solution (at $d[\mathrm{P}]/dt = 0$) of this differential equation is

$$[\mathrm{P}] = \frac{L}{z(\sigma + \rho)} \tag{8.2}$$

where L is areal P loading rate and z is lake mean depth. On analysis of the data pool, Vollenweider (1976) noted that σ may be approximated by

$$\sigma = \frac{10}{z} \tag{8.3}$$

since $\rho = 1/T_w$ (where T_w = water residence time in year), L is described as

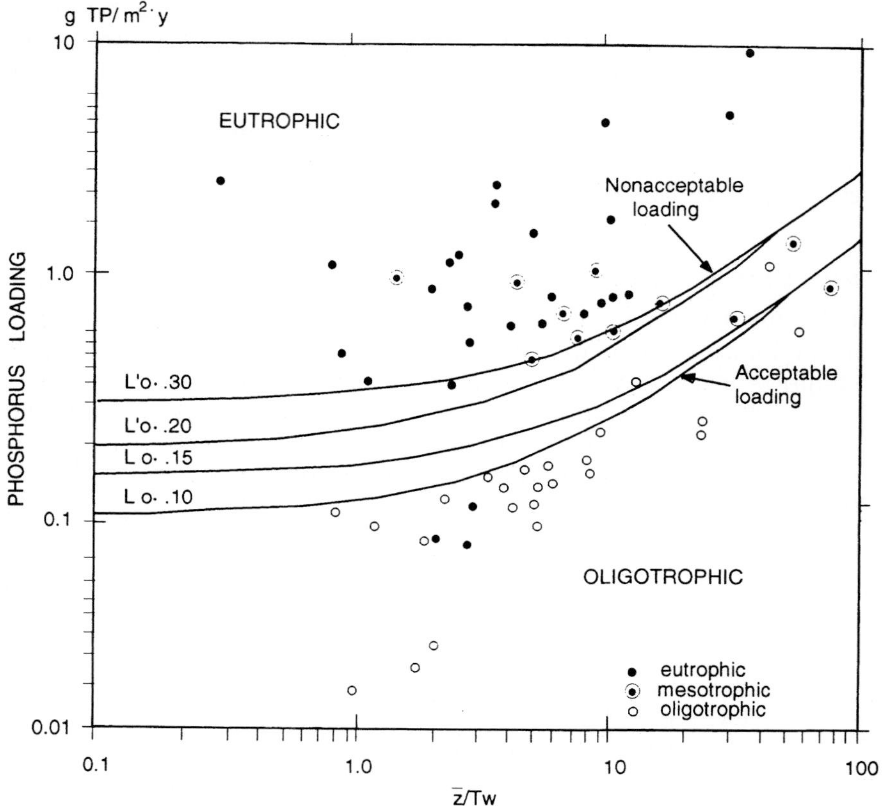

Figure 8.11 Relationship between phosphorus loading and mean depth z/hydraulic residence time (T_w) (Vollenweider, 1976).

$$L \quad (\text{mg/m}^2\cdot\text{yr}) = [\text{P}]\left[\left(\frac{z}{T_w}\right) + 10\right] \tag{8.4}$$

Figure 8.11 shows the plots of L-vs.-z/T_w data from more than 60 lakes (Vollenweider, 1976). The plots for eutrophic lakes were distributed above the L-vs.-z/T_w curve at [P] = 20 mg/m^3 (L_0, L at $z/T_w = 0$ is 0.2 g P/m^2·yr) and the data for oligotrophic lakes, below that at (L_0 is 0.1 g P/m^2·yr). In his initial work, Vollenweider (1969) proposed a concept of permissible and dangerous loading as a threshhold level for eutrophication and suggested the critical values as a function of mean depth. The two lines representing 20 and 10 mg/m^3 in Fig. 8.11 indicate nonacceptable and acceptable loading, respectively, reflecting a revised presentation of the original dangerous and permissible loading.

From Eq. (8.2) and empirical analysis, Vollenweider (1976) developed a new model to estimate critical area loading L (transitional loading between oligotrophic and mesotrophic states) (Reckhow, 1979) as a function of critical in-lake P concentration $[P_c]$, q_s, and z:

$$L_c\ (\text{mg/m}^2\cdot\text{yr}) = [P_c]q_s(1 + (\sqrt{z/q_s})) \tag{8.5}$$

where q_s (m/yr) is the annual areal hydraulic load (= Q/A) and A is lake surface area. Since $z/q_s = T_w$, writing Eq. (8.5), [P] is described as a function of the annual areal P loading L, annual areal hydraulic load (q_s), and water residence time T_w:

$$[P] = \frac{L}{q_s}\left(\frac{1}{1 + \sqrt{T_w}}\right) \tag{8.6}$$

Around the same time as Vollenweider (1976), Dillon (1975) considered the importance of hydraulic flushing in determining the in-lake P concentration and provided a model to describe [P] as a function of P loading and P retention coefficient R (Dillon and Rigler, 1975):

$$[P] = L\ \frac{1 - R}{z} \tag{8.7}$$

According to Vollenweider (1976), R is defined as $\sigma/(\sigma + \rho)$. Because a direct measurement of σ is generally difficult, especially in collecting reliable data in many lakes (Dillon and Rigler, 1975; Larsen and Mercier, 1976), R has generally been calculated as

$$R_{\text{exp}} = 1 - \frac{\text{outflow load}}{\text{inflow load}} \tag{8.8}$$

Because outflow and inflow load are measurable, the in-lake P concentration [P] can be predicted from L using Eq. (8.7).

Larsen and Mercier (1976) noted that in-lake TP concentration [P] at steady state can be described as a function of the average influent TP concentration [p] and R_{exp}, which ranged from 0.00 to 0.95 in low productive lakes in North America:

$$[P] = [p](1 - R) \tag{8.9}$$

The graph [Fig. 1 in Larsen and Mercier (1976)] showing the [p]-vs.-R relationship with the declination lines (10 and 20 mg P/m^3) for oligotrophic-mesotrophic and mesotrophic-eutrophic lakes clearly identified the trophic state of surveyed lakes, suggesting its usefulness for prediction of the trophic state.

Because of their simplicity and general applicability, the mass-balance models described by a simple equation have been widely employed for prediction and assessment of water quality (Ahl, 1975; Forsberg et al., 1978; Benndorf et al., 1981; Faafeng and Nielsen, 1981). In particular, the OECD models developed from Vollenweider's model using numerous data from international monitoring surveys have been used as the global model with a general applicability for environmental assessment of many lakes in the world. One example OECD model is presented in Fig. 8.12.

Although the mass-balance model has general applicability, it does have some limitations. Studies on water quality of the inland waters in Australia indicated that the OECD model was not useful in Australian reservoirs (Cullen and Smalls, 1981). There was overestimation of [P] due to a high proportion of particulate P entering the lakes and its deposition on the lake bed. Limited light availability by phytoplankton due to high turbidity of the lake water may be respon-

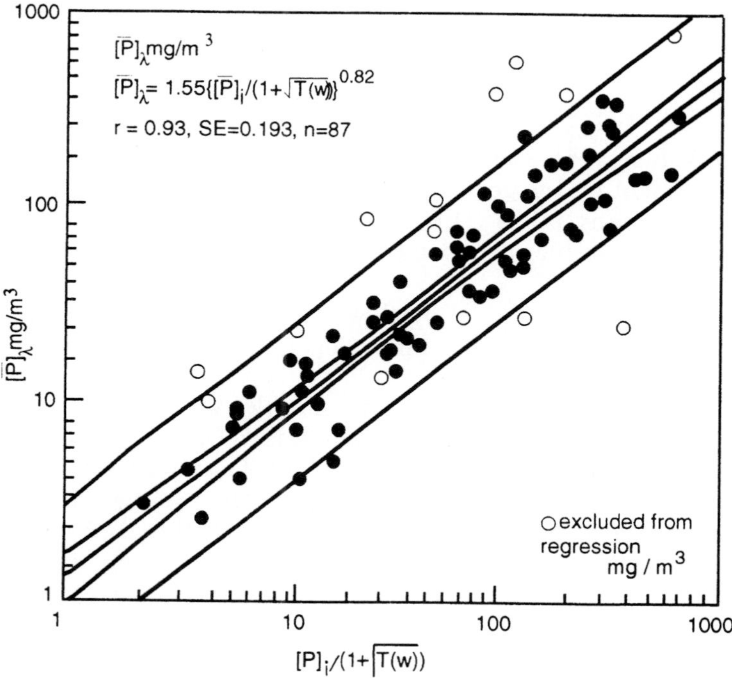

Figure 8.12 Annual average in-lake TP concentrations $[P]_\lambda$ as a function of annual average inflow TP concentration ($[P]_j/1 + \sqrt{T_w}$) (OECD, 1982).

sible for this low applicability. The fact that a major part of aquatic production was due to the presence of macrophytes could also account for the limited usefulness of Vollenweider's models in Australian lakes. A steady state through a time period such as a year is assumed in Vollenweider's mass-balance model. A variable T_w could exert significant influence on the calculated results. In northern Irish lakes with T_w<1 year, Foy (1992) found that the OECD model underestimated the in-lake P concentration. In the lakes with shorter T_w, short-term changes of both external P loading and hydraulic loading may violate Vollenweider's (1976) steady-state assumption and give lower estimates of [P].

From studies on the lake rehabilitation process after P reduction, much evidence has been presented on the slow response of water quality to reduced P loading such as sewage diversion (Larsen et al., 1981; Boers et al., 1993). Continued internal P loading from the bottom sediments was definitely responsible for the observed slow recovery of water quality. Sas (1990) noted that the exponent of the recovery trajectory of the in-lake P concentration as a function of reduced loading was significantly different from those in the Vollenweider-OECD model, because of remaining old P stock and continued release of P from the sediments.

Another problem associated with the use of a general "global" model is that the trophic-state criteria were different for temperate lakes and tropical lakes. As described before, in temperate lakes, if [P] is less than 10 mg P/m^3, the lake should be oligotrophic; if greater than 20 mg P/m^3, it will be eutrophic. In tropical and subtropical lakes, the corresponding values are around twofold or threefold larger (Thornton, 1987; Salas and Martino, 1991). This might be more or less due to greater metabolic activities and higher nutrient circulation rates in tropical and subtropical regions.

Dynamic models

Although much experience has shown that simple models can offer sufficient information for most management decisions as exemplified by the OECD regression models, the complex dynamic models play a valuable role in improving our understanding of eutrophication (Rast and Kerekes, 1981). In the dynamic models, manipulating the state variables can provide valuable insight on the control factors of eutrophication. They can simulate the complex processes (simulation models) of an aquatic ecosystem and provide essential information on the process to control a specific issue in a lake ecosystem such as cyanobacterial bloom.

There has been considerable development of ecosystem modeling for the past 20 years, and a large number of complex models on eutrophication have been presented. According to the overview of eutrophication models reported after the 1970s (Jørgensen, 1989), there was a different level of complexity in eutrophication models from a biologically simple system of no breakdown in trophic levels in the hydrodynamically undifferentiated lakes to biologically complex models, including several species at various trophic levels in two- or three-dimensional layers with horizontal and vertical gradients.

Some case studies using the complex models for predicting the changes of water quality in eutrophied lakes are reviewed here. The SALMO model was developed for scenario analysis to select an optimal environmental control strategy in both reservoirs and lakes in Germany (Benndorf et al., 1981; Recknagel and Benndorf, 1992). In the hypertrophic Beeiloch Reservoir, where eutrophication could not be controlled by P load reduction, examination of the response of several control measures showed that the reduced internal P load can exert no effect on phytoplankton because of the small area of hypolimnetic sediments but that the reduction of zooplankton mortality to 10 percent by biomanipulation would lead to a significant decrease of algal biomass.

In shallow lakes in the Netherlands, eutrophication has been a severe problem in water-quality management because of high algal biomass, dominance of cyanobacteria, and loss of emergent plants. Increased nutrient loading was caused by inflow of domestic sewage, industrial wastewater, and agricultural runoff as well as nutrient loading from surrounding countries via the River Rhine (Boers et al., 1993). Two-thirds of the annual freshwater supply originated from the River Rhine, which carried large amounts of nutrients to lakes. The bottom muds also supply considerable amounts of nutrients, because most lakes are shallow and unstratified (Los et al., 1984). P and N removal by wastewater-treatment plants and restriction of agricultural manure supply to fields was programmed with 50 percent reduction in 1995 and 70 to 75 percent in 2000. Dredging of the bottom muds was undertaken in some lakes to reduce internal loading (Boers et al., 1993). The effects of these measures were predicted using mathematical models (Los et al., 1984; Brinkman et al., 1988; Los and Brinkman, 1988; Janse and Aldenberg, 1991; Boers et al., 1993). The revised phytoplankton model, BLOOM II, was employed in combination with a chemical model and a water-quality model to reproduce alga dynamics depending on nutrient dynamics. Application of this model to Lake Veluwe and Lake Wolderwijd in 1975–1983 indicated the agreement of the computed and measured Chl values in both lakes. It predicted the dominance of cyanobacteria *Oscillatoria agardhii* in all cases. Similar analysis using BLOOM II coupled with other models in lake systems in Friesland (Brinkman et al., 1988) indicated that flushing could cause the shift of cyanobacteria to green algae and sanitizing domestic sewage disposals could reduce P concentrations by at least 75 percent.

In a eutrophication model employed for eutrophic Loosdrecht Lakes (Janse and Aldenberg, 1991), all biotic and abiotic components were modeled in carbon and phosphorus, bringing the number of state variables to 18. The upper sediment and the water just above the sediment were included in modeling the P cycle. It showed that the dynamics of algal P/C ratios could be an important factor in the ecosystem resilience to the changes of P loading. It also showed no further changes in the system without further P reduction.

Seip (1990) noted that food-web models can allow prediction of the timing of algal blooms, specific composition, and average biomass over a given growth period, whereas a regression model usually predicts only the average biomass for the growth period. By applying the NORECO models reflecting food webs and the OECD regression model to eight Norwegian lakes, the food-web models were developed and provided better and more information on lake behaviors than did the regression models. This suggests that dynamic models can be of wider applicability to many lakes while a general eutrophication model applicable for all lakes, such as the regression model, is nonexistent in the complex dynamic model (Rast and Kerekes, 1981).

Strategy for Lake Recovery: Toward Sound Management of the Lake-Watershed System

N and P loading as the cause of eutrophication

The present review of lake eutrophication clearly indicates that accelerated eutrophication of freshwater lakes in the world was caused by an increase in supply of P and N and resulting enhanced algal growth. P was the major limiting nutrient in a larger percentage of temperate lakes and reservoirs. N was a factor limiting algal production in Lake Tahoe and Lake Titicaca as well as in tropical lakes. In high eutrophic lakes, N also often played an important role as a major limiting nutrient because of internal load of phosphorus and denitrification loss of N at bottom sediments. When TN/TP ratios were larger than 10 and smaller than 17, the colimitation by both P and N was documented.

For this reason, the major effort on eutrophication control in the temperate lakes has been given to P reduction from the sewage and industrial effluents or complete diversion of P containing sewage from the lakes. However, this remediation measure is not always applicable to all eutrophic lakes. There are many naturally eutrophic lakes in fertile lowlands. In these naturally eutrophic lakes, P reduction is meaningless. P reduction can be applied only to the lakes where the P level of lake water is higher than the natural level (as expected from natural loading controlled by the characteristics of soils in the basin and atmosphere supply) and often causes serious impairment of water uses.

Many nutrient sources located inside and outside the watershed can contribute to P loading into lakes. For rehabilitation of eutrophied lakes, it is necessary to quantify their contributions prior to the application of control measures to eutrophied lakes. The external nutrient sources are grouped into two sources: point sources and nonpoint sources. *Point sources* are locally concentrated sources such as municipality sewage and their industrial effluents. Their contributions to P loading are relatively easy to evaluate, and it is also easy to collect them in sewer systems. Reviews on this subject indicate that in lakes located along urbanized regions, especially with higher population densities, the contribution of domestic sewage and industrial wastewater to total P and N loading is considerable. Discharge of raw sewage and secondary treated sewage rich in nutrients have increased P and N loading levels in these lakes and caused considerable algal bloom, resulting in severe water-quality changes such as decreased hypolimnetic oxygen concentration and increased liberation of ferrous and manganous ions.

As *nonpoint* (diffuse) *sources* of nutrient supply, atmospheric sources, such as dry fallout and rainwater as well as terrestrial sources such as discharge from urban and rural areas and from agricultural lands and forested area, are of vital importance (Fricker, 1980). Atmospheric deposition of N and P has ranged from 1 to 19 g N m^{-2} yr^{-1} and 0.1 to 0.9 g P m^{-2} yr^{-1} (Jansson, 1979; Mosello et al., 1990; Dillon et al., 1991), although gradual increase of N inputs is expected as a result of anthropogenic pollution of rain water. The observed atmospheric inputs were at a level roughly similar to that of N and P loading from terrestrial sources into oligotrophic lakes with a mean depth $>$10 m (Vollenweider, 1968), suggesting that the atmospheric impacts were less significant in mesotrophic and eutrophic lakes. Conversely, the contribution on nonpoint sources on the watershed is generally considerable there.

The minimal recovery of water quality of the lakes after P reduction of sewage or sewage diversion was due mostly to continued nutrient supply from nonpoint sources on the watershed or from internal sources. Vollenweider (1968) and Reckhow et al. (1980) provided extensive information on export coefficients of N and P from various land uses. A considerable range of variations were detected in their export from agricultural fields, pastures, manure storage areas, and urbanized areas. The variation seems to be due to differences in agriculture practices, including fertilization and manure handling; and different effects of climate, topography, and soils; and anthropogenic disturbances of the watershed (Reckhow et al., 1980).

P and N releases from the bottom sediments are significant internal sources that greatly affect increased algal production (Ryding and Forsberg, 1976; Larsen et al., 1981). Recent efforts to reduce P loading in shallow eutrophic lakes often resulted in little decline of in-lake P concentration (Boers et al., 1993). In lakes with a long history of high external P loading, even after reduction of external loading, the sediments continued P release, especially at anoxic conditions. Bengtsson (1975) observed a P release of up to 22 to 40 mg P m^{-2} d^{-1} from a highly eutrophic lake sediment. This internal load corresponded to 90 percent of the total P load. The published values of P release in anoxic sediments ranged from 0.8 to 34 mg P m^{-2} d^{-1} (Nürnberg, 1984). For 23 stratified lakes with anoxic hypolimnion, the P release from the sediments contributed 39 percent on average to the total P load (Nürnberg and Peters, 1984). These observations clearly indicated a significant importance of internal loading of P from the bottom sediments in eutrophication and therefore its remediation.

Strategy of lake restoration

Because P and/or N is the dominant factor in eutrophication, the relevant control measure is to decrease P and N concentrations in lake water below the permissible level. The permissible level could vary with the effects of changed water quality on human use of water bodies and biotic communities in the lakes. Considering the limnological aspects of the trophic state, Vollenweider (1976) subjectively defined 10 mg P/m^3 as the permissible level of P concentration as a boundary value between oligotrophic and mesotrophic lakes in temperate regions. In OECD criteria, this permissible level was applied to the trophic state of lakes as defined using the measures of Chl, TP, and Tr. Berge (1990) defined the acceptable TP level in terms of predominance of cyanobacteria and provided an inverse relationship between acceptable TP level and mean lake depth, with the lowest value of 7 mg P/m^3 in lakes deeper than 15 m. This suggests that the permissible level of eutrophication would differ with one's view on the water quality and also differ between temperate and tropical regions as described in the section on mass-balance eutrophication models (above).

Several techniques have been employed to decrease P and N levels in lake water. They are divided into two groups. First is the external control of nutrient inputs from outside the lakes. The techniques of sewage diversion and sewage treatment at sewage-treatment plants with P and N reduction are most effective in reducing nutrient loading. Lake recovery in Lake Washington, the Laurentian Great Lakes, and many European lakes has demonstrated the effectiveness of this remediation measure [see section on world lake and reservoir eutrophication case studies (above)].

However, even after sewage diversion and P reduction from sewage and wastewater, many reports indicated only minimal recovery because of continued nutrient release from bottom sediments or continued nutrient inputs from diffused sources on the watershed. To reduce these continued nutrient inputs, two methods have been applied: (1) decreasing in-lake nutrient concentration by internal control and (2) reducing external nutrient inputs by watershed management.

Cooke (1983) and Cooke et al. (1986) reviewed various in-lake restoration techniques of eutrophication. Dilution and flushing, phosphate inactivation and precipitation, sediment removal, hypolimnetic withdrawal, and hypolimnetic aeration have been employed for reduction of in-lake P concentration. Successful lake restoration by sediment removal in Lake Trummen was described earlier in the world lake-reservoir eutrophication case-studies section. For details, see USEPA (1980), Cooke (1983), and Cooke et al. (1986). Biological controls to reduce biomass of algae and macrophyte by lake-level drawdown, sediment cover, and harvesting are direct procedures destined to have a cosmetic effect on the lake. Although successful trials of vegetation control were reported, the vegetation may be soon recovered without continued controls.

Two new techniques are cited here as promising remediation measures of lake eutrophication: (1) biomanipulation of algae and nutrients and (2) nutrient removal by wetland. Both methods are good use of the biological processes of aquatic organisms in and along the lakes. In terms of sound conservation of lake ecosystems, these techniques seem to make a greater contribution to sustainable management of lake environments.

Lake restoration by biomanipulation was first reported by Shapiro and Wright (1984). Lake biological manipulation is based on the theory that increased piscivore abundance will result in decreased planktivore abundance and increased zooplankton abundance, followed by increased zooplankton grazing pressure, which can lead to reduction in phytoplankton abundance and improved water clearness. After elimination of planktivorous and benthivorus fishes, a high-density piscivore population was restocked in a small lake and the changes of phyto- and zooplankton abundance were studied over a 2-year period (Shapiro and Wright, 1984; Wright and Shapiro, 1984). Consistent low abundance of phytoplankton and zooplankton was recorded. Their studies were followed by a number of experimental examinations. Fifty papers have been presented on food-web manipulation from 1961 to 1989 (DeMelo et al., 1992). In a small lake in Norway (Faafeng and Brabrand, 1990), removal of fish caused mass occurrence of filter-feeding zooplankton, and led to a dramatic reduction in both algal biomass and percentage of cyanobacteria despite high levels of dissolved P and N in the lake. Reduction of P in the lake with high fish biomass led to linear decrease of phytoplankton, but in a year with low fish density a greater decrease of algal density was observed. Even with a small population of fish, the fish reduced herbivore zooplankton and allowed phytoplankton to utilize the available nutrients to maximum growth. Although a consistent response of biomanipulation has not been obtained in all studies, it was suggested that food-web control could have a significant effect on phytoplankton biomass in eutrophic water. Reynolds (1994) pointed out the importance of experimental conditions for successful biomanipulation. Further studies are needed for reproducible responses of biomanipulation before it can be applied for eutrophication control.

Aquatic macrophytes and emergent plants are characterized by greater biomass during the vegetation season and high contents of N and P in their plant bodies. If vegetation is removed during the vegetation season, it provides a good opportunity to reduce the N and P in inflowing water through vegetation removal. Kis-Balaton, by Lake Balaton in Hungary, was originally natural wetland through which the River Zara drained into the lake under considerably accelerated eutrophication (Herodek et al., 1988). In Lake Balaton, at least half of the nutrients were supplied from diffuse sources, so they could not be removed before entering the river by the traditional method (Pomogyi, 1989). In order to restrict the transport of plant nutrients at the mouth of the river flowing in Lake Balaton, the embankment with a bridge construction between Kis-Balaton and Lake Balaton isolated the Kis-Balaton as a shallow reservoir (Herodek et al., 1988; Dióos, 1989). After flooding the reservoir, very rapid changes in macrophyte communities occurred. Around half of the reservoir was covered with Scirpo-Phragmitetum and the *Polygonetum amphibi* communities. This vegetation accumulated 120 t of N and 20 t of P in 1987. Harvesting of these plants removed 48 t of N and 0.7 t of P, which corresponded to 20 percent of the total external and internal loading (Pomogyi, 1989).

In the Chaohu Lake catchment in China, there are 150 reservoirs and several canals together with thousands of ponds. Reservoirs, ditches, and ponds from complicated irrigation systems support the high yields of rice, wheat, and other crops. The studies of nutrient budget on the rainfall events revealed that multipond system had a trapping efficiency of 99 percent for N and 98 percent for P (Yin et al., 1991, 1993). The system was less efficient when rainfall was intense and especially when some ponds were full, causing water overflow.

For control of non-point-source pollution from urban and rural areas, increasing efforts have been made to construct wetlands (Brix, 1994; Mitsch, 1994). Many reports have been presented on the role of natural wetlands along and in rivers in improving water quality by processes such as sedimentation, denitrification, and biological uptake (Peterjohn and Correll, 1984; Mitsch et al., 1989; Fleischer et al., 1991; Mitsch, 1994; Yan and Zhang, 1994). Studies for several sites, including natural instream wetland and constructed riverine marsh in the United States, indicated a considerable retention of P and sediments with comparable retention rates in both natural and constructed systems in wetlands [wetland, 1 to 4 g P m^{-2} yr^{-1}; sediments, 200 to 2000 P m^{-2} yr^{-1} (Mitsch, 1994). These studies suggest very effective role of wetlands for controlling non-point-source pollution at concentrations much lower than in the common wastewater-treatment plants.

Not all lakes are associated with wetlands around lakes. Construction of wetlands with aquatic vegetation at or near the mouth of inflowing rivers and streams might be, if hydrologically applicable, an appropriate control measure for reducing nutrient inputs from nonpoint sources. There is a limitation of nutrient and sediment removals by wetlands. Frequent refreshment and management to maintain healthy wetland systems are needed. A considerable supply of nutrients from nonpoint sources is associated mostly with unwise management of agricultural lands, manure storage, and wastewater from urbanized area. Development of a sustainable agricultural production system, coupled with recycling of agricultural waste and manure, as well as establishment of new lifestyle in urbanized areas with a focus on nutrient recycling are of paramount importance for successful management of eutrophication through lake-watershed ecosystems.

Acknowledgments

This review was undertaken in cooperation with the International Lake Environment Committee (ILEC), Kusatsu, Japan, under the editorial supervision of Prof. A. K. Biswas, which is gratefully acknowledged. Sincere thanks are given to Prof. T. Kira and Dr. G. Ando for their kind support of this review; Dr. S. Ide and the staff of ILEC for their help in providing published information and in handling the manuscript; and the librarians of the University of Shiga Prefecture for research help.

References

Introduction

ASLO (American Society of Limnology and Oceanography) (1972). Likens, G. E. (ed.). *Nutrient and Eutrophication: The Limiting-Nutrient Controversy. Special Symp.* no. 1. American Society of Limnology and Oceanography.

National Academy of Sciences, Washington, D.C. (1969). Rohlich, G. A. (ed.). *Eutrophication: Causes, Consequences and Correctives. Proc. of Symp.* National Academy of Sciences, Washington, D.C.

OECD (Organization for Economic Cooperation and Development) (1970). *Eutrophication in Large Lakes and Impoundments.* Uppsala Symposium, OECD.

OECD (Organization for Economic Cooperation and Development) (1982). *Eutrophication of Waters. Monitoring of Inland Waters: Eutrophication Control.* Environmental Directorate, OECD, Paris.

Ryding, S.-O., and Rast, W. (1989). *The Control of Eutrophication of Lakes and Reservoirs.* Man and the Bioshere Series 1. UNESCO, The Parthenon Publishing Group Ltd.

Vollenweider, R. A. (1968). *Scientific Fundamentals of the Eutrophication of Lakes and Flowing Waters, with Particular Reference to Nitrogen and Phosphorus as Factors in Eutrophication.* OECD, DAS/CSI/68.27.

Vollenweider, R. (1989). Global problems of eutrophication and its control. In Salánki, J., and Herodeck, S. (eds.). *Conservation and Management of Lakes. Symp. Biol. Hung.* **38:** 19–41.

Specific features of eutrophication studies

Ahlgren, I. (1978). Response of Lake Norrviken to reduced nutrient loading. *Verh. Internatl. Verein. Limnol.* **20:** 846–850.

Barica, J. (1989). Recovery of the Laurentian Great Lakes, 1970–1985: eutrophication aspects. (1989) In Salánki, J., and Herodeck, S. (eds.). *Conservation and Management of Lakes. Symp. Biol. Hung.* **38:** 43–59.

Barica, J. (1990). Long-term trends in Great Lakes nutrient concentrations: a summary. *Verh. Internatl. Verein. Limnol.* **24:** 369–370.

Bengtsson, L. (1975). Phosphorus release from a highly eutrophic lake sediment. *Verh. Internatl. Verein. Limnol.* **19:** 1107–1116.

Boers, P. C. M., van der Does, J., van der Molen, D. T., and Roip, W. J. (1993). Eutrophication control in the Netherlands by reduction of P loadings: approach, results, challenges for the future. In Giusani, G., and Callieri, C. (eds.). *Proc. 5th Internatl. Conf. Conservation and Management of Lakes,* pp. 50–53.

Clasen, J., and Bernhardt, H. (1979). *Final Report of Regional Project, Shallow Lakes and Reservoirs,* vol. 1. Cooperative Programme for Monitoring of Inland Waters, OECD.

Cullen, P., and Forsberg, C. (1988). Experience with reducing point sources of phosphorus to lakes. *Hydrobiologia* **170:** 321–336.

Dillon, P. J., and Rigler, F. H. (1975). The chlorophyll-phosphorus relationship in lakes. *Limnol. Oceanogr.* **19:** 767–773.

Edmondson, W. T. (1969). Cultural eutrophication with special reference to Lake Washington. *Mitt. Internatl. Verein. Limnol.* **17:** 19–32.

Edmondson, W. T. (1972a). Nutrient and phytoplankton in Lake Washington. *Limnol. Oceanogr. Spec. Symp.* **1:** 172–193.

Edmondson, W. T. (1972b). The present condition of Lake Washington, *Verh. Internatl. Verein. Limnol.* **18:** 284–291.

Edmondson, W. T., and Lehman, J. T. (1981). The effect of changes in the nutrient income on the condition of Lake Washington. *Limnol. Oceanogr.* **26:** 1–29.

Faafeng, B. A., and Nilssen, J. P. (1981). A twenty-year study of eutrophication in a deep, soft-water lake. *Verh. Internatl. Verein. Limnol.* **21:** 412–424.

Fricker, Hj. (1980). *OECD Eutrophication Program—Regional Project Alpine Lakes.* Swiss Federal Board for Environmental Protection, Bern.

Hillbricht-Ilkowska, A., and Zdanowski, B. (1983). Sensitivity of lakes to inorganic enrichment stress—some results of experimentally induced fertilization. *Internatl. Rev. Ges. Hydrobiol.* **68:** 153–174.

Jansson, M. (1978). Experimental lake fertilization in the Kuokkel area, northern Sweden: Budget calculations and the fate of nutrients. *Verh. Internatl. Verein. Limnol.* **20:** 857–862.

Jansson, M. (1979). Nutrient budgets and the regulation of nutrient concentrations in a small sub-arctic lake in northern Sweden. *Freshwater Biol.* **9:** 213–231.

Jensen, J. P., Kristensen, P., and Jeppesen, E. (1990). Relationships between nitrogen loading and in-lake nitrogen concentrations in shallow Danish lakes. *Verh. Internatl. Verein. Limnol.* **24:** 201–204.

Jørgensen, S. E. (1989). Eutrophication models: The state-of-the-art. In Salánki, J., and Herodek, S. (eds.). *Conservation and Management of Lakes. Symp. Biol. Hung.* **38:** 609–621.

Klein, G., and Chorus, I. (1991). Nutrient balances and phytoplankton dynamics in Schlachtensee during oligotrophication. *Verh. Internatl. Verein. Limnol.* **24:** 873–878.

Lake Biwa Research Institute (1984). *Data Book of World Lakes*. The Secretariat, LECS'84, Otsu, Japan.

Larsen, D. P., Schults, D. W., and Malueg, K. W. (1961). Summer internal phosphorus supplies in Shagawa Lake, Minnesota. *Limnol. Oceanogr.* **26:** 740–753.

Larsen, D. P., Maluueg, H. K. W., Schults, D. W., and Brice, R. M. (1975). Response of eutrophic Shagawa Lake, Minnesota, U.S.A., to point-source, phosphorus reduction. *Verh. Internatl. Verein. Limnol.* **19:** 884–892.

Lee, G. F., and Jones, R. A. (1984). Summary of U.S. OECD eutrophication study. Results and their application to water quality management. *Verh. Internatl. Verein. Limnol.* **22:** 261–267.

Los, F. J., De Rooij, M., and Smits, G. C. (1984). Modeling eutrophication in shallow Dutch lakes. *Verh. Internatl. Verein. Limnol.* **22:** 917–923.

Los, F. J., and Brinkman, J. J. (1988). Phytoplankton modeling by means of optimization: a 10 year experience with BLOOM II. *Verh. Internatl. Verein. Limnol.* **23:** 790–795.

Lundgren, A. (1978). Experimental lake fertilization in the Kuokkel area, northern Sweden: changes in sestonic carbon and the role of phytoplankton. *Verh. Internatl. Verein. Limnol.* **20:** 863–868.

Møller Andersen, J. (1974). Nitrogen and phosphorus budgets and the role of sediments in six shallow Danish lakes. *Arch. Hydrobiol.* **74:** 528.

NORDFORSK (Scandinavian Council for Applied Research) (1980). *Monitoring of Inland Waters*. OECD Eutrophication Programme; the Nordic Project. NORDFORSK, Secretariat of Environmental Sciences Publication 1980:2.

OECD (Organization for Economic Cooperation and Development) (1982). *Eutrophication of Waters. Monitoring of Inland Waters: Eutrophication Control.* Environmental Directorate, OECD, Paris.

Oglesby, R. T. (1976). Relationships of fish yield to lake phytoplankton standing crop, production, and morphoedaphic factors. *J. Fish. Res. Board Can.* **34:** 2271–2279.

Pace, M. L. (1986). An empirical analysis of zooplankton community size structure across lake trophic gradients. *Limnol. Oceanogr.* **31:** 45–55.

Persson, G. (1978). Experimental lake fertilization in the Kuokkel area, northern Sweden: the response by the planktonic rotifer community. *Verh. Internatl. Verein. Limnol.* **20:** 875–880.

Reynolds, C. S. (1984). Phytoplankton periodicity: the interactions of form, function and environmental variability. *Freshwater Biol.* **14:** 111–142.

Rigler, F. R. (1973). Nutrient kinetics and the new typology. *Verh. Internatl. Verein. Limnol.* **19:** 197–210.

Ryding, S.-O., and Forsberg, C. (1980). Short-term load-response relationships in shallow, polluted lakes. In Barica, J., and Mur, L. R. (eds.). *Hypertrophic Ecosystems*, pp. 95–103.

Sakamoto, M. (1966) Primary production by phytoplankton community in some Japanese lakes and its dependence on lake depth. *Arch. Hydrobiol.* **62:** 1–28.

Salas, H. J., and Martino, P. (1991). A simplified phosphorus trophic state model for warm-water tropical lakes. *Water Res.* **25:** 341–350.
Schindler, D. W. (1974). Eutrophication and recovery in experimental lakes: implications for lake management. *Science* **184:** 897–899.
Schindler, D. W. (1977). Evolution of phosphorus limitation in lakes: natural mechanisms compensate for deficiencies of nitrogen and carbon in eutrophied lakes. *Science* **195:** 260–262.
Schindler, D. W., and Fee, E. J. (1974). Eutrophication Lakes area: whole-lake experiments in eutrophication. *J. Fish Res. Board Can.* **31:** 937–953.
Schindler, D. W., and Nighswander, J. E. (1970). Nutrient supply and primary production in Clear Lake, Eastern Ontario. *J. Fish. Res. Board Can.* **28:** 189–201.
Schindler, D. W., Armstrong, F. A. J., Holmgren, S. K., and Brunskii, G. J. (1971). Eutrophication of Lake 227, Experimental Lakes Area, northwestern Ontario, by addition of phosphate and nitrate. *J. Fish. Res. Bd. Canada.* **28:** 1763–1782.
Schindler, D. W., Kling, H., Schmidt, R. V., Prokopowich, J., Frost, V. E., Reid, R. A., and Capel, M. (1973). Eutrophication of Lake 227 by addition of phosphate and nitrate: the second, third and fourth years of enrichment, 1970, 1971 and 1972. *J. Fish. Res. Board Can.* **30:** 1415–1440.
Sommer, U. (1981). The role of r- and K-selection in the succession of phytoplankton in Lake Konstance. *Acta Oceanologia* **2:** 327–342.
Sommer, U., Gliwicz, Z. M., Lampert, W., and Duncan, A. (1986). The PEG-model of seasonal succession of planktonic events in fresh waters. *Arch. Hydrobiol.* **106:** 433–471.
Tilmann, D. (1977). Resource competition between planktonic algae: an experimental and theoretical approach. *Ecology* **58:** 338–348.
van Liere, L., van Tongen, O. F. R, Breebaart, L., and Kates, W. (1990). Trends in chlorophyll and total phosphorous in Loosdrecht Lakes, The Netherlands. *Verh. Internatl. Verein. Limnol.* **24:** 707–710.
Varis, O. (1990). Development of phytoplankton community in a restored eutrophic lake—a multivariate analysis. *Verh. Internatl. Verein. Limnol.* **24:** 693–697.
Vollenweider, R. A. (1968). *Scientific fundamentals of the eutrophication of lakes and flowing waters, with particular reference to nitrogen and phosphorus as factors in eutrophication.* OECD, DAS/CSI/68.27.
Vollenweider, R. A. (1969). Möglichkeiten und Grenzen elementarer Modelle der Stoffbilanz von Seen. *Arch. Hydrobiol.* **66:** 1–36.
Vollenweider, R. A. (1976). Advances in defining critical loading levels for phosphorous in lake eutrophication. *Mem. Inst. Itali. Idrobiol.* **40:** 1–24.

Trophic state and eutrophication

Definition

Johannessen, M., Lande, A., and Rognerud, S. (1984). Fertilization of 6 small mountain lakes in Telemark, Southern Norway. *Verh. Internatl. Verein. Limnol.* **22:** 673–678.
Lindeman, R. L. (1942). The trophic dynamic aspects of ecology. *Ecology* **23:** 399–418.
Naumann, E. (1919). Några synpunkter angående planktons ökologi. *Svenske Botanisk Tidskript.* **13:** 129–158.
Rast, W., and Kerekes, J. J. (1981). *Proc. International Workshop on the Control of Eutrophication.* UNESCO, IIASA, and OECD.
Stockner, J. G. (1981). Whole-lake fertilization for the enhancement of sockeye salmon (*Oncorhynchus nerka*) in British Columbia, Canada. *Verh. Internatl. Verein. Limnol.* **21:** 293–299.
Stockner, J. G., and Shortreed, K. S. (1985). Whole-lake fertilization experiments in coastal British Columbia lakes: empirical relationships between nutrient inputs and phytoplankton biomass and production. *Can. J. Fish. Aquat. Sci.* **42:** 649–658.
Thienemann, A. (1913–1915). Physicalische und chemische untersuchungen in den Maaren der Eifel. *Verhandl. Nat. Ver. Preuss. Rheinlande u Westfalens:* **70:** 249–302, **71:** 274–389.

Thienemann, A. (1921). Seetypen. *Naturwissenschaften* **9:** 343–346.
Thienemann, A. (1925). *Die Binnengewasser Mitteleuropas. Eine limnologische Einführung.* Die Binnengewässer 1, Stuttgart.
Thienemann, A. (1926). Der Nahrungskreislauf im Wasser. *Verhandl. Deutsch. Zool. Gesell.* **31:** 20–79.
Vollenweider, R. (1989). Global problems of eutrophication and its control. In Salanki, J., and Herodek, S. (eds.). *Conservation and Management of Lakes, Symp. Biol. Hung.* **38:** 19–41.

Criteria and indicator of trophic state

Anderson, N. J., Rippey, B., and Gabson, C. E. (1993). A comparison of sedimentary and diatom-infered phosphorus profiles; implications for defining predisturbance nutrient conditions. *Hydrobiologia* **253:** 357–366.
Bays, J. S., and Crisman, T. L. (1982). Zooplankton and trophic state relationships in Florida Lakes. *Can. J. Fish. Aquat. Sci.* **40:** 1813–1819.
Beaver, J. R., and Crisman, T. L. (1990). Use of microzooplankton as an early indicator of advancing cultural eutrophication. *Verh. Internatl. Verein. Limnol.* **24:** 532–537.
Best, E. P. H., de Vries, D., and Reins, A. (1984). The macrophytes in the Loosdrecht Lakes: a story of their decline in the course of eutrophication. *Verh. Internatl. Verein. Limnol.* **22:** 868–875.
Brook, A. J. (1965). Planktonic algae as indicators of lake types, with special references to the desmidiaceae. *Limnol. Oceanogr.* **10:** 403–411.
Brooks, J. L. (1969). Eutrophication and changes in the composition of the zooplankton. In Rohlich, G. A. (ed.). *Eutrophication: Causes, Consequences, Correctives,* pp. 236–255. National Academy of Sciences, Washington, D.C.
Brungum, R. B. (1979). A reevaluation of the Araphidineae/Centrales index as an indicator of lake trophic status. *Freshwater Biol.* **9:** 451–460.
Brungum, R. B., and Patterson, C. (1983). The A/C (Araphidineae/Centrales) ratio in high and low alkalinity lakes in eastern Minnesota. *Freshwater Biol.* **13:** 47–55.
Carlson, R. E. (1977). A trophic state index for lakes. *Limnol. Oceanogr.* **22:** 361–369.
Carpenter, S. R. (1982). Submerged macrophyte community structure and internal loading: relationship to lake ecosystem productivity and succession. *Proceedings of the Second Annual Conference on Lake Restoration, Protection, and Management*, pp. 105–122. North American Lake Management Society/U.S. EPA. EPA 440/5-83-001.
Cooke, G. D. (1981). Perspectives in lake management and restoration. Paper presented at UNESCO/IIASA Meeting on Control of Eutrophication. Laxenburg, Austria, Oct. 1981.
Forsberg, C., and Ryding, S.-O. (1980). Eutrophication parameters and trophic state indices in 30 Swedish waste-receiving lakes. *Arch. Hydrobiol.* **89:** 189–207.
Hall, D. J., Cooper, W. E., and Werner, E. E. (1970). An experimental approach to the productive dynamics and structure of the freshwater animal communities. *Limnol. Oceanogr.*
Hall, R. I., and Smol, J. P. (1992). A weight-averaging regression and calibration model for inferring total phosphorus concentration from diatoms in British Columbia (Canada). *Freshwater Biol.* **27:** 417–434.
Hillbricht-Ilkowska, A. B. (1984). The indices and parameters useful in the evaluation of water quality and the ecological state of temperate, lowland lakes, connected with their eutrophication. *Proc. Shiga Conf.* (1984) *Conservation and Management of World Lake Environment,* pp. 55–69. Shiga Prefectural Government, Otsu, Japan.
Howard-Williams, C., and Allanson, B. R. (1981). An integrated study of littoral and pelagic primary production in a South African coastal lake. *Arch. Hydrobiol.* **92:** 507–534.
Jønasson, P. M. (1969). Bottom fauna and eutrophication. In Rohlich, G. A. (ed.). *Eutrophication: Causes, Consequences and Correctives,* pp. 50–64. Washington, D.C. National Academy of Sciences.
Jønasson, P. M. (1984). Decline of zoobenthos through five decades of eutrophication in Lake Esrom. *Verh. Internatl. Verein. Limnol.* **22:** 800–804.

Lang, C. (1984). Eutrophication of lakes Léman and Neuchatel (Switzerland) indicated by oligochaetae communities. *Hydrobiologia* **115:** 131–138.

Lang, C. (1985). Eutrophication of Lake Geneva indicated by the oligochaete communities of the profundal. *Hydrobiologia* **126:** 237–243.

Lang, C. (1990). Quantitative relationships between oligochaete communities and phosphorus concentrations in lakes. *Freshwater Biol.* **24:** 327–334.

Lang, C., and Reymond, O. (1992). Reversal of eutrophication in Lake Geneva: evidence from the oligochate communities. *Freshwater Biol.* **28:** 145–148.

Lang, C., and Raymond, O. (1993). Recovery of Lake Neuchatel (Switzerland) from eutrophication indicated by the oligochaete communities. *Arch. Hydrobiol.* **128:** 65–71.

Liu, H. (1990). Classification of trophic types of lakes in China. In Liu, H., Zhang, Y., and Li, H. (eds.). *Lake Conservation and Management. Proc. 4th Internatl. Conf. Conservation and Management of Lakes* ("Hangzhou '90"), pp. 24–40. Chinese Research Academy of Environmental Sciences.

Lods-Crozet, B., and Lachavanne, J.-B. (1994). Changes in the chironomid communities in Lake Geneva in relation with eutrophication, over a period of 60 years. *Arch. Hydrobiol.* **130:** 453–471.

Lorenzen, M. W. (1980). Use of chlorophyll-Secchi disk relationships. *Limnol. Oceanogr.* **25:** 371–377.

Megard, R., Settles, J. C., Boyer, H. A., and Combs, W. S. (1980). Light, Secchi disks, and trophic states. *Limnol. Oceanogr.* **25:** 373–366.

Millard, E. S., and Johnson, M. G. (1986). Effects of decreased phosphorus loading on primary production, chlorophyll-a and light extinction in the Bay of Quinte, Lake Ontario, 1972–82. In Minns, C. K., Hurley, D. A., and Nicholls, K. H. (eds.). *Project Quinte. Point Source Phosphorus Control and Ecosystem Response in the Bay of Quinte, Lake Ontario. Can. Spec. Publ. Fish Aquat. Sci.* **86:** 100–113.

Nygaard, G. (1949). Hydrobiological studies of some Danish ponds and lakes II. The quotient hypothesis and some new or little known phytoplankton organisms. *Kgl. Danske Videnskab. Seleskab. Biol. Skrifer.* **7:** 293.

OECD (Organization for Economic Cooperation and Development) (1982). *Eutrophication of Waters. Monitoring, Assessment and Control.* OECD Cooperative Programme on Monitoring of Inland Waters (Eutrophication Control). Environmental Directorate, OECD, Paris.

Pearsall, W. H. (1920) The aquatic vegetation of the English Lakes. *J. Ecol.* **8:** 163–199.

Pearsall, W. H. (1932). Phytoplankton in the English Lakes II. The composition of the phytoplankton in relation to dissolved substances. *J. Ecol.* **20:** 241–262.

Rast, W., and Kerekes, J. J. (1981). *Proceedings of the International Workshop on the Control of Eutrophication.* UNESCO, IIASA, OECD.

Rodhe, W. (1969). Crystalization of eutrophication concepts in Northern Europe. In Rohlich, G. A. (ed.). *Eutrophication: Causes, Consequences, and Correctives,* pp. 50–64. National Academy of Sciences, Washington, D.C.

Ryding, S.-O., and Rast, W. (1989). *The Control of Eutrophication of Lakes and Reservoirs.* Man and the Biosphere Series 1. UNESCO, The Parthenon Publ. Group Ltd.

Sakamoto, M. (1966). Primary production by phytoplankton community in some Japanese lakes and its dependence on lake depth. *Arch. Hydrobiol.* **62:** 1–28.

Stockner, J. G. (1971). Preliminary characterization of lakes in the Experimental Lakes Area, northwestern Ontario using diatom occurrence in sediments. *J. Fish. Res. Board Can.* **28:** 265–275.

Stockner, J. G. (1972) Paleolimnology as a means of assessing eutrophication. *Verh. Internatl. Verein. Limnol.* **18:** 1018–1030.

Thienemann, A. (1925). Die Binnengewasser Mitteleuropas. *Eine limnologische Einführung.* Die Binnengewässer 1, Stuttgart.

Thunmark, S. (1945). Zur Soziologie des Susswasserplanktons. *Folia Limnol. Scand.* **3:** 66.

USEPA (U.S. Environmental Protection Agency) (1975). *The Relationship of Phosphorous and Nitrogen to the Trophic State of Northeast and North Central Lakes and Reservoirs. National Eutrophication Survey,* U.S. Environmental Protection Agency, Working Paper no. 23.

Westlake, D. F. (1980). Primary production. In Le Cren, E. D., and Lowe-McConell, R. H. (eds.). *The Functioning of Freshwater Ecosystems,* pp. 141–240. Cambridge University Press. Cambridge, U.K.

Wetzel, R. G. (1975). *Limnology.* Saunders, Philadelphia.

Whitmore, T. J. (1989) Florida diatom assemblages as indicators of trophic state and pH. *Limnol. Oceanogr.* **34:** 882–895.

Willén, T. (1975). Biological long term investigations of Swedish lakes. *Verh. Internatl. Verein. Limnol.* **19:** 1117–1124.

Yoshimura, S. (1935). The concentrations of dissolved nitrogenous compounds in the harmonic lake waters of Japan. *Bull. Jpn. Soc. Sci. Fish.* **4:** 183–189.

Factors involved in the control of eutrophication

Ahl, T. (1972). Plant nutrients in Swedish lake and river waters. *Verh. Internatl. Verein. Limnol.* **18:** 362–369.

ASLO (American Society of Limnology and Oceanography) (1972). Likens, G. E. (ed.). *Nutrient and Eutrophication: The Limiting-Nutrient Controversy. Spec. Symp.* 1 American Society of Limnology and Oceanography.

Berman, T. (1970). Alkaline phosphatase and phosphorus availability in Lake Kinneret. *Limnol. Oceanogr.* **15:** 663.

Best, E. P. H., de Vries, D., and Reins, A. (1984). The macrophytes in the Loosdrecht Lakes: a story of their decline in their course of eutrophication. *Verh. Internatl. Verein. Limnol.* **22:** 868–875.

Brylinsky, M., and Mann, K. H. (1973). An analysis of factors governing productivity in lakes and reservoirs. *Limnol. Oceanogr.* **18:** 1–14.

Campos. H. (1984). Limnological study of Araucanian lakes (Chile). *Verh. Internatl. Verein. Limnol.* **22:** 1319–1327.

Carpenter, S. R. (1982). Submerged macrophyte community structure and internal loading: relationship to lake ecosystem productivity and succession. *Proc. 2nd Annual Conf. Lake Restoration, Protection, and Management,* pp. 105–122. North American Lake Management Society, USEPA. EPA 440/5-83-001.

Dillon, P. J., and Rigler, F. H. (1974). The chlorophyll-phosphorus relationship in lakes. *Limnol. Oceanogr.* **19:** 767–773.

Dodds, W. K., and Priscu, J. C. (1990). Comparison of methods for assessment of nutrient deficiency of phytoplankton in a large oligotrophic lake. *Can. J. Fish. Aquat. Sci.* **47:** 2328–2338.

Downing, J. A., and McCauley, E. (1992). The nitrogen:phosphorus relationship in lakes. *Limnol. Oceanogr.* **37:** 936–945.

Eriksson, C., and Pettersson, K. (1984). The physiological status of spring phytoplankton in Lake Erken—field measurements and nutrient enrichment experiments. *Verh. Internatl. Verein. Limnol.* **22:** 743–749.

Faafeng, B. A., and Hessen, D. O. (1993). Nitrogen and phosphorus concentrations and N:P ratios in Norwegian lakes: perspectives on nutrient limitation. *Verh. Internatl. Verein. Limnol.* **25:** 465–469.

Fee, E. J.. (1979). A relation between lake morphometry and primary productivity and its use in integrating whole-lake eutrophication experiments. *Limnol. Oceanogr.* **24:** 401–416.

Fitzgerald, G. P., and Nelson, T. C. (1966). Extractive and enzymatic analysis for limiting or surplus phosphorus in algae. *J. Phycol.* **2:** 32–37.

Forsberg, C., and Ryding, S.-O. (1980). Eutrophication parameters and trophic state indices in 30 Swedish waste-receiving lakes. *Arch. Hydrobiol.* **89:** 189–207.

Forsberg, C., Ryding, S.-O., Claesson, A., and Forsberg, A. (1978). Water chemical analyses or algal assay? Sewage effluent and polluted lake water studies. *Mitt. Internatl. Verein. Limnol.* **21:** 352–363.

Goldman, C. R. (1960). Primary productivity and limiting factors in three lakes of the Alaska Peninsula. *Ecol. Monogr.* **30:** 207–230.

Goldman, C. R. (1964). Primary productivity and micro-nutrient limiting factors in some North American and New Zealand lakes.*Verh. Internatl. Verein. Limnol.* **15:** 365–374.

Goldman, C. (1981). Lake Tahoe: two decades of change in a nitrogen deficient oligotrophic lake. *Verh. Internatl. Verein. Limnol.* **21:** 45–70.

Goldman, C. R. (1988). Primary productivity, nutrients and transparency during the early onset of eutrophication in ultra-oligotrophic Lake Tahoe, California-Nevada. *Limnol. Oceanogr.* **33:** 1321–1333.

Goldman, C. R., and Armstrong, R. (1969). Primary productivity studies in Lake Tahoe, California. *Verh. Internatl. Verein. Limnol.* **17:** 49–71.

Gorham, E., Lund, J. W. G., Sanger, J. E., and Dean, W. E., Jr. (1974). Some relationships between algal standing crop, water chemistry and sediment chemistry in the English Lakes. *Limnol. Oceanogr.* **19:** 601–617.

Gray, C. B., Dickman, M., Krushelnicki, B., and Cromie, V. (1990). A survey of Canadian Lakes. In Liu, H., Zhang, Y., and Li, H. (eds.). *Lake Conservation and Management. Proc. 4th Internatl. Conf. Conservation and Management of Lakes* ("Hangzhou '90"), pp. 41–68.

Hammer, U. T. (1964). The succession of "bloom" species of blue green algae and some causal factors. *Verh. Internatl. Verein. Limnol.* **15:** 829–836.

Healey, F. P. (1978). Physiological indicators of nutrient deficiency in algae. *Mitt. Internatl. Verein. Limnol.* **21:** 34–41.

Healey, F. P., and Hendzel, L. L. (1979). Indicators of phosphorus and nitrogen deficiency in five algae in culture. *J. Fish. Res. Board Can.* **36:** 1364–1369.

Healey, F. P., and Hendzel, L. L. (1980). Physiological indicators of nutrient deficiency in lake phytoplankton. *Can. J. Fish. Aquat. Sci.* **37:** 442–453.

Hecky, R. E., and Kilham, P. (1988). Nutrient limitation of phytoplankton in freshwater and marine environments: a review of recent evidence on the effects of enrichment. *Limnol. Oceanogr.* **33:** 796–822.

Istvánovics, V., Pettersson, K., Pierson, D., and Bell, R. (1992). Evaluation of phosphorus deficiency indicators for summer phytoplankton in Lake Erken. *Limnol. Oceanogr.* **37:** 890–900.

Kalff, J. (1971). Nutrient limiting factors in an arctic tundra pond. *Ecology* **52:** 655–659.

Köhler, J. (1992). Influence of turbulent mixing on growth and primary production of *Microcystis aeruginosa* in the hypertrophic Bautzen Reservoir. *Arch. Hydrobiol.* **123:** 413–429.

Kopáček, J., Procházková, Stuchlík, E., and Blazka, O. (1995). The nitrogen-phosphorus relationship in mountain lakes: influence of atmospheric input, watershed, and pH. *Limnol. Oceanogr.* **40:** 930–937.

Lam, C. W. Y. (1979). Phosphate requirements of Anabaena oscillarioides and its ecological implications. *Hydrobiologia* **67:** 89–96.

Lean, D. R. S., and Pick, F. R. (1981). Photosynthetic response of lake plankton to nutrient enrichments. A test for nutrient limitation. *Limnol. Oceanogr.* **26:** 1001–1019.

Lehman, J. T., and Sangren, G. D. (1978). Documenting a seasonal change from phosphorus to nitrogen limitation in a small temperate lake, and its impact on the population dynamics of *Asterionella. Verh. Internatl. Verein. Limnol.* **20:** 375–380.

Levine, S. N., and Schindler, D. W. (1989). Phosphorus, nitrogen, and carbon dynamics of experimental lake 303 during recovery from eutrophication. *Can. J. Fish. Aquat. Sci.* **46:** 2–10.

Lin, C. K., and Schelske, C. L. (1981). Seasonal variation of potential nutrient limitation to chlorophyll production in southern Lake Huron. *Can. J. Fish. Aquat. Sci.* **38:** 1–9.

Miller, M. C., Kannan, M., and Colinvaux, P. A. (1984). Limnology and primary productivity of Andean and Amazonian tropical lakes of Ecuador. *Verh. Internatl. Verein. Limnol.* **22:** 1264–1270.

OECD (Organization for Economic Cooperation and Development) (1982). *Eutrophication of Waters: Monitoring, Assessment and Control.* OECD Cooperative Programme on Monitoring of Inland Waters (Eutrophication Control). Environmental Directorate, OECD, Paris.

Pettersson, K. (1980). Alkaline phosphatase activity and algal surplus phosphorus as phosphorus-deficiency indicators in Lake Erken. *Arch. Hydrobiol.* **84:** 54–87.

Pettersson, K., Istvánovics, V., and Pierson, D. (1990). Effects of vertical mixing on phytoplankton phosphorus supply during summer in Lake Erken. *Verh. Internatl. Verein. Limnol.* **24:** 236–241.
Powers, C. F., Schultz, D. W., Malueg, K. W., Brice, R. M., and Schuldt, M. D. (1972). Algal response to nutrient additions in natural waters. II. Field experiments. *Limnol. Oceanogr. Spec. Symp.* **1:** 141–156.
Rawson, D. S. (1939). Some physical and chemical factors in the metabolism of lakes. *Am. Assoc. Adv. Sci.* **10:** 9–26.
Rawson, D. S. (1955). Morphometry as a dominant factor in the productivity of large lakes. *Verh. Internatl. Verein. Limnol.* **12:** 164–175.
Reynolds, C. S. (1984). Phytoplankton periodicity: the influence of form, function and environmental variability. *Freshwater Biol.* **14:** 111–142.
Ryding, S.-O., and Rast, W. (1989). *The Control of Eutrophication of Lakes and Reservoirs*. Man and the Biosphere Series 1. UNESCO/Parthenon Publ. Group.
Sakamoto, M. (1966a). Primary production by phytoplankton community in some Japanese lakes and its dependence on lake depth. *Arch. Hydrobiol.* **62:** 1–28.
Sakamoto, M. (1966b). Primary production and trophic degree of lakes (in Japanese with English summary). *Jpn. J. Limnol.* **23:** 73–85.
Sakamoto, M. (1971). Chemical factors involved in the control of pytoplankton production in the Experimental Lakes Area, northwestern Ontario. *J. Fish. Res. Board Can.* **28:** 203–213.
Sakamoto, M. (1984). Eutrophication and its causative agents, phosphorus and nitrogen, in Japanese lakes. *Proc. Shiga Conf.* (1984) *Conservation and Management of World Lake Environment,* pp. 80–85. Shiga Prefectural Government, Otsu, Japan.
Schelske, C. L., and Stoermer, E. F. (1972). Phosphorus, silica and eutrophication of Lake Michigan. *Limnol. Oceanogr. Spec. Symp.* **1:** 157–171.
Schindler, D. W. (1971). Carbon, nitrogen and phosphorus and the eutrophication of freshwater lakes. *J. Phycol.* **7:** 321–329.
Schindler, D. W. (1974). Eutrophication and recovery in experimental lakes: implications for lake management. *Science* **184:** 897–899.
Schindler, D. W. (1975). Whole-lake eutrophication experiments with phosphorus, nitrogen and carbon. *Verh. Internatl. Verein. Limnol.* **19:** 3221–3231.
Schindler, D. W. (1977). Evolution of phosphorus limitation in lakes: natural mechanisms compensate for deficiencies of nitrogen and carbon in eutrophied lakes. *Science* **195:** 260–262.
Schindler, D. W. (1978). Factors regulating phytoplankton production and standing crop in the world's freshwaters. *Limnol. Oceanogr.* **23:** 478–486.
Schindler, D. W., and Fee, E. J. (1974). Eutrophication Lakes area: whole-lake experiments in eutrophication. *J. Fish. Res. Board Can.* **31:** 937–953.
Schindler, D. W., and Nighswander, J. E. (1970). Nutrient supply and primary production in Clear Lake, Eastern Ontario. *J. Fish. Res. Board Can.* **28:** 189–201.
Schindler, D. W., Armstrong, F. A. J., Holmgren, S. K., and Brunskill, G. J. (1971). Eutrophication of Lake 227, Experimental Lakes Area, northwestern Ontario, by addition of phosphorus and nitrate. *J. Fish. Res. Board Can.* **28:** 1763–1782.
Schindler, D. W., Kling, H., Schmidt, R. V., Prokopowich, J., Frost, V. E., Reid, R. A., and Capel, M. (1973). Eutrophication of Lake 227 by addition of phosphate and nitrate: the second, third and fourth years of enrichment, 1970, 1971 and 1972. *J. Fish. Res. Board Can.* **30:** 1415–1440.
Schindler, D. W., Fee, E. J., and Ruszczynski, T. (1977). Phosphorus input and its consequences for phytoplankton standing crop and production in the Experimental Lake Area and in similar lakes. *J. Fish. Res. Board Can.* **35:** 190–196.
Serruya, O., and Berman, T. (1975). Phosphorus, nitrogen and the growth of algae in Lake Kinneret. *J. Phycol.* **11:** 155–162.
Smith, V. H. (1982) The nitrogen and phosphorus dependence of algal biomass in lakes: an empirical and theoretical analysis. *Limnol. Oceanogr.* **27:** 1101–1112.
Smith, V. H. (1983). The nitrogen and phosphorus dependence of blue-green algal dominance in lakes. In *Restoration of Lakes and Inland Waters: Internatl. Symp. Inland Waters and Lake Restoration,* pp. 237–241. USEPA. 440/5-83-001.
Smith, V. H. (1986). Light and nutrient effects on the relative biomass of blue-green algae in lake phytoplankton. *Can. J. Fish. Aquat. Sci.* **43:** 148–153.

Straškrava, M., and Straškrabová, V. (1969). Eastern European lakes. In Rohlich, G. A. (ed.). *Eutrophication: Causes, Consequences and Correctives,* pp. 65–97. National Academy of Sciences, Washington, D.C.

Takamura, N., Iwakuma, T., and Yasuno, M. (1985). Photosynthesis and primary production of *Microcystis aeruginosa* Kütz, in Lake Kasumigaura. *J. Plankton Res.* **7:** 303–312.

Talling, J. F. (1971). The underwater light climate as a controlling factor in the production ecology of freshwater phytoplankton. *Mitt. Internatl. Verein. Limnol.* **19:** 214–243.

Thienemann, A. (1925). Die Binnengewässer Mitteleuropas. *Eine limnologische Einführung*. Die Binnengewässer 1, Stuttgart.

Wurtsbaugh, W. A., Vincent, W. F., Tapia, R. A., Vincent, C. L., and Richerson, P. J. (1985). Nutrient limitation of algal growth and nitrogen fixation in a tropical alpine lake, Lake Titicaca (Peru/Bolivia). *Freshwater Biol.* **15:** 185–195.

Yin, C., Lan, Z., and Zhao, M. (1994). Factors limiting algal growth in eutrophic Chaohu Lake, China. *Mitt. Internatl. Verein. Limnol.* **24:** 213–218.

Eutrophication of world lakes and reservoirs

Ahl, T. (1975). Effects of man-induced and natural loading of phosphorus and nitrogen on the large Swedish lakes. *Verh. Internatl. Verein. Limnol.* **19:** 1125–1132.

Ahlgren, I. (1972). Changes in Lake Norrviken after sewage diversion. *Verh. Internatl. Verein. Limnol.* **18:** 355–361.

Ahlgren, I. (1978). Response of Lake Norrviken to reduced nutrient loading. *Verh. Internatl. Verein. Limnol.* **20:** 846–850.

Ahlgren, G. (1978). Response of phytoplankton and primary production to reduced nutrient loading in Lake Norrviken. *Verh. Internatl. Verein. Limnol.* **20:** 840–845.

Alvarez, C. M. A., Munoz, R. O., Rubino, O. A., and Prat, F. N. (1992). Current state of eutrophication in Spanish inland waters. *Eur. J. Water Pollution Control* **2:** 27–32.

Ambühl, H. (1981). The status of eutrophication and its control in Alpine lakes. *Water Qual. Bull.* **6:** 113–156.

Andersson, G., Berggren, H., and Hambrin, S. (1975). Lake Trummen restoration project. III. Zooplankton, macrobenthos and fish. *Verh. Internatl. Verein. Limnol.* **19:** 1097–1106.

Baker, L. A., Brezonik, P. L., and Kratzer, C. R. (1981). Nutrient loading—trophic state relationships in Florida lakes. Univ. Florida Water Resources Research Center Publ. 56, Gainesville.

Barica, J. (1987). Water quality problems associated with high productivity of prairie lakes in Canada: a review. *Water Qual. Bull.* **12:** 107–114.

Barica, J. (1989). Recovery of the Laurentian Great Lakes, 1970–1985: eutrophication aspects. In Salánki, J., and Herodeck, S. (eds.). *Conservation and Management of Lakes. Symp. Biol. Hung.* **38:** 43–59.

Barica, J. (1990). Long-term trends in Great Lakes nutrient concentrations: a summary. *Verh. Internatl. Verein. Limnol.* **24**: 369–370.

Bengtsson, L., Fleischer, S., Lindmark, G., and Ripl, W. (1975). Lake Trummen restoration project. 1. Water and sediment chemistry. *Verh. Internatl. Verein. Limnol.* **19:** 1080–1087.

Björk, S., Bengtsson, L. Berggren, H. Cronberg, G. Digerfeldt, G., Fleischer, S., Gelin, C., Lindmark, G., Malmer, N., Plejmark, F., Ripl, W., and Swanberg, P. O. (1972). Ecosystem studies in connection with the restoration of lakes. *Verh. Internatl. Verein. Limnol.* **18:** 379–387.

Brylinsky, M., and Mann, K. H. (1973). An analysis of factors governing productivity in lakes and reservoirs. *Limnol. Oceanogr.* **18:** 1–14.

Bugenyi, F. W. B., and Balirwa, J. S. (1989). Human intervention in natural processes of the Lake Victoria ecosystem, the problem. In Salánki, J., and Herodeck, S. (eds.). *Conservation and Management of Lakes. Symp. Biol. Hung.* **38:** 311–339.

Calderoni, A., Bernardi, R., and Ruggiu, D. (1993). The changing trophic state of Lago Maggiore. In Giussani, G., and Calleri, C. (ed.). *Strategies for Lake Ecosystems beyond 2000. Proc. 5th Internatl. Conf. Conservation and Management of Lakes,* pp. 58–61.

Campos, H. (1984). Limnological study of Araucanian lakes (Chile). *Verh. Internatl. Verein. Limnol.* **22:** 1319–1327.
Clasen, J., and Bernhardt, H. (1979). *Final Report of Regional Project, Shallow Lakes and Reservoirs,* vol. 1. Cooperative Programme for Monitoring of Inland Waters, OECD.
Cronberg, G., Gelin, G., and Larsson, K. (1975). Lake Trummen restoration project. II. Bacteria, phytoplankton and phytoplankton productivity. *Verh. Internatl. Verein. Limnol.* **19:** 1088–1096.
Cullen, P. (1991). Response to changing nutrient inputs over a twenty year period in Lake Burley Griffin, Australia. *Verh. Internatl. Verein. Limnol.* **24:** 1471–1476.
Cullen, P., and Smalls, I. (1981). Eutrophication in semi-arid areas. The Australian experience. *Water Qual. Bull.* **6:** 79–91.
Dobson, H. F. H. (1981). Trophic conditions and trends in the Laurentian Great Lakes. *Water Qual. Bull.* **6:** 146–158.
Edmondson, W. T. (1969). Cultural eutrophication with special reference to Lake Washington. *Mitt. Internatl. Verein. Limnol.* **17:** 19–32.
Edmondson, W. T. (1972a). Nutrient and phytoplankton in Lake Washington. *Limnol. Oceanogr. Spec. Symp.* **1:** 172–193.
Edmondson, W. T. (1972b). The present condition of Lake Washington. *Verh. Internatl. Verein. Limnol.* **18:** 284–291.
Edmondson, W. T., and Lehman, J. T. (1981). The effect of changes in the nutrient income on the condition of Lake Washington. *Limnol. Oceanogr.* **26:** 1–29.
Environmental Agency of Japan (1987). *National Report on Lake Environments.* Third National Environmental Survey for Nature Conservation (in Japanese).
Environmental Agency of Japan (1993). *Quality of the Environments in Japan.*
Faafeng, B. A., and Nielsen, J. P. (1981). A twenty-year study of eutrophication in a deep, soft-water lake. *Verh. Internatl. Verein. Limnol.* **21:** 412–424.
Faafeng, B. A., and Hessen, D. O. (1993). Nitrogen and phosphorus concentrations and N:P ratios in Norwegian lakes: perspectives on nutrient limitation. *Verh. Internatl. Verein. Limnol.* **25:** 465–469.
Forsberg, C., and Ryding, S.-O. (1980). Eutrophication parameters and trophic state indices in 30 Swedish waste-receiving lakes. *Arch. Hydrobiol.* **89:** 189–207.
Forsberg, C., Ryding, S.-O., Forsberg, A., and Claesson, A. (1978). Research on recovery of poluted lakes. I. Improved water quality in Lake Boren and Lack Ekoln after nutrient reduction. *Verh. Internatl. Verein. Limnol.* **20:** 825–832.
Fricker, Hj. (1980). OECD Eutrophication Program—Regional Project Alpine Lakes. Swiss Federal Board for Environmental Protection, Bern.
Gaggino, G. F., and Provini, A. (1988). Eutrophication of lakes and reservoirs in Italy. *Verh. Internatl. Verein. Limnol.* **23:** 553–557.
Galazy, G. I. (1989). State of the ecosystem of Lake Bailkal and its catchment area: problems of conservation and rational use of resources. In Salánki, J., and Herodek, S. (eds.). *Conservation and Management of Lakes. Symp. Biol. Hung.* **38:** 349–359.
Galazy, G. I., and Tarasova, E. N. (1993). Baikal and the problem of eutrophication. In Giussani, G., and Callieri, C. (ed.). *Strategies for Lake Ecosystems beyond 2000. Proc., 5th Internatl. Conf. Conservation and Management of Lakes,* pp. 96–99.
Ghilarov, A. M. (1982). Report on eutrophication studies in the U.S.S.R. *Water Res.* **17:** 607–611.
Giesen, W. (1994). Indonesia's major freshwater lakes: a review of current knowledge, development processes and threats. *Mitt. Internatl. Verein. Limnol.* **24:** 115–128.
Goldman, C. (1981). Lake Tahoe: two decades of change in a nitrogen deficient oligotrophic lake. *Verh. Internatl. Verein. Limnol.* **21:** 45–70.
Goldman, C. R. (1988). Primary productivity, nutrients and transparency during the early onset of eutrophication in ultra-oligotrophic Lake Tahoe, California-Nevada. *Limnol. Oceanogr.,* **33:** 1321–1333.
Goldman, C. R., and Armstrong, R. (1969). Primary productivity studies in Lake Tahoe, California. *Verh. Internatl. Verein. Limnol.* **17:** 49–71.
Goldman, C. R., and de Amezaga, E. (1975). Spatial and temporal changes in the primary productivity of Lake Tahoe, California-Nevada between 1959 and 1971. *Verh. Internatl. Verein. Limnol.* **19:** 812–825.

Goldman, C. R., and de Amezaga, E. (1984). Primary productivity and precipitation at Castle Lake and Lake Tahoe during twenty-four years, 1959–1982. *Verh. Internatl. Verein. Limnol.* **22:** 591–599.

Gray, C. B., Dickman, M., Krushelnicki, B., and Cromie, V. (1990). A survey of Canadian Lakes. In Liu, H., Zhang, Y., and Li, H. (eds.). *Lake Conservation and Management. Proc. 4th Internatl. Conf. Conservation and Management of Lakes* ("Hangzhou '90"), pp. 41–68.

Hajós, B. (1992). Development of water quality and economics at Lake Balaton. *Eur. J. Water Pollution Control* **2:** 20–25.

Harper, D. M., Phillips, D. G., Chilvers, A., Kitaka, N., and Mavuti, K. (1993). Eutrophication prognosis of Lake Naivasha, Kenya. *Verh. Internatl. Verein. Limnol.* **25:** 861–865.

Hecky, R. E., and Bugenyi, F. W. B. (1992). Hydrology and chemistry of the African Great lakes and water quality issues: problems and solution. *Mitt. Internatl. Verein. Limnol.* **23:** 45–54.

Hecky, R. E., Coulter, G. W., and Spegel, R. H. (1991). The nutrient regime. In *Lake Tanganyika and its Life.* British Museum and Oxford University Press.

Henry, R., and Tundishi, J. G. (1983). Response of the phytoplankton community of a tropical reservoir (Sáo Paulo, Brazil) to the enrichment with nitrate, phosphate and EDTA. *Internatl. Rev. Ges. Hydrobiol.* **68:** 853–862.

Henry, R., Hino, K., Gentil, J. G., and Tundisi, J. G. (1985). Primary production and effects of enrichment with nitrate and phosphate on phytoplankton in the Barra Bonita Reservoir (State of Sáo Palo, Brazil). *Internatl. Rev. Ges. Hydrobiol.* **70:** 561–573.

Herodek, S. (1984). The eutrophication of Lake Balaton: measurements, modeling and management. *Verh. Internatl. Verein. Limnol.* **22:** 1087–1091.

Herodek, S., Istvánovics, V., and Zlinsky, J. (1988). Phosphorus metabolism and eutrophication control of Lake Balaton. *Verh. Internatl. Verein. Limnol.* **23:** 517–521.

Holtan, H. (1978). Eutrophication of Lake Miøsa in relation to the pollutional load. *Verh. Internatl. Verein. Limnol.* **20:** 734–742.

Holtan, H. (1979). The Lake Miøsa story. *Arch. Hydrobiol. Beih.* **13:** 242–258.

Holtan, H. (1981). Eutrophication of Lake Miøsa and its recovery. *Water Qual. Bull.* **6:** 99–107.

ILEC (International Lake Environmental Committee) (1987–1989). *Data Book of World Lake Environments. A Survey of the State of World Lakes.* International Lake Environmental Committee, Japan.

Janus, L. L., and Vollenweider, R. A. (1981). *The OECD Cooperative Programme on Eutrophication.* Canadian contribution, Scientific Series no. 131. Canada Center for Inland Waters. Burlington, Ontario, Canada.

Jin, X. (1990). Analysis of the characteristics of eutrophication of China's lakes. In Liu, H., Zhang, Y., and Li, H. (eds.). *Lake Conservation and Management. Proc. 4th Internatl. Conf. Conservation and Management of Lakes* ("Hangzhou '90"), pp. 110–121.

Jin, X. (1994). An analysis of lake eutrophication in China. *Verh. Internatl. Verein. Limnol.* **24:** 207–211.

Jones, R. A., and Lee, G. F. (1982). Recent advances in assessing impact of phosphorus loads on eutrophication-related water quality. *Water Res.* **16:** 503–515.

Kalff, J. (1983). Phosphorus limitation in some tropical African lakes. *Hydrobiologia* **100:** 101–112.

Kalff, J. (1991). The utility of latitude and other environmental factors as predictors of nutrients, biomass and production in lakes worldwide: problems and alternatives. *Verh. Internatl. Verein. Limnol.* **24:** 1235–1239.

Kalff, J., and Brumelis, D. (1993). Nutrient loading, wind speed and phytoplankton biomass in a tropical African lake. *Verh. Internatl. Verein. Limnol.* **25:** 860.

Kilham, S. S., and Kilham. P. (1990). Tropical limnology: do African lakes violate the "first law" of limnology? *Verh. Internatl. Verein. Limnol.* **24:** 68–72.

Kira, T. (1984). Lake Biwa: a case history of contracts between water and life of the Japanese. *Proc., Shiga Conf.* (1984) *Conservation and Management of World Lake Environment,* pp. 38–51. Shiga Prefectural Government, Otsu, Japan.

Lake Biwa Research Institute (1984). *Data Book of World Lakes*. The Secretariat, LECS '84, Otsu, Japan.

Lean, D. R. S., Neilson, M. A., Stevens, J. J., and Mazumder, A. (1990). Response of Lake Ontario to reduced phosphorus loading. *Verh. Internatl. Verein. Limnol.* **24:** 420–425.

Lee, G. F., and Jones, R. A. (1984). Summary of U.S. OECD eutrophication study. Results and their application to water quality management. *Verh. Internatl. Verein. Limnol.* **22:** 261–267.

Liu, H. (1990). Classification of trophic types of lakes in China. In Liu, H., Zhang, T., and Li. H. (eds.). *Lake Conservation and Management Proc. 4th Internatl. Conf. Conservation and Management of Lakes* ("Hangzhou '90"), pp. 24–40. Chinese Research Academy of Environmental Sciences.

Matsumura-Tundisi, T., Hino, K., and Claro, S. M. (1981). Limnological studies at 23 reservoirs in southern part of Brazil. *Verh. Internatl. Verein. Limnol.* **21:** 1040–1047.

Miller, M. C., Kannan, M., and Colinvaux, P. A. (1984). Limnology and primary productivity of Andean and Amazonian tropical lakes Ekuador. *Verh. Internatl. Verein. Limnol.* **22:** 1264–1270.

Morioka, T. (1984). *Water Quality Management in Lake Biwa Basin: Non-point Source Loading and Pollution Control. Proc. Shiga Conf. Conservation and Management of World Lake Environment,* pp. 167–182. Shiga Prefectural Government, Otsu, Japan.

Mosello, R., and Ruggiu, D. (1985). Nutrient load, trophic conditions and restoration prospects of Lake Maggiore. *Internatl. Rev. Ges. Hydrobiol.* **70:** 63–75.

Moss, B. (1969). Limitation of algal growth in some central African waters. *Limnol. Oceanogr.* **14:** 591–601.

Nontji, A. (1994). The status of limnology in Indonesia. *Mitt. Internatl. Verein. Limnol.* **24:** 95–113.

NORDFORSK (The Scandinavian Council for Applied Research) (1980). *Monitoring of Inland Waters.* OECD Eutrophication Programme; the Nordic Project. Secretariat of Environmental Sciences, NORDFORSK. Pub. 1980:2.

Oritz, C. J. L., and Pena, M. R. (1984). Applicability of the OECD eutrophication models to Spanish reservoirs. *Verh. Internatl. Verein. Limnol.* **22:** 1521–1535.

Quiros, R. (1991). Empirical relationships between nutrients, phyto- and zooplankton and relative fish biomass in lakes and reservoirs of Argentina. *Verh. Internatl. Verein. Limnol.* **4:** 1198–1206.

Rast, W., and Lee, G. F. (1978). *Summary Analysis of the North American* (U.S. portion) *OECD Eutrophication Project: Nutrient Loading-Lake Response Relationships and Trophic State Indices.* Ecological Research Series. EPA. 600/3-78-008, USEPA, Environmental Research Laboratory, Corvallis, Oreg.

Ruttner, F. (1931). Hydrographisce und hydrochemische Beobachtungen auf Java, Sumatora und Bali, *Arch. Hydrobiol. Suppl.* **8:** 197–454.

Sakamoto, M. (1984). *Eutrophication and Its Causative Agents, Phosphorus and Nitrogen in Japanese Lakes. Proc., Shiga Conf.* (1984) *Conservation and Management of World Lake Environment,* pp. 80–85. Shiga Prefectural Government, Otsu, Japan.

Salas, H. J., and Martino, P. (1991). A simplified phosphorus trophic state model for warm-water tropical lakes. *Water Res.* **25:** 341–350.

Schanz, F., and Thomas, E. A. (1981). Reversal of eutrophication in Lake Zurich. *Water Qual. Bull.* **6:** 108–156.

Schindler, D. W. (1978). Factors regulating phytoplankton production and standing crop in the world's freshwaters. *Limnol. Oceanogr.* **23:** 478–486.

Schindler, D. W., Kalff, J., Welch, H. E., Brunskill, G. J., Kling, H., and Kritsch, N. (1974). Eutrophication in the high arctic-Meretta Lake, Cornwallis Island (75°N lat.). *J. Fish. Res. Board Can.* **31:** 647–662.

Shiga Prefectural Government (1995). *Quality of the Environment in Shiga Prefecture* (in Japanese).

Straškrava, M., and Straškrabová, V. (1969). Eastern European lakes. In Rohlich, G. A. (ed.). *Eutrophication: Causes, Consequences and Correctives,* pp. 65–97. National Academy of Sciences, Washington, D.C.

Talling, J. F. (1966). The annual cycle of stratification and phytoplankton growth in Lake Victoria (east Africa). *Internatl. Rev. Ges. Hydrobiol.* **51:** 421–621.

Talling, J. F. (1969). The incidence of vertical mixing, and some biological and chemical consequences, in tropical African lakes. *Verh. Internatl. Verein. Limnol.* **17:** 998–1012.

Talling, J. F., and Talling, I. B. (1965). The chemical composition of African lake waters. *Internatl. Rev. Ges. Hydrobiol.* **50:** 421–463.

Thienemann, A. (1931). Tropisceh See und Seetypenlehre. *Arch. Hydrobiol. Suppl.* **9:** 205–231.

Thomas, E. A. (1969). The process of eutrophication in Central European Lakes. In Rohlich, G. A. (ed.). *Eutrophication: Causes, Consequences, Correctives,* pp. 29–49. National Academy of Sciences, Washington, D.C.

Thornton, J. A. (1987). Aspects of eutrophication management in tropical/sub-tropical regions: a review. *J. Limnol. Soc. S. Afr.* **13:** 25–43.

Toerien, D. F., and Steyn, D. (1975). The eutrophication levels of four South African impoundments. *Verh. Internatl. Verein. Limnol.* **19:** 1947–1956.

Toerien, D. F., and Walmsley, R. D. (1978). Man-made stress patterns in two impoundments. *Verh. Internatl. Verein. Limnol.* **20:** 1690–1694.

Tóth, L. G., and Padisák, J. (1984). A study on the eutrophication of Lake Balaton: a multivariate analysis on the variables of phyto- and bacterioplankton. *Verh. Internatl. Verein. Limnol.* **22:** 1092–1096.

Tundisi, J. G. (1981). Typology of reservoirs in Southern Brazil. *Verh. Internatl. Verein. Limnol.* **21:** 1031–1039.

Tundisi, J. G. (1984). Tropical limnology. *Verh. Internatl. Verein. Limnol.* **22:** 60–64.

Tundisi, J. G., and Matsumura-Tundisi, T. (1984). Comparative limnological studies at three lakes in tropical Brazil. *Verh. Internatl. Verein. Limnol.* **22:** 1310–1314.

Tundisi, J. G., Matsumura-Tundisi, T., Calijuri, M. C., and Novo, E. M. L. (1991). Comparative limnology of five reservoirs in the middle Tiete River, S. Paulo State. *Verh. Internatl. Verein. Limnol.* **24:** 1489–1496.

Vighi, M., and Chiaudani, G. (1985). A simple method to estimate lake phosphorus concentrations resulting from natural, background, loadings. *Water Res.* **19:** 987–991.

Viner, A. B. (1973). Response of a tropical mixed phytoplankton population to nutrient enrichment of ammonia and phosphate, and some ecological implications. *Proc. Roy. Soc. Lond.* **B183:** 351–370.

Vollenweider, R. A. (1976). Advances in defining critical loading levels for phosphorus in lake eutrophication. *Mem. Inst. Itali. Idrobiol.* **40:** 1–24.

Vollenweider, R. (1989). Global problems of eutrophication and its control. In Salánki, J., and Herodek, S. (eds.). *Conservation and Management of Lakes. Symp. Biol. Hung.* **38:** 19–41.

Vörös., L., and Göde, P. N. (1993). Long-term changes of phytoplankton in Lake Balaton (Hungary). *Verh. Internatl. Verein. Limnol.* **25:** 682–686.

Wallsten, M. (1978). Situation of twenty-five lakes now and 40 years ago. *Verh. Internatl. Verein. Limnol.* **20:** 814–817.

Willén, T. (1975). Biological long-term investigations of Swedish lakes. *Verh. Internatl. Verein. Limnol.* **19:** 1118–1124.

Winderholm, T. (1978). Long-term changes in the profundal benthos of Lake Mälaren. *Verh. Internatl. Verein. Limnol.* **20:** 815–824.

Yamaguchi, T. (1984). Environmental conservation in Lake Biwa. *Proc. Shiga Conf.* (1984) *Conservation and Management of World Lake Environment,* pp. 301–305. Shiga Prefectural Government, Otsu, Japan.

Eutrophication models

Ahl, T. (1975). Effects of man-induced and natural loading of phosphorus and nitrogen on the large Swedish lakes. *Verh. Internatl. Verein. Limnol.* **19:** 1125–1132.

Benndorf, J., Uhlman, D., and Pötz, K. (1981). Strategies for water quality management in reservoirs in the German Democratic Republic. *Water Qual. Bull.* **6:** 68–90.

Biswas, A. K. (1976). Mathematical modeling and water-resources decision making. In Biswas, A. K. (ed.). *Systems Approach to Water Management,* pp. 398–414. McGraw-Hill, New York.

Boers, P. C. M., van der Does, J., van der Molen, D. T., and Roip, W. J. (1993). Eutrophication control in the Netherlands by reduction of P loadings: approach, results, challenges for the future. In Giusani, G., and Callieri, C. (eds.). *Proc. 5th Internatl. Conf. Conservation and Management of Lakes,* pp. 50–53.

Brinkman, J. J., Groot, S., Jos, F. J., and Griffioen, P. (1988). An integrated water quality- and water quantity models as a tool for water management. Application to the Province of Friesland, the Netherlands. *Verh. Internatl. Verein. Limnol.* **23:** 1483–1494.

Canfield, D. E., Jr. (1983). Prediction of chlorophyll a concentration in Florida lakes: the importance of phosphorus and nitrogen. *Water Resources Bull.* **19:** 255–262.

Cullen, P., and Smalls, I. (1981). Eutrophication in semi-arid areas. The Australian experience. *Water Qual. Bull.* **6:** 79–91.

Dillon, P. J. (1975). The phosphorus budget of Cameron Lake, Ontario: the importance of flushing rate to the degree of eutrophy of lakes. *Limnol. Oceanogr.* **20:** 28–39.

Dillon, P. J., and Rigler, F. H. (1974). The chlorophyll-phosphorus relationship in lakes. *Limnol. Oceanogr.* **19:** 767–773.

Dillon, P. J., and Rigler, F. H. (1975). A simple method of predicting the capacity of a lake for development based on lake trophic status. *J. Fish. Res. Board Can.* **32:** 1519–1531.

Dillon, P. J., Nicholls, K. H., Locke, B. A., de Grosbois, E., and Yan, N. D. (1988). Phosphorus-phytoplankton relationships in nutrient-poor soft-water lakes in Canada. *Verh. Internatl. Verein. Limnol.* **23:** 258–264.

Faafeng, B. A., and Hessen, D. O. (1993). Nitrogen and phosphorus concentrations and N:P ratios in Norwegian lakes: perspectives on nutrient limitation. *Verh. Internatl. Verien. Limnol.* **25:** 465–469.

Faafeng, B. A., and Nielsen, J. P. (1981). A twenty-year study of eutrophication in a deep, soft-water lake. *Verh. Internatl. Verein. Limnol.* **21:** 412–424.

Forsberg, C., and Ryding, S.-O. (1980). Eutrophication parameters and trophic state indices in 30 Swedish waste-receiving lakes. *Arch. Hydrobiol.* **89:** 189–207.

Forsberg, C., Ryding, S.-O., Forsberg, A., and Claesson, A. (1978). Research on recovery of polluted lakes. I. Improved water quality in Lake Boren and Lake Ekoln after nutrient reduction. *Verh. Internatl. Verein. Limnol.* **20:** 825–832.

Foy, R. H. (1992). A phosphorus loading model for northern Irish lakes. *Water Res.* **26:** 633–638.

Janse, J. H., and Aldenberg, T. (1991). Modelling the eutrophication of the shallow Loosdrecht Lakes, *Verh. Internatl. Verein. Limnol.* **24:** 751–757.

Jørgensen, S. E. (1989). Eutrophication models: the state-of-the-art. In Salánki, J., and Herodek, S. (eds.). *Conservation and Management of Lakes. Symp. Biol. Hung.* **38:** 600–621.

Jørgensen, S. E. (1990). Application of models in limnological research. *Verh. Internatl. Verein. Limnol.* **24:** 61–67.

Jørgensen, S. E., and Mejer, H. (1979). A holistic approach to ecological modelling. *Ecol. Model.* **7:** 169–189.

Jørgensen, S. E., Jørgensen, L. A., Kamp-Nielsen, L., and Mejer, H. F. (1981). Parameter estimation in eutrophication modelling. *Ecol. Model.* **13:** 111–129.

Larsen, D. P., and Mercier, H. T. (1976). Phosphorus retention capacity of lakes. *J. Fish. Res. Board Can.* **33:** 1742–1750.

Larsen, D. P., Schults, D. W., and Malueg, K. W. (1981). Summer internal phosphorus supplies in Shagawa Lake, Minnesota. *Limnol. Oceanogr.* **26:** 740–753.

Los, F. J., and Brinkman, J. J. (1988). Phytoplankton modelling by means of optimization: a 10-year experience with BLOOM II. *Verh. Internatl. Verein. Limnol.* **23:** 790–795.

Los, F. J., de Rooij, N. M., and Smiths, J. G. C. (1984). Modelling eutrophication in shallow Dutch lakes. *Verh. Internatl. Verein. Limnol.* **22:** 917–923.

Mazumder, A. (1994a). Phosphorus-chlorophyll relationships under contrasting hervivory and thermal stratification: predictions and patters. *Can. J. Fish. Aquat. Sci.* **51:** 390–400.

Mazumder, A. (1994b). Phosphorus-chlorophyll relationships under contrasting zooplankton community structure: potential mechanisms. *Can. J. Fish. Aquat. Sci.* **51:** 401–407.

McCauley, E., Downing, J. A., and Watson, S. (1989). Sigmoid relationships between nutrients and chlorophyll among lakes. *Can. J. Fish. Aquat. Sci.* **46:** 1171–1175.

Molot, L. A., and Dillon, P. J. (1991). Nitrogen/phosphorous ratios and the prediction of chlorophyll in phosphorus-limited lakes in central Ontario. *Can. J. Fish. Aquat. Sci.* **48:** 140–145.

Nicholls, K. H., and Dillon, P. J. (1978). An evaluation of phosphorus-chlorophyll-phytoplankton relationships for lakes. *Internatl. Rev. Ges. Hydrobiol.* **63:** 141–154.

OECD (Organization for Economic Cooperation and Development) (1982). *Eutrophication of Waters. Monitoring, Assessment and Control.* OECD Cooperative Programme on Monitoring of Inland Waters (Eutrophication Control). Environmental directorate, OECD, Paris.

Prairie, Y. T., Duarte, C. M., and Kalff, J. (1989). Unifying nutrient-chlorophyll relationships in lakes. *Can. J. Fish. Aquat. Sci.* **46:** 1176–1182.

Rast, W., and Kerekes, J. J. (1981). *Proc. Internatl. Workshop on the Control of Eutrophication.* Laxenburg, Austria, Oct. 1981. UNESCO and IIASA.

Reckhow, K. H. (1979). *Quantitative Techniques for the Assessment of Lake Quality.* USEPA, Office of Water Planning and Standards. EPA-440/5-79-015.

Recknagel, F., and Benndorf, J. (1982). Validation of the ecological simulation model "SALMO." *Internatl. Rev. Ges. Hydrobiol.* **67:** 113–125.

Ryding, S.-O., and Forsberg, C. (1976). Sediments as a nutrient source in shallow polluted lakes. In Golterman, H. L. (ed.). *Interaction between Sediments and Freshwater. Proc. Internatl. Symp.,* pp. 227–234. Amsterdam, Sept. 1976. Dr. W. Junk, The Hague.

Ryding, S.-O., and Rast, W. (1989). *The Control of Eutrophication of Lakes and Reservoirs.* Man and the Biosphere Series 1. UNESCO, The Parthenon Publ. Group Ltd.

Sakamoto, M. (1966). Primary production by phytoplankton community in some Japanese lakes and its dependence on lake depth. *Arch. Hydrobiol.* **62:** 1–28.

Salas, H. J., and Martino, P. (1991). A simplified phosphorus trophic state model for warm-water tropical lakes. *Water Res.* **25:** 341–350.

Sas, H. (1990). Lake restoration by reduction of nutrient loading: expectations, experiences, extrapolations. *Verh. Internatl. Verein. Limnol.* **24:** 247–251.

Scavia, D., and Lang, G. A. (1987). Dynamics of Lake Michigan plankton: a model evaluation of nutrient loading, competition, and predation. *Can. J. Fish. Aquat. Sci.* **45:** 165–177.

Seip, K. L. (1990). Simulation models for lake management—how far do they go? *Verh. Internatl. Verein. Limnol.* **24:** 604–608.

Smith, V. H. (1982). The nitrogen and phosphorus dependence of algal biomass in lakes: an empirical and theoretical analysis. *Limnol. Oceanogr.* **27:** 1101–1112.

Smith, V. H., and Shapiro, J. (1980). A retrospective look at the effects of phosphorus removal in lakes. In *Restoration of Lakes and Inland Waters,* pp. 73–77. EPA 440/5-91-010.

Thornton, J. A. (1987). Aspects of eutrophication management in tropical/sub-tropical regions: a review. *J. Limnol. Soc. S. Afr.* **13:** 25–43.

Vollenweider, R. A. (1969). Möglichkeiten und Grenzen elementarer Modelle der Stoffbilanz von Seen. *Arch. Hydrobiol.* **66:** 1–36.

Vollenweider, R. A. (1976). Advances in defining critical loading levels for phosphorus in lake eutrophication. *Mem. Inst. Itali. Idrobiol.* **40:** 1–24.

Watson, S., and McCauley, E. (1992). Sigmoid relationships between phosphorus, algal biomass, and algal community structure. *Can. J. Fish. Aquat. Sci.* **49:** 2605–2610.

Strategy for lake recovery

Bengtsson, L. (1975). Phosphorus release from a highly eutrophic lake sediment. *Verh. Internatl. Verein. Limnol.* **19:** 1107–1116.

Berge, D. (1990). FOSRES—a phosphorus loading model for shallow lakes. *Verh. Internatl. Verein. Limnol.* **24:** 218–223.

Boers, P. C. M., van der Does, J., van der Molen, D. T., and Roip, W. J. (1993). Eutrophication control in the Netherlands by reduction of P loadings: approach, results, challenges for the future. In Giusani, G., and Callieri, C. (eds.). *Proceedings of the 5th International Conference on the Conservation and Management of Lakes,* pp. 50–53.

Brix, H. (1994). Constructed wetlands for municipal waste water treatment in Europe. In Mitsch, W. J. (ed.). *Global Wetlands: Old World and New,* pp. 325–333. Elsevier Science Publishers.

Chang, W. Y. B. (1994). Management of shallow tropical lakes using integrated lake farming. *Verh. Internatl. Verein. Limnol.* **24:** 219–224.

Cooke, G. D. (1983). Review of lake restoration techniques and an evaluation of harvesting and herbicides. In *Lake Restoration, Protection, and Management. Proc. 2d Annual Conf. North American Lake Management Society,* pp. 257–266. U.S. Environmental Protection Agency, Washington, D.C.

Cooke, G. D., Welch E. G., Peterson, S. A., and Newroth, P. R. (1986). *Lake and Reservoir Restoration.* Butterworth, Boston.

DeMelo, R., France, R., and McQueen, D. J. (1992). Biomanipulation: hit or myth? *Limnol. Oceanogr.* **37:** 192–207.

Dillon, P. J., Molot, L. A., and Scheider, W. A. (1991). Phosphorus and nitrogen export from forested catchments in central Ontario. *J. Environ. Qual.* **20:** 657–664.

Diósi, A. (1989). Post-project analysis of the reconstructed Kis-Balaton. In Salanki, J., and Herodek, S. (eds.). *Conservation and Management of Lakes. Symp. Biol. Hung.* **38:** 489–503.

Faafeng, B. A., and Brabrand, A. (1990). Biomanipulation of a small, urban lake. Removal of fish exclude bluegreen blooms. *Verh. Internatl. Verein. Limnol.* **24:** 597–602.

Faafeng, B. A., and Nielsen, J. P. (1981). A twenty-year study of eutrophication in a deep, softwater lake. *Verh. Internatl. Verein. Limnol.* **21:** 412–424.

Fleischer, S., Stieb, L., and Leonardson, L. (1991). Restoration of wetlands as a means of reducing nitrogen transport to coastal waters. *Ambio* **20:** 271–272.

Fricker, Hj. (1980). *OECD Eutrophication Program—Regional Project Alpine Lakes.* Swiss Federal Board for Environmental Protection, Bern.

Herodek, S., Istvánovics, V., and Zlinsky, J. (1988). Phosphorus metabolism and eutrophication control of Lake Balaton. *Verh. Internatl. Verein. Limnol.* **23:** 517–521.

Jansson, M. (1979). Nutrient budgets and the regulation of nutrient concentrations in a small sub-arctic lake in northern Sweden. *Freshwater Biol.* **9:** 213–231.

Larsen, D. P., Schults, D. W., and Malueg, K. W. (1981). Summer internal phosphorus supplies in Shagawa Lake, Minnesota. *Limnol. Oceanogr.* **26:** 740–753.

Larsen, D. P., Maluueg, H. K. W., Schults, D. W., and Brice, R. M. (1975). Response of eutrophic Shagawa Lake, Minnesota, U.S.A., to point-source, phosphorus reduction. *Verh. Internatl. Verein. Limnol.* **19:** 884–892.

Mitsch, W. J. (1994). The nonpoint source pollution control function of natural and constructed reparian wetlands. In Mitsch, W. J. (ed.). *Global Wetlands: Old World and New,* pp. 351–361. Elsevier Science Publishers.

Mitsch, W. J., Reeder, B., and Klarer, D. (1989). The role of wetlands for the control of nutrients with a case study of western Lake Erie. In Mitsch, W. J., and Jørgensen, S. E. (eds.). *Ecological Engineering: An Introduction to Ecotechnology,* pp. 129–158. Wiley, New York.

Mosello, R., Marchetto, A., Boggero, A., and Tartari, G. A. (1990). Relationships between water chemistry, geographical and lithological features of the watershed of alpine lakes located in NW Italy. *Verh. Internatl. Verein. Limnol.* **24:** 155–157.

Nürnberg, G. (1984). The prediction of internal phosphorus load in lakes with anoxic hypolimnion. *Limnol. Oceanogr.* **29:** 155–157.

Nürnberg, G., and Peters, R. H. (1984). The importance of internal phosphorus load to the eutrophication of lakes with anoxic hypolimnia. *Verh. Internatl. Verein. Limnol.* **22:** 190–194.

Nürnberg, G., Shaw, M., Dillon, P. J., and McQueen, D. J. (1985). Internal phosphorus load in an oligotrophic Precambrian Shield lake with an anoxic hypolimnion. *Can. J. Fish. Aquat. Sci.* **43:** 574–580.

OECD (Organization for Economic Cooperation and Development) (1982). *Eutrophication of Waters. Monitoring, Assessment and Control.* Final Report, OECD Cooperative Programme on Monitoring of Inland Waters (Eutrophication Control). Environment Directorate, OECD, Paris.

Peterjohn, W. T., and Correll, D. J. (1984). Nutrient dynamics in an agricultural watershed: observations on the role of a riparian forest. *Ecology* **65:** 1466–1475.

Pomogyi, P. (1989). Macrophyte communities of the Kis-Balaton reservoir. In Salánki, J., and Herodek, S. (eds.). *Conservation and Management of Lakes. Symp. Biol. Hung.* **38:** 505–515.

Reckhow, K. H., Beaulac, M. N., and Simpson, J. T. (1980). *Modeling Phosphorus Loading and Lake Response under Uncertainty: A Manual and Compilation of Export Coefficient.* U.S. Environmental Protection Agency, Clean Lakes Section. EPA-440/5-80-011.

Reynolds, C. (1994). The ecological basis for the successful biomanipulation of aquatic communities. *Arch. Hydrobiol.* **130:** 1–33.

Ryding, S.-O., and Forsberg, C. (1976). Sediments as a nutrient source in shallow polluted lakes. In Golterman, H. L. (ed.). *Interaction between Sediments and Freshwater. Proc. Internatl. Symp.,* pp. 227–234. Amsterdam, Sept. 1976. Dr. W. Junk, The Hague.

Ryding, S.-O., and Forsberg, C. (1980). Short-term load-response relationships in shallow polluted lakes. In Baric, J., and Mur, L. R. (eds.). *Hypertrophic Ecosystems,* pp. 95–103. Dr. W. Junk, BV Publishers, The Hague.

Shapiro, J., and Wright, D. I. (1984). Lake restoration by biomanipulation: Round Lake, Minnesota, the first two years. *Freshwater Bio.* **14:** 371–383.

USEPA (1980). *Clean Lakes Program Guidance Manual.* EPA 440/5-81-003. U.S. Environmental Protection Agency, Office of Water Regulations and Standards. Washington, D.C.

USEPA (1981). *Restoration of Lakes and Inland Waters.* EPA 440/5-81-010. U.S Environmental Protection Agency, Office of Water Regulations and Standards. Washington, D.C.

Vollenweider, R. A. (1968). *Scientific Fundamentals of the Eutrophication of Lakes and Flowing Waters, with Particular Reference to Nitrogen and Phosphorus as Factors in Eutrophication.* OECD, DAS/CSI/68.27.

Vollenweider, R. A. (1976). Advances in defining critical loading levels for phosphorus in lake eutrophication. *Mem. Inst. Itali. Idrobiol.* **40:** 1–24.

Wright, D., and Shapiro, J. (1984). Nutrient reduction by biomanipulation: an expected phenomenon and its possible cause. *Verh. Internatl. Verein. Limnol.* **23:** 518–524.

Yan, J. S., and Zhang Y. S. (1994). How wetlands are used to improve water quality in China. In Mitsch, W. J. (ed.). *Global Wetlands: Old World and New,* pp. 369–376. Elsevier Science Publishers.

Yin, C., Jin, W., Lan, Z., and Zhao, M. (1991). Chaohu Lake eutrophication and nutrient load reduction. *Verh. Internatl. Verein. Limnol.* **24:** 1086–1087.

Yin, C., Zhao, M., Jin, W., and Lan, Z. (1993). A multi-pond system as a protective zone for the management of lakes in China. *Hydrobiologia* **251:** 321–329.

Chapter

9

Wastewater Reuse

Takashi Asano
Department of Civil and Environmental Engineering
University of California at Davis
Davis, California

Wastewater reuse is one element of water-resource development and management which provides innovative and alternative options for traditional water supply. The water-pollution control efforts in many countries have made available treated effluent that may be an economical augmentation to existing water supply when compared to increasingly expensive new water-resource development. However, wastewater reuse is only one alternative in planning to meet the water resources needs of a region. Water conservation, water recycling, efficient management and use of existing water supplies, and new water-resource development are examples of other alternatives.

Wastewater reuse involves considerations of public health and also requires close examination of facilities planning, wastewater-treatment plant reliability, economic and financial analyses, and water-utility management. Whether wastewater reuse will be appropriate depends on economic considerations, potential uses for the reclaimed water, the degree of stringency of waste-discharge requirements, and public policy in which the desire to conserve rather than develop available water resources may override economic and public-health considerations.[1,2] Today, technically proven wastewater treatment or purification processes exist to provide water of almost any quality desired. Thus, wastewater reuse has a rightful place and an important role in optimal planning and more efficient management and use of water resources in many countries.[3–7]

The specific topics covered in this chapter are (1) overview of wastewater reuse, (2) wastewater-reuse planning, (3) types of water-reuse applications, and (4) microbial health-risk estimation for wastewater reuse. The focus of this chapter is the unique aspects of wastewater reuse and concentrates on the *big-picture* developments and technologies which contribute more to the water-supply benefits than traditional water-pollution control measures discussed in many sanitary- and environmental-engineering textbooks.

Overview of Wastewater Reuse

Terms related to wastewater reuse that are important in understanding the concepts discussed in this chapter are highlighted in the following paragraphs.

Definitions

Wastewater reclamation is the treatment or processing of wastewater to make it reusable, and *wastewater reuse* or *water reuse* is the utilization of this water for a variety of beneficial uses. In addition, either reclamation and reuse of water frequently implies the existence of a pipe or other water-conveyance facilities for delivering the reclaimed water. In contrast to reuse, *wastewater recycling* or *water recycling* normally involves only one use or user, and the effluent from the user is captured and redirected back into that use scheme. In this context, the term *wastewater recycling* is applied predominantly to industrial applications such as in the stream-electrical, manufacturing, and minerals industries.

Water potentially available for reuse includes discharge from municipalities (sewage, municipal wastewater), industries (industrial wastes), agricultural return flows, and storm water. Of these, return flows from agricultural irrigation and storm water are usually collected and reused without further treatment.

Reclaimed wastewater can be used directly (*direct reuse*) without passing through a natural body of water for such applications as agricultural and landscape irrigation or indirectly (*indirect reuse*) such as in groundwater recharge. The topics covered in this chapter are related to deliberate or *planned reuse* as described above. It is recognized that indirect reuse, through discharge of a treated effluent to a receiving water for assimilation and withdrawal downstream, while important, does not normally constitute planned reuse. For example, the diversion of water from a river downstream of a discharge of treated wastewater constitutes an incidental or *unplanned reuse*. However, the distinction between the various types of water reuse are arbitrary, and, in fact, every degree of wastewater reuse exists.

Another distinction in wastewater reuse is *potable-water reuse* in which highly treated reclaimed water is used to augment drinking-water supplies and *nonpotable-water reuse*. Direct potable reuse is the incorporation of reclaimed water into a potable-water supply system, without relinquishing control over the resource. Indirect potable reuse incorporates reclaimed waters into a raw-water supply, but control of reclaimed water is surrendered to allow mixing and dispersion by discharge into an impoundment or receiving water.[2,8]

The distinction between inadvertent or unintentional and indirect reuse is, after all, one of intention or attention.[3] It is no longer realistic that a community dumping its wastewater in a river not know where the wastewater is going or that the community taking the river water downstream not know that it contains wastewater.[3,4]

Categories of wastewater reuse

Unplanned indirect wastewater reuse, through effluent disposal to streams and groundwater basins, has been a long-accepted practice throughout the world. Many communities situated at the end of major waterways—New Orleans, United States (the Mississippi River); London, England (the River Thames); and the cities and towns in the Rhine River Valley, Germany and Osaka, Japan (the Yodo River)—ingest water that has already been used many times through repeated river withdrawal and discharge. Argo and Rigby[9] estimated that up to 82 percent of the water in the Santa Ana River in southern California originated from wastewater discharged by upstream municipal wastewater-treatment plants. Also, a review of the data revealed that the Ohio River at Cincinnati contained more than 19 percent wastewater during the seven-consecutive-day minimum flow condition for the river, which occurs once every 10 years.[9] Similarly, riverbeds or percolation ponds may recharge underlying groundwater aquifers with waste-containing water, which, in turn, is withdrawn by subsequent communities for domestic water supply.

Indirect water reuse can also be planned. For example, municipal wastewater from Stevenage, north of London, is treated to remove nitrates before being discharged into the River Lea, a tributary of the River Thames. This is done so that London, which withdraws 20 percent of its water from the Lea, will not have too high a level of nitrates in its drinking water.[3] Similarly, as part of its responsibilities, the Los Angeles County Flood Control District in California has implemented an artificial groundwater-recharge program using reclaimed water since 1962.[10,11] In general, however, water-reclamation permits issued by the California Regional Water Quality Control Boards stipulated that not more than 20 percent of the water applied to surface spreading area should be reclaimed water during a 12-month period.[12]

Because of the health and safety concerns, direct nonpotable-water reuse has become a major option for planned reuse for supplementing public water supplies in the United States and elsewhere in the world.[13] Some communities are, however, developing or implementing plans for direct potable reuse where no other economic possibilities exist for supplemental freshwater supplies.[8,14,15] While the quantities involved in these projects are small, some of the salient features are discussed in this chapter because of their technological and public-health interest. The attractiveness of wastewater reuse for non-potable purposes is summarized in Table 9.1.

Nonpotable reuse projects were, until recently, initiated locally without much stimulation from regulatory or water-management authorities. Many projects were located in areas close to the waste-water-treatment plants where additional treatment and extensive transmission pipelines were unnecessary. Thus, wastewater reuse was more of an opportunistic nature than the result of a well-planned program to supplement or replace the use of potable water for non-

TABLE 9.1 Attractiveness of Wastewater Reuse for Nonpotable Purposes

Attractiveness	Reasons
Resourceful	Nonpotable reuse is often the only feasible way of supplementing water resources in a community where additional freshwater resources are not available
Economical	Nonpotable reuse is often the least expensive option for increasing water resources in a community
Serviceable	Nonpotable reuse often offers an expedient approach to pollution abatement, and treatment for nonpotable-water reuse may be less costly than treatment or discharge to a receiving body of water
Feasible and convenient	The technology for nonpotable reuse is relatively well established, and experimental investigations are not required to establish the design criteria, except to improve or optimize the practice
Suitable	The public-health problems associated with nonpotable reuse are more easily addressed than for potable reuse and do not require research in each instance
Dependable	Customers are pleased with the availability of reclaimed water because it is generally provided at a significantly lower cost than freshwater and the supply may be more reliable in periods of water shortage
Acceptable	Where nonpotable reuse is practiced with proper guidelines for reliability and monitoring, and such practice is now widespread, the public has been enthusiastic in its endorsement
Good public policy	Water conservation and reclamation and the efficient use of water are all in the public's mind, and wastewater reuse is one of the major policy issues debated in arid and semiarid regions of the country

SOURCE: Developed in part from Ref. 13.

TABLE 9.2 Categories of Municipal Wastewater Reuse

1. Agricultural	2. Landscape irrigation
Crop irrigation	Park
Commercial nurseries	Schoolyard
Commercial aquaculture	Freeway median
3. Industrial reuse	Golf course
Cooling	Cemetery
Boiler feed	Greenbelt
Process water	Residential
Heavy construction	4. Groundwater recharge
5. Recreational and environmental	Groundwater replenishment
Lakes and ponds	Saltwater intrusion control
Marsh enhancement	Subsidence control
Streamflow enhancement	6. Nonpotable urban
Fisheries	Fire protection
Snowmaking	Air conditioning
7. Potable reuse	Toilet flushing
Blending in municipal water-supply reservoir	
Pipe to pipe supply	

potable purposes.[13] This began to change, however, about two decades ago as water demands increased and, at the same time, highly treated effluents were becoming available from many municipal wastewater-treatment plants in the United States. Thus, treated effluent is perceived as a valuable resource and disposing of it into the ocean, for example, is considered to be a waste of natural resources. In some cases, treated effluent is the only affordable source of additional water available, and therefore, wastewater reuse has become more of a necessity than a means of effluent disposal.

Seven categories of municipal wastewater reuse are listed in Table 9.2. These categories are arranged in descending order of anticipated volume of use. Major factors affecting the implementation of wastewater reuse in different categories are (1) water-quality requirements for intended wastewater-reuse applications, (2) existing or proposed wastewater-treatment facilities, (3) requirements for degree of treatment process reliability, (4) potential health-risk mitigation, and (5) public perception and acceptance.

Potential and status of wastewater reuse

The second National Water Assessment by the U.S. Water Resources Council[16] provided, for the first time, nationally consistent current and projected water use and supply information by regions and subregions for the entire United States. Table 9.3 shows estimates of freshwater withdrawals and wastewater recycling and reuse in the years 1975, 1985, and 2000. The data in this summary indicate that the quantity of water recycling and reuse is expected to increase over

TABLE 9.3 Estimated Freshwater Withdrawals and Wastewater Recycling and Reuse in the United States in the Years 1975, 1985, and 2000

Category	Quantity ($\times 10^6$ m^3/d) 1975	1985	2000
Total freshwater withdrawals	1373	1349	1253
Wastewater recycling (industrial)	527	1464	3276
Steam-electrical	215	—	1958
Manufacturing	231	—	1197
Minerals	79	—	121
Wastewater reuse (municipal)	3	8	18

SOURCE: Ref. 16 provided the 1975 data, while the data for the years 1985 and 2000 are the projected values from Ref. 20.

twofold during the next 15 years. These data also help somewhat in placing municipal wastewater reuse in perspective on a national scale. The relatively small municipal wastewater-reuse fraction compared to the water recycling in 1985 is expected to remain about the same in the future even though the actual quantity of wastewater reuse will increase significantly. Wastewater reuse has their greatest impacts in the arid West in such states as Arizona, California, Colorado, and Texas; although the humid Southeast, such as Florida, is implementing wastewater reuse because of rapidly growing water demands and their associated water-pollution problems.

Potentially large quantities of reclaimed water can be used in four sectors of activities:

1. *Irrigation.* On the basis of quantity of use, irrigation is undoubtedly ranked highest. Overall irrigation use, consisting of both agriculture and landscape applications, is projected to account for 54 percent of total freshwater withdrawals by the year 2000 in the United States.[17]

2. *Industry.* The second major potential user of reclaimed water is industry, primarily for cooling and process needs. However, industrial uses vary greatly, and additional wastewater treatment beyond secondary treatment is usually required.

3. *Groundwater recharge.* The third major potential reuse application is groundwater recharge by means of either spreading basins or direct injection into aquifers. Groundwater recharge with reclaimed water involves storage of the mixture of groundwater and treated wastewater. The point of recharge and the points of withdrawal of the water for use are usually some distance apart. Time-in-storage and separation-in-space factors are important public-health considerations, providing time for sampling and testing of water from the recharged aquifer before the groundwater is used. It also provides time for pathogen die-off and biodegradation of organic substances.

Perhaps, the greatest single advantage of including groundwater recharge in any program of wastewater reuse, especially potable reuse, is the loss of identity that groundwater recharge seems to provide for reclaimed water.[18,19]

4. *Miscellaneous.* A fourth sector of activity is characterized as miscellaneous subpotable uses, such as for recreational lakes, aquaculture, and toilet flushing. These subpotable uses are projected to be minor wastewater-reuse applications, accounting for less than 5 percent of total wastewater reuse. However, in selected instances, they may provide an ideal means of effluent disposal or a supply adequate for low-quality needs.

The Federal Clean Water Construction Grant Program provided most construction costs of wastewater reclamation and reuse projects. However, because of the limited financial resources available in recent years, only wastewater-reclamation projects which also proved to be cost-effective water-pollution control options have been eligible for funding.[21] Because the need to expand the use of reclaimed water is evident, several federal and state financial-assistance programs have now been developed, such as water-reclamation loan programs in California, as well as local and regional financing of wastewater-reclamation and -reuse projects.

Data from the survey conducted in California in 1987 indicated that over 329×10^6 m^3/yr of municipal wastewater was reclaimed by 200 wastewater-treatment plants that supplied reclaimed water to more than 854 use areas.[22] Much of the volume of reclaimed wastewater (63 percent) was used for agricultural irrigation. An important use in recent years is irrigation of golf courses, other turfgrass, and landscape areas (about 13 percent).[25] Similar trends have been observed in other states such as in Arizona, Colorado, Florida, and Texas, where significant wastewater reuse projects have been implemented.[20–22]

Wastewater-Reuse Planning

A number of factors affect the planning and implementation of municipal wastewater-reuse projects. Impetus for wastewater reuse has, in general, resulted from four motivating factors: (1) availability of high-quality effluents; (2) increasing cost of freshwater development; (3) desirability of establishing comprehensive water-resource planning, including water conservation and wastewater reuse; and (4) avoidance of more stringent water-pollution control requirements such as needs for advanced wastewater-treatment facilities.

A common misconception in planning for wastewater reuse is that reclaimed water represents a low-cost new water supply. This assumption is valid only when wastewater reclamation facilities are conveniently located near large industrial or agricultural users, and when additional wastewater treatment is not required. The conveyance and distribution systems for reclaimed water represent the principal cost of most wastewater-reuse projects. Recent experience in California indicates that approximately \$2.5 in average capital costs are required for making each cubic meter per year of reclaimed municipal wastewater available for wastewater reuse.[23] Assuming a facilities life of 20 years and a 9 percent interest rate, the amortized cost of this reclaimed water is in the neighborhood of $\$0.24/m^3$, excluding operation and maintenance (O&M) costs.

The optimum wastewater-reuse project is best achieved by integrating both wastewater-treatment and water-supply needs into planning. Thus, the facilities planning for wastewater reuse should consist of (1) wastewater treatment and disposal needs assessment, (2) water supply and demand assessment, (3) detailed reclaimed-water market analysis, (4) engineering and economic analyses of alternatives, and (5) implementation plan with financial analysis. These planning steps are described in more detail in this section, and important factors to consider in planning and implementation are highlighted.

Planning basis

The typical framework for analysis is first to establish clearly defined objectives and identify whether a project is intended as primarily single- or multiple-purpose, that is, to serve two or more basic functions.[23]

Single- and multiple-purpose projects. Generally wastewater-reclamation projects serve the functions of either water-pollution control or water supply. Because most public-works agencies, or subdivisions of agencies, are established as single-purpose entities, planning for wastewater-reclamation projects is usually initiated with a single purpose in mind. For example, a city wastewater department is confronted with the need to meet more stringent effluent-discharge requirements and will investigate wastewater reclamation and reuse as one of the pollution-control options. On the other hand, a water department may be faced with a falling groundwater table and regard wastewater reuse as a means of satisfying some of the water demands by supplementing existing water supply.

In recent years, however, at least two simultaneous trends in the United States should force U.S. citizens to view wastewater reuse as fundamentally serving multiple purposes: (1) standards for the discharge of wastewater are becoming increasingly more stringent, and (2) freshwater resources are becoming increasingly stressed to meet growing and competing water demands.[23]

Many projects intended originally as single-purpose inevitably have spillover benefits. If these would be recognized at the outset of project planning, the scope could be expanded. Because of their multiple benefits and beneficiaries, there are additional options available in terms of sharing project responsibility and costs and achieving the optimum balance of benefits (i.e., realizing maximum net benefits). The point of emphasizing the multiple-purpose concept is that the traditional perspective of a single-purpose agency and funding program is often becoming outmoded and a disservice to meeting increasingly complex needs of society with an environmentally conscious public.

Project study area. The project study area is another critical planning issue. The typical approach is to equate the study area with the project sponsor's jurisdictional boundaries. However, this approach can have serious pitfalls. The project study area should include all the area that can potentially benefit from reuse of effluent from a particular wastewater-treatment plant. Because water supply is typically dependent on water resources outside the project study area, it is essential to look beyond the local area to obtain an understanding of the water-resource situation. For example, overdrafted groundwater basins may have the most serious impact on communities great distances from the local area. Thus, implementing wastewater reuse in the project area could result in a water-supply savings in another area.

Wastewater-reuse planning. Planning for wastewater reuse typically evolves through three stages: (1) concept-level planning, (2) preliminary feasibility investigation, and (3) facilities planning. At the concept-level planning, a potential project is sketched out, rough costs estimated, and a potential reclaimed water market identified. On the basis of the preliminary feasibility investigation, if wastewater reuse appears to be viable and desirable, then detailed planning can be pursued, refined facilities alternatives developed, and a final facilities plan proposed.

Reclaimed-water market assessment

A key task in planning a water-reclamation project is to find potential customers who are capable and willing to use reclaimed water. The approach to take in marketing the reclaimed water depends on two factors:

1. *Project purpose.* Is the intent solely to treat and dispose of the wastewater or also to obtain optimum water-supply benefit?
2. *User option.* Will the use of reclaimed water be voluntary or mandatory?

If the primary purpose is to treat and dispose of wastewater on land, then planners usually seek land application sites where water can be applied at high rates, usually in excess of optimum crop-uptake rates, at least cost. Unless the system is designed with backup wastewater-disposal methods, the users will have to make a long-term commitment to accept the treated effluent and may not have full control over the quantities of water delivered. If users cannot be found to accept treated effluent on a voluntary contractual basis, the wastewater agency will itself have to purchase wastewater application sites and apply the reclaimed water or lease the land to a private farmer.

Projects designed with the primary purpose of water supply can usually be operated more flexibly if alternate disposal, such as stream discharge, is available to dispose of effluent that cannot be reused. The reclaimed water can be marketed on a voluntary basis. However, if water supply is critical, the managing agency may elect to impose the use of reclaimed water in place of freshwater where it is environmentally and humanly safe to do so.

Whether a user is capable of using reclaimed water depends on the quality of effluent available and its suitability for the type of use involved. Willingness to use reclaimed water depends on whether its use is voluntary and, if it is, on how well reclaimed water competes with freshwater with respect to cost, quality, and convenience. It is essential to have a thorough knowledge of the water-supply situation, especially if reclaimed water is to be marketed on a voluntary basis.

Market assessment. The wastewater-reuse market assessment consists of two parts: determination of background information and a survey of potential reclaimed water users and their needs, as shown in Table 9.4. Important water-supply information includes a complete background on all the wholesale and retail water agencies in the planning area, including their boundaries, quality of water served, prices charged, and willingness to allow reclaimed water use in their jurisdictional areas. Because the introduction of reclaimed water could reduce freshwater revenues, at least in the short run, there might be resistance to implement wastewater reuse by some agencies. There should be the willingness to consider the freshwater revenue impacts in the analysis and appropriate revenue and cost sharing to obtain the full cooperation of all affected agencies.[23]

Without much investigation it is possible to list most of the potential reclaimed water use categories in the study area, e.g., landscape irrigation, industrial cooling, and irrigation of food crops. On the basis of the use categories, health and water-pollution-control regulatory authorities should be consulted to obtain their respective requirements. These would include wastewater-treatment requirements, on-

TABLE 9.4 Reclaimed-Water Market Assessment: Background Information and Survey

Inventory potential users and uses of reclaimed water

Determine health-related requirements regarding water quality and application requirements (e.g., treatment reliability, backflow prevention, use-area controls, irrigation methods) for each type of application of reclaimed water

Determine regulatory requirements to prevent nuisance and water-quality problems, such as restrictions to protect groundwater

Develop assumptions regarding probable water quality that would be available in the future with various levels of treatment and compare those to regulatory and user requirements

Develop an estimate of future freshwater supply costs to potential users of reclaimed water

Survey potential reclaimed-water users, obtaining the following information:

- Specific potential uses of reclaimed water
- Present and future quantity needs
- Timing and reliability of needs
- Quality needs
- On-site facility modifications to convert to reclaimed water and meet regulatory requirements for protection of public health and prevention of pollution problems from reclaimed water
- Capital investment of the user for on-site facility modifications, changes in operational costs, desired payback period or rate of return, and desired water cost savings
- Plans for changing use of site in future
- Preliminary willingness to use reclaimed water now or in the future

Inform potential users of applicable regulatory restrictions, probable water quality available with different levels of treatment, reliability of the reclaimed water, future costs, and quality of freshwater compared to reclaimed water

SOURCE: Adapted from Ref. 23.

site facility modifications (e.g., backflow-prevention devices), and use-area controls (e.g., no irrigation in areas of direct human contact). Technical experts, such as farm advisors, can be consulted to determine acceptable water quality for various use categories.

Identification of users. It is then possible to begin identifying and contacting individual potential users of reclaimed water. Access to records of water retailers can be especially helpful. Several years of actual water-use records are helpful to ensure that planning is not misled by data from unusually wet or dry precipitation years. It is important to obtain actual prices paid for water or, if a user has its own supply, its fixed and variable costs. Potential users should be contacted and the reuse sites visited to determine potential site problems or on-site water-system modifications needed to accommodate use of reclaimed water. These factors have cost implications which must be assessed in the planning stage. The concern, needs, and financial expectations of users must be identified. Group presentations with potential users may be useful to disseminate information and make technical experts accessible to respond to questions.

Monetary analyses

While technical, environmental, and social factors are considered in project planning, monetary factors[23] tend to override in the key decisions of whether and how to implement a wastewater-reuse project. Monetary analyses fall into two categories: economic analysis and financial analysis, the distinction between which is critical. The economic analysis focuses on the value of the resources invested in a project to construct and operate it, measured in monetary terms and computed in the present value. On the other hand, the financial analysis is based on the market value of goods and services at the time of sale, incorporating any particular subsidies or monetary transfers which may exist. Whereas economic analysis evaluates wastewater-reuse projects in the context of impacts on society, financial analysis focuses on the local ability to raise money from project revenues, government grants, loans, and bonds to pay for the project.

The basic result of the economic analysis is to address the issue of whether a reuse project *should* be constructed. Equally important, however, is to determine whether a reuse project *can* be constructed. Both orientations, therefore, are necessary. However, only wastewater-reuse projects which are viable in the economic context should be given further consideration for a financial analysis.[23,24]

Economic analysis. The role of an economic analysis is to provide a basis for justifying a wastewater-reuse project in monetary terms. A project is considered justified if its total benefits exceed its total costs. If several alternatives can meet the same objective, then the alternative providing the maximum net benefit is the economically justifiable project. While the cost/benefit ratio is a common measure of economic justification, it is not the best measure to determine the optimum project size.[24,25]

An important aspect of the economic analysis is that it takes into consideration all costs and benefits associated with the alternatives under consideration, placing all alternatives on equal footing for comparison. Also, this analysis is completely independent of financing considerations. To identify all costs and benefits it is essential to look beyond the boundaries of the agency doing the planning. For example, an agency may be seeking a new source of water supply outside its boundaries. To perform an economic comparison it would be necessary to identify the construction, operation, and maintenance costs of this supply.

Cost and price of water. Another important aspect of an economic analysis is that it considers only the future flow of resources invested in or derived from a project. Past resource investments are considered sunk costs that are irrelevant to future investment decisions. Thus, debt service on past investments is not included in an economic analysis. A common error in this respect is to confuse water price with water cost. *Water price* is the purchase price paid to a water wholesaler or retailer to purchase water and usually reflects a melding of current and past expenditures for a combination of projects, as well as water-system administration costs, which are generally fixed costs. The costs of relevance to an economic analysis would be only costs for future construction, operation, and maintenance. If a wastewater-reuse project were to be compared to a particular new water-supply development, the relevant costs would be the future stream of costs to (1) construct new freshwater facilities and (2) operate and maintain all the facilities needed to treat and deliver the new increment of water supply developed. This stream of costs may bear no resemblance to the present and future price, at the wholesale or retail level, charged for water.

In contrast, *water prices* embody debt service on existing facilities, and future projections are an average price to recover costs for both existing facilities and future additions. Typically, water price will be much lower than the marginal cost of developing a new water supply, because the cost of each new source of supply is increasingly expensive as a result of inflation and the greater difficulty in developing new supplies.

Financial analysis. The financial analysis addresses the question whether a wastewater-reuse project is financially feasible. The project sponsor will need a source of capital and sources of revenue to pay for debt service and operational costs for both the proposed reuse project and any existing facilities. Fixed costs for existing facilities, while irrelevant in the economic analysis, must be considered in a financial analysis if they are a continuing financial obligation.

The wastewater-reuse project sponsor is not the only important party to consider in a financial analysis. Of particular importance is the participation of the user of the reclaimed water. The user will expect a net cost of reclaimed water that is no more than it would have paid for freshwater. For example, a reclaimed-water customer may have to invest in piping modifications or a dual water-supply system to accommodate the reclaimed water. On the other hand, a farmer may be able to save on fertilizer costs by taking advantage of nitrogen and phosphorus contained in reclaimed water. A prospective user will expect the difference in price between freshwater and reclaimed water to reflect any added costs or savings.

Because the sale of reclaimed water may reduce revenue from freshwater sale, there may be a need to evaluate the effect on the freshwater retailer and freshwater prices. It may be necessary to allocate some of the reclaimed-water revenue to compensate for the freshwater revenue loss. On the other hand, if reclaimed water offsets the purchase or development of more-expensive freshwater, then it may be appropriate for freshwater revenue to be used to subsidize the wastewater-reuse project.

Potential users not uncommonly have different sources of water or different rate schedules. It is important, therefore, not to assume that there is an average price that all users are paying for freshwater. Failure to take into account the financial situation of each user could result in the loss of key reclaimed-water customers. The initial market assessment should have included these financial data. In conclusion, there should be flexibility in tailoring a financial scheme to fit each situation best.

Other planning factors

A number of factors[23] besides the monetary aspects have to be evaluated during the planning for a wastewater-reuse project, such as environmental impacts. However, factors of particular significance in project development are related to engineering and public health.

Engineering involves more than water-distribution system design. A wastewater-reuse project is a relatively small-scale water-supply project with considerations of matching supply and demand, appropriate levels of wastewater treatment, reclaimed-water storage, and supplemental or backup freshwater supply.

Water-demand characteristics. In freshwater systems, water demands are first projected and then water supplies are developed to meet the demand. The reverse procedure is often applied in wastewater-reuse system planning. The wastewater supply rate is accepted as a given and the reclaimed-water demand is added to the system until the economic optimum is met. For example, landscape irrigation demand in California is seasonal. However, wastewater production is nearly constant on a year-round basis. Reclaimed-water supply may be sufficient to meet annual demands, but only if seasonal storage is available. Seasonal storage, however, is costly and, in urban settings, even impossible to site.

Another option is to include fewer users in the system such that the peak demands can be met entirely by the reclaimed-water supply without seasonal storage. This could, however, result in the waste of as much as 40 percent of the available reclaimed water. What will probably be the optimum situation is to add users in excess of supply and meet the peak demands with supplemental freshwater. There may still be some supply that cannot be used or economically stored during low-water-demand periods, but this lost supply can be reduced substantially because of the availability of a supplemental supply in the peak season. Some projects have incorporated an added benefit by utilizing a poor-quality water supply unsuitable for potable use, such as an abandoned groundwater basin, to supplement reclaimed water.

Supplemental-water supply. Supplemental freshwater can be blended with reclaimed water in the distribution system or on the user's site. Because of public-health concerns about potential cross connection or backflow of reclaimed water into potable-water-supply systems, it may be necessary to provide an airgap between the supplemental supply and the reclaimed system.

Even if supplemental water is not needed to meet demands, there may still be a need to provide an emergency backup water supply during periods of treatment plant upset or equipment failure. Because a backup water supply would be utilized in place of, rather than simultaneously with, reclaimed water, there are more options for introducing it into the system. With appropriate backflow prevention, the reclaimed-water distribution system can be connected to the potable system during the emergency period. It should be noted that with the availability of a backup water supply, there is less need for equipment redundancy in the reclaimed-water system to ensure 100 percent reliability.

Water quality. If a significant market could be added by upgrading reclaimed-water quality, project alternatives should be developed for various treatment levels. The levels of wastewater treatment and water quality for landscape and agricultural irrigation uses are normally governed by health-related regulations, although crop sensitivity to effluent constituents such as salts and boron should be investigated.[26]

Industrial users will have more stringent physical and chemical water-quality requirements that will affect levels for wastewater treatment. Generally, it is impracticable to serve more than one quality of water. Thus, the level of wastewater treatment provided may be higher than many of the users actually require. If there is a reclaimed-water market for two levels of water quality, it should be considered whether the distribution system can be separated so that the higher and more-expensive treatment can be sized to serve only those users needing such higher water quality.

Planning report

The results of the completed planning effort should be documented in a facilities planning report on wastewater reclamation and reuse. A suggested outline is shown in Table 9.5, which also serves as a check-

TABLE 9.5 Outline for Wastewater-Reuse Facilities Plan

Study-area characteristics: geography, geology, climate, groundwater basins, surface waters, land use, population growth

Water-supply characteristics and facilities: agency jurisdictions, sources and qualities of supply, description of major facilities, water-use trends, future facilities needs, groundwater management and problems, present and future freshwater costs, subsidies, and customer prices

Wastewater characteristics and facilities: agency jurisdictions, description of major facilities, quantity and quality of treated effluent, seasonal and hourly flow and quality variations, future facilities needs, need for source control of constituents affecting reuse, description of existing reuse (users, quantities, contractual and pricing agreements)

Treatment requirements for discharge and reuse and other restrictions: health- and water-quality-related requirements, user-specific water-quality requirements, use-area controls

Potential water-reuse customers: description of market analysis procedures, inventory of potential reclaimed-water users, and results of user survey

Project alternative analysis: capital and O&M costs, engineering feasibility, economic analysis, financial analysis, energy analysis, water-quality impacts, public and market acceptance, water-rights impacts, environmental and social impacts, comparison of alternatives, and selection

- Treatment alternatives
- Alternative markets: based on different levels of treatment and service areas
- Pipeline route alternatives
- Alternative reclaimed water-storage locations and options
- Freshwater alternatives
- Water-pollution control alternatives
- No project alternative

Recommended plan: description of proposed facilities, preliminary design criteria, projected cost, list of potential users and commitments, quantity and variation of reclaimed water demand in relation to supply, reliability of supply and need for supplemental or backup water supply, implementation plan, operational plan

Construction financing plan and revenue program: sources and timing of funds for design and construction; pricing policy of reclaimed water; cost allocation between water-supply benefits and pollution-control purposes; projection of future reclaimed water use, freshwater prices, reclamation project costs, unit costs, unit prices, total revenue, subsidies; sunk costs, and indebtedness; analysis of sensitivity to changed conditions

SOURCE: Data from Refs. 21 and 23.

list for planning considerations. All the items listed have been found at one time or another to affect the evaluation of a water-reuse project. Thus, while all the factors shown do not deserve an in-depth analysis, they should at least be considered. The overall level of detail should be commensurate with the size and complexity of the proposed project.

While the emphasis on the wastewater or water-supply aspects will vary depending on whether the purpose of the project is single or multiple, the nature of wastewater reuse is such that both aspects must at least be considered. Even if it is determined that a wastewater-reuse project is not feasible at the conclusion of the study, it is still advisable to publish the information and data collected and the analysis performed to arrive at this conclusion. Wastewater-reuse projects are good public policies in appropriate situations, and public interest in them will continue to recur as long as water-supply needs are perceived to be critical, such as in drought years. Documentation of even unsuccessful reuse planning is helpful in responding to public inquiry and in orienting future planning efforts.

User contracts

Facilities design is the next logical step after a positive recommendation resulting from the planning effort to implement a wastewater-reuse project. Equally important, however, is securing users to take the reclaimed water. There are two approaches to achieve this: mandatory and voluntary. Where the conditions and political climate have been appropriate, some jurisdictions such as water districts in California have mandated the acceptance of reclaimed water on the basis of the "waste and unreasonable use of water" doctrine when there is considered to be no harm to the potential user. The more general case, however, is that reclaimed-water customers are elicited on a voluntary basis.

Before an agency or a sponsor embarks on the significant cost of a wastewater-reuse project, it may wish to ensure participation of potential users through contractual agreements. The level of effort needed to negotiate such agreements may seem to be cumbersome and establish an adversarial relationship with potential users. Experience has shown, however, that potential users quite easily express positive interest in using reclaimed water and yet still resist final acceptance when the facilities are built. Examples of reasons for such resistance are given in Table 9.6. The contract-negotiating process can also be an educational process to win the support and confidence of potential customers and bring out hidden issues much earlier in the project-development process.

TABLE 9.6 Potential Obstacles to Securing Reclaimed-Water User Commitments

Concern of users regarding detrimental effects of reclaimed water on industrial process, landscaping, or crops
User has its own water supply at less cost than either municipal water supply or offered reclaimed-water price
Disagreement over acceptable reclaimed-water price
User unwilling to pay or incapable of paying for extra costs for pipelines or on-site water-system modifications
User lies outside project proponent's boundaries, requiring negotiations with other jurisdictions
Disapproval of local or state health department because of health risks

SOURCE: Data from Ref. 23.

TABLE 9.7 Desirable Provisions of Reclaimed-Water User Contracts

Contract duration: term, conditions for termination
Reclaimed-water characteristics: source, quality, pressure
Quantity and flow variations
Reliability of supply: potential lapses in supply, backup supply provisions
Commencement of use: when user can or will begin use
Financial arrangement: pricing of reclaimed water, payment for facilities
Ownership of facilities, rights of way: responsibility for operation and maintenance
Miscellaneous: liability, restrictions on use, right of purveyor to inspect site

SOURCE: Data from Ref. 23.

Some of the factors to address in a user contract are shown in Table 9.7. The contracts need to address the concerns of the purveyor and the user and to clearly establish financial and operational responsibility and legal liability.

Water-Reuse Applications

The purpose of this section is to discuss an overview of water-reuse applications; water-quality requirements, including health aspects of wastewater reuse; and several case studies in (1) agricultural and landscape irrigation, (2) industrial recycling and reuse, (3) groundwater recharge, and (4) potable reuse.

Preapplication treatment of wastewater is necessary to (1) protect public health, (2) prevent nuisance conditions during application and storage, and (3) prevent damages to crops and soils. Municipal wastewater treatment consists of a combination of physical, chemical, and biological processes and operations to remove solids, organic matter, pathogens, and sometimes nutrients from wastewater. General terms used to describe different degrees of treatment, in order of increasing treatment level, are preliminary, primary, secondary, and advanced treatment. A disinfection step to remove pathogens usually follows the last treatment step. A generalized wastewater-treatment diagram is shown in Fig. 9.1

Because various industrial pretreatment programs have been implemented by the publicly owned treatment works (POTWs), toxic industrial wastes in municipal wastewater have been increasingly reduced. Both prohibited discharge and categorical pretreatment standards established by the U.S. Environmental Protection Agency (EPA)[27] are being incorporated in POTWs discharge limitations for their industrial users.

Agricultural and landscape irrigation

The total national rate of withdrawal of both surface water and groundwater in 1980 was approximately 1.7×10^9 m^3/d.[17] The quantity of water withdrawn for irrigation in 1980 was estimated at about 568×10^6 m^3/d. The water was used on approximately 234×10^3 km^2 of farmland. This represents an increase in both water use and irrigated area of about 7 percent from the 1975 estimate.

The nine western (U.S.) water-resource regions accounted for 91 percent of the total water withdrawn for irrigation in 1980. In the eastern regions, most of the water used for irrigation was in the South Atlantic–Gulf and Lower Mississippi regions. California was by far the largest user of irrigation water, withdrawing about 140×10^6 m^3/d, 25 percent of the national total, which is more than the next two largest users, Idaho and Colorado, combined.

Irrigation of crops developed along with the settlement of the arid West because most years farmers needed to irrigate to raise any crops. In the humid eastern states, irrigation was used to supplement natural rainfall in order to increase the number of plantings per year and yield of crops, and to reduce the risk of crop failure during drought periods. Irrigation also is used to maintain recreational lands such as parks and golf courses.

Characteristics of irrigation with reclaimed water. Although irrigation with municipal wastewater is in itself an effective form of wastewater treatment, normally wastewater treatment must be provided before such water can be used for agricultural or landscape irrigation (see Fig. 9.1). The design approach for irrigation with reclaimed municipal wastewater depends on whether emphasis is placed on wastewater treatment or on water supply. The term *slow-rate land treatment* is

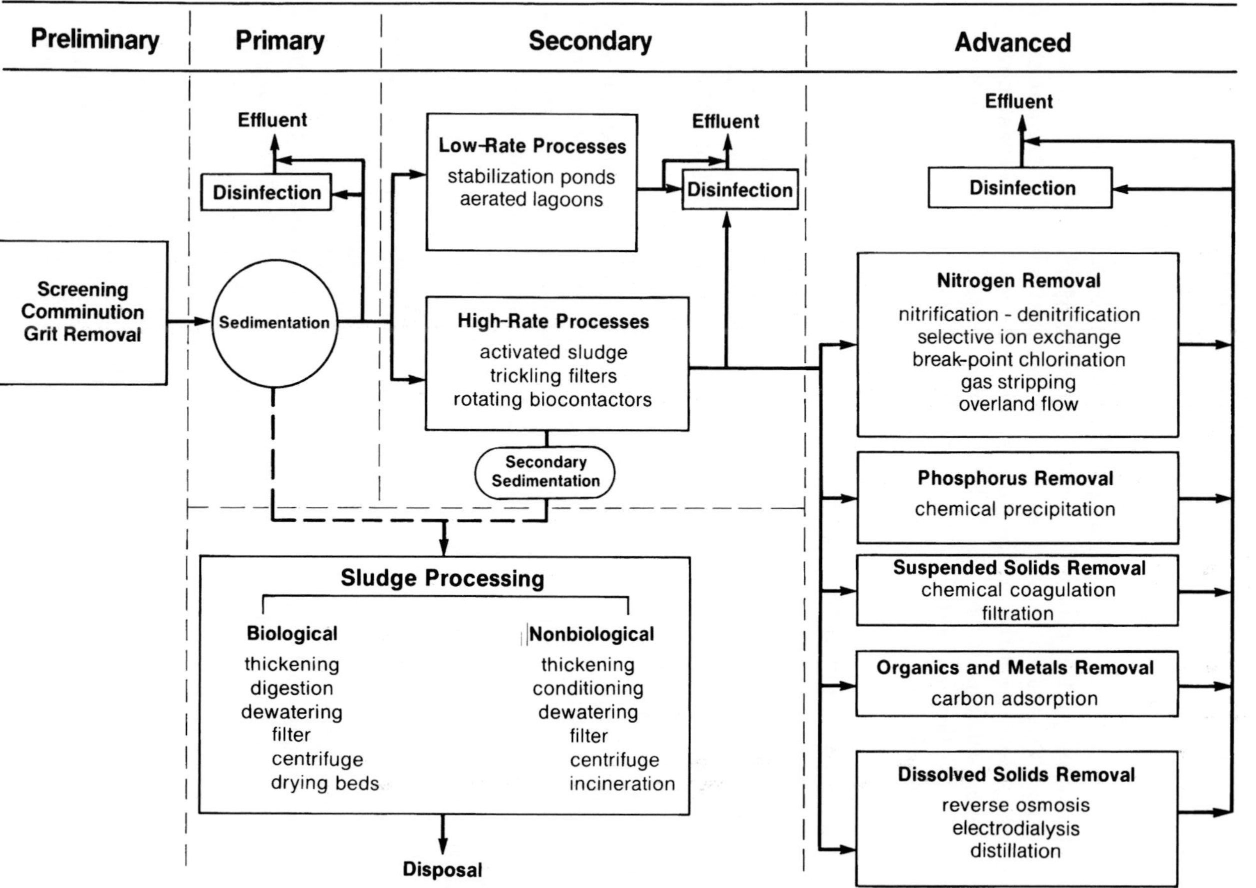

Figure 9.1 Generalized flow diagram of wastewater treatment and water reuse. (*Adapted from Pettygrove and Asano.*[26])

generally used to focus attention on wastewater treatment and disposal, while the terms *agricultural reuse* and *wastewater irrigation* imply that reclaimed water is used for irrigation substituting freshwater. An important use of reclaimed water in recent years is irrigation of golf courses, other turfgrass, and landscaped areas in the urban environment.

The objective of this section is to discuss the unique aspects of irrigation with reclaimed municipal wastewater that contribute more to water-supply benefits than pollution-control purposes. Substantial amounts of reclaimed water are presently used for agricultural and landscape irrigation.

Evaluation of irrigation-water quality. The primary factor in evaluating water quality for irrigation is the quantity and kind of salts present in irrigation water. As salinity increases in the reclaimed water used for irrigation, the probability for certain soil, water, and cropping problems increases. These problems are related to the total salts content, to one or more types of salts, or to excessive concentrations of one or more trace elements.[26,34] The problems, however, are no different from those caused by salinity or trade elements in freshwater supplies and are of concern only if they restrict the use of the water or require special management to maintain acceptable yields.

A number of different water-quality guidelines are related to irrigation. The guidelines presented in this section were originally developed by the University of California Committee of Consultants and were subsequently expanded by Ayers and Westcot in two publications.[26,30] The guidelines shown in Table 9.8 stress the management needed to successfully use water of a certain quality. These guidelines are intended to cover the wide range of conditions encountered in California's irrigated agriculture. Several basic assumptions have been used to define the range of usability for these guidelines. If the water is used under greatly different conditions, the guidelines may need to be adjusted. Wide deviations from the assumptions might result in wrong judgments on the usability of a particular water supply, especially if it is a borderline case. Where sufficient experience, field trials, research, or observations are available, the guidelines may be modified to more closely fit local conditions.

Yield potential. Full production capability of all crops, without the use of special practices, is assumed when the guidelines indicate no restrictions on use. A "restriction on use" indicates that there may be a limitation such as choice of crop or the need for special management in order to maintain full production capability, but a "restriction on use" does not indicate that the water is unsuitable for use.

TABLE 9.8 Guidelines for Interpretation of Water Quality for Irrigation

Potential irrigation problem	Units	Degree of restriction on use: None	Slight to moderate	Severe
Salinity (affects crop water availability)				
EC_w[a]	dS/m or mmho/cm	<0.7	0.7–3.0	>3.0
TDS	mg/L	<450	450–2000	>2000
Permeability (affects infiltration rate of water into the soil; evaluate using EC_w and SAR[b] together)[c,d]				
SAR = 0–3	and EC_w	= >0.7	0.7–0.2	<0.2
= 3–6		= >1.2	1.2–0.3	<0.3
= 6–12		= >1.9	1.9–0.5	<0.5
= 12–20		= >2.9	2.9–1.3	<1.3
= 20–40		= >5.0	5.0–2.9	<2.9
Specific-ion toxicity (affects sensitive crops)				
Sodium (Na)				
Surface irrigation	SAR	<3	3–9	>9
Sprinkler irrigation	mg/L	<70	>70	
Chloride (Cl)[e,f]	mg/L	<140	140–350	>350
Surface irrigation	mg/L	<100	>100	
Boron (B)	mg/L	<0.7	0.7–3.0	>3.0
Miscellaneous effects (affects susceptible crops)				
Nitrogen (total N)[g]	mg/L	<5	5–30	>30
Bicarbonate (HCO_3) (overhead sprinkling only)	mg/L	<90	90–500	>500
pH		Normal range 6.5–8.4		
Residual chlorine (overhead sprinkling only)	mg/L	<1.0	1.0–5.0	>5.0

[a]Electrical conductivity of the irrigation water, reported in millimhos per centimeter or decisiemens per meter.

[b]Total dissolved solids, reported in milligrams per liter.

[c]Sodium-adsorption ratio. SAR is sometimes reported as R_{Na}. At a given SAR, the infiltration rate increases as salinity (EC_w) increases. Evaluate the potential permeability problem by SAR and EC_w in combination (see Ayers and Westcot,[30] and Pettygrove and Asano[26]).

[d]For wastewater, it is recommended that the SAR be adjusted to include a more correct estimate of calcium in the soil water following an irrigation. A procedure is given in Ayers and Westcot[30] and Pettygrove and Asano.[26] The adjusted sodium adsorption ratio (adj R_{Na}) calculated is to be substituted for the SAR value.[31]

[e]Most tree crops and woody ornamentals are sensitive to sodium and chloride; use the values shown. Most annual crops are not sensitive.

[f]With overhead sprinkler irrigation and low humidity (<30%), sodium or chloride greater than 70 or 100 mg/L, respectively, have resulted in excessive leaf absorption and crop damage to sensitive crops.

[g]Total nitrogen should include nitrate-nitrogen, ammonia-nitrogen, and organic nitrogen. Although forms of nitrogen in wastewater vary, the plant responds to the total nitrogen.

SOURCE: Adapted from Ayers and Westcot (1985).[30] The basic assumptions of the guidelines are listed in the text.

Site conditions. Soil texture ranges from sandy-loam to clay with good internal drainage. Rainfall is low and does not play a significant role in meeting crop water demand or leaching. In the Sierra and extreme northern coastal areas of California, where precipitation is high for part or all of the year, the guideline restrictions are too severe. Drainage is assumed to be good, with no uncontrolled shallow water table present.

Methods and timing of irrigations. Normal surface and sprinkler irrigation methods are used. Water is applied infrequently as needed, and the crop utilizes a considerable portion of the available stored soil water (50 percent or more) before the next irrigation. At least 15 percent of the applied water percolates below the root zone [leaching fraction (LF) >15 percent]. The guidelines are too restrictive for specialized irrigation methods, such as drip irrigation, which result in near daily or frequent irrigations. The guidelines are not applicable for subsurface irrigation.

Water uptake by crops. Different crops have different water-uptake patterns, but all take water from wherever it is most readily available within the root zone. Each irrigation leaches the upper root zone and maintains it at a relatively low salinity. Salinity increases with depth and is greatest in the lower part of the root zone. The average salinity of the soil solution is about three times that of the applied water.

Salts leached from the upper root zone accumulate to some extent in the lower part but eventually are moved below the root zone by sufficient leaching. The crop responds to average salinity of the root zone. The higher salinity in the lower root zone becomes less important if adequate moisture is maintained in the upper, "more active" part of the root zone.

Four categories of potential water quality-problems—(1) salinity, (2) toxicity, (3) water-infiltration rate, and (4) miscellaneous—are used to evaluate the suitability of irrigation water. These guidelines emphasize the long-term influence of water quality on crop production, soil conditions, and farm management, and are applicable to both freshwater and reclaimed water. Referring to Table 9.8, the following potential problems and solutions are of special interest.[26,30]

Salinity. Salinity, measured by electrical conductivity, is the single most important parameter in determining the suitability of a water for irrigation. A salinity problem exists if salt accumulates in the crop-root zone to a concentration that causes a loss in yield. In irrigated areas, these salts often originate from a saline water, waters of high water table, or salts in the applied water. Plant damage from both salinity and specific ions is usually tied closely to an increase in salinity. Establishing a net downward flux of water and salt through the root zone is the only practical way to manage a salinity problem. Under such conditions, good drainage is essential in order to allow a continuous movement of water and salt below the root zone. Long-term use of reclaimed water for irrigation is seldom possible without adequate drainage.

Leaching of salts is done by applying sufficient water to allow a portion of water to percolate through and below the entire root zone, carrying with it a portion of the accumulated salts. The fraction of applied water that passes through the entire rooting depth and percolates below is called the *leaching fraction* (LF):[26]

$$\mathrm{LF} = \frac{\text{depth of water leached below the root zone}}{\text{depth of water applied at the surface}} \tag{9.1}$$

A high leaching fraction (LF = 0.5) results in less salt accumulation in the root zone than does a lower leaching fraction (LF = 0.1). If the salinity of irrigation water ($\mathrm{EC_w}$) and the leaching fraction are known, salinity of the drainage water that percolates below the rooting depth can be estimated as follows:

$$\mathrm{EC_{dw}} = \frac{\mathrm{EC_w}}{\mathrm{LF}} \tag{9.2}$$

where $\mathrm{EC_{dw}}$ = salinity of the drainage water percolating below the root zone (equal to salinity of soil-water, $\mathrm{EC_{sw}}$)
$\mathrm{EC_w}$ = salinity of irrigation water
LF = leaching fraction

Specific-ion toxicity. Toxicity due to a specific ion occurs when that ion is taken up by the plant and accumulates in the plant in amounts that result in damage or reduced yield. The ions of most concern in wastewater are sodium, chloride, and boron. The most prevalent toxicity from the use of reclaimed municipal wastewater is from boron. The source of boron is usually household detergents or discharges from industrial plants. Chloride and sodium contents also increase during domestic usage, especially with the use of household laundry water softeners. For sensitive crops, toxicity is difficult to correct, short of changing the crop or the water supply. The problem is usually accentuated by severe (hot and dry) climatic conditions due to high evapotranspiration rates.[30]

Water-infiltration rate. In addition to Na effects on the plant, sodium in irrigation water may affect soil as well as reduce soil aeration. If the infiltration rate is greatly reduced, it may be impossible to supply the crop or landscape plant with enough water for good growth. In addition, irrigation systems with reclaimed municipal wastewater are often located on less-desirable soils or those already having soil-permeability and soil-management problems, which accentuate the problem. It may be necessary in these cases to modify soil profiles by engineering means by excavating and rearranging the affected land.

Water-infiltration-rate problems usually occur in the top ~10 cm of the soil and are related mainly to the structural stability of the surface soil. In the past, several procedures have been used to predict a potential infiltration problem. The most commonly used method to evaluate the infiltration problem is the *sodium-adsorption ratio* (SAR),[7,26,30,31] which is given as

$$\text{SAR} = \frac{\text{Na}}{[(\text{Ca} + \text{Na})/2]^{1/2}} \tag{9.3}$$

where all cations are expressed in milliequivalents per liter.

At a given SAR, the infiltration rate increases as salinity increases or decreases as salinity decreases. Therefore, SAR and electrical conductivity (EC_w) of irrigation water should be used in combination to evaluate the potential permeability problem (see Table 9.8).

Reclaimed municipal wastewater is normally high enough in both sodium and calcium, and there is little concern for the water dissolving and leaching too much calcium from the surface soil. However, reclaimed water is sometimes high in sodium; the resulting high SAR is a major concern in planning irrigation projects with reclaimed water in such a case.

Nutrients. The nutrients in reclaimed municipal wastewater provide fertilizer value to crop or landscape production but, in certain instances, are in excess of plant needs and cause problems related to excessive vegetative growth, delayed or uneven maturity, or reduced quality. Nutrients in reclaimed water which are important to agriculture and landscape management include nitrogen (N), phosphorus (P), and occasionally potassium (K), zinc (Zn), boron (B), and sulfur (S). The mostly beneficial but frequently excessive nutrient in reclaimed municipal wastewater is nitrogen.

Miscellaneous problems. Clogging problems with sprinkler and drip-irrigation systems have been reported, particularly with primary and oxidation pond effluents. Growth (slimes and bacteria) in the sprinkler head, emitter orifice, or supply line causes plugging, as do heavy concentrations of algae and suspended solids. The most frequent clogging problems occur with drip-irrigation systems. From the standpoint of public health such systems are, however, often considered ideal, as they are totally enclosed systems and minimize the problems of worker exposure to reclaimed water or spray drift.

In the disinfection step by chlorine in the municipal wastewater treatment, a residual chlorine of less than 1 mg/L does not affect plant foliage, but excessive residual chlorine causes plant damage. When chlorine residual is in excess of 5 mg/L, severe plant damage can occur if reclaimed water is sprayed directly on foliage.[26]

Health and regulatory requirements. In every wastewater reclamation-reuse operation, there is some risk of human exposure to pathogenic organisms, but the health concern is in proportion to the degree of human contact with the reclaimed water and the adequacy and reliability of the treatment processes. The contaminants in reclaimed water that are of health significance may be classified as biological and chemical agents. For most of the uses of reclaimed water, however, pathogenic organisms pose the greatest health risks, and water-quality standards and criteria are properly directed at these agents.[26,38] Bacterial pathogens, helminths, protozoa, and viruses are removed in wastewater-treatment processes in varying degrees. The most important treatment processes from the standpoint of pathogen removal and/or destruction are sedimentation and disinfection.

To protect public health without unnecessarily discouraging wastewater reclamation and reuse, many regulations include water-quality standards and requirements for treatment, sampling and monitoring, treatment-plant operations, and treatment process reliability. To minimize health risks and aesthetic problems, tight controls are imposed on the delivery and use of reclaimed water after it leaves the treatment facility. Regulations for a specific irrigation use are based on the expected degree of contact with the reclaimed water and the intended use of the irrigated crops.[1,34–36]

Although there is no uniform set of federal standards for wastewater reuse, several states have developed wastewater-reclamation criteria, often in conjunction with regulations on land treatment and disposal of wastewater. California, for example, requires that reclaimed water used for landscape irrigation of areas with unlimited public access be "adequately oxidized, filtered, and disinfected prior to use";[35] in other words, it requires the wastewater treatment processes including biological secondary treatment and tertiary treatment consisting of filtration followed by disinfection. Table 9.9 contains a summary of irrigation requirements excerpted from the State of California *Wastewater Reclamation Criteria*.[35] These criteria are health regulations, intended to assure an adequate degree of health protection from disease transmission, and thus do not specifically address the treat-

TABLE 9.9 California Wastewater Reclamation Criteria for Irrigation and Recreational Impoundments

Reclaimed-wastewater applications	Description of minimum wastewater characteristics			
	Primary effluent*	Secondary and disinfected	Secondary, coagulated, filtered,† and disinfected	Median total coliform per 100 mL
Crop irrigation				
Fodder crops	X			NR‡
Fiber	X			NR
Seed crops	X			NR
Produce eaten raw, surface irrigated		X		2.2
Produce eaten raw, spray irrigated			X	2.2§
Processed produce, spray irrigated		X		23¶
Landscape irrigation				
Golf courses		X		23
Freeways		X		
Landscape irrigation				
Parks			X	2.2
Playgrounds			X	
Recreational impoundments				
No public contact		X		23
Boating and fishing only		X		2.2
Body contact (bathing)			X	2.2

*Effluent containing not more than 0.5 mL L^{-1} h^{-1} of settlable solids. No primary effluent is used presently in California, and this category will be deleted in future regulation.

†This requirement is often referred to as the "Title 22 requirement" or the "Title 22 treatment process." Effluent does not exceed an average of two nephelometric turbidity units (NTU) and does not exceed 5 NTU more than 5 percent of the time during any 24-h period.

‡No requirement.

§The median number of coliform organisms in the effluent does not exceed 2.2/100 mL and the number of coliform organisms does not exceed 23/100 mL in more than one sample within any 30-day period.

¶The median number of coliform organisms in the effluent does not exceed 23/100 mL in seven consecutive days and does not exceed 240/100 mL in any two consecutive samples.

SOURCE: Adapted from Ref. 35.

ment technology or the potential effect of reclaimed water on the crops or soil. The median number of total coliform density is used for the assessment of reliability of treatment in the California regulations. The State of Arizona's wastewater-reuse regulations contain enteric viruses limits [e.g., not to exceed one plaque-forming unit (PFU) per 40 L] for the most stringent reclaimed-water applications such as spray irrigation of food crops.[37]

Other safety measures for nonpotable-water reuse applications include (1) installation of separate storage and distribution systems of potable water, (2) use of color-coded labels to distinguish potable and nonpotable installation of the pipes, (3) backflow prevention devices, (4) periodic use of tracer dyes to detect the occurrence of cross-contamination in potable-water supply lines, and (5) irrigation during off hours to further minimize the potential for human contacts.

Comparison of various water-reuse criteria. Since major issues surrounding wastewater reclamation and reuse are often related to health, considerable amounts of research have been conducted in the general area of public health. The major microbiological health hazards associated with water consumption and contact with reclaimed wastewater originate from fecal contamination.

Reclaimed-water quality criteria for protecting health in developing countries must be established in relation to the limited resources available for public works and other health-delivery systems that may yield greater health benefits for the funds spent. Confined sewage-collection systems and wastewater treatment are often nonexistent, and reclaimed wastewater often provides an essential water-resource and fertilizer source. Thus, for most developing countries, the greatest concerns for the use of wastewater for irrigation are caused by the enteric helminths such as hookworm, *Ascaris* species, *Trichuris* species, and under certain circumstances, the beef tapeworm.[34,36] These pathogens can damage the health of both the general public consuming the crops irrigated with untreated wastewater and sewage farm workers and their families.[38–42] Table 9.10 compares the most cited microbiological quality guidelines: the *California Water Reclamation Criteria*[35] and the World Health Organizations' health guidelines for the use of wastewater in agriculture and aquaculture[38–40] which were recommended for most developing countries.

A case study for agricultural irrigation: the *Monterey Wastewater Reclamation Study for Agriculture.*[28] The combination of fertile soils and long growing season makes the lower Salinas Valley in northern Monterey County, California, a rich agricultural region. Artichokes are a major crop, but a variety of annual crops is also grown: broccoli, cauliflower, celery, and lettuce are grown throughout the region. It became evident during the early 1970s that the groundwater supply was decreasing because of extensive withdrawal of groundwater for irrigation. This overdraft lowered the water tables and created an increasing problem of saltwater intrusion from the Pacific Ocean. At the same time, wastewater-treatment facilities were reaching full capacity, requiring expansion to meet the growing needs of the region. The water-quality-management plan recommendations recognized that wastewater reclamation had to be proved safe before regional

TABLE 9.10 Summaries of Recommended Microbiological Quality Guidelines and Criteria by the World Health Organization[38–42] and the State of California's *Wastewater Reclamation Criteria*[35]

Category	Reuse conditions	Intestinal nematodes*	Fecal† or total‡ coliforms	Wastewater treatment expected to achieve microbiological quality
WHO	Irrigation of crops likely to be eaten uncooked, sports fields, public parks	≤ 1/L	≤ 1000/100 mL	A series of stabilization ponds or equivalent treatment
Calif.	Spray and surface irrigation of food crops, high exposure landscape irrigation such as parks	No standard recommended	≤ 2.2/100 mL	Secondary treatment followed by filtration and disinfection
WHO	Irrigation of cereal crops, industrial crops, fodder crops, pasture, and trees	≤ 1/L	No standard recommended	Retention in stabilization ponds for 8–10 days or equivalent removal
Calif.	Irrigation of pasture for milking animals, landscape impoundment	No standard recommended	≤ 23/100 mL	

*Intestinal nematodes (*Ascaris* and *Trichuris* species and hookworms) is expressed as arithmetic mean number of eggs per liter during the irrigation period.

†WHO recommends more stringent guideline (≤ 200/100 mL fecal coliforms) for public lawns, such as in hotels, with which the public may come into direct contact.

‡The California Wastewater Reclamation criteria are expressed as the median number of total coliforms per 100 mL, as determined from the bacteriological results of the last 7 days for which analyses have been completed.

implementation could be considered. This provided the impetus for the Monterey Wastewater Reclamation Study for Agriculture (MWRSA), which was a 10-year, $7.2 million (1987 dollars) demonstration project designed to assess the safety and feasibility of agricultural irrigation with reclaimed municipal wastewater.

Planning for the project was begun in 1976 by the Monterey Regional Water Pollution Control Agency (MRWPCA), the regional agency responsible for wastewater collection, treatment, and disposal. Full-scale field studies began in 1980 and continued through May 1986. During these 5 years, a perennial crop of artichokes was grown along with rotating annual crops of celery, broccoli, lettuce, and cauliflower. Extensive sampling and analysis of water, soils, and plant tissues were conducted.

Project description. The site for the MWRSA field operations was a farm in Castroville, California. The existing 1500-m^3/d Castroville Wastewater Treatment Plant was selected for modification and upgrading to be used as the pilot tertiary reclamation plant for MWRSA. A portion of the secondary effluent was diverted to a new pilot tertiary treatment plant which consisted of two parallel treatment process trains. The first process, Title-22 (T-22), conformed strictly to the requirements of the California *Wastewater Reclamation Criteria*[39] for irrigating food crops that may be consumed without cooking (see Table 9.9). The second process produced a treated wastewater designated as filtered effluent (FE). This is a wastewater treated less extensively than T-22 effluent via direct filtration of secondary effluent. Well water from local wells was used as the control for the study.[28,43,44]

Agricultural operation. The 12-ha field site was divided into two parts: demonstration fields and experimental plots. Large demonstration fields were established because farm-scale feasibility of using reclaimed water is of special importance to the growers, farm managers, and operators responsible for day-to-day farming practices.

To investigate the large-scale feasibility of using reclaimed wastewater, two 5-ha plots were dedicated to reclaimed-water irrigation, using the FE flow stream. On one plot, artichokes were grown; on the other plot, a succession of broccoli, cauliflower, lettuce, and celery was raised during the first 3 years of the field investigation. The crops were observed carefully for appearance and vigor.

Split-plot design. A split-plot design was chosen for the experimental plots. This design allowed the use of two treatment variables: water type and fertilization rate. Twelve main plots were established, divided into four blocks or replicates. Three water types were used in each replicate of three main plots: T-22 and FE effluents and well water.

These three water types were assigned randomly to main plots within each block or replicate to achieve a randomized complete block (i.e., each block contained all three of the main water-type treatments). Each main plot was then divided into four subplots, each of which was randomly assigned a different fertilization-rate treatment: the full amount of nitrogen fertilizer used by local farmers (3/3), two-thirds the full rate (2/3), and one-third the full rate (1/3), and no fertilizer (0/3). The full design thus had 48 plots. This process was performed for artichokes and repeated for annual row crops, for a total of 96 plots which occupied 1.2 ha. This experimental design allowed comparison of both irrigation with different water types and the effect of varying fertilization rates. The fertilization rates were designed to elucidate the value of the two effluents as a supplement to fertilization.

Results of public-health studies[28]

1. *Virus survival.* Monitoring for the presence of naturally occurring animal viruses showed that the influent to the two pilot processes contained measurable viruses in 53 of the 67 samples taken. The median concentration of virus was two plaque-forming units per liter (2 PFU/L): 90 percent of the samples contained less than 28 PFU/L. During the 5-year period, no in situ viruses were recovered from the chlorinated tertiary effluent of either process. No viruses were recovered from any of the crop samples. This was also the case for the soil irrigated with the reclaimed water.

2. *Bacteria and parasites.* All three types of water, including the well-water control, periodically exhibited high coliform levels. No salmonellae, shigallae, *Ascaris lumbricoides, Entamoeba histolytica,* or other parasites were ever detected in any of the irrigation waters. The levels of total and fecal coliform in soils and plant tissue irrigated with all three types of water were generally comparable. No significant difference attributable to water type was observed.

3. *Aerosols.* Aerosol-carried microorganisms from FE sprinklers were not significantly different from those generated by well-water sprinklers. This finding was verified through replications in both daytime and nighttime operations to account for die-offs or organisms caused by natural ultraviolet radiation.

4. *Health of field workers.* The health status of each person assigned to the field tasks in MWRSA was monitored regularly through frequent questionnaires and thorough initial and exit medical examinations administered by qualified medical professionals. One hundred questionnaires were completed by personnel during the 5 years. No complaints could be related to contact with treated wastewater effluents. No formal epidemiological investigation was deemed appropriate or necessary for the purposes of MWRSA.

Results of agricultural studies

1. *Heavy metals.* None of the nine heavy metals studied (Cd, Zn, Fe, Mn, Cu, Ni, Co, Cr, and Pb) manifested any consistently significant difference in concentration among plots irrigated with different water types. Input of zinc and other heavy metals from the commercial chemical fertilizer impurities is far greater than from any of the water sources and accounts for the large concentration differences observed at the three soil depths sampled. These differences have occurred over many decades of continuous farming with regular application of chemical fertilizers. No consistently significant difference in heavy-metal concentrations was observed between plants irrigated with either effluent and with well water.

2. *Water-infiltration rate.* Infiltration rates in the lettuce field were highest in those plots irrigated with well water, but these levels were not significantly different because of the great variation of infiltration rates within each water type. Infiltration rates in the artichoke field was higher than in the lettuce field, probably because the artichoke field receives less irrigation water and is less frequently compacted by equipment used for field preparation.

3. *Crop yields.* Artichoke yields were similar for all three water types; in the first 2 years, the different fertilization rates had no effect on yield. In the last 3 years, a significant effect of fertilization became apparent. All three fertilization rates showed significantly higher yields than did the unfertilized plots.

Field-quality assessments and shelf-life measurements revealed no differences between produce irrigated with reclaimed water and that irrigated with well water. Visual inspection of artichoke plants in the field showed no differences in appearance or vigor of plants irrigated with different water types.

Shelf life and quality of row crops were similar for all water types. No problems with increased spoilage of produce irrigated with effluents were encountered.

Conclusions. The following conclusions were made in the *Monterey Wastewater Reclamation Study for Agriculture:*[28]

1. On the basis of virological, bacteriological, and chemical results from sampled vegetable tissues, no differences were found among irrigation with filtered effluent or T-22 effluent and with well water.

2. Few statistically significant differences in measured soil or plant parameters were attributable to the different water types. None of these differences had important implications for public health.

3. No in situ virus was detected in any of the reclaimed waters sampled, although it was often detected in the secondary effluent.
4. The T-22 process was somewhat more efficient than the FE process in removing the vaccine-strain polio virus when influent water was inoculated at extremely high rates. Both treatment trains can remove more than five logs of virus (i.e., removal to below 1/100,000 of the seeded concentration).
5. Marketability of produce was not expected to be a problem according to the market survey.
6. The cost of producing filtered effluent (FE process after secondary treatment) was estimated to be 6¢/m^3, excluding conveyance and pumping costs.

Industrial water recycling and reuse

Industrial recycling and reuse are practiced with (1) reuse of municipal wastewater for an industrial process, (2) cascading use of industrial process water between successive processes within an industry, and (3) agricultural reuse of industrial-plant effluent. Among them, three categories of industrial water use are of particular interest because they are high-volume uses with excellent prospects for using reclaimed municipal wastewater: (1) cooling-tower makeup, (2) once-through cooling, and (3) process water. The remaining categories have less potential for reuse in significant quantities and are not considered further.

Cooling-tower makeup water. Cooling-tower makeup water represents a significant water use for many industries. For industries such as electrical-power-generating stations, oil refining, and many types of chemical and metal plants, one-quarter to more than one-half of a facility's water use may be cooling-tower makeup. Because a cooling tower normally operates as a closed-loop system isolated from the process, it can be viewed as a separate water system with its own specific set of water-quality requirements, which are largely independent of the particular industry involved.[45]

There are significant variations among large industrial cooling systems. The range includes once-through noncontact cooling such as at large power facilities or refineries near the ocean: direct-contact cooling of relatively inert material being processed, as in quenching in the primary-metals industry; and noncontact recirculating cooling at large inland industries with limited water resources.

Once-through cooling transfers process heat to water which is then wasted. Recirculating systems go one step further and transfer the heat from the warmed water to the air so that the water can again be used to absorb process heat. The heat is transferred from water to air primarily through evaporation.[46] Cool, dry outside air is pulled into the sides of the tower, at a rate of 600 to 750 m^3 air/m^3 recirculating water, up through and out the top by a fan rotated at the top. Warmed water from an industrial heat exchanger is pumped into the top of the tower and allowed to fall down through the upcoming airstream. Packing inside the tower breaks up the water into droplets to allow efficient air-water contact. A fraction of the water evaporates and leaves the tower as vapor. The cooled remainder falls into a collecting basin at the base of the tower, and is again ready for process cooling.[46,47]

Industrial cooling-tower operations face four general water-quality problems: (1) scaling, (2) corrosion, (3) biological growths, and (4) fouling in heat exchanger and condensers. Both freshwater and reclaimed municipal wastewater contain contaminants which can cause these problems, but their concentrations in reclaimed municipal wastewater are generally higher. In addition, since the possibility of exposure to aerosols exists, a high degree of disinfection may be required to ensure the protection of the neighboring public as well as plant operators.

In most cases, disinfected secondary effluent is supplied to noncritical, once-through cooling. For recirculating cooling-tower operation, it still contains constituents which, if not removed, would limit industries to very low cycles of concentration in their cooling towers. Additional treatment generally includes lime clarification, alum precipitation, or ion exchange. Internal chemical treatment involves the addition of acid for pH control, biocide, scale, and biofoul inhibitors. In many cases, the water-quality requirements for the use of reclaimed municipal wastewater are the same as those for freshwater.

Use of reclaimed municipal wastewater in a nuclear power plant.[48,49] The Arizona Nuclear Power Project's Palo Verde Nuclear Generating Station (PVNGS) is located about 80 km west of Phoenix, near the town of Wintersburg, Arizona. The PVNGS has three 1270-MW pressurized-water reactor units. The resulting total generating-capacity output is 3810 MW, making Palo Verde the largest nuclear power plant in the United States.

It is located in the desert far from lakes, rivers, and oceans and requires an average of 79×10^6 m^3 of water annually for the process of cooling and condensing steam. An on-site advanced wastewater-reclamation plant satisfies this demand by treating secondary effluent pumped from the cities of Tolleson and Phoenix through a 60-km pipeline. The on-site advanced wastewater-treatment processes consist of four stages: (1) biological nitrification, (2) lime and soda-ash addition, (3) chlorination, and (4) filtration. From the gravity filters, the treated water flows into a storage reservoir consisting of 32 ha surface area and 9 m depth. It holds 2.5×10^6 m^3, which is at least a 7-day supply of cooling-tower makeup water. The purpose of the advanced wastewater treatment at the PVNGS is to reduce corrosion and scaling in the cooling-tower systems.

Water-quality requirements. Effluent quality from the advanced wastewater-reclamation plant is as follows:[48]

Parameter	Maximum value
Suspended solids (total)	10
Turbidity	15 NTU
PO_4 (as P)	0.5 mg/L
Ca (as $CaCO_3$)	70 mg/L
Alkalinity (as $CaCO_3$)	100 mg/L
Si (as SiO_2)	10 mg/L
Mg (as $CaCO_3$)	10 mg/L
NH_3-N	5 mg/L

The PVNGS's cooling-tower operation is to maintain 15 cycles of concentration. The initial operations of the water-reclamation plant are reported to be in compliance with the design specifications satisfying the cooling-tower makeup requirements at the PVNGS.

Groundwater recharge

Groundwater recharge with reclaimed water is an approach to wastewater reuse that results in the planned augmentation of groundwater. The purposes of artificial recharge of groundwater have been to (1) reduce, stop, or even reverse declines of groundwater levels; (2) protect underground freshwater in coastal aquifers against saltwater intrusion from the ocean; and (3) store surface water, including flood or other surplus water, and reclaimed wastewater for future use.[50,51] Groundwater recharge is also incidentally achieved in land treatment and disposal of municipal and industrial wastewater through percolation and infiltration.

There are several advantages to storing water underground: (1) the cost of artificial recharge may be less than the cost of equivalent surface reservoirs; (2) the aquifer serves as an eventual distribution system and may eliminate the need for surface pipelines or canals; (3) water stored in surface reservoirs is subject to evaporation, potential taste and odor problems due to algae and other aquatic productivity, and pollution—these may be avoided by underground storage; (4)

suitable sites for surface reservoirs may not be available or environmentally acceptable; and (5) the inclusion of groundwater recharge in a water-reuse project may also provide psychological and aesthetic secondary benefits as a result of the transition between reclaimed water and groundwater. This aspect is particularly significant when it is possible to augment substantial portions of potable-water supplies during wastewater reclamation and reuse.[18,19,51–55]

Techniques of groundwater recharge. Two types of groundwater recharge are commonly used with reclaimed municipal wastewater: (1) surface spreading or percolation and (2) direct injection.

In *surface spreading,* recharge water such as reclaimed municipal wastewater percolates from spreading basins through the unsaturated zone. The advantage of groundwater recharge by surface spreading is that groundwater supplies may be replenished in the vicinity of metropolitan and agricultural areas where groundwater overdraft is severe, with the added benefits of the treatment effect of soils and transporting facilities of aquifers.

In *direct injection,* treated wastewater is pumped under pressure directly into the groundwater zone, usually into a well-confined aquifer. Groundwater recharge by direct injection is practiced, in most cases, where groundwater is deep or where the topography or existing land use makes surface spreading impractical or too expensive. This method of groundwater recharge is particularly effective in creating freshwater barriers in coastal aquifers against intrusion of saltwater from a sea.[50,51]

In both surface spreading and direct injection, locating the extraction wells as great a distance as possible from the spreading basins or the injection wells increases the flow path length and residence time of the recharged groundwater. These separations in space and in time contribute to the mixing of the recharged water and the other aquifer contents, and the loss of identity of the recharged water originated from wastewater.[18,19,55]

Pretreatment requirements. Pretreatment requirements for groundwater recharge vary considerably, depending on the sources of reclaimed water, recharge methods, and location. Inasmuch as recharged groundwater may be an eventual source of potable-water supply, water reclamation for groundwater recharge may involve treatment beyond the secondary level. For surface-spreading operations, common wastewater-reclamation processes include tertiary treatment of chemical coagulation and filtration followed by chlorine disinfection. Reclaimed water from both San Jose Creek and Whittier Narrows Wastewater Treatment Plants operated by the County Sanitation

Districts of Los Angeles County, California, is used as recharge water as well as storm water.

Groundwater recharge with reclaimed municipal wastewater in southern California. Southern California, like many semiarid regions of the United States, does not receive sufficient water from local sources to support the considerable population of the area. Almost two-thirds of the water supply is imported 300 to 800 km from the point of use. The remainder is derived from local groundwater basins. In some areas, the occurrence of overdraft conditions and saltwater intrusion has led to the adjudication of groundwater extractions and the implementation of artificial groundwater replenishment. Water sources used for groundwater replenishment include storm runoff, imported water, and, in some cases, reclaimed municipal wastewater.

There is considerable uncertainty at this time regarding the sufficiency of water supplies for the future water needs of the area. Population-growth projections coupled with reductions in imported-water deliveries indicate that, by the early 21st century, water needs may exceed available supplies. These water-shortage predictions have stimulated regional planning activities aimed at optimizing available water supplies through conservation efforts and developing new local sources of supply through conjunctive groundwater storage and wastewater reclamation and reuse. Of all the reclamation projects under consideration, groundwater recharge represents the largest and most economical use of reclaimed water in the region.[10]

Despite these economic incentives, implementation of proposed groundwater-recharge projects is constrained by concerns over the potential health impacts of indirect reuse for potable purposes. The existing groundwater-recharge projects in Los Angeles and Orange Counties provided an opportunity to gather the data needed to evaluate the health significance of water reuse by groundwater recharge. Foremost among these is the Whittier Narrows groundwater-recharge project located in the Montebello Forebay area of Los Angeles County. Planned replenishment of groundwater with reclaimed water has been practiced in this location since 1962.

Whittier Narrows Groundwater Recharge Project, Los Angeles County, California.[10,11,19] Under natural conditions, groundwater in the Montebello Forebay was replenished by (1) percolation from the channels of the San Gabriel River and Rio Hondo and other streams crossing the area, (2) subsurface flow from the adjacent groundwater basin, and (3) percolation of local precipitation.

As part of its responsibilities, the Los Angeles County Flood Control District has implemented an artificial replenishment program in the Montebello Forebay to augment recharge that occurs naturally. This program combines the elements of water conservation and water-resource management. First, the Flood Control District has constructed surface storage facilities for storm runoff in the mountainous area tributary to the San Gabriel River and Rio Hondo. Captured stormflows, which otherwise would have been wasted to the ocean, are released at rates compatible with the downstream percolation capacity. Second, the Flood Control District has constructed special spreading areas designed to increase the indigenous percolation capacity. Specifically, this activity has consisted of modifications to the San Gabriel River channel and construction of off-stream spreading basins adjacent to the Rio Hondo and San Gabriel River.

The Rio Hondo spreading facility is the largest recharge location, having a total of 1.8 km^2 of wetted areas available for spreading. The San Gabriel River spreading basins have approximately 0.5 km^2 wetted areas available for recharge, with an additional 0.5 km^2 of river bottom that can be used when needed. The Rio Hondo and San Gabriel River spreading grounds are subdivided into individual basins that range in size from 1.6 to 8.1 ha.[19]

Recharge operation. Under normal operating conditions, batteries of the basins are rotated through a 21-day cycle consisting of (1) a 7-day flooding period during which the basins are filled to maintain a constant 1.2 m depth, (2) a 7-day draining period during which the flow to the basins is terminated and the basins are allowed to drain, and (3) a 7-day drying period during which the basins are allowed to thoroughly dry out. This wetting-drying operation serves several purposes, including maintenance of aerobic conditions in the upper soil strata and vector control.

Infiltration rates during the flooding period average about 0.6 m/d. The capacity of the spreading grounds during normal operation is about 8.5 m^3/s. During the winter storm period, when all the basins are in use, the capacity increases to 27 m^3/s.[19]

Source of recharge water. The available percolation capacity of the recharge facilities is utilized during only a small portion of the year for spreading storm runoff, thus allowing for replenishment by water from other sources. This practice first began in the 1950s with the purchase of Colorado River water imported by the Metropolitan Water District of Southern California.

The other water source used for replenishment in the Montebello Forebay has been reclaimed municipal wastewater purchased from the County Sanitation Districts of Los Angeles County. Reclaimed water was first made available in 1962 following the completion of the Whittier Narrows Water Reclamation Plant and later in 1973 when the San Jose Creek Water Reclamation Plant was placed in service. Reclaimed water has also been incidentally supplied by the

Districts' Pomona Water Reclamation Plant. Effluent from the Pomona plant that is not put to beneficial use is discharged into San Jose Creek, a tributary of the San Gabriel River, and ultimately becomes a source of recharge for the Montebello Forebay. As the Pomona effluent becomes fully utilized for irrigation and industrial applications within the Pomona area, this source of recharge is expected to diminish.

Through 1977/78, the reclaimed water used for replenishment was disinfected activated sludge secondary effluent. Since that time, all three treatment plants have been upgraded to include either dual-media filtration (Whittier Narrows and San Jose Creek Water Reclamation Plants) or activated-carbon filtration (Pomona Water Reclamation Plant) prior to chlorine disinfection. The final effluents produced by each treatment facility comply with primary drinking-water standards and meet average coliform and turbidity effluent discharge requirements of less than 2.2 MPN (most potable number) per 100 mL and less than 2 NTU, respectively.[19,35,56]

During the 20-year period following the completion of the Whittier Narrows Water Reclamation Plant, over 542×10^6 m^3 of reclaimed water has infiltrated into the Montebello Forebay. On an annual basis, the amount of reclaimed water entering the Forebay averages about 33×10^6 m^3, or 16 percent of the total inflow to the basin. An arbitrary upper limit for reclaimed water of 40×10^6 m^3/yr has been established on the basis of historical spreading operations. The Los Angeles Regional Water Quality Control Board has recommended that the current level of use of reclaimed water for groundwater recharge in this location be expanded to a maximum of 62×10^6 m^3/yr, or approximately 30 percent of the total inflow to the Montebello Forebay.[10,19]

To reduce the mining of groundwater and to provide a secure water supply over a long term, the community is looking at many water sources. A unique wastewater reclamation–groundwater recharge project has been in operation since June 1985 at El Paso, Texas, as described in the following example.

Advanced wastewater treatment and groundwater recharge by direct injection at El Paso, Texas.[19,57] El Paso is located in an arid area with a water-supply problem. It is of a long-range nature since current mining of groundwater reserves is meeting present demands. At the 38×10^3 m^3/d Fred Hervey Water Reclamation Plant, municipal wastewater is treated in a state-of-the-art advanced wastewater-reclamation system. The system consists of 10 separate treatment steps before reclaimed water is returned to the aquifer via a system of ten

240-m-deep recharge wells. Ultimately, the reclaimed municipal wastewater returns to the city's potable-water system via groundwater supply.

Purposes of groundwater recharge. Groundwater recharge was chosen for El Paso to (1) solve existing wastewater problems, (2) add to the reliable drinking-water supply, and (3) be a prototype for larger-scale wastewater reclamation and reuse which would provide more than one-fourth of El Paso's water needs over the next 70 years. The water-supply benefits of wastewater reclamation and reuse are judged by the public to warrant increased dollar costs, resource use, and environmental impact because alternative water-supply projects, such as importation, are even more expensive and are expected to result in significant resource use and environmental impact.

Since the reclaimed water is being added to a protected water-supply source, meeting drinking water standards is crucial and the treatment process must have a high factor of safety for removal of those pollutants which are difficult to identify, particularly toxic materials, trace organics, and viruses. To accomplish these objectives, a multistage powdered activated-carbon system was chosen in conjunction with lime treatment, recarbonation, sand filtration, ozone disinfection, and granular activated-carbon adsorption.

Operating costs of advanced treatment and groundwater recharge. In 1986 a total of 5.3×10^6 m^3 of reclaimed water meeting current drinking-water standards was returned to the Hueco Bolson aquifer by the Fred Hervey Water Reclamation Plant. Power and labor costs were nearly equal and made up a total of 66 percent of the overall costs. Chemical costs made up another 18 percent, with powdered carbon representing 5 percent of the chemical costs. Total operating costs for the 38×10^3 m^3/d Fred Hervey Water Reclamation Plant were 41¢/m^3 in 1986.[57]

The major areas of cost increase have been power, chemicals, and maintenance. It is reasonable to expect some increase in maintenance costs as equipment warranties expire and costs are shifted from the supplier to the owner.

Since the Fred Hervey Water Reclamation Plant was placed in service in 1985, the facility has consistently met the operational priority established with aquifer recharge goals. However, the amount of water recharged into the potable-water system to date is minor in terms of injected volume and displacement volumes involved in traveling to the production wells. The processes in the treatment train have proved their ability to accomplish the required treatment reliability, at a cost commensurate with the strict water-reclamation requirements imposed on them.

Another example of groundwater recharge with municipal wastewater involves an interregional scheme for the collection, purification, and reclamation of the municipal wastewater in Israel. Three major issues have been dealt with in the project described below:[58,59]

1. The degree of treatment to be provided prior to groundwater recharge
2. The need and desirability of recharging the effluent to a high-quality potable aquifer
3. Conversion of the existing single-supply network to a dual-supply system, conveying potable water and high-quality nonpotable water.

Dan Region Project in Israel.[59,60] Israel, at present, utilizes practically all its freshwater potential of 4.1×10^6 m^3/d. Of this quantity, approximately 75 percent is supplied to agriculture and 25 percent is used for domestic and industrial purposes. The Dan Region Project provides for advanced wastewater treatment, groundwater recharge, and indirect reuse of municipal wastewater from the largest urban area of the country—the Tel Aviv metropolitan area. The first stage of the project consists of biological treatment in two parallel series of facultative oxidation ponds and chemical treatment by the high-lime-magnesium process, followed by detention of the high-pH effluent in polishing ponds, mainly for free-ammonia stripping and natural recarbonation. Most of the recharged effluent, after additional treatment and prolonged detention in the soil-aquifer system, is pumped up via the recovery well. The pumped groundwater is supplied to nonpotable uses mainly for unrestricted irrigation of agricultural crops.

Recharge site. The recharge site is located in an area of rolling sand dunes near the Mediterranean coast, which lies above the central part of the coastal aquifer. This aquifer is composed mainly of calcareous sandstone and is divided into subaquifers by silt and clay layers.

The climate of the area is typically Mediterranean. Summers are warm and dry and winters are mild, with rainy spells. The average annual precipitation is 500 to 550 mm. The average temperatures range between 20 and 30°C in summer and 10 and 20°C in winter.

Purification of the recharged effluent by soil-aquifer treatment. In addition to its major purification effect, groundwater recharge, as practiced in the Dan Region Project, fulfills a series of additional functions; it provides seasonal and multiyear storage, serves as a safety barrier against any unpredictable quality deterioration, increases the system's reliability, and has an important psychological effect, since the consumers will be supplied from groundwater wells and not from the treatment-plant outlet.

The scope of the stage two project is to collect all the municipal wastewater from the Dan Region, and to renovate the effluent to the highest level achievable by an advanced biological treatment process. The final water reclamation, storage, and reuse of the renovated water through the stage 2 facilities are part of separate projects currently in the advanced stages of planning and design.[60]

Guidelines for groundwater recharge with reclaimed water. Groundwater recharge with reclaimed municipal wastewater in groundwater basins that serve as sources of domestic water supply present a wide spectrum of health concerns. Reports by water-quality and public-health experts have been prepared to provide information needed to assess health issues and establish criteria for groundwater recharge with reclaimed municipal wastewater.[11,12,19,56] These reports did not establish specific guidelines but provided assessments regarding risks and comparisons of reclaimed-water quality with other sources of water supply that have been historically acceptable.

It is essential that water extracted from a groundwater basin for domestic use be of acceptable physical, chemical, microbiological, and radiological quality. Main concerns governing the acceptability of groundwater-recharge projects are that adverse health effects could result from the introduction of pathogens or trace amounts of toxic chemicals into groundwater that is eventually consumed by the public. Because of the increasing concern for long-term health effects, every effort should be made to reduce the number of chemical species and concentration of specific organic constituents in the applied water. A source-control program to limit potentially harmful constituents entering the sewerage system must be an integral part of any recharge project. Extreme caution is warranted because of the difficulty in restoring a groundwater basin once it is contaminated. Additional cost would be incurred if groundwater quality changes resulting from recharge necessitated the treatment of extracted groundwater and/or the development of additional water sources.

The level of municipal wastewater treatment necessary to produce suitable reclaimed water for groundwater recharge depends on the groundwater-quality objectives, hydrogeologic characteristics of the groundwater basin, and the amount of reclaimed water and percentage of reclaimed water applied. Major considerations are (1) the total amount and types of recharge water available for recharge on an annual basis, (2) size of the groundwater basin and probability of dilution with natural groundwaters, (3) soil types, (4) depth to groundwater, (5) method of recharge, (6) and the length of time the reclaimed water is retained in the basin prior to withdrawal for domestic use. These factors must be evaluated in establishing criteria for groundwater recharge with reclaimed water.[12,19]

In the United States, federal requirements for groundwater recharge in the context of wastewater reuse have not been established. Because of this, wastewater-reuse requirements for groundwater recharge are presently regulated by the state agencies with a case-by-case determination.

Proposed California groundwater-recharge regulations. The State of California's proposed regulations for groundwater recharge with reclaimed municipal wastewater rightly reflect a cautious attitude toward such short-term as well as long-term health concerns. The proposed regulations[12] are shown in Table 9.11. The regulations rely on a combination of controls intended to maintain a microbiologically and chemically safe groundwater-recharge operation. No single method of control would be effective in controlling the transmission and transport of contaminants of concern into and through the environment. Therefore, source control, wastewater-treatment processes, treatment standards, recharge methods, recharge area, extraction-well proximity, and monitoring wells are all specified.

TABLE 9.11 Proposed Requirements for Groundwater Recharge with Reclaimed Municipal Wastewater[12]

Treatment and recharge-site requirements	Project category			
	Surface spreading			Direct injection
	I	II	III	IV
Level of wastewater treatment				
Primary and secondary	X	X	X	X
Filtration	X	X		X
Organics removal	X			X
Disinfection	X	X	X	X
Maximum allowable reclaimed wastewater in extracted well water, %	50	20	20	50
Depth to groundwater, m, at initial percolation rate of				
50 mm/min	3	3	6	NA*
80 mm/min	6	6	15	NA
Retention time underground, months	6	6	12	12
Horizontal separation,† m	150	150	300	300

*Not applicable.
†From the edge of the groundwater-recharge operation to nearest potable-water-supply well.

The requirements listed in Table 9.11 are specified by "project category" identifying a set of conditions that constitute an acceptable project. An equivalent level of perceived risk is inherent in each project category when all conditions are met and enforced. The main concerns governing the acceptability of groundwater-recharge projects with reclaimed municipal wastewater are that adverse health effects could result from the introduction of pathogens or trace amounts of toxic chemicals into groundwater that is eventually consumed by the public.

Microbiological considerations. Of the known waterborne pathogens, enteric viruses have been considered most critical in wastewater reuse in California because of the possibility of contracting disease with relatively low doses and difficulty of routine examination of reclaimed wastewater for their presence. Thus, essentially virus-free effluent via the full treatment process (primary and secondary, coagulation and flocculation, clarification, filtration, and disinfection; see Fig. 9.1) is deemed necessary by the California Department of Health Services for reclaimed-wastewater applications with higher potential exposures, e.g., spray irrigation of food crops eaten raw, or most of groundwater-recharge applications (project categories I, II, and IV in Table 9.11).

The wastewater treatment requirements in Table 9.11 are designed to provide assurance that reclaimed water is essentially pathogen-free prior to extraction from the groundwater. In addition to the treatment processes, passage through an unsaturated zone of significant depth (>3 m) reduces organic constituents and pathogens in treated effluents. At low infiltration rates (<5 m/d) in sands and sandy loams, the rates of virus removal are approximated by a semilogarithmic plot ($k = -0.007$ log/cm) against infiltration rates, resulting in approximately 99.2 percent or 2.1-log removal for 3-m-deep soils. The overall estimates for the removal of enteric viruses by the treatment processes, unsaturated zone, and horizontal separation (retention time in groundwater) as specified in the proposed criteria are in the range of 13- to 17-log removal.[12,44]

Trace-organics removal. The regulations intend to control the concentration of organics of municipal wastewater origin as well as anthropogenic chemicals that have an impact on health when present in trace amounts. Thus, the dilution requirements and the organics removal specified in project categories I and IV in Table 9.11 is to limit the average concentration of unregulated organics in extracted groundwater affected by the groundwater-recharge operation. The concentration of unregulated and unidentified trace organics is of great concern since other constituents and specific organics are dealt with through the established maximum contaminant levels and action levels developed by the California Department of Health Services. Approximately 90% by weight of the organics constituting the total organic carbon (TOC) in treated municipal wastewater is unidentified.[12] One health concern related to the unidentified organics is that an unknown but small fraction of them are mutagenic.

Regulating the presence of trace amounts of organics in reclaimed water can be accomplished by dilution using surface water or groundwater of a less-contaminated source. When reclaimed water makes up more than 20 percent of the water reaching any extraction well for potable-water supply, treatment to remove organics must be provided. Because of lack of ideal measure for trace amount of organics in reclaimed water as well as in the affected groundwater, TOC was chosen, as a surrogate, to represent the unregulated organics of concern.[12] Although TOC is not a measure of specific organic compounds, it is considered to be suitable measure of gross organics content of reclaimed water as well as groundwater for the purpose of determining organics removal efficiency in practice. However, there is insufficient basis for the establishment of a gross organics standard for the recharge water that protects public health. The proposed regulations shown in Table 9.11 require that the groundwater recharge projects by surface spreading resulting in a 20 to 50 percent reclaimed wastewater contribution at any extraction well (category I), and the recharge project by direct injection resulting in a 0 to 50 percent contribution (category IV), provide organic removal step sufficient to limit the TOC concentration of wastewater origin in extracted water to 1 mg/L.

Table 9.12 shows the maximum TOC concentration that may be allowed in the reclaimed water, for a given percent reclaimed water

TABLE 9.12 Maximum Allowable TOC Concentration in Reclaimed Wastewater (Recharge Water) Where Organics Removal to Achieve 1 mg/L TOC in Extracted Well Water Is Required[12]

Percent reclaimed wastewater in extracted water	Maximum allowable TOC, mg/L, in recharge water	
	Surface spreading (category I)	Direct injection (category IV)
0–20	20	5
21–25	16	4
26–30	12	3
31–35	10	3
36–45	8	2
46–50	6	2

contribution, to achieve no more than 1 mg/L TOC of wastewater origin in the extracted water.[12] The numbers in Table 9.12 assume a 70 percent reduction through the unsaturated zone and no TOC removal in the aquifer. The numbers associated with the direct injection were derived by dividing 1 mg/L of TOC concentration by the fractional contribution of reclaimed water to native groundwater at the extraction point. Thus, the numbers for the direct injection are 30 percent of those for the groundwater recharge by surface spreading. In addition, direct-injection projects would have to achieve a 70 percent TOC reduction to compensate for the lack of unsaturated zone in the overall soil-aquifer treatment capability.

Inorganic chemicals. Inorganic chemicals, with the exception of nitrogen in its various forms, are adequately controlled by meeting all maximum contaminant limits (MCLs) in the reclaimed wastewater. By limiting the concentration of total nitrogen in the reclaimed water, detrimental health effects such as methemoglobinemia can be controlled. In those recharge operations where adequate nitrogen removal cannot be achieved by treatment processes or passage through an unsaturated zone, the criteria provide the alternative method such as wellhead treatment to meet the allowable total nitrogen concentration of 10 mg/L as N.[12]

Potable-water reuse

A repeated thesis for the last two decades has been that advanced wastewater treatment provided a water of such high quality that it should not be wasted but put to further use. This conviction, when joined with increasing problems of water shortage, provided a realistic atmosphere for considering wastewater reclamation and reuse in the United States. As discussed earlier, in the section on wastewater-reuse categories, direct nonpotable-water reuse has become a major option for planned reuse for supplementing public water supplies in the United States and elsewhere in the world.

Because of the health and safety concerns, there has been a very cautious attitude toward crossing the threshold of potable-water reuse. Nevertheless, some communities are developing or implementing plans for direct or indirect potable reuse where no other possibilities exist for supplemental freshwater supplies.[8,17,18] Although the quantities involved in potable reuse are small, the technological and public-health interests are the greatest, and hence considerable debate has been directed toward possible formulation of potable-reuse criteria and standards.

Planned indirect potable-reuse systems in operation today include such groundwater-recharge operations as in the Whittier Narrows groundwater-recharge project in Los Angeles County, California, and in El Paso, Texas, where reclaimed waters are percolated or injected into groundwater aquifers. Another example is the Occoquan Reservoir in northern Virginia. Effluent from the 57×10^3 m^3/d Manassas advanced wastewater-treatment plant operated by the Upper Occoquan Sewage Authority is directly discharged into the Occoquan Reservoir, a principal drinking-water reservoir for more than 660,000 people.[61,62] To date, no systems in the United States have been planned to deliver reclaimed municipal wastewater pipe to pipe or immediately added to the drinking-water supply of the community.

Emergency use of reclaimed municipal wastewater for potable supply. The first well-documented case of direct potable reuse of treated municipal wastewater occurred at Chanute, Kansas. The resulting scientific paper[8] constituted a major demonstration of short-term use of reclaimed municipal wastewater for potable supply. During the severe drought of 1952–1957, the city of Chanute was intermittently plagued by water shortages. Normally, this town of 12,000 persons, which at the time obtained its water supply from Neosho River, used an average of 5.3×10^3 m^3/d and a maximum of about 7.6×10^3 m^3/d. In the summer of 1956, the situation became progressively worse and the Neosho River ceased to flow:[8] "On October 14, 1956, without fanfare, the city opened the valve which permitted mixing of treated municipal wastewater with water stored in the river channel behind the water treatment dam." During the ensuing 5 months, chlorinated secondary effluent was collected behind the dam on the river and used as intake water for the city's water-treatment plant. During the period of water reuse, the sewage-treatment plant removed, on the average, 86 percent of the BOD and 76 percent of the COD. The impounding reservoir served very effectively as a waste-stabilization pond.

The clinical observations by local physicians and the laboratory data were in agreement as to the safety of the water. The tap water met the drinking-water standards for bacteriological quality during the entire recirculation period. It was felt, however, that the margin of safety was uncomfortably narrow. On the basis of chloride data and other methods of estimating the number of cycles, recirculation of the reclaimed water occurred 8 to 15 times during the ensuing 5 months.

The treated water had a pale-yellow color and an unpleasant musty taste and odor. It foamed when agitated and contained undesirable quantities of dissolved minerals and organic substances. Consumer acceptance of water was poor, and many people obtained their drinking water from other sources.[8]

Findings. Several important conclusions were drawn from this well-documented episode:[8]

1. The reuse of sewage-treatment-plant effluent to supplement deficient water supplies should not be considered or permitted except under the most severe emergency conditions. It is certainly a last resort, to be used only after all other possible sources of supply have been fully investigated. With the degree of treatment provided at Chanute, it is doubtful that the process could have been continued for more than a few weeks beyond the 5-month period.
2. The most serious problem was that of public acceptance or, more accurately, public rejection of the water, although there were also serious technical difficulties to be overcome. People were willing to pay many times the normal cost for safe and palatable water.
3. The accumulation of nitrogen made free residual chlorination impractical and rendered tests, odor control, and color removal difficult. The studies demonstrated the effectiveness of chlorine as a disinfectant, even under very adverse conditions. Bacteriological quality as judged by the coliform test was excellent.
4. A rapid increase occurred in the concentrations of dissolved salts and organic materials, many of which were not amenable to removal by ordinary treatment processes.
5. There are many unresolved questions concerning the safety of recycled water from a public-health standpoint, despite the apparently favorable results obtained in the Chanute study. Standard techniques available to water-plant laboratories are not adequate to detect the presence of some pathogens which may be present in heavily polluted waters, and the health significance of some of the organic chemical contaminants which may be present is not known.[8]

Since the Chanute episode, considerable research and technological advancement have been made in both water and wastewater fields which have direct impact on potable reuse. However, experience with full-scale operation of the processes is still limited, particularly when the different unit processes and operations are used in combination in a complete wastewater-reclamation scheme.

Thus, in proposing potable-water reuse, serious consideration must be given to whether the need is for a short-term emergency as seen in the preceding example or for normal use over a prolonged period. Regardless of the length of use, however, the major emphasis placed on potable-water reuse today concerns the chronic health problems that might result from ingesting the mixture of inorganic and organic contaminants that remain in water, even after subjecting it to the most advanced treatment methods.[61,63] In many cases, alternative sources of water have been developed in ensuing years and urgency to adopt pipe-to-pipe potable-water reuse has been avoided.

Direct potable-water reuse in Windhoek, Namibia. The only full-scale direct potable-reuse facility in operation is at the 4500 m^3/d potable-reuse plant in Windhoek, the capital of Namibia. The plant was inaugurated in 1969, following pilot-scale research by the National Institute for Water Research of the South African Council for Scientific and Industrial Research.[67] The decision to build the plant was precipitated by severe water shortages which faced Windhoek, and the absence of alternative sources of supply at that time. Although alternative sources have been developed since then, the plant is being operated intermittently today according to need.[15] Microbiological and epidemiological study at Windhoek to assess the health effects of reclaimed-water consumption have been conducted. Analysis of more than 15,000 episodes of diarrheal illness during the period August 1976 to March 1983 showed that the consumption of reclaimed water does not increase the risk resulting from waterborne infectious agents.[64,65]

No water-reclamation schemes for potable water have been established in South Africa since the commissioning of the original Windhoek plant in 1969. The reason is simply one of relative urgency. Windhoek is located in a very arid region, and at the time the wastewater-reclamation plant was commissioned, the city could no longer meet its water demands from the available conventional supplies. No city has yet reached the same state of urgency, but in a few areas the economically feasible alternatives are rapidly being exhausted.[67] In the decade following the inauguration of the Windhoek reclamation plant, the further development and evaluation of physicochemical technology for potable reuse continued at the 4500 m^3/d Stander demonstration plant in Pretoria and other locations in South Africa. The state of the art was drawn together in guidelines published by the Water Research Commission.[66]

While the implementation of direct potable use of reclaimed municipal wastewater is obviously limited to extreme situations, research relating to potable reuse has continued in several locations. As the proportional quantities of treated wastewater discharged into the nation's waters increase, as seen in New Orleans on the Mississippi River, much of the research which addressed only potable reuse is becoming equal in relevance to unplanned indirect potable reuse.

It has been argued that there should be a single water-quality standard for potable water and that if reclaimed water can meet this standard, it should be acceptable. However, it must be recognized that current drinking-water standards have been derived on the basis of experience with water supplies from relatively unpolluted freshwater sources. While great advances have been made in analytical methods to identify contaminants in water, only a small fraction of the contaminants present can be identified. This has frustrated attempts to develop acceptable quality criteria for drinking water from highly contaminated sources. In assessing water being considered for potable reuse, comparison should be made with the highest-quality water locally available.[61]

The Potomac Estuary Experimental Water Treatment Plant Study. In the projects related to the future water-supply needs of the Washington, D.C., metropolitan area, a study to determine the feasibility of using the Potomac estuary waters as a source of water supply was conducted by the U.S. Army Corps of Engineers.[69,70] In the review of the study by the National Academy of Science/National Academy of Engineering, a panel on quality criteria for water reuse was appointed and charged to advise on the criteria needed for determining the suitability of water supplies produced from unacceptable or polluted sources, such as wastewater. In its 1982 report,[61] the panel attempted to offer the best practical scientific statement concerning health-effects criteria for the evaluation of reclaimed water intended for human consumption:

> The Panel...strongly endorses the generally accepted concept that drinking water should be obtained from the best quality source available. Because the costs of wastewater treatment for potable use are high, it is anticipated that reuse would be contemplated only in the few locations where alternatives are nonexistent or even more costly.
>
> Even though there are precedents that have been cited as evidence that indirect planned potable reuse has been accepted in some locations, there is inadequate information from which to judge the safety of such a practice. In the panel's opinion, U.S. drinking water regulations were not established to judge the suitability of raw water supplies heavily contaminated with municipal and industrial wastewaters. Thus, criteria to judge the relative safety of using heavily contaminated water supplies as part of the potable water supply—direct or indirect, planned or unplanned—need to be developed.[61]

Following a decade of experimentation and the piloting of many unit processes and operations, potable-water-reuse research has been under way at the Denver Water Department's 3.8×10^3 m^3/d Potable Water Reuse Demonstration Plant.[71–73] The plant has been operated since 1984, and various unit processes have been examined to eliminate unreliable or unnecessary treatment steps without jeopardizing product water quality or consumer safety. As a result of the analysis, the final plant configuration was adopted in 1988 for the health-effects-study portion of the project.

For a period of 2½ years, concentrates of product water from the Denver Potable Water Demonstration Plant will be compared to the Denver's drinking-water concentrates in chronic and acute toxicity testing, reproductive effects, histopathological examination, and tumor-generation analysis. It is hoped that the relative risk between the existing drinking-water supply and the reclaimed water produced by this plant can be determined in the manner consistent with the future water-supply situations in Denver, Colorado.[72]

Microbial Health-Risk Estimation for Wastewater Reuse

Despite a long history of wastewater reuse in many parts of the world, the issue of *safety* of wastewater reuse is still difficult to define and *acceptable* health risks have been hotly debated. In this section, a comparative assessment of the safety of wastewater reuse is discussed with reference to two recent publications on enteric virus risk assessment. Furthermore, some observations are presented pertaining to the proposed revision of the State of California's *Wastewater Reclamation Criteria*.[35]

When treated municipal wastewater effluents are used in the urban environment, with the resulting possibility of direct contact with humans, considerable health concerns may be justified. These health concerns are specifically directed, in industrialized countries with high health standards, to the presence of enteric viruses because of their low-dose infectivity, long-term survival in the environment, difficulty of monitoring, and low removal efficiency in conventional wastewater treatment. The potential health hazards inherent in the use of reclaimed municipal wastewater must always be guarded against in such wastewater-reuse applications.

Enteric-virus concentration in wastewater

The enteric-virus data assembled included 424 unchlorinated secondary effluent samples in which 283 samples (67 percent) were virus-positive and 814 chlorinated tertiary (filtered) effluent samples with 7 positive samples (1 percent).[74] The database was obtained from the published reports from the water and wastewater agencies in California. Quantifying the virus concentration [expressed as viral units (vu) per liter] in the treated effluent is the first step for estimating the risk of virus infection on exposure to reclaimed municipal wastewater. The statistical model used was the lognormal distribution and the goodness of fit of the hypothesized distribution was evaluated using the nonparametric Kolmogorov-Smirnov test.[75]

The virus concentrations vary in a wide range in the unchlorinated secondary effluents. Furthermore, the virus concentrations among different treatment plants show distinctively different characteristics; for example, the geometric means ranged four orders of magnitude (10^{-4} to 10^{0} vu/L) and the spread factors range from 4 to 115. The geometric-mean values of unchlorinated secondary effluent samples ranged from 0.0002 to 2.3 vu/L, and 90 percentile concentrations ranged from 0.34 to 29 vu/L. The virus concentration of 0.01 vu/L is considered to be the reasonable estimate of limit of detection. Thus, characterizing the variability of enteric-virus concentrations in unchlorinated secondary effluent is extremely important in virus-risk assessment.

Two sets of virus concentrations were used for the risk analysis: the data derived from the unchlorinated activated-sludge effluents and the chlorinated tertiary filtration effluents. For the first run, the geometric mean and the 90 percentile values for enteric viruses found in unchlorinated activated-sludge effluents were used, and 5-log removal (99.999 percent) of viruses was assumed in tertiary filtration and chlorine disinfection. For the second run, two computer simulations used the virus concentrations of 0.01 and 1.11 vu/L from the chlorinated tertiary filtration effluents, which are the reasonable estimate of limit of detection for enteric viruses and maximum concentration found in tertiary effluents.

Virus risk analysis

To assess potential risks associated with the use of reclaimed wastewater, the following exposure scenarios are developed for landscape irrigation, spray irrigation of food crops, recreational impoundments, and groundwater recharge[74] and summarized in Table 9.13.

The beta-distributed probability model based on Haas' work[76] was chosen for use in risk calculations because it best represented the frequency distribution of virus infection. Infectious models based on echovirus 12 and polioviruses 1 and 3 were used by Asano et al.,[74] and the rotavirus model based on Rose and Gerba[77] was used in the study by Tanaka et al.[75]

Results of the annual risk calculations are shown in Table 9.14, using the virus concentrations of 0.01 and 1.11 vu/L from the chlorinated tertiary filtration effluents, which are the reasonable estimate of the limit of detection for enteric viruses and the maximum concentration found in tertiary effluents in California.

The estimates of risk of infection presented in Table 9.14 show the range of risks associated with annual exposures encountered in different wastewater-reuse situations. The overall probability of infection due to ingestion of viruses is a combination of virus removal and

TABLE 9.13 Summary of Exposure Scenarios Used in Risk Analysis[74]

Reuse application	Risk receptor	Exposure frequency	Amount of water ingested in a single exposure, mL	Decay or fate in the environment
Scenario 1: golf-course irrigation	Golfer	Twice per week	1	Irrigation one day before playing
Scenario 2: food-crop irrigation	Consumer	Every day	10	Stop irrigation before harvest and shipment; viral reduction due to sunlight
Scenario 3: recreational impound-ments	Swimmer	40 days per year—sum-mer season only	100	No virus reduction
Scenario 4: groundwater recharge	Ground-water consumer	Every day	1000	3-m vadose zone and 6-month retention in aquifer

TABLE 9.14 Annual Risk of Contracting at Least One Infection from Exposure to Reclaimed Wastewater Based on Two Different Enteric-Virus Concentrations[74]

	Exposure scenarios			
Viruses	Landscape irrigation for golf courses	Spray irrigation for food crops	Unrestricted recreational impoundments	Groundwater recharge
Maximum Virus Concentration of 1.11 vu/L Enteric Viruses in Chlorinated Tertiary Effluent				
Echovirus 12	1E-03	4E-06	7E-02	6E-08
Poliovirus 1	3E-05	2E-07	3E-03	5E-09
Poliovirus 3	3E-02	1E-04	8E-01	2E-08
Minimum Virus Concentration (Concentration Limit of Detection) of 0.01 vu/L Enteric Virus in Chlorinated Tertiary Effluent				
Echovirus 12	9E-06	4E-08	7E-04	5E-10
Poliovirus 1	3E-07	1E-09	2E-05	5E-11
Poliovirus 3	2E-04	1E-06	2E-02	2E-10

inactivation by wastewater treatment, die-off in the environment, and dose/response ratios. For each exposure scenario presented, the range of risks covers two to three orders of magnitude. This reflects the differences in infectivity among different viruses. For groundwater recharge with reclaimed wastewater, with an effluent virus concentration of 0.01 vu/L, the annual risk of infection ranges from 5×10^{-10} to 5×10^{-11}. When virus concentration is increased to 1.11 vu/L, which is the maximum virus concentration found in the chlorinated tertiary effluent, the risk of infection increased by two to three orders of magnitude (6×10^{-8} and 5×10^{-9}). Similar trends are noted in the other exposure scenarios.

Of the remaining three categories, the least-restrictive use of reclaimed wastewater is for recreational impoundments where water-contact sports such as swimming take place. In all cases, regardless of the starting virus concentration, the use of reclaimed wastewater for unrestricted recreational impoundments allows the most liberal exposure to enteric viruses according to the risk assessment. The relatively high probability of infection is attributed to the fact that no dilution or virus die-off in the environment were included in the risk calculations, assuming the worst possible case.

Landscape irrigation for golf courses posed the second most exposure to reclaimed wastewater, and spray irrigation of food crops ranked third, being two orders of magnitude lower in relative risk. The lower risks of infection in the cases of spray irrigation of food crops and groundwater recharge can be attributed to environmental factors such as use-area controls. In both cases, the exposure scenarios developed include virus die-off in the environment. These risk analyses, however, do not account for the variability of enteric viruses in the environment. Seasonal fluctuations in the endemic virus populations will affect the quantity and species present in the wastewater at any given time.

The lack of positive samples indicates that the chlorinated tertiary effluent is essentially virus-free; however, none of the monitoring has produced enough positive samples to establish a measure of process reliability with respect to virus removal. Since a wastewater-treatment process rarely produces a constant-quality effluent, because of daily and seasonal water-quality variations, flow fluctuations, or process variability, the effluent produced should be expected to vary.

In the 7 positive samples out of 841 samples analyzed (roughly 1 percent positive), the virus concentration ranged from 2 to 111 vu/100 L.[74] Therefore, it might be assumed that the wastewater-treatment process produced a virus-free effluent 99 percent of the time. In this circumstance, the assumption of 5-log removal of viruses in tertiary treatment needs close examination in light of treatment-process reliability. If a treatment process for wastewater reuse has a reliability of 99 percent, the process is expected to meet the performance requirements 99 percent of the time. Thus, the permit limit of virus-free effluent is expected to be exceeded 1 percent of the time, or three to four times per year. The issue in risk management in wastewater reuse, then, becomes one of determining whether the presence of enteric viruses in the concentration range of 2 to 111 vu/100 L in approximately 1 percent of the time in tertiary-treated reclaimed wastewater is significant to protect public health. Further research is needed to characterize treatment-process reliability that contributes to the overall reduction of infectious risks in wastewater reclamation and reuse.

The goal of essentially virus-free reclaimed wastewater contained in California's *Wastewater Reclamation Criteria*[35] should not be interpreted to mean that the practice of using such water is risk-free. As Table 9.14 clearly shows, there is always some risk of infection due to exposure to reclaimed wastewater. However, this does not mean that the practice of wastewater reclamation and reuse is unsafe. The *safety* of wastewater reclamation and reuse practice is defined by the acceptable level of risks developed by the regulatory agencies responsible for risk management and endorsed by the public.

References

1. Ongerth, H. J., and J. E. Ongerth: "Health Consequences of Wastewater Reuse," *Ann. Rev. Public Health* **3:** 419–444, 1982.
2. *Water Reuse,* Manual of Practice SM-3, Water Pollution Control Federation, Washington, D.C., 1989.
3. Dean, R. B., and E. Lund: *Water Reuse: Problems and Solutions,* Academic Press, Orlando, Fla., 1981.
4. *Reuse of Sewage Effluent,* Proceedings of the International Symposium, Thomas Telford, London, 1985.
5. Neis, U. (ed.), *Water Reuse,* Institut für Wissenschaftliche Zusammenarbeit, Tubinngen, Federal Republic of Germany, 1984.
6. Hahn, H. H., et al. (eds.), *Recycling in Chemical Water and Wastewater Treatment,* Institut für Siedlungswasserwirtscaft, Universität Karlsruhe, Karlsruhe, Germany, 1986.
7. Reed, S. C., and R. W. Crites, *Handbook of Land Treatment Systems for Industrial and Municipal Wastes,* Noyes Publications, Park Ridge, N.J., 1984.
8. Metzler, D. F., et al.: "Emergency Use of Reclaimed Water for Potable Supply at Chanute, Kansas," *J. Am. Water Works Assoc.* **50**(8):1021, 1958.
9. Argo, D. G., and M. G. Rigby: "Water Reuse—What's the Big Deal?" in *Proceedings of the 1986 Conference,* American Water Works Association, Denver, 1986.
10. Nellor, M. H., et al.: "Health Aspects of Groundwater Recharge," in T. Asano (ed.), *Artificial Recharge of Groundwater,* Butterworth Publishers, Boston, 1985.
11. Nellor, M. H., et al.: "Health Effects of Indirect Potable Water Reuse," *J. Am. Water Works Assoc.* **77**(7):88, 1985.
12. State of California, *Policy and Guidelines for Ground Water Recharge with Reclaimed Municipal Wastewater,* draft edition, Department of Health Services and State Water Resources Control Board, Sacramento, Calif., 1989.
13. Crook, J., and D. A. Okun: "The Place of Nonpotable Reuse in Water Management," *J. Water Pollution Control Fed.* **59**(5):236, May 1987.
14. Lauer, W. C., et al.: "Denver's Potable Water Reuse Project—Current Status," in *Proceedings of the Water Reuse Symposium III, Future of Water Reuse,* vol. 1, p. 316, AWWA Research Foundation, Denver, 1985.
15. Odendaal, P. E., and W. H. Hattingh: "The Status of Potable Reuse Research in South Africa," in *Proceedings of Water Reuse Symposium IV, Implementing Water Reuse,* p. 1339, AWWA Research Foundation, 1988.

16. U.S. Water Resources Council: *The Nation's Water Resources, the Second National Assessment,* Washington, D.C., 1978.
17. Solley, W. B., et al.: *Estimated Use of Water in the United States in 1980,* Geological Survey Circular 1001, U.S. Geological Survey, Alexandria, Va., 1983.
18. Culp, R. L.: "Selecting Treatment Processes to Meet Water Reuse Requirements," in T. Asano and P. V. Roberts (eds.), *Water Reuse for Groundwater Recharge,* California State Water Resources Control Board, May 1980.
19. Asano, T. (ed.), *Artificial Recharge of Groundwater,* Butterworth Publishers, Boston, 1985.
20. *Culp / Wesner / Culp: Water Reuse and Recycling,* Office of Water Research and Technology, U.S. Department of the Interior, July 1979.
21. Asano, T., and R. S. Madancy: "Water Reclamation Efforts in the United States," in E. J. Middlebrooks (ed.), *Water Reuse,* Ann Arbor Science Publishers, Ann Arbor, Mich., 1982.
22. State of California: *California Municipal Wastewater Reclamation in 1987,* State Water Resources Control Board, Office of Water Recycling, June 1990.
23. Asano, T., and R. A. Mills: "Planning and Analysis for Water Reuse Projects," *J. Am. Water Works Assoc.* **82**(1):38–47, 1990.
24. State of California: *Interim Guidelines for Economic and Financial Analyses of Water Reclamation Projects,* prepared by Ernst & Ernst for State Water Resources Control Board, Sacramento, Calif., Feb. 1979.
25. James, L. D., and R. R. Lee: *Economics of Water Resources Planning,* McGraw-Hill, New York, 1971.
26. Pettygrove, G. S., and T. Asano (eds.): *Irrigation with Reclaimed Municipal Wastewater—a Guidance Manual,* Lewis Publishers, Chelsea, Mich., 1985.
27. *Environmental Regulations and Technology, the National Pretreatment Program,* EPA/625/10-86/005, Office of Water Enforcement and Permits, Office of Water, U.S. Environmental Protection Agency, July 1986.
28. *Monterey Wastewater Reclamation Study for Agriculture,* prepared for Monterey Regional Water Pollution Control Agency, prepared by Engineering-Science, Berkeley, Calif., April 1987.
29. *The Porter-Cologne Water Quality Control Act and Related Code Sections* (including 1986 amendments), California State Water Resources Control Board, Sacramento, Calif., Jan. 1987.
30. Ayers, R. S., and D. W. Westcot: *Water Quality for Agriculture,* FAO Irrigation and Drainage Paper 29, revision 1, Food and Agriculture Organization of the United Nations, Rome, Italy, 1985.
31. Suarez, D. L., "Relation between pH_c and Sodium Adsorption Ratio (SAR) and Alternative Method of Estimating SAR of Soil or Drainage Waters," *Soil Sci. Soc. Am. J.* **45:**469, 1981.
32. Ayers, R. S., and K. K. Tanji: "Agronomic Aspects of Crop Irrigation with Wastewater," *Water Forum '81,* vol. 1, pp. 578–586, American Society of Civil Engineers, 1981.
33. Asano, T., and G. S. Pettygrove: "Using Reclaimed Municipal Wastewater for Irrigation," *Calif. Agric.* **41**(3–4):15–18, 1987.
34. Crook, J.: "Health and Regulatory Considerations," in G. S. Pettygrove and T. Asano (eds.), *Irrigation with Reclaimed Municipal Wastewater—a Guidance Manual,* Butterworth Publishers, Chelsea, Mich., 1985.
35. State of California: *Wastewater Reclamation Criteria,* California Administrative Code, Title 22, Division 4, Environmental Health, Department of Health Services, Sacramento, Calif., 1978.
36. Shuval, H. I.: "Health Risks Associated with Water Recycling," *Water Sci. Technol.* **14:**6E1, 1982.
37. Bouwer, H., and W. L. Chase: "Water Reuse in Phoenix, Arizona," in *Future of Water Reuse, Water Reuse Symposium III,* vol. 1, pp. 337–353, AWWA Research Foundation, Denver, 1985.
38. Shuval, H. I., et al.: *Wastewater Irrigation in Developing Countries—Health Effects and Technical Solutions,* World Bank Technical Paper 51, The World Bank, Washington, D.C., 1986.

39. Feachem, R. G., et al.: *Sanitation and Disease—Health Aspects of Excreta and Wastewater Management,* World Bank Studies in Water Supply and Sanitation 3, published for the World Bank by Wiley, Chichester, U.K., 1983.
40. *Human Wastes: Health Aspects of Their Use in Agriculture and Aquaculture,* International Reference Centre for Waste Disposal, Duebendorf, Switzerland, May 1988.
41. Biswas, A. K., and A. Arar (eds.): *Treatment and Reuse of Wastewater,* Butterworths, London, 1988.
42. Pescod, M. B., and A. Arar (eds.): *Treatment and Use of Sewage Effluent for Irrigation,* Butterworths, London, 1988.
43. Kirkpatrick, W. R., and T. Asano: "Evaluation of Tertiary Treatment Systems for Wastewater Reclamation and Reuse," *Water Sci. Technol.* **18**(10), 83, 1986.
44. Asano, T., and R. Mujeriego: "Pretreatment for Wastewater Reclamation and Reuse," in H. H. Hahn and R. Klute (eds.), *Pretreatment in Chemical Water and Wastewater Treatment,* Springer-Verlag, Berlin, Federal Republic of Germany, 1988.
45. *Industrial Water Recycling,* Office of Water Recycling, California State Water Resources Control Board, Sacramento, March 1979.
46. *Evaluation of Industrial Cooling Systems Using Reclaimed Municipal Wastewater: Applications for Potential Users,* Office of Water Recycling, California State Water Resources Control Board, Sacramento, Nov. 1980.
47. Burger, R.: *Cooling Tower Technology: Maintenance, Upgrading and Rebuilding,* Cooling Tower Institute, Houston, Tex., 1979.
48. Blackson, D. E.: "Treated Municipal Sewage as a Source of Power Plant Cooling Tower Makeup Water," in *Proceedings of the Symposium on Water Use Data for Water Resources Management,* p. 571, American Water Resources Assoc., 1988.
49. *Water Reuse,* Arizona Nuclear Power Project, Phoenix, Ariz., 1985.
50. Todd, D. K.: *Groundwater Hydrology,* 2d ed., Wiley, New York, 1980.
51. Bouwer, H.: *Groundwater Hydrology,* McGraw-Hill, New York, 1978.
52. Freeze, R. A., and J. A. Cherry: *Groundwater,* Prentice-Hall, Englewood Cliffs, N.J., 1979.
53. Huisman, L., and T. N. Olsthoorn: *Artificial Groundwater Recharge,* Pitman Books, London, 1983.
54. Ward, C. H., et al. (eds.): *Ground Water Quality,* Wiley, New York, 1985.
55. Roberts, P. V.: "Water Reuse for Groundwater Recharge: An Overview," *J. Am. Water Works Assoc.* **72**(7):375, 1980.
56. County Sanitation Districts of Los Angeles County: *Health Effects Study,* Final Report, prepared by M. H. Nellor, R. B. Baird, and J. R. Smyth, March 1984.
57. Knorr, D. B., et al.: "Wastewater Treatment and Groundwater Recharge: A Learning Experience at El Paso, TX," in *Implementing Water Reuse, Proceedings of the Water Reuse Symposium IV,* p. 211, AWWA Research Foundation, Denver, 1988.
58. Idelovitch, E., and M. Michail: "Groundwater Recharge for Wastewater Reuse in the Dan Region Project: Summary of Five-Year Experience, 1977–1981," in T. Asano (ed.), *Artificial Recharge of Groundwater,* Butterworth Publishers, Boston, 1985.
59. Idelovitch, E., et al.: "The Role of Groundwater Recharge in Wastewater Reuse: Israel's Dan Region Project," *J. Am. Water Works Assoc.* **72**(7):390, 1980.
60. Farchill, D. J.: "The Dan Region Wastewater Project—Stage II: An Advanced Single Stage Biological Wastewater Treatment and Renovation System," in *Proceedings of the Water Reuse Symposium, Water Reuse—from Research to Application,* March 25–30, 1979, p. 830, Washington, D.C., AWWA Research Foundation, 1980.
61. *Quality Criteria for Water Reuse,* Panel on Quality Criteria for Water Reuse, National Research Council, National Academy Press, Washington, D.C., 1982.
62. Robbins, M. H., Jr.: "Operation and Maintenance of the UOSA Water Reclamation Plant," paper presented at the 57th WPCF Annual Conference, New Orleans, La., 1984.
63. *Protocol Development: Criteria and Standard for Potable Reuse and Feasible Alternatives,* Report of Workshop Proceedings, July 29–31, 1980, Warrenton, Va., U.S. Environmental Protection Agency, Dec. 1982.

64. Isaacson, M., et al.: *Studies on Health Aspects of Water Reclamation During 1974–1983 in Windhoek, South West Africa/Namibia,* WRC-Report no. 38/1/87, Water Research Commission, Pretoria, South Africa, 1987.
65. Isaacson, M., and A. R. Sayed: "Human Consumption of Reclaimed Water—the Namibian Experience," in *Proceedings of Water Reuse Symposium IV, Implementing Water Reuse,* p. 1047, AWWA Research Foundation, Denver, 1988.
66. Water Research Commission: *A Guide for the Planning, Design and Implementation of a Water Reclamation Scheme,* prepared by P. G. J. Meiring and Partners, Pretoria, South Africa, 1982.
67. Odendaal, P. E., and L. R. van Vuuren: "Reuse of Wastewater in South Africa—Research and Application," in *Proceedings of Water Reuse Symposium: Water Reuse—From Research to Application,* vol. 2, p. 886, AWWA Research Foundation, Denver, 1980.
68. World Health Organization, *Health Aspects of Treated Sewage Re-Use,* report on a WHO Seminar, EURO Reports and Studies 42, Regional Office for Europe, Copenhagen, Denmark, 1980.
69. *The Potomac Estuary Experimental Water Treatment Plant, a Review of the U.S. Army Corps of Engineers, Evaluation of the Operation, Maintenance and Performance of the Experimental Estuary Water Treatment Plant,* National Academy Press, Washington, D.C., 1984.
70. *Water for the Future of the Nation's Capital Area 1984, a Review of the U.S. Army Corps of Engineers Metropolitan Washington Area Water Supply Study,* National Academy Press, Washington, D.C., 1984.
71. Linsted, K. D., and M. R. Rothberg: "Potable Water Reuse," in E. J. Middlebrooks (ed.), *Water Reuse,* Ann Arbor Science Publishers, Ann Arbor, Mich., 1982.
72. Rogers, S. E., et al.: "Organic Contaminants Removal for Potable Reuse," *J. Water Pollution Control Fed.* **59**(7):722, 1987.
73. Rogers, S. E., and W. C. Lauer: "Disinfection for Potable Reuse," *J. Water Pollution Control Fed.* **58**(3):193, 1986.
74. Asano, T., L. Y. C. Leong, M. G. Rigby, and R. H. Sakaji: "Evaluation of the California Wastewater Reclamation Criteria Using Enteric Virus Monitoring Data," *Water Sci. Technol.* **26**(7–8):1523–1524, 1992.
75. Tanaka, H., T. Asano, E. D. Schroeder, and G. Tchobanoglous: "Estimating the Reliability of Wastewater Reclamation and Reuse Using Enteric Virus Monitoring Data," paper presented at the 66th Annual Conference & Exposition, Oct. 3–7, 1993, Water Environment Federation.
76. Haas, C. N.: "Estimation of Risk Due to Low Doses of Microorganisms: A Comparison of Alternative Methodologies," *Am. J. Epidemiol.* **118**(4): 573.
77. Rose, J. B., and C. P. Gerba: "Use of Risk Assessment for Development of Microbial Standards," *Water Sci. Technol.* **24**(2):29–34, 1991.

Chapter

10

Inland Fisheries

U. Barg
Food and Agriculture Organization of the United Nations
Fisheries Department
Rome, Italy

I. G. Dunn
Aquatic Biological Consultancy Services Ltd.
Chatham, Kent, England

T. Petr
Food and Agriculture Organization of the United Nations
Fisheries Department
Rome, Italy

R. L. Welcomme
Food and Agriculture Organization of the United Nations
Fisheries Department
Rome, Italy

Introduction

The purpose of this chapter is to present a short overview of major issues in inland fisheries, including aquaculture, in relation to availability and uses of water resources. FAO, in its report on the state of food and agriculture in 1993, has examined current issues and policies relevant to global water use and agriculture (FAO, 1993a). Water is an increasingly scarce and valuable resource. Of principal concern is our failure to recognize and accept that there is a finite supply of water. Growing water scarcity, misuse of freshwater, and competition among agriculture, industry, and cities for limited water supplies is already constraining development efforts in many countries. As populations expand and economies grow, the competition for limited supplies will intensify and so will conflicts among water users.

Despite water shortages, misuse of water is widespread. Small communities and large cities, farmers and industries, developing countries, and industrialized economies are all mismanaging water resources. Water quality is deteriorating in key basins from urban and industrial wastes. Because large segments of the population in developing countries do not have access to treated water for drinking, preparing food, and bathing, polluted water poses a significant public-health problem (WRI, 1990). Decreasing water flows are reducing hydroelectric-power generation, pollution assimilation, and fish and wildlife habitats. An attempt is made to highlight constraints, opportunities, and prospects for development of freshwater fisheries in the context of increased competition for and degradation of aquatic resources. Emphasis is given to inland fisheries development issues in developing countries.

Fisheries exploit living aquatic resources. In general, the term *fisheries* includes both capture fisheries and various forms of aquaculture, or fish farming. Capture fisheries, for the most part, take fish from waters where the overall production is controlled by natural climatic events and the impacts of environmental factors on the water body and its catchment area. Aquaculture or fish-farming activities involve the raising of fish and other aquatic organisms, under controlled conditions where the final yield is determined by the inputs of the operator. Aquaculture activities, when carried out in inland waters, can be described as livestock raising in an aquatic environment. Other more specific definitions of fisheries and aquaculture are available (FAO, 1990a; FAO/FIRI, 1995b). Since both activities exploit aquatic environments, generally, they are very susceptible to changes in water quality and supply, and depend to varying degrees on integrity, diversity, and functions of aquatic ecosystems and habitats.

Status of Inland Fisheries and Aquaculture

According to FAO (FAO, 1995a), world fish production in 1989 reached 100.3 million t (metric tons). Production declined in 1990 and 1991 to 97 million t, but increased to 98.7 million t in 1992, and has been estimated to be 101.3 million t in 1993. The increase in total production between 1992 and 1993 came almost entirely from aquaculture. Both marine and freshwater capture fisheries have slightly declined since the major drop of 3.5 million t between 1989 and 1990, while aquaculture has been increasing by over 1 million t per year since 1991. Total fish production in 1992 resulting from inland waters through aquaculture and capture fisheries was 14.9 million t.

It should be noted that trends in inland fish production are difficult to quantify. Food fisheries, although the basis of regular statistical collections and analysis, tend to ignore the portion of the production not passing through regular markets or to ignore "minor" and subsistence fisheries altogether. Typically, the catches from subsistence fisheries, prevalent in many developing regions, and recreational fisheries, which often also contribute to food supply in low-income groups, tend to be unreported in official catch statistics provided to FAO. Actual catches may be at least twice the reported figures. Notwithstanding these weaknesses, the FAO nominal statistics are the only consistent body of data available against which to assess the trends in fisheries and aquaculture (FAO, 1994a; FAO/FIDI, 1994).

Capture fisheries

The yields from inland capture fisheries, as reported, increased steadily until 1989, when they peaked at approximately 7 million t. Since this time the catches have stabilized or even slightly declined to 6.5 million t as recorded for 1992. Inland capture-fishery yields provide the equivalent of some 1 kg per caput of the world's population. Inland fisheries exploit a great diversity of natural and artificial human-made water bodies. Catchments which furnish high concentrations of dissolved nutrients to the drainage will have a high potential primary productivity and consequently high production of the target resource. Production, and therefore potential sustainable yields, will be greater in the tropics with high ambient temperatures and long growing periods. Tropical inland fisheries provide the greater proportion of the total world catches from freshwater environments. Between 1970 and 1992, records (FAO/FIRI, 1995a) indicate that about 80 percent of all inland capture-fishery yields were derived from tropical and subtropical zones. In 1992, capture fisheries in Asia contributed 57 percent and Africa 27 percent to the total inland fish production in the world.

It appears that most major inland capture fisheries are currently being exploited at close to, or above, their sustainable levels, although catches will vary from season to season, reflecting the quantitative and qualitative fluctuations of the target populations. These natural fluctuations result from changing climatic conditions and the complex population dynamics of interacting groups of organisms. This lack of stability is particularly obvious in tropical inland fisheries, which typically exploit a mix of species groups.

Virtually all inland fish are used for human consumption. Catches are most significant for rural areas, both commercially and for subsistence, although there are some export markets for high-value fish. For poorer and isolated communities, subsistence fishing is very important to their food supplies, and exploitation is usually at low levels of efficiency with catches which seem to stay in balance with the available resource. Inland fish also provide a basis for valuable and widespread recreational fisheries.

Aquaculture

Total inland aquaculture production in 1992 was estimated by FAO (FAO/FIRI, 1995b) to be 8.4 million t, contributing more than 60 percent of the world's total aquaculture production and more than 56 percent of total inland fish production. In 1992, about 84 percent of total aquaculture yield was produced in Asian countries, particularly China and India. Since 1984 yearly growth, in global yields, including coastal aquaculture, has been at a compound rate of 9 percent, although in recent years the rate of increase has reduced to 6.5 percent. For 1992, the value of total animal production through aquaculture has been estimated to be U.S. $27.6 thousand million.

Environmental Threats to Living Aquatic Resources

The character of aquatic resources and the hydrologic regime of a river or lake basin is determined by natural factors such as climate, geomorphology and geology, and vegetation cover. However, the totality of human activities exerts a complex pressure on living and nonliving aquatic resources. Aquatic environments are increasingly subject to effects of water abstraction and diversion, aquatic pollution, and physical modifications of aquatic habitats.

Water bodies are being utilized as sinks for waste substances or are recipients of impacts stemming from all land-use activities which may be present in catchment areas supplying water to rivers, lakes, and eventually, the sea (Martin et al., 1981; GESAMP, 1987). Further, not only is the quality of these waters changed in terms of their physicochemical characteristics, but other important components in aquatic ecosystems are affected, such as the organisms and the substrates and sediments of rivers and lakes. Human interventions, including fisheries, also affect the structure and function of aquatic communities and habitats. In short, aquatic ecosystems have to assimilate the consequences of ecological alterations resulting from chemical inputs, physical modifications, and direct biological interventions due to land and water uses in regions as large as any catchment area. However, assimilative capacities are limited and, when exceeded, lead to rapid degradation in water quality.

Generally, modification of inland waters often has negative consequences for inland fisheries (FAO, 1991a, 1991b), which in many cases are adversely affected by construction of dams, channelization and regulation of rivers, water diversion and extraction, lowering lake water levels, drainage of wetlands, and water pollution.

Physical modifications

The range of human activities with potential to damage and degrade river-basin systems can be categorized (Boon, 1992a; Arthington and Welcomme, in press) as including (1) supracatchment effects such as acid deposition and interbasin transfers, (2) catchment land-use change (afforestation and deforestation, urbanization, agricultural development, land drainage, and flood protection), (3) corridor engineering (removal of riparian vegetation, flow regulation via dams and weirs, channelization, dredging, and mining), and (4) in-stream impacts (chemical and thermal pollution, water abstraction, navigation, exploitation of native species, and introduction of exotic species).

Schemes which transfer large quantities of water between natural river basins (interbasin transfers) may present serious problems in terms of water balance, water quality, the spread of pests and diseases, alterations to natural biotic distribution patterns, and disruption of significant ecological processes. On the other hand, such schemes may assist in maintaining a healthy environment in the recipient river in which substantial irreversible uptake of water, such as for irrigation, takes place.

Deforestation, agricultural practices, urbanization, land-drainage and flood-control measures, and other land uses in catchment areas have cumulative geomorphological, hydrologic, and ecological impacts on rivers (Alabaster, 1985; Welcomme, 1985; Dunn, 1989). Primary effects of land-use activities in large rivers are the reduction of the physical heterogeneity of the channel network, a decrease in the lateral boundary between the terrestrial and aquatic environment, and disconnection of the floodplain from the main channel.

Deforestation of drainage basins usually enhances the rate of water and sediment delivery to the channel, accentuating high- and low-flow conditions, increasing temporal variation of abiotic factors, and often degrading water and stream habitat quality. More unpredictable, spiky flood regimes, faster rise and fall of water levels, and lower dry-season flows are potentially detrimental to many species of fish adapted to particular flow regimes. Freshwater ecosystems may be adversely affected by agricultural land use, through removal or reduction of catchment and riparian vegetation, increased soil erosion and siltation, elevated nutrient levels, and influx of a wide range of agricultural chemicals. Removal of riparian vegetation can result in loss of its important buffering role and increased bank instability. Overgrazing in riparian corridors due to high cattle density may degrade streamside vegetation, particularly ground cover, and cause soil compaction and erosion of riverbanks, nutrient, and soil inputs.

High silt loads from badly managed agriculture, afforestation, and urban development also tend to accelerate the natural erosion-deposition processes which conserve the dynamic equilibrium of floodplain systems, and this increased instability affects aquatic plant communities, invertebrates, and fish. Suspended solids directly interfere with light penetration and may reduce the depth to which phytoplankton can develop or shade out aquatic macrophytes, resulting in reduced plant biomass and the availability of shelter and food for aquatic biota. Suspended matter may also reduce the visibility of pelagic food resources, smother benthic substrates and spawning sites, and clog fine structures such as fish-gill rakers and filaments, with consequences for fish-breeding success, egg and larva survival, growth, population size, and age structure.

Corridor engineering implies a degree of control over discharge and channel form and may involve the barring of the channel either transversely by weirs or dams or longitudinally by levees, bankline revetment, dikes, and other structures. Widening, straightening, and deepening of the channel may also be involved. The impoundment of rivers and flow regulation are major disturbing influences on lotic (running water) systems, and the scale of global impact is enormous (UNESCO, 1990; Arthington and Welcomme, in press). By the year 2000, more than 60 percent of the world's streamflow will be regulated (Petts, 1989) and many rivers may be reduced to cascades of artificial lakes.

From a fisheries ecologist's perspective, impoundment has three significant categories of effect (Petts, 1989): (1) the isolation of river channels from their alluvial floodplains, eliminating access to backwater habitat, lakes, and marshes, with severe effects on biotic diversity, river productivity, and fish populations; (2) the introduction of barriers to migration, causing declines in fish populations, and (3) changes to in-stream habitats involving interactions of geomorphology, hydrology, and water quality. These changes have major impacts on biological systems, including increased autotropic production, disruption of invertebrate and fish reproductive cycles, and increased juvenile mortality due to loss of habitat, shelter, and exposure when water levels fluctuate rapidly. There may also be water-quality effects with severe consequences on downstream fish populations due to release of anoxic, hypolimnetic water from deep-release reservoirs, gas supersaturation below hydroelectric-power dams, and lowering of water temperatures over considerable distances downstream. Waters flowing out of reservoirs not only have reduced concentrations of suspended solids and nutrients but often are more saline, which may have adverse effects on downstream agriculture and fisheries. Other impacts include the overall loss of lotic (running water) habitat and the inability of many riverine species to benefit from increased lentic (stagnant water) habitat. Habitat loss takes many forms, such as the loss of heterogeneity when braided and branching channels are replaced by a single channel of reduced dimensions and habitat quality, or channel degradation below dams. Major effects of hydraulic works on river fish communities have been summarized by Welcomme (1985, 1989).

Drinkwater and Frank (1994) reviewed the effects of river regulation and diversion on marine fish and invertebrates, specifically the response of river-regulation projects on the Nile, Indus, and rivers flowing into the Black Sea, San Francisco Bay, and James Bay. They found that a decline in some coastal fisheries with an overall impact on the biota is generally associated with reductions in freshwater flow.

Aquatic pollution

Aquatic-pollution problems are due to the addition of deoxygenating substances, eroded sediments, and chemical contaminants arising from salinization and acidification of soils, urban sewage, agriculture, wastes from industrial activities, radionuclides, and heated effluents from power plants. Recent assessments by the UNEP/WHO Global Environmental Monitoring System (GEMS) have identified the main water pollutants including sewage, nutrients, and toxic metals as well as agricultural chemicals (UNEP, 1991). The most common water pollutant is organic material from domestic sewage, municipal waste, and agroindustrial effluent (Lean et al., 1990). Acid rain and the acidification of surface waters has caused severe problems in industrialized areas of temperate regions. Thermal pollution stems largely from electricity-generating stations that often produce discharges 6 to 16°C warmer than ambient levels.

Discharges of urban sewage and industrial and agroindustrial wastewaters rich in organic matter, such as effluents derived from pulp and paper production, palm-oil extraction, and sugar-beet processing containing high levels of decomposable organic substances, can lead to reduction of the oxygen content of recipient waters and create anoxic conditions in poorly mixed waters. The release of human and animal organic wastes carries the risk of high microbial loading, which may lead to bacterial and viral infections.

Hazardous, sometimes toxic, trace elements are accidentally or deliberately dumped into water bodies. For example, substances such as mercury may be released in industrial wastewaters from chloralkali plants, arsenic, and cadmium in mining and smelter waters, and lead in urban runoff. Such contaminants may also leach from solid-waste deposits. Continuous exposure to elevated lead concentrations has led to fish death. An increasing number of organic micropollutants are disposed of in sewers and domestic, urban, and industrial effluents or may be leaching from solid- and liquid-waste dumps. Pesticides and herbicides in drainage water and agricultural runoff from fields are becoming a widespread source of aquatic pollution, and enclosed water bodies such as lakes, reservoirs, small water bodies, and ponds are particularly at risk. Organochlorine compounds are of major concern because of their persistence in the aquatic environment. Most toxic trace elements and many organic pollutants (e.g., PCBs, DDT) of low solubility or which are fat-soluble are predominantly adsorbed by particles of inorganic or biological origin. Fish species which are top predators in the food chain may accumulate high levels of toxic compounds in their body tissues, which may well present a health hazard to humans if consumed in sufficient quantity. Sediment-bound polychlorinated aromatic hydrocarbons have been shown to have carcinogenic effects in fish species in lakes affected by their proximity to urban centers.

Acidification of inland waters may be due to mining or to indirect inputs through acidic atmospheric deposition, mainly as nitric and sulfuric acids resulting from motor exhausts and fossil-fuel combustion (Jorgensen, 1993). The acidification of temperate lake waters in industrialized regions, together with the acid leaching of lake basin soils, resulted in the release of trace elements that are toxic to aquatic organisms. Aluminum has proved to be dangerous for fish, with asphyxiation following the deposition of aluminum oxide on the gill filaments.

Freshwater environments may suffer from salinization resulting from the release of certain salts in mining and industrial wastewaters, or from evaporation, in the case of lakes and reservoirs in semiarid or arid regions.

Eutrophication is increasing in many lakes and reservoirs because of high inputs of nitrogen and phosphorus (Jorgensen and Vollenweider, 1988; Seko, 1994). Generally, the increase in primary productivity and biomass production is desirable to a certain degree, since it may enhance fisheries. However, the decay and deposition of phytoplankton can lead to oxygen depletion in the hypolimnion, remobilization of nutrients and metals such as iron and manganese, and generation of gases from bottom sediments, including ammonia, which is toxic to fish.

Kira (1993) summarized the most widespread environmental disruptions in the world's lakes: (1) lowering the lake level due to overuse of water, (2) rapid siltation caused by accelerated soil erosion in catchment areas, (3) acidification of lake water due to acid precipitation, (4) contamination with toxic chemicals, (5) eutrophication, and (6) disintegration of aquatic ecosystems and resultant loss of diversity of lacustrine biota. It is generally recognized that large lakes such as the Caspian Sea, the Aral Sea, the African Great Lakes, and the North American Great Lakes are particularly sensitive to human activities.

Accelerated eutrophication has proved to be less of a problem in many rivers since natural processes result in the downstream accumulation of organic matter within the system (Arthington and Welcomme, in press). Furthermore, the floodplain acts as a sink for carbon and nutrients, and, provided it is still connected to the river, plays a considerable role in the dynamics of nutrient flux. Nevertheless, the tolerance of river systems can be exceeded, especially in the tropics, where excessive nutrient loading can cause severe anoxia in floodplain lagoons and even in static waters in the main channel. This can cause local fish kills, especially of migratory species with high dissolved-oxygen requirements.

Generally, the first severe consequences of pollution in rivers and lakes are the elimination of the most sensitive species, and as the loading increases, it leads to the formation of large sections of rivers or parts or layers in lakes that do not support fish. However, the sublethal and chronic effects of pollutants on fish should be considered (Müller and Lloyd, 1994), since even low pollution levels may influence fish production in numerous ways: reduced number of offspring and lowered growth rates; increased susceptibility to diseases; gradual decline, or changes in species composition of fish communities; and, in some cases, reduction of stocks by mass mortalities. It should be noted that deterioration of quality of water supplies is becoming a major constraint to freshwater aquaculture development.

Environmental impacts of inland fisheries and aquaculture

Fishery resources are not inexhaustible. Overexploitation of fish populations can result from excessive levels of fishing effort which may be due to a high number of small-scale fishermen, or intensive fishing by large vessels equipped with modern gear (Scudder and Conelly, 1985). Potential consequences of overfishing may include changes in the abundance, dominance, and reproduction patterns of fish species. Predator-prey relationships and species composition within fish communities may be altered with consequent changes of food web structures. Increasing seasonal and long-term variability of fish-stock sizes may well be due to overfishing possibly enhancing natural fluctuations.

The use of explosives and poisons as a method of fishing, although not widespread, is of particular concern since it kills eggs, larvae, and adult stages of targeted or nontargeted species, and is especially devastating when applied in more confined circumstances of small water bodies.

Nevertheless, inland fisheries, by nature, do not tend to have a high impact on the environment as a whole. In most inland fisheries, fishing effort is characterized by individuals or groups of individuals who rarely use mechanized or industrial gears. Very often, these small-scale or artisanal fishermen, have seasonal alternative employment, which enables them to respond to changes in the population of their "prey" organisms. When the target fish populations begin to decrease, as a result of fishing predation or a natural decline, the fishing activity will be reduced in response to the necessity of increasing effort for smaller returns.

There is little evidence of gross and permanent change in fish populations due to removal of target species. Catch records indicate that the effects of fishing simply distort the biological equilibrium and recovery takes place when fishing pressure ceases. Permanent, irreversible changes in inland fishery ecosystems result from chemical and physical changes in the environment and, normally, not from fishery-induced perturbations in fish populations. There are, however, worrying signs that the impact of overfishing is becoming severe in some areas of the world, such as China and Bangladesh and in some rivers of Africa, where levels of exploitation have apparently exceeded the resilience of the fish population.

There are as yet relatively few environmental problems resulting from most tropical inland aquaculture practices (Beveridge and Phillips, 1993; Pullin et al., 1993; FAO/NACA, in press), although trends to intensify aquaculture production may lead to increased environmental hazards. Evidence from temperate regions, however, has shown that there is a range of potential impacts of intensive (feed-lot) aquaculture farms (Alabaster, 1982; NCC, 1990; Martinez-Espinosa and Barg, 1993; Pullin, 1993). The freshwater environment may be affected by the release from fish farms of uneaten food; feces and dissolved excretory products; high microbial loads, parasites, disease organisms, and vectors; and aquaculture chemicals. The chemistry of the recipient water bodies and their bottom sediments can be changed by an increase in suspended solids, biochemical and chemical oxygen demands, and increased nutrient loading, particularly nitrogen and phosphorus. Thus quantitative and qualitative changes in the biota (bacteria, protozoa, plankton, benthos, and fish) of recipient waterbodies are likely.

Nutrient and organic enrichment may lead to local eutrophication and hypoxia or anoxia (Beveridge, 1984), although the fertilization of some oligotrophic waters may increase fish production. Aquaculture chemicals and their residues may cause sublethal and lethal effects on wild aquatic organisms, depending on their potential for bioaccumulation, toxicity, and characteristics of physicochemical persistence. Excessive use of drugs may generate drug-resistant pathogens. Such impacts on rivers and freshwater lakes also depend on the residence time and amount of water flowing through the aquaculture installations and on the kinetics of the recipient waters. Here, for instance, seasonal changes in river flow, lake flushing, and water space around cages may well influence the dilution and distribution of contaminated waters.

Inland fisheries can be affected by intensive freshwater aquaculture. Abstraction of surface and groundwater, diversion of water courses, dam and pond construction, and setting up of fish pens and cages in open-water areas can have implications for aquatic wildlife. Wild fish populations may be threatened by diseases and parasites emanating from high levels of infection in aquaculture installations. The feeding and breeding habitats of many species could be disturbed. With aquatic life cycles being disrupted and recruitment reduced, overall productivity may be lowered (Dunn, 1989).

Aquaculture is a net consumer of water (Phillips et al., 1991; Beveridge and Phillips, 1993; Beveridge et al., 1994; Yoo and Boyd, 1994). However, there is a wide range of aquaculture systems, for example, ponds, tanks, raceways, all with varying water demands. Rainfed and low-flow, extensive, or semi-intensive aquaculture types, defined here as systems in which water is added simply to counteract evaporative and seepage losses, predominate in the tropics. Evaporative and seepage losses in pond culture systems can be considerable in more-arid zones. Intensification of production requires increased water use in order to maintain adequate water quality and for disposal of wastes. Intensive, high-flow freshwater systems do occur in the tropics, but are rare.

Introduction of exotic species and genetically modified organisms

Fish species with good characteristics for aquaculture and stocking and good marketability have been and will continue to be introduced to new habitats. However, introduced exotic species and genetically modified organisms may alter and impoverish local aquatic biodiversity and genetic resources, as they may affect endemic species via competition, predation, destruction of habitats, transmission of parasites and diseases, and interspecific breeding (Welcomme, 1988). Tropical waters support a number of exotic plants, many originating from South America. The free-floating water hyacinth (*Eichhornia crassipes*) and salvinia (*Salvinia molesta*) create serious socioeconomic problems by impeding water flows and water transport, reducing water quality, blocking waterways, and providing shelter and habitat for insect vectors of diseases (schistosomiasis and malaria).

Responses of fish communities to stress

Fish communities subjected to increasing stress tend to behave in a characteristic manner (Rapport et al., 1985). Initially, assemblages consist of the full range of species and sizes characteristic of the undisturbed ecosystem. As stress is applied, the large, long-lived individuals and species tend to disappear, and as stress increases, a progressive replacement of the largest remaining elements by small, short-lived species takes place. The evolution of the composition of the assemblage is similar whether the stress is environmental or fishery-based (Welcomme, in press). The two types of stress may be additive or even synergistic, and it is thus difficult to determine to what degree a community response can be traced to excessive fishing or environmental degradation. This makes it extremely difficult to assign the responsibility for change in systems suffering from more than one type of stress. Furthermore, natural fluctuations in the environment, such as the Sahelian drought of the 1980s, may further stress aquatic ecosystems, exaggerating the effects of mismanagement.

Regional Issues of Environmental Degradation and Fisheries

It is generally recognized that environmental stress on living aquatic resources and related constraints to inland fish production are increasing worldwide. However, it is important to note that there are local and regional differences in the patterns of multiple water uses and their implications for fisheries. Such differences exist, for example, between developing and industrialized countries and between freshwater environments and river-drainage basins exploited in the various continents. In addition to environmental considerations, management responses to multiple water uses need to take account of a variety of factors, including development constraints, social and economic conditions and needs, and population growth. A general overview by continent follows, emphasizing specific fisheries issues in relation to the status of rivers, lakes, and reservoirs. Issues relevant to aquaculture are summarized for Africa, Asia, South and North America, Australia, and Europe.

Africa

Most African rivers in general remain among the least modified of any region of the world. Most rivers still have their floodplains, but some of these have been adversely affected by continued drought conditions since the 1970s. Recent droughts have resulted in decline in fish production in many water bodies, including large reservoirs (Kariba, Volta) and numerous small dams. The impact of major dams, built on the Niger, Zambezi, Volta, and Nile systems, has been relatively slight (Arthington and Welcomme, in press). Although modifying the original riverine fish community, artificial lakes have developed significant fisheries [see also Bernacsek (1984)]. African reservoir fisheries are estimated to contribute 150,000 t or about 10 percent of the overall inland fishery yield of the continent (Kapetsky and Petr, 1984; Knaap, 1994; Petr, 1994). Fishery resources in small water bodies, especially reservoirs built for water-supply purposes, seem generally to be underexploited, and the possibilities for enhancing fisheries are much greater in these small water bodies than in larger ones (Giasson and Gaudet, 1989; Marshall and Maes, 1995).

The natural lakes of eastern Africa, which are among the largest lakes in the world, support very important fisheries (Greboval et al., 1994). Many lakes have naturally very clear water accompanied by high fishery yields based on a deep euphotic zone of primary production. Serious decline in their fisheries could be triggered by increased turbidity resulting from external impacts, in particular from loadings with suspended matter. Contaminants can accumulate in deeper waters because of the stratification of lake waters through the formation of thermoclines. Accumulation of organic matter can lead to oxygen depletion in deeper waters, and fish kills can occur as a result of periodic mixing processes. Lake Victoria suffers from symptoms of eutrophication which include deoxygenation of bottom waters, blooms of blue-green algae, fish kills, and the rapidly spreading problem of water hyacinth, *Eichhornia crassipes*. The eutrophication of lakes and reservoirs seems likely to become an increasingly serious problem in many other areas of the continent (Marshall, 1994). The problem is mostly caused by the discharge of sewage, or sewage effluents, into the streams that flow into lakes and reservoirs (Calamari et al., 1994).

The growth of urban populations and extensive, unimproved farming practices mean that sedimentation and pollution are becoming increasingly important constraints to fish production. Contamination of inland waters is still relatively low (Calamari and Naeve, 1994), with the exception of some hot spots, where urban development, industrial-waste discharge, and use of certain pesticides are the main sources of pollution (West and Biney, 1991). The CIFA Working Party on Pollution and Fisheries, on the basis of its own work and the work of others, recognized pollution by organic matter, causing eutrophication and anoxia as a major threat to fisheries (Alabaster, 1981; Symoens et al., 1981; Calamari, 1985; Biney et al., 1987, 1994; Dejoux, 1988; Saad et al., 1990; Osibanjo et al., 1994).

Generally, urbanization, erratic rainfall over the drier parts of the continent, increasing catchment-area degradation, and excessive fishing effort in some inland waters may limit significant fisheries development potentials.

Asia

Tropical and subtropical Asia are naturally dominated by large river basins, and lakes are relatively small. Asian rivers are heavily fished with intensive management through mixed natural and aquaculture techniques involving stocking, cage culture, and extraction of juvenile fish for rearing in ponds. However, overfishing in some areas is affecting migratory species as well as reservoir fish stocks (Bhukaswan, 1985; De Silva, 1985; Petr, 1995).

Efforts in the Indian subcontinent, Southeast Asia, and China have been directed to controlling flooding and harnessing river flows to provide for irrigated agriculture, hydroelectric-power development, urban water supply, and navigation (Arthington and Welcomme, in press). Plans exist for further containment of river systems through dams on the Irrawaddy, Mekong, and Yangtze Rivers, and the Mekong in particular is destined to become a series of impounded reaches, especially in the middle course tributaries.

In arid countries of Asia, up to 95 percent of the total river discharge may be taken away for irrigation purposes, leaving some river stretches dry, terminal lakes drying out, and with increasing salinities. Concentrations of agrochemicals in water bodies receiving drainage water and elevated salinities cause the disappearance of some fish species and lead to accumulation of toxic elements in other fish. As a consequence, the fish stocks become degraded in quantity and quality (Petr and Mitrofanov, 1995). The impact of irreversible water uptake in arid countries of the world needs urgently to be evaluated and strategy developed for implementing the best remedial measures.

The extensive development of large reservoirs has contributed to important fisheries managed generally through stocking (Baluyut, 1983; Petr, 1994). In India, stocking reservoirs with three major Indian carps (*Labeo rohita, Catla catla,* and *Cirrhinus mrigala*) is a widely used management method, which ensures that all the niches in the reservoir are utilized with good economic results (Jhingran, 1992). In China, high yields of fish production are achieved in reservoirs, based on a wide range of species and methods which include "stock and take" fishing; "cove and cage" culture; integrated fish farming of fish, livestock, poultry, and agriculture in small reservoirs; and application of inorganic fertilizers to enhance fish yields in small reservoirs (Lu, 1986, 1992).

General pollution problems in Southeast Asia include (Alabaster, 1986) (1) absence of sewerage, leading to deoxygenation of rivers and fish kills; (2) suspended solids from land erosion and mining tailings; (3) industrial wastes with significant concentrations of heavy metals; and (4) use of pesticides. Agriculture is the major water user in the region, and the demand for regulating water deliveries for irrigation has had a major impact on water resources (IPFC, 1994). Other activities, such as deforestation and mining, have contributed to the gradual degradation of water resources, as has urban pollution.

More specifically, the densely populated river basins of the Ganga and Mekong, and several Chinese rivers, have been found to be subject to heavy pressure due to aquatic pollution, resulting in significantly reduced yields in certain fisheries (Lu, 1994). Dudgeon (1992) identified three major threats to the conservation of river ecosystems in Asia: (1) degradation of drainage basins (particularly through deforestation and overgrazing), leading to increased suspended sediment loads; (2) river regulation and control; and (3) river pollution due to untreated sewage from densely populated areas, industrial effluents, and mining wastes. For example, chloride levels in the Ganga (Ganges) River have increased two to three times over three decades, usually an indication of increased domestic pollution (Natarajan, 1989). The combined effect of abstracting large volumes of water for irrigation and other purposes, and the need to dilute effluent is likely to be an increasing problem.

The major impact on natural tropical lakes is that of water withdrawal in its catchment (Petr, in press). This has led to a gradually shrinking water surface and decreased depth of some lakes, with several accompanying negative features: siltation, overgrowing with aquatic plants, penetration of settlements and agriculture into lake margins, and increase in water temperature and turbidity during critically low water levels. While increasing fish production in eutrophic, highly productive shallow lakes may be limited, there is potential for increasing fish production in deeper, less-productive lakes. This could be achieved, for example, through cage culture.

Indian floodplain lakes constitute an important fisheries resource and as an integral component of the Ganga and Brahmaputra basins, these versatile lakes still offer promising expansion possibilities to integrate both capture and culture fisheries if habitat deterioration due to siltation, eutrophication, and industrial pollution could be reduced (Natarajan, 1989; IPFC, 1994). Urban development in Papua

New Guinea, however, affected the sediments of a floodplain lake (Polunin et al., 1988).

Laguna de Bay in the Philippines supports a significant fishery (and cage culture) which has possibly to some extent been enhanced by massive inputs of nutrients. However, the situation is further aggravated by inputs of industrial effluents from numerous factories situated on this lake (Tamayo-Zafarwalla, 1983). In Malaysia, fish caught in rice fields for subsistence and export purposes are exposed to excessive pesticide application, and there is a potential hazard of the pesticide bioaccumulating at the end of the food chain in humans (Yap, 1992).

Both the Caspian and Aral Seas have shown marked environmental deterioration over the past three decades as a consequence mainly of industrial and agricultural intensification, and water abstraction for irrigation. Area and depth of the Aral Sea have diminished drastically, with a consequent increase in salinity which none of the indigenous fish could tolerate. Desiccation of the Aral Sea is considered to be a major management disaster for the countries of the former Soviet Union (Micklin, 1988). In the 1960s and 1970s significant fisheries declined dramatically in the Caspian Sea, affected by a prolonged period of decline in water level until 1977, and leading to desiccation of large areas of shallows in the Volga delta, with its richest breeding, nursery, and feeding ground. The subsequent rise in the water level of the Caspian Sea, still in progress, on the other hand, has led to the rehabilitation of this situation.

Aquaculture operations are increasingly subject to problems of water availability and consequences of water pollution and habitat destruction (FAO/NACA, in press). However, cage-culture systems appear to be more vulnerable to environmental changes than pond-culture systems, which are usually more isolated from surrounding aquatic ecosystems. Still, major effects on aquaculture can include lack of water for aquaculture development, reduction in dry-season water supply, reduced quality of water supplies, siltation and increase in suspended solids in farming units, aquatic weeds infestation, contamination of produce with toxic chemicals, and fish kills. Effects on wild stocks may also affect aquaculture with reduction in seed supply, loss of genetic resources, and potential species for aquaculture.

In general, regulation of water resources is progressing with increasing needs for water supply and hydropower and requirements for water-control structures for irrigation and intensification of agriculture. The impoundment of rivers and floodplains has involved both loss and gain in fisheries. Catchment-area degradation, resulting in high silt loads, and industrial and domestic pollution at present appear secondary or localized problems.

Latin America

South America contains two of the world's largest river basins, the Amazon and the La Plata, as well as two others of considerable significance, the Magdalena and the Orinoco. Riverine fish resources are still little exploited (Welcomme, 1990). Although the Magdalena River has been fished intensively, recently some intensification of the Amazonian and Orinoco fisheries has occurred, with signs of local overfishing around major urban centers. River fisheries are usually very selective; however, some of the major targeted fish stocks in the Amazon appear to be declining (Bayley and Petrere, 1989).

Water quality in the Amazon and the Orinoco is generally unaltered, however, with some localized pollution around main urban centers. Both rivers lack mainstream dams, but barrages have been placed across some of the tributaries. In the La Plata system, the Paraguay River still retains extensive pristine floodplains and wetlands, whereas the Uruguay and Parana Rivers have been modified (Quiros, 1993), in some parts with cascades of reservoirs. For example, the construction of 45 reservoirs have transformed the main tributaries of the Parana, Grande, Paranaiba, Tiete, Paranapanema, and Iguacu into a succession of lakes in which only 483 km of the original 809 km of the river is flowing, and the construction of the Porto Primavera reservoir is expected to reduce the proportion flowing to less than 50 percent (Agostinho et al., 1994).

Reservoir fisheries are well developed, particularly in Cuba, northeastern Brazil, and Mexico, with systematic stocking and improved fishing techniques, applied in many large and small reservoirs. The potential for development and intensification of fisheries in reservoirs may be considerable (Petr, 1994; Quiros, 1994).

Wastes generated by mining have resulted in the bioaccumulation of copper, zinc, and cadmium in fish of Lake Poopo in Bolivia (Beveridge, 1983). A major concern is contamination with mercury resulting from extraction of metals, especially gold. There is a risk that mercury, accumulating in migratory fish species, is being distributed throughout large tracts of some river basins (Lacerda and Salomons, 1991).

The high content of deoxygenating organic substances in domestic and industrial effluents has caused periodic events of anoxia or even permanent anoxic conditions in some parts of the rivers Cauca, Tuy, Medellín, and Bogota (Escobar-Ramirez and Barg, 1990). An investigation in Argentina showed that fish catches in lakes open to human influences initially might rise possibly as a result of enrichment with untreated sewage, but that in the medium to long term, catches declined significantly as pollution overwhelms the lakes (Quiros, 1993).

A recent survey has shown that agriculture is seen as a source of potential pollution of water bodies resulting from input of silt, fertilizers, and pesticides (Chanduvi and Prieto-Celi, 1993). However, fertilizer consumption within Latin America is relatively stagnant and is not expected to contribute as a major source of water pollution (Joly, 1993). Local effects of high levels of pesticides, or oil spills and emissions of oil residues into water bodies, have been reported (Escobar-Ramirez and Barg, 1990).

Damming and pollution of rivers have compound effects, and three examples are given here. A dam has been placed across a tributary of the Magdalena, the river Cauca, and its downstream floodplain is being affected by land reclamation and pollution to a point where fish stocks have been diminished (Valderrama and Zarate, 1989). The river Tiete, after receiving the untreated sewage of São Paolo (14 million inhabitants), passes through a cascade of six dams. While the first one can be anoxic, there is a general amelioration down the chain as suspended solids are progressively deposited and water is forcibly aerated over spillways (Petrere, 1994). The mineralization of organic matter results in artificial enrichment of the water with increased fish production. Fish diversity and stock density increase down the chain, progressively supporting intensive sport and artisanal fishing. The last reservoir of the chain is reported to contain a fauna representative of natural conditions. Quiros (1993) found that increased reservoir storage capacity may have had a net positive effect on fish production, while industrial pollution has had a negative effect.

Generally, fisheries in Latin America are influenced mainly by impoundments, and locally by increasing pollution resulting from urban sewage, mining, and industrial activities.

Australia, Europe, and North America

In Australia, the hydrologic consequences of river regulation and abstraction on the river Murray-Darling system are pronounced (Arthington and Welcomme, in press). High sediment and nutrient loads, sewage, organochlorine pesticide, and industrial effluent contamination are causing significant water-quality problems. River Murray fishes and commercial fishing have been affected, and large fish kills have occurred.

European rivers are extensively polluted by industrial wastewater and enriched to the point of eutrophication by domestic and rural effluents. Water quality in eastern European rivers such as the river Vistula in Poland is extremely poor (Backiel and Penczak, 1989; Kajak, 1992; Dill, 1990, 1993). Channelization and bank reconstruction of the Rhine and most other rivers have resulted in the decline in abundance or disappearance of fish species. Only the Danube and the Rhone and a few other rivers retain traces of their former floodplains. The large rivers of the Russian Federation are impounded, and the Volga has been transformed in a cascade of large reservoirs (Pavlov and Vilenkin, 1989).

Pollution in North America is not as generalized or severe as in other regions, although severe water-quality problems persist around the most heavily industrialized areas, in spite of massive expenditure since 1970 on sewage treatment. It is being recognized that heavily polluted systems have a very long response time to improved conditions. For example, the concentrations of hydrophobic organic chemicals [polychlorinated biphenyls (PCBs), polycyclic aromatic hydrocarbons (PAHs), dichlorodiphenyltrichloroethylene (DDT), etc.] in long-lived fish in the North American Great Lakes peaked in the 1970s, underwent a rapid decrease in concentration over the next decade, and now have leveled off at values that still exceed guidelines for consumption (Eisenreich, 1993).

North American rivers have been extensively modified by straightening, channelization, and bank protection as well as by the construction of navigation locks and impoundments (Arthington and Welcomme, in press). The Mississippi system has been reduced to a cascade of locks, and the large west- and north-flowing rivers in the United States and Canada are barred by dams for power generation [see various references cited in Dodge (1989)]. River degradation has had severe effects on fish yields.

Option for Conservation and Enhancement of Fisheries

Preventing and reversing environmental degradation of inland waters will help sustain existing fisheries. In addition, there are opportunities to further increase fisheries yields through enhancement measures.

Future fisheries development

The opportunities for future fisheries development will be determined by the perceived need, capability, and political will to implement a set of actions which may be guided by following general approaches: protection, rehabilitation, mitigation, and intensification (Welcomme, in press).

Protection. Conservationists usually advocate the protection of aquatic environments, in particular those which have not yet been strongly affected by development. There are conservation measures which are technically feasible, and which may aim at the conservation of whole river basins and the restoration of riparian vegetation, trophic state of lakes, physical characteristics and flow regimes of rivers, and some degree of ecological integrity of these aquatic systems. However, environmental protection as a management option needs careful evaluation in view of national economic development needs as well as of opportunity costs for local communities.

Where it is determined that there is a net benefit in conserving aquatic environments and resources, protection efforts should discourage physical modifications such as the building of dams, leveling of river channels, and the reveting of lacustrine shorelines. Access to water and fishery resources should be limited, and levels of exploitation of the fishery, extraction of the water, and loading with nutrients and pollutants should be strictly controlled, and other related measures should be enforced, in particular to prevent long-term and irreversible effects. Where basin-wide conservation measures are difficult to implement, increasing emphasis is given to protection and conservation of riparian vegetation and to exclusion of further developments within specified riparian zones of rivers or lakes (Arthington and Welcomme, in press).

However, this approach may require considerable research to define the costs of full or partial conservation measures or to determine permissible levels of exploitation. It also requires a greater integration of fisheries management into more general frameworks of development planning and management at the rural, national, and basin levels. Pressures for the intensification of land and water use arising from human population growth will ensure that many of the existing areas, particularly in the tropics, such as wetlands and floodplains, will either be reclaimed through land-drainage schemes or submerged under reservoirs.

Rehabilitation. The restoration of already modified ecosystems or stocks of fish is advocated where the pressures producing the modification have eased or where new technology can be introduced to reduce stresses (Welcomme, in press; Regier et al., 1989). In the temperate zone, there have been successful attempts to rehabilitate polluted rivers and lakes by the installation of sewage and industrial effluent treatment plants (Arthington and Welcomme, in press). Other interventions involve the manipulation of lake biota and habitats specifically to reduce algal biomass (Gophen, 1990; Lammens et al., 1990; Shapiro, 1990; Starling and Rocha, 1990; Harper, 1992).

Concepts for the physical rehabilitation of water courses modified in the nineteenth century for navigation and flood control appear to be popular in Europe, North America, and Australia, and attempts are increasingly being made to rehabilitate smaller systems and to restore some of the large rivers such as the Mississippi, the Murray-Darling river system in Australia, and major European rivers (Brookes, 1988; Boon et al., 1992). Rehabilitation measures may include interventions such as modification of the morphology of rivers, riparian retirement, setting back the levees from the main channel, removal of revetments and wing dikes from the banks, restoring portions of the floodplains by local piercing of the dikes, and other measures aiming principally to reestablish the river's historical flow characteristics and its connectivity to the floodplain. The EIFAC Working Party on the Effects of Physical Modifications of the Aquatic Habitat on Fish Populations has produced guidelines for the improvement of riverine habitats for European inland fisheries (EIFAC, in press). Rehabilitation measures usually require a single major expenditure, and there may be considerable social and economic implications of reconversion of large tracts of land with the transfer of population and changes in property rights that these imply.

Mitigation. Because many of the changes to inland aquatic systems are irreversible and because of the greater value attached to alternative uses, by far the most common tendency today is to attempt to live with modified systems through a series of interventions and management strategies designed to lessen the impact of stress. Mitigation usually involves recurring expenditure, which should be considered part of the cost of the major modifying use of the system (Welcomme, in press). Most mitigation measures attempt to retain something of the original diversity of the ecosystems. Actions can range from installation of fish passes in dams, the creation of artificial habitat, including zones of spawning substrate for key fish species (e.g., salmonids), arranging for flood-simulating water releases, or systematic stocking to maintain populations of fish which have no alternative source of recruitment.

Intensification. Intensification strategies imply the modification of the aquatic resource aiming specifically at the enhancement of fisheries. A range of tools may be deployed to this end, including introduction and systematic stocking with desired species, elimination of unwanted species, fertilization of the water body, and physical modification of the system to facilitate exploitation. In the following paragraphs the term *enhancement* is used in the sense of interventions on the biotic or abiotic component of the fishery ecosystem in order to increase biological production, total yield, or value beyond the levels corresponding to existing environmental conditions. Enhancement may imply a process involving progression from simple capture fisheries through supplementary stocking, maintenance stocking, nutrient enrichment, and mixed culture-capture systems to intensive aquaculture.

Fisheries enhancement by modifying the aquatic environment

The environmental conditions in any body of water may be found to be below the optimum for achieving maximum levels of primary production and the production of other organisms at other levels of the food chain. The controlled input of nutrients into freshwater, particularly reservoirs and lakes, up to a certain limit, can enhance biological production and fishery yields (Ryding and Rast, 1989). However, in some cases nutrient inputs may exceed carrying capacities and serious eutrophication problems may result. Creation of turbulence as an aid to maintaining sound levels of dissolved oxygen, such as applied in some reservoirs and lakes (Fast et al., 1982) and in waters with high organic loading, will assist in maintaining optimum conditions and prevent the loss of fish through oxygen starvation.

Water regimes may be controlled to the benefit of fisheries. It can be generally assumed that the shallower water body with more frequent mixing will have higher levels of production. Bays and shallows will also have higher productivities than will the open water mass of large reservoirs and lakes. Water level control can be a minimal intervention by the retention of wet-season (flood) waters into the dry season to allow for a longer period of fish growth followed by planned harvesting. Where it is feasible to control the flow of water into and out of a basin, it is possible to allow the water level to rise or fall at a rate which maximizes the growth of rooted plants or provides optimum conditions for the reproduction of fish or other organisms. For example, in irrigation networks fish production can be optimized (Redding and Midlen, 1991) and, especially in the tropics, environmental health problems caused by waterborne vectors of diseases minimized by simple alterations in water regimes. As mentioned above, fish populations are very dependent on seasonal flow patterns. Where the other uses of the impounded waters allow, it should be possible to initiate water-release programs intended to enhance fish stocks downstream, thus restoring a part of the natural productivity that existed before the impoundment of the river.

Modifying the physical features of impoundments, such as "sculpting" or landscaping reservoirs or canals, can create highly productive areas of shallow water. Where poor-quality water is discharged from reservoirs, the provision of barriers to create pools and turbulence will increase the carrying capacity of the water and enhance fish production.

Installation of fish passes or lifts at barrages and dams will improve capabilities of fish passing to or from their feeding or breeding grounds (Pavlov, 1989; Clay, 1995). A new type of fishways designed in Australia (Mallen-Cooper, 1994) for migratory non-salmonid fish has been successfully tested in pilot experiments in Bangladesh (Bernacsek, personal communication). The results from Bangladesh show that this type of fishpass may represent a solution to this problem for tropical and subtropical countries.

Fisheries may be enhanced by facilitating access to fishable stocks and by making fishing conditions easier. The density of fish may be increased by fish-aggregating devices (FADs), such as the brush parks (acadjas) commonly used in western Africa or Southeast Asia (Welcomme and Kapetsky, 1981), consisting of cut branches, bamboo, etc. set in shallow waters to act as shade, shelter, and feeding areas for many fish species which are caught when concentrating around these devices. In artificial lakes and reservoirs there is often much scope for preparing the area before flooding to ease fishing activities. For example, access to fish stocks can be enhanced by providing areas free of drowned brush or forest. Conversely, areas deliberately left as difficult to fish by leaving obstructions may provide "refuges" for the fish population and prevent overexploitation.

Fisheries enhancement by introductions and stocking

Biological manipulations of aquatic ecosystems involving introductions of species or strains and stocking of artificially raised juvenile fish into inland waters are particularly promising. *Introductions* refer to the introduction of a species into a location where it has never existed or a reintroduction into a location from which it has disappeared. *Stocking* refers to the continuous, seasonal addition of young fish to a water body to replace or supplement the natural breeding capacity of the stocks.

Concerns and recommendations related to the conservation and utilization of aquatic genetic resources have been formulated at various expert consultations dealing with genetic resources of fish, stock enhancement, and aquaculture genetics (FAO/UNEP, 1981; EIFAC, 1982; Chevassus and Coche, 1986; FAO, 1993b). Efforts by international expert groups led to the formulation of related codes of practice and protocols (EIFAC, 1984; FAO, 1986a; Turner, 1988). FAO is currently preparing an International Code of Conduct for Responsible Fisheries which contains provisions aiming at the conservation of biological diversity of aquatic ecosystems (FAO, 1994b). Many countries have enacted legislation to regulate and control the movement of eggs, larvae and juveniles, and adult stages of exotic fish species, often combined with compulsory certification of stocks to be free of certain diseases and banning of all movements of diseased stocks (Van Houtte et al., 1989).

Introductions are effectively "one-off" activities which are usually irreversible. Once the introduced species is successfully established, it will usually require no further intervention to maintain the stocks. If well planned and managed, introductions can become a very cost-effective way of providing significant fishery benefits. Introductions have been undertaken for a number of reasons, including the development of large-scale subsistence or commercial fisheries, particularly in developing countries, for sport and recreation, aquaculture, and disease and/or pest control.

Introductions make most sense where a "vacant" or "underutilized" ecological niche occurs. Although the concept of a vacant niche is controversial, from a fisheries point of view such opportunities clearly exist. Occupying an unoccupied trophic niche or replacing a species considered of lesser value (economically or as a food) can make a great contribution to an existing fishery or create one where it did not exist before. It should be noted that pollution and habitat degradation have resulted in shifts in natural fish populations, thereby creating such vacancies. Species that are able to withstand these changes in habitat and water quality may be introduced for the improvement of fisheries production.

There are many examples for successful introductions (Welcomme, 1988), and in terms of food production, many of these have been spectacular. The introduction of tilapias into reservoirs and other small water bodies in Sri Lanka resulted in the development of a very productive fishery with a total yield of approximately 27,000 to 38,000 t/yr (Amarasinghe, 1992). The Amu Darya River fishery, especially in the created reservoirs and large irrigation canals, has been enhanced significantly through the introduction of Chinese carps (Welcomme, 1985). Small pelagic zooplankton-eating fish introduced into a number of African lakes and reservoirs support important artisanal fisheries. Considerable increase in fish catch has been achieved through the introduction of the Nile perch (*Lates niloticus*) into Lake Victoria and the freshwater herring *Limnothrissa miodon* into the Cahora Bassa reservoir and Lakes Kivu and Kariba (Marshall and Mubamba, 1993; Petr, 1994). In the Kariba reservoir on the Zambezi River the kapenta fishery now produces about 20,000 t annually (Lupikisha, 1993). In Latin America, introduced tilapias have become the dominant fish in reservoirs of Cuba, Mexico, and Brazil (Juarez-Palacios and Olmos-Tomassini, 1992). Where there is a high production of fish with a relatively low value, the introduction of a high-value predator may convert the low-value fish into a high-value resource, as experienced with the introduction of the Nile perch into Lake Victoria (Reynolds and Greboval, 1988), or the reintroduction of salmon into the Laurentian Great Lakes. Fish have also been introduced to control unwanted organisms (Welcomme, 1988), and may play a significant role in combating the spread of aquatic weeds as well as vectors and waterborne diseases (Petr, 1987, 1993). Such measures not only rehabilitate lacustrine environments but also protect human health and enhance fisheries in rural areas.

Stocking of fish into natural temporary or permanent small water bodies and the large artificial impoundments will probably provide the greatest future potential for enhancing fishery yields on a global scale. In contrast to fish introductions, stocking is a continual or repetitive process of releasing seed fish which are obtained through breeding and raising of juveniles in hatcheries. Fisheries relying on stocking practices are commonly termed *culture-enhanced fisheries,* since the fingerlings to be stocked are usually produced through aquaculture techniques, whereas the adults are the target of capture fisheries. Stocking can be a beneficial intervention in any body of water that is known to have a high production potential but low breeding and recruitment potential for the fish which are considered a valued yield.

There are several advantages of improving production through culture-enhanced fisheries, including

Existing water resources are used.

Low-resource-input systems are involved.

Using species low in the food chain can help maximize the biological efficiency of production.

Increases in production do not require any major technological changes.

Management requirements are well known and not technically complex.

There are no pollution risks.

Exploitation of the resource as a capture fishery results in increased participation of people in rural communities.

Beneficiaries are often low-income, resource-poor communities.

Probably the most common reason for stocking a water body is the lack of natural breeding areas for the fish community because of human modification of the environment (e.g., blocking access to spawning grounds due to construction of weirs or pollution of access waters, and removal of ecological conditions required in breeding areas). This is often the case with stocking activities in China and India. For example, a stocking program has been initiated in India to restore the once important hilsa (*Hilsa ilisha*) fishery upstream of the Farakka barrage on the Ganga River (Jhingran, 1992). The fecundity of breeding adults has a finite limit, and with too few adult broodstock it may be impossible to produce enough young to ensure adequate recruitment to the adult stock. Such a shortage of breeding adults, through either overfishing or environmental disturbances, is also another common reason for stocking programs. An example of the potential for enhancing temporary waters, or perennial waters that undergo a regular seasonal cycle of flood and drawdown, is the program for stocking and physically managing oxbow lakes ("beels") of the Brahmaputra floodplain system in Assam, India (Jhingran, 1992). Here significant increases in production are achieved by the construction of hatcheries, undertaking water-control works, and considering the social and economic needs of the communities which exploit the stocks.

Significant increases in catches from inland waters can be expected by applying and improving culture-based enhancement techniques. These stocking activities offer particular promise for smaller water bodies and reservoirs and are already contributing to a major proportion of the catch from inland waters in many regions, particularly in Asia (Lu, 1986, 1992; Moreau and De Silva, 1991; Petr, 1994). It is estimated that in sub-Saharan Africa, the potential fisheries yield from small water bodies is in the order of 1 million metric tons per year (Townsley, 1992).

From a fisheries point of view, it can be expected that new reservoirs will present additional opportunities for increased production. For example, in tropical and subtropical Asia, the total area covered by the larger reservoirs is in the order of 86,000 km^2, with this total expected to rise to some 200,000 km^2 by the year 2000 (Costa-Pierce and Soemarwoto, 1987). Reservoir surface area in Asia will thus exceed the surface of natural inland waters (lakes, marshes, floodplain lakes, and freshwater swamps) which cover approximately 185,000 km^2.

In the long term, progress in the ability to economically mass-produce sterile fish (e.g., through triploidy) will increase the feasibility of stocking of such fish in circumstances where uncontrolled reproduction in the wild is a problem. Further, the development of techniques to mark mass-produced fish, such as using genetic techniques ("genetic tags"), will encourage private-sector involvement in stocking activities, since it will be possible to discriminate hatchery contributions to the stock and hence ownership or rights to the final product.

Enhancement by management of the fishery

Last but not least, conventional fisheries management intervention, whether traditional or based on analysis of scientific information, can be considered as enhancement techniques. Ideally, fishery management intervention should be based on information about the status and dynamics of fish populations. The fish stocks should be categorized by the location of the stocks at the different seasons of the year, the total biomass, the size and age groups of the individual fish, growth rates, reproductive capacity, ecological requirements for breeding, and nursery functions. In practice, however, only part of this information will be available.

Several management interventions may be undertaken to maintain or enhance fishery yields. Regulation of fishing effort is generally aimed at restricting the entry of individuals or fishing units into a fishery. Use of fishing gear and equipment may be controlled, for example, by regulating size, length, and mesh size of nets, or the power of fishing boats. Fishing activity may be restricted to certain periods or areas. Any fishery management intervention, however, will have to take due account of social and economic circumstances prevailing in fishing communities. In practice it is for the short term and immediate future that regulations will have to be implemented and accepted by fishermen, particularly if they are required to restrain their effort, reduce their catches, and accept less income and/or food.

Prospects for Fisheries

In many countries, environmental impacts arising from other sectors are the major constraint to sustaining, or increasing, inland fish production. It can be expected that degradation of aquatic environments will—to a varying degree—continue in some parts of the world, despite the growing awareness of the need for conservation of water resources. Inland fisheries cannot be seen as an economic sector in isolation, and the competition for resources with other sectors is likely to increase in view of population growth and the demand for food and energy.

Too often it is the discrepancy in distribution of economic power among the various social groups in a country which leads to environmentally degrading practices. On one hand, poverty and marginalization force subsistence farmers to unsustainable land use (Lundqvist et al., 1985); on the other hand, considerable short-term gains from trading in international markets often cause national and foreign entrepreneurs to embark on environmentally unacceptable land and water usage.

The prospects for future development of inland fisheries will be different in the various regions and countries. Prevailing social and economic conditions will influence the choice of options for development and environmental management of water resources. In many countries, priorities will focus on food security, generation of employment and income for growing populations, and foreign-exchange earnings, although environmental considerations currently are being included in most land- and water-development efforts (FAO, 1995b, 1995c).

As control over water as a resource is intensifying worldwide, so, too, is the control over the aquatic environment and its use for fisheries. At a recent regional meeting, Asian experts discussed possible management strategies for increasing the efficiency of inland fisheries under current constraints (IPFC, 1994). It was concluded that while there is still potential for increasing fish harvest from some underexploited water bodies, the main increase from inland fish-capture fisheries will have to be achieved through enhancement by stocking in a large diversity of water bodies, such as small multipurpose storages, derelict and saline waters, and medium and large reservoirs constructed for hydroelectricity production and/or irrigation purposes. It was felt that for rehabilitation of inland fisheries, some measures of environmental quality with respect to fish requirements need to be maintained by the users responsible for environmental degradation. It was recommended that development of strategies for rehabilitation of the environment be included among government priorities.

In Africa and Latin America, inland fisheries yields may be maintained if trends of increasing catchment-area degradation, including urban and industrial pollution, could be halted. There is potential to increase inland fish production in African small water bodies and Latin American reservoirs. Some fish stocks in some Latin American rivers are underexploited.

Inland fisheries in Australia, Europe, and North America may benefit from growing interest in protecting and rehabilitating aquatic environments.

The growth of aquaculture is expected to continue. Aquaculture is expected to make a major and increasing contribution to total world-food fish supply (FAO, 1995a; FAO/FIRI, 1995b). Still, pollution of water resources has proved to be an increasing problem for aquaculture developments, and, in some countries, the availability of land and water is likely to set limits to the growth of inland aquaculture in the future. Fortunately, carp production in Asia, which currently accounts for almost half of the global production of aquaculture (excluding plants), is through long-standing extensive and semi-intensive polyculture systems which are extremely efficient in the use of water. Competition with agriculture for resources is not likely in the near term because of the integration of activities at the farm level, provided future development and expansion is based on similar systems. However, aquaculture production processes are being intensified in some regions, and the release of wastes may cause water pollution.

Integrated Management of Resources

For many years, fishery experts have generally been very supportive of approaches aiming at the integrated management of land and water resources in catchment areas or river and lake basins (see, e.g., Kapetsky, 1981; Petr, 1983). There are guidelines available on the planning and development of inland fisheries in the context of multiple use of resources (Dill et al., 1975; Petr, 1985; Scudder and Conelly, 1985; Vanderpuye, 1985; Baluyut, 1986; Dunn, 1989; Boon, 1992b; Burbridge, 1994). Bernacsek (1984) provided guidelines for dam design and operation to optimize fish production in impounded river basins. Options for integrated resource management for sustainable aquaculture have been described or are being advocated by AIT (1994), Chang (1989), Chen et al. (1995), Edwards (1993), FAO/NACA (in press), Pullin (in press), and many others.

FAO has also published numerous reports and guides on land use planning and use of land resources (FAO, 1993c; Sombroek and Sims, 1995), rural-area development planning (Bendavid-Val, 1990), integrated plant nutrition (Dudal and Roy, 1995), integrated pest management and distribution and use of pesticides (FAO, 1990b, 1994c), forest harvesting (Dykstra and Heinrich, 1995), integrated watershed management and natural-resource conservation (FAO, 1986b; Gregersen et al., 1988; Quesada Mateo, 1993). Practices and concepts of integrated rural water management have been addressed (FAO, 1993d), water-resource management policies have been formulated (World Bank, 1993), experiences with river-basin developments have been reviewed (Scudder, 1994), and general frameworks for water-sector policy review and strategy formulation are being prepared (FAO/UNDP/World Bank, in press). As mentioned before, FAO is assisting Member States in the formulation of an international Code of Conduct for Responsible Fisheries, which aims at the worldwide promotion of appropriate and sustainable fishery resource-management measures (FAO, 1994b).

It is hoped that efforts which promote integrated resource management at the various levels of intervention will be continued. It is also hoped that fishery interests will be considered in planning and management of resources utilization. The successful implementation of such management approaches will help in sustaining and enhancing inland fish production.

Acknowledgments

This contribution is based partly on three background and discussion papers prepared for FAO: "Inland capture fisheries and enhancement: status, constraints and prospects for food security," by D. Coates; "Enhancement of inland fisheries and future management prospects," by I. G. Dunn; and "Environmental constraints and management of the environment for enhancement of inland fisheries," by A. I. Payne and S. A. Temple.

References

Agostinho, A. A., Julio, H. F., and M. Petrere, 1994. Itaipu reservoir (Brazil): impacts of the impoundment on the fish fauna and fisheries. In Cowx, I. G. (ed.), *Rehabilitation of Freshwater Fisheries,* pp. 171–184. Fishing News Books, Blackwell, Oxford.

AIT (Aquaculture), 1994. *Partners in Development: The Promotion of Sustainable Aquaculture.* Asian Institute of Technology, Bangkok.

Alabaster, J. S., 1981. *Review of the State of Aquatic Pollution in East African Inland Waters.* CIFA Occasional Paper no. 9.

Alabaster, J. S., 1982. *Report of the EIFAC Workshop on Fish-Farm Effluents.* EIFAC Technical Paper no. 41, FAO, Rome.

Alabaster, J. S. (ed.), 1985, Habitat modification and freshwater fisheries. In *Proceedings of a symposium of the European Inland Fisheries Advisory Commission* (EIFAC). FAO, Rome and Butterworths, London.

Alabaster, J. S., 1986. *Review of the state of aquatic pollution affecting inland fisheries in Southeast Asia.* FAO Fisheries Technical Paper no. 260. FAO, Rome.

Amarasinghe, U.S., 1992. Recent trends in the inland fishery of Sri Lanka. In Baluyut E. A. (ed.), *Indo-Pacific Fishery Commission, Country Reports Presented at the Fifth Session of the Indo-Pacific Fishery Commission Working Party of Experts on Inland Fisheries* (Bogor, Indonesia, June 24–29, 1991) and *Papers Contributed to the Workshop on Tilapia in Capture and Culture-Enhanced Fisheries in the Indo-Pacific Fishery Commission Countries* (Bogor, Indonesia, June 27–29, 1991), FAO Fisheries Report no. 458, suppl. FAO, Rome.

Arthington, A. H., and R. L. Welcomme, in press. The condition of large river systems of the world. Paper presented to the World Fisheries Congress, Athens, Greece, May 1992.

Backiel, T., and T. Penczak, 1989. The fish and fisheries in the Vistula river and its tributary, the Pilica river. In D. P. Dodge (ed.), *Proceedings of the International Large River Symposium (LARS)*, pp. 488–503. Canadian Special Publications on Fisheries and Aquatic Science no. 106.

Baluyut, E. A., 1983. *Stocking and Introduction of Fish in Lakes and Reservoirs in the ASEAN Countries.* FAO Fisheries Technical Paper no. 236. FAO, Rome.

Baluyut, E. A., 1986. *Planning for Inland Fisheries under Constraints from Other Uses of Land and Water Resources: General Considerations and the Philippines.* FAO Fisheries Circular no. 798. FAO, Rome.

Bayley, P. B., and M. Petrere, 1989. Amazon fisheries: assessment methods, current status and management options. In D. P. Dodge (ed.), *Proceedings of the International Large River Symposium (LARS)*, pp. 385–408. Canadian Special Publications on Fisheries and Aquatic Science no. 106.

Bendavid-Val, A., 1990. *Rural Area Development Planning: A Review and Synthesis of Approaches.* FAO/ESP/TMAP no. 21. FAO, Rome.

Bernacsek, G. M., 1984. *Guidelines for Dam Design and Operation to Optimize Fish Production in Impounded River Basins (Based on a Review of the Ecological Effects of Large Dams in Africa).* CIFA Technical Paper no. 11. FAO, Rome.

Beveridge, M. C. M., 1983. *Un Estudio de los Niveles de Metales Pesados en el Lago Poopo, Bolivia.* University of Stirling, Stirling, Scotland.

Beveridge, M. C. M., 1984. *Cage and Pen Fish Farming. Carrying Capacity Models and Environmental Impact.* FAO Fisheries Technical Paper no. 255. FAO, Rome.
Beveridge, M. C. M., and M. J. Phillips, 1993. Environmental impact of tropical inland aquaculture. In R. S. V. Pullin, H. Rosenthal, and J. L. MacLean (eds.), *Environment and Aquaculture in Developing Countries,* pp. 213–236. ICLARM Conference Proceedings no. 31.
Beveridge, M. C. M., L. G. Ross, and L. K. Kelly, 1994. Aquaculture and biodiversity. *Ambio* **23**(8): 497–502
Bhukaswan, T., 1985. The Nam Pong Basin (Thailand). In Petr, T. (ed.), *Inland Fisheries in Multi-purpose River Basin Planning and Development in Tropical Asian Countries: Three Case Studies,* pp. 55–90. FAO Technical Paper no. 265.
Biney, C. et al., 1987. Scientific bases for pollution control in African inland waters. *Chem. Ecol.* (3): 49–74.
Biney, C. et al., 1994. Review of heavy metals in the African aquatic environment. *Ecotoxicol. Environm. Safety* (28): 134–159.
Boon, P. J., 1992a. Essential elements in the case of river conservation. In P. J. Boon, P. Calow, and G. E. Petts (eds.), *River Conservation and Management,* pp. 11–33. Wiley, Chichester, U.K.
Boon, P. J., 1992b. Channeling scientific information for the conservation and management of rivers. *Aquat. Conserv. Marine Freshwater Ecosyst.* **2:** 115–123.
Boon, P. J., P. Calow, and G. E. Petts (eds.), 1992. *River Conservation and Management.* Wiley, Chichester, U.K.
Brookes, A., 1988. *Channelized Rivers: Perspectives for Environmental Management.* Wiley, Chichester, U.K.
Burbridge, P. R., 1994. Integrated planning and management of freshwater habitats, including wetlands. *Hydrobiology* (285): 311–322.
Calamari, D., 1985. *Review of the State of Aquatic Pollution in West and Central African Inland Waters.* CIFA Occasional Paper no 12.
Calamari, D., and H. Naeve (eds.), 1994. *Review of Pollution in the African Environment.* CIFA Technical Paper no. 25. FAO, Rome.
Calamari, D., M. O. Akech, and P. B. O. Ochumba, 1994. *Pollution of Winam Gulf, Lake Victoria, Kenya. CIFA Seminar on African Inland Fisheries, Aquaculture and the Environment* (Harare, Zimbabwe, Dec. 5–7, 1994). Committee for Inland Fisheries in Africa, CIFA/94/Seminar A-25. FAO, Rome.
Chanduvi F., and M. Prieto-Celi, 1993. Results of a questionnaire filled out by participating countries from Latin America and the Caribbean. In *Prevention of Water Pollution by Agriculture and Related Activities. Proceedings of the FAO Expert Consultation* (Santiago, Chile, Oct. 20–23, 1992), pp. 11–26. FAO Water Reports no. 1. FAO, Rome.
Chang, W. Y. B., 1989. Integrated lake farming for fish and environmental management in large shallow Chinese lakes: a review. *Aquacult. Fish. Manag.* (20): 441–452.
Chen, H., B. Hu, and A. T. Charles, 1995. Chinese integrated fish farming: a comparative bioeconomic analysis. *Aquacult. Res.,* (26): 81–94.
Chevassus, B., and A. G. Coche (eds.), 1986. *Report of the Symposium on Selection, Hybridization and Genetic Engineering in Aquaculture of Fish and Shellfish for Consumption and Stocking* (Bordeaux, France, May 27–30, 1986). EIFAC Technical Paper no. 50. FAO, Rome.
Clay, C. H., 1995. *Design of Fishways and Other Fish Facilities,* 2d ed. Lewis Publishers, Boca Raton, Fla.
Costa-Pierce, B. A., and O. Soemarwoto, 1987. Proliferation of Asian reservoirs: the need for integrated management. *NAGA* **10**(1): 9–10.
Dejoux, C., 1988. *La Pollution des Eaux Continentales Africaines.* Travaux et documents no. 213. ORSTOM, Paris.
De Silva, S., 1985. The Mahaweli Basin (Sri Lanka). In T. Petr (ed.), *Inland Fisheries in Multipurpose River Basin Planning and Development in Tropical Asian Countries: Three Case Studies,* pp. 91–166. FAO Technical Paper no. 265. FAO, Rome.
Dill, W., 1990. *Inland Fisheries of Europe,* IEFAC Technical Paper no. 52. FAO, Rome.
Dill, W., 1993. *Inland Fisheries of Europe,* EIFAC Technical Paper no. 52, suppl. FAO, Rome.

Dill, W., D. W. Kelley, and J. C. Fraser, 1975. *Water- and Land-Use Development and the Aquatic Environment—Problems and Solutions.* FAO Fisheries Technical Paper no. 141. FAO, Rome.

Dodge, D. P. (ed.), 1989. *Proceedings of the International Large River Symposium (LARS).* Canadian Special Publications on Fisheries and Aquatic Science no. 106.

Drinkwater, K. F., and K. T. Frank, 1994. Effects of river regulation and diversion on marine fish and invertebrates. *Aquat. Conserv. Freshwater Marine Ecosyst.* (4): 131–151.

Dudal, R., and R. N. Roy, 1995. *Integrated Plant Nutrition Systems.* FAO Fertilization Plant Nutrition Bulletin no 12. FAO, Rome.

Dudgeon, D., 1992. Endangered ecosystems: a review of the conservation of tropical Asian rivers. *Hydrobiology* (248): 167–191.

Dunn, I. G., 1989. *Development of Inland Fisheries under Constraints from Other Uses of Land and Water Resources: Guidelines for Planners.* FAO Fisheries Circular no 826. FAO, Rome.

Dykstra, D. P., and R. Heinrich, 1995. *FAO Model Code of Forest Harvesting Practice.* FAO, Rome.

Edwards, P., 1993. Environmental issues in integrated agriculture—aquaculture and wastewater-fed fish culture systems. In R. S. V. Pullin, H. Rosenthal, and J. L. MacLean (eds.), *Environment and Aquaculture in Developing Countries,* pp. 139–170. ICLARM Conference Proceedings no. 31.

EIFAC (European Inland Fisheries Advisory Commission), 1982. *Report of the Symposium on Stock Enhancement in the Management of Freshwater Fisheries* (Budapest, Hungary, May 31–June 2, 1982; held in conjunction with the 12th session of the European Inland Fisheries Advisory Commission). EIFAC Technical Paper no. 42. FAO, Rome.

EIFAC, 1984. *Report of the EIFAC Working Party on Stock Enhancement* (Hamburg, Germany, May 16–19, 1984). EIFAC Technical Paper no. 22. FAO, Rome.

EIFAC, in press. *Guidelines for the Rehabilitation of Riverine Habits in Fish Communities.* FAO, Rome.

Eisenreich, S. J., 1993. The fate of in-lake micropollutants: response times of ecosystem change. In G. Giussani and C. Callieri (eds.), *Strategies for Lake Ecosystems beyond 2000. Proceedings of the 5th International Conference on the Conservation and Management of Lakes* (Stresa, Italy, May 17–21, 1993), pp. 9–12. International Lake Environmental Committee Foundation.

Escobar-Ramirez, J. J., and U. Barg, 1990. *La Contaminacion de las Aguas Continentales de Bolivia, Colombia, Costa Rica, Ecuador, Panama, Peru y Venezuela.* COPESCAL Technical Document no. 8. FAO, Rome.

FAO, 1986a. *Report of the Fourteenth Session of the European Inland Fisheries Advisory Commission* (Bordeaux, France, May 27–June 3, 1986). FAO Fisheries report no. 364. FAO, Rome.

FAO, 1986b. *Strategies, Approaches and Systems in Integrated Watershed Management.* FAO Conserv. Guide no. 14. FAO, Rome.

FAO, 1990a. *The Definition of Aquaculture and Collection of Statistics.* FAO Aquaculture Min. no. 7.

FAO, 1990b. *International Code of Conduct on the Distribution and Use of Pesticides.* FAO, Rome. (This code is supported by additional guidelines and manuals.)

FAO, 1991a. *Environment and Sustainability in Fisheries. Information Paper for the Nineteenth Session of the FAO Committee of Fisheries.* COFI/91/3. FAO, Rome.

FAO, 1991b. *Fish for Food and Development. Strategy and Action Programme for Fisheries.* FAO, Rome.

FAO, 1993a. *The State of Food and Agriculture 1993.* FAO Agriculture Series no. 26. FAO, Rome.

FAO, 1993b. *Report of the Expert Consultation on Utilization and Conservation of Aquatic Genetic Resources* (Grottaferrata, Italy, Nov. 9–13, 1992). FAO Fisheries Report no. 491.

FAO, 1993c. *Guidelines for Land-Use Planning.* FAO Development Series no. 1.

FAO, 1993d. *Integrated Rural Water Management. Proceedings of a Technical Consultation* (Rome, March 9–13, 1993). FAO, Rome.

FAO, 1994a. *FAO Yearbook: Fishery Statistics. Catches and Landings 1992*. Vol. 74, FAO Fisheries Series no. 43; FAO Statistical Series no. 120. FAO, Rome.

FAO, 1994b. *Draft Code of Conduct for Responsible Fisheries* (revised Secretariat Draft; COFI/95/2, Nov. 1994), prepared for the Twenty-First Session of the FAO Committee on Fisheries (Rome, Italy, March 10–15, 1995).

FAO, 1994c. *Sustainable Agriculture through Integrated Pest Management. Twenty-second Regional Conference for Asia and the Pacific* (Manila, Philippines, Oct. 3–7, 1994). APRC/94/3. FAO, Rome.

FAO, 1995a. *The State of World Fisheries and Aquaculture*. FAO, Rome.

FAO, 1995b. *Food, Agriculture and Food Security: The Global Dimension. Historical Development, Present Situation, Future Prospects*. WFS 96/TECH/1 (advance unedited version). FAO, Rome.

FAO. 1995c. *Water Development for Food Security*. WFS 96/TECH/2 (advance unedited version). FAO, Rome.

FAO/FIDI, 1994. *Aquaculture Production, 1986–1992*. FAO Fishery Information, Data and Statistics Service. FAO Fisheries Circular no. 815, rev. 6. FAO, Rome.

FAO/FIRI, 1995a. *Review of the State of World Fishery Resources: Inland Capture Fisheries*. FAO Inland Water Resources and Aquaculture Service (FIRI). FAO Fisheries Circular no. 885. FAO, Rome.

FAO/FIRI, 1995b. *Review of the State of World Fishery Resources: Aquaculture*. FAO Inland Water Resources and Aquaculture Service (FIRI). FAO Fisheries Circular no. 886. FAO, Rome.

FAO/NACA, in press. *Regional Study and Workshop on Environmental Assessment and Management of Aquaculture Development* (Bangkok, Thailand, Feb. 21–26, 1994). FAO Project TCP/RAS/2253. FAO and Network of Aquaculture Centres in Asia-Pacific (NACA).

FAO/UNDP/World Bank, in press. *Water Sector Policy Review and Strategy Formulation: A General Framework*. FAO, Rome.

FAO/UNEP, 1981. *Conservation of the Genetic Resources of Fish: Problems and Recommendations. Report of the Expert Consultation on the Genetic Resources of Fish* (Rome, June 9–13, 1980). FAO Fisheries Technical Paper no. 217.

Fast, A. W., L. H. Bottroff, and R. L. Miller, 1982. Largemouth bass, Micropterus salmoides, and bluegill, Lepomis macrochirus, growth rates associated with artificial destratification and threadfin shad, Dorosoma petenense, introductions at El Capitan reservoir, California. *Calif. Fish Game* **67:** 4–20.

GESAMP (IMO/FAO/UNESCO/WMO/WHO/IAEA/UN/UNEP Joint Group of Experts on the Scientific Aspects of Marine Pollution), 1987. *Land/Sea Boundary Flux of Contaminants: Contributions from Rivers*. Report on Studies by GESAMP no 32. UNESCO, Paris.

Giasson, M., and J. L. Gaudet, 1989. *Summary of Proceedings and Selected Papers. Symposium on the Development and Management of Fisheries in Small Water Bodies*. FAO Fisheries Report no. 425. FAO, Rome.

Gophen, M. 1990. Biomanipulation: retrospective and future development. *Hydrobiology* (200/201): 1–11.

Greboval, D., M. Bellemans, and M. Fryd, 1994. *Fisheries Characteristics of the Shared Lakes of the East African Rift*. CIFA Technical Paper no. 24. FAO, Rome.

Gregersen, H. M., K. N. Brooks, J. A. Dixon, and L. S. Hamilton, 1988. *Guidelines for Economic Appraisal of Watershed Management Projects*. FAO Conservation Guide no. 16. FAO, Rome.

Harper D., 1992. *Eutrophication of Freshwaters. Principles, Problems and Restoration*. Chapman and Hall, London.

IPFC (Indo-Pacific Fishery Commission), 1994. *Report of the Sixth Session of the IPFC Working Party of Experts on Inland Fisheries* (Bangkok, Thailand, Oct. 17–21, 1994) and *Report of Regional Symposium on Sustainable Development of Inland Fisheries Development under Environmental Constraints*. FAO Fisheries Report no. 512. FAO, Rome.

Jhingran, A. G., 1992. Inland fisheries management in India: Development potential and constraints, pp. 1–24. In E. A. Baluyut (ed.), *Indo-Pacific Fishery Commission, Country Reports Presented at the Fifth Session of the Indo-Pacific Fishery*

Commission Working Party of Experts on Inland Fisheries (Bogor, Indonesia, June 24–29, 1991) and *Papers Contributed to the Workshop on Tilapia in Capture and Culture-Enhanced Fisheries in the Indo-Pacific Fishery Commission Countries* (Bogor, Indonesia, June 27–29, 1991). FAO Fisheries Report no. 458, suppl. FAO, Rome.

Joly, C., 1993. Plant nutrient management and the environment. In *Prevention of Water Pollution by Agriculture and Related Activities. Proceedings of the FAO Expert Consultation* (Santiago, Chile, Oct. 20–23, 1992), pp. 223–245. FAO Water Reports no. 1. FAO, Rome.

Jorgensen, S. E. (ed.), 1993. *Guidelines of Lake Management,* vol. 5, *Management of Lake Acidification.* International Lake Environment Committee Foundation and United Nations Environment Programme. Shiga, Japan.

Jorgensen, S. E., and R. A. Vollenweider (eds.), 1988. *Guidelines of Lake Management,* vol. 1, *Principles of Lake Management.* International Lake Environment Committee Foundation and United Nations Environment Programme. Shiga, Japan.

Juarez-Palacios, J. R., and M. E. Olmos-Tomassini, 1992. Tilapia in capture and culture-enhanced fisheries in Latin America. In E. A. Baluyut (ed.), *Indo-Pacific Fishery Commission, Country Reports Presented at the Fifth Session of the Indo-Pacific Fishery Commission Working Party of Experts on Inland Fisheries* (Bogor, Indonesia, June 24–29, 1991) and *Papers Contributed to the Workshop on Tilapia in Capture and Culture-Enhanced Fisheries in the Indo-Pacific Fishery Commission Countries* (Bogor, Indonesia, June 27–29, 1991), pp. 244–273. FAO Fisheries Report no. 458, suppl. FAO, Rome.

Kajak, Z., 1992. The river Vistula and its floodplain valley (Poland): its ecology and importance for conservation. In P. J. Boon, P. Calow, and G. E. Petts (eds.), *River Conservation and Management,* pp. 35–50. Wiley, Chichester, U.K.

Kapetsky, J. M. (ed.), 1981. *Seminar on River Basin Management and Development* (Blantyre, Malawi, Dec. 8–10, 1980). CIFA Technical Paper no. 8. FAO, Rome.

Kapetsky, J. M., and T. Petr (eds.), 1984. *Status of African Reservoir Fisheries.* CIFA Technical Paper no. 10. FAO, Rome.

Kira, T., 1993. Major environmental problems in world lakes. In G. Giussani and C. Callieri (eds.), *Strategies for Lake Ecosystems beyond 2000. Proceedings of the 5th International Conference on the Conservation and Management of Lakes* (Stresa, Italy, May 17–21, 1993), p. 3. International Lake Environmental Committee Foundation.

Knaap, M. van der, 1994. *Status of Stocks and Fisheries in Thirteen Medium-Sized African Reservoirs.* CIFA Technical Paper no. 26. FAO, Rome.

Lacerda, L. D., and W. Salomons, 1991. *Mercury in the Amazon: A Chemical Time Bomb?* Report sponsored by the Dutch Ministry of Housing, Physical Planning and Environment.

Lammens, E. H. R. R. et al., 1990. The first biomanipulation conference: a synthesis. *Hydrobiology* (200/201): 619–627.

Lean, G., D. Hinrichsen, and A. Markham, 1990. *Atlas of the Environment.* Prentice-Hall, Englewood Cliffs, N.J.

Lu, X., 1986. *A Review of Reservoir Fisheries in China.* FAO Fisheries Circular no. 803.

Lu, X., 1992. *Fishery Management Approaches in Small Reservoirs in China.* FAO Fisheries Circular no. 854.

Lu, X., 1994. *A Review of River Fisheries in China.* FAO Fisheries Circular no. 862.

Lundqvist, J., U. Lohm, and M. Falkenmark (eds.), 1985. *Strategies for River Basin Management (Environmental Integration of Land and Water in a River Basin).* Reidel, Dordrecht, Netherlands.

Lupikisha, J. M. C., 1993. Catch trends of *Limnothrissa miodon* in Lake Kariba. In B. E. Marshall and R. Mubamba (eds.). *Papers Presented at the Symposium on Biology, Stock Assessment and Exploitation of Small Pelagic Fish Species in the African Great Lakes Region* (Bjumbura, Burundi, Nov. 25–28, 1992), pp. 51–67. CIFA Technical Paper no. 19. FAO, Rome.

Mallen-Cooper, M., 1994. How high can a fish jump? *New Sci.* (April 16): 32–37.

Marshall, B. E., 1994. *Eutrophication in African Lakes and Its Impact on Fisheries. CIFA Seminar on African Inland Fisheries, Aquaculture and the Environment*

(Harare, Zimbabwe, Dec. 5–7, 1994). Committee for Inland Fisheries in Africa. CIFA/94/Seminar A-14. FAO, Rome.

Marshall, B. E., and B. Maes, 1995. *Small Water Bodies and Their Fisheries in Southern Africa.* CIFA Technical Paper no. 29. FAO, Rome.

Marshall, B. E., and R. Mubamba (eds.), 1993. *Papers Presented at the Symposium on Biology, Stock Assessment and Exploitation of Small Pelagic Fish Species in the African Great Lakes Region* (Bjumbura, Burundi, Nov. 25–28, 1992). CIFA Technical Paper no. 19. FAO, Rome.

Martin, J. M., J. D. Burton, and D. Eisma (eds.), 1981. *River Inputs into Ocean Systems.* UNEP and UNESCO, Paris.

Martinez-Espinosa, M., and U. Barg, 1993. Aquaculture and management of freshwater environments, with emphasis on Latin America. In R. S. V. Pullin, H. Rosenthal, and J. L. MacLean (eds.), *Environment and Aquaculture in Developing Countries,* pp. 42–59. ICLARM Conference Proceedings no. 31.

Micklin, P. P., 1988. Desiccation of the Aral Sea: a water management disaster in the Soviet Union. *Science* **241:** 1170–1176.

Moreau, J., and S. S. De Silva, 1991. *Predictive Fish Yield Models for Lakes and Reservoirs of the Philippines, Sri Lanka, and Thailand.* FAO Fisheries Technical Paper no. 319. FAO, Rome.

Müller, R., and R. Lloyd (eds.), 1994. *Sublethal and Chronic Effects of Pollutants on Freshwater Fish.* FAO and Fishing News Books, Oxford.

Natarajan, A. V., 1989. Environmental impact of Ganga Basin Development on gene pool and fisheries of the Ganga river system. In D. P. Dodge (ed.), *Proceedings of the International Large River Symposium (LARS),* pp. 545–560. Canadian Special Publications on Fisheries and Aquatic Science no. 106.

NCC (Nature Conservancy Council), 1990. *Fish Farming and the Scottish Freshwater Environment.* Nature Conservancy Council, Edinburgh, Scotland.

Osibanjo, O. et al., 1994. Review of chlorinated hydrocarbon substances in the African aquatic environment. In *Committee for Inland Fisheries in Africa. Report of the Fourth Session of the CIFA Working Party on Pollution and Fisheries,* pp. 7–45. FAO Fisheries Report no. 502. FAO, Rome.

Pavlov, D. S., 1989. *Structures Assisting the Migrations of Non-salmonid Fish: USSR.* FAO Fisheries Technical Paper no. 308. FAO, Rome.

Pavlov, D. S., and B. Y. Vilenkin, 1989. The present state of the environment, biota and fisheries of the Volga river. In D. P. Dodge (ed.), *Proceedings of the International Large River Symposium (LARS),* pp. 504–514. Canadian Special Publications on Fisheries and Aquatic Science no. 106.

Petr, T. (ed.), 1983. *Summary report and selected papers presented at the IPFC Workshop on Inland Fisheries for Planners* (Manila, Philippines, Aug. 2–6, 1982). FAO Fisheries Report no. 288.

Petr, T. (ed.), 1985. *Inland Fisheries in Multi-Purpose River Basin Planning and Development in Tropical Asian Countries: Three Case Studies.* FAO Technical Paper no. 265.

Petr, T., 1987. Food fish as vector control, and strategies for their use in agriculture. In *Effects of Agricultural Development on Vector-Borne Diseases. Papers presented to the 7th Annual Meeting of the Joint WHO/FAO/UNEP Panel of Experts on Environmental Management for Vector Control* (Rome, Sept. 7–11, 1987), pp. 87–92. AGL/MISC/12/87. FAO, Rome.

Petr, T., 1993. Aquatic weeds and fisheries production in developing regions of the world. *J. Aquat. Plant Manag.* **31:** 5–13.

Petr, T., 1994. Intensification of reservoir fisheries in tropical and subtropical countries. *Internatl. Rev. ges. Hydrobiol.* **79**(1): 129–136.

Petr, T., 1995. The present status of inland fisheries development: Southeast Asia. In T. Petr and M. Morris (eds.), *Proceedings of the IPFC Regional Symposium on Sustainable Development of Inland Fisheries under Environmental Constraints.* FAO Fisheries Report no. 512, suppl. FAO, Rome.

Petr, T., in press. The future of fisheries in natural lakes under environmental stress in tropical Asia. Paper presented at the International Conference on Fisheries and Environment: Beyond 2000 (Kuala Lumpur, Malaysia, Dec. 6–9, 1993).

Petr, T., and M. P. Mitrofanov, 1995. Fisheries in arid countries of Central Asia and Kazakhstan under the impact of irrigated agriculture. In T. Petr and M. Morris (eds.), *Proceedings of the IPFC Regional Symposium on Sustainable Development of Inland Fisheries under Environmental Contraints.* FAO Fisheries Report no. 512, suppl. FAO, Rome.

Petrere, M., 1994. Sintesis sobre las pesquerias de los grandes embalses tropicales de America del Sur. Paper presented at Simposio Regional sobre Manejo de la Pesca en Embalses en America Latina (La Habana, Cuba, Oct. 24–28, 1994). Comision de Pesca Continental para America Latina (COPESCAL). FAO, Rome.

Petts, G. E., 1989. Perspectives for ecological management of regulated rivers. In J. A. Gore and G. E. Petts (eds.), *Alternatives in Regulated River Management,* pp. 3–24. CRC Press, Boca Raton, Fla.

Phillips, M. J., M. C. M. Beveridge, and R. M. Clarke, 1991. Impact of aquaculture on water resources. In D. E. Brune and J. R. Tomasso (eds.), *Advances in World Aquaculture,* vol. 3, pp. 568–591. World Aquaculture Society, Baton Rouge, La.

Polunin, N. V. C., P. L. Osborne, and R. G. Totome, 1988. Environmental archive: tropical urban development reflected in the sediment geochemistry of a floodplain lake. *Arch. Hydrobiol.* **114**(2): 199–211.

Pullin, R. S. V., 1993. An overview of environmental issues in developing country aquaculture. In R. S. V. Pullin, H. Rosenthal, and J. L. MacLean (eds.), *Environment and Aquaculture in Developing Countries,* pp. 1–19. ICLARM Conference Proceedings no. 31.

Pullin, R. S. V., in press. Aquaculture, integrated resources management and the environment. Paper presented at the International Workshop on Integrated Fish Farming (Wu-Xi, China, Oct. 11–15, 1994), ICLARM contribution no. 1096.

Pullin, R. S. V., H. Rosenthal, and J. L. MacLean (eds.), 1993. *Environment and Aquaculture in Developing Countries.* ICLARM Conference Proceedings no. 31.

Quesada Mateo, C. A., 1993. *Debt-for-Nature Swaps to Promote Natural Resource Conservation.* FAO Conservation Guide no. 23. FAO, Rome.

Quiros, R., 1993. Inland fisheries under constraints by other uses of land and water resources in Argentina. In *Prevention of Water Pollution by Agriculture and Related Activities. Proceedings of the FAO Expert Consultation* (Santiago, Chile, Oct. 20–23, 1992), pp. 29–44. FAO Water Reports no. 1. FAO, Rome.

Quiros, R., 1994. *Intensificacion de la pesca en los pequenos cuerpos de agua en America Latina y el Caribe.* COPESCAL Documento Ocasional no. 8. FAO, Rome.

Rapport, D. J., H. A. Regier, and T. C. Hutchinson, 1985. Ecosystem behavior under stress. *Am. Natur.* **125:** 615–640.

Redding, T. A., and A. B. Midlen, 1991. *Fish Production in Irrigation Canals. A Review.* FAO Fisheries Technical Paper no. 317. FAO, Rome.

Regier, H. A., R. L. Welcomme, R. J. Steedman, and H. F. Henderson, 1989. Rehabilitation of degraded river ecosystems. In D. P. Dodge (ed.), *Proceedings of the International Large River Symposium (LARS),* pp. 86–97. Canadian Special Publications on Fisheries and Aquatic Science no. 106.

Reynolds, J. E., and D. F. Greboval, 1988. *Socio-economic Effects of the Evolution of Nile Perch Fisheries in Lake Victoria: A Review.* CIFA Technical Paper no. 17. Committee for Inland Fisheries of Africa. FAO, Rome.

Ryding, S. O., and W. Rast, 1989. The possibilities for reuse of nutrients. In *The Control of Eutrophication of Lakes and Reservoirs,* pp. 213–230 (Chap. 10). Man and Biosphere Series, vol. 1. Parthenon Publishing Group and UNESCO, Paris.

Saad, M. A. H. et al., 1990. Scientific bases for pollution control in African inland waters—domestic and industrial organic loads. In *Committee for Inland Fisheries in Africa. Report of the Second Session of the CIFA Working Party on Pollution and Fisheries,* pp. 6–24. FAO Fisheries Report no. 437. FAO, Rome.

Scudder, T., 1994. Recent experiences with river basin development in the tropics and subtropics. *Natural Resources Forum* **18**(2): 101–113.

Scudder, T., and T. Conelly, 1985. *Management Systems for Riverine Fisheries.* FAO Fisheries Technical Paper no. 263.

Seko, I., 1994. Survey on the state of world lakes. *UNEP-IETC News.* (Fall 1994): 1–2. UNEP International Environmental Technology Centre.

Shapiro, J., 1990. Biomanipulation: the next phase—making it stable. *Hydrobiology* (200/201): 13–17.

Sombroek, W. G., and D. Sims, 1995. Planning for sustainable use of land resources: towards a new approach. Background paper to FAO's Task Managership for Chap. 10 of UNCED's Agenda 21. FAO, Rome.

Sreenivasan, A., 1986. *Inland Fisheries under Constraints from Other Uses of Land and Water Resources: Indian Subcontinent and Sri Lanka.* FAO Fisheries Circular no. 797. FAO, Rome.

Starling, F. L. R. M., and A. J. A. Rocha, 1990. Experimental study of the impacts of planktivorous fishes on plankton community and eutrophication of a tropical Brazilian reservoir. *Hydrobiology* (200/201): 581–591.

Symoens, J. J., M. Burgis, and J. J. Gaudet (eds.), 1981. *The Ecology and Utilization of African Inland Waters,* UNEP Report Study no. 1. UNEP, Nairobi, Kenya.

Tamayo-Zafarwalla, M., 1983. Conflicts between fisheries and mining, industrialization and pollution of inland waters in the Philippines. In T. Petr (ed.), *Summary Report and Selected Papers presented at the IPFC Workshop on Inland Fisheries for Planners* (Manila, Philippines, Aug. 2–6, 1982), pp. 83–87. FAO Fisheries Report no. 288.

Townsley, P., 1992. *Rapid Appraisal for Small Water Bodies.* FAO, Aquaculture for Local Community Development Programme. ALCOM Report no. 11.

Turner, G. E., 1988. *Codes of Practice and Manual of Procedures for Consideration of Introductions and Transfers of Marine and Freshwater Organisms.* EIFAC Occasional Paper no. 23. FAO, Rome.

UNEP, 1991. *Freshwater Pollution.* UNEP/GEMS Environmental Library, no. 6. UNEP, Nairobi, Kenya.

UNESCO, 1990. *The Impact of Large Water Projects on the Environment. Proceedings of an International Symposium Convened by UNESCO and UNEP and Organized in Cooperation with IISA and IAHS* (UNESCO Headquarters, Paris, Oct. 21–31, 1986). UNESCO/UNEP, Paris.

Valderrama, M., and M. Zarate, 1989. Some ecological aspects and present state of the fishery of the Magdalena river basin, Colombia, South America. In D. P. Dodge (ed.), *Proceedings of the International Large River Symposium (LARS)*, pp. 409–421. Canadian Special Publications on Fisheries and Aquatic Science no. 106.

Van Houtte, A. R., N. Bonucci, and W. R. Edeson, 1989. *A Preliminary Review of Selected Legislation Governing Aquaculture.* ADCP/REP/89/42. FAO, Rome.

Vanderpuye, C. J., 1985. *Evaluation Guidelines for Rational Planning and Management of Tropical and Subtropical Inland Fisheries under Constraints from Other Uses of Land and Water Resources: Africa.* FAO Fisheries Circular no. 789. FAO, Rome.

Welcomme, R. L., 1985. *River Fisheries.* FAO Fisheries Technical Paper no. 262. FAO, Rome.

Welcomme, R. L., 1988. *International Introductions of Inland Aquatic Species.* FAO Fisheries Technical Paper no. 294. FAO, Rome.

Welcomme, R. L., 1989. Floodplain fisheries management. In J. A. Gore and G. E. Petts (eds.), *Alternatives in Regulated River Management,* pp. 209–233. CRC Press, Boca Raton, Fla.

Welcomme, R. L., 1990. Status of fisheries in South American rivers. *Interciencia* (15): 337–345.

Welcomme, R. L., in press. Status and trends of global inland fisheries. Paper presented to the World Fisheries Congress, Athens, Greece, May 1992.

Welcomme, R. L., and J. Kapetsky, 1981. Acadjas: the brush park fisheries of Benin, West Africa. *ICLARM Newsl.* **4**(4): 3–4.

West, W. Q. B., and C. A. Biney, 1991. *African Fisheries and the Environment.* RAFR Publication FI/91/1. FAO Regional Office for Africa, Accra, Ghana.

World Bank, 1993. *Water Resource Management. A World Bank Policy Paper.* The World Bank, Washington, D.C.

WRI (World Resources Institute), 1990. Freshwater. In *World Resources 1990–1991,* pp. 161–177. Oxford University Press.

Yap, S. Y., 1992. Inland capture fisheries in Malaysia. In E. A. Baluyut (ed.), *Indo-Pacific Fishery Commission, Country Reports Presented at the Fifth Session of the*

Indo-Pacific Fishery Commission Working Party of Experts on Inland Fisheries (Bogor, Indonesia, June 24–29, 1991) and *Papers Contributed to the Workshop on Tilapia in Capture and Culture-Enhanced Fisheries in the Indo-Pacific Fishery Commission Countries* (Bogor, Indonesia, June 27–29, 1991), pp. 25–46. FAO Fisheries Report no. 458, suppl. FAO, Rome.

Yoo, K. H., and C. E. Boyd, 1994. *Hydrology and Water Supply for Pond Aquaculture*. Chapman and Hall, New York.

Chapter

11

Aquatic Weeds

P. R. F. Barrett
Consultant in Aquatic Plant Management
Oxford, England

Introduction

One definition of a *weed* is that it is a plant growing in the wrong place. However, water plants perform a number of valuable functions and so are not necessarily weeds. They help stabilize riverbanks and riverbeds, provide a habitat for fish and fish-food organisms, improve poor water quality, and provide environmental and conservational value to most waters. In many waters, aquatic plants cause no problems and are definitely not weeds. In other situations, however, they can cause problems, particularly if the quantity of plant material becomes excessive. Therefore, this rather general definition that a weed is a plant in the wrong place does not fully describe most aquatic plants. A better definition of aquatic weeds would be that aquatic plants become weeds when their growth becomes excessive. How much growth is "excessive" depends heavily on the type of weed and the place in which it is growing. For example, a small amount of weed in an irrigation channel may be excessive while the same amount in a large river may have no adverse effect at all. Similarly, the presence of single-celled alga in a hydroelectric reservoir would not impede the turbines, but the same alga in a potable supply reservoir could have very serious effects on the filtration process. However, many waters are becoming multifunctional, being used for water supply, navigation, fishing, irrigation, and many other applications, and it is increasingly likely that the presence of aquatic plants will have an adverse effect on at least one of these functions. Where a water body has a number of functions, the identification that a weed problem exists and the solution to that problem must take into account the whole range of interests and uses of the water, even if some of the uses are not adversely affected by the weeds.

Extent of Weed Problems

There are thousands of species of aquatic plants and algae which have evolved to exploit virtually all aquatic habitats from hot springs to ice-covered ponds. Even the rarest and least aggressive of these plants can become a weed problem locally when conditions are particularly suitable for them. However, the major aquatic weed problems are caused by the more dominant and aggressive species which do not have very specific growth requirements and, so, can exploit a wide range of situations. The factors which encourage the growth of aquatic plants are similar to those required by terrestrial plants. Aquatic plants need light, nutrients, water, and carbon dioxide. One major difference between aquatic and terrestrial plants is that aquatic plants are less tolerant of desiccation, and so aquatic weed problems occur most frequently in permanent water bodies.

One characteristic of aquatic plants which gives them the potential to become troublesome is their high vegetative growth rate and their ability to reproduce asexually by, for example, budding off daughter plants (*Eichhornia crassipes*), regrowing from detached fragments (*Elodea canadensis*), or producing tubers and turions (*Hydrilla verticillata* and *Potamogeton pectinatus*). The survival and reproductive strategies adopted by individual weed species can have an important impact on the success of different methods of control, and water managers would be well advised to identify the species of weed and to learn about their growth habits before deciding on the most appropriate method and timing of control techniques.

Aquatic weed problems can arise simply because the native weed species grow to such an extent that they cause adverse effects. This usually arises because a change has occurred in the water body which has benefited plant growth. Such changes are frequently brought about, either directly or indirectly, by human actions.

Human activities have altered the ecology of the natural environment. Rivers have been impeded by dams, weirs, and obstructions; nutrients are released into the water as a result of agriculture and industry; and both land drainage and irrigation have altered water velocity and depth. Artificial (human-made) channels, created especially for irrigation and drainage, tend to be shallow with relatively slow flows and are an ideal habitat for aquatic plants.

As the human population increases, there is a growing demand for water and water space so that less weed can be tolerated. Water is needed for domestic use, industry, and irrigation. Water is used for fishing, navigation, and other commercial and recreational uses. All these activities can be adversely affected by aquatic weeds. In the past, it was sometimes possible to avoid weed-infested areas and obtain water and water space elsewhere. With growing demand for these resources, the need to maintain all available water in a usable condition is growing.

A number of aquatic plants have been transported, either accidentally or deliberately, away from their native habitats, where their growth is controlled by natural pests and predators, into countries where there are no natural control agents and their growth has become explosive. *Eichhornia crassipes* is an example of a plant which originates in South America, where it is of little consequence, but which has been introduced into many tropical and subtropical countries and is sometimes described as the world's worst aquatic weed (Cook, 1990a).

Weed problems occur worldwide, and, while it is tempting to say that the worst problems occur in tropical regions with year-round aquatic plant growth, problems can be just as severe, although of shorter duration, during the summer period in temperate countries.

Problems Caused by Aquatic Weeds

Aquatic weeds can cause a wide range of problems to humans and other components of the ecosystem. However, the majority of these problems can be summarized under the following headings:

1. *Impedance to flow.* The most widespread and frequent problem associated with aquatic weed growth is their ability to impede the flow of water in channels. Dense weed growth extending from bed to water surface can, in some instances, reduce the discharge capacity of a channel almost to zero. The extent to which individual plant species impede flow depends on the morphology of the weed and the density of its growth. Individually, the stiff, erect stems of some emergent aquatic plants impede flow and further increase hydraulic resistance by trapping detritus and other weeds carried by the current. In contrast, the individual stems of the shorter, more-flexible submerged weeds do not impede flow as much but, because they form dense masses which fill the channel, are collectively probably more troublesome in impeding flow than the stiffer emergent weeds. Clearly, the denser a weed bed becomes, the more it impedes water movement, and one of the effects of cutting weed beds is to synchronize their regrowth to ensure that all the weed beds reach their maximum bio-

mass at the same time and so create greater impedance, albeit temporarily, than would occur if the plants had been allowed to mature with a natural variation in growth rates. The long-term effects of cutting also vary with species; some are stimulated to regrow more vigorously while others are unaffected or may gradually reduce in quantity (Westlake and Dawson, 1982, 1986).

2. *Interference with fishing, navigation, and recreation.* Weeds trap fishing nets and lines, tangle around propellers, and can prevent the passage of boats. Severe infestations, particularly of floating weeds which can become interlocked so that they form floating islands, can prevent the movement of even large boats (Pieterse, 1990a, 1990b).

3. *Blocking of pumps, intakes, filters, and irrigation systems.* Weeds can be drawn into pumps and intakes, where they cause damage. While weed screens can be used to keep out the larger weeds, small fragments and algae pass through these screens and block filters and pumps.

4. *Water loss.* Emergent and floating weeds increase the rate of water loss by evaporation. Estimates of the rate of loss compared with open water have varied considerably depending on weed species and temperature, but it would not be unrealistic to say that dense infestations of weed can double the rate of water loss by evapotranspiration.

5. *Harboring pests and diseases.* Aquatic plants are a natural habitat for snails and other invertebrate animals. Snails act as the host for a number of parasites which attack humans and livestock as part of their life cycles. Various fly larvae, parasitic worms, and protozoa also live on, or are protected by, aquatic plants during part of their life cycles (Gangstad and Cardarelli, 1990).

6. *Taint and odor.* Some aquatic plants can create taste and odor problems in potable water. Many species of weed can cause this problem, but it is most commonly associated with blooms of algae, particularly diatoms.

7. *Deoxygenation.* Although aquatic plants add oxygen to the water during the day, they take it up at night. Normally, the two processes are sufficiently balanced throughout the 24-h cycle that fish and other aquatic organisms can survive the low point around dawn; however, under certain conditions—usually of high temperature and low rainfall—submerged weeds and algae can cause deoxygenation which results in fish kills. Floating plants also cause deoxygenation by covering the water surface and preventing light from reaching the submerged plants. These die, and the loss of photosynthesis, coupled with oxygen demand of the bacteria on the rotting vegetation, deoxygenates the water.

8. *Poisonous plants.* A number of poisonous plants grow in or near water. These are generally avoided by livestock but can become

palatable when they are cut or sprayed with a herbicide. Some species of cyanobacteria (blue-green algae) can also produce toxins and livestock and domestic pets have been killed by drinking water containing dense blooms of these algae (National Rivers Authority, 1990).

9. *Damage to structures and banks.* When large quantities of weed are washed downstream, they can build up against structures such as bridges, dams, and piers. Once the process starts, the weeds act as filters, trapping more weed and detritus which becomes bonded and compressed by water pressure. The damming effect of the mass of compressed weed can result in large forces which damage the structures holding the weed. In flowing systems, weeds can also divert and accelerate the current close to the banks, leading to erosion and collapse.

10. *Conservation and loss of habitats.* While a diversity of plant species provides a range of habitats for aquatic fauna, dense monocultures of weed destroy habitats, leading to loss of species. Many species of waterfowl and fish require a proportion of open water for feeding and reproductive purposes. Dense weed infestations are harmful to these species.

11. *Amenity values.* Water bodies can have a high amenity value for local residents as well as tourists. Excessive weed growth can destroy amenity values, leading to loss of income for the local tourist industry and residents.

12. *Impact on local industry.* Fish farms, rice paddies, and harvests of various aquatic plants can be adversely affected by weed growth. Industries which depend on a supply of high-volume and/or high-quality water can also be damaged by excessive weed growth.

Types of Aquatic Weed

Before considering the ways in which weeds can be controlled, it is useful to discuss the types of weed which can cause problems in water. This is not a botanical classification but is simply a way of describing the various types of weed in terms of their control. Most water managers describe aquatic weeds as being emergent, floating, or submerged. Algae are usually considered in a category separate from that for the vascular plants and are subdivided into unicellular or filamentous forms. Managing water bodies also involves controlling weeds which grow in the bankside. These bankside plants may be semiaquatic plants but can also include terrestrial species.

1. *Emergent weeds.* These are plants which grow with much of the stem, leaves, and flowers above the water surface. The leaves and flower stalks are sufficiently rigid to remain erect without the sup-

port of the surrounding water. They are usually rooted in the bottom sediment, although, occasionally, they can form floating islands growing on a mat of dead aquatic vegetation which provides support for the root-rhizome system. Normally, they occur in shallow water up to about 1.5 m deep, although floating mats can become detached and drift into deeper water. Most species are perennial, growing from rhizomes buried in the sediment. An example is the *common reed* (*Phragmites australis*), which is found throughout the world in both freshwater and estuarine situations. Some emergent aquatic weeds are poisonous to livestock.

2. *Floating plants.* These are plants whose leaves float on the water surface as a result of the buoyancy of the stems and leaves. Sometimes, particularly when growing in dense masses, the stems and leaves may stand slightly above water level, often supported by their floating neighbors. There are two types of floating plant; one is rooted in the bottom sediment, and the other floats freely on the water surface with the roots hanging down into the water. Rooted floating plants such as the water lilies (*Nuphar* and *Nymphaea* spp.) grow in shallow water up to about 2 m deep. Free-floating plants such as *Salvinia* species can survive in any depth of water. The distinction between free-floating and rooted floating plants is important when considering control methods because of the ability of free-floating plants to drift rapidly back into areas which have recently been cleared.

3. *Submerged plants.* The majority of the stems and leaves of these plants are submerged below the water surface. These plants are buoyant because of oxygen bubbles trapped within the plant and will usually float to the surface if cut or detached from the anchoring root system. Occasionally, the tips of the stems and the flowers may reach the water surface or even extend above it for a few centimeters, but the main part of the plant is under water. A few submerged plants are rootless (e.g., *Ceratophyllum* spp.) and drift freely in the water, but most species are rooted in the bottom sediment (e.g., *Hydrilla verticillata*). The depth of water in which they can grow is limited by the clarity of the water. In muddy, turbid water, few species can obtain enough light to survive and photosynthesize below about 2 to 3 m but, in very clear water, extensive growth can occur down to 10 m or more. The depth range of these species can sometimes be increased where water levels fluctuate. The plants become established when water levels are low at depths below their normal tolerance limits and can grow upward as the level rises, maintaining sufficient leaf area in the photic zone.

4. *Algae.* There are many thousands of species of algae, and these primitive plants grow in a number of forms. The two most commonly troublesome forms are filamentous and free-floating unicellular algae.

Filamentous algae commence growth as spores attached to the bottom or to submerged plants and, when the mass of filaments becomes thick enough, trap bubbles of oxygen and become buoyant, rising to the surface and covering the water with a floating mat or blanket. These mats can drift or be carried by currents and become entangled with other plants. Dense mats of filamentous algae can impede flow in channels and block filters and abstraction points. Unicellular algae can form blooms which make the water appear green and turbid. These algae seldom impede flow in channels but can block filters, cause taint and odor problems, and, under some conditions, cause deoxygenation of the water. Many species of the toxic blue-green algae are either unicellular or composed of such fine filaments that they behave as unicellular algae. These algae can be particularly dangerous in waters used for potable supply or livestock watering.

5. *Bankside weeds.* These are semiaquatic and terrestrial species which grow on the banks and batters of watercourses and lakes. They can interfere with access to the water and may impeded flow in channels when water levels are high. Their presence can interfere with or impinge on the methods used to control aquatic weeds within the channel.

While these categories give a general description of the growth form of different types of weed, some aquatic plants are very plastic in their growth and produce different-shaped leaves as the growing season progresses or when conditions change. For example, some submerged weeds can grow for short periods as emergent plants on damp soil above water level when the water level falls; others change the shape of their leaves depending on water level or velocity and may produce emergent, floating, or submerged leaves depending on the conditions. If aquatic plants are growing temporarily in an unusual form, they can sometimes be wrongly identified and incorrect control techniques can be adopted. For this reason, it is always advisable to identify a weed accurately to species level to ensure that the most appropriate method of control is being selected. A useful book on the identification of water plants was published by Cook (1990b).

Methods of Controlling Aquatic Weeds

In general terms, there are four techniques of controlling aquatic weeds: mechanically, chemically, biologically, and environmentally. Under each of these headings, a range of methods have been developed, some of which can be applied to almost any water body while others are more specifically designed for particular conditions. The following sections describe the various methods available under the four headings and outline some of the advantages and disadvantages of each method.

Mechanical control

This involves some form of physical action by which the weeds are cut, raked, dredged, or damaged by hand or machine. The simplest and, generally, fastest approach to mechanical control is to cut the weed. The slowest and usually most expensive form of mechanical control is to remove the weeds by dredging the channel. However, dredging is rarely recommended unless the benefits of weed control form part of an overall need to clean, enlarge, or restructure the channel. This section deals mainly with methods of weed cutting.

The types of weed which can be controlled by cutting are emergent, floating weeds which are rooted and most submerged weeds. Free-floating weeds and filamentous algae are seldom affected by cutting, although they can sometimes be raked out, particularly where they can be captured by floating booms. It should also be remembered that cut weed can cause many problems if left in the water, and it is important to emphasize that mechanical control normally involves two processes: *cutting,* which is relatively easy as most aquatic plants are soft and easily cut; and *removal of the cut weed,* which is usually more difficult and time-consuming than the cutting operation.

Mechanical-control methods and equipment

Hand cutting and pulling. Weeds in shallow water can be cut manually using scythes or sickles. Hand cutting is slow and limited to situations in which the water is no more than about 1 m deep. In many countries, it is used only in small areas where machinery cannot be used or where the high conservation value dictates the use of very precise, localized cutting.

In shallow water, weeds can sometimes be pulled and uprooted by hand. Hand pulling has the advantage over cutting that it can remove the roots and rhizomes of shallow or weakly rooted weeds so that regrowth is delayed until recolonization occurs. However, hand pulling is generally slower than cutting and has little advantage over cutting if the weeds are strongly rooted as the plants break off, leaving the roots and rhizomes to regrow.

Where labor is cheap, manual weed control is used more extensively (Wade, 1990) but is inefficient and exposes the workers to the risk of disease where they are required to work in water contaminated by toxins or pathogens.

Manual cutting can also be undertaken in small channels using chain scythes. These consist of a series of linked cutting blades with a rope at each end. The operators stand on opposite banks, pulling the blades backward and forward across the channel bed.

Where the channel is too deep for wading and too wide for chain scythes, hand cutting from a boat is sometimes practiced. A pair of scythe blades is attached to a pole to form a trailing V shape, which is pulled in a series of jerks by the operator standing in the back of a boat.

Filamentous algae are sometimes removed by hand using rakes or long-handled grabs. The process is inefficient because many filaments remain in the water and algal mats regrow rapidly from these. Also, the disturbance of floating mats releases the trapped oxygen bubbles so that the mats sink temporarily, giving the impression that the weed has been removed, but refloat again once photosynthesis has generated new bubbles and buoyancy is restored.

Small free-floating weeds, such as *Lemna* species, are also difficult to remove manually because they can escape through the tines of conventional rakes and scoops. Fine-mesh netting can be placed over the rake to enhance the rate of capture. Floating nets or booms may also help hold and concentrate the weed in a small area from which it can be pulled by rake or net.

Weed-cutting boats. Numerous sizes and shapes of weed-cutting boat have been designed. Some are commercially produced machines, but others have been built by water authorities for their own use. With a few, mainly experimental, exceptions, they employ one of two cutting methods. The simpler of the two designs uses a mechanical version of the trailing scythe blades described above. Two pairs of V-shaped blades, attached by a chain to a rotating cam at the rear of a boat, are pulled in a series of jerks by the cam as the boat travels forward. The blades are 1 to 1.5 m long and set at an angle of about 50°, with the interblade distance just wide enough to allow the two pairs of blades to meet at the tip, forming a VV shape. The blades are constructed of mild steel and the leading edges are sharpened sufficiently to ensure that they cut the vegetation. The length of the chain can be adjusted so that the two heavy blades lie almost flat on the channel bed or even sink slightly into the sediment. These blades cut very close to the channel bed and may even cut into the sediment damaging the plant roots and rhizomes. Because the blades are of robust steel construction, they are not easily damaged by underwater obstructions and can be sharpened when they become blunt. This type of weed cutter is most effective against submerged or rooted floating-leaved weeds but less effective against emergent weeds. Dense infestations of emergent weeds impede the passage of the boat and are bent over by the boat so that the blades can slide over them without cutting effectively.

Other weed boats use reciprocating knives similar to those on some agricultural mowers. The toothed blades move in opposing directions usually powered by a hydraulic motor. Various shapes of cutter exist but, most commonly, the cutting blades are in the form of either a U or an inverted T. The depth of the cutting blade is usually controlled hydraulically, most machines cutting to a depth of about 1.5 m. Reciprocating knives are more easily damaged than trailing V blades and, so, are usually operated a few centimeters above bed level to avoid damage from unseen obstructions. In some cutting bars the individual knife blades are bolted to the cutter bar; in others they are attached by rivets. Operators generally prefer the bolted knives because these can be more easily removed and replaced if they are broken or become blunt. The U- or inverted-T-shaped cutters are usually mounted at the bows of the boat so that the weed is cut before the boat passes over it. For this reason, emergent weeds can be cut more effectively than with the V-shaped trailing blades. Floating mats of weed can also be cut by the upright parts of the blade, provided the floating mat has not become too thick, and this aids the passage of the boat through dense masses of weed and allows the floating mats to be cut into small sections which can be towed or allowed to float to the shore where they can be removed.

The speed at which weed-cutting boats can operate effectively is limited, especially in small channels, by the *Venturi effect,* which causes the water to flow backward past the boat as it moves forward; this water movement can cause the weeds to lie flat on the bed of the channel so that the weed-cutting machinery passes over the top. The larger the boat is in proportion to the cross-sectional area of the channel, and the faster it moves, the worse this problem becomes. Boats with reciprocating knives are adversely affected more than those with trailing V blades.

Some weed boats also mount a hydraulically controlled and powered side-cutting arm which can be used to cut emergent and bankside weeds or weeds in water too shallow for the boat to operate directly over the weeds.

Most weed boats are propelled by either paddles or by an Archimedes screw. Unlike conventional boat propellers, these do not become choked or entangled by thick weed beds or by the weed released by the cutter.

In flowing water, both forms of weed-cutting boat work more effectively if operated in an upstream direction so that the cutting blades move underneath the weed and are assisted by the water current to cut the weed. In static water, the blades cut equally well in either direction.

Weed boats provide the fastest form of mechanical weed control. They are the preferred method in channels which are deep and wide enough for the boat and where there are few low bridges or overhead obstructions to impede the passage of the boat.

On some weed boats, the cutting equipment can be replaced with a form of weed rake which is used to catch and collect the weed after it has been cut and has floated to the surface. The weed is then transported to the bank, where it can be removed mechanically or by hand. In large water bodies, experience now suggests that it is more efficient to operate a second boat solely as a harvester, and a number of purpose-built harvesting boats are available with hydraulically controlled baskets which catch and lift the cut weed onto the bank. Weed harvesters are useful in lakes and channels where there is insufficient flow to carry the cut weed downstream. In faster-flowing water, a floating boom can be anchored diagonally across the stream so that cut weed, carried downstream by the current, is captured by the boom and forced to one side where it is removed mechanically or by hand.

Weed buckets. These are best described as modified dredging buckets. They consist of a steel cage in a shape which is similar to but usually wider but shallower than a dredging bucket, with a reciprocating cutter blade attached to the leading bottom edge. They are normally mounted on a hydraulic arm on an excavator or large tractor, although some models can be used on a dragline excavator. The cutter blade is powered by a hydraulic motor using flexible hydraulic hoses connected to the excavator. As the bucket is drawn through the water, the weed is cut and captured. Water drains out as the bucket is removed from the water, and the weed is deposited on the bank. Weed-cutting buckets are slower than weed boats but have the advantage that the weed is cut and removed in the same operation. Weed-cutting buckets range in width from about 1 to 4 m; the smaller ones are more suitable for tractor-mounted operations in small watercourses. The largest excavators, fitted with long-reach booms, can reach about 17 m from the bank so that, if access is available along both banks, a channel of approximately 34 m wide can be cleared of weed. Weed buckets can be used in any depth of water, and skilled operators can cut closer to bed level than is possible with a weed boat so that weed control usually lasts longer.

Another advantage of the weed bucket over the weed boat is that it is possible to cut bankside vegetation as well as aquatic weeds growing in the channel so that narrower channels can be cleared from bank top to bank top in a single operation. However, for maximum efficiency, the excavator needs good access along the bank and obstructions such as trees, fences, and side channels, which have to be circumvented, all reduce working efficiency. Crops growing close to the channel margin may also restrict access or limit the seasonal timing of weed control.

In some situations, the weed bucket can be replaced by a simple weed rake, consisting of a series of slightly curved tines sometimes joined at the bottom by a simple static blade. Sturdily constructed week rakes have the advantage over weed buckets that they can be used to remove soft silt as well as dragging out some of the rooted weed. However, they remove less weed, in a single pass, than a weed bucket and so may be slower in heavily infested channels.

Both weed buckets and weed rakes can be used to remove free-floating weeds and filamentous algae, although this tends to be a slow and inefficient process when compared with chemical and other methods of control. The smaller free-floating weeds such as *Lemna* and *Azolla* species pass easily through the tines of weed rakes and buckets but can be captured if the bucket is fitted with a fine-mesh wire or plastic net. However, the ability of these small, free-floating weeds to spread rapidly into cleared areas makes the process inefficient unless the floating weeds can be contained by movable floating booms which can concentrate the weed around the removal point.

Dredging. Although dredging is much slower than other mechanical-control methods, it has the advantage of removing not only the shoots and leaves of aquatic weeds but also the roots, rhizomes, seeds, and other vegetative reproductive structures buried in the sediment. Dredging can also remove the soft substrate into which the plants root; therefore, it can produce the longest lasting of any form of mechanical control of rooted aquatic weeds. However, it is very slow and expensive when compared with other methods, and it is seldom cost-effective to use dredging solely as a means of weed control. The removal of competition from the rooted weeds coupled with the release of nutrients from the sediment caused by the disturbance during dredging can sometimes result in rapid growth of algae or free-floating plants. When this occurs, methods other than mechanical control may be needed.

Dredging can be used to alter the channel shape and depth and, thus, the growing conditions for aquatic plants. This can have long-term implications for the types of weed that are able to colonize the channel and so affect the needs for weed control. This is discussed further under the section on environmental control.

Advantages and disadvantages of mechanical control. The following generalizations about the advantages and disadvantages of mechanical control when compared with chemical, biological, and environmental methods can be made.

Advantages

- Cutting has an immediate effect. The hydraulic resistance of the weed and any other adverse effects are reduced as soon as the weed is removed.
- There is no pollution or residue in the form of herbicide molecules, biological agents, or environmental changes which could adversely affect or produce long-term changes to the ecosystem (except when major dredging and restructuring is carried out).
- Cutting requires relatively little knowledge of aquatic weeds or operator training.

Disadvantages

- Weeds recover rapidly after cutting, and several cuts may be needed each year.
- The weed must be removed as part of, or after, the cutting operation.
- Removal and disposal can increase costs considerably, even if the cut weed can be used in some way.
- Cutting never produces a long-term cure for a weed problem, whereas herbicides, biological-control agents, and environmental control can (sometimes) eradicate a troublesome weed species or reduce the growth of aquatic weeds to tolerable levels for long periods.
- Access for cutting and dredging equipment is often more limited than for other forms of control.
- Mechanical control affects only those weeds directly in the path of the equipment and, unlike some chemical, biological, and environmental techniques, produces no spread of effect beyond the point of application. Thus it must be applied uniformly and effectively to all areas requiring treatment.

Chemical control

Chemical control of aquatic weeds started in the nineteenth century, when it was discovered that heavy-metal salts such as copper sulfate could control algae. Other chemicals, such as sodium azide and sodium arsenite, were used in the early part of the twentieth century as aquatic weed-control agents. These chemicals are used very little now because it is recognized that they combine two highly undesirable properties. They produce very persistent residues in water and are highly toxic to the aquatic and terrestrial fauna which use the water and to the operators who apply them. Some inorganic chemicals are still used in water for specialized uses. For example, Kay et al. (1982) showed that hydrogen peroxide had algicidal properties and could be used to control unicellular green algae and cyanobacteria in fish tanks. Subsequently, Fowler and Barrett (1986) showed that it can control filamentous algae. It has been used in large raw-water reservoirs, intended for potable-water supply, to control cyanobacteria (Barrett, unpublished data). Other inorganic chemicals are used to control algae either by flocculation of the cells or by precipitating phosphates to render them unavailable for algal growth. However, the majority of chemicals used in aquatic weed control are organic molecules which have been discovered and manufactured as herbicides since 1950.

Herbicides. Worldwide there are about 35 herbicides used for aquatic weed control, although within individual countries the choice is usually more restricted (Pieterse and Murphy, 1990). The susceptibility of weed species to specific herbicides and local legislation are the main factors which determine the range of chemicals available to users. This section describes the more commonly used herbicides and the benefits and disadvantages of their use.

Most of the herbicides which are currently used for aquatic weed control were developed originally for terrestrial use. Their properties and behavior in soil were well known before additional research established their efficacy against aquatic weeds, their safety in aquatic situations, and that any adverse effects could be minimized by appropriate precautions. Prior to the 1980s, many herbicides were tested for use in aquatic situations. In recent years, the high cost of the research necessary to satisfy increasingly strict safety and efficacy regulations imposed by many countries precludes the development of new chemicals for use in the relatively small aquatic market. It is unlikely that new products will become available in the foreseeable future. Nevertheless, there is no doubt that the biological effects of herbicides in aquatic ecosystems have been investigated in far more detail than the effects of mechanical control, and, to date, there is no evidence that they are more harmful to the environment and may sometime be less harmful than some forms of mechanical control. However, they can produce longer-lasting control and, so, can lead to changes to the animal community because species dependent on large masses of weed decline, while those which require more open water increase in abundance.

To ensure that herbicides are used safely and correctly, it is important that operators have a thorough knowledge of the herbicides which they are using. They need to know the correct dose, timing, and method of application and must be able to identify the weeds that are susceptible to the chosen product. They also need to have an understanding of the persistence of the herbicide and any potential risks to nontarget organisms so that appropriate precautions can be taken to prevent or minimize adverse effects. There have been examples, in both terrestrial and aquatic situations, of herbicides being wrongly selected or applied, with the result that little weed control was achieved, there was avoidable damage to nontarget organisms, and much money was wasted.

The choice of which herbicide to use depends on the target weed species and the conditions under which they are growing. The method of application of the selected herbicide will be determined by the properties and formulation of a particular product. When used to control aquatic weeds, some herbicides are applied as a spray directly onto exposed foliage, while others are added to the water and subsequently absorbed by the submerged foliage or roots. The following section describes some of the more commonly used aquatic herbicides and the weeds on which they are used.

Control of emergent and floating-leaved plants by direct foliar spray. Aquatic plants which produce floating or emergent leaves can be controlled by spraying a herbicide onto the exposed foliage. For this to be effective, a sufficient proportion of the plant's foliage must be at or above the water surface at the time of spraying for the leaves to absorb a lethal dose of the herbicide. Floating leaves must be dry on the upper surface so that the spray droplets are retained and absorbed by the leaves. Most floating-leaved, emergent, and bankside species can be controlled by the application of an appropriate herbicide, although, where plants of different heights are present in the same location, smaller plants may be screened from the spray droplets by their taller neighbors, and a second application may be needed to control them after the taller plants have died.

Herbicides commonly used to control bankside, emergent and floating weeds are shown in Table 11.1. These herbicides seldom have any effect on submerged weeds or algae at the doses recommended for use against emergent of floating-leaved weeds.

The conditions under which herbicides can be applied are similar to those for terrestrial spraying. Plants should not be sprayed if there is a likelihood of rain falling and washing the herbicide off the leaves before it has been adequately absorbed. Usually, a few hours is sufficient to allow absorption to take place. To minimize spray drift, spraying should be undertaken only under calm conditions or when there is no more than a gentle breeze. In aquatic situations, especially in narrow drainage or irrigation channels with adjacent crops, it is generally preferable to use a low-pressure, coarse-spray nozzle which pro-

TABLE 11.1 Herbicides Applied as Sprays for the Control of Bankside, Emergent, and Floating-Leaved Plants

Herbicide name	Target weeds
Asulam	Broad-leaved weeds on banks
Dalapon (2,2-dichloropropionic acid)	Grasses, rushes, reeds, and sedges
2,4-D-Amine (and related alkanoic herbicides)	Broad-leaved bankside, emergent, and floating-leaved weeds
Fosamine ammonium	Trees and shrubs on banks
Glyphosate	Almost all bankside, emergent, or floating-leaved plants
Diquat	Free-floating weeds

duces relatively large spray droplets which do not drift. The volume of water in which the concentrated herbicide is diluted before spraying should be selected so that the leaves are adequately covered but are wetted so that the spray droplets coalesce and run off the leaves.

The quantity of chemical (dose) varies with both the herbicide (and regulations which govern its use) and the target weed. Doses of herbicides applied as sprays directly onto foliage are expressed in terms of weight of active ingredient (or volume of product) per unit area of land or water surface. For example, the recommended dose for glyphosate in Britain is 1.8 kg/ha to control most emergent weeds and at 2.2 kg/ha on many floating weeds. However, in other countries and on different weed species, the dose may be higher or lower. The concentration of active ingredient (herbicide) within individual products can vary, and this should always be checked before calculating the dilution and preparing spray solutions. Most manufacturers provide information on the product label, or in leaflets, on the dose, timing, dilution, and other factors which determine when and how the chemical should be applied. These recommendations are the result of careful research and should normally be observed. It is sometimes tempting to increase the dose to improve efficacy. This seldom works because doses which are too high may scorch the exposed leaves and can reduce uptake and translocation. Reducing the dose, for reasons of economy, below that recommended by the manufacturer increases the risk of incomplete control and the development of resistance in plants which have been repeatedly treated with sublethal doses.

Herbicides which are applied as sprays for the control of emergent or floating weeds can be applied by conventional knapsack or motorized spraying equipment operated either on the bank or in a boat. However, two problems can arise. First, with a conventional handheld spray lance, it is difficult to reach far enough from the bank or from a boat to cover the weeds adequately. Second, the passage of a boat submerges floating and emergent plants and so washes off any spray which has fallen on them. The first problem may be overcome by using specially constructed long-lance sprayers or by mounting long spray booms onto boats or vehicles operating from the bank. The second problem is minimized by steering the boat between the weed beds as far as possible or, where it is impossible to avoid the weed beds, by treating any surviving plants with a second spray application. In some instances, herbicides can be applied by aircraft. This overcomes both of the problems associated with conventional spraying equipment but is expensive and is usually worthwhile only when very large areas are involved.

Control of submerged weeds and algae by herbicide application to water. Herbicides commonly used to control submerged weeds and algae are shown in Table 11.2. Some of these herbicides may also control floating and emergent weeds as well as submerged species, and this can be an advantage where all these types of weed are growing together and causing problems. Submerged plants and algae are generally most susceptible to herbicides when they are young and growing actively and least susceptible later in the season when they are nearing the end of their growth cycle, especially when they are dying back in the autumn or are in a moribund or inactive state. Submerged weeds which are coated with epiphytes, sediment, or calcium marl tend to be less susceptible than those with clean foliage. These deposits accumulate as the plants age, and this reinforces the need to treat young, actively growing plants.

TABLE 11.2 Herbicides Applied Directly into Water to Control Submerged Weeds, Algae, and Some Floating Weeds

Herbicide name	Target weeds
Diquat (and paraquat)	Submerged weeds and algae
Terbutryn (and other triazines)	Submerged weeds and algae
Dichlobenil (2,6-dichlorobenzonitrile)	Submerged weeds
Endothal	Submerged weeds
Acrolein*	All weeds
Fluridone	Submerged weeds
Diuron	Submerged weeds and algae
Copper* salts	Submerged weeds and algae

*Acrolein and copper are highly toxic to fish and freshwater invertebrates. Acrolein is also explosive and toxic to mammals. Acrolein is used in only a few specialized situations, such as irrigation channels, where its environmental impacts are considered to be acceptable.

When a herbicide is applied directly to water, even static water, it disperses and becomes diluted so that the submerged plants are exposed to a much lower concentration of herbicide than occurs in the spray droplets applied to emergent or floating-leaved plants. Also, the herbicide in spray droplets becomes more concentrated with time as the water evaporates from the spray droplet, whereas a herbicide in water is continually diluted and dispersed as time passes. In flowing water, the problem of achieving sufficient uptake by the plants may be worse because the herbicide can be washed away from the target weeds before sufficient uptake has taken place. Some herbicides are adsorbed and inactivated by mud or clay particles in the water and sediment. The diluting effect of the water may also reduce the ability of plants to absorb the chemical. For these reasons, many of the herbicides developed for use against submerged weeds are specially formulated to hold the herbicide close to the plants and to aid absorption. The formulation exploits the properties of a particular herbicide by, for example, controlling the release rate of the chemical so that the plants can absorb it effectively. The formulation may also aid application, ensuring that the product is dispersed as evenly as possible over the surface of the water or holding it within the formulation until it has sunk down to the weeds. Thus, for example, Dichlobenil is formulated as a granule which carries the active ingredient down to the plant roots, where it is absorbed (Tooby and Spencer-Jones, 1978). Diquat, which is absorbed by the stems and leaves of submerged plants but is inactivated on contact with hydrosoil, has been formulated as a thick viscous gel (diquat alginate) which sinks rapidly and sticks onto submerged foliage, thus holding the herbicide close to the site of absorption (Barrett, 1978).

The simpler herbicide formulations allow the herbicide to disperse throughout the water. In static water, the appropriate dose is calculated by measuring the volume of water to be treated and applying a dose which will achieve the recommended concentration. For example, terbutryn is applied at a concentration of 0.005 to 0.1 g^{-3}, which is equivalent to 0.05 to 0.1 parts of active ingredient per million parts of water (ppm). In flowing water, some herbicides are applied through a metered delivery system so that a concentration is maintained in the water flowing past the input point. In this instance, the discharge in the channel is measured and the metered flow adjusted accordingly and maintained long enough for the plants downstream to absorb the herbicide (Bowmer and Smith, 1984).

Other herbicides, such as diquat alginate or the granular formulation of Dichlobenil, are formulated so that they are applied in an undiluted form to the water and sink rapidly onto the weeds or bottom sediment. For these products, the depth and, therefore, the volume of water are less important than the area, and the dose is calculated on an area basis.

Two factors which affect the performance of these herbicides are the water velocity and water quality. Most products carry a maximum velocity limit above which they become ineffective and should not be used. Water quality, particularly the concentration of suspended mud, clay, or organic matter, also affects performance, and the dose may need to be increased, or an alternative herbicide considered, if the levels of these substances are too high.

Application methods for the herbicides applied to water depend on the formulation. Granular formulations are applied by hand (for small-scale treatments), by spinning disk, or by motorized air blower. Spinning disks operate by allowing a stream of granules to fall onto a disk which is rotating rapidly so that the granules are flung out in a circle or arc. The disk can be turned by hand or driven electrically. Granules can also be applied using motorized air blowers which consist of a small petrol engine (often worn as a backpack) which drives a fan. The granules fall into the stream of air and are blown out of a flexible hose which can be directed as required. The viscous diquat alginate formulation is applied either by a hand-operated or motorized pump through a flexible lance which delivers a pencil jet of the formulation over a range of about 5 m. The strings of diquat alginate are directed so that they fall onto the water surface with a space of about 30 cm between each line of alginate. When they hit the water surface, the strings break up into small droplets which sink onto the submerged weed. Liquid herbicide formulations are sometimes applied using trailing hoses. These are mounted on a boom behind a boat so that the liquid is distributed via the boom down each hose which is weighted so that the end of the hose sinks into the weed bed.

Advantages and disadvantages of herbicides. The advantages and disadvantages of the use of herbicide, compared with mechanical, biological, or environmental control, can be generalized as follows.

Advantages

- Herbicides can be applied more quickly than most other control methods.
- Herbicides usually control the target weed for longer than most forms of mechanical control, although some forms of biological and environmental control may last longer.
- It is unnecessary to remove weeds after they have been treated with a herbicide, unlike herbicide treatment in mechanical control.
- Careful selection and application of an appropriate herbicide may allow selective control of only the target weed species and/or very

precise localized control of weeds in predetermined areas. Biological control can be equally or more selective but can seldom be localized.

Disadvantages

- Some herbicide residues may adversely affect the use of water for a period after application.
- Oxygen concentrations in the water may be seriously depleted when large masses of weed, particularly submerged weed, are killed by a herbicide.
- Appropriate selection and use of herbicides requires training, knowledge, and skill by the user. This can also apply to other methods of control, but the risks resulting from misuse of herbicides can be greater.
- The range of times and conditions when herbicides can be used effectively may be more limited than for most other forms of control.
- Long-term control of a weed species allows time for other weed species to colonize and, potentially, cause a new problem for which different control methods may be required. This also applies to selective forms of biological control.

Selective control of algae. Problems caused by algae are increasing worldwide. One reason for this is the relatively rapid eutrophication caused by agricultural runoff and industrial discharges into watercourses which has occurred in recent years. Nutrients, particularly nitrate and phosphate, are rapidly absorbed by algae, which have a higher growth rate than most other aquatic plants. The dense blooms of unicellular algae or the floating mats of filamentous algae compete with and suppress the growth of submerged vascular plants, which further reduces competition and, so, allows even more algal growth to occur. Another reason why algal problems have increased is because more effective forms of vascular plant control have been developed. By controlling the vascular plants, the natural succession of plants, in which algae are replaced by vascular plants, has been pushed backward in favor of algae. It is not surprising, therefore, that when the two processes of eutrophication and effective weed control occur together in the same water body, algal problems develop.

Algal problems are difficult to control by conventional methods. Mechanical control is of little value because unicellular algae are too small to be removed, except by filtration. Filamentous algae can sometimes be raked out but so many filaments remain, and growth is so fast, that recovery is rapid and little long-term benefit is obtained.

Some unicellular algae are controlled biologically by zooplankton, but it is seldom possible to apply biological control except in carefully controlled conditions. Prevention of nutrient inputs, which is a form of environmental control, is the long-term solution to algal problems, but this is extremely difficult to achieve, especially as artificial fertilizers are used increasingly in agriculture and phosphate-rich detergents are used in domestic and industrial situations.

Herbicides, particularly the triazine molecules such as terbutryn, are effective algicides. However, they control most species of submerged plants as well as algae. Once the herbicide has degraded, algae tend to recolonize the water more rapidly than do vascular plants and benefit from the reduced competition. The effect of using nonselective herbicides to control algae can, therefore, make algal problems worse in the long term.

There are very few selective algicides. Copper salts were probably the most extensively used algicidal compounds, although hydrogen peroxide (referred to earlier) has also been used to a limited extent. Copper salts are toxic to fish and invertebrate animals at, or just above, the concentration required to control algae and they accumulate in the sediment. For these reasons, the use of copper salts has been banned, or restricted, in many countries.

Recent research has shown that selective control of algae can be achieved by adding barley or other straw to water. The mode of action is still not fully understood, but it has been established that most alga species are controlled by an unknown substance released from straw as it rots in water (Barrett, 1994). Successful experiments have been carried out in many countries and on a wide range of algae. The efficacy of straw as a selective algicide is now generally accepted and is being used in widely differing water bodies, including garden ponds, canals, lakes, reservoirs, rivers, and streams. A recent study has shown that it is a cost-effective way of controlling algae in potable reservoirs (Barrett et al., in press). A summary of the technique is given below.

Types of straw. Barley, wheat, maize, rice, linseed, lavender, and other straw types have all shown some activity against algae. However, of the more commonly available straw types, barley was the most active for the longest period. Where barley straw is unavailable, wheat straw is probably the second-best option. Hay and straws with a very low lignin content are less likely to be effective.

Mode of action. When straw rots in water, an unknown factor which inhibits the growth of algae is produced. This factor is produced only if the straw is maintained in an oxygenated condition, usually by the movement of oxygenated water through the straw mass. The straw does not become active immediately when added to water and may take 1 to 4 weeks after immersion depending on water temperature and other factors which are probably related to the rate at which the straw decomposes. Once production of the antialgal factor starts, the straw continues to generate it until decomposition of the straw is almost complete. In one carefully controlled experiment, straw remained active for approximately 8 months (Ridge and Barrett, 1992), but in other trials, straw may have remained active for up to 12 months. The period of activity probably depends on the quantity of straw and how tightly it is packed and on conditions such as water velocity, temperature, and nutrient availability.

Algal cells take up the antialgal factor rapidly, and growth stops within 24 to 48 h after the cells have come into contact with the factor. The factor appears to be algistatic rather than algicidal because algae can recover and start to grow again if removed from contact with the factor. However, if the straw remains active, the algal cells are unable to grow and divide and, so, die when they reach the end of their life cycle. Short-lived algae (mainly unicellular species and cyanobacteria) may die 3 to 6 weeks after the straw has been applied. Filamentous algae may survive for several weeks or months, depending on their longevity, although growth stops within a few days of exposure.

Timing of treatments. Straw can be applied at any time of year either to treat an existing infestation or as a preventive treatment. Generally, the latter is preferable since it prevents a problem from occurring and, where season algal growth occurs, straw should be applied well before algal growth starts. This allows time for the straw to become active and so prevent the formation of algal blooms. Because algal growth can recover rapidly after the straw has decomposed, it is advisable to apply fresh straw at least one month before the old straw has decomposed. In temperate regions, a cycle of biannual treatments has been found to be effective with applications in autumn, which inhibit algal growth in the following spring and a spring application to maintain algal control throughout the summer (Newman et al., 1994).

Application methods for straw. Straw floats when first put into the water and, in a loose form, can be blown by wind or carried by currents away from the target area. Therefore, the straw should be retained in nets or applied as bales which can be anchored or attached to posts. Bales are tightly compressed and do not allow water to move easily through the straw to keep it well oxygenated. If bales are used, they should be small (not more than ~20 kg), loosely packed, and should be used only where there is sufficient wave action or current to ensure that oxygenated water is forced through the bale. The preferred method of applying straw, particularly in static, sheltered waters, is to loosen the straw from a bale and retain it in netting. Straw can also be applied in wire cages (gabions) or in post and wire enclosures. These enclosures, which can be constructed using three or four posts driven into the sediment and surrounded by wire mesh, will retain loose straw and have the advantage over nets or cages that more straw can be added as required without necessitating removal of the old net or container.

Required dose of straw. Much of the initial research on the quantity (or dose) of straw required was carried out under laboratory conditions and was calculated on the basis of weight of straw to volume of water. The minimum effective dose of barley straw has been found to be approximately 2.5 g of straw per cubic meter (Newman and Barrett, 1993) but a higher dose ($\geq$5.0 g/m^3) is normally recommended under field conditions and has produced consistent results in numerous treatments. However, it has also been noted in field trials that the depth of water appears to have little effect on the performance of barley straw as an algicide and that the area of water rather than the volume is the more important criterion in determining the required dose. This may be because much of the algal growth takes place in the top meter and the depth of water below this is immaterial. A weight of 50 kg/ha is equivalent to 5 g/m^3 in water one meter deep, and this has been effective even in water several meters deep. In some heavily infested waters when straw has been used to treat an existing algal problem, higher doses (100 to 200 kg/ha) were used for the initial treatment, but these can usually be reduced in subsequent treatments once the algae have been controlled.

Placement of straw. The antialgal factor produced by the straw is rapidly absorbed by algae but is equally rapidly adsorbed by mud or organic material. Straw is most effective if it is held in the surface layers where most of the algal growth occurs and away from the bottom sediment. In order to prevent the straw from sinking to the bottom once it becomes waterlogged, the bales or nets should be attached to floats so that they remain near the surface even when waterlogged.

It is preferable to use numerous small quantities of straw rather than a few very large masses, and the amount of straw in each net or bale should not exceed 10 to 20 kg. In static waters, these nets or bales should be spaced in the water body so that the average distance between each straw mass is not more than about 70 m and they should be located where they are exposed to wind and wave action so that water is circulated through the straw, keeping it well oxygenated and distributing the antialgal factor through the water body. In flowing waters, straw can be anchored along either side of the main channel, with a distance of about 70 m between each net or bale. It is important to ensure that bales or nets are firmly anchored and will not be washed away by currents or dislodged by boat traffic even when the straw has started to decompose and the straw mass becomes soft. Where possible, anchoring sites should be chosen where the current will distribute the antialgal factor across the whole width of the channel and not direct it along the margins where it will be inactivated by contact with the sediment and bank.

Additional information. Although the nature of the antialgal factor has not yet been fully established, experiments have shown that higher plants are not affected by it and their growth may increase as a result of straw applications, probably because of the reduction in competition from algae. In some lakes, temporary suppression of the algae has allowed vascular plants to recolonize the water so that their dominance was restored and further straw treatments were unnecessary. Invertebrate animals often colonize and breed in the straw, producing very large numbers which eventually become food for fish and wildfowl. Straw is being used in a number of potable supply reservoirs, and there is no evidence so far of an adverse effect on water quality. In one experiment, the water-treatment process has been shown to benefit because of the reduction in algal cell numbers so that filters worked more efficiently and taste and odor problems were reduced (Barrett et al., in press).

Biological control

Many water plants are attacked by pathogenic fungi and bacteria or are eaten by invertebrate animals, fish, birds, and mammals. A close examination of almost any water plant will show some degree of damage by a biological agent. In many instances, the damage is small and the growth rate of the plant is sufficient to compensate for the damage so that the plant survives and reproduces. Sometimes, the damage is sufficient to retard or kill the plant so that the population is kept in check. Where this occurs naturally, the fact that biological control is preventing a weed problem from developing may pass unnoticed.

The existence of a weed problem shows that natural biological control is not occurring or has been unable to keep pace with the growth rate of the plants. This may be because there are no natural control agents present or because the conditions favor the plant more than the biological agent so that it is unable to suppress growth adequately. When this occurs, it may be possible to introduce a control agent or to aid an existing biological agent to ensure that the weed is controlled.

The definition of *biological control* given by the report on terminology in weed science (Coordinatiecommissie Onkruidonderzoek, 1984) is "a control agent based upon a living organism or virus."

Pieterse (1990b) identified three approaches to biological control of aquatic weeds:

1. The use of selective organisms, i.e., organisms which attach one or only a few species.
2. The use of nonselective organisms, i.e., organisms which attack all (or nearly all) weed species present.
3. The use of competitive plant species, i.e., plants which compete with, and suppress, the weed species but are not themselves regarded as weeds.

Competitive plants achieve control of another species by producing allelopathic chemicals (Elakovich and Wooten, 1989) or by altering the growth conditions to the disadvantage of other plants. Although there is some evidence that allelopathy may occur in aquatic plants, there are no generally recognized techniques by which this property can be used as a weed-control technique. Introduction of a plant species which competes aggressively with the target weed can result in the latter's decline and ultimate replacement by the introduced species. This may arise from competition for nutrients or light. In the latter case, bankside vegetation may be managed so as to shade the water and reduce aquatic plant growth. It can be argued that this is really environmental—rather than biological—control of the weed (see later). Certainly, there are nonbiological means of nutrient removal or shading and the use of shade and other environmental-control methods is described later. This section describes control by approaches 1 and 2 (above): selective and nonselective organisms.

The use of selective organisms for biological control has produced some highly successful results. Wapshere (1982) defined two methods of selective control:

1. *Classic biological control.* This generally involves the importation and release of an exotic pest or predator to control an introduced exotic weed species which is not prone to attachment by organisms native to its new environment. The biological agent is released, usually in small quantities and sometimes on repeated occasions, and allowed to reproduce and spread throughout its target's range where it reduces and maintains the target's population at a low level that can be tolerated. It seldom, if ever, eradicates the weed.

2. *Inundative biological control.* This approach is a strategy akin to herbicidal control of weeds. Large quantities of the biological agent

are applied over the whole of the target infestation. The intention is to induce a rapid kill of the target and establish control throughout the season. The agents used are generally indigenous and do not survive for long, at least in useful quantities, after the demise of the target. Thus, fresh applications are normally made each year or when the weed reestablishes itself. This strategy is well suited to the use of microorganisms as the control agent, although it can also be used with weed-eating fish on a put-and-take basis.

Classic biological control. This has produced some spectacular successes. Biological agents which are specific to a single species or a very small group of plants have been isolated from the region of origin of the target weed and, after very careful research to ensure that they will not harm any other native species, introduced into the region where control is required. Usually, the introduced control agent breeds and spreads so that a relatively small inoculum is required and the process can be left to continue without further effort. Sometimes, the control agent may require periodic reintroductions to ensure that adequate distribution is achieved or that numbers are maintained. Harley and Forno (1990) give examples of classic biological control using insects on some emergent and floating weeds.

Alternanthera philoxeroides (alligator weed), which originates in South America, has been successfully controlled by the alligator flea beetle *Agasicles hygrophila* in a number of countries, including the United States, Australia, Thailand, and the warmer parts of New Zealand. Colonies of this beetle are available from the Commonwealth Scientific and Industrial Research Organisation (CSIRO) Australia, National Biological Research Centre (NBRC) Thailand, or the U.S. Department of Agriculture (USDA).

Salvinia molesta, which originates in southeastern Brazil, has been controlled in Australia, Papua New Guinea, India, and Namibia by the weevil *Cyrtobagous salviniae,* collected from the native range of the weed. It is predicted that this control agent will successfully control *S. molesta* in most tropical and subtropical areas of the world. Colonies of the weevil are available from CSIRO Australia.

Eichhornia crassipes (water hyacinth) is a native of Brazil and has now spread widely to almost every tropical and subtropical region. Biological control has been partly successful with a number of insects. Of these, the weevil *Neochetina eichhorniae* has been the most successful so far and has controlled this weed in the United States, Australia, and parts of Africa, although an interval of about 5 years may be necessary between the release of the insect and any noticeable reduction in the growth of *E. crassipes*. The opportunities for biological control of this weed throughout its introduced range are thought to be good. Colonies of *N. eichhorniae* are available through the Commonwealth Institute of Biological Control or from CSIRO, Australia, and the USDA.

Pistia stratiotes (water lettuce) probably originates in Central America although it is widely distributed throughout tropical regions. Control of this weed has been obtained in Australia using *Neohydronomus pulchellus* (order Coleoptera; family Curculionidae), although some resurgence occurred each summer (Harley et al., 1984) and it has been introduced into Papua New Guinea, South Africa, and the United States. Harley and Forno (1990) considered that considerable potential for control by this agent exists in other parts of the world. Colonies are available from CSIRO Australia. *Pistia stratiotes* was also successfully controlled in Zimbabwe by the weevil *Neohydronomus affinis,* which reduced the cover of water lettuce by at least 80 percent after 8 to 15 months (Chikwenhere, 1994).

The control of submerged weeds, such as *Hydrilla verticillata* and *Myriophyllum spicatum,* by insects is being studied in many countries, and some promising experimental results have been obtained (e.g., Creed and Sheldon, 1994). However, further research is required before the techniques are developed sufficiently to recommend their use.

Biological control by plant pathogens. Charudattan (1990) reviewed the potential for biological weed control using fungi. This research is still in a relatively early stage and, although a number of plant pathogens which show promise as aquatic weed-control agents have been identified, none have so far been developed fully. Some emergent and floating-leaved plants have been controlled successfully with plant pathogens, either alone or in combinations with insect agents, but control of submerged weeds does not appear to be feasible at present.

Inundative biological control. This involves the use of a range of biological agents including microorganisms, invertebrates, and vertebrates. They are generally native species in the area of use, although exotic vertebrates (e.g., grass carp) can be used. Although there has been some success in developing microbial (plant pathogenic fungi) agents for the control of terrestrial weeds, their use against aquatic weeds remains generally experimental. One example is the commercial wettable powder formulation of *Cercospora rodmanii* developed in the United States for the control of *E. crassipes*. After field trials, this product has not progressed for various reasons; principally, the weed outgrows the disease. It is clear that the fungus needs to be made more virulent (by genetic or selection means), or it needs to be aided by abiotic agents such as low doses of herbicide or used in conjunction with another biological agent (Charudattan, 1991).

Invertebrate and vertebrate agents can be very effective but are generally less selective to individual species of weed than are microorganisms. They may be selective for groups of weed such as submerged weeds or unicellular algae. If a weed problem exists, it is clear that these native agents are not present in sufficient numbers and will require some form of assistance to increase their numbers or potency in order to achieve effective weed control. In the simplest form, this can sometimes be achieved by a captive breeding program. Examples include enhancing the populations of various native weed-eating fish species or introducing colonies of invertebrate animals such as the water flea, *Daphnia species,* which graze on unicellular algae.

The problem with this approach is that the biological-control agent will almost certainly be part of a food chain and be prey to some carnivore further up the food chain. If its numbers are increased artificially, it will have a temporarily beneficial effect on the weed problem but the increased population will also enable predator numbers to increase so that the situation reverts to a state in which the populations of the biological control agent are too small to be effective. Restocking of the biological-control agent has progressively less effect on the target weed as the numbers of predators are maintained and increased by a constant source of food. Frequently, the way to enhance the populations of a native biological-control agent is to control its predators so that numbers can increase naturally. This technique is sometimes called *biomanipulation* rather than biological control. An example of this approach was demonstrated by Giles et al. (1990), who wished to increase the number of invertebrate animals in a pit formed by gravel extraction as food for ducklings. They removed the coarse fish from a gravel pit, thus allowing the number of algal grazing invertebrates to increase. One result of this treatment was that algal populations were controlled so that the water cleared and vascular plant growth increased. A similar experiment was carried out by Van Donk (1990) in which a lake dominated by unicellular algae and cyanobacteria was cleared by removing the coarse fish populations. The reverse process has also been observed. Domaniewski (personal communication) has observed that stocking lakes with fish species, particularly carp and bream, has resulted in an almost total loss of submerged macrophytes within one season and an increase in turbidity caused by the feeding behavior of the fish and by blooms of unicellular algae. This form of biomanipulation may have applications to slow-flowing drainage and irrigation channels, especially those with soft, silty bottoms in which the benthic feeding habits of the fish cause turbidity.

To be effective, biomanipulation requires a detailed knowledge of the interrelationships between the target weed, the biological-control agent, and the predators which control the agent. While this approach has considerable promise, it is still in an early stage of development. Biomanipulation techniques must be tailored to suit exactly the ecosystem presented in each water body because of the complexity of the biological and environmental relationships. Other factors may also influence their use. For example, the removal of fish stocks to enhance invertebrate populations can have a major impact on other aspects of the ecosystem as well as on the human population who may be dependent on the fish as a source of protein.

Nonselective biological control. It is perhaps misleading to suggest that any control agent (whether biological, chemical, or mechanical) is totally nonselective and will control any and every type of water plant. There is always some degree of selectivity in any method of weed control. The majority of nonselective biological-control agents are animals, and the range of aquatic plants which can be controlled by any individual biological agent may be limited by factors such as the plant's size or roughness, palatability, or location within the water body. Nevertheless, nonselective biological control agents will attack a wide range of species. Frequently, however, their ability to control a particular species will vary with the growth stage of the plant. For example, young tender shoots may be eaten in preference to old, tough ones.

Many animal species live exclusively or largely on water weed. Warm-blooded animals which could be regarded as biological-control agents of aquatic weeds include the water buffalo, capybara, and manatee as well as birds such as ducks, geese, and swans. Cold-blooded, nonselective biological-control agents are mainly various freshwater fish, although some invertebrate animals, including snails and crayfish, have been used locally. Their ease of handling and control, reproductive rates, and resistance to disease and predation influence their suitability, especially when they are being considered for importation to regions where they do not occur naturally. Of those animals studied to date, the most widely used is the herbivorous fish *Ctenopharyngodon idella.* This is known as the *Chinese grass carp* or the *white amur,* as it originates from rivers in the Amur basin in northeastern China and adjacent parts of Siberia. It eats a wide range of aquatic plants, particularly submerged vascular plants, some filamentous algae, and also some floating weeds. When submerged and floating weeds have been controlled and food is scarce, grass carp have been observed to pull emergent weeds and bankside grasses down into the water and consume them. These fish have been imported into many countries and are widely used as a nonselective biological-control agent.

The use of phytophagous fish, including grass carp, was reviewed by van der Zweede (1990). Grass carp become almost totally herbivorous once they reach a length of about 3.5 cm but have a poor food-conversion ratio (about 30 g of plant food to 1 g of fish weight). This makes them useful as a weed-control agent because of the large quantity of weed eaten by the fish. Much of the cell-wall material is undigested by the fish and is released into the water as feces. The fish can tolerate a wide range of water temperatures and quality but require very specialized conditions for breeding. This is considered to be an advantage because it reduces the risk that natural breeding will lead to overstocking and eradication of all aquatic plant growth, which might be damaging to native fish and wildlife stocks.

The feeding rate of the fish is dependent on the temperature of the water. Feeding starts when the water temperature reaches about 12°C and increases until water temperatures are at or above 16°C. In temperate climates, where water warms up slowly in the spring or where summer temperatures remain below the optimum feeding temperature of the fish, weed growth can exceed grazing rates so that control is poor. Where high water temperatures remain throughout the year and where rapid increase in temperature occurs in the spring, effective control is obtained.

The stocking rate necessary to achieve effective control within one year has been found to be about 200 to 250 kg of fish per hectare. However, when planning a stocking policy, it is important to remember that the balance between growth rate and mortality can vary considerably with the age of the fish and with external factors. Young fish can double their weight annually so that the growth rate usually exceeds mortality for several years and the biomass of fish increases. Therefore, if fish are stocked initially at around 200 kg per hectare, they can produce a satisfactory level of control in the first year but will be overstocked in the second and subsequent years. Culling may then be necessary to ensure survival of the fish, to preserve the habitat for other species, and to protect banks and beds from erosion caused by overgrazing of the weed. Alternatively, a lower stocking rate (perhaps 75 to 120 kg/ha) may be used so that, although the level of control is reduced in the first year, it is not necessary to cull the fish in the second or, perhaps, subsequent years. As the fish age, mortality begins to exceed growth and the grazing pressure is reduced so that restocking may be necessary after several years to maintain adequate weed control. The presence of predators may alter the ratio of growth to mortality, and it is sometimes necessary to increase the frequency of restocking, particularly in waters which contain predatory fish species, because grass carp are sluggish and easily taken by them. Grass carp are widely considered to be a palatable species and are deliberately grown for food in China and a number of other countries. In some instances, they have been caught and eaten when they were introduced for weed-control purposes, so the benefits were lost. While these fish can live for 30 years or more and reach a weight of 30 to 40 kg, it is clear that they will be an effective and acceptable form of biological control only if managed regularly.

In many countries, there are strict controls on the release of grass carp to ensure that fish diseases and environmental damage do not result from their release. Often their use is restricted to enclosed waters or to channels where they can be contained by fences or barriers. It is doubtful that their release into large, uncontrolled river systems would be effective for a number of reasons. One of the main reasons is that the fish have preferences in the species of weed which they eat. When enclosed, they eat the preferred species first, followed by less-palatable weeds, until effective control of most species is achieved. Given complete freedom of movement, they may migrate once they have eaten the preferred weed species, leaving the other weeds to recolonize the cleared area. Unless the stocking rate throughout the entire river system was such that the fish were forced to eat less-palatable weeds, no effective control would be achieved. The tendency of the fish to migrate upstream at breeding time, even if no breeding took place, and to move downstream subsequently, would mean that control was not maintained consistently throughout the system. It would also be virtually impossible to manage the stocking density of fish through an entire river system.

Despite these limitations, grass carp have been imported to much of Europe, America, and Asia, where they have been used effectively for weed control in enclosed waters and drainage systems and have been the subject of an intensive breeding and stocking program in Egypt. In the right conditions, they are a useful way of controlling aquatic weeds and of converting unwanted plant material into edible food for human consumption.

Environmental control

As the term implies, this form of control is achieved by changing the environment in or around a water body to make it less suitable for weed growth. A weed problem exists because the conditions are favorable for a plant or range of plants to grow, and the objective of environmental control is to reduce or remove one or more of the factors which favor plant growth to the point where growth is no longer troublesome. It has the advantage over other methods of control that it targets the cause of the problem rather than the symptoms and, therefore, can have a much longer-lasting effect.

The factors which are essential for plant growth are light, nutrients (particularly nitrate and phosphate), carbon dioxide, water, and, for most aquatic plants, a substrate in which they can root. By reducing or altering the availability of one or more of these factors, growth can be suppressed. At first sight, it may seem an impossible task to reduce the availability of any of these components because they all occur naturally within the water body, or enter it from external sources over which the water manager has little or no control. However, various environmental-control techniques which have a useful and sometimes profound effect on weed growth have been developed.

Environmental-control techniques

Light limitation. All green plants, including aquatic species, require light to photosynthesize, although there is considerable variation between species as to the quantity and quality of light required for growth. Some require direct sunlight, while others can tolerate degrees of shade. In general, emergent and floating-leaved plants require more sunlight than do submerged species because the latter are adapted to the lower light levels found underwater. Water, and suspended particles within it, absorb light so that the light level falls rapidly as the depth increases. Even so, in clear water, submerged plants can grow in depths of 10 m or more, although troublesome levels of growth usually occur only in waters less than about 5 m deep.

Pitlo (1978) described a number of ways of reducing light levels in water. One way is to shade the water surface by planting trees or encouraging tall emergent vegetation along the bank (Dawson and Haslam, 1983). This method is applicable to narrow watercourses and is a well-established technique which has been practiced for centuries. In recent years, the benefit of tree shading in reducing aquatic weed growth has sometimes been forgotten. Trees require regular management, and, for reasons of short-term economy, engineers have sometimes removed the trees with the result that weed growth increased and management costs escalated. Because newly planted trees may take a number of years to grow sufficiently to produce effective shade, other weed-control methods may need to be maintained in the early period, and it is sometimes preferable to plant the trees along one bank, leaving the other accessible to weed-cutting machinery until the trees have become established. It may then be possible to plant the other bank, but the need to maintain access for periodic dredging and other essential maintenance operations should be considered. The degree of control depends on the species of tree and the density of the leaf cover. Even partial shading will reduce the growth of many aquatic weeds, but some plants (e.g., *Hydrilla verticillata*) require only 0.5 percent of full sunlight to fall onto the water surface to survive (Bowes et al., 1977). In tropical regions, where the sun is almost directly overhead at midday, adequate shading is more difficult to achieve than in temperate regions, where a line of trees along the equatorial side of the channel is adequate to reduce weed growth. In very narrow channels, even the presence of tall herbs or reeds along the bank can reduce the growth of the more-troublesome submerged weeds. Tree species suitable for shade purposes should produce dense canopies and deep-rooting systems which help stabilize the banks. They should be tolerant of flooding and should not shed branches easily but should be tolerant of regular coppicing or pollarding, which are required when they become too large. The wood produced can be a valuable by-product.

If light cannot be prevented from falling onto the water surface, it can sometimes be prevented from penetrating below the surface. This can be effective in reducing the growth of submerged weed and algae. Surface shade has been induced artificially by floating black polyethylene sheeting or other floating materials on the surface. Although this can reduce or eliminate submerged weeds, generally, there are disadvantages which outweigh any advantages. Anchoring the floating material in place is extremely difficult; the technique impedes many uses of the water and is damaging to wildlife.

Another form of artificial shade is produced by adding a vegetable dye, Aquashade, which absorbs light at the wavelengths used by aquatic plants for photosynthesis (Barltrop et al., 1982). The dye is nontoxic and gives the water an attractive blue color. It has been used effectively in small static water bodies where it persists sufficiently to kill submerged plants by preventing photosynthesis. However, it is unlikely to be cost-effective in flowing-water situations where the dye would be continuously diluted and washed away.

Floating-water plants such as water lilies produce large floating leaves anchored in place by long, slender stalks. They shade the water surface so that little submerged weed grows beneath them but have relatively little impedance to water flow (Pitlo, 1978; Dawson and Haslam, 1983). In some parts of Britain, they are deliberately encouraged to grow along the margins of larger watercourses where they will grow in water up to about 2 m deep and reduce the growth of submerged weeds. In the center of these channels, the water is often too deep for submerged weed to become established, so that little weed control is required. If the water lily spreads too far into the channel, it is easily controlled by cutting or by spraying the floating leaves with glyphosate. Free-floating weeds such as *E. crassipes, S. molesta,* and *Lemna* species have the same effect on submerged-weed growth but have a number of disadvantages. They are too aggressive and can cover the water surface so effectively that many problems arise. These plants can produce a very large biomass which impedes fishing and navigation. They shade the water so effectively that all light is excluded and the water becomes deoxygenated, killing fish and other aquatic organisms. Free-floating weeds can also be washed away in fast-flowing channels, causing problems downstream and allowing light back into the channel. For these reasons, it is important to select and preserve only those weed species which have the right properties. These are

1. Preferably native species which will grow well in situations where weed control is required and that are easy to establish.
2. Plants which produce an effective degree of shade without growing so aggressively that they become a problem.
3. Plants which are easily controlled when necessary.

Another means of light reduction is to increase the load of suspended silt. Suspended-silt particles not only absorb light but also settle out onto plant surfaces, reducing photosynthetic efficiency. This form of environmental control is limited to slow-flowing or static waters which have a very fine, easily suspended sediment. Silts can be brought into suspension mechanically, for example, by the turbulence produced by boat propellers. Although some weed control is achieved by the chopping action of the propeller, the majority of control is achieved by the reduction in available light caused by the suspended silt. This is why canal systems which are shallow and experience numerous boat movements seldom have weed problems (Murphy and Eaton, 1983).

Turbidity is also caused by the feeding behavior of various species of bottom-feeding fish. Species such as carp and bream feed in the bottom sediment, taking mouthfuls of mud and filtering out invertebrate animals. The mud is ejected through the gills, and submerged weeds are suppressed by the turbidity produced by this suspended material. Some weed is also uprooted during the feeding activity of the fish so that they produce a combination of biological and environmental control. Lakes which have been stocked with these fish can lose most of their submerged-weed growth within one year (Domaniewski, personal communication). This method of control may result in the submerged weeds being replaced by blooms of unicellular algae. Their growth is probably encouraged by the loss of competition from the higher plants and by the nutrients released from the mud by the disturbance caused by the fish. These blooms of unicellular algae also increase the turbidity of the water, thus adding to the shading effect produced by the suspended sediment. However, in some experiments, this process was reversed by removing the benthic-feeding fish. The result was that turbid lakes dominated by blooms of unicellular and blue-green algae reverted to clear water with healthy growth of submerged vascular plants. Fish also feed on the invertebrate animals which help control unicellular algae, so that the manipulation of fish stocks to determine the type of weed is partly a biological and partly an environmental form of control.

Another method of reducing light penetration is to raise the water level so that a greater percentage of the bottom is below the photic zone (Nichols, 1984). This approach can have important implications for the design of irrigation and drainage channels. Although it is seldom possible to define exactly how deep water must be to prevent weed growth, a study of local conditions and an assessment of the likely turbidity of water in the channel may help in designing channels which do not support extensive weed growth and, therefore, have lower management costs.

Nutrients. The two plant nutrients which are most important in determining the quantity and type of aquatic weed growth are nitrate and phosphate. High levels of nitrate generally produce heavy infestations of vascular plants and green filamentous algae. Phosphates are also necessary for the growth of vascular plants and green algae but are particularly important in encouraging the growth of blue-green algae or cyanobacteria. These organisms are capable of fixing atmospheric nitrogen and, provided phosphate levels are adequate, can thrive in waters where the nitrogen level is too low to sustain significant quantities of vascular plant growth.

Nitrates are highly soluble and are washed into watercourses as runoff and drainage from agricultural land. Excessive use of nitrate fertilizers has contaminated groundwater in some countries and has resulted in dangerously high levels of nitrate in potable water abstracted from aquifers or from rivers. Because the movement of these nitrate-rich waters through soil is slow, it might take 30 years for nitrate levels to fall after any reduction in fertilizer use. However, there is evidence that at least some of the nitrate draining off agricultural land can be removed by maintaining a margin of herbs and trees about 20 m in width along the banks of a watercourse. Some of the nitrogen is absorbed by the plants, while more is lost as gaseous nitrogen produced by denitrifying bacteria living in the conditions created by the plant roots (Vought et al., 1994). There is also evidence that the emergent and submerged macrophytes in wetlands, through which water percolates before entering the river or drainage system, may remove nitrates (Weisner et al., 1994).

Phosphate levels are increasing rapidly in many waters, leading to problems of excessive algal growth. Phosphates have limited solubility and do not drain as readily as do nitrates from agricultural land. However, they enter water from point sources of release such as sewage and industrial effluent outfalls. Modern, tertiary sewage treatment removes much of this phosphorus, which originates principally from detergents and other industrial chemicals. Less-efficient treatment processes allow release into the water. Once phosphates enter a water body, they can remain for many years, recycling from the sediments as soluble phosphates which are absorbed by the algae and, when they die, being returned to the sediment along with the dead algal cells. While the surface of the sediment remains well oxygenated, the phosphates remain in the sediment in insoluble forms, but if the sediment becomes anaerobic, the phosphate becomes soluble and diffuses into the water to be absorbed again by algae. In temperate regions, algal dieback takes place mainly in autumn and, under the colder conditions of autumn and winter, the surface layers of the sediment remain well oxygenated. In the spring, as water temperatures rise and microbial activity increases, the surface of the sediment becomes anaerobic and the phosphates are released in time to fertilize the new growth of algae.

Phosphate concentrations can be reduced in several ways. In potable-water supply reservoirs, ferric salts are metered into the water as it enters the reservoir. These precipitate the phosphates onto the bottom of the reservoir. This treatment is very expensive, and the high level of iron in the water may be environmentally damaging. In relatively small ponds and lakes, phosphates can be kept in an insoluble form and, therefore, unavailable to algae, by artificially inducing a current of oxygenated water over the sediment. These currents are produced by compressed air or by pumps, sometimes powered by wind-generated equipment. In rivers and channels, there is usually sufficient current to keep the surface of the sediment oxygenated so that the phosphates remain insoluble. However, during periods of low rainfall, velocities may decrease to the point where sediments become anaerobic and phosphates are released. One way of maintaining a good velocity over the sediment and reducing the release of phosphates, as well as producing other beneficial effects, is the use of two-stage channels. These are discussed later.

Carbon dioxide. It is very difficult to influence the amount of carbon dioxide available to aquatic plants. Emergent and floating plants absorb much of the carbon dioxide which they need for photosynthesis, directly from the atmosphere, and, for most of their life cycles, their growth is unaffected by levels of dissolved carbon dioxide in the water.

Submerged vascular plants and algae absorb carbon dioxide from the water, where carbon dioxide is always present as dissolved gas or as bicarbonate ions. Gaseous carbon dioxide dissolves readily in water from the atmosphere and is generated during respiration, by aquatic fauna and microorganisms in the water and sediment. Carbon dioxide is absorbed by plants during photosynthesis and released during respiration. In daylight, when plants are growing actively, carbon dioxide uptake exceeds production so that the dissolved carbon dioxide concentration falls while pH and dissolved-oxygen levels rise. At night, plant and animal respiration continues but photosynthesis stops and dissolved carbon dioxide levels rise and the pH of the water falls. When plants are dying back at the end of their growing period, or when they are stressed for other reasons, respiration may exceed photosynthesis, leading to high carbon dioxide levels and low pH values.

In hard waters containing calcium carbonate, there is a chemical reaction between calcium carbonate and carbon dioxide in which calcium bicarbonate is formed. This reaction buffers the pH of the water, preventing it from rising above pH $\sim$ 8.5. Many submerged aquatic plants are able to use bicarbonate ions as their source of carbon. During this process, the remaining calcium carbonate, which is less soluble than calcium bicarbonate, is precipitated onto the plant surface, leaving a crusty marl on the leaves.

There is no practical method of removing sufficient quantities of carbon dioxide from water except by allowing aquatic plants to photosynthesize. Dense stands of *Hydrilla verticillata* have been shown to absorb carbon dioxide so rapidly that photosynthesis is limited to the first few hours of daylight and less-competitive plants are suppressed (Van et al., 1976, 1978). Even with this limited period of photosynthesis, dense growth of the dominant weed occurs so that the only effect is to reduce species diversity without reducing the quantity of weed. If the biomass of *H. verticillata* were to be reduced by control measures, the amount of available carbon dioxide would be increased proportionally so that the growth rate of the remaining plant material would increase, leading to rapid recovery.

Water. It may seem inappropriate to include water as one of the factors which could be limited because, by definition, it must be present in a water body. However, there are a number of ways in which water is used to alter or control the growth of aquatic plants.

It has already been stated that most aquatic plants are limited to relatively shallow water where sufficient light reaches the bottom to allow plants to grow. This depth, which is called the *photic zone,* varies with the clarity of the water and the amount of light falling on the surface. However, submerged-weed problems occur mainly in shallow water, usually less than 2 to 3 m deep, and seldom more than 4 to 5 m deep. Emergent weeds seldom grow in water more than about 1 to 1.5 m deep. Therefore, a deep, steep-sided channel will grow fewer weeds than will a wide, sloping-sided channel which allows emergent weeds to colonize the margins and submerged weeds to establish across the bed of the channel. In ponds and lakes, weed problems have sometimes been alleviated simply by obstructing the outflow so that the water level was raised.

Few water plants grow to problem proportions in fast-flowing rivers. The high velocities damage or uproot the plants, and the soft silts into which most aquatic plants prefer to root tend to be washed away and deposited along the margins and in slow-flowing reaches downstream. Free-floating weeds are seldom a problem in fast-flowing waters because they are flushed away before they can develop to nuisance levels. Therefore, a narrow, high-velocity channel tends to have fewer weed problems than does a wide, slow-flowing one. Some plants, such as *Ranunculus* species, are adapted to growth in fast-flowing waters and are unable to tolerate slow flows. These plants are common in some of the fast-flowing chalk streams in southern England, but they showed a marked decline during the drought period of 1989–1992, when river discharges were low. It is possible that, where water velocity can be controlled by sluices and weirs, periodic fluctuations in velocity might be used to limit the growth of both slow- and fast-flowing water species.

One way in which both water depth and water velocity have been incorporated into channel design so as to reduce management costs is the construction of two-stage channels. These are used particularly in rivers and drainage systems where the discharge varies considerably with rainfall. The first-stage channel is designed to carry only the summer or low-flow discharge with a sufficient depth and velocity to discourage the growth of excessive amounts of weed and to prevent the deposition of silt. The second stage, sited on one or both banks, is above summer flow levels and is dry for most of the year. It does not grow aquatic plants, and the terrestrial species which grow on it can be managed by cutting or by grazing by livestock. During periods of high discharge, the water level rises and the second stage comes into use, increasing the channel capacity sufficiently to carry the designed flood discharge. The width of the second-stage channel can vary from a few meters up to several kilometers. An example of these large second-stage channels is seen in situations such as the washes in parts of eastern England which are grazed by cattle in summer but provide a large storage capacity for water during the winter or under flood conditions.

In regions where winter temperatures are consistently below zero and where water depth can be controlled, winter drawdown is an effective way of controlling weeds, particularly in channels and around the margins of deep lakes. The water level is reduced in autumn, exposing the margins and shallow areas to frost action. The roots, rhizomes, and overwintering organs of aquatic plants do not generally tolerate exposure to frost and thus are controlled. In spring, the water level is raised back to normal levels. However, this technique should not be used where winter temperatures do not fall below zero long enough to freeze the exposed sediment. This is because winter drawdown, in the absence of freezing temperatures, will not control dormant weeds buried in the exposed sediment and, under mild conditions, may also enable overwintering fragments of weed to become established in water which, because of the reduced depth, would normally be beyond their depth range. Once established, these fragments grow, as the water level rises in spring, to form weed beds in previously uncolonized areas.

In regions where long periods of consistently hot, dry weather can be predicted, weeds in channels can be controlled by draining the channel and allowing the aquatic plants to desiccate. This can kill the stems and leaves of submerged and floating aquatic plants rapidly, although control of buried vegetative organs and seeds takes considerably longer. Emergent plants are generally more tolerant of short periods of desiccation but will eventually die if the sediment dries out. Normally, summer drawdown should not be attempted in water bodies which cannot be fully drained because, while the water level is low, aquatic plants can become established in water beyond their normal depth range, thus spreading the weed problem.

Both summer and winter drawdown can be damaging to fish and other aquatic fauna, and these techniques should not be used in waters where conservation of these species is important.

In large water bodies, especially those exposed to strong winds, wave energy may control some species of aquatic weed. The most susceptible weeds are the small free-floating species such as *Salvinia* and *Lemna* species, which appear to have little tolerance to the inundation caused by waves. Larger free-floating plants such as *Eichhornia crassipes* and the emergent and submerged weeds may be damaged by wave action but are seldom controlled.

Sediments. Most aquatic weeds depend on sediment as a rooting medium, partly for nutrients but mainly for anchorage. The majority of these plants grow best in soft sediments found mainly in slow-flowing or static waters. As these plants grow, they impede water movement, causing more silt to settle around them. Their roots and rhizomes grow into the new sediment, binding and holding it so that more plants become established and the process is perpetuated. Gradually, the silt beds become shallower and larger until, eventually—unless remedial action is taken—channels and lakes can become so silted that their functions are impaired and large parts become dry land.

Removal of silt by dredging is an essential maintenance operation in many water bodies, and it has several weed-control benefits. Dredging removes not only the stems and leaves of aquatic plants but also the roots and rhizomes so that the whole plant is controlled. Dredging deepens the water and, in some instances, can make it too deep for submerged plants to reestablish themselves, at least until the sediments build up again and the water becomes shallow enough for light to penetrate sufficiently for plant growth. It also removes the nutrients, particularly phosphates, stored in the sediment. Against this, dredging is an expensive operation and is undertaken only when absolutely necessary. It is also damaging to the aquatic ecosystem, destroying habitats and whole communities of plants and animals.

In the long term, poorly undertaken dredging in channels can encourage weed growth. In long-established channels, a stone-and-gravel bed may be exposed as the silts are washed away. These hard surfaces are less easily colonized by aquatic plants and thus tend to have fewer weed problems. However, when these channels are dredged to the extent that the gravel bed is removed and the underlying silt or clay is exposed, aquatic plants become established more easily and weed problems develop. Anther adverse effect of dredging has become apparent in recent years. In the 1950s and 1960s, when powerful and efficient dredging equipment became readily available, engineers were encouraged to dredge watercourses to a channel capacity greater than was needed for the normal range of discharges, presumably in the belief that this would reduce the long-term management costs. The result was that, in all except the most severe flood conditions, the channel was wide, shallow, and slow-flowing. These were ideal conditions for aquatic weeds which rapidly colonized the channels. The beds of weed restricted the flow, causing accelerated velocities between the weed beds, which scoured the soft exposed silt, and caused silt beds to accumulate in the slow-flowing areas within the weed beds. Eventually, these became shallow enough for emergent weeds to become established, further restricting the channel capacity. The remedy, in many instances, has been to dredge only part of the channel to create the two-stage channel design referred to earlier. The deeper water and higher velocity in the first stage of these channels discourages weed growth and provides sufficient velocity to prevent the buildup of silt beds in the main channel, allowing much of the suspended silt to deposit in the slower-velocity areas, particularly among the weeds growing on the second-stage channel during periods of high discharge.

TABLE 11.3 Environmental Control* of Aquatic Weeds

Technique	Types of weed				
		Floating			
	Emergent	Rooted	Free-floating	Submerged	Algae
Light reduction					
Shade (trees or tall herbs)	H	H	H	H	H
Shade (floating plants)	NA	NA	NA	H	H
Turbidity (fish or boats)	NA	S	S	H	U
Nutrient reduction					
Nitrate	S	U	U	U	S
Phosphate	NA	NA	S	S	H
Water					
Steep banks	H	NA	NA	NA	NA
Deep water	NA	H	S	U	U
High velocity	S	U	H	U	H
Winter drawdown	H	H	S	U	NA
Summer drawdown	U	H	H	H	H
Wave action	S	S	U	NA	NA

*Levels of control: H = highly effective control; U = useful level of control; S = some control possible; NA = not applicable or ineffective.

The various methods of environmental control and their potential effect on the different types of aquatic weed are summarized in Table 11.3. The scoring system assumes that appropriate conditions for each type of environmental control exists and indicates the degree of control which may be achieved.

Advantages and disadvantages of environmental control. The advantages of environmental control compared with mechanical, chemical, and biological control are

- The cause, rather than the symptoms, of the weed problem is removed or alleviated.
- It produces very long-term control usually lasting many times longer than alternative methods.
- Management costs can be significantly less than alternative methods.

The disadvantages of environmental control are

- The techniques are site-specific and very limited in the situations in which they can be applied.
- They often involve an initial high capital cost.

- They are slow to produce an effect.
- They can be damaging to nontarget organisms which are affected by the altered environment.

Integrated Weed Control

It has already been stated that no single method of weed management will control all types of water weed and, in very few situations, will a single weed-control operation eradicate a weed problem permanently. It follows from this that several methods of weed control may be needed in a water body to achieve a satisfactory level of overall weed management. These weed-control operations can be totally independent of one another so that neither operation benefits or harms the other, or they can be integrated so that there is an overall net benefit. The objective of integrated weed control is to reduce overall management costs and/or to increase the range and effectiveness of weed control by combining two or more operations.

Most of the techniques which have been described previously can be used in combination with each other, although timing and sequence of application can be important. There are also some instances in which the integration of two techniques can be inappropriate. For example, it is counterproductive to use mechanical equipment to cut submerged weeds if, in the same location, emergent or floating weeds are present which are to be sprayed with glyphosate soon before or after the cutting operation, since this will prevent translocation of the herbicide to the root system and totally negate the effect of the herbicide. On the other hand, some submerged and floating weeds are controlled more effectively if they are cut immediately prior to treatment with Dichlobenil, as this stimulates their regrowth and thus increases their susceptibility to the herbicide.

When considering the use of integrated control, it is first necessary to identify which methods would be effective against the problem weeds in the water body and then to consider how each method might be used to enhance the performance of the other. It is also necessary to ensure that neither technique will reduce the performance of the other as would occur in the example involving cutting and glyphosate given above. There are too many options for integrated control to list them all, and the following are given as examples only. Examples of integrated control are

1. *Grass carp used with either herbicides or mechanical control.* The weeds are first cut or treated with a herbicide to reduce the quantity. Provided this treatment still leaves enough of the weed for the grass carp to survive on, this allows a lower stocking rate to be

effective. The fish will control the target weeds for a number of years, although some control of nonsusceptible weeds (e.g., emergent and bankside weeds) by either cutting or herbicide spraying may be necessary. However, the reduced stocking rate of the fish and the lower overall biomass of weed which needs to be controlled by chemical or mechanical means may reduce management costs in the longer term.

2. *Environmental control* (e.g., *tree shade*) *combined with mechanical* (*or chemical and biological*) *control.* When first planted, trees are too small to have any effect of weed growth in a channel which must be managed by other methods for several years. The trees will also need management, especially to prevent weed growth around the base of the trees to allow them to become established. Initially, the weeds in the channel can be cut and used as a mulch around the base of the trees. As the trees grow, they shade the water, reducing the frequency with which weed cutting is necessary. Alternatively, if the main weed problem in the channel is caused by emergent and floating weeds, they can be controlled with glyphosate, which can also be used to spray off weed growth around the base of the trees until they are fully established. Once the trees are established and large enough to be safe from grazing livestock, weed growth on the banks can be controlled by allowing livestock to graze on the emergent and bankside vegetation. Weed growth in the channel will be reduced, but occasional cutting may be necessary, and the cut weed can be fed to the livestock.

3. *Algal control with straw combined with herbicides or mechanical control.* In waters containing a mixture of algae and other weeds, control can be difficult because of the mass of weed which interferes with mechanical techniques and reduces the efficacy of some herbicides. By using barley straw to control the algae, both the frequency and the cost of controlling other weeds mechanically or chemically can be reduced. In some situations, drainage ditches have been treated with a herbicide, such as terbutryn, to control both algae and submerged weeds because rapid and total control was required. At the same time, straw was introduced to prevent return of the algae. The slower rate of recolonization by vascular plants meant that no other weed control was needed for about 5 years instead of the annual weed control required previously (Newman et al., 1994).

4. *Dredging combined with spraying emergent and bankside weeds.* When a watercourse is dredged, the roots and rhizomes of emergent and bankside weeds are moved with the spoil and may become established on adjacent agricultural land. By spraying these weeds in time to let them die off before the dredging operation, the operator can ensure that they are not spread onto adjacent land, and the dredging operation may be easier because of the reduced mass of material

which has to be moved and spread and because the operator has a clearer view of the channel.

5. *Two-stage channels combined with other weed-control techniques.* The advantages of two-stage channels in reducing the need for weed control has already been mentioned. Even so, some weed control is likely to be necessary, and the presence of a dry second stage at the time of year when most aquatic weed growth occurs provides an adjacent platform for weed-control operations adjacent to the watercourse without the need to gain access through crops.

In many instances, water managers use integrated control techniques, without identifying the significance of the process, because they recognize that no single technique will solve all the weed problems. Even where the weed problem is of a nature where a single technique produces adequate results, it is advisable to have an alternative approach ready for use to aid, augment, or replace the existing technique because changes in weed species or breakdown of equipment can suddenly render existing methods unusable.

Uses of Aquatic Weed

In can be argued that, if a use were to be found for a weed, it would cease to be a weed and would become a crop which could be usefully harvested. Various species of water plant have been used productively for thousands of years. Rushes and reeds have been used for thatching, baskets or mats, and boat manufacture.

Papyrus provided what was probably the earliest form of paper. Other water plants are eaten by humans or fed to livestock. They have been used as fuel, paper and building materials, compost, and mulch. New uses are being discovered, and there is no doubt that many species of water plant could be put to useful purposes if the necessary time, money, and research were to be invested.

Why, if all these uses exist, are water plants still weeds? Essentially, water plants are usually more difficult to harvest than comparable terrestrial plants and, because of their high water content, are more expensive and difficult to transport and use. When the relative costs of harvest and transport are compared, the economic value of water weed is generally much lower than that of terrestrial alternatives. Thus, aquatic weeds have not been exploited widely, and on a global scale, it is extremely unlikely that their potential uses will ever require even a small fraction of the total production or remove the need for weed control. If an economically viable use could be found for an aquatic plant, it is likely that it would be grown as a crop by farmers who would supply it more cheaply and at a higher and more uniform quality than would be possible by harvesting wild populations growing in diffuse and widespread areas. Nevertheless, it is worth considering the potential uses of water weeds which are a local problem and which might be turned into an asset while, at the same time, reducing weed-control costs. Potential uses for aquatic plants were reviewed by Little (1968) and the National Academy of Sciences (1976). The following section summarizes and updates their conclusions.

Herbivorous animals

Fish. A number of fish species live totally or mainly on aquatic plants. Of these, *Ctenopharyngodon idella* (Chinese grass carp) and various species of *Tilapia* are best known. Grass carp are fast-growing fish which can grow to over 30 kg. They can reach 1 kg in their first year and grow subsequently at 2 to 3 kg per year in temperate regions and 4 to 5 kg per year in tropical regions. They are considered to be excellent food with a firm, flaky, white flesh containing about 18 percent protein on a fresh-weight basis.

There are several species of herbivorous *Tilapia* which are an important food source. They are found mainly in Africa and the Middle East, where they feed on a range of aquatic plants and algae. Some species are farmed as a source of food. In other situations they have been stocked into waters where they control weed problems and help reduce insect problems by consuming the larvae.

Other species of fish which eat aquatic plants and algae include *Hypophthal michthys molitrix* (silver carp), *Cyprinus carpio* (common carp), and *Mylossoma argenteum* (silver-dollar fish). All these are reported to be edible and have, in the right circumstances, demonstrated an ability to achieve useful levels of weed or algae control.

Marine fish stocks are being depleted and the need for fish as a source of protein is increasing. The culture of freshwater fish which feed directly on aquatic plants and, to a lesser extent, fish which consume organisms in the food chain which feed on weeds, are potentially one of the main uses of aquatic plants.

Crayfish. There are some 300 species of crayfish, crawfish, or freshwater lobster mostly in the genera *Orconectes, Procambarus,* and *Cambarus*. They are native to most continents except Africa and grow to around 10 to 18 cm long with large edible tail muscles. They are a highly prized food in many countries and are farmed commercially as a delicacy. They are mainly omnivorous scavengers but will eat a wide range of aquatic weeds if other sources of food become scarce. While their ability to control aquatic weed problems has not been researched fully, they are known to have produced useful levels of control in some enclosed waters.

Ducks, geese, and swans. These and other species of waterfowl eat water weed, foraging on submerged and floating weeds and filamentous algae. They will also eat emergent plants, especially when the shoots are young and tender. Swans are not easy to domesticate, but both ducks and geese can be farmed. Clipping the flight feathers on one wing prevents the birds from escaping. Some food supplement may be necessary to keep the birds in good condition if the grazing is poor, but the birds need relatively little management and provide a valuable source of high-quality meat.

Other herbivorous animals. In various parts of the world, there are animals which graze on water weed to some extent. Many of these are useful, or could be useful, to humans as a source of protein, fur, or power or for other uses. In Asia, the Mediterranean basin, South America, and the Caribbean, the water buffalo provides meat, milk, hides, and power. These animals graze on coarse, low-value semi-aquatic grasses such as *Imperata cylindrica* and on floating weeds, including water hyacinth, and will also feed underwater on submerged weeds.

In South America, the capybara (*Hydrochaerus hydrochaeris*), which grows to about 60 kg, and the nutria (or coypu; *Myocastur coypus*), which grows to about 8 kg, are rodents which feed on both aquatic and marsh plants. The capybara is edible, although not considered a delicacy, but the hides of the capybara and of the nutria are valuable.

Processed animal feeds. Domestic animals, including cattle, sheep, pigs, and goats, can be fed on aquatic plants. However, it is usually necessary to process the plants, or to add supplements, to ensure that the animals receive a balanced diet. Water plants contain up to 95% water when harvested, which means that animals eating freshly harvested weeds have to process a very large bulk of material and excrete large quantities of water. If the weeds are partially dried before feeding, the problem is reduced, but there is a danger, particularly in humid regions, that decomposition will start before the weeds are sufficiently dry, and this can create a risk to the animals. Surprisingly, the level of protein and salts in many aquatic plants, as a percentage of dry weight, is as high or higher than in some forage crops. In some instances, livestock may receive all the salts required in their diet when fed on a mixture containing only 10 to 20 percent aquatic plants. Higher percentages of aquatic weed in the diet can lead to an imbalance of salts.

Various ways of processing aquatic plants to improve the feeding value and the storage have been tested. Machines which press and squeeze the plants can reduce the water content, but if the plants are overpressed, valuable cell contents are lost and the resulting liquor is a potential danger if it is allowed to flow back into the watercourse. Silage production may prove to be a useful alternative, as this can be achieved by a simple wilting process before the plants are put into the silage clamps. Again, liquor draining from the silage clamps is a potential hazard to fish and other aquatic organisms.

The value of water weeds as animal foods varies with the condition of the weed when harvested. Young, actively growing weeds with plenty of green leaves and little root are generally best. Older plants tend to have a lower feed value and may become encrusted with epiphytes and carbonate deposits.

Compost and soil conditioning. Water weeds can be composted, used as a mulch or as a green manure. When composted, the plants are left on the bank for a day or two to wilt and are then heaped with additions of soil, ash, and a little animal manure. The resulting compost enhances water retention in sandy soils and improves soil structure. As a mulch, water weeds suppress weed growth and reduce water loss from the soil. When used as a green manure, the weeds can be spread on the soil surface or, preferably, dug in. The disadvantage of using water weeds as mulches or green manure is that large quantities of vegetation have to be transported and spread onto the soil surface. This is a slow and, potentially, expensive operation.

Paper, pulp, and fiber. Tall emergent grasslike aquatic plants, including reeds (*Phragmites australis*), bulrushes (*Typha* spp.), clubrushes (*Schoenoplectus* or *Scirpus* spp.), papyrus (*Cyperus papyrus* and *C. antiquorum*), and various aquatic grasses have been harvested for centuries for food, fiber, and building material. The earliest form of paper was made by the Egyptians from papyrus stems. Reed boats are also known to have existed since prehistoric times. There is a large commercial industry in the Danube delta using reeds to make paper, cellophane, cardboard, and other products. In other parts of the world, aquatic plants are used locally to make roofing material, matting, baskets, bags, and other products. However, the quantities used for any of the products are seldom likely to have a major impact on the amount of weed available or the problems which it causes.

Energy. Water hyacinth and other aquatic weeds can be fermented anaerobically to produce biogas. This is composed of approximately 70% methane and 30% carbon dioxide. It will burn readily and can be used for cooking, for heating, and as a source of power. However, the weed must be chopped or crushed to allow bacterial decomposition to take place, and oxygen must be excluded. This presents a problem if continuous production is to be achieved because oxygen is likely to be introduced into the fermentation vessel when weed is added. This usually stops production of the biogas until the oxygen has been used up by aerobic bacteria which produce carbon dioxide and reduce the calorific value of the biogas.

Wastewater treatment and removal of inorganic and organic pollutants. Aquatic and semiaquatic riparian plants can be used to improve water quality in several ways. The use of wetlands and of buffer strips along banks to remove inorganic nutrients, especially nitrate, has already been mentioned. Much of the nitrate removed by these ecosystems is lost as gaseous nitrogen into the atmosphere, but the use of shallow lagoons has been shown to be an effective form of tertiary treatment for sewage and industrial effluents. The weeds can be harvested and composed so that the valuable nitrates and phosphates are recovered instead of polluting the watercourses and leading to further weed problems. Specially constructed reed beds are being used as an alternative to more traditional forms of sewage and industrial-effluent treatment. Organic pollutants are absorbed and decomposed by the microflora associated with the root structures of the reed.

Conclusions and Discussion

No doubt, aquatic weed have troubled humans since prehistoric times when the earliest *Homo sapiens* tried to catch fish or cross rivers. More recently, the development of agriculture led to the construction of drainage and irrigation channels. The ability of weeds to impede flow in these channels is one of the principal reasons why weed control is now an essential part of land management. If an irrigation channel is blocked by weeds, the result is a drought; if a drainage channel is blocked by weeds, the result is a flood. In either situation, there can be losses of crops, livestock, and human life. The more intensive agriculture becomes, and the more urbanization which takes place, particularly in floodplains, the greater is the risk from weed growth and the greater the need to manage the weed. The more use that is made of water and water bodies, the more opportunities are created for weeds to cause problems. Water abstraction for potable supply usually involves the creation of reservoirs in which weeds and algae grow. They cause taint and odor problems in the water and block filters and pumps. Irrigation reservoirs experience the same problems of blockage to filters and pipelines. Hydroelectric dams become covered with floating mats of weed which block and damage turbines. The greater the demand for water and for water space, the less weed can be tolerated.

Two other factors are exacerbating the problem. First, the need to increase food production has increased the use of agricultural fertilizers, some of which inevitably enter the watercourse and fertilizes the growth of more water weed. In many countries, industry also contributes to the level of nutrients in the water, either directly in the form of effluents from factories or indirectly as products such as detergents which find their way into the water. As well as causing eutrophication, agriculture and industry create a demand for water which then has to be managed to ensure that supplies are constant and available in sufficient quantities. This means that flow has to be controlled and the water held back, thus creating more opportunities for weed growth. Second, weeds have been moved from their country of origin into new areas where no natural control agents exist. There are many examples of introduced plants becoming troublesome weeds. An early example was the accidental introduction into Europe of *Elodea canadensis* (Canadian waterweed), probably brought in with shipments of timber from North America, which is now a widespread problem. Similarly, *Myriophyllum spicatum* (Eurasian water milfoil) was introduced into North America from Europe where it has also become a major problem. Many other weeds have been moved, either accidentally or deliberately, into new regions. The author has seen ornamental jars containing plants of *E. crassipes, S. molesta,* and *P. stratiotes* on sale in Europe. In temperate regions, these plants may not survive if released into the wild because they are not tolerant of low temperatures, but they have already been released into many countries where they not only survive but grow vigorously and present serious problems.

Another example is the deliberate sale of *Crassula helmsii* (Australian swamp stonecrop) in garden centers and aquarist shops. This plant is highly invasive and has already become established in ponds and lakes in Britain, where it dominates and suppresses almost all native aquatic vegetation. Recently, it has been observed in some drainage channels where its dense, compact growth habit is likely to impede flow more than any native species. If it continues to spread, it is likely to become a major weed species in many temperate regions. There are many aquatic plant species which, so far, have remained in their native habitats but which could become new weed species if wrongly introduced into new areas away from their natural control agents.

It seems unlikely that the demand for water and water space, the need for drainage and irrigation, the accidental or deliberate introduction of fertilizers into water, or the introduction of new species will decrease in the foreseeable future. If so, the need to manage and control aquatic weeds will continue and may increase. If a weed problem exists in a water body, it is because the conditions are suitable for the plant species which is causing the problem. Even if that plant can be eradicated, others may soon take its place. In existing water bodies, this means that regular and continuing management will be required. It is only in the design of new, artificial water bodies, or in the reconstruction of old ones, that the opportunity exists to create conditions where weed growth is limited to tolerable levels and where little management is required. So far, there is little evidence that the design and construction of new water bodies has included any consideration of the likelihood of weed growth and the results have been the loss of valuable crops and an ever increasing cost of management.

References

Barltrop, J., Martin, B. B., and Martin, D. F. (1982). Response of Hydrilla to selected dyes. *J. Environ. Sci. Health* **A17**(5), 725–735.

Barrett, P. R. F. (1978). Some studies on the use of alginate for the placement and controlled release of diquat on submerged plants. *Pest. Sci.* **9,** 425–433.

Barrett, P. R. F. (1994). Field and laboratory experiments on the effects of barley straw on algae. In BCPC Monograph no. 59: *Comparing Glasshouse and Field Pesticide Performance* pp. 191–200.

Barrett, P. R. F., Curnow, J., and Littlejohn, J. W. in press. The control of diatom and cyanobacterial blooms in reservoirs using barley straw. In *Proceedings of EWRS 9th Symposium on Aquatic Weeds.*

Bowes, G., Van, T. K., Garrard, L. A., and Haller, W. T. (1977). Adaptation to low light levels by Hydrilla. *J. Aquatic Plant Manag.* **15,** 32–35.

Bowmer, K. H., and Smith, G. H. (1984). Herbicides for injection into flowing water: acrolein and endothal-amine. *Weed Res.* **24,** 201–211.

Charudattan, R. (1990). Biological control of aquatic weeds by means of fungi. In *Aquatic Weeds,* A. H. Pieterse and K. J. Murphy, eds. Oxford Scientific Publications, pp. 186–201.

Charudattan, R. (1991). The mycoherbicide approach with plant pathogens. In *Microbial Control of Weeds,* D. O. TeBeest, ed. Chapman and Hall, London, pp. 24–57.

Chikwenhere, G. P. (1994). Biological control of water lettuce in various impoundments of Zimbabwe. *J. Aquatic Plant Manag.* **32,** 27–29.

Cook, C. D. K. (1990a). Origin, autecology and spread of some of the world's most troublesome aquatic weeds. In *Aquatic Weeds,* A. H. Pieterse and K. J. Murphy, eds. Oxford Scientific Publications, pp. 31–38.

Cook, C. D. K. (1990b). *Aquatic Plant Book.* SPB Academic Publishing, The Hague.

Coordinatiecommissie Onkruidonderzoek (1984). *Report on Terminology in Weed Science in the Dutch Language Area* (2d ed.). Report, Coordinatiecommissie Onkruidonderzoek, NRLO, Netherlands.

Creed, R. P., and Sheldon, S. P. (1994). The effect of two herbivorous insect larvae on Eurasian Water Milfoil. *J. Aquatic Plant Manag.* **32,** 21–26.

Dawson, F. H., and Haslam, S. M. (1983). The management of river vegetation with particular reference to shading effects of marginal vegetation. *Landscape Plan.* **10,** 147–169.

Elakovich, S. D., and Wooten, J. W. (1989). Allelopathic potential of sixteen aquatic and wetland plants. *J. Aquatic Plant Manag.* **27,** 78–84.

Fowler, M. C., and Barrett, P. R. F. (1986). Preliminary studies on the potential of hydrogen peroxide as an algicide on filamentous species. In *Proceedings of EWRS/AAB 7th Symposium on Aquatic Weeds,* pp. 113–118.

Gangstad, E. O., and Cardarelli, N. F. (1990). The relation between aquatic weeds and public health. In *Aquatic Weeds,* A. H. Pieterse and K. J. Murphy, eds. Oxford Scientific Publications, pp. 85–90.

Giles, N., Street, M., Wright, R., Phillips, V., and Traill-Stevenson, A. (1990). A review of the fish and duck research at Great Linford 1986–1990. In *The Game Conservancy Review of 1990,* 129–133.

Harley, K. L. S., Forno, I. W., Kassulke, R. C., and Sands, D. P. A. (1984). Biological control of water lettuce. *J. Aquatic Plant Manag.* **22,** 101–102.

Harley, K. L. S., and Forno, I. W. (1990). Biological control of aquatic weeds by means of arthropods. In *Aquatic Weeds,* A. H. Pieterse and K. J. Murphy, eds. Oxford Scientific Publications, pp. 177–186.

Kay, S. H., Quimby, P. C., and Ouzts, J. D. (1982). A potential algicide for aquaculture. In *Proceedings of 35th Southern Weed Science Society Annual Meeting.* Georgia, pp. 275–289.

Little, E. C. S., ed. (1968). *Handbook of Utilization of Aquatic Plants,* Food and Agriculture Organization of the United Nations.

Murphy, K. J., and Eaton, J. W. (1983). Effects of pleasure boat traffic on macrophyte growth in canals. *J. Appl. Ecol.* **20,** 713–729.

National Academy of Sciences (1976). *Making Aquatic Weeds Useful; Some Perspectives for Developing Countries.* Report of an Ad Hoc panel of the Advisory Committee on Technology Innovation Board on Science and Technology for International Development Commission on International Relations.

National Rivers Authority (1990). *Toxic Blue-Green Algae,* Water Quality Series no. 2.

Newman, J. R., and Barrett, P. R. F. (1993). Control of *Microcystis aeruginosa* by decomposing barley straw. *J. Aquatic Plant Manag.* **31,** 203–206.

Newman, J. R., Barrett, P. R. F., and Cave, T. G. (1994). The use of barley straw to control algae in drainage ditches. In *Conference Proceedings—Nature Conservation and the Management of Drainage System Habitats* (Nottingham University, Sept. 22–25, 1994).

Nichols, S. A. (1984). Macrophyte community dynamics in a dredged Wisconsin lake. *Water Resources Bull.* **20,** 573–576.

Pieterse, A. H. (1990a). Introduction. In *Aquatic Weeds,* A. H. Pieterse and K. J. Murphy, eds. Oxford Scientific Publications, pp. 3–16.

Pieterse, A. H. (1990b). Introduction to biological control of aquatic weeds. In *Aquatic Weeds,* A. H. Pieterse and K. J. Murphy, eds. Oxford Scientific Publications, pp. 174–176.

Pieterse, A. H., and Murphy, K. J. (1990). Herbicides used in fresh waters. In *Aquatic Weeds,* A. H. Pieterse and K. J. Murphy, eds. Oxford Scientific Publications, app. C. pp. 443–445.

Pitlo, R. H. (1978). Regulation of aquatic vegetation by interception of daylight. In *Proceedings of EWRS 5th Symposium on Aquatic Weeds,* pp. 91–99.

Ridge, I., and Barrett, P. R. F. (1992). Algal control with barley straw. In *Aspects of Applied Biology,* vol. 29, *Vegetation Management in Forestry, Amenity and Conservation Areas,* pp. 457–462.

Tooby, T. E., and Spencer-Jones, D. H. (1978). The fate of the aquatic herbicide dichlobenil in hydrosoil, water and roach (Rutilus rutilus L.) following treatment of areas of a lake. In *Proceedings of EWRS 5th Symposium on Aquatic Weeds,* pp. 323–331.

Van, T. K., Haller, W. T., and Bowes, G. (1976). Comparison of the photosynthetic characteristics of three submerged aquatic macrophytes. *Plant Physiol.* **58,** 761–768.

Van, T. K., Haller, W. T., and Bowes, G. (1978). Some aspects of the competitive biology of *Hydrilla.* In *Proceedings of EWRS 5th Symposium on Aquatic Weeds,* pp. 117–126.

van der Zweede, W. (1990). Biological control of aquatic weeds by means of phytophagous fish. In *Aquatic Weeds,* A. H. Pieterse and K. J. Murphy, eds. Oxford Scientific Publications, pp. 201–221.

van Donk, E. (1990). Necessity for aquatic plant management after lake restoration by biomanipulation. In *Proceedings of EWRS 8th Symposium on Aquatic Weeds,* pp. 91–96.

Vought, L. B-M., Dahl, J., Pederson, C. L., and Lacoursiere, J. O. (1994). Nutrient retention in riparian ecotones. *Ambio* **23** (Sept. 6), 342–348.

Wade, P. M. (1990). Physical control of aquatic weeds. In *Aquatic Weeds,* A. H. Pieterse and K. J. Murphy, eds. Oxford Scientific Publications, pp. 93–135.
Wapshere, A. J. (1982). Biological control of weeds. In *Biology and Ecology of Weeds,* W. Holzner and N. Numala, eds., pp. 47–56.
Weisner, S. E. B., Eriksson, P. G., Graneli, W., and Leonardson, L. (1994). Influence of macrophytes on nitrate removal in wetlands. *Ambio* **23**(Sept. 6), 363–366.
Westlake, D. F., and Dawson, F. H. (1982). Thirty years of weed cutting on a chalk stream. In *Proceedings of EWRS Symposium on Aquatic Weeds,* pp. 132–140.
Westlake, D. F., and Dawson, F. H. (1986). The management of Ranunculus calcareus by pre-emptive cutting in Southern England. In *Proceedings of EWRS/AAB 7th Symposium on Aquatic Weeds,* pp. 395–400.

Recommended Reading

Cook, C. D. K. (1990). *Aquatic Plant Book.* SPB Academic Publishing, The Hague (228 pp., ISBN 90-5103-043-6).
Little, E. C. S. (1979). *Handbook of Utilization of Aquatic Plants.* Food and Agriculture Organization of the United Nations (176 pp., ISBN 92-5-100825-6).
NAS (1976). *Making Aquatic Weeds Useful: Some Perspectives for Developing Countries.* National Academy of Sciences, Washington, D.C. (175 pp., Library of Congress Catalog no. 76-53285).
Pieterse, A. H., and Murphy, K. J., eds. (1990). *Aquatic Weeds: The Ecology and Management of Nuisance Aquatic Vegetation.* Oxford Scientific Publications (593 pp., ISBN 0-19-854181-3).
RSPB, NRA, and RSCN (1994). *The New Rivers and Wildlife Handbook.* Royal Society for the Protection of Birds (426 pp., ISBN 0-903138-70-0).

Chapter

12

Institutional Principles for Sound Management of Water and Related Environmental Resources

Harald D. Frederiksen
World Bank
Washington, D.C.

The Nature of Water Resources and Environmental Institutions

General

A nation's institutions—its form of government, laws, customs, organizations, and all that is associated with these institutions—constitute the framework within which society functions. They constitute the framework for every action from group relations to commercial activities. This framework, is, by nature, constantly subjected to pressures for change as a nation alters its economic and social objectives and the available resources and means to satisfy them change. The adequacy of a country's institutions, more than any other factor, determines its ability to guide its destiny. This applies most directly to managing its water resources and the related environment.

The adequacy of a nation's institutions becomes even more important as the quantity of uncommitted resources diminishes; no less than the social stability and economic well-being of the country and the health of its environmental surroundings are vulnerable to misguided or ineffective water-resource management. The limited success of concerted efforts to improve water-resource management over the last few decades confirms that the institutions of many countries are severely limited. Governments proposing changes—institutional, technical, or policy—to address critical conditions, particularly governments facing dire conditions, must recognize that they cannot risk failure. Proven approaches must be selected—untested concepts and unrefined theories must be avoided as time constraints permit only limited experimentation.

So what is the specific evidence of conditions suggesting institutional inadequacies? Today conditions in Asia, not just Africa and the Middle East, have degenerated to the point where countries encounter water shortages under conditions of normal precipitation. The pervasive long-term, if not permanent, pollution of the groundwaters supply a major portion of almost every metropolitan center, with many in overdraft, is in many ways even more serious. Countries lack plans and facilities to deal with the calamity that will result from the next widespread regional drought, and the affected public doesn't even conceive of the impending disaster. These conditions exist with the world population growing by 1 billion in the next 10 years and 2 billion in 20 years. The time for action is extremely short.

The dilemma involves choosing among the physical alternatives and devising measures that are technically, politically, procedurally, and financially feasible and can be executed in time to ameliorate the crisis. The term *ameliorate* is used, as for many countries it is too late to avoid the serious repercussions of mismanaged resources. The remedy includes creating a comprehensive set of interlinked institutions that will facilitate resolution of the issues and the effective application of the measures. This chapter presents an outline of institutional principles for consideration by governments undertaking significant institutional modifications for managing their water and related environmental resources. Although the principles for the most part are universally applicable, this discussion is oriented toward governments that have need for substantial and early change. Fortunately, the situation in many of these also offers a greater degree of flexibility to create a broader set of interlinking laws, regulations, and procedures than found among the more mature governments.

The principles outlined are drawn from several sources, mostly by direct observation of government and private entities in the course of executing their assigned responsibilities. The observed governments represent a range of physical and institutional settings with difficult water-management issues. Several books, some of which are noted in the reference list, have been written about the subject, including specific efforts in various countries. Of necessity, this short chapter is not an in-depth treatment of the subject, but rather constitutes a synopsis that can serve as background for individuals and agencies grappling with this complex and often emotional subject. It is hoped that the chapter's shortcomings and omissions are seen in this context.

Evolution of water resources and environmental institutions

Water resources development and management institutions have evolved over centuries. During this period, villages and urban centers have developed and managed their water supply and distribution. The necessity to provide the services was felt directly by all inhabitants, and, until recent times, it was equally evident that they could not expect higher levels of government to provide for their needs. As a result, both the principle that local villages and cities should provide these services and the institutions to implement the service evolved in a similar manner in many countries under a wide range of conditions.

As cities grew, more substantial facilities for storm drainage and flood control became necessary and viable. Many received support from higher levels of government for the larger infrastructure as central government had an obvious interest to reduce the potential for extensive economic losses caused by major inundations while local entities constructed the immediate works. Unfortunately, because the damage from inadequate waste management was not as obvious, this aspect of resource management has received minimal attention. Many urban centers did not construct the means to provide this service, while central government did not feel any pressures to support a program. As a consequence, waste management has only gained widespread attention in the more recent decades and in large regions it remains the most serious water resource and environmental management problem. Today, local government is assuming greater responsibility for both protection from inundation and waste management. And as government budgets tighten, it will fall on local institutions to provide the services through financially self-sufficient entities.

Historically, farmers have developed and managed much of the basic water facilities for irrigation and agricultural drainage. The local group of irrigators established administrative and operating procedures, individual rights within the group's service area, the responsibilities of members, payment methods, and the means for enforcing the system rules. Informal group rights to primary water supply, usually determined by the sequence of development, were recognized among the various diverters drawing from the same source. The provision of drainage service has a similar history where beneficiaries, cultivators, and villages, organized to drain their lands and build levees to control river floods and high seas—some systems now 8 to 10 centuries old. It is ironic that the establishment of this form of service entity has just recently come into vogue—some, unfamiliar with its history, viewing it as a recent invention.

Regional or national rulers developed large irrigation schemes beginning many centuries ago. Immense additional areas came under irrigation in the 1800s and early 1900s to alleviate widespread hunger and provide farming opportunities in sparsely settled arid regions. Only governments could devote the immense funds required for these large, long-maturing undertakings. Some countries required farmers, villages, and cities to form legal entities with the power of taxation to assume responsibility for operating and maintaining the distribution facilities, pay for the bulk water service from the government, and repay portions of the capital costs as a precondition for constructing new projects.

However, it was during the rapid expansion from the 1950s into the 1980s, to increase food production and improve rural incomes, that governments in developing countries undertook the massive developments. The enthusiasm of governments and lending agencies fueled programs for constructing and operating facilities from source to the farmer's gate. Similar programs were launched for massive drainage and flood-control systems. Many of the flood-control programs focused more on the urban centers rather than rural areas. The beneficiaries of major flood-control schemes were even more widely favored with few, even industrial and urban areas, assessed for the services. These policies and actions that installed governments as generous service providers at all levels produced some of the most vexing institutional issues countries now face.

The institutions now grouped under the environmental banner vary from country to country. In many instances, they are an assemblage of the former public-health, water, and land-use regulatory entities. The function of the original entities largely was to protect the health of the people. The expanded functions evolved as the public became more concerned about protecting the environment in which they live and equally, the health of the natural habitat.

The desire to expand the protective function began in earnest near the turn of the century. Forest management with a conservation component was an early effort. Large tracts of land were withdrawn from economic development and placed in the public domain. Indeed, European countries launched major programs in the 1700s. Soil-conservation measures were promoted for decades in the 1800s, but gained momentum in the 1920s and 1930s. However, except for land set aside, most of the initial efforts outside public health and safety were more of an advisory nature. Governments, through financial incentives, encouraged proper husbandry by the property owners. The administrative regulatory approach to protecting the environment, other than water, waste, and public health, didn't gain widespread importance until the 1960s through the 1980s.

Early environmental regulations set forth in-stream water quality and minimum discharge standards to sustain the biosystem. Measures to protect endangered species, restore wetlands, further restrain pollution, and elevate standards for public water supply were enacted in the developed and most developing countries. Some actions depended on administrative action, while others depended on incentives to alter behavior. Almost universally, legislation was enacted that required an assessment of potential environmental impacts of all government and private-sector actions. These were used both by the regulatory entities to execute their responsibilities and to inform the public, which, in turn, exerted its views through the political process.

Institutions for water-resource and environmental management will have to continue to evolve the conditions. Population and economic growth, particularly in number and size of urban centers, will force reallocation of land and water and alter present services in those areas. It will affect the configuration and operation of irrigation, drainage, and flood-control facilities. Single-purpose irrigation projects will become multipurpose, providing bulk supply to the villages and urban centers with inadequate sources or where groundwater overdraft or pollution occurs. This will dictate adjustments in the operation of the main facilities, while the operation of the irrigation and village distribution systems may remain unaltered. The private sector will become more engaged in the water supply systems. Investment in major facilities as well as purchase of entire systems will be seen. And management contracts with government entities to operate existing systems may proceed even more rapidly. Although groundwater extraction is largely by individuals, the recharge and maintenance of long-term storage for droughts will have to be managed by a responsible water-management agency. Privatization of urban water treatment will largely be limited to new facilities where cities cannot self-finance. Local government will retain its responsibilities for sewerage and solid-waste management. Sewerage collection will likely remain under a government entity; however, the sewerage treatment plants as well as solid-waste collection and disposal may well be privatized.

The future modifications in drainage and flood-control services will be more subtle. As villages and urban areas expand within farm areas, higher levels of drainage and flood-control services will set criteria for design and operation of the drainage conveyance and disposal facilities. Capital investment, O&M (operation and maintenance) costs, classes of beneficiaries and customer responsibilities will change accordingly.

The trends in providing the services in the future will clearly alter the point of decisions and the administration, O&M, and financial management. These have implications in the matters of ownership, cost recovery, and the government and customer roles.

Politics and institutional change

The role of government and even the form of government are being greatly altered in several countries, while the approach to resource management is undergoing technological and philosophical changes. Yet, the need to better match institutions to the evolving conditions, even within a limited field, is not always recognized as a positive opportunity. Nor is it seen that the measures can be formulated through rational endeavor. Instead, too many adjustments are narrowly focused. Sometimes incidental to major legislation. And sometimes it results as a compromise based on an incomplete assessment of the broader sectorwide needs and impacts.

Indeed, aspects of the water-resource institutions are under debate by the political bodies and nongovernment and government groups alike. Environmental concerns have brought intense and continuous public scrutiny of water development relating to both existing and new facilities to meet the needs of the future. The public is questioning many existing laws, regulations, and practices involving water quality and allocations. And water shortages, exacerbated by rapid population and economic growth, are forcing governments to seek new directions in water management. The conflicts among social, regional development, economic, environmental, and security objectives are complicating the modernization of water-resource institutions.

Institutional evolution in many countries, nevertheless, has not kept pace, often resulting in delays with serious implications for the country's future. Special interests and long-established customs are powerful constraints. Political will to change and strong leadership to carry it through are essential. But the ease and pace by which sound comprehensive institutions can be introduced depend heavily on the perceived need for change and sometimes on the relative powers of the legislative and the executive branches of government. In many countries, a weak conviction that change is needed coupled with the desire by both bodies of government to retain power sidetracks reform in the short term while conditions worsen.

However, no matter where the political power lies, the politics of introducing institutional reform must be placed at the fore by anyone seriously wanting results. People who wish to reform institutions must compose a blend of practical politics with sound institutional principles. Often independent boards or commissions composed of nongovernment and nonpolitical figures appointed from the community can be the most effective vehicle for engaging the political leaders in the process. This provides a forum that brings equals together over a period of time. It avoids political posturing and allows them to consider matters in depth. At the same time, public hearings can be conducted to also bring the public into the discussions.

The most important point to recognize in this entire chapter is that decisions and programs in the water-resource sector evolve through the political process. The process cannot be circumvented or dictated to by the ideal. And this is as it should be. A positive perspective on this fact will advance sound institutional strengthening most rapidly.

The form of government, decentralization, and public participation

It should be noted that the term *public* is used in most of this discussion instead of the currently popular term *stakeholder,* because it has served citizens well and has been understood by them whenever national issues have arisen; and resource management today is of national and international importance. However, there is a much more important reason. The water, land, and related resources are owned by all the citizens of a country, and they, as a body, should husband the basis for their national well-being. At the local level, the local citizens should participate. The use of the term *stakeholders* has proved effective to allow groups to determine who should and—by virtue of that decision—who should not participate in debate, who should decide, and who should be excluded. It is the public that should participate in policy, program, and project discussion, not selected groups within the public! And government and nongovernment entities should strive to engage the public and make it aware of problems and alternative solutions in an unbiased, unrestricted manner.

The form of government—federal or unitary—usually determines the level in a country's governmental structure that retains the ownership of the resources; the authority for resource planning, development, and management; the sources of program funding; and the administration of regulatory programs that guide management.

Similar to a unitary form, typically, the federal government has the responsibility and power to assure a measure of consistency among all states and provinces in meeting national objectives. In the water-resource areas, minimum standards for drinking water, instream water quality, pollution control, and waste management are mandated by central government in compliance with national legislation. To be effective, this usually includes monitoring, enforcement, and levying penalties for violations. This does not preclude local government from setting higher or additional standards within their jurisdiction.

Central agencies under all forms of government will deconcentrate (locating central government offices as needed throughout the country) to the extent most effective to fulfill their mandate. The nature and magnitude of the function, degree of access, and reliability of communications will dictate this decentralization. Often the local offices of the national government, through close relationships, serve as direct local government support. Decentralization (delegation of responsibilities and powers to lower levels of government) usually increases as local capability matures.

The effect has been that much of the day-to-day management is increasingly carried out at the local level under both federal and unitary governments. However, where decentralization results in the delegation of power in areas key to the national goals, oversight usually remains at the center, while decentralization provided by the constitution may be subject to only partial oversight by the center.

There are pluses and minuses to decentralization. The degree to which decentralization of resource management can be effective depends heavily on the consistency in decentralizing several other component functions of government—political, administrative, and financial. Many countries are or have decentralized, through their constitution or special legislation, much of the political power and, with it, certain authority to act. Some have included powers to raise funds to finance programs; however, others have largely retained these powers at the center. Under the latter provisions, local jurisdictions are actually restrained from encumbering themselves to meet what they perceive to be their priorities in resource management. Centrally controlled budget allocations effectively ensure central direction, and even a proportionate distribution of untied central funds will limit the freedom of local decisions. To avoid conflicts in some resource-management areas, some countries have assigned the functions as comanagement responsibilities of higher and lower levels of government.

Another arrangement that cripples decentralized actions arises when central governments enact programs and regulatory measures that local jurisdictions must implement, yet do not provide the necessary funds. The conflicts are most frequently found in the regulatory area, particularly in countries with significant differences in the conditions and means of regional government. One legitimate fear of central government may underlie the tendency to retain central control through revenue constraints and budget allocations. This is the real or perceived risk that powerful local interests will overwhelm the local entities' efforts to manage the nation's resources in a manner that will meet the country's objectives. Decentralization and what is often defined as *stakeholder participation* can lead to results quite different from what is ideally assumed.

The risk lies mostly in the regulatory field, where the uses and abuses of resource management are defined and penalties for violations are set forth. The tendency for local jurisdictions to ignore resource-management plans or be lax in applying regulations may arise from the desire to create advantageous economic and employment conditions compared to other localities. Industrial and commercial developments are located inconsistent with pollution- and supply-management regulations and associated infrastructure. Broader land-use-related problems arise from granting development before the necessary supply-, waste-, and flood-management facilities are in place. In practice, many locally formulated and enforced plans do not incorporate best management and are merely an illusion that there is order. Indeed, if left to the local entities' decision, many would not enact the regulations to begin with. This is confirmed by the absence of regulations until forced by the central government. This does not mean that central authorities never violate regulations for gain.

Before deciding on the degree of decentralization of specific resource management functions, one should examine the environmental conditions and the adequacy of services in the water sector found under various central and local government arrangements in different countries. Some of the poorest results are found in countries with the most extensive and longest period of decentralization—political, administrative, and financial. Some of the best, though, are found in countries with highly centralized, almost dictatorial, control. It seems that there is a degree of decentralization of selected management functions where better management is found—provided there is a fully informed and vigilant public.

Thus, the degree of decentralization and powers of the public reflected in the sum of the nation's institutions must be tailored carefully to the situation:

How knowledgeable and informed are the citizens?

How effective are the national and local political mechanisms in expressing dissatisfaction?

Are there adequate sources of funds for the local unit?

Are penalties for violation of regulations severe enough to deter violators?

Are the means in place to effect a timely halt to undesirable actions?

Is there local unbiased transparent oversight with unquestioned power to remedy noncompliance with regulations?

And is there effective audit from higher levels?

In sum, how mature and how effective are the existing institutions? For local corruption is common even in every developed country!

The point to recognize is that decentralization and public participation are not the same. Local government can refine the national programs to best meet their particular needs, while higher levels can control mismanagement. Public participation through local and national hearings can be a very effective means to assure satisfaction while the management or regulatory control remains above the local level. More basic mechanisms for public participation are discussed later.

Common institutional deficiencies

Conditions generic to many countries define fundamental problems requiring institutional actions to improve resource management. Foremost is the absence of clear resource-management goals, objectives, and policies. Often land-use policies, which dictate water use and pollution, are inconsistent with water-management policies. The linkages to land-use objectives and geographic location within basins are not defined. Many governments set general water-allocation priorities, but these general priority statements neither clarify the allocations among specific users nor set priorities under long-term and emergency shortages. Water-quality considerations are absent in criteria governing the use of an allocation.

Countries lack the water-rights systems to record allocations protecting the investors and the affected public. Indeed, few effective allocation mechanisms are in place in the developing countries other than what government does or does not construct. Even then, government does not define firm project water rights, although the undertaking should have a 50-year economic life and form a much longer base for development of the affected region. In spite of legislated instream water-quality standards and restrictions on waste disposal, the quality of surface supplies is deteriorating because of the discharge of untreated urban and industrial waste into waterways. There is no enforcement of the pollution charges or waste standards, nor does local government construct the necessary waste-handling facilities.

The primary reason for increased flood damage is not unheard-of hydrologic events; it's the habitation and economic activities expanding onto flood-prone lands. Similarly, sites for future storage and conveyance facilities, essential to managing a nation's resources efficiently, are being permanently lost to other uses. Enormous future costs will accrue to society by not reserving these sites now when disruptions are relatively manageable. The electric-power-generation companies often operate large reservoirs as essentially single-purpose projects. Without basinwide multipurpose operations rules, maximum benefits cannot be realized—neither water supply, flood protection, nor in-stream conditions can be optimized. Many local water agencies do not apply sound business practices. Deficiencies in mid- and long-term planning, budgeting, accounting, and financial control prevent them from becoming effective, self-sufficient entities that can efficiently meet their service responsibilities and maintain their assets.

Financial responsibilities are integral to a country's institutions. What aspects of resource development and management should society pay for? To what extent should activities be subsidized? If the beneficiaries should pay, what facilities and responsibilities should government relinquish to them? Can for-profit privatization, as applied in some developed countries, be adapted to developing countries?

These example problems are only a few of those confronting governments in managing their water resources and the related environmental aspects. Only a continuing and objective assessment of the situation with a fresh perspective leading to concrete actions can ensure that a country's institutions are equipped to meet the ever-growing needs.

Mechanisms and considerations in evaluating institutional modification

As inferred, most issues that may warrant modifying a country's institutions are generic to the sector. One might intuitively expect that institutions devised specifically for the development and management of water resources would be essentially the same in all countries, leading to the view that one format fits all. After all, water-resource management involves a substance of given physical characteristics, obtained from similar sources and applied to similar uses for the same economic, social, and environmental purposes. This is not the case. To emphasize an earlier admonition, goals and practices are set, resources allocated, and programs formulated through the political process. Likewise with institutions.

It is a complex political process for which the legislative and executive bodies have ultimate responsibility. They formulate alternatives and debate the options and eventually enact the measures. Commissions composed of government and public figures and outside experts, as are discussed in subsequent sections, can play a pivotal role. The function of the "experts" is to offer specific expertise to the political bodies by helping devise feasible alternatives and by presenting realistic, unbiased assessments of the consequences of proposed actions. The bureaucracies' role, in addition to being one source of expertise, is to implement the political decisions. Thus, it must be recognized that line agencies also inevitably represent their clientele—whether urban, agriculture, industry, energy, fisheries, or forestry. Hence, the contributions of line agencies must be seen in total, in part through the planning and programs of a water-resource-sector agency assigned responsibility for sectorwide management.

Countries throughout the world have confronted the same issues, and many have devised solutions that give insight into principles that others might adopt. Indeed, although institutions differ from country to country, fundamental institutional principles are common under good water management, particularly countries nearing full utilization of their resources. Four general principles apply when modifying institutions:

1. Ideally, institutional modifications to any aspect of water-resource development and management should be comprehensive and made in the context of devising effective management of all government and nongovernment functions in the entire water-resource sector. This ideal should prevail at least in major subsectors even if implemented in discrete steps.
2. The institutional framework to guide modifications should be consistent. Deficiencies arise from inconsistencies in the interlinked legislation, the organizational structure, the rules and procedures, and the exercise of public and governmental responsibilities in the various subsectors. Partial measures, no matter how sound, risk conflict.
3. Water-resource institutions should be dynamic and must continuously adjust, even if slightly, so that they will best match the evolving conditions. This should hold true at every level, particularly within each agency. Proposals for institutional change of any type, at any point in time, should be viewed in that context.
4. Any modifications must be politically practical under the situation. Changes must have the strong support of key leaders and be acceptable to the public and an adequate number of the dominant players.

An examination of the institutional principles common to most successful institutional arrangements might best begin by noting the reasons for the differences encountered when examining how different countries manage their resources. First, the relative availability of water has a predominant influence. A variety of institutions may be found in countries with ample water, particularly if the countries are still in a "development" phase. Loose arrangements suffice under conditions of surplus when conflicts and environmental concerns are minimal. As countries mature, however, problems of scarcity or quality force institutional modifications that follow similar paths. A second influence on the institutions is the rate at which the related resource problems have arisen. Has it been at a rate to which the institutions could adjust in an orderly manner or one that occurs at such a pace that institutions are overwhelmed and adjustments have been made in response to crises?

A third influence is the climate, primarily the amount, pattern, and nature of precipitation. This variable largely determines the extent of major commitments to flood control, irrigation, and drainage, usually determining the extent of large investments and their consequences on sustaining the country's basic economy and social well-being and the power of beneficiary groups. Finally, the form of the government and the legal system create the basic structure for water- and land-use legislation. Nations display differences caused by the extent of state and provincial autonomy and decentralization of resource ownership and management.

Before discussing institutions, it is important to identify the primary areas of government actions: legislative, operational, regulatory, and financing. It is in these areas, at every level of government but particularly at the level of central government, where institutional efforts usually must be applied to improve resource-management capability. The functions are treated in the appropriate sections that follow, but will be defined briefly.

In addition to the constitution, the country's legislation creates the legal basis for all activities, and its history reflects the country's response to changing conditions. It includes legislated policies, regulations, and authorization and funding of basic programs and projects. The identified need for much of the legislation may originate in the government bureaucracy, and much of the legislation's content may be formulated there as well. But ultimately, the content and fate of programs are determined through legislation and budget formulation and, hence, the political process. The bureaucracy cannot undertake programs and projects unless the majority of the legislature wants it to—and the bureaucracy cannot halt programs or projects unless the legislators want them halted!

The operational area includes data collection, planning, design, construction, operation and maintenance (O&M), and public support. The primary actions to utilize and manage the resources are executed in accordance with legislative authorization. In most developing countries, governmental line agencies at various levels dominate the operational area. In others, the nongovernmental entities play expanding roles as the involved activities approach the O&M phase.

The regulatory area constitutes the framework for guiding the behavior of government and nongovernment entities and individuals and measuring results in conformity with set objectives, often expressed as standards, particularly in the environmental area. It includes monitoring activities and conditions and enforcement of established laws, regulations, and specific-purpose agreements bearing on resource use. A majority of the resource-management issues debated today directly or indirectly relate to the regulatory area.

The government function of financing services and programs for the benefit of or to cover the costs resulting from the actions of nongovernment or individuals is the least understood by the public. It entails direct loans, guarantees of the payment of loans or bonds by lower-level government entities, and grants and subsidies for programs and services. It is the functional area most intimately linked to politics.

Time may seem an odd consideration in formulating institutions, yet today, as with financing policy, it is an essential. Many countries must make significant and prompt changes in their institutions if they are to maintain a healthy environment and sustain their economic and social well-being. And neither time nor funds exist to experiment with untried approaches, a luxury that the developing countries in particular cannot afford. The real or perceived uncertainty about how to solve most problems confronted by developing countries in particular is not a reason to "pilot" yet another concept without thoroughly researching and understanding what others have tried in a like situation. The similarities of the resources and of the peoples' aspirations in different countries have led to the evolution of several principles proved sound.

The Foundation for Resource Management

Management goals and resource ownership, allocation, and rights

Relationship of national goals, water allocation, and rights. Uppermost in today's water debate are the issues associated with supply and the water-related environment. In the debate, the relationships among a nation's goals, its water-resource-allocation objectives, its water-allocation mechanisms, and its water-rights system for documenting the allocations are not recognized. Allocation mechanisms are being proposed without reference to national goals. There is no understanding that some allocation mechanisms will yield results contrary to the most basic objectives, particularly the social and environmental. Water-rights systems are proposed without deciding which combination of allocation mechanisms should be adopted to best meet the nation's objectives. Not only are allocation objectives, allocation mechanisms, and rights systems related; they should be determined in that sequence. And obviously, all must conform to the nation's constitution.

Goals and objectives. Every society has set goals for its nation. These may be explicit, often in the constitution, but usually are reflected in its body of legislation and government programs with limited documentation. The goals may be grouped as national security, social well-being, economic betterment, equality of opportunity, and a healthy environment. More specific to resource management are the objectives such as sustainable national economic growth, balanced regional development, stability of rural regions, and ensuring a truly healthy environment throughout all regions and for the benefit of all citizens. A country should clarify its goals and the entire set of specific objectives to fulfill these goals before it decides on matters regarding resource allocations, mechanisms, and rights. The goals and objectives must be described in clear terms so an enlightened public can debate the alternatives, and subsequently debate the allocation issues. Their selection is too important for limited, biased, or distorted information or for sloganeering.

The nation's land and water resources constitute the base for its economic, social, and environmental well-being. Although water receives most attention, land and water-resource management must be joint and truly multipurpose and directed in a coordinated manner to meet the nation's goals and objectives. Linkages of water-allocation decisions, mechanisms, and rights to the land-use objectives, use planning, and use controls should be explicit, particularly as they affect the balancing of water-supply priorities, controlling pollution, and meeting environmental objectives.

Water allocation, allocation mechanisms, and rights. Regardless of the form of government or the underlying political system, the constitution of most countries retains national ownership of the water resources, and some even the land, with government administration on behalf of all its citizens. "Use" rights are issued to individuals and service entities that may be treated as quasi-private ownership to the extent that their use complies with the conditions of their issuance. Usually, this links water to specific land uses. Indeed, water-use rights of limited or open duration are the basic form of water rights used most commonly today. The principle that society as a whole should retain control of water use to meet its national goals within bounds that protect long-term investments is increasingly being adopted. This principle is made most effective through the adoption of limited-term water-use rights with review and renewal opportunity at stated times (10 to 20 years are ample under some situations, while those uses entailing substantial long-term investments may be for 50 to 75 years' duration).

A limited number of formal water-allocation mechanisms dominate in the world. A common objective when water is abundant is to allocate water to develop a region's lands and other resources. The mechanism to meet the allocation objective was simple—issuing formal water rights for uses meeting the general measures of "beneficial purposes." The type of rights involved, designated use (by legislation), and "first-in-time and first-in-right" (appropriative) outlined the general conditions of permitted use and set forth the priority among users according to the date acquired. (It should be noted that the system of *riparian* rights under which land owners along natural bodies of water have certain minimum rights serves as one allocation mechanism linking land use to water use.)

The most common water allocation mechanism in developing countries is through legislative authorization of major water-related developments. This has long been used in developed countries, most recently to allocate water for environmental purposes. Indeed, project legislation incorporates the allocation objective and serves as the allocation mechanism: a focused allocation mechanism available through the political process. (Unfortunately, as discussed later, many countries have not set out or recorded a project and customer water-use right for the development.) The situation in many developing countries, coupled with the size of farms and villages and their near subsistence conditions, precludes any type of allocation at this time other than government-guided development. Modifying the original project allocations may inflict severe social conflict and at best be of questionable benefit.

Less-formal allocation mechanisms, not linked to firm individual rights or project authorizations, also are found in developing countries. Some simply set priorities for whatever water is available at a given time. Others may dedicate a proportion of the available water for each category of use (agricultural, urban, industrial, etc.), with more elaborate allocation mechanisms for the users within each category. Or there may be a changing share among users as resource deficiencies occur—long-term, annually, and seasonally (features of this concept may also be found in formal allocation and rights systems). Usually a regional or basin entity administers the real-time distribution. A very serious flaw of this ad hoc basis for allocation, rather than a rigorously enforced water-rights system, is the tendency for development to exceed the available water resources. Through political pressures, governments often proceed with projects that in total far exceed supply, wasting huge investments, both public and private.

Environmental allocations are made through legislative appropriation in developed (and developing) countries often independent of the allocation mechanisms applied to the traditional users. Usually seasonal flow and quality levels are set. Currently, allocations to this purpose do not have to meet economic criteria and often do not compete in a priority ranking with other uses for the water except in times of extreme drought. In spite of stream-quality legislation, few developing countries have allocated the water to meet their standards, blaming the deterioration of quality on pollution, which the government also fails to remedy.

Under all existing allocation mechanisms, an overriding priority of use usually governs actual allocations during critical drought shortages. Other mechanisms are under trial in developed countries in an attempt to improve management during limited periods of scarcity. "Water banking" is practiced, which allows water-use-rights holders who have stored water within a basin to exchange water during shortages. Most exchanges are limited to systems in jointly owned reservoirs. Also, government agencies serving as brokers have set firm prices for all sellers and buyers as a means to free up water for high-priority users during drought emergencies. These schemes depend on a complex set of factors: regulatory capabilities of enforced, measurable, recorded rights; reliable facilities to measure water; an independent broker; and extensive interlinking conveyances, thus essentially limiting their consideration to developed countries.

Some countries are testing free-market mechanisms for reallocating water-use rights. To date, no large-scale free markets with bidding among "systems" have evolved in the developed countries. The major direct impacts on third parties, issues of rural versus urban economies, and conformance to land-use objectives, as well as other objectives, constrain its application. Adverse third-party impacts must be fully compensated, a reasonable stipulation but difficult to comply with. This is at the heart of the problem; by definition, there is no free market in which third-party impacts arise. But perhaps the most serious limitation to the use of private markets—reallocation and private ownership of this basic resource by the highest bidder—will not ensure meeting the fundamental national goals, particularly the key social and environment issues. The appeal of private ownership is that it avoids the difficult political and social questions of how to allocate existing resources and whether to develop additional ones.

One variation allows a free market of members' share of water and the rental of the associated use rights among irrigators and village users within the system service-area boundary. Two features are important in this case. The water in question is imported from another basin, and the service entity retains ownership of all return flows; thus there are no third-party rights to return flows and, by definition, no impacts. Legal provisions retain the resource within the system bounds following a policy to sustain the local area's economy.

Even large urban jurisdictions do not maintain markets for their customers under conditions of shortages. Indeed, they limit every individual's allocation consistent with their social and equity objectives, regardless of the customers' willingness to buy. Some jurisdictions have retracted supply allotments from public uses, forcing parks and golf courses to secure reclaimed wastewater, even though the golf courses offered to accept greatly increased rates. Almost universally, urban supplies are allocated to customers through a combination of service limits and the application of a tiered rate of service charges.

Again, any allocation mechanism is appropriate only if the results satisfy the nation's goals and all the country's water-allocation objectives. The features of the mechanisms should ensure transparency of results in meeting the allocation objectives. They should make obvious any linkages to non-water-related consequences such as third-party rights and land-use management. Administrative practicality, a means to establish rights of the investor, user, and third parties and a provision for future change should be inherent in provisions of the mechanisms.

As noted, once allocations have been made, most countries successful in water-resource management issue formal rights to its use primarily to record the individual acts, many described in terms of allocation mechanisms. But protecting the investments of government, water utilities, and individuals and equitable real-time water allocation under actual hydrologic events requires definitive rights and effective administration of those rights. A country's water-rights system should apply to surface water, groundwater, and coastal water. It should stipulate the category of use (agricultural, urban, industrial, environmental, etc.); class of use (consumptive, nonconsumptive, and polluting), quantity and quality implications; priority, time, and duration; and administrative procedures. It should define source of water, geographic restrictions on use (including linkages to land use), limitations on class of use, and quality restrictions on source and return flows.

Water management through land use. Increasingly, land-use management is used to meet water-resource management objectives. A rapidly growing trend is to impose land-use restrictions to control water pollution, and to a larger extent to protect the environment. Example actions are restrictions on the application of agricultural chemicals, the density of livestock feeder and fattening operations, mineral extraction, siting of industries and power plants and urban configuration as affecting the quality and quantity of watershed runoff; and retention of groundwater-infiltration sites. Regulators and the public realize that these are the only effective means to control major pollution sources and maintain supplies, and public support for these measures is growing.

A second land-use restriction helps ensure water availability to meet regional priority purposes during short periods of scarcity. When the total quantity of water allocated to high-priority demands approaches the total supply under normal hydrologic conditions, the volume allocated to low-priority use available to transfer to priority needs during droughts shrinks. Indeed, the full use of a region's normal water supplies by domestic users and critical industries, particularly if all the water savings and "efficiency" measures are in place, assures economic disaster during prolonged shortages. Countries should retain a mix of low- and high-priority uses drawing on a given water source to assure the flexibility to manage water allocations in times of drought. The water supplies above the critical drought supply quantities should be allocated only to lower-priority uses such as annual field cropping and park lands.

As a principle, countries should formulate land-use objectives and land-use rights in parallel with water-allocation objectives and water rights. The land-use rights should contain explicit statements concerning any reserved water rights, conditions of surface-water and groundwater use, and conditions of land uses that affect water-resource management. They should address management aspects from quality impacts to floodplain reservation.

Standards, regulations, and administrative rules

The regulation of government and nongovernment activities is a critical aspect of long-term and real-time development and management of land and water resources. The actions usually regulated pertain to physical and quality impacts of any water use, broader environmental impacts, facilities configuration, public safety, and the use-rights provisions discussed in the previous section. Other regulatory functions pertain to the obligations, quality of services, and financial operations of water-service entities. All users—government, nongovernment entities, and individuals—are responsible for complying with regulations.

The formulation of regulations and rules to administer the standards entails a range of actions from clarifying legislation to preparing manuals. What constitutes a violation and the associated penalties must be clearly stated; known to all, particularly the public; and enforced promptly and uniformly. The mechanisms to monitor activities, the triggering values signifying noncompliance, and the enforcement actions should be clear. The triggering conditions should be set forth at locations where real-time conditions, particularly water quality, hydrologic, and land use can be observed and are measured. The administrative measures should allow for all actions that may help meet the objectives of the standards. For example, a system of land-use rights is inherent in land-use zoning for controlling development in the most large urban centers. The administration falls under the municipal government, usually part of the construction and property approval and evaluation programs. The issuance and administration of land-use rights for purposes of water-resource management, however, remains a vague concept. And if not directly coordinated with the water-management units, it will not be applied as effectively as provided by the law.

Likewise, real-time administration of water rights entails the ongoing monitoring and enforcement of rights among diverters consistent with current hydrologic conditions in the source river, reservoir, or groundwater, particularly if the resources are fully committed. Practicality of the monitoring and enforcement means and openness to public scrutiny are essential and require a range of measures.

The government is responsible to assure the safety of government and nongovernment facilities, including dams. Almost all countries have enacted some legislation. However, too many developing countries have not effected a comprehensive dam safety assurance program to implement the law. Regulations and rules may cover technical aspects, but only minimal rules exist for administering the standards. The owner's legal and financial responsibilities for adequate maintenance should be obvious under law, but liability for remedial work and conditions under which the government can assume control and assess full costs of corrective action to the owner help emphasize the seriousness of this matter.

Effective service standards and regulations, promptly administered, are as essential to sustaining water-service systems as the physical facilities. This applies equally to supply, drainage, flood control, and waste collection and disposal. The quality of water services greatly influences the economic potential for customers, and, in the case of domestic use, their well-being. It is a proven principle that where comprehensive service standards are enforced, service quality is high and the beneficiaries are more able and willing to pay. Where service standards are not met by the operating entity, beneficiaries will resist charges even to meet O&M costs.

Standards for urban water-service quality and waste collection and disposal are long-proven, and numerous examples are available from international agencies. Standards for irrigation delivery and farmer actions are peculiar to each scheme. Indeed, standards are often lacking—inevitably in problem systems—including those under government operation. The generic service problems on government-owned systems caused by illegal canal breaches and offtakes and the excessive "headender" diversions demonstrate that neither the standards nor the rules to administer the scheme regulations exist. And there is no facilities design that can overcome actions of undisciplined farmers or dishonest operators. Although drainage and flood-control services are not as easily defined, there also should be standards for performance, beneficiary actions, and rules to enforce them.

Beneficiary-owned and local-government service entities are commonly created under legislation with legal provisions for their management and financing. The entities are responsible to the customers and to the lenders for operational efficiency and fiscal integrity. Most failures of such entities to deliver services, or even to survive where established, result from unsound financial decisions either in the day-to-day operation or in system expansion. Governments should devise detailed performance standards and regulations for management and accounting and audit rules for fiscal activities specific to service entities—whether independent or subunits of larger government agencies. Independent oversight units should administer the program with strong enforcement powers and the staff and means to match the task.

Responsibilities and Organization of Government and Nongovernment Entities

General

Having set the objectives and policies for resources utilization and standards with procedures for their administration, a country can more rationally organize and assign responsibilities for resource management among the government and nongovernment sectors. Of course, every country has arrangements in place, and improvements depend on a great many factors. However, institutional change arises from new conditions and pressures by the public as it becomes more knowledgeable and capable of assuming a role. An understanding of some proven principles should equate to improved results. But first a general statement about the participants and their roles to preface the subsequent discussion.

There are three main participants in resource development and management: government, nongovernment entities, and the individual. The government controls overall development and management of the resources for the benefit of society, undertakes programs in all phases of their development and management, and, through its subdivisions, provides various public services in the sector (Fig. 12.1). Nongovernment entities (not-for-profit beneficiary-owned or investor-owned companies) may develop and manage resources for their members' or customers' benefits within bounds established by legislation. Individuals' activities parallel those of the nongovernmental entities, except for scale. Advocacy organizations, particularly in the environmental and the customer-protection fields, constitute an expanding group of nongovernmental entities joining the longer-established beneficiary, industry, and professional organizations. Regardless of what level water-resource management activities one examines, however, government capability is critical to the end result.

The application of the term *government* in Fig. 12.1 includes the legislative, line, and regulatory functions, as described earlier. The following sections focus on principles applicable to the line and the regulatory.

Functional linkages for effective resource management

Linkage of water- and land-use management. The interdependence of water rights and land-use rights has been discussed. The mechanics of linking rights can be debated, but the need to reflect this linkage in the government's organizational structure, in assigning responsibilities, and in government programs should not await resolution of rights legislation. Most countries still manage water and land use in isolation and without recognizing the interdependence of their impacts. Plans committing the resources are made by different agencies, often by several different entities.

Countries should fully integrate all land use and water use in their planning, in their regulatory functions, and, as appropriate, in the management and operations functions through assignment of organizational responsibilities. The resource rights and the regulatory controls and enforcement of water and land use should rigorously conform to the resource plans—national, basin, and local—and be administered or overseen by one regulatory agency at the respective levels. The owning and operating entities, government and private, should be held accountable for compliance.

Linkage of water quantity and quality management. Governments clearly define and assign water-quantity management to specific agencies: planning, operations and services, and regulatory. Yet, water-quality management responsibilities are often vague and most commonly viewed only as regulatory. Seldom do national investment policies and

Figure 12.1 Functions of the central government in water-resource management.

Viet Nam, through its constitution, various legislation, and proclamations by its leaders, has created a framework of objectives to guide management of its water resources. Simply stated, the nation's resources, which are owned by its people, should be committed to uses that best meet the social, economic, environmental, and security goals of the country. A balanced development and management of its water resources in support of the nation's well-being, considering all the many interests of society, will be the underlying objective on which the government will focus its actions.

Central government has two roles to help attain these goals: (1) the focused programs and assistance of line agencies in the individual water-resource subsectors and (2) the overall multipurpose resource management and water-resource-sector-wide oversight isolated from subsector activities. Governments initiated programs and assistance in the subsectors—agriculture, navigation, city development, storm drainage, and local flood control—essentially from the time central governments evolved. These line functions dominated, indeed, were the only civil functions, until recent times. The ample resources, small populations, and absence of conflicts among users of the water resources precluded the need for coordinated management or regulatory intervention.

The government's role of multipurpose management and oversight is just the opposite of its subsector roles. This broader role requires independence from the pressures of specific groups of users. It must be positively focused on resource planning and operation of multipurpose programs to meet all objectives from the entire nation's perspective.

The water-resource-sectorwide role may be assigned to an agency with a multidisciplinary capability: one that can be held accountable. Or it can be assigned to a body that coordinates single-purpose roles of several different subsector line agencies. Most countries and their local jurisdictions have established a single agency to assume these responsibilities—usually a ministry or department of water resources. The use of coordinating bodies inevitably has led to distortions as subsector pressures and strengths continuously shift with a lack of the critically important, consistent long-term policies and programs. The lack of clear accountability with control over the "program staff" causes costly delays and places water-resource-sectorwide programs hostage to subsector participants.

The subsectors of the water-resource sector typically assigned to or guided by separate line agencies include

1. Irrigation and agricultural drainage
2. Hydroelectric power
3. Environment
4. Urban water and sewerage
5. Navigation
6. Public health

The multipurpose functions typically assigned to or guided by a broader water-resource-sector agency(ies) include

1. Hydrologic data collection
2. National, regional, and basin* planning
3. Major multipurpose reservoirs
4. Regional flood protection
5. Bulk water supply*
6. Bulk storm drainage*
7. Stream management
8. Regulatory

*May be local or central agencies, depending on magnitude and complexity.

(*Continued*)

Many assign all the multipurpose functions to a single ministry of water resources. In this case, function 8 may be directed by an appointed board and provided administrative support only from the ministry. In some countries function 1 is created by combining several existing entities, but still placed under the ministry of water resources for administrative support. The remaining functions are interdependent, with multipurpose objectives, and hence, combined into one entity.

Figure 12.1

program priorities reflect the interrelationship. They do not take advantage of the cost tradeoffs between pollution control and prevention, water-supply treatment, and developing new sources in the same basin. Pollution actions are focused on meeting standards. The options for using a given water source depend directly on its quality and the maintenance of that quality. In turn, the specific use of a water source determines the quality of return flows and, hence, the potential for subsequent reuse. The quantity of in-stream flow required to meet quality objectives may be dictated in large part by waste discharges that must be diluted.

It has proved feasible to consolidate responsibilities for water quality (affected by both solid and liquid wastes) and quantity in the same functional units in the areas of planning, operations and services, and regulatory, particularly in water supply and waste treatment. The service functions may be combined in the local "utility" and in governmental agencies providing bulk services to local entities. Formal coordination and strict review and approval procedures among existing single-purpose entities may be suitable until more extensive changes are warranted. But consolidation in the planning and regulatory areas should proceed immediately.

Linkage of surface and groundwater management. Another much-espoused, but seldom-applied concept is the conjunctive management of surface and groundwater. The separation of management responsibilities for these two water resources gives rise to overly optimistic resource-availability projections, conflicting exploitation policies and projects, ineffective control of groundwater buildup, inefficient operations, and wasteful investments. The increased resources potential from short-term and long-term conjunctive management is not realized. Countries should consolidate responsibilities for surface water and groundwater and assign to the same functional units in the planning, operations and services, and regulatory areas with equal attention to this principle at all levels of government. This consolidation is usually easier and yields more immediate benefits than do many other institutional changes.

Linkage of agency jurisdiction to geographic and political bounds. The organizational arrangements of some water-management functions, primarily at the basin and local levels, may influence management efficiency. Inconsistencies of political and hydrologic boundaries present few obstacles to broad planning, data-collection, and regulatory activities. On the other hand, the detailed planning and the regulation of water and land use at the local level may warrant greater emphasis on the selection of the responsible agency's boundaries. Difficulties arise when jurisdictions violate hydrologic boundaries at the operational levels of services (water supply, water distribution, sewerage, drainage, and flood protection).

The O&M function, critical to sustainability, dictates that beneficiaries of a given system be included within the bounds of the responsible entity. The service area of large schemes can be subdivided in a manner so that the subareas constitute hydrologic units. Government line agencies, such as irrigation and public-works departments; and administrative government units, such as counties and municipalities, can readily meet this principle by creating single-purpose service utilities complying with hydrologic boundaries for the services under their responsibility. Political jurisdictions should join in forming umbrella service entities where the service area overlaps more than one.

Functional independence for regulation and specialization

Separation of line (operations) and regulatory functions. Many governments in developing countries do not separate responsibilities for providing services from the responsibilities for regulating actions pertaining to the services. This lack of division creates adversarial relationships between agency subunits and conflicting objectives for agency managers. Some government water agencies determine the allocation of water among projects—their own and others—with resulting overcommitment and overconstruction. Water-supply agencies also may be responsible for enforcing (as opposed to managing) water quality. Industry ministries with powers to promote industrial development regulate land use and siting and sometimes even act as the regulators to control industrial waste disposal and groundwater extraction. Owners of major water facilities are the sole judge of the safety and the adequacy of maintenance of their own facilities. Environmental protection activities, which are regulatory, are inconsistently assigned to executing agencies.

The principle of explicitly keeping agencies responsible for operating functions independent of those with regulatory responsibilities is perhaps one of the most essential institutional principles in government. This is evident in the countries having greatest success in managing their water resources. To assume that service organizations will administer regulatory standards consistently and unbiased without independent oversight is not realistic.

Separation of functions in the environmental area. The institutional principles that apply to the broad regulatory functions generally apply in the environmental area. Consistent with these principles, some traditional water-related regulatory activities, particularly water quality and solid-waste disposal, are increasingly grouped in "environmental" regulatory agencies. Indeed, "environmental" agencies in developed countries perform a regulatory function.

Yet, even in these countries some resource agencies retain seemingly dual functions. These apparent violations of the principle must be examined more closely to understand the reasons and the risks. Typically, the environmental units formulate general standards and monitor the actions of government and nongovernment entities for compliance and enforcement. But the fisheries and forestry-resource agencies typically refine the applicable standards and regulations and perform the enforcement function of their own regulations. This procedure is common where control over resource use is the primary means of protecting that resource's feature of the environment. The risk lies in the power of the agency's clientele. Agencies managing the watershed resources (forest and mineral) and those managing freshwater recreational fisheries or commercial ocean fisheries that have a strong clientele, particularly if the agency's funds come from exploitation licenses, face that dilemma and the environment often suffers.

Also, today the principle of separating operations from the regulatory area is often violated when external bodies introduce new "environmental" programs in developing countries. Too often operational and regulatory responsibilities are intermixed under vaguely defined programs and placed under a single unit. Indeed, the indiscriminate application of the term "environmental activities" complicates effective assessment and resolution of environmental problems and issues. Urban waste treatment, land drainage, watershed management, and groundwater management should be referred to by the traditional, more descriptive terms rather than "environmental" projects. The institutional arrangements and responsibilities to set criteria and standards, monitor conditions, introduce improved operational procedures, enforce regulations, and execute remedial programs associated with protecting the environment could then be set forth in these same terms. They would be clearly understood by the public and the agencies. And the established line units can execute those tasks within their functional areas while the appropriate independent regulatory units can assure compliance.

Separation of the line functions. Comprehensive data collection and dissemination, resource planning, design, construction, O&M, and—where explicitly provided—financing are the six primary line functions in the water-resource sector. Many levels of government have specialized units in each functional area to provide expertise and establish performance accountability. Nevertheless, some countries still retain all-purpose units that handle several functions with whatever staff is available, shifting work as budgets dictate. Some separate project operations from project maintenance—a step that assures failure of an O&M program. Others have several parallel units performing the same function in the same geographic area, making it impossible to develop true centers of expertise or efficient services. Program continuity, staff currency with evolving technology, program quality, and performance accountability are sacrificed.

The program of organizing along functional lines is probably the second most important, yet often violated, principle for effective formulation and execution of water-resource programs. Planning requires considerable knowledge of country and sector policies, governmentwide programs, and budgets. It requires a multidisciplinary team. Design entails higher technology, knowledge of latest materials and methods, and experience in specifying the work. Construction must oversee the fieldwork and ensure quality, being intimately familiar with standards, contract administration, and construction methods. While O&M entails service operations, maintenance of the facilities, exchanges with the customers, and, more than any other government function, the expertise and discipline of managing a responsive, self-sufficient permanent "business."

Quality assurance at each stage of an undertaking requires clear separation of functions so that units can be held accountable for performance. This is most evident in the design, construction, and O&M phases of a project. For example, any deficiencies in construction should be remedied by the construction organization from their capital construction budget and not passed on to the inevitably inadequate O&M budget. Each functional area should be assigned to a specialized unit of a size and capacity to match the technology and workload. Staff tenure, training, and compensation policies should encourage retention of capable, experienced personnel, which such structuring facilitates.

Magnitude of change. Government must adjust its structure to the maturing of its stages in resource development and to the advancing

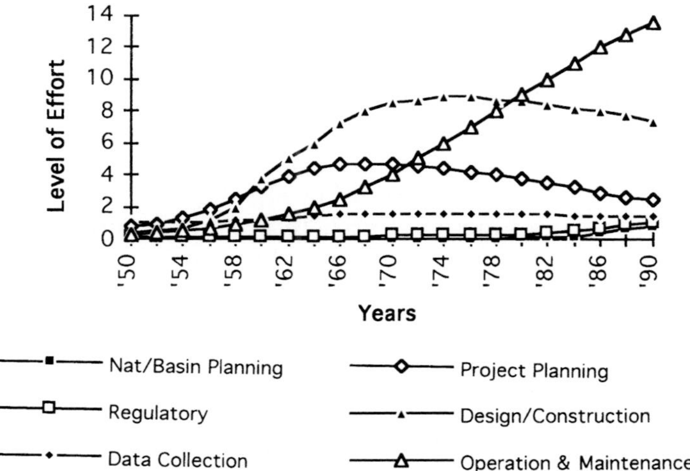

Figure 12.2 Relative government effort by function.

technology available in the field if it is to manage resources effectively. The nature of change and the accompanying staff adjustments, however, generate opposition—sometimes sufficient to abort the effort. Indeed, it is the private sector's flexibility, particularly in personnel actions, that allows it to rapidly adjust and thus remain efficient. Although rarely addressed directly, the future workload should be a major consideration when structuring and staffing agencies. Figure 12.2 illustrates the relative changes in effort required of a hypothetical group of government agencies during a given development period. Many countries are currently completing their development stage: a time when organizational and staffing flexibility is essential. Natural attrition reduces staff rapidly, if allowed, and eases the difficulties of consolidation. During periods of shifting workloads, the employment of consultants can prove effective for accommodating peak loads and highly specialized tasks. Limited-term employment can serve in temporary units.

Improvement of a government's organizational structure should commence by carefully clarifying the present and the future mandate of the government and the composition of government and nongovernment activities in the water-resource sector. Then the nature and magnitude of government programs in the previously listed functional areas should be determined, existing units evaluated, and overlapping and new responsibilities assigned to the ongoing units. Subsequently, the adjustments to specific responsibilities and authority, internal structuring, staffing, specialization, procedures, work, and support will fall in place.

Location of specialties. Success in meeting its responsibilities depends on a nation's management capability at each level of government and effectiveness of communications between the responsible units, the activity site, and those affected. Certain skills can be sustained only in central units. Sound quality control demands a central programming together with technical and budget review. But excessive centralized direction can frustrate local input in tailoring the program to the need. As a principle, governments should locate activities in the affected area consistent with sound government. It should decentralize as appropriate by assigning line responsibilities to units at the lowest level where quality work can be attained, incorporating measures to facilitate any necessary higher-level guidance for consistency among all functional areas. Some functions may necessitate that central line agencies retain responsibilities. Then deconcentration of central units by assigning subunits to the locality may be appropriate, particularly where a few highly skilled staff are required or the local government units lack the necessary ability.

Government functions, assignment of responsibilities, and structure of entities

As has been noted, government participates in water development and management through six primary functions—five line and one regulatory. The scope of each function is adequately defined by its title for purposes of discussing principles of organizational structure and agency responsibilities. The more-successful governments use well-crafted line and regulatory agencies together with advisory entities in the execution of these functions.

Advice and oversight bodies. The participation of individuals outside government in advice and oversight roles is a tradition in the water-resource sector. Indeed, some government functions can be performed best by such special-purpose entities. Neither the legislative bodies nor the bureaucracy may have the required expertise or blend of views or can act free from pressures. The public and beneficiaries serve on government agency boards, experts serve on technical committees, and public figures serve on policy and oversight commissions.

The use of limited-term commissions to help articulate national objectives in the water-resource areas was noted earlier. They have been established from time to time at every level of government to formulate proposals to resolve particularly contentious water-use or land-use issues. The traditional government entities are not well suited for these tasks. Greater public understanding and support of government actions are an important result, essential to instituting new approaches or altering established programs—particularly if the existing program is popular with the public.

Permanent commissions are particularly effective to oversee, if not have, final approval powers of planning, regulatory, and resource-allocation matters. They have become almost standard adjuncts to most regulatory agencies, including environmental-protection agencies. The advisory commissions are effective in reviewing the development and management programs of resource agencies. They are the mechanism of choice to provide oversight of land-use plans and the administration of these plans. Permanent commissions in this manner provide a timely, unbiased reaction from an outside perspective.

Most commissions do not have approval powers, leaving final decisions to the elective body or line agency which they serve. But they often hold public hearings to receive comments from outside interests. Commissions delegated powers of approval over basic policies and defined levels of action may have substantial staff to support their work with members serving either part or full time. The qualification of commission members are set in the authorizing legislation or acts creating the body. The individuals are typically selected by the executive branch and approved by the legislative—often the candidates are selected by each of several groups to be represented. The mechanisms described have long served key aspects of the currently popular "stakeholder" participation and a key forum in the political process.

Governments would gain much by generous use of this type of entity in managing its resources. Their mere existence will improve the bureaucracy's configuration of programs and projects. They will keep the public appraised and give them greater assurance that the programs and projects have received careful consideration. This will reduce the time and effort and build positive support for actions. The adopted mechanisms should ensure that all segments are represented: recognized community leaders, the public at large, advocacy groups, and the professions. This principle could begin at the national and regional levels with technical advisory committees and with oversight commissions in the policy, planning, and regulatory areas. It should be encouraged at the local level in planning, regulatory enforcement, and operations.

Assigning data-collection functions. The adequacy of the data-collection efforts for use in water development and management varies greatly among countries. Deficiencies stem largely from the dispersed responsibilities and in part from inadequate budget. Hydrologic units invariably have inadequate funds for gathering information from an inadequate network of poorly maintained sites. Development agencies obtain data for use in project formulation and design, terminated on completion. O&M agencies gather information only as essential to their seasonal and day-to-day operations. Regulatory agencies should collect highly specialized and location specific information, but gather little due to limited funds, staff, and facilities. Characteristic of dispersed efforts, data quality suffers from the use of inexperienced personnel, inconsistent procedures, and poorly maintained with records either lost or not processed.

Successful countries have adopted the principle that this is a specialty field justifying an independent entity serving the common needs of all government agencies in the sector (and making information available to the private sector). Planning, design, operating, and regulatory entities select the level of detail desired of the agency. Line and regulatory units may augment the information of special need to their work with their own internal efforts. The data unit should have authority to coordinate data-collection standards and review data-collection programs of others to ensure quality information. Deconcentration of the central data agency often results in higher-quality programs than can be offered by units of lower levels of government. The data users, including the regulatory agencies, should still retain responsibility to set their requirements and, of course, the use of the data.

Assigning planning responsibilities. Comprehensive water-resource planning should be assigned to the appropriate levels in government, reflecting the purpose of planning and the decisions to be taken. Unfortunately, aspects of water-resource planning are often scattered among several line agencies with only minimal coordination by the finance ministry through the budgeting process. The relative powers of the individual line agencies determine the government's thrust. Some countries have national plans, but these also may suffer from the dominant-agency syndrome or are one-time efforts by consultants with limited budgets, inadequate dialog with government leaders, and insufficient data. Consequently, too few countries update their plans at regular intervals.

The degree of detail necessary in resource planning depends on the maturity of resource development, magnitude, and purpose of the facilities and the management mechanisms already in place. Water- and land-use framework planning with 10-, 25-, and 50-year time horizons (followed by refined plans issued at 5-year intervals) should be conducted at the levels where resource ownership resides and allocation is made: the national and state levels. Local government (central, if local government is weak) should prepare detailed plans for each river basin consistent with the framework plans to guide water and land use in their jurisdiction using 10-, 25-, and 50-year time horizons. This will force coordination among the line and service agencies and land-use management agencies to better assure sensible land acquisition and development. Line agencies and service entities must plan in the detail required for long-term development scheduling to meet their responsibilities and for the project-specific programs consistent with the national, state, and basin plans.

Most successful countries create "water resources" entities (ministries, departments, etc.) to conduct resources planning efforts together with other multipurpose activities. When guided by a "commission" composed of representatives from all affected agencies and the public, they remain effectively unbiased. This allows easy access to a reservoir of technical support by ready linkage to the staff engaged in multipurpose programs. The unit should (1) compile information on present and projected demands on the resource and formulate alternatives to best satisfy them, (2) maintain the plan document reflecting decisions, and (3) offer advice to legislative bodies and policymakers on the consequences of actions they or others may propose. Plan units should keep these different roles clearly in mind and should not play an advocacy role.

The local jurisdictions' planning unit similarly should report to the local administrative head with an external oversight body. Local planning responsibilities should match the assigned management and enforcement responsibilities, including land use, waste management, water utilization, and environmental protection. Line agencies with development and project responsibilities should conduct development and project planning in their respective geographic and subject areas of operation. This requires coordination among all agencies, but the national and state, basin, and local plans should be the primary reference for the line agencies' programs.

Assigning design and construction functions. During the development phase in any service area, from flood control to irrigation, governments often established large design units. This has proved wise as long as the workload continues. A minimum engineering capacity is required to serve the larger service entities even during the O&M phase. However, the use of consultants to carry short-term needs and specialty work is common. And the centralization of engineering within the larger units is essential if a high-quality capability is to be maintained. The risk is that governments do not anticipate the downturn in their programs and reduce their design organizations accordingly.

Most countries today use private firms to construct their facilities. This policy should be adopted by all, as this activity requires technical and business skills that can be applied in several sectors with resulting efficiencies. Only fully employed construction firms can afford the timely introduction of new methodologies and equipment. And competitive bidding among private firms ensures cost-effective construction. No government at any level should maintain construction units.

Assignment of basin management. It is increasingly recognized that resource management—long-term planning and real-time operations—can best be executed within the context of the river basin, a long-proven concept. Basin planning, as discussed earlier, may be conducted at various levels of government, although best locally if talent exists. However, services and operations, by nature, must be local. An assortment of different basin O&M entities are found to be successful in countries throughout the world, influenced by the physical situations, preexisting agencies, and the underlying institutions. A few have mandates to foster development of all the basin's resources with authority to undertake most government responsibilities and services. Too often these have performed poorly in the developing countries. The complexity of managing divergent "businesses" with conflicting purposes and demands on staff and budget has been overwhelming. To further complicate their existence, all-purpose authorities overlap with and antagonize the established line agencies.

Strong, more-narrowly focused river basin entities to operate mainstream facilities, assume water-quality management responsibilities, or generate power have been more successful. In developing countries, it may be best to initially assign the basin responsibilities for bulk services to a modified unit within established provincial agencies. This would avoid creating new independent authorities in each basin that would overlap and compete for budget, staff, and power with the existing agencies. Such a model, found in several countries, builds on the existing entities while consolidating functions as appropriate. New authorities that focus on bulk services have been successful in other instances.

The resulting basin service-operations entity is responsible for the O&M of all water-resource services: bulk water supply, bulk storm drainage, primary flood-protection facilities, and the river channel and floodplain. This increases efficiencies and presents one entity for customer entities to work with and be served by. The manager of the basin O&M entity may report on technical matters to a provincial or central engineering department, as appropriate, at the same level or at a level different from the level at which basin planning is assigned. An oversight board or committee would guide the basin O&M entity and be composed of officials of the government agencies involved, customer-beneficiary entities served by the basin entity, and public leaders from the immediate basin community. Importantly, this board or committee also should be the same board or committee that oversees resource planning and allocation. The importance of having a basin committee to not just provide oversight but also to isolate the basin O&M entity from outside pressures for water allocation and operations cannot be overstressed.

Under such an arrangement, the primary responsibilities and functions of existing local government and customer-owned systems who would purchase services (bulk supply, storm drainage, flood protection, etc.) from the basin entity would remain untouched. Most would continue to function as now, with the exception of activities within the main channels. But no single basin resource-management structure satisfies all situations. While basin-specific factors influence the choice, recent trends in basin management should be examined closely before new agencies are established.

Assigning service responsibilities and application of the utilities concept. Power generation and distribution, water supply, waste disposal, irrigation supply, and—to a lesser extent—navigation, drainage, and flood-protection services are measurable. The first four are almost always operated as utilities in developed (and many developing) countries. The entity provides and is responsible for the quality of a readily defined service to the customer, with no peripheral support or unrelated activities. It owns assets, conducts O&M, procures new facilities and equipment, finances capital improvements, and charges for services. Where beneficiary organizations own the system, the management board consists of customers, similar to an investor-owned entity. Where a unit of government, usually at the local level, functions as an independent service "utility" within government, the budgeting and accounting are carefully isolated from the parent organization. The objective of "sustainability" requires that the service function be managed as a business, without other influences or activities that divert attention or create a maze within which accountability is lost. As a utility, it isolates a function that a broad-based government line agency is ill-suited to perform unless the activity is organizationally separate. And this is the key. Water-service systems constructed by farmers are operated with all the cited characteristics of an independent utility at no expense to government. But those constructed by governments in developing countries have usually remained government operations, blended into the array of agency programs, susceptible to the priority dictates of the agency's overall budget.

The adoption of the "utility" form of organization to provide water services is one of the most important principles found in effective water management. Countries should structure all government water-service entities at every level as self-sufficient utilities with rigid programming, budgeting, financing, and public accountability. The utility should have clearly defined, unquestioned rights to the resources (if involved). Services should be easily defined and measured. Beneficiaries should participate as appropriate in project formulation, contribute to the investment, and pay the cost of services through a combination of fees and taxes. Any subsidies to government-owned entities should be public and transparent.

Local services by local government entities. Typically, the distribution of bulk water deliveries in urban areas or villages is made by the local government unit. The entity should be structured as a utility similar to the bulk supply-service unit described previously. And in most successful urban arrangements, sewage collection and treatment are combined with the water-distribution service and provided by the same unit. The service area, the customers, the maintenance of the physical works, and—to a large extent—the location and operation of the supply and sewerage systems are the same, similar to the situation warranting the combining of irrigation and agricultural drainage services. All charges and taxes collected should be deposited to the account of the supply or sewerage utility, to be applied as deemed most effective. (Unfortunately, some local governments view service fees as a general source of revenues for all uses, violating the principle of equating the charges to the service and destroying the basis for sustainability and for customer trust.)

In effect, most urban water services are under direct beneficiary control. Municipal water supply and waste collection and treatment are typically provided by subunits of local government; and under local government, they report to the local people. In most cases, the local political process should allow direct influence on both long-term and daily operations. (Assignment of the services and operations to beneficiaries is treated under the following section.)

Assigning regulatory functions. Effective regulatory administration requires that the responsibility to comply with the standards be assigned to the operating entities, whether public or private. They must be held accountable. Special-purpose units should not be created to remedy a given "environmental" problem. Indeed, negligent agencies must apply their own resources to such work, so they do not believe that a cleanup effort by an "environmental" unit will rectify their errors or even their willful omissions.

The monitoring and enforcement of all water-quality regulations, water-quantity allocation, related environmental regulations, and interlinked land use should be assigned at the level of resource ownership. Some powers are properly delegated to lower levels if capacity exists, but the level of government that owns the resource still retains prime responsibility.

Since financially self-sufficient service entities inevitably must engage in financing, an oversight entity should review the viability of their plans, the cost-recovery scheme, the application of charges, and fiscal operations, including disposition of revenues. In many countries, a public-utilities commission oversees nongovernment utilities and approves the financing and repayment plan before the action may proceed. This step is particularly important where financial instruments are offered on the market. A ministry such as "home affairs," "interior," or "finance" may have this function for government entities and, in some countries, for customer-owned entities. In every regulatory area, central government should have jurisdiction where states have inadequate capability and on matters where national objectives dictate a uniform application of standards.

Role of Nongovernment Entities and the Public

Nongovernment organizations (NGOs)

General. Nongovernment entities have always had a major part in managing water resources, playing both advocacy and service-operations roles. They are not a new vehicle for providing services or for influencing decisions on resource development and management. Accordingly, it is important that those who seek vehicles for public expression and participation in resource management to investigate past experiences and understand how undertakings were conceived, decisions were really reached, and actions were executed.

Advocacy NGOs. In the past, as today, most nongovernment organizations were established to influence government actions. NGOs, as advocacy groups are now usually labeled, have long been a dominant force in guiding the development and management of resources in developed countries—in many, the most powerful single force. In some developed countries, the largest NGO role has been the active pressure groups composed of local residents striving to secure government action. Contrary to commonly expressed views today, it was the sustained and vocal pressures from local NGOs that brought about some of the largest water-resource developments. Client pressures on local and national legislators produced the authorization and subsequently the funds. This is not to say that bureaucracies did not support these efforts. These projects, indeed, were built on local participation, local voice, and local support—a confirmation of the power of local segments of the public.

Well-organized national-advocacy NGOs also evolved in many developed countries. Rather than being formed to promote local interests, these promote government programs at the national level. These have been particularly active in support of programs in urban water supply, urban waste collection and treatment, irrigation, rural electrification, and flood control. The organization and general purposes of these NGOs mimic the broad array of non-water-related urban and industrial advocacy groups also prevalent for influencing government programs in their fields of interest.

In recent years, additional advocacy NGOs have formed. The objectives have been mostly to seek more-comprehensive assessments of proposed programs and projects in the water-resource sector and give greater consideration to environmental objectives. These groups have been instrumental in changing the attitude of the public and legislatures. They helped evolve standards and administrative measures in new areas such as in-stream water quality, water habitats, wetlands, and land use adjacent to water bodies. But a few NGOs seek to muster outright opposition to all water-resource development and to alter the existing development. Examples of ill-conceived undertakings have galvanized these efforts at the international level, too, a new phenomenon.

As with all advocates, advocacy NGOs can foster or frustrate rational debate. There is no better evidence that politics determines actions in the water-resource sector than a review of group actions and the resulting legislation over the past several decades. Responsible NGOs have contributed and can contribute immensely to improving resource management. Only by the diligent efforts of government agencies and all NGOs to provide comprehensive unbiased information and to promote calm debate of all views by the public can sensible solutions evolve and the best interests of society be realized. How to assure the public full information, rigorous analysis, and rational debate—this is the important challenge to everyone engaged in institutional change and officials in every branch of government. Government should take the initiative, working with responsible NGOs and providing educational information for schools and, as it evolves, information to the general public on current topics of debate.

Industry and professional NGOs. Professional and water-industry NGOs play a major role in providing technical innovation, a vehicle for exchange of information and forums for establishing standards. New planning, design, construction, and operations concepts and methodologies have been instrumental in advancing the quality of water utilization and management. The standardization of materials has protected customers and greatly reduced the costs of facilities. The lack of industry associations and weakness of professional NGOs in many developing countries should be a concern of government. Unfortunately, rather than encouraging these organizations and participating in their work, many governments have restricted their formation and even hindered exchanges between the NGOs and their staff.

Beneficiary entities providing services. Nongovernment not-for-profit entities have developed and today operate many times more water-service enterprises than do investor-owned companies and, in some countries, more than even the government. Usually they are overseen by a board of directors composed of beneficiaries. But many are overseen and even administered by a local government board or even the courts. The latter arrangements are usually found where the entities are organized for agricultural drainage or flood control because the direct service and benefits for individuals are not readily evident, yet funds must be collected, maintenance performed, and debt repaid.

An extension of the argument for establishing self-sustaining water-service "utilities" as subunits of the responsible government agencies is to ultimately transfer the government utility to the direction and responsibility of the service beneficiaries. It has an important characteristic when considering institutional change—it is a well-proved mode for providing services with many examples in almost every country. For new water-service projects outside local government jurisdiction, beneficiaries should be required to organize as a public corporation with specified legal powers, including those of taxation, before the project is finalized or any construction initiated. The transfer of existing government-operated facilities to local entities can proceed in parallel under a program of system repair and management advice. Beneficiaries under both approaches should actively participate in project formulation and assume a significant part of the investment financing so that they are satisfied with the concepts and quality in line with their acceptance of O&M responsibility while restraining excessive construction expenditures.

The adoption of policies for full cost recovery, and hence increasing charges, is the main stimulus for seeking customers' support for transfers. By assuming O&M responsibilities, beneficiaries can reduce their costs through the lower unit wages of local hires and their own direct labor. They also can realize substantial gains from the improved efficiencies of locally directed efforts. And finally, the customers will, with time, be better able to determine the level of investment in improvements and changes in O&M practices that will yield the greatest return to their personnel operations that depend on the service.

For these and other reasons, customers should assume responsibility for as many of the O&M tasks as they can manage effectively as a group. However, they should not overreach, as accepting even a small portion of the scheme presents enough problems. The prevailing conditions in each case will determine the scope of responsibilities for operations and maintenance. As competence grows, the beneficiary entity can expand by either merging with other entities or creating umbrella entities.

Irrigation services are readily suited to beneficiary ownership and operation, as noted earlier. The changing conditions of supply shortages and growing villages and cities within the service areas are complicating both operations and ownership of existing irrigation systems. But these changes will require only a change in the governing board membership and modified operating policies—not in the entity's structure. The collection and disposal of localized drainage in small areas and the collection of agricultural drainage in service areas within large projects are not unlike irrigation-distribution services. The primary difference is that the benefits from drainage are not as immediate or obvious as those derived from irrigation. Farmers can maintain their on-farm works and join in clearing larger collector channels. The entity can employ private-sector contractors for heavy maintenance and major cleaning of channels. And government policy should encourage the provision of both distribution and drainage-collection services within a service area by the same entity.

A more advanced organizational form can be adopted when customers wish to assume responsibility for larger systems or major components thereof. "Umbrella" organizations constituting a federation of smaller beneficiary entities and/or local government entities can take responsibility for some of the higher-level government functions cited above. Such an entity has the political, technical, and financial power to meet the members' objectives which the individual entities alone cannot accomplish. By federating, beneficiary entities can join in solving their common supply problems and avoid costly conflicts and duplication of facilities. By remaining independent of the activities of the member entities, the federation can provide a highly efficient and cost-effective service. This form of organization has been successful in supplying bulk water within large metropolitan areas governed by numerous city governments and in supplying bulk water to groupings of irrigation entities. Under these arrangements, each member entity retains full responsibility for all services within its boundaries, including any independent supplies it may already have developed. The management of the umbrella entity reports to a board of directors composed of representatives of the member entities. Each member pays its share of the costs to the federation on behalf of its service area.

Examples of the arrangements described here can be found in practice, and anyone considering forming such entities should visit successful operations. Specific legislation presenting the legal format and, equally important, the government's regulatory rules can be examined. Established operations offer much that theoretical analysis cannot match.

The role of investor-owned private companies. Investor-owned companies have long provided support in the development and management of water resources. By far the most common have been to furnish materials, construct facilities, and maintain works. Indeed, government should secure all materials and construction through competition from private firms and increasing design and basic maintenance as discussed earlier. Private firms have developed and managed recreational activities utilizing the opportunities provided by water-resource developments.

Recently, private service firms have engaged more directly in water-resource management, including some form of collection and meter reading; management contracts; build, operate, and transfer of ownership of facilities to the government (BOT); and full ownership of the system. The willingness to go beyond facilities management and assume a BOT role in a specific situation depends on the extent of uncertainties about the condition, the adequacy of basic infrastructure, the cooperation of government agencies, political stability, and obligations for maintenance and system expansion. Water- and waste-treatment plants and new network systems offer more straightforward arrangements for BOT contracts. These facilities comprise a well-defined scope of activity with manageable risk, far different from the distribution or collector systems. Contract management services is perhaps the most straightforward. These may vary from limited-duration concessions that include construction activities to fixed-fee management contracts with performance bonuses. Although more simple, the government still must have a strong independent audit and regulatory capability.

The issues in privatization are encountered most often when contemplating having investor-owned entities provide the entire primary water-related service. Both the inefficiencies displayed by government operations and the complications and risks in having private firms provide services arise because the function is a monopoly activity. The desired responsibility must be clearly defined, the selection transparent, and competition open. Subsequently the firm's operation must be subject to full scrutiny. The only protection to the public from unscrupulous private firms in these situations is an active, strong, unbiased regulatory entity independent of special interests or political favors. It is the difficulty of creating such capacity that precludes privatization in many countries, particularly those without a regulatory tradition and lacking mature institutions. The extent to which these private-service firms are appropriate in developing countries will be learned, but the government oversight essential to assure responsible private-sector utilities has proved to be stretched even in countries with long-established regulatory and public oversight institutions.

A government contemplating any level of privatization of urban services should conduct a thorough investigation of the practices and experiences in other countries. This is the type of task that government should assign to a well-chosen independent study commission with ample time and resources to secure the advice of consultants and officials from other countries. Until new approaches are fully proved, countries should not reduce the use of traditional approaches to render their water and sewerage services. Government utilities provide efficient quality services in many countries, and delays in constructing water and sewerage services are costly.

There appears to be no significant role for investor-owned companies to provide water-related services in the nonurban area. In the past, major irrigation undertakings failed as a result of extended droughts when revenues disappeared and severe depressions when farmers were unable to make payments to any creditors or service providers. Flood control and drainage, for slightly different reasons, offer little opportunity to attract investor-owned entities.

Financial Aspects of Resource Management

Financing and institutions

This chapter treats financing to a very limited extent. However, financing cannot be fully divorced from a discussion of institutions for managing water and related environmental resources. Linked financial and institutional issues include water-allocation mechanisms, pricing of services, structuring agencies, and assignment of responsibilities. The reallocation of water, whether legislated or by markets, has major financial and economic impacts on the involved individuals and existing service entities. Third-party impacts of reallocations include the financial health of the affected villagers and the local governments. The policies on what constitutes public good and what constitutes defined beneficiaries of services, and hence cost recovery, are inherently dependent on the nation's goals and government programs. All organizations, including their programs and behavior, reflect financial policies.

One of the more obvious linkages arise when contemplating the transfer of service responsibilities. It is critical that customers understand that after the transfer, there will be charges for any remaining government services, that the basis for those charges is clear, and that the amounts can be forecasted. A breakdown of the cost of bulk and distribution services of proposed transfers will help clarify the financial benefits for the beneficiaries. If there is no logical way to evaluate the cost impacts of a transfer, there can be little incentive for customers to act.

Cost recovery and institutional implications

A government's policies on cost recovery and subsidy payments determine the financial viability of service entities. They relate to the broader government policies on subsidizing all government services and bear directly on the success of transfer programs. Unfortunately, the government's policies and programs for financing services to urban government, industry, agriculture, and individuals are some of the least understood by the public. Indeed, the uniformed views regarding payments and the recipients of subsidies are due largely to the efforts of special interests who wish to keep their own subsidy benefits out of the public's scrutiny by protesting that of others.

Many countries have initiated charges—partial or full—for all identified government services. This is more equitable and indicates a return to the period when beneficiaries constructed and operated facilities to provide all these services, charging the customers at rates to recover the investment, replacement, and O&M costs. Some countries now charge for the issuance and renewal of all user licenses, usually to cover the cost of administering the licensing and monitoring program.

Customer-owned systems, some of which have existed for decades and even centuries, prove that municipal water-supply, drainage, and irrigation schemes can be physically and financially self-sustaining. They follow common principles that are important to their success. The system is designed as an affordable facility consistent with the benefits, the construction is of a type and quality that results in affordable maintenance, the beneficiaries invest in the facilities, government does not guarantee rehabilitation when deficiencies or failures occur, the service is reliable as measured by the rules, and a disciplined operation is supported by strong beneficiary enforcement. Cultural and economic conditions in most developing countries justify the fact that the beneficiaries pay for services, unless government has undertaken costly or unsuitable projects or is not providing a reliable service. Unfortunately, these conditions are too common and first must be rectified.

The provision of subsidies is of increasing concern as government revenues shrink, demands on the budget increase, and O&M budget allocations lurch from year to year. Excessive costs and inefficient maintenance result. Traditionally, flood protection of major urban, industrial, and agricultural areas has been fully subsidized by government. Subsidies were used to support regional or subsector development such as irrigation. The subsidy program was expanded to improve public health by constructing treatment and network facilities to provide safer urban and village drinking water. Now subsidies to clean up waste and pollution from urban dwellers and industry and remedy adverse environmental impacts are major budget items. As should be readily apparent, the issue of cost-recovery policies can be rationally discussed only if all subsidies are clearly identified, fully costed, and open to full public debate.

Governments will have to tighten their cost-recovery policies if they are to expand service and maintain what is built. The financial crisis dictates such action. Importantly, for reasons of equity and to facilitate the application of cost recovery, its policy should be rigorously consistent for every type and level of service and every class of beneficiary, with the exception of internal system support to the very poor.

Basis for service charges

The cost of providing the service is the most common basis used worldwide for determining service charges to the beneficiary or customer. The costs of services represent all investment, debt, and O&M costs of the local single-purpose service facilities plus the same categories of costs allocated from any multipurpose projects providing bulk services to the local single-purpose scheme. These costs, in turn, are allocated to each beneficiary or customer in proportion to the service or benefits received. The costs are measurable and can be judged as fair by beneficiaries and the public alike. Differing from system to system, this basis better matches the services that the beneficiaries actually receive and what they decide they want to invest in capital and O&M for the given service. The principle is straightforward; is obvious to the public, which has a direct interest; and is easily examined.

Allocating service charges to beneficiaries or customers of water supply and irrigation are relatively easy applications as most services can be directly or indirectly measured. Assessing the benefits of drainage and flood-control services is more complicated. People view flood-control and storm-drainage benefits as uniform in relatively flat terrain. The costs can be allocated directly in proportion to area or value of property. But the benefits may differ significantly in an area of undulating terrain depending on a property's location relative to the natural drainageways. However, a methodology for determining the "zones of benefit"—maybe three to five zones in a service area—was developed long ago for this purpose.

Recently, water-quality degradation has been charged as an operating cost, since downstream users incur direct expenses by the polluters' action. This most obviously applies to urban and industrial dischargers. Theoretically, it could apply to agricultural wastes when caused by chemicals or livestock operations.

Opportunity costs have been suggested as the basis for charging customers of water-supply services—irrigation and urban. But the mechanics of calculating this cost and the changes in this component seasonally and from year to year as affecting an investor's plans, deciding how to free it of political influence, and applying it to other water uses such as hydroelectric power, navigation, recreation, and environmental protection and enhancement are not resolved. This, just as "free market" pricing, would function largely as an allocation mechanism and is appropriate only for countries where the overriding national objective is to allocate water to the highest calculated direct economic use. (This, however, does not preclude the calculation of opportunity costs as one item to consider in resources planning.) Since opportunity cost pricing is not applied in the developed countries, it should remain under study for application in the developing countries.

Marginal costs have also been suggested as the basis for charging customers for water services. This method is not found in the water sector for reasons related to the nature of the activity. The policy of both government and nongovernment entities is to not discriminate against new customers within a defined service area. This fairness doctrine is stipulated in laws or ordinances that designate service areas and grant the associated water rights to the service provider. Expansion costs are melded into the total costs, and all customers share. However, some use the mechanism of escalating hookup fees and surcharges to pass higher marginal costs of these components on to the individual demanding the expansion.

Service-charge mechanisms

Several factors influence the selection of service-charge mechanisms, the means to assess charges against the customer or beneficiary. The factors include the nature of the service, annual variation in costs, annual and seasonal variation in services, costs relative to benefits, water-conservation incentives, water-quality-control incentives, the extent of subsidies, and other policies to address poverty, equity, and ability-to-pay issues. A variety of customer service-charge mechanisms are in use. Examples are quantity charges for urban and irrigation supply, standby charges for services that enhance the property value or for fire protection and zoned property-tax assessments to beneficiaries and the adjacent public for all types of services.

Usually, a mix of mechanisms best provides funds for municipal supply, irrigation, and sewerage services. A component property tax may cover a portion of the allocated service costs and fund the entity through years of drought, while quantity charges reflect actual services derived by the customer in a given period. A tiered or escalating rate structure encourages water conservation in municipal systems. Minimal rates apply to the system's low-income group cross-subsidized within the system or subsidized by direct government payments to the service entity. Experience shows that drainage and flood-control costs are most effectively and equitably charged through zoned property taxes. Several countries are experimenting with new waste-charge mechanisms. These should be viewed from the standpoint of adequacy of recovery, equity, and effectiveness, if serving as a pollution management tool. But enforcing criteria of land-use and water-use licenses may be far more effective.

Service charges should be calculated independently for each individual system to reflect its peculiarities, the level of services to the beneficiaries, and the basic principle of no cross-subsidizing by beneficiaries among different systems. All should apply in an open, easily monitored manner. The collected funds should be rigorously accounted for and dedicated to their purpose and promptly deposited in the account of the service entity.

Financing capital investment

Government at every level will have to secure financing for capital investment from other than government budgets. All countries are encountering demands far exceeding budget revenues. Funding available from international institutions is diminishing in real terms and even more in terms of total government needs. Capital investment financing will have to be sought to an increasing extent from (1) an increase in service fees dedicated to this purpose, (2) issuance of obligation and revenue bonds on the financial markets, and (3) private sector investment.

Customers and beneficiaries will have to be informed of the urgency of securing new sources of funds to finance infrastructure rehabilitation and construction. Service tariffs will have to include additional amounts for this use. There is no alternative. And there is no painless way to introduce this change. Political will by all parties is the only solution. The brevity of this paragraph does not relate to the importance of this action—it is critical now.

Instruments from financial markets are the dominant sources used in the developed countries to finance major rehabilitation and new construction, and are used an increasing extent by developing countries. These means are employed by government, customer-owned, and investor-owned entities alike. Without such sources of financing most countries will not be able to expand infrastructure to meet even the most basic needs of their citizens. However, this calls for the borrowing entity to establish a detailed history of financial prudence and performance. It requires services to be operated as a "business," not as unaudited, unaccountable entities relying on the whim of a government budget as their primary source of funds—hence customers must pay for the cost of services. It also requires efficient organizations directed by professionals, and government regulatory bodies to monitor and report on the service and financial performance of every service entity. These efforts should receive urgent support from government officials and political leaders should become thoroughly familiar with these means. Governments now using these financial sources can readily compile the information necessary to formulate an approach and a program suitable for a given country.

Private sector investment was mentioned earlier. Its primary application will be for major works where revenues can be guaranteed by the government. As may be observed, water and waste treatment facilities and electrical generation are the obvious, and have proven to be essentially the only, investments suitable from the investor's perspective. Even for these, political and financial risks have resulted in high prices and funding of only a very minor percentage of needs. Few if any opportunities are evident to attract private investment in irrigation, drainage, flood control, waste collection, or water distribution. Besides the cited risks for the investor, the government must create a monitoring and regulatory mechanism in which both the investor and the public will have confidence. The opportunities for distortions without strong controls only lend support to the interests that wish to retain inefficient government operations.

Funding O&M expenditures and government guarantee

Governments have readily assumed responsibility for extensive water development. They have allocated the resources, planned the developments, and executed the work, but usually without considering funding of the inherent O&M obligation. Indeed, not even the beneficiaries have been engaged in devising the development program. Unfortunately, the condition of facilities today confirms the result.

Legislatures should guarantee O&M funding of government-owned facilities at a level to sustain the facilities in a condition to fully provide the design services on into the future. As a part of this, a reserve fund for emergency repair should be incorporated in the O&M budget. The O&M funding needs must override all other agency budget demands, particularly any construction of new works. The principle followed in bond financing of such facilities should be rigorously applied by the government and any lending agencies that support construction of such facilities. The adequacy of the O&M effort should be verified by an outside review unit, and, if satisfactory, any revenues surplus to O&M could be applied to other budget items.

Indeed, full charges should be agreed to with the beneficiaries before any new facilities are constructed. The beneficiaries should be compelled to form legal organizations to facilitate assessing charges, as outlined in an earlier section. Simultaneously, the government should introduce, as part of any facilities rehabilitation program, full payment of O&M and at least a portion of investment costs by the beneficiaries.

These principles should never be violated, even in the short term. There is no other way that a country can reconcile its wishes with its means. Nor is there any other action that assures sound, affordable investment in resource development. The application of these policies should encourage the political leaders and the bureaucracies to introduce consistent sectorwide service charges.

Closing

A country's institutions and financing constitute the means, in every sense of the word, for managing its water, land, and related environmental resources. The state of a nation's resources directly confirms the adequacy of its means and the extent and urgency for introducing modifications. When changes seem necessary, the undertaking should be pursued comprehensively, not ad hoc, formulating a framework for sectorwide management with realistic objectives. The urgency of the situation will determine the schedule for change. The immediate modifications should be geared to produce results; rarely can a country afford experimentation of unproven concepts while conditions deteriorate further or spend 20 years to introduce the more ideal. In these instances, it is well to determine why new unproven concepts are not widely found in successful countries. Fortunately, there is ample information in the world on a great many approaches—both successes and failures.

The most important step for those considering institutional changes is to thoroughly understand the basic principles found consistently in effective institutions. This should commence with an examination of how institutions in water, land, and environmental-resources management evolved under various conditions. The country will realize handsome returns from investing substantial effort to obtain in-depth information on how successful governments manage their resources today. And this should include countries with different demographics, different levels of industrialization, and at different stages in the development of their resources.

Institutional reform is the type of assignment where study and advisory commissions reporting to the most senior government leaders can prove extremely effective. Constituted of members who as a group are independent of bureaucratic or political pressures, these commissions can examine and formulate alternatives for government leaders to consider. Such bodies can provide the broad perspective and independence essential to help in this task while giving credibility to the process. Of course, the final changes will be selected by the nation's leaders through the political system.

Bibliography

Legal aspects

Smith, Turner T. and Pascale Romarek (1987). *Understanding U.S. and European Environmental Law: A Practitioner's Guide*, Graham & Trotman/Martinus Nijhoff, London.

Hayton, Robert D. and Albert E. Hutton. *Transboundary Ground Waters: The Bellagio Draft Treaty,* International Transboundary Resources Center, University of New Mexico.

California Water Code (1994). Shepard's/McGraw-Hill, Colorado Springs.

Attwater, William R. and James Markle (1988). "Overview of California water rights and water quality law," *Pacific Law Journal*, Vol. 19.

Water Resources and Other Related Laws of Illinois, 1990 ed. Illinois Department of Transportation.

Gould, George A. (1988). "Water rights transfers and third party effects." *Land and Water Law Review*, Vol. 23, No. 1.

Teerik, J. R. and M. Nakashima (1993). *Water Allocation, Rights and Pricing: Examples from Japan and the United States,* World Bank Technical Paper No. 198.

Government and agency organization

Wilson, James Q. *Bureaucracy: What Government Agencies Do and Why They Do It*, HarperCollins, New York.

March, James G. and Johan P. Olsen. *Rediscovering Institutions: The Organizational Basis for Politics*, Macmillan.

Shafritz, Jay M. and Albert C. Hyde. *Classics of Public Administration,* Wadsworth.

Delaware River Basin Compact, United States Public Law 87-328, 1961; Delaware Laws Chapter 71, 1961; New Jersey Laws of 1961 Chapter 13; New York Laws of 1961 Chapter 148; Pennsylvania Acts of 1961, Act No. 268.

Agency histories

Warne, William E. *The Bureau of Reclamation*, Praeger, New York.

Lilienthal, David E. *TVA Democracy on the March*, Harper & Brothers, New York.

Van de Ven, G. P. *Man-made Lowlands: History of Water Management and Land Reclamation in the Netherlands,* Uitgeverij Matrijs, Ultrecht.

Barnes, Dwight H. (1987). *The Greening of Paradise Valley: The First 100 Years of the Modesto Irrigation District,* Crown Printing, Fresno.

Shady, Aly M. (1989). *Irrigation Drainage and Flood Control in Canada.*

Rangely, Robert (1994). *International River Basin Organizations in Sub-Saharan Africa,* World Bank Technical Paper No. 250.

Parker, Dennis, J., and W. R. D. Sewell (1988). "Evolving water institutions in England and Wales: An assessment of two decades of experience." *Natural Resources Journal,* Vol. 28, Fall.

Principles and practice

Ali, M., G. Radosevich, and A. Khan (1987). *Water Resources Policies for Asia,* Balkema, Rotterdam.

Burchi, Stefano (1994). *Preparing National Regulations for Water Resources Management: Principles and Practice.* U.N. Food and Agriculture Organization.

Making Development Sustainable (1994). World Bank.

Frederiksen, H. D. (1992). *Water Resources Institution,* World Bank Technical Paper No. 191.

Frederiksen, H. D., J. Berhoff, and W. Barber (1993). World Bank Technical Paper No. 212.

Frederiksen, H. D. (1994). "Considerations in the transfer responsibilities for services in the water resources sector." *Water Reports 5: Irrigation Management Transfer,* selected papers of the International Conference on Irrigation Management Transfer, Wuhan, China. Published by the U.N. Food and Agriculture Organization and the International Irrigation Management Institute.

Chapter

13

Economic Mechanisms for Managing Water Resources: Pricing, Permits, and Markets

K. William Easter, Nir Becker, and Yacov Tsur
Department of Applied Economics
University of Minnesota
St. Paul, Minnesota

In this chapter we consider the different economic instruments that are usually suggested as means to deal with water-allocation and water-quality problems. The popular administrative method of water allocation through a central governmental authority is used as the reference point for comparison with other methods. Prices, markets, taxes, and permits are then considered as a means to improve water allocation and water quality. This is followed by two sections that are concerned with the practical problems of developing water markets and implementing water-pricing systems.

Government Involvement in Water Allocation

Water is a scarce resource, primarily because it is usually in the wrong place or available at the wrong time relative to demand. When water resources are scarce, there is an important role for economics in determining the best way to allocate a given amount of water among different users. Two major problems exist in water allocation, but both are interconnected and tied to agriculture. In most countries, the agricultural sector consumes a large share of the water resources. However, this is also the sector which can pay the least for water, relative to other major consumers such as municipal and industrial users (Gibbons, 1986). In addition, low water charges are one of the ways governments subsidize the agricultural sector.

The first allocation problem occurs because the low fees or prices charged for water encourage agricultural water users to consume more water than is socially optimum for crop production. This, in turn, can cause the degradation of the water quality as well as increase pumping costs (Rogers, 1986; Becker, 1995). When water is inadequate to meet demands at these low prices, a solution often suggested is to invest in new high-cost sources of water that are subsidized by government.

The second allocation problem arises, not from overuse in general, but from a misallocation of water among users since the amount that each user receives is the result of administrative quantity rationing or some historically established nontradable water right. Greater economic benefits could be obtained from the water if it could be traded by users. This, however, is not allowed in most countries since it would mean that farmers could sell subsidized water and appropriate most of the economic rents generated by the sales. Thus, the two problems are interconnected: low water prices often cause governments to restrict water trading (Anderson, 1983; Frederick, 1986).

Besides helping the agricultural sector, there are several other reasons why governments around the world have been directly involved in the water sector. The first reason is the lack of well-defined property rights, especially for groundwater. Without property rights, groundwater tends to be overpumped. This is the open-access problem where if "you do not use the resource, someone else will." (See App. A for a more detailed discussion of the open-access problem.) A second, but related, reason for public intervention arises when open-access resources provide services that are nonrival or nonsubtractable (see Fig. 13.1). In other words, another user can enjoy the resource at no additional cost. These resources are called *public goods* and are not profitable for the private sector to supply.

The third reason why governments often play a role in the water sector is that water-development projects entail increasing returns to scale. That is, the per unit cost of supplying water declines as projects become large. This, in turn, could lead to a natural monopoly in which only one firm controls the water supply (lowest-cost supplier) and charges high water prices (Mead, 1952; Lauria et al., 1977). In addition, in many developing countries, private capital is not adequate for

Since many water investments involve *large capital investments* with *long periods* before *payoff,* it has been difficult to attract private investors, particularly in developing countries. Moreover, water supply and irrigation projects typically exhibit *increasing returns to scale,* and therefore, are prone to natural monopolies. Consequently, without government intervention, there may be underinvestment in water supply and monopoly pricing. Furthermore, many water resources and water-resource investments produce *joint projects,* such as recreation, electric power, and irrigation, which complicate water pricing and allocation decisions.

Inadequate information concerning water supply and demand, which can vary widely within and between years, as well as information concerning who actually receives water and how much is used, has in many cases, made it difficult to effectively manage and price water.

Some of the services provided by water are of a *public good* nature, where one person's use of a good or service does not decrease or *subtract* from its value to others who use the same good or service. Since public goods are not consumed (in the traditional sense) when used, they can continue to provide the same benefits to everyone, as long as they are not damaged or *congested.* As long as there is no cost to society for the added use of or *subtraction* from benefits to other consumers (marginal cost of serving another user is zero), increased use adds to total economic welfare.

A second characteristic of public goods is the inability or higher cost of excluding consumers from using the resource. *Excludability* refers to the cost or difficulty involved in preventing a consumer from using a resource who does not meet the conditions set by the supplier. These two characteristics (excludability and subtractability) can be used to set up a broad two-way classification of goods and services (a) *public goods* (low excludability and subtractability); (b) *price goods* (high excludability and subtractability); (c) *toll goods* (low subtractability and high excludability); and (d) *open-access goods* (high subtractability and low excludability). Private firms do not engage in activities with low excludability because it is difficult to get consumers to pay. Where low subtractability exists, market forces will not produce an optimal level of investment. For large-scale irrigation projects in South Asia, excludability is low because of large numbers of farmers involved and the difficulty of monitoring each farm plot. Subtractability is low for flood-control and navigation projects. Goods or services classified in any one of these four categories cannot be considered fixed for all time. As technology, institutions, or other conditions change, a resource may move from one category to another. For example, developments in tube-well technology have reduced the economies of scale in tube-well irrigation, so that it now fits in the private goods category, even for relatively small-scale farmers.

A closely related problem involves the provision of public goods in large "lumpy" units. In such cases, the consumer does not have the opportunity to purchase different quantities of the goods or services and cannot be charged on the basis of quantity consumed. Flood control is an example of such lumpy goods that are also public goods. It is also likely to be difficult and expensive to determine how much each individual benefits or to *exclude* people from enjoying flood protection.

Unregulated market systems may generate outcomes that do not satisfy environmental concerns or a country's *social goals* in terms of poverty alleviation, food security, income distribution, and public health. In cases where water resources are *transnational,* or involve transnational environmental effects, water-allocation decisions are necessarily made by transactions among governments.

Figure 13.1 Characteristics of water-resource development.

such large "lumpy" investments. To take advantage of these economies to scale while keeping prices low, governments have built and operated many water projects. These government investments have helped overcome capital constraints, but in the process they have created a number of other problems such as the misallocation of investment funds, rent seeking, and an overextension of governmental agencies (Easter and Hearne, 1995).

Still, governments are now and will continue to be an inevitable part of future water-allocation decisions, at least in setting the "rules of the game." The issue is how government can improve water use without exerting direct management control. To address this issue, an important first step is to define an economically efficient allocation of water resources.

The Efficient Allocation of Water

Although the focus is on the quantitative aspects of water allocation, quality concerns can be analyzed in the same manner. Because, in many cases, water is considered a renewable resource (except for some groundwater stocks), the problem of allocating water between generations is less important relative to nonrenewable resources such as oil or natural gas. Thus, for simplicity, the efficiency criterion is confined to one time period, or one generation (see Apps. B and C for discussion of the value of groundwater over time).

The economic *efficient allocation* of water is determined by that amount of water which, if allocated to each user in the basin, results in the highest return for the amount of water available. How does one achieve an efficient (sometimes referred to as the optimal) allocation? To answer that question, one needs to understand how aggregate water demand is calculated for a given region.

The aggregate demand is based on the sum of water demands from individual users. To begin with, assume that an individual user who owns some land has different water quantities available for use. How much is the user willing to pay for one unit of water? This depends on how much a unit of water contributes to the use that the consumer intends for the water, e.g., the price (value) of that use (good) times the output. The value of a marginal (small) change in water use is called the *value of marginal product* and can be written as

$$\text{VMP}_w = P_{x_i}\,\text{MP}_x^w \tag{13.1}$$

where VMP_w = value of marginal product
P_{x_i} = price of commodity x_i
MP_x^w = marginal product of one more unit of water applied to commodity x

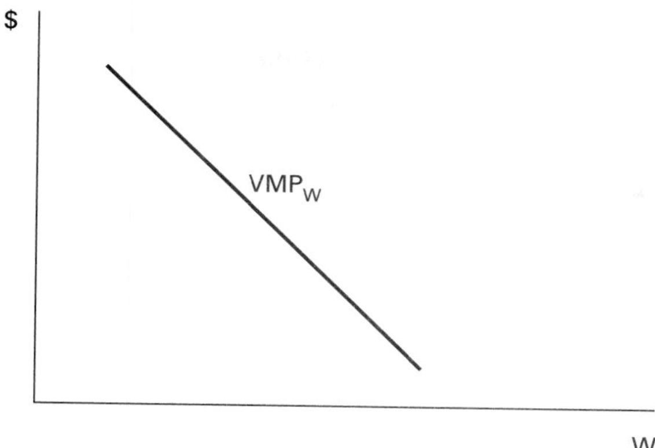

Figure 13.2 Value of marginal product for water.

An important characteristic of the MP_x^w is that as more and more water is used, the MP_x^w declines because of the law of diminishing marginal productivity. Since P_x is assumed constant, VMP_w will be a decreasing function of W (Fig. 13.2). The demand function for an individual user is given by the VMP_w curve. It shows how much the user is willing to pay for varying quantities of water.

A number of factors will affect the demand for water. Among them are prices and availability of other factors of production, the technology available, and the prices of the different commodities for which water is used. But all these factors will not change the conclusion that the demand function for water is downward-sloping as shown in Fig. 13.2.

Now that the individual-demand curve is defined, the next step is to determine the regional-demand curve. While the individual-demand curve is used to define individual water use, a regional demand can be used to help determine regional water use. Within a region there are a variety of uses for water, and all of them have legitimate competing claims. It is, therefore, necessary to find a mechanism that will allocate the water among the different users so that benefits are maximized for a basin or region.

For simplicity, assume that there are two users. The results can later be generalized to more users. The regional-demand curve for water is based on how much will be demanded, at each price, by both users. To estimate this, the demands of the two users are added horizontally (see Fig. 13.3). In this example, user B is willing to use less water than user A at each price. At price P per unit of water, the overall water demanded will be W, which is composed of the quantity W_A demanded by A and the quantity W_B demanded by B. Is this price economically efficient? Since both users have the *same* VMP (which is

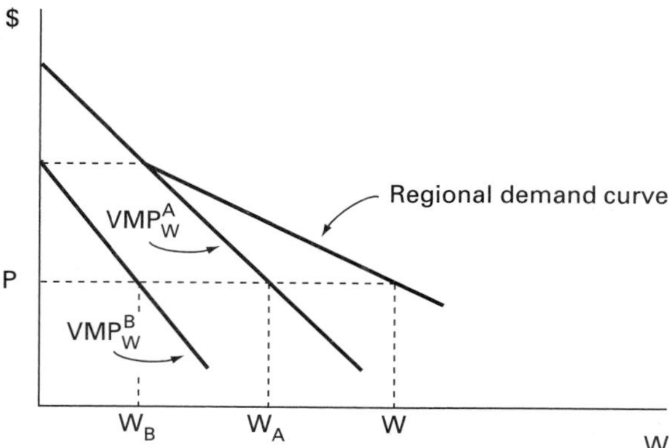

Figure 13.3 A regional-demand curve for water.

equal to the water price charged) and changing the allocation would reduce the welfare more for one than it would raise it for the other, the allocation is efficient. In other words, in order to achieve economic efficiency without rationing, all users have to be charged the *same* price for the same quality of water. This is one of the most important conditions for creating an efficient allocation of water (Howe et al., 1986).

If the net basin supply of water is *W*, then economic efficiency dictates a water charge of *P*. In practice, the regional-demand curve may be unknown, yet an efficient allocation can be reached if a price is selected that makes the total quantity demanded equal to *W* (this ignores the impact of uncertainty in water supply and other factors in the system). The key to an efficient allocation is that both users face the same price.

The other half of the allocation equation is supply. The *supply curve* is usually upward-sloping from left to right. This means that an increasing amount of water can be supplied only at an increasing cost and, therefore, a higher price. For simplicity, however, assume a vertical supply function. For this special short-run case, it means that no matter what price is charged for the water, the same amount is available (and it equals the basin's net supply). To further simplify things, suppose that the production cost of water is zero, which is depicted in Fig. 13.4. If the net basin supply is S_0 and the regional-demand curve is given by D_0, then the equilibrium point is *E*, where *W* units of water are used and the price at equilibrium is P^*. Even though there is no cost involved in water production, in equilibrium, the price is positive. This reflects the scarcity value of the resource itself and is sometimes called *scarcity rent* (or the shadow price).

Without this signal of scarcity, the aggregate water demand in the basin would be W_1, yet there is not enough water to satisfy W_1, so water prices have to be raised in order to induce conservation. Prices

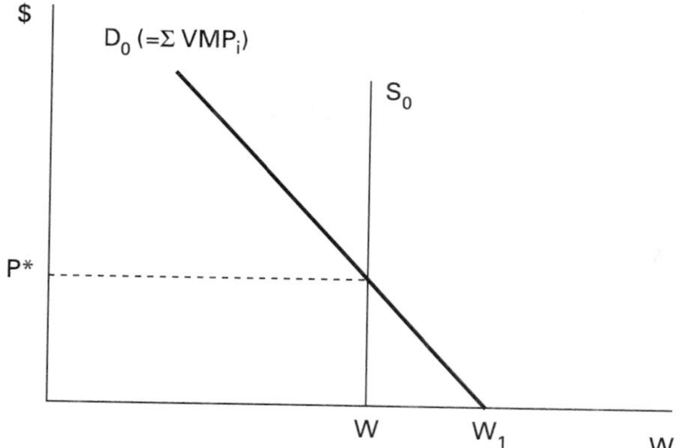

Figure 13.4 Equilibrium of demand and supply in a representative basin.

would have to increase until the price P^* is reached and the total demand exactly equals the net basin supply. The other alternative is to establish quotas for all users. Ideally the quotas would be based on the efficient allocation determined by charging price P^*.

When the costs of supplying water are added, they should include fixed and variable costs, as well as indirect costs. *Variable costs* usually include the operation and maintenance cost of a project as well as the energy needed to divert a unit of water from the supply source to the point of use. The *fixed-cost* component includes the capital cost, which is the opportunity cost of the capital invested in the hardware component of the network. The indirect cost component includes "third-party impacts," especially environmental impacts. These are the spillovers that harm parties not directly involved in the water exchange and include damage to fisheries and aquatic ecosystems.

In order to achieve efficiency in any reasonably well-functioning market, the following condition must be met:

$$P_W = \mathrm{MC}_x^w \tag{13.2}$$

where MC_x^w is the marginal cost of using one more unit of water. Yet, as discussed above, even with zero costs and, therefore, zero marginal costs, there is still an equilibrium price that includes the scarcity value or rent. This means that the efficient price of water should include the scarcity rent λ:

$$P_w^* = \mathrm{MC}_w + \lambda \tag{13.3}$$

If the planning authority tries to cover only the operating and capital costs, and excludes the scarcity rent, the price signal will be too low, and users will be encouraged to use excessive quantities of water.

The result could be a decline in water availability, higher pumping costs, and a degradation in water quality. Administratively determining the efficient allocation is not an easy task and requires extensive knowledge of both supply and demand for water. This is the task we turn to next.

Estimating demand

In general terms, water uses can be categorized into irrigation, industry use, hydroelectric-power generation, domestic (municipal) use, and recreation. In principle, the input derived demands can be used to estimate water demands. In actual practice, this approach may not be feasible, in many cases. Water used for domestic and recreation purposes does not produce marketable outputs. Thus output prices for these sectors do not exist, and the input derived-demand approach breaks down. Industrial processes are often too complex to possess well-defined production function representations, thus rendering the input derived-demand approach inappropriate. For the irrigation and hydropower sectors, the derived-demand approach is feasible. However, prices of agricultural commodities and of electricity—the output prices corresponding to irrigation and hydropower—are, many times, obtained in distorted and regulated markets, respectively, and hence do not reflect true output values. Thus, assessing water demands according to these prices would bias the value of marginal productivity of water and result in an inefficient allocation. Nevertheless, it is possible to evaluate the derived demand with respect to the "correct" output prices and to obtain genuine demand estimates. This section briefly discusses various methods for valuing or estimating demand for the five sectors; valuation methods are normative in nature and are based on calculating the value of water in production while estimation methods are positive in nature and use actual price and quantity data.

Irrigation. Because prices of irrigation water are typically rigid, direct estimations of water demand for irrigation using price and quantity data are uncommon [see Griffin and Perry (1985) for an example]. Valuation methods, advanced initially by Young and Gray (Young et al., 1972), include *farm-crop budget analysis,* the *linear programming approach,* and a combination of the two. A summary of these methods is given in Gibbons (1986) and Colby (1989).

In the input derived-demand approach, one first estimates a production function and then calculates the derived demand for water as the water input that maximizes profit at each input/output price configuration. However, these production function estimates are dependent on the availability of crop water-response data (Gibbons, 1986;

Colby, 1989). Although constrained by these data limitations, the derived-demand approach often has been applied to estimate irrigation water demand (see the discussion below for possible biases in this approach).

Another approach uses farmland sales data to infer a value for irrigation water (Colby, 1989). If farmland prices reflect the present value of future rents to be derived from a plot of farmland, and if these rents measure the return to the "free" (or underpriced) factors of production, then it is possible to determine the value of irrigation water from farmland prices. The condition set by the two "if"s are quite restrictive (e.g., farmland prices may reflect speculative motives in the demand for farmland or false expectations regarding commodity programs); thus irrigation water prices based on increased land values are questionable.

A few additional qualifications need to be borne in mind when using estimates of irrigation water demand:

1. All estimates are made with respect to the prevailing commodity prices and prices of other inputs of production (e.g., fertilizers). Some of these prices may be distorted as a result of government intervention and hence do not reflect their economic values. This directly affects the demand for water as well as indirectly through its effect on the demand for other inputs such as land.
2. Because of the seasonality in crop production, there is a considerable difference in the demand for irrigation water across seasons (water available during the peak of the irrigation season has a much higher value than water during the off season). Water available at one particular time is really a different commodity than water available at some other time.
3. Many crops exhibit high sensitivity to differences in water quality (particularly salinity). Water of different qualities should be considered as different commodities. A water commodity is thus distinguished by both time and quality, while a well-defined demand schedule for a given water commodity is based on its time and quality parameters.
4. Some courses of action are available in the long run, but not in the short run. In the long run, farmers can change the crop mix or switch to using water-saving irrigation techniques such as drip irrigation. Consequently, the demand for irrigation water is different depending on the time frame used (in the short run vs. the long run).

Municipalities. There are two basic approaches to estimate the value of an additional water unit for municipal use. The first utilizes price and quantity data to estimate a demand equation by specifying an econometric model and applying regression techniques to estimate the model's parameters. The second approach uses contingent valuation techniques based on survey questionnaires directly asking households or individuals to state their willingness to pay for additional water. Numerous studies have estimated water demand for municipal use from price and quantity data, starting with the seminal work of Howe and Linaweaver (1967), and more recently, Carver and Boland (1980). They differ according to the data used (time series, cross section, or panel) and the level of aggregation employed (Gibbons, 1986).

Price-elasticity estimates exhibit intraseasonal variation as well as spatial and intertemporal (long-run vs. short-run) variations. For example, price elasticities estimated for Tucson, Arizona, range between −0.23 and −0.7 between the winter and summer, respectively; in Raleigh, North Carolina, elasticity estimates varied between −0.3 and −1.38 (Gibbons, 1986). Carver and Boland (1980) estimated water-demand elasticities in the Washington, D.C., metropolitan area using cross section, time series, and panel data. The panel data allowed Carver and Boland to separate short-run from long-run elasticities. Elasticity estimates range between −0.05 (Nov.–April, cross section, short run) and −0.7 (Nov.–April, cross section, long-run). Thomas and Syme (1988) estimated demand elasticities for the Perth, Australia, metropolitan area that ranged from −0.58 to −0.1, depending on different specifications of the demand equation.

Contingent valuation techniques were originally developed to value extra market benefits from improved environmental qualities (water quality or a scenic preservation). Thomas and Syme (1988) employed this technique to evaluate willingness to pay for municipal water in Perth, western Australia, and compared the contingent valuation results with those obtained by econometric estimation. The similarity of the results suggest that contingent valuation methods can replace econometric estimation whenever data problems exist.

As in the case of irrigation water, it is important to distinguish by quality. Important water-quality variables include palatability (taste, odor, color), softness, corrosiveness, temperature, and the content of bacteria, organic matter, and dissolved salts and acids (Hirshleifer et al., 1960). Treatment of these different characteristics requires different methods which vary in their costs. The cost of treating domestic water would thus depend on the treatment required and the extent to which it is applied. Water softening costs in the United States range from $10 to $23 per acre-foot (Hirshleifer et al., 1960). Filtration costs in the Cannonsville-Hudson study amount to about 20 percent of annual operating costs (Hirshleifer et al., 1960). Over a 40-year life of the Hudson project, operating and maintenance costs were estimated at $306 per million gallons, of which filtration costs were $177 per million gallons ($100 per million gallons is equivalent to $32.59 per acre-foot). These figures are given in 1950 dollars and are calculated at a 10 percent interest rate (Hirshleifer et al., 1960).

Raw water and treated water, therefore, should be considered as different commodities. Within treated water, there are different methods and levels of treatment, which, if they vary in cost, render further refinement necessary. Thus, the quality of water is an important feature in analyzing urban water demand. In contrast, time is generally not very important unless the system does not have sufficient supply or capacity to meet water demands on a continuous basis.

Urban water demand grows rapidly over time, as the population expands and the standard of living improves. This raises the need to consider future demands in present decisions, which, in turn, requires using present and past patterns of demand to forecast future demands (Gardiner and Herrington, 1986). However, these data may be seriously limited by a lack of price variation, especially the lack of higher prices.

Industry. Industrial processes use water mainly for cooling, condensation (stream electricity-generation plants), washing of raw materials, equipment, and products, and as a direct input (the beverage industry). Water is also used in waste disposal and assimilation. The largest share of industrial water intake is for cooling and condensation. In general, water costs are only a small fraction of production costs in most industrial processes. Thus industrial water demand tends to be inelastic. This complicates the task of estimating a demand relation from price and quantity data (since the variability in quantity data is small), but implies that in the short run, the demand for water is almost fixed and hence easy to predict. In the long run, firms can use various techniques to recycle water and to decrease their consumptive use of water. Estimates of long-run demands, thus, are based on equating the value of water with the least-cost method of recycling and conserving water. The cost of recycling provides an upper bound on the (long-run) willingness to pay for an additional unit of water in industry (Young et al., 1972; Gibbons, 1986).

Hydropower. Hydropower production technologies typically exhibit constant returns to scale with respect to the water input in the short run; multiplying the rate of water flow over a given head will multiply the power generation. Thus, in the short run, the marginal cost of water in hydropower generation is constant for all output levels that do not exceed the generators' capacity. Obtaining the value of water would require assigning the value of the kilowatthour produced by hydropower. Because the electric-utility industry is heavily price-regulated, the kilowatthour price does not reflect its economic value to society, undermining the use of the input derived-demand approach. An alternative approach is to value hydropower water according to the alternative cost of power generated by other means: "The value imputed to the water used for hydropower is the difference between the alternate cost of electricity and the cost of hydropower generation" (Gibbons, 1986, p. 89). Suppose a kilowatthour's worth of water is removed from a river and the electricity is generated via some alternate method (e.g., steam-electricity plant or gas turbine). The value of the lost water is defined as the marginal cost of making up the kilowatthour through the lowest-alternate-cost method, less the marginal cost of generating the kilowatthour at the hydroelectric facility (which is no longer incurred). Marginal water values are denoted as peak or baseload, depending on whether the alternate generation method is used to meet peak or baseload demand, respectively. Estimates of the marginal value of water range between $18 per acre-foot (short-run, baseload) in the Columbia River, to $80 per acre-foot (short-run, peak) in the Snake River (Gibbons, 1986).

There are reliability and environmental quality advantages of hydropower compared to other electricity-production methods, which should be accounted for in the cost calculations (Colby, 1989). This would increase the value of water for hydropower generation as long as the environmental costs of alternative methods for generated electricity have not been included in their marginal costs.

Recreation. The demand for recreation has been increasing as the population expands and the desire for outdoor activities grows. Direct evidence on willingness to pay for recreational water uses is rare, and methods to elicit such information are based on survey questionnaires or indirect techniques. The techniques that have been devised for placing a value on water used in recreation are typically used to evaluate changes in quality rather than quantity. Contingent valuation involves asking a question such as "How much would you be willing to pay for...?" Contingent-valuation studies can use survey methods or referendum data (McConnell, 1990). Another approach entails using travel-cost data and discrete-choice analysis. Travel-cost data are used to input a value to water quality by measuring how far people will travel to obtain water of improved quality (Feather, 1992). In principle, the methods discussed above can also be used to value the differences in quantity flows, i.e., the willingness to pay for a marginal increase in instream flows. This subject, however, has attracted less attention than the concerns about water quality (Daubert and Young, 1981; Loomis and Ward, 1985).

The value of recreation water includes, in addition to the user value discussed above, nonuser values; these are option value, existence value, and bequest value. *Option value* represents willingness to pay to have the option of using the resource at a later time (Hanemann, 1989). *Existence value* represents the benefit consumers derive from knowing that the resource exists although they may have no intention of ever visiting it (Brookshire et al., 1986). *Bequest value* represents willingness to pay for saving the source for future generations. These nonuser values can be a nonnegligible component of the demand for water (Walsh et al., 1984).

Summary of water demand. The demand for a commodity, as it appears in standard textbooks, is the willingness to pay for the last unit of the commodity. In the case of water, the complexity of this concept is exacerbated by a number of factors:

1. Water is not a homogeneous commodity. Different users will tolerate water quality differently, and this will influence their willingness to pay for water of different qualities.
2. The willingness to pay for water varies seasonally and annually (water demand for irrigation changes during the growing season; urban water demand changes over time). Hence, water demand should be defined for each time period (e.g., season) as well as for a particular water quality.
3. There are nonuser values attached to water, which are potentially sizable. The option, existence, and bequest values were mentioned above in the context of recreation demand for water. But similar concepts apply to irrigation and domestic use of water. Views of irrigated green fields and grazing herds have aesthetic value for others beyond the direct benefit accrued to the user (farmer). Likewise, green lawns and public parks in urban areas provide indirect benefits for nonusers.
4. Some sectors require a large quantity of water but consume only a small fraction of it. Water used to generate hydroelectricity, for example, can be used by downstream industries and irrigators. The same is true for water used for cooling and condensation in industrial processes, although the water discharged may be of lesser quality (high temperature) and hence is excluded from the supply of water of the original quality but adds to the supply of water of lesser quality.

Estimating supply

A country, a region, or a state has a given amount of water resources endowed to it. Some of the supply is derived from recurring annual flows (such as rainfall and streamflows), while others are available as a result of past stock accumulations (such as the stocks in an aquifer, a reservoir, or a lake). The use of recurring flows that are independent of water stocks does not affect the available supply of water in the future, whereas drawing down stocks does. Thus, water derived from recurring flows affects only present production, while water derived from stock sources also affects future supplies. Accordingly, water derived from recurring flow sources and water derived from stock sources should be considered as different commodities.

As discussed above, the same water may be used for more than one purpose; water used to generate electricity may be available downstream for industries or irrigation. In such a case, firms that share the water do not compete, but rather complement each other. If the same cubic meter of water was used first to generate electricity and then to irrigate a crop, the contribution of this cubic meter of water is the sum of the contributions in these two activities. The use of water for electricity has not eliminated it from the supply available for irrigation. The two sectors compete only for water used for hydroelectricity generation during periods when demand for irrigation water is essentially zero or for water that could otherwise be stored and used for future irrigation. Such timing conflicts may be hard to resolve. Irrigation is season-specific, unlike electricity, which is produced throughout the year. Even during the growing season, irrigators may prefer to irrigate during the day, while hydroelectricity may be used extensively during peak demand in the afternoon, so that the water used in its production is available to downstream growers at night when it is not wanted.

Water sources differ also in the quality and the degree of uncertainty regarding their availability, both spatially and temporally. Rainfall, for example, may fluctuate substantially from year to year, within a year, and across regions. The stock of an aquifer may be imperfectly known. The recharge rate of an aquifer is a result of ground flows and surface-water percolation, both of which cannot be perfectly predicted. Shallow aquifers underlaying cultivated areas absorb fertilizers, pesticides, and salts. Deep fossil aquifers in arid regions contain water of different degrees of salt content. Thus, groundwater from shallow aquifers and groundwater from deep aquifers usually differ in salinity level, and the quality of surface streamflows is likely to differ from both.

Since water derived from recurring sources that do not contribute to water stocks has no effect on future supplies, its marginal cost consists of the cost of processing and delivering the last unit. This, in turn, depends on the existing technology and facilities used to process and deliver the water. Water for domestic use should pass certain quality standards, thus may require processing in addition to delivery. The same is true for water used to irrigate crops which are sensitive to water salinity. The supply curve of water derived from recurring sources is thus given by the marginal cost of delivering and processing the water. The short-run problem is to evaluate the existing water delivery and processing technologies and to estimate their associated marginal cost functions. A long-run analysis should consider all the available technologies. Examples of such studies include the works of Teeples and Glyer (1987) and Swallow and Martin (1988).

The marginal cost (supply) curve of water depends on the source from which the water is derived. The marginal cost of water derived from recurring sources, which are independent of water stocks, again, consists of the extraction costs of delivering and processing the last unit. The cost of water that draws down an available stock (or is derived from a flow that would otherwise contribute to a stock source) includes an intertemporal scarcity value or rent. Scarcity values depend on the extraction technology and cost and on the demand for water in the present and in the future. They are sensitive to the various types of future uncertainty. The cases of future demand and uncertainty and their impact on the scarcity values are discussed in Apps. A and B.

Efficient Pricing and Water Permits

Once the basin's supply and demand have been estimated, the optimum water allocation can be determined. Again, assume that two users have different demands for water. This can be thought of as two industries or two sectors that have different value marginal products (VMPs) for water. The situation is depicted in Fig. 13.5. In the short run, the overall supply of water is assumed to be fixed at W. The two VMP curves for users 1 and 2 are drawn in opposite directions, while the length of the vertical axis in Fig. 13.5 is determined by the total amount of water W in the basin.

The efficient allocation occurs at point 0, where the VMPs are equal and Q^* is allocated to user 1 and $W - Q^*$, to user 2. However, efficient allocations are the exception rather than the rule. To illustrate, assume a historical water allocation of $\overline{Q}$ units to user 1 and $W - \overline{Q}$ to user 2. The two users can be regarded as the agricultural and the nonagricultural sectors. The allocation between the two sectors does not involve the inefficiency of total water use since it is assumed that total consumption is W units.

An efficient allocation can be achieved if the price is set at P^* (or a tax P^*) because at that price, user 1 will demand Q^* while user 2 will demand $(W - Q^*)$. In addition, both users' VMPs are equal to P^*. This achieves the two conditions necessary for efficiency: the total water

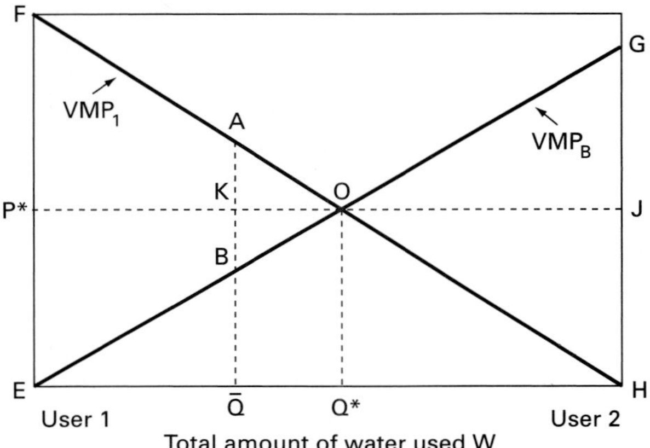

Figure 13.5 Water use by two individuals.

demand equals the net basin supply and the VMPs of all users are equal. The gain in efficiency can be measured by the change in producer benefits or the area under the VMP curves. With an efficient allocation, the benefits would be FOQ^* and GOQ^* for users 1 and 2, respectively. Under the inefficient historical allocation, the benefits would be $FA\overline{Q}$ and $GB\overline{Q}$. The net gain from the efficient allocation is AOB. Although society and user 1 are better off, user 2 is not.

The users will also experience a loss from paying higher prices. Again, this loss can be calculated by looking at the changes in producer benefits. With the price increase, the net benefits for users 1 and 2 would be only FOP^* and GOJ because part of the benefits would be collected by the government through the higher prices (P^*OQ^* and JOQ^* collected from users 1 and 2, respectively). When the purpose of the low water prices is to subsidize these sectors, it is not surprising that governments resist raising water prices. While efficiency dictates that the optimal price include the scarcity value of water in addition to its marginal cost, there is likely to be resistance to higher administrative prices. The strong agricultural lobby which exists in many countries has helped prevent efficient water pricing.

Another alternative is to use tradable water permits or rights. A *transferable* water permit or right is permission to use a previously specified amount of water and the right to sell it at a price which is determined by the buyer and seller in a transaction. Permits and rights can be allocated in a number of different ways, yet two of the most common ways are by auction and by historical use.

With an *auction* of W permits, the equilibrium price will be exactly the same as the efficient price shown in Fig. 13.5, that is, P^*. Only for this permit price does the amount supplied exactly equal the quantity demanded. When the permits are allocated by *historical use,* user 1 holds $\overline{Q}$ permits while user 2 has $W - \overline{Q}$ permits. Since these permits are tradable, there will be a sale of permits (assuming no transaction costs). User 1 will purchase $Q^* - \overline{Q}$ units of water at P^*, which results in a total cost shown by the area $KOQ^*\overline{Q}$. The added value of that water to user 1 is given by the area $AOQ^*\overline{Q}$. Thus user 1 has a net gain of AKO. The value of the water sold by user 2 is given by the area $BOQ^*\overline{Q}$, while the payment received from user 1 is $KOQ^*\overline{Q}$. This leaves user 2 with net gain of KOB. Both buyer and seller gain from the trade!

The original allocation determines the payment from user 1 to user 2. Any other initial allocation will result in the same final allocation, namely, Q^* for user 1 and the rest for user 2 (assuming that there are no income constraints for the buyer and zero transaction costs). The initial allocation, however, determines who pays what to whom.

Receiving permits based on historical use are typically preferred by users over higher water prices or an auction. The question of how the permits are actually allocated is not an important allocation issue under the assumptions described above, because any initial allocation will move to an efficient allocation provided users are free to trade permits. How the permits are allocated does affect income distribution and is, therefore, an important politicoeconomic question. In addition, if transaction costs are high, trading will be constrained and not lead to an efficient allocation. Permits can be traded between two users in the same industry (e.g., agriculture) or trade between two industries (e.g., from agriculture to urban use). Permits can also be used to deal with water-quality problems.

Water-quality management

When considering permits or prices (taxes) to deal with water-quality problems, two issues deserve special attention: (1) the nonuniformity in the impacts of water use and (2) non-point-source pollution. The issue of nonuniformity is important in water quality because of the varying impacts of different emitters on the water resource. If a polluter is emitting closer to a receptor (a river or lake), then the impact will likely be greater than the impact of another emitter who is located farther from the receptor. To achieve efficiency, a program to implement either pollution taxes or tradable pollution permits would have to take into account the differences in impacts.

To illustrate, assume that there is only one receptor indicated by R and Z_R is the pollution measured in that receptor. Each source (or producer i) emits pollution defined by E_i. The transfer coefficient Tc_i, translates the amount of emission discharged by producer i into the amount of pollution that actually reaches R. Because of the nonuniformity, the transfer coefficients will be different for each user. For example, the amount of soil reaching a river from a given field will decline, the farther the field is from the river (assuming constant slope, soil type, and vegetative cover).

The concentration level at R is related to emissions from all sources as follows:

$$Z_R = \sum_i Tc_i E_i \tag{13.4}$$

The policy target is to bring Z_R to a prespecified level of concentration. Since the impact is not uniform among polluters, the economic incentives to control the pollution will have to differ among polluters.

The optimal tax (price) for each user is given by

$$t_i = Tc_i\, U \tag{13.5}$$

where t_i is the tax paid by polluters i, which will differ among polluters depending on their transfer coefficient. The marginal cost of concentration reduction U should be equal among polluters for the last unit of concentration reduction at the receptor. Otherwise, the pollution standard could be attained at a lower cost if emissions were reallocated among sources. Since any U can be efficient, U is chosen so that it is consistent with the overall desired level of concentration at the receptor (Tietenberg, 1992).

The same efficient reduction in pollution can be achieved with a system of tradable permits. A permit system needs to be designed differently depending on the type of permit being issued. The permit, in this example, is to allow the polluter to change the concentration level at the receptor by one unit. This is called an *ambient permit* and is different from using emission permits which do not require transfer coefficients. Assume that there are two emitters and there is a predetermined concentration at the receptor. To assure that the same level of concentrations will exist after the trade, the following conditions must be satisfied:

$$Tc_1\, \Delta E_1 = Tc_2\, \Delta E_2 \tag{13.6}$$

$$\Delta E_1 = \frac{Tc_2}{Tc_1}\, \Delta E_2 \tag{13.7}$$

If Tc_2 is smaller than Tc_1, emitter 2 has less of an impact on the receptor than emitter 1. Therefore, emitter 1 is allowed lower levels of emissions than emitter 2. So the amount permitted is a function of the transfer coefficient. The emitter that is closer to the source (a higher *Tc*) is allowed lower levels of emissions on the *same* permit as the emitter farther away.

The second problem associated with water-quality control is non-point-source pollution. This is a more difficult problem because information about sources of pollution as well as transfer coefficients is hard to obtain. Even when mechanisms are implemented to control nonpoint pollution, it is difficult to detect violators and their impacts.

The major sources of nonpoint pollution are agricultural activities, individual disposal systems, and urban runoff. Because these kinds of pollution are difficult to control, the government adjusts by using several different programs. These include programs which subsidize farmers to leave the land fallowed or adopt conservation practices and ones that tax or use quantitative restrictions to control polluting inputs such as fertilizers and pesticides. However, efforts to control these nonpoint sources have been small compared to those directed at controlling point-source pollution. Thus, although it is not clear how efficient the different mechanisms have been in controlling either type of pollution, it appears that more needs to be done to control non-point-source pollution.

Experience with tradable permits

Much of the research done on tradable permits has been on air-pollution permits. Yet, water-quality permits have been tried in several cases. One of the first studies which compared various economic mechanisms for controlling water pollution was done on the Delaware River Basin (Kneese, 1977). Although not a large river, it serves a large population. The study showed that the ratio of pollution-control cost decreased by 3 (!) if economic incentives such as taxes or permits were used.

Another important study was done on the Fox River Basin in northern Wisconsin (Ehart et al., 1993). There were about 10 pulp and paper mills as well as four municipalities polluting the river. The study revealed that there was a significant difference in the transfer coefficients between the different sources. Also, the abatement cost differed between sources. This indicated a large potential gain from use of economic instruments, and, as the study showed, the control cost could be reduced by almost 50 percent with the use of tradable pollution permits.

Finally, there were two recent studies on tradable permits. One was done in California (Howitt, 1994) and the other, in Israel (Becker, 1995). Both of these studies concluded that permits could increase efficiency by about one-third and there would be budget savings due to a reduction in the high water subsidies to agriculture (water is transferred out of agriculture). The negative side effect of the policy would be a large decrease in intensive cultivation in a number of irrigated areas.

Implementing water pricing

The method used to deliver water to the final consumer affects the variable cost, as well as the feasible pricing schemes. Water may be delivered in a continuous flow or be available only at certain periods of time. The conveyance system may consist of open channels or closed pipes. Often the water in a country is delivered by more than one method, depending on tradition, physical conditions, water infrastructure, and institutional arrangements.

Similarly, water-pricing practices are influenced by factors such as historical arrangements, political pressures, and cost-recovery considerations, all of which can lead to deviations from efficient (marginal) pricing based on supply and demand. In addition, marginal cost pricing may entail a multiplicity of rates that are costly to monitor and implement. Deviation from marginal pricing, in some cases, may therefore be justified on "efficiency" grounds as well as equity (Seagraves and Easter, 1983).

When water is plentiful and the value of an additional unit of water is low, a simple or indirect method of charges or taxes may be the best method of paying for water. When the supply of water is more limited and the value of an additional unit of water is high, more sophisticated methods of allocating water become important. More elaborate water-pricing schemes, therefore, are more critical when value of water is high and the cost of measurement is low (Young, 1986). New technology has already reduced measurement costs while the growing world population is demanding more and better quality water.

At any given point in time, the water price that leads to efficient allocation of water of a particular quality is the price that equates the marginal cost of supplying the overall demand. The total water supplied is allocated between users such that the marginal productivity of water is equal across sectors. The marginal cost consists of the marginal transportation and processing cost plus scarcity cost (plus the cost of foregone nonuse values, if relevant). The price of water of a particular quality at a particular point in time should vary across users only to the extent that the marginal costs of delivering and processing the water vary across users; block-rate pricing represents such variability across users. The prices of water of a particular quality to a group of users, which are identical in their marginal cost of supply, should vary over time to the extent that their demands (and possible supply) vary over time; peak-load pricing represents such variability over time. Water supplies of different qualities should be priced according to the demand and marginal cost of supplying each quality. A perfect differentiation of water prices according to marginal costs (supply) and benefit (demand) requires excessive monitoring. The resources needed to implement such a pricing practice may exceed the gains, as compared to simpler methods. A possible remedy for the high transaction costs associated with marginal cost pricing is offered by markets for water rights, which are discussed in a later section.

Irrigation charges

As emphasized above, the agricultural sector is the largest consumer of water, and the value of water in crop production is low relative to other uses. In addition, it is not easy to measure the water delivered to individual farmers when there are a large number of small farming units and the water flows are highly variable. Yet, because agriculture is such a large consumer of water, pricing is important, both for allocation among sectors and for efficient water use within the sector.

There are four general types of water charges used in irrigation: (1) direct charges based on the volumes delivered, (2) direct charges per share of the flow in a river or canal, (3) direct charges per acre irrigated, and (4) indirect charges on crops marketed or purchased inputs such as fertilizer. Volumetric charges are used primarily in developed economies and are best suited for cases in which the value of water is high.

One major reason for the reluctance of developing countries to use volumetric charges has been the measurement cost, including enforcement costs. Developing countries have tended to use fixed charges per acre or per share of a canal flow. In the latter case, farmers are charged according to the length of time water flows into their fields. Charges based on the flow are easy to apply if farmers are on a fixed rotation with a set time allotment. In this case, the charge will not influence the quantity demanded since farmers cannot alter the length of the delivery time or the water charge. If water is supplied only on demand, then charges based on time will influence farmers' water use and encourage conservation.

The fixed charge per hectare is the most common method used in developing countries. The charge may vary with the type of crop, the method of irrigation, or the season of the year. In many countries, the water charges are higher when there are large investments in infrastructure (reservoirs and lined canals) than when the water comes from a simple diversion directly from a river. The charges for govern-

TABLE 13.1 Water Charges for Surface Irrigation Water in Maharashtra, India, 1980

Crop	Water rate ($/ha)
Sugar cane	50
Other perennial crops	33
Kharif-season crops (summer)	3
Rabi-season crops (fall and winter)	5
Hot-season crops (spring)	10
Crops for seed production in kharif season	7
Crops for seed production in rabi season	10
Cotton (spring)	16–26
Groundnut (spring)	13–20
Postseason water (summer)	1
Postseason water (winter)	2

Seasons: winter, December–February; spring, March–May; summer, June–September; fall, October–November.

ment-supplied groundwater are usually higher than those for large gravity-flow systems. A schedule of water charges can be quite complex, as shown in Table 13.1. In this example, charges are varied by crop and by season. Crops using the most water, such as sugar cane, have the highest charge ($50/ha), and the charges are also high in the hot (spring) season, when evaporation is high and water supplies are low. This system of water charges, at least, relates the charge to the amount of water likely to be used. Yet, the charges will influence water use only by altering what crops are grown and how and when they are grown.

A two-part charge could be used to combine a fixed charge per year with a volumetric charge. The fixed charge may be the same for all users, or it could be varied according to farm or firm size. The pricing method shown in Table 13.1 is advocated primarily for situations in which a public utility produces at a marginal cost below its average cost and must cover total costs (variable and fixed). They can still base charges on marginal costs while using the fixed charge to make up the difference between total costs and sales revenues.

The rates charged can also be varied by volume used or by hectares irrigated. Such a tiered or block pricing method could prescribe multiple rates based on the amount of water delivered to a user during a set time period. For example, the rate for the first 1000 m^3 per month could be lower than for the second 1000 m^3, and so on. Similarly, a tiered pricing system could be designed on the basis of acres irrigated. Such a system might be used to subsidize small farms. A low rate could be offered for the first 2 ha irrigated, then raised progressively as the acreage irrigated increased. The higher fixed charges based on firm size might simply reflect the greater volume of water used on larger farms and, therefore, be designed to reduce their subsidy.

When there are a large number of small farms and metering at each farm is not cost-effective or enforceable, the village could be selected as the control point. Meters would be placed at the village or subvillage level. The village would then be given the power to deliver the water to the farmers and to collect the fees. The village might even contract with the irrigation agency for a given water supply and price. The control over water delivery is critical because without it, the village might have a difficult time collecting water fees (Easter, 1975). This type of system would be a good alternative for large surface irrigation systems in many parts of Asia.

When cost recovery is the main reason for water charges, the ease of collection becomes the prime concern. In such situations, the *betterment levy* has been used. This is a water charge that is generally added to and collected as part of a land tax. It is a per area charge based on the increase in land value resulting from the irrigation.

Indirect methods of charging for water are also used when the major issue is cost recovery. Since irrigation should increase crop production and input use, these indirect charges are levied as taxes on inputs and crop outputs that farmers buy and sell. Since there are a limited number of outlets for these inputs and products, it is, generally, easier to collect these taxes than direct water charges. Although the cost of collection is lower, indirect taxes and betterment levies do not give farmers any incentive to use their water more efficiently.

Bos and Walters (1990) investigated farmers representing 12.2 million ha of irrigated farms worldwide, and found that in more than 60 percent of the cases, water charges are levied on a per unit area basis. In less than 15 percent of the irrigation projects, the water fees are based on a combination of area and volume. In about 25 percent of the cases studied, the charges are based on the volume delivered.

Municipal and industrial charges

Water charges for domestic and industrial use take much the same form as those used in agriculture. Both fixed and volumetric charges are widely used. The major difference is that the marginal value of water is, generally, higher and the cost of metering is lower relative to the value of the water.

In developed countries, volumetric charges are being used more widely while the fixed charge is common in developing countries. When fixed charges are used by utilities that must cover operating costs, they are generally set high enough to cover, in a normal year, the total operating and maintenance (O&M) costs and possibly some of the capital costs. No meters are required with fixed charges, but they provide no incentive to conserve water. Financial autonomous water utilities can have financial problems in dry years since they have less water to sell and may not be able to raise prices enough to make up the difference. To prevent such problems, they should be allowed more price flexibility and the option of carrying over surplus revenues from good years.

The most frequently used volumetric method is a flat rate per unit delivered to the consumer. Meters and monitoring are required, and all users face the same constant marginal cost for water. This is the simplest volumetric charge, and all users face the same positive incentive to conserve water. Volumetric charges can also be tiered or block rates such that charges depend on the quantity used per unit of time. If it is an increasing block rate (the higher the rates, the more you use), the marginal cost to the user increases, as does the incentive to conserve water (Fig. 13.6). The opposite is true for a decreasing block rate. The former is appropriate in areas with scarce water supplies and the need to subsidize low-income families who do not use much water. The latter or decreasing rate may be justified when the cost of delivering water to certain large-scale consumers, such as food processors, is much lower than for others. It also may be used to attract certain industries into an area.

Finally, peak-load or seasonal pricing can be employed to reduce water use during periods of high (peak) demand. Peak-load pricing has frequently been used for energy pricing but not much for water pricing. On the other hand, seasonal pricing has been used for water, particularly in urban areas where there is high water use for car washing, lawns, golf courses, and swimming pools. Higher prices in the summer can reduce the overall quantity demand and significantly reduce the storage capacity needed to meet peak urban water demands. Both peak-load pricing and seasonal pricing account for the variation in water scarcity at different points in time. How much conservation higher prices will induce depends on the elasticity of demand. As pointed out above, the price elasticities for water are much higher in summer, suggesting that seasonal prices should be used more widely, especially in areas with scarce water supplies.

Issues in water pricing

Several problems continue to plague water agencies and public utilities in their efforts to collect water fees. One continuing problem is the failure to raise charges over time to keep up with inflation. Water rates that may have been reasonable when they were originally set may be totally inappropriate after 10 years with no increases. Another related issue is the failure of most water agencies and public

Water and sewerage in Istanbul are managed by an autonomous agency, the Istanbul Water and Sewerage Authority (ISKI). Only very limited subsidized financing is available for water and sewerage investments from central government agencies and the municipality. The ISKI has been obliged, therefore, to finance most of the investment program, as well as the operation of the system, from charges to customers. The issue has been to raise the necessary funding, while assuring that the basic water needs of the population are met, including, in particular, those of lower-income households (about 18 percent of the population is estimated to have incomes below the poverty threshold). ISKI has used the practice of increasing block volumetric water charges for households to meet these dual objectives. Charges were set in 1987 as follows:

Customer category	Basic rate per 3 m^3 (U.S. $)	Percent of total consumption
Households		
0–7.5 m^3/month	0.26	4
>7.5 m^3/month	0.53	45
Office	0.88	12
Industry	1.24	38

Households are thus cross-subsidized by industry since water charges up to a minimum to meet basic health requirements (7.5 m^3 per month), were kept particularly modest. Affordability calculations indicate that low-income households would spend about 2.7 percent of average monthly income on water and sewerage charges, compared with 3.4 percent for median-income households. Water and sewerage charges have been raised several times since 1987, and the principle of full payment by consumers for investment and operating costs has been accepted. This system has worked because consumers are metered, and vandalism of meters has not been a serious problem. Still, ISKI had major problems in billing and collection, with billed water amounting to about 65 percent of water consumed. ISKI has also faced rising investment costs, and sites originally planned for sewage treatment have been found unacceptable by local communities because of proposed high fees. Despite these difficulties, the ISKI model has been adopted by four other large municipalities in Turkey, all of which have introduced the increasing block volumetric water charges.

Figure 13.6 Increasing block volumetric water charges in Istanbul.

utilities to collect enough in fees to cover operations and maintenance costs, let alone contribute to repaying capital costs. A third problem in some countries is that the fees collected go into general government revenues. Thus, there is no relationship between fees collected and the amount spent on the water system. This cuts the vital link between fees paid and performance of the water-management unit. Finally, more research is needed on the responsiveness of water users to different systems of water charges.

Market Approaches to Water Management

There is growing evidence that water markets can be an effective mechanism for allocating water resources (Chang and Griffin, 1992;

Hearne and Easter, 1995). In areas with limited water supplies and inefficient water allocation, markets are being evaluated as one possible mechanism for improving water use and allocation. One advantage of water markets over administrative pricing and allocation is that market prices are based on individual user assessments of the value of water. These assessments are made with information that is available to water users but expensive for government agencies to collect. Without markets, it is expensive for government agencies to determine the value of water to individual water users and to allocate water to those with the highest marginal value product.

Water markets not only provide a mechanism to address the water misallocation problem but also give water owners an incentive to move water to its highest-valued uses. Because sellers are compensated, they have an incentive to sell their water whenever the market price minus transaction costs exceeds the net marginal value in its existing use. The market introduces an opportunity cost which encourages users to transfer water to its highest-valued uses and make long-term investments that improve water-use efficiency. By adopting new technology such as drip irrigation, farmers can either expand their area irrigated or sell the water saved to the highest bidder.

Unfortunately, efficient water markets do not occur naturally. The mobility and liquidity of water makes it difficult to establish secure water rights without collective action. In addition, the bulky nature of water makes it expensive to move large quantities upstream or outside the river basin. This, combined with the heterogeneous character of the commodity, prevents the development of large national water markets, except for bottled water. These characteristics, combined with some of the market factors discussed above, mean that a number of institutional and organizational arrangements are needed before local or regional water markets can function efficiently.

In establishing water markets, at least six arrangements should be in place to ensure that markets can be efficient, equitable, and sustainable. First, institutional arrangements must be created to establish tradable water rights that are separate from land. A second important ingredient is an organization or management unit to implement the trades. A third component for an effective water market, particularly for irrigation, is a flexible infrastructure (adjustable gates and interconnecting canals) that allows sellers to transfer water (at a reasonable cost) to the buyers. This may be less of an issue if large quantities of water are being transferred to high-valued urban uses that can pay for a new pipeline. Fourth, institutional arrangements need to be in place that protect third parties who may be damaged by water trades because of the interdependent nature of water uses. Fifth, mechanisms are needed to help resolve conflicts over water rights and third-party interests. Finally, if equity is important, then the initial distribution of water rights should be based on an assessment of society's basic water needs, especially for poor families, as well as past water uses and overall concerns about future development.

Water rights

Once a decision is made to allow markets to allocate water, the next step is to determine how water rights will be defined. (These rights can be granted as either the actual water right or as a use right; the use right allows government to maintain the fiction that water belongs to all its citizens.) Rights can be defined in terms of a share of an uncertain streamflow, or a share of an aquifer or reservoir. Alternatively, rights can be defined in actual quantity terms for a given time period, e.g., 20,000 m^3 annually. If water rights are denoted in quantity terms, some agreed-on method will be needed for sharing water during periods when not enough water is available to fully supply all the rights. Two alternatives are generally used. First, each right can be given a priority on the basis of some characteristic such as the date the right was issued, the location of the right, or the type of water use (irrigation vs. domestic consumption). When water is in short supply, only high-priority water rights would be fully served (Becker et al., 1996). In the western United States, this is done by dating each water right according to when it was first established or appropriated. Under these "appropriated water rights," the oldest or most senior water-rights holders receive their full share before the more junior rights holders receive water.

A second approach is to reduce all rights proportionately according to how large the shortage is expected to be. Sometimes a combination of the two approaches is used. In this case, certain priority uses such as domestic supplies are reduced less or not at all while others, such as industrial or agricultural uses, are reduced the most. These priorities will vary among countries depending on what they determine to be their most critical water uses. In Chile, water-use rights are generally specified as a share of a section of the river's flow. In most cases, the total quantity of shares available for allocation are based on a river's flow that is estimated to be available 85 percent of the time. Thus, in 15 percent of the years, there will be a shortage and all agricultural shares will be reduced proportionately to make up the difference.

The advantage of the appropriative water rights in the western United States is that one does not need as much historical information concerning water availability. During times of drought, only senior water-rights holders receive water while in time of surplus flows, even the most junior rights holders will receive water. Granting another junior water right does not reduce the amount available for other more senior water-rights holders. In Chile, they have created a special class of water rights for water that is available when the supply exceeds the normal flows (which have already been allocated to permanent water-rights holders, based on the 85 percent availability rule). These water rights, called *contingent water rights,* are contingent on high streamflows. This was done because granting additional permanent water rights would have reduced the flows available to all other permanent rights holders during times of shortage.

To establish rights on the basis of shares or volumes without establishing priorities requires historical information concerning the probabilities that certain quantities or flows of water will be available at different times during a year. Without this information, it is difficult to determine how many water rights can be distributed. This information will also improve the efficiency of the water market by reducing the uncertainty regarding how much water will be received under a given water right. Storage capacity, in either a reservoir or an aquifer, will also reduce uncertainty and improve market efficiency. In fact, water rights can be established in terms of a share of reservoir capacity as has been done in Australia, or as a share of aquifer storage (Dudley and Musgrove, 1988, 1991). A rights holder receives a percentage of whatever volume of water is in the reservoir during a given time period. For an aquifer, the rights could be based on the average annual recharge. This would allow some use of the aquifer stock during drought periods that would be recharged in high-rainfall years.

It is important to establish water rights for both surface water and groundwater. In many cases, surface and groundwater supplies are directly interconnected. For such cases, establishing only surface-water rights may leave out much of the water supply. Furthermore, the surface-water rights are not secure if anyone can install a well next to a canal or river and draw out water that "belongs" to owners of surface-water rights. The lack of compatible surface and groundwater rights has caused serious water-management problems in a number of areas such as Arizona and is likely to cause problems in other areas (Charney and Woodard, 1990; Hearne, 1995).

Once the water rights are established, independent of land, they need to be recorded and enforced. This is generally an important role for government, but strong water-user associations can record and help enforce the water rights. Having the rights recorded should improve water-market efficiency by reducing uncertainty and making information about the supply of water rights more accessible to potential buyers.

Management of the system

For water exchange beyond one's neighbors, some management unit will be needed to change gates and assure the interested parties that the correct amount of water reaches each new owner. In some countries, this can be the organizations currently operating the water system such as a government water agency, a water-user association, or some form of utility (public or private). In other cases, managers who have been hired by the water users to manage the water system can implement the transfer. However, an active water market will require more effective management than most system operators have exercised in the past, particularly for government-managed irrigation systems. Thus, at a minimum, the staff may need training to improve their management skills and responsiveness to user requests.

An effective management unit, such as the water-users association in the Limarí Valley of Chile, can implement trades as well as provide information concerning who wants to buy or sell water. By improving the flow of information about water availability, demand, and water prices, the association can improve the market efficiency. In the Limarí Valley, water users are well informed about the prices of water and how much is available for a given season. Thus, they have an active water market (Hearne and Easter, 1995).

Infrastructure

The other component necessary to actually implement a trade includes the canals, pipes, and gates that allow operators to change the water flows. The more flexible the system, the easier it is to make the necessary changes and the lower will be the transaction costs for making a trade. In fact, if the infrastructure is inflexible and the transaction costs are too high (i.e., if new canals or gates are required), many agricultural trades will be blocked. These costs appear to be an important constraint for canals with fixed gates. The cost of alternating the canals and gates reduces the price of the water offered to the seller and raises the cost to the buyer (assuming that the buyer pays for changing the structures). Thus, few trades have been consummated in Chile when canals have fixed gates (Hearne, 1995).

A water system that includes reservoir storage also appears to facilitate water trades, particularly if water rights can be translated into a volume of water in the reservoir. This, then, allows anyone in the system to buy from anyone else, assuming that the reservoir services the whole area. In a water system with storage and a large number of water users, there is the potential for an efficient market with numerous buyers and sellers.

Third-party impacts

Institutional arrangements are generally needed to protect third parties who may be damaged by a trade. Because of return flows, withdrawals or sales upstream can reduce the quantity and quality of water availability downstream. Even changing the type of water use may change the return flow. To mitigate this problem, most water sales in the western United States are of the consumptive portion of the water right. In addition, water sales, especially those to users outside a water district or basin, are highly regulated in most western states (Colby et al., 1993).

An equally serious issue is the discharge of pollutants into water bodies that affects other users of the same water source. "In addition to pollution that is visible and degradable, new types of pollution have arisen involving small quantities of nondegradable synthetic chemicals that are invisible, toxic, persistent, and difficult and costly to treat" (World Bank, 1993, p. 32). These pollutants can have a negative impact on all downstream users. Since water rights have been specified only in quantitative terms, this means that governments must establish overall water-quality regulations or impose emission charges on polluters. If this is not done, water markets will be constrained because of the uncertainty concerning water quality. Reductions in water quality can essentially render water unfit for certain water uses, and thus decrease the value of water rights.

Another possible third-party impact may arise from changes in economic activities and jobs. For areas that are buying the water, these economic impacts will be positive. However, when water is transferred to another region or river basin and agricultural land goes out of production or into dry-farming input suppliers, processors, and farm workers are likely to see a drop in their business, income, and employment. These are primarily local impacts, and from a national perspective, the gains in the area receiving the water are likely to be as large as, or larger than, the losses in the exporting area.

Resolving conflicts

A formal mechanism must be developed to help resolve disputes over water rights and to bring third-party concerns into the decision-making process. A commission could be established to act as a clearinghouse for water sales. They would receive complaints regarding proposed sales and have the authority to block or delay sales if large negative third-party impacts are likely. In some countries, this could be done by a government water agency and the court system, while in others with strong water-user organizations (WUOs), the water users working through their WUO, may be able to resolve most of the water-rights disputes.

In the western United States, the court system along with water agencies or the state engineer have been effective in resolving most disputes. Colby (1990, p. 1191) argues that policy-induced transaction costs in the western United States play an important role of accounting for the social costs of water trades.

> Transaction costs generated by state policies are a reasonable means to account for social costs of water transfers, given the immense centralized information requirements for an optimal tax on transfers. The present institutional structure gives affected parties, each weighing their own costs and benefits, an incentive to generate information on transfer impacts and to negotiate transfer conditions and mitigation of externalities. State water transfer criteria are not arbitrary hindrances imposed upon the marketplace. These policies protect existing investments by water rights holders and, as they are broadened to include other interests, they can reflect public good values affected by water use and transfer.

The positive role of water policy is illustrated in Fig. 13.7, where policy-induced transaction costs reduce the size of a water transfer and move the water allocation closer to the social optimum. Assume that Q_1 is the initial allocation of water to the urban sector, while the rest is used by agriculture. Once water becomes tradable, farmers begin to sell water to urban users. The farmers' (private) supply curve PSC, measures the benefits farmers forego when they sell water. In the new market allocation, without transaction costs, the urban demand D_1 equals PSC at Q_2. Under this market allocation, the urban sector obtains an additional $Q_2 - Q_1$ units of water. Yet this is not optimum for society since third-party impacts are excluded. When the third-party impacts are included, the supply curve becomes SSC (socially appropriate supply curve). The difference between PSC and

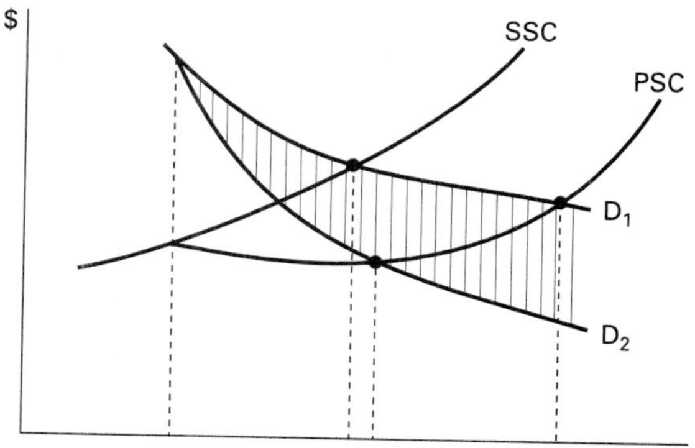

Figure 13.7 Transaction costs and the market allocation of water between sectors. [*Source: Adapted from Colby (1990).*]

SSC measures the negative impact of the transfer on third parties. The social optimum allocation is, therefore, Q^*, where SSC equals D_1, which is less than private optimum Q_2.

If the transaction costs of negotiating and implementing the trades are included, the urban demand drops to D_2, which is the urban demand minus all transaction costs (this assumes that the buyer pays the transaction costs). The private allocation, with transaction costs included, becomes Q_3, where PSC equals D_2. In this example, transaction costs move the private market solution closer to the social optimum Q^*. It is also quite possible that the transaction costs could be high enough to move the urban sector's share to less than the social optimum (to the left of Q^*) and even close to the starting allocation Q_1. In such cases, the government should take steps to reduce transaction costs and encourage more trades.

The distribution of rights

The key equity issue involved with water markets is deciding who gets the initial rights. In terms of economic efficiency, the distribution is not important as long as the rights are freely tradable and transaction costs are not too high. When trading is allowed, water will move to its highest-valued uses. However, the distribution of the benefits from such trades and the initial use of the water will be directly tied to whomever initially captures the water right. When the water has been used in the past, this past use will need to be given some weight in the allocation of water rights. If there is additional water available after water has been allocated to past users, the rest can be auctioned off to the highest bidder or sold on a first-come first-served basis.

If there is a plan to redistribute land resources within a country, then the water rights can be redistributed in a similar manner. Where the land resources on which most of the water is used are held by a few individuals and there is no plan for land redistribution, then the water rights should be auctioned off. This will create an efficient allocation of the water and allow some of the economic rents created by public water projects to be recovered and used to pay for the projects.

Closely related to the issue of who gets the rents is the allocation of water rights for "societal water uses" (uses with significant public-goods characteristics). This involves in-stream water uses to preserve fisheries and aquatic ecosystems and to prevent saltwater intrusion. For example, during low-flow periods, California irrigation has dried up several rivers with drastic impacts on fisheries and aquatic ecosystems. At the time water rights were allocated in these rivers, the state government failed to recognize that in-stream water flows provide numerous social benefits. They took the position, with the support of farmers and community leaders, that all water should be used and none should be allowed to flow into the ocean. The state is now faced with legal requirements under the *public-trust doctrine,* which requires the government to hold certain water rights in public trust to protect in-stream uses, to set aside water for these basic in-stream uses. This means that California has to take back water rights that have already been allocated to and used by agriculture. Other countries can learn from this experience and remember that it is important to consider basic in-stream uses when allocating water rights.

Alternatives to the sale of water rights

Two other alternatives that the government could use to improve water-use efficiency are to sell water options or lease water. These might be used in cases where the government is reluctant to allocate tradable water rights or as a way for the government to operate within a water market. Options would be for a given volume of water or share and would be used primarily by firms that have to maintain a constant amount of water even during periods of severe shortage. An option could also be for a specific time period so that firms might protect time-dependent water uses. The options would reduce the water-supply uncertainty and should encourage investment in production capacity because of the reduced uncertainty.

Leases would serve much the same purpose as water-use rights, except that they would be shorter in duration and may not be as easy to renew. Here, governments are faced with a tradeoff. The longer the lease and the easier the renewal, the more limited will be the government's control; in contrast, the shorter the lease, the smaller the incentive for firms to invest in water-using activities. For example, with a one year lease, farmers might grow small grains with a simple flood-irrigation system. With a 20- or 30-year lease, they might grow vegetables with a sprinkler system.

Government action and water rights

Establishing water markets necessitates taking concrete actions. Many governments fail to take the necessary steps either because they are loath to give up some of their control over the water resource, particularly if it has been developed with public funds, or they are concerned about the potential for monopoly control over water and the possible third-party impacts. Countries seem to overcome these fears either when they are restructuring the role of government in the economy (Chile, Mexico, and Peru) or when water scarcity and water misallocation become too severe (California and Texas). Yet even after markets are introduced, these concerns still remain. For example, Chile and Mexico are granting water-use rights for a given period of time. This allows the country to retain the actual water right and the rights to reallocate water during times of drought or major changes in the economy.

Establishing water rights and their distribution is likely to be the most important issue facing countries that want to introduce water markets. The distribution issue is particularly contentious when farmers are currently using water that has been provided by the government at a highly subsidized rate. The question is "What should water users give up to be able to trade the water?" In many cases, these trades create large economic rents for both the buyer and the seller. Should some of these rents accrue to government to help pay for past subsidies? If the government tries to capture most of the rents generated by the sale, this will eliminate the incentives to trade water.

As the recognized uses for water expand, it will become more difficult to agree on who is entitled to an initial water allocation. There will also be issues concerning how much flexibility a water-rights owner should be given to move the water or change its use. Furthermore, as rights are expanded to include the infrastructure such as dams and canals, who will own the excess capacity in these structures? Can markets for water options be used to help reduce the need for excess capacity in the system? Finally, the introduction of water markets can result in some lumpy changes, particularly if large amounts of water are traded between sectors or regions. Are there more effective ways to smooth out these changes, and reduce the third-party impacts?

Conclusions

As new sources of water supply become difficult to find and expensive to develop, economic mechanisms for managing water resources will come to the fore. Water prices, markets, and permits can all improve water allocation and give users a stronger incentive to conserve water. As the electric utilities in the United States have already learned, conservation is often a cheaper source of supply than building a new power plant. The same is true for water resources where many countries have already developed the low-cost sources of water supply and their sources of new supplies are much more expensive to develop (World Bank, 1993).

Economic tools will also be more widely used to help deal with water-quality problems. Water users who have to pay to discharge polluted water will either find low-cost methods to reduce the amount of pollution or use less water. The particular economic tools used will vary depending on the particular institutional and resource setting.

However, the clear overriding principle must be that polluters pay for their contamination of water resources.

This means that the price users pay for water, either in the market or through administered prices, should include the cost of providing the water, a scarcity cost or value, and a charge for any contamination caused by the discharge of water. For stock water resources such as groundwater that are not being recharged, the scarcity costs would include an intertemporal charge because the water will not be available for future use. The cost of providing the water should also include a charge for replacing the infrastructure. Finally, the charge for water contamination should include all the costs imposed on downstream users, including damage to aquatic ecosystems.

Appendix A The Open-Access Problem in Water-Resource Management

The *open-access problem* is defined as the overuse of a natural resource due to a lack of well-defined property rights. This is not the same problem as inefficient water pricing, although there is a similarity between the two in that the wrong price signal is received by users. In open-access resources, the user does not include the cost that this individual use imposes on all other users. This appendix describes a simple open-access problem and demonstrates the inefficiency that results from the lack of private property rights.

Assume that there are N users on a single lake or aquifer. The production cost (total cost) of the water for one user is given by

$$\mathrm{TC}_i(w_i, W, S) = w_i \cdot \mathrm{AC}_i(w_i, W, S) \tag{13.8}$$

where w_i = water extracted by user i
W = total amount of water used (by all the users)
S = stock of water in the aquifer
$\mathrm{AC}_i(\cdot)$ = average cost of water

Further, assume that

$$W = w_i + \sum_{\substack{j=1 \\ j \neq i}}^{n} W_{-i}$$

that is, W_{-i} is the amount of water used by all other users except user i. The cost of pumping water for an individual i is a function not only of what that person uses but also what others are using, because each user affects, to some extent, the stock of water and therefore, the pumping costs in the basin.

If there is only one product being produced and its price is given by P, the private profit function to be maximized for any individual can be written as

$$\max_{[w_i]} \pi = w_i P - w_i \mathrm{AC}_i(w_i, W, S) \quad \text{s.t.} \quad w_i + W_{-i} \leq S, \quad w_i \geq 0 \qquad (13.9)$$

That is, each user tries to maximize profit with a nonnegativity restriction on water use and another one that limits the total water use to the stock in the aquifer. In order to solve this problem, an assumption has to be made regarding expectations of each user concerning the other users. The most common expectations assumption is the Courno-Nash, which can be written as

$$\frac{\partial W_{-i}}{\partial w_i} = 0$$

That is, small changes in the behavior of user i do not change the behavior of the other users.

A necessary condition for solving the maximization problem is given by

$$P = \mathrm{AC}_i(\cdot) + w_i \frac{\partial \mathrm{AC}_i(\cdot)}{\partial w_i} \qquad (13.10)$$

The price is equal to the average cost plus any increase in average cost caused by user i's consumption of another unit of water w_i. If all the users are assumed to be identical in their characteristics, $w_i = W/N$, Eq. (13.10) can be rewritten as

$$P = \mathrm{AC}_i(\cdot) + \frac{W}{N} \frac{\partial \mathrm{AC}(\cdot)}{\partial w_i} \qquad (13.11)$$

The right-hand side of this equation is the marginal cost of extracting one unit of water for user i. It is composed of the average cost $\mathrm{AC}_i(\cdot)$ plus the increase in cost for all inframarginal units W. However, the ith user bears only $1/N$ of this inframarginal cost while imposing an external cost on the other users equal to

$$\frac{N-1}{N} W \frac{\partial \mathrm{AC}(\cdot)}{\partial w_i}$$

The cost of pumping depends on the aggregate water use and hence on the number of users that exploit the aquifer. The following decomposition of Eq. (13.11) illustrates how the difference between marginal social cost and the private cost increases as the number of users N increases:

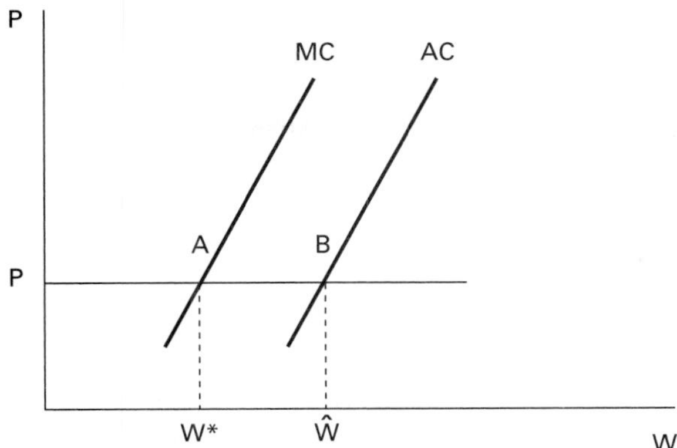

Figure 13.8 The marginal cost and the average cost of water.

$$P = \frac{1}{N}\left[\mathrm{AC}_i(\cdot) + W\,\frac{\partial \mathrm{AC}_i\,(\cdot)}{\partial W}\right] + \frac{N-1}{N}\,\mathrm{AC}_i(\cdot) \qquad (13.12)$$

The marginal cost is a weighted average of the marginal social cost and the average cost. The size of W is a function of N. When $N = 1$, there is sole ownership of the aquifer and Eq. (13.12) reduces to an efficient marginal condition; in other words, the marginal social cost is equal to the price (the last term in the equation becomes zero). If $N \rightarrow \infty$, profits are driven to zero since prices are equated to the average cost (the first term on the right-hand side is zero).

In the intermediate case in which the ownership is limited to a finite number of firms, aggregate use would be between these two polar cases, and the perceived marginal cost will shift toward the marginal social cost as N gets smaller (see Fig. 13.8). At point A, where the marginal-cost curve (MC) intersects the price line P, there is an efficient use of water in the basin W^*. When property rights are not well defined and there are a large number of water users, the average cost curve AC, which intersects with the price line at B, determines the overall water use in the basin, $\hat{\mathbf{W}}$. This results in an overuse of water equal to $\hat{\mathbf{W}} - W^*$.

Appendix B Scarcity and Intertemporal Allocation

The intertemporal scarcity value of water represents competition between present and future users. Therefore, it must be determined within an intertemporal allocation framework. Let $Y(S_t + g_t)$ be the overall water revenue obtained by using $s_t + g_t$ units of water allocated efficiently among all sectors (as defined above), of which s_t (for surface) is water derived from recurring flow sources, g_t (for groundwater) is water derived from a stock source (e.g., an aquifer), and $t = 1,2,\ldots$ represents the time period.

Let G_t denote the aquifer's stock at time $t = 1,2,\ldots$ over the indefinite horizon and G_0 the initial stock. Then

$$G_t = G_{t-1} - g_t + R(G_{t-1}) \qquad t = 1,2,\ldots$$

where $R(G_t)$ is the natural recharge rate. Let $z(G_t)$ be the unit extraction and delivery cost when the aquifer's stock is at the level G_t. Extracting the amount g_t generates the instantaneous profit $Y(s_t + g_t) - z(G_t - 1)g_t$, where it is assumed for simplicity that the supply of s_t is costless. The intertemporal objective function is the present value of the stream of profits:

$$\sum_{t=1}^{\infty} [Y(s_t + g_t) - z(G_{t-1})g_t](1 + r)^{-t}$$

where r is the rate of discount. The dynamic allocation problem can be formulated as

$$V(G) = \max_{\{g_t \geq 0\}} \sum_{t=1}^{\infty} [Y(s_t + g_t) - z(G_{t-1})g_t](1 + r)^{-t}$$

subject to $G_t = G_{t-1} - g_t + R(G_t) \geq 0$, $t = 1,2,\ldots$, and $G_0 = G$ [see Tsur and Graham-Tomasi (1991) for a continuous-time formulation]. Applying dynamic programming techniques gives the Bellman optimality condition:

$$rV(G) = \max_{g \geq 0} \{Y(s + g) - z(G)g + V'(\tilde{G})[R(G) - g]\}$$

where $g = g_1$, $V'(G) = \partial V(G)/\partial G$, and $\tilde{G}$ is a point between G and $G - g + R(G)$. If G is large relative to $R(G) - g$, $\tilde{G}$ is close to G and $V'(\tilde{G})$ can be approximated by $V'(G)$. Undertaking the maximization on the right-hand side, the optimal extraction $g(G)$ given that the stock level is G, satisfies

$$Y'[s + g(G)] = z(G) + V'(G)$$

The term on the left-hand side is the value of marginal productivity of water, which constitutes the (inverse of the) derived demand for water in each time period. The right-hand side is the marginal cost, which consists of the extraction and delivery component $z(G)$ and the scarcity value $V'(G)$.

If a system of future markets were properly operated, the price of g_t determined in this system would equal $z(G_{t-1}) + V'(G_{t-1})$, $t = 1,2,\ldots$. Lacking such market systems, $V'(G_t)$ must be calculated by solving the dynamic programming problem given above. Analytical solutions are rare, but given the rapid advances in computation technology, numerical solutions are feasible in many cases.

Consider, for example, the special case where both $z(G)$ and $R(G)$ are constants independent of the stock G. Observe first that if $Y'(s) \le z$, then the aquifer cannot be profitably exploited; that is, the cost of extracting groundwater always exceeds the benefit. Suppose that $Y'(S) > z$, so that it pays (at least initially) to extract groundwater. Then, it can be shown that (1) the scarcity value $V'(G_t)$ takes a particularly simple form

$$V'(G_t) = C_0(1 + r)^t$$

where $C_0 = V'(G_0)$ is the initial scarcity value and (2) the aquifer will eventually be depleted. Given these two conditions, one can calculate C_0 and hence the entire time path of V_t' [see example in Tsur and Graham-Tomasi (1991)].

When the cost of extraction $z(G)$ varies with G, one can view the different layers of the aquifer as different water sources. The extraction cost of each layer is constant. When moving from one layer to the next, extraction costs jump. Thus $z(g)$ can be approximated by a step function which is constant within a certain G interval (layer) and jumps between intervals. A backstop source (e.g., seawater or transporting recurring water supplies from other regions) can also be included as a final water source, with its associated extraction costs (desalination cost or transportation cost). In this case, the scarcity value for each water source raises exponentially over time, although the initial value C_0 may vary from source to source, and hence the time path of the scarcity value V_t' may exhibit kinks when extraction switches from one source to another. Moncur and Pollock (1988) applied this procedure to the case of two sources: an aquifer and a backstop source (seawater). Extensions to multiple source situations are straightforward.

Appendix C Uncertainty and Intertemporal Allocation

Whenever planning concerns future events, uncertainty is unavoidable. In the present context, uncertainty may involve the recurring supplies (rainfall, streamflows), the available stock supplies (the aquifer's stock), or both. Rainfall uncertainty implies that s_t is uncertain. The objective function with uncertainty is specified as the expectation of the present stream of profit and takes the form

$$V(G) = \max_{\{g_t \geq 0\}} E\left\{ \sum_{t=1}^{\infty} [Y(s_t + g_t) - z(G_{t-1})g_t](1 + r)^{-t} \right\}$$

where the expectation is taken with respect to the random s_t and assumes different meanings depending on whether the decision on g_t is made before or after the realization of rainfall is observed (Tsur and Graham-Tomasi, 1991). The scarcity value of g_t is still represented by $V'(G_{t-1})$, but $V(\cdot)$ takes a form different from that specified in the certainty case, and hence the scarcity value will be different, with obvious effects on the temporal allocation of the stock supplies.

In the case of stock uncertainty, the random variable is the depletion time. Accordingly, the objective function takes the form

$$V(G) = \max_{\{g_t \geq 0\}} E\left\{ \sum_{t=1}^{T} [Y(s_t + g_t) - z(G_{t-1})g_t](1 + r)^{-t} \right\}$$

where the expectation is taken with respect to the random depletion time T and G_{t-1} measures cumulated extractions as of time t (not including time t), rather than the (unknown) remaining stock.

The effect of stock uncertainty is twofold: (1) it changes the form and values of the scarcity value $-V'(G_t)$ and (2) it changes the allocation rules. The marginal cost associated with the extraction of g_t consists of the extraction cost, $z(G_{t-1})$, the scarcity value, $-V'(G_{t-1})$, and an additional cost which accounts for the fact that extracting gt changes the probability of depletion. If this last component is denoted by $K(G_{t-1})$, the extraction decisions (the allocation rule), given that the cumulated extraction as of time t is G, takes the form

$$Y'[s_t + g(G)] = z(G) + [-V'(G)] + K(G)$$

where $K(G) = \lambda(G)\ [V(G) - V_T(G)]$, $\lambda(G) =$ the probability that extraction occurs in the next time instant given that it has not occurred yet and $V_T(G)$ is the postdepletion value function [see Tsur (1992) for details]. The full marginal cost of water in this case consists of the three terms on the right-hand side.

When the uncertainty involves the recharge rate, the stock process

$$G_t = G_{t-1} - g_t + \tilde{R}(G_{t-1})$$

is random according to the random recharge rate $\tilde{R}(G_{t-1})$. The random recharge rate includes randomness in the future stock levels and thereby in the value function. The allocation problem takes the form

$$V(G) = \max_{\{g_t \geq 0\}} E\left\{\sum_{t=1}^{\infty} [Y(s_t + g_t) - z(G_{t-1})g_t](1 + r)^{-t}\right\}$$

subject to $G_t = G_{t-1} - g_t + \tilde{R}(G_{t-1}) \geq 0$, $t = 1,2,\ldots$, and $G_0 = G$, where the expectation is taken with respect to the randomness in future stocks G_t induced by the random recharges $\tilde{R}(G_t)$. The allocation rule in this case is of the form

$$Y'[s + g(G)] = z(G) + V'(G)$$

which means that the marginal cost of groundwater consists of the extraction costs $z(G)$ and the scarcity value $V'(G)$. The temporal behavior of $V'(G)$, though, will differ substantially from its certainty counterpart, with obvious implications for the temporal allocation of groundwater.

References

Anderson, T. L. (1983). *Water Crisis: Ending the Policy Drought,* Cato Institute, Washington, D.C.

Becker, N. (1995). "Value of moving from central planning to a market system: lessons from the Israeli water sector," *Agricultural Economics* **12:** 11–21.

Becker, N., N. Zeitouni, and M. Shechter (1996). "Reallocating water resources in the Middle East through market mechanisms," *International Journal of Water Resources Development* **12**(1): 17–32.

Bos, M. G., and W. Walters (1990). "Water charges and irrigation efficiencies," *Irrigation and Drainage Systems* **4:** 267–278.

Brookshire, D. S., L. S. Eubanks, and C. F. Sorg (1986). "Existence values and normative economics: implications for valuing water resources," *Water Resources Research* **22**(11): 1509–1518.

Carver, P. H., and J. J. Boland (1980). "Short-run and long-run effects of price on municipal water use," *Water Resources Research* **16**(4): 609–616.

Chang, C., and R. Griffin (1992). "Water marketing as a reallocative institution in Texas," *Water Resources Research* **28:** 879–890.

Charney, A., and G. Woodard (1990): "Socioeconomic impacts of water farming on rural areas of origin in Arizona," *American Journal of Agricultural Economics* **72:** 1193–1199.

Colby, B. (1989). "Estimating the value of water in alternative uses," *Natural Resources Journal,* **29:** 511–527.

Colby, B. (1990). "Transaction costs and efficiency in western water allocation," *American Journal of Agricultural Economics* **72:** 1184–1192.

Colby, B. G., K. Crandall, and D. B. Bush (1993). "Water rights transactions: market values and price dispersions," *Water Resources Research* **29**(6): 1565–1572.

Daubert, J. T., and R. A. Young (1981). "Recreational demands for maintaining instream flows: a contingent valuation approach," *American Journal of Agricultural Economics* **63:** 666–676.

Dudley, N. J., and W. F. Musgrove (1988). "Capacity sharing of surface water reservoirs," *Water Resources Research* **24**(5): 649–658.

Dudley, N. J., and W. F. Musgrove (1991). "The economics of irrigation water allocation under conditions of uncertain supply and demand," in *The World Congress on Water Resources* (Rabat, Morocco, May 12–13, 1991), p. 8.

Easter, K. W. (1975). "Field channels: a key to better Indian irrigation," *Water Resources Research* **11**(3): 392.

Easter, K. W., and R. Hearne (1995). "Water markets and decentralized water resources management: international problems and opportunities," *Water Resources Bulletin* **31:** 9–20.

Ehart, J., W. E. Downey Brill, Jr., and R. M. Lyon (1993). "Transferable discharge permits for control of BOD: an overview," in E. F. Joeres and M. H. David (eds.), *Buying a Better Environment: Cost Effective Regulation through Permit Trading.*

Feather, P. M. (1992). "Valuing water quality using discrete choice models," Ph.D. dissertation, University of Minnesota.

Frederick, K. D. (1986). *Scarce Water and Institutional Change,* Washington, D.C.: Resources for the Future.

Gardiner, V., and P. Herrington (eds.) (1986). *Water Demand Forecasting,* proceedings of a workshop sponsored by the Economic and Social Research Council, Short Run Press, Ltd., Exeter, U.K.

Gibbons, D. (1986). *The Economic Value of Water,* Resources for the Future, Washington, D.C.

Griffin, R. C., and G. M. Perry (1985). "Volumetric pricing of agricultural water supplies: a case study," *Water Resources Research* **21**(7): 944–950.

Haneman, W. M. (1989). "Information and the concept of option value," *Journal of Environmental Economic Management* **16:** 23–37.

Hearne, R. (1995). "The market allocation of natural resources: transactions of water-use rights in Chile," Ph.D. dissertation, University of Minnesota.

Hearne, R., and K. W. Easter (1995). *Water Allocation and Water Markets: An Analysis of Gains-from-Trade in Chile,* World Bank Policy Paper, Washington, D.C.

Hirshleifer, J., J. C. Dehaven, and J. W. Milliman (1960). *Water Supply, Economics Technology and Policy,* University of Chicago Press.

Howe, C. W., and F. P. Linaweaver, Jr. (1967). "The impact of price on residential water demand and its relation to system design and price structure," *Water Resources Research* **3**(1): 13–32.

Howe, C. W., D. R. Schurmeier, and W. D. Shaw (1986). "Innovative approaches to water allocation: the potential for water markets," *Water Resources Research* **22:** 439–445.

Howitt, R. E. (1994). *Empirical Analysis of Water Market Institutions: The 1991 California Water Market,* Working Paper 13.94, Foundazione Eni Enrico Mattie.

Kneese, A. V. (1977). *Economics and the Environment,* Penguin Books, New York.

Lauria, D. T., D. L. Schlenger, and R. W. Wentworth (1977). "Models for capacity planning of water systems," *Journal of Environmental Engineering Division* **103:** 273–291.

Loomis, J., and F. Ward (1985). *The Economic Value of Instream Flow: An Assessment of Methodology and Benefit Estimates,* draft paper from the U.S. Fish and Wildlife Service, Ft. Collins, Colo.

McConnell, K. E. (1990). "Models for referendum data: the structure of discrete choice models for contingent valuation," *Journal of Environmental Economics and Management* **18:** 19–34.

Meade, J. (1952). "External economies and diseconomies in a competitive situation," *Economic Journal* **62:** 54–67.

Moncur, J. E. T., and R. L. Pollock (1988). "Scarcity rents for water: a valuation and pricing model," *Land Economics* **64**(1): 60–72.

Rogers, P. (1986). "Water: not as cheap as you think," *Technology Review* (Nov./Dec.): 31–43.

Seagraves, J. A., and K. W. Easter (1983). "Pricing irrigation water in developing countries," *Water Resources Bulletin* **19**(4): 663–672.

Swallow, S. K., and C. M. Marin (1988). "Long-run price inflexibility and efficiency loss for municipal water supply," *Journal of Environmental Economics and Management* **15:** 233–247.

Teeples, R., and D. Glyer (1987). "Production functions for water delivery systems: analysis and estimation using dual cost function and implicit price specifications," *Water Resources Research* **23**(5): 765–773.

Thomas, J. F., and G. J. Syme (1988). "Estimating residential price elasticity of demand for water: a contingent valuation approach," *Water Resources Research* **24**(11): 1847–1857.

Tietenberg, T. H. (1992). *Environmental and Natural Resource Economics,* 3d ed., Harper Collins, New York, chap. 14.

Tsur, Y. (1992). "The exploitation of aquifers of unknown stock," discussion paper, Department of Agricultural and Applied Economics, University of Minnesota.

Tsur, Y., and T. Graham-Tomasi (1991). "The buffer value of groundwater with stochastic surface water supplies," *Journal of Environmental Economics and Management"* **21:** 201–224.

Walsh, R. G., J. B. Loomis, and R. A. Gilman (1984). "Valuing option, existence and bequest demands for wilderness," *Land Economics* **60**(1): 14–29.

World Bank (1993). *Water Resources Management* (A World Bank policy paper), Washington, D.C.

Young, R. A. (1986). "Why are there so few transactions among water users?" *American Journal of Agricultural Economics* **68**(5): 1145.

Young, R. A., S. L. Gray, R. B. Held, and R. S. Mack (March 1972). *Economic Value of Water: Concepts and Empirical Estimates,* Technical Report of the National Water Commission, NTIS, no. PB210356, National Technical Information Service, Springfield, Va., pp. 231–234.

Chapter

14

Social Impacts

Thayer Scudder
California Institute of Technology
Pasadena, California
and Institute for Development Anthropology
Binghamton, New York

Introduction

This volume is appearing at a time of increasing criticism of major water-resource development projects and especially of projects involving dams, canals, and embankments that modify river systems in a major way. Such criticism is overdue and welcome because benefits have often been inflated and costs underestimated. Moreover, diminishing freshwater resources (Gleick, 1993), coupled with population increase, will require difficult choices to be made between alternative procurement techniques, agricultural, industrial, and residential uses, and, in regard to transnational river basins, allocational issues.

To date critiques have emphasized such underestimated environmental costs as reduction of highly productive wetlands and biodiversity, and increased flooding, especially in Europe and North America. Aside from a three-volume series on the social and environmental effects of large dams (Goldsmith and Hildyard, 1984, 1986; Goldsmith et al., 1992), and the growing literature on the relocation experience [which is assessed in a separate chapter (Chap. 15) in this volume], much less has been written on social impacts. Well-designed long-term research is urgently needed. Because of its absence, the arguments presented in this chapter will have to rely to a large extent on case studies. What is known from them indicates that, whether short-term or cumulative, adverse social impacts, as with environmental ones, have been seriously underestimated. When combined with adverse health impacts (Hunter et al., 1993), it is clear that large-scale water-resource development projects unnecessarily have lowered the living standards of millions of local people.

Because of its importance as a model emulated by other countries, the Tennessee Valley Authority (TVA) provides a cautionary tale of how projects intended by their sponsors (in this case President Franklin Roosevelt and first chairman A. E. Morgan) to provide major social benefits can be implemented in ways that leave many local people worse off. Long considered one of the most successful large-scale attempts at integrated river-basin development, early in the TVA's history the agency became primarily an electricity-generating company because of the dominant influence of director David E. Lilienthal (Wheeler and McDonald, 1986, pp. 7–8). The authors further note that "In the short term TVA's 'social experiment' was a failure—a failure brought about by the lack of a coherent and unified grassroots policy; by conflicting dictatorial roles and administrative structures; and by highly variant and persistent conceptual views held by people within TVA of the agency's aims and goals" (ibid., p. 264). It was not meant to be that way; TVA's intent under Roosevelt's New Deal was to raise the living standards of low-income people. Although reasons vary, such a narrowing of initial goals during implementation has been a rather frequent occurrence in the history of large-scale river-basin development projects. The main risk-takers and the main losers have been project-affected people.

In TVA's case, few detailed studies had been completed until the 1980s that dealt with the social and environmental impacts of specific TVA projects. In their *TVA and the Disposed,* which deals with Norris Dam relocation, McDonald and Muldowny (1982) note that aside from Selznick's (1949) *TVA and the Grass Roots,* there is a "dealth" of TVA literature that assessed impacts from below as opposed to the extensive literature "that has provided a view of TVA from above" (McDonald and Muldowny, 1982, p. 6). Part of the reason was that while TVA espoused a "grassroots orientation," actually TVA had maintained that posture "by guiding state and local agencies to do its will" (ibid., p. 15).

In his *The Myth of TVA,* Chandler (1984) attempts to compare TVA's impact on basin states with development trends in surrounding but unaffected states. As indicators he uses per capita income, industrialization, changes in employment, population shifts, and rural electrification. He concludes that "no persuasive evidence supports the notion that TVA significantly improved the economy of the Tennessee Valley." There are methodological problems with Chandler's study. He also pays scant attention to the influence TVA has had on land reclamation, reforestation, and nonstructural approaches to floodplain management (White, 1985); to energy conservation when prices per kilowatthour began to escalate (Freeman, 1985, p. 684); and on small holder agriculture, engineering design, and standards; and a range of benefits "not amenable to quantitative analysis" (Wheeler and McDonald, 1986, p. 217). Furthermore, Chandler does not deal with changes in TVA policies through time or with political constraints that interfered with TVA's realization of its broader development goals. Nonetheless, Chandler's conclusions require a more careful consideration than they have received, especially because so much of the literature on TVA has been written by TVA officials or advocates. Never a general agency for the integrated development of the Tennessee Valley, what TVA benefits have occurred were often skewed, with Chandler pointing out that over 80 percent of flood-control benefits, for example, can be attributed to the city of Chattanooga, with urban areas and urban residents in general benefiting more than rural residents.

Such was especially the case with approximately 120,000 people (Gaventa, 1982, p. 983) relocated from TVA's several dozen reservoir basins, especially those farming the rich bottom lands that were inundated and an unknown number of farm laborers and sharecroppers whose relocation was not required but whose employment was dependent on those lands. Adverse impacts on black Americans were analogous to those involving Native Americans elsewhere in the United States and Canada and ethnic minorities elsewhere in the world. According to Grant, racial discrimination within the Tennessee Valley Authority itself significantly reduced both the number of black Americans and their upward mobility. As for people relocated, "studies showed that families forced by TVA to move were worse off than they had been before the coming of TVA" and "the tenant and poor landowner bore the brunt of readjustment" (Grant, 1990, p. 75).

While black families experienced "particular difficulty in relocation" (ibid., p. 77), "TVA regional planners did not significantly aid any relocated families, regardless of color. TVA did not keep records on the fate of its reservoir removal families; there were no follow-up studies to measure adjustment and living conditions after removal" (ibid., p. 83).

Definitions

Generally speaking, environmental-impact analysis has dealt with impacts on the physical and nonhuman biotic components of ecosystems, while social-impact analysis has dealt with impacts on sociocultural systems (SCOPE, 1972). In this chapter, an equally broad definition of social impacts is used, with particular emphasis on the impacts of large-scale river-basin development projects on the lifestyles of affected communities, households, and individuals.

The topic is a vast one since it includes, in effect, impacts on entire societies. To make it manageable, some restrictions are necessary. The emphasis throughout the chapter is on the low-income rural majority that lived in subtropical and tropical river basins prior to being impacted on by one or more water-resource development projects. Furthermore, analyses in this chapter and Chap. 15 deal mainly with the impacts of dams on major rivers on three categories of people. These include those who must be relocated because of project works and future reservoir-basin inundation (the relocatees), those whose communities must receive relocatees (the host population or hosts), and those other project-affected people (OPAP) who are neither relocatees nor hosts.

Because environmental- and social-impact assessments, as well as supervisory and project-completion reports, have tended to ignore project impacts on OPAP, they are dealt with in some detail. They include three major types: (1) those who live in the vicinity of project works, including not just the dam site and township but also access roads and transmission lines; (2) those who live in communities within reservoir basins that do not require relocation nor incorporation of relocatees; and (3) numerically tending to be by far the most numerous, people who live below dams whose lives are affected, for example, by the implementation of irrigation projects or changes in the annual regime of rivers in whose basins they reside. Taken together, such other project-affected people usually outnumber relocatees and hosts. Therefore, failure to assess impacts on them can be expected to distort feasibility-study results.

Regrettably, the restrictions outlined above do not include several important categories of people, including immigrants from without the river basin, and inhabitants of cities, mining townships, and other major industrial complexes both within and without a river basin. Since these categories—as opposed to local rural communities—tend to be the major beneficiaries (in terms of rising living standards resulting from increased supplies of industrial and residential electricity and water) of water resource development, it is important for readers to realize that this chapter is biased toward those who, to date, are most apt to be adversely affected. While some effort is made to correct this bias, it is intentional in order to emphasize two points:

1. The worldwide tendency to ignore (in the case of OPAP) or underestimate (in the case of relocatees and hosts) the costs of major water-resource development projects to large numbers of people has inflated their benefits to the extent that insufficient attention has been paid to other alternatives.

2. There are ways, to be outlined later, for increasing the likelihood of making a larger proportion of all categories of project-affected people beneficiaries in cases where future river-basin development projects are selected for implementation.

Numbers and Knowledge

Numbers

Little accurate data exists on the total number of people affected by even a single major project, let alone water-resource development projects completed during a single year or over a 10-year period. The major exception is where a project impacts on an easily defined area with a relatively small population whose number is already known. Examples include Quebec's large-scale James Bay Project, whose components will have a direct impact on approximately 11,000 Cree Indians and an equal number of other residents; and Botswana's Southern Okavango Integrated Water Development Project, the social impacts of which would be largely restricted to about 50,000 people. In both cases the number of relocatees and hosts would constitute only a small minority of the total, a relatively small number of people lived below dam sites, and no major cities would be involved. In contrast, where projects impact upon large numbers of downstream residents—as with the Kariba Dam on the Zambezi, the Kainji Dam on the Niger, and the Aswan High Dam on the Nile, and the Gezouba and Three Gorges Dams on the Yangtze—numbers impacted on can run into the millions.

Although underestimations are common, the most accurate information pertains to relocatees. According to the World Bank (1994, pp. 1, 3), "The displacement toll of the 300 large dams that, on average, enter into construction every year is estimated to be above 4 million people," with at least 40 million so relocated over the past 10 years. In India alone, it is estimated that at least 10 million people have been relocated during that time period because of water-resource projects, many of whom are of tribal origin. Counts for specific projects are relatively accurate where governments attempt to calculate numbers for compensation and other purposes. To date the largest number of relocatees from a single project was 383,000 in connection with China's Danjiangkou Dam on a Yangtze tributary. Completion of the Three Gorges Dam on the main Yangtze will require the resettlement of over a million people, as will Bangladesh's Flood Action Plan as originally conceived.

Although donors may emphasize the importance of involving the host population in project benefits, few planning documents attempt to calculate the size of the host population. Aside from time, personnel, and financial constraints to enumeration, there is a problem of defining who is a host. The simplest solution would be to restrict enumeration to inhabitants of settlements that physically receive relocatees, but that would place too much emphasis on the housing component as opposed to the impact of those resettled on the arable and grazing lands and other natural resources, employment opportunities, and social services of the recipients.

Resettlement always reduces access to land of a recipient population; hence all such recipients should be considered hosts. On the other hand, well-planned and well-implemented resettlement theoretically could increase employment opportunities and improve access to social services. The latter is the commonest output with both relocatees and hosts not infrequently benefiting from improved educational and medical facilities as well as improved communications. The former is more common in regard to urban as opposed to rural relocation, with the major exception of reservoir fisheries, where carefully designed training programs can benefit both hosts and relocatees.

Analysis of one project will demonstrate the difficulties of estimating numbers. This is the binational Highlands Water Project (LHWP), which involves both the Kingdom of Lesotho and the Republic of South Africa. It is used as an example not just because it is current and one of the world's five largest water-resource projects, but also because it is a World Bank–assisted project. Though insufficient because of insufficient attention paid to OPAP and because they do not insist on relocatees becoming project beneficiaries, the Bank nonetheless has the best guidelines of any donor in regard to social impacts (World Bank, 1990, 1994). Furthermore, provided plans are adequately implemented, LHWP has the potential to make a positive contribution to the economies of both countries and to a majority of project-affected people. Little information is available, however, on the number of such people.

According to the 1986 treaty between Lesotho and South Africa, the LHWP has three major purposes. As water is the most crucial scarce natural resource in South Africa, the first is to transfer up to 70 m^3/s of water to the Vaal Triangle, which includes the Johannesburg-Pretoria area. Of high quality, that water is designated for commercial, industrial, and residential use, as opposed to its sale to commercial farmers growing field crops. Properly used through the new South Africa's Reconstruction and Development Programme, it could raise the living standards of hundreds of thousands of low-income urban residents and workers. Revenue received will become a major source of income for Lesotho.

En route to South Africa, sufficient electricity will be generated from the transferred water to replace imports from South Africa and to ensure, for the present, Lesotho self-sufficiency in electricity. That is the second purpose of LHWP. The third purpose is to benefit people directly impacted on by project infrastructure and construction activities in a fashion that is environmentally, economically, and institutionally sustainable.

As conceived in the 1980s, LHWP is to be implemented in four phases that would involve the construction of six dams. Construction is currently under way on the first two as well as on the water-transfer tunnels (Phase 1A), while contracts have been let for advanced infrastructure serving the third (Phase 1B). By international standards, planning is of a relatively high order. Yet, to date even the number of potential relocatees affected by the first phase has been consistently underestimated, while virtually no assessments exist in either Lesotho or South Africa of the numbers, let alone the social impacts, that would be affected by project implementation.

The Lahmeyer-MacDonald Consortium (1986) joint feasibility study estimated Phase 1A relocatees at only 51 persons in the mid-1980s. By the mid 1990s that number had increased by a factor of nearly 10 to nearly 500 (Maema and Reynolds, in press). Not only was the 1986 figure a serious underestimate, but current estimates for Phase 1B would also appear to be underestimated by a factor of at least 5. Aside from the difficulty of enumerating dispersed rural populations in which a significant portion of the male labor force is absent as migrant laborers, the main reason for the Phase 1A underestimate was the failure to count households whose physical relocation would be required because of the construction of hundreds of kilometers of roads and transmission lines. Indeed, the total of such relocatees now exceeds those who must move because of inundation of the two reservoir basins.

The major reason for suspecting the Phase 1B estimate to be too low is that it is based more on loss of housing than on loss of livelihood due to inundation of arable land. Worldwide, relocation planners continue to base relocation estimates more on loss of residence than on loss of the means of production (especially in relationship to access to arable land, grazing and other natural resources). Inundation caused by Phase 1B construction will inundate approximately 25 percent of the arable land in the local catchment basin, including the best concentration of fertile bottom land. One village, sited well above the planned reservoir, will lose approximately 50 percent of its arable land. Because of the potential loss of livelihood, it is not unlikely that up to 2000 of the 7000 plus people currently residing within the local catchment will require relocation, if their living standards are not to deteriorate, as opposed to the 591 persons noted in the latest LHWP estimates (Tshabalala, 1994).

In the local catchments of the three dams, the host and other project-affected people categories are rarely the same people. This is because tenure to grazing land, fuel, thatch, and other natural resources is based on limited-access communal tenure. Those numbers can be roughly extrapolated from national census figures. Such is not the case, however, with downstream communities. Nor is it the case with the urban recipients of water in South Africa and of electricity in Lesotho. As with other major water-resource development projects, to date little information, reliable for planning purposes, is available on the social impacts of the LHWP on downstream communities along the ≥1000-km length of the Orange River between the lowest dam in Lesotho and the Atlantic Ocean. Studies currently under way suggest that those impacts may be sufficiently great as to question the wisdom of proceeding, for example, with Phases 3 and 4.

Estimates of impacts on urban residents in both countries will depend on investment strategies yet to be determined. Take, for example, Johannesburg and such surrounding satellite communities as Soweto and Empangeni. The social impacts of LHWP water on low-income communities will be influenced by whether water and delivery facilities will allow periurban residents to undertake smallholder irrigated agriculture on household plots. On the basis of utilization of household plots around Harare, Zimbabwe; Maputo, Mozambique; and Maseru, Lesotho, such activities could provide higher living standards to thousands of households while providing low-income consumers with improved supplies of vegetables, fruits, and small animals. The necessary policy decisions, however, have yet to be taken. As for the industrial development that will be facilitated by increased water availability, again different but yet-to-be-defined strategies relating to reliance on technology versus labor will influence the numbers of other project-affected beneficiaries.

Knowledge

Resettlement of relocatees is dealt with in a separate chapter (Chap. 15) not just because of the importance of the topic but also because of the relatively large number of studies, including several longitudinal ones, of the resettlement process. The same cannot be said of the other categories of project-affected people. Indeed, there is a shocking lack of knowledge about the more general longer-term and cumulative impacts—environmental, economic, and political—of water-resource development projects. While comparative studies exist on the impact of irrigation-based land-settlement schemes (World Bank, 1985) and of relocation (Scudder and Colson, 1982; Cernea, 1988; Cernea and Guggenheim, 1993; World Bank, 1994) on participants, there are no broader comparative studies dealing with the totality of project-affected people.

This deficiency is especially serious in regard to large-scale river-basin development projects. Notwithstanding the fact that such projects frequently are the largest projects in national development plans in terms of financial costs and impacts on national heartlands, there are few, if any, projects for which implementing agencies or donors have completed a detailed, long-term evaluation. Biswas makes the same point in regard to irrigation projects, stating that "one would be hard pressed to use all the fingers of one hand to count the number of projects in developing countries which have been reliably and objectively evaluated five, ten and 20 years after their completion" (Biswas, 1987, p. 47).

This omission is especially surprising in the case of the World Bank, which continues to be the major international financier of large-scale river-basin development projects and which has, in its Operations Evaluation Department, a time-tested monitoring and evaluation capacity. While the Bank regularly issues project completion reports (now relabeled *project-implementation reports*) after the completion of disbursements and the project cycle (as defined by the Bank), no long-term studies that assessing project impacts at various intervals over several generations have been completed. Granted the fact that the impacts of major water-resource development projects tend to be ongoing over an extended time period, and that the productivity of innumerable project-affected people has tended to decline over time, for most major projects it is impossible to assess the extent to which they have met defined goals.

The necessary studies will involve methodological difficulties, especially in the assessment of cumulative impacts (Dickson et al., 1994). Problems relate not just to changes that would have occurred among affected people without a project but also, in regard to longer-term cumulative impacts, to factoring out the relevance of other postproject influences. Such problems must not be allowed, however, to put off further the necessary studies since the few long-term studies that have been undertaken, by academics largely, agree that longer-term project-induced social impacts have been seriously underestimated (see, e.g., Colson, 1971; Bartolome and Barabas, 1990; Scudder, 1993a).

A major problem concerns the inadequacy of preproject benchmark studies against which subsequent impacts can be assessed. Even where such studies are undertaken, their initiation and completion tends to be out of phase with the engineering cycle. Hence, in the case of the Lesotho Highlands Water Project, the completion of both epidemiological and socioeconomic benchmark studies occurred after construction activities had begun. The same was the case with detailed socioeconomic studies carried out in connection with System B of Sri Lanka's Accelerated Mahaweli Project.

The only solution to this situation is to initiate the necessary preproject benchmark studies much earlier in the planning process. Much of the necessary data is required by the World Bank's Guidelines, which state that borrowers must present a resettlement plan before project appraisal. Gathering data for such a plan provides the opportunity for integrating the necessary benchmark studies into the planning process. Once gain, however, all too often construction activities are apt to proceed more rapidly. LHWP is a case in point.

While there is no alternative to undertaking detailed preproject benchmark studies among potential relocatees, one reason why there are so few follow-up studies relates to arguments among social scientists and statisticians regarding methodologies for monitoring and evaluation. Ideally, it would be advantageous to follow a statistically significant random sample drawn from the original benchmark studies. This, however, is seldom done because of time, personnel, and financial constraints. These appear to be insurmountable in part because the delay in providing results, a delay which can involve years as opposed to months, makes such follow-up studies of little use for dealing with key issues as they arise.

One cost-effective solution is to select a small, carefully stratified subsample of no more than 50 to 100 households, reinterviewed over an extended time period once or twice annually. Such a methodology can provide planners with a problem-oriented report within a 6- to 8-week period. Skepticism continues, however, as to the reliability of the data collected. Certainly the results of such studies need be periodically checked by more detailed sample surveys. Nonetheless, a methodology combining participatory rapid survey with periodic reinterviewing of the same households has already proved its worth. Undertaken between 1979 and 1989, periodic reinterviewing of a stratified opportunity sample that grew to only 40 households amongst tens of thousands of Mahaweli settler households, reached accurate and policy-important conclusions within a 6-week period (Scudder, 1993b). Furthermore, none of the major conclusions has subsequently been found to be in error. This is largely because of the longitudinal nature of the monitoring process. Use of indicators relating, for example, to housing, water supply and sanitation, fuel and lighting, and production technology, linked with questions relating to income and expenditures, allows the interviewer to conclude within an hour as to whether living standards are improving, remaining the same, or deteriorating. Complementary questionnaires, following by group-focused interviews, can then be expected to identify the causes for change or lack of change.

Immigrants

Immigrants, both temporary and permanent, from without a particular river basin are major beneficiaries of river-basin development projects. The largest number of temporary immigrants are associated with a project's construction phase. While employment on advanced infrastructural and the main civil-engineering contracts are often listed as a benefit for local people, local workers in fact tend to be only a small minority of the labor force. For example, during construction of the first stage (La Grande) of the James Bay Project in Quebec, less than 5 percent of the labor force were Cree Indians in spite of the existence of available and otherwise unemployed workers. There are several reasons for such a situation. Most important is the fact that local people seldom have the skills required by contractors, with crash training programs seldom bringing skills up to the desired level—as the LHWP found out in regard to local training efforts. Moreover, contractors often bring a significant portion of their labor force with them or recruit immigrant workers with previous construction experience. On completion of China's Gezouba Dam, for example, many of the 40,000 workers stayed in the area in the expectation of joining the labor force for the construction of the Three Gorges Dam.

Although a small minority of temporary workers choose to remain in the area after their contracts end, most permanent immigrants come after the completion of the construction phase. Seeking the new opportunities created by the project, they frequently are able to outcompete local people. Two major rural benefits of large-scale river-basin development projects are fisheries associated with new reservoirs and irrigation. Although both involve project affected people as well as immigrants, the latter tend to dominate unless a special effort is made to increase the competitive abilities of local people.

Case Studies of Other Project-Affected People (OPAP)

Introduction

Large-scale river-basin development projects accelerate the incorporation of all project-affected people, including relocatees and hosts, within wider political economies. Theoretically that should be a "plus" in terms of potential national development. What studies have been completed among affected rural populations suggest, however, that such incorporation over the longer term is more apt to reduce than improve the living standards of a majority. This certainly has been the case with Kariba, the first mainstream dam on the Zambezi. Today, over 30 years after project completion, the majority of over

100,000 project-affected people in the reservoir basin are worse off as a result of environmental degradation influenced by resettlement-related increases in population densities, higher adult mortality rates, reduced incomes, and social disorganization. Hundreds of thousands of downstream riparian residents are also worse off because of the adverse effects of dam-altered Zambezi flows, and especially reduced annual flooding, on their production systems. The time factor has been important, with conditions worsening as the years pass.

On the other hand, millions of downstream rural residents between the Aswan High Dam and the Mediterranean are probably better off today than they were before the dam because of improved irrigation, flood control, and rural electrification (see, e.g., White, 1988). But are those benefits sustainable over the longer term? In an analysis of the impact of the Aswan High Dam and other factors such as rising sea level on Nile Delta erosion, a recent conclusion is "no" (Stanley and Warne, 1993). Egypt's breadbasket, the delta, is by far the nation's most important agricultural resource. There, however, "human intervention...has caused northern Egypt to cease as a balanced delta system." After postulating a range of interventions for reversing declining conditions, caused, among other factors, by loss in the High Dam reservoir of the Nile's high sediment load, the authors conclude that "at current levels of population growth...these measures will be inadequate" (ibid., p. 634).

Downstream project-affected people

Introduction. There have been very few detailed studies of the impacts of river-basin development projects on the multimillions of people living below dams, under the assumption that flood-control benefits would more than compensate for any disbenefits. Worldwide, informed people are beginning to realize, for several reasons, just how wrong such an assumption is. As in the Mississippi and Rhine River basins, such previous development initiatives as levees and other flood-control mechanisms have, in conjunction with urbanization, led to increased flooding by reducing farmland and water-absorptive wetlands and channeling more and more water into a river's primary channel. That is one reason.

Another reason is the growing realization of the very high productivity of riverine habitats, wetlands in particular (IUCN, ongoing), and the need to enhance rather than reduce their extent and productivity. Related to that reason, a third reason—of primary concern here—is the fact that millions of contemporary people are dependent on the productivity of those wetlands and floodplains. where dams, irrigation canals, and embankments reduce them, local people are impoverished. Developers and donors rarely take that into consideration. If they had in the past, and if the necessary broader feasibility studies had been carried out, many completed projects would have been shown to be uneconomic.

Floodplain utilization has played a major role in the formation of city-states and of civilization in Africa, Asia, the Middle East, and the Americas as shown by archaeological studies. Granted population increase, such utilization may well support more people today than at any time in the past, but far less sustainably. The floodplains of the Brahmaputra-Ganges system in Bangladesh is one Asian example. Others are the floodplains of the Mekong in Vietnam and of the Yangtze in China. In Africa, the floodplains of all the major rivers are of crucial importance to millions of people—whether the Nile, the Zambezi, or the Niger.

Oddly, few socioeconomic studies have dealt in detail with the contemporary utilization of floodplains. Those that have emphasize the extent to which floodplains constitute by far the most important resource in local production systems (Scudder, 1962, 1980, 1991; Horowitz et al., 1990). Annual flooding is critical not only for maintaining that resource but also for the survival of dependent communities. As floods recede, for example, communities throughout the arid and semiarid lands of Africa practice flood-recession agriculture. As the dry season progresses, the much higher carrying capacity of floodplain grazing allows cattle and small ruminants to survive until the coming of the rains, while the flood cycle itself is necessary for sustaining productive fisheries and recharging aquifers.

Dam construction is apt to impoverish downstream riverine communities in one of two ways: (1) through alteration of the annual regime of dammed rivers or (2) through attempts by governments and immigrants to displace resident households and communities from their land.

Dams and flood regularization. Aside from run-of-the-river installations, a major function of dam construction is to regularize a river's annual regime by augmenting low-flow periods and greatly reducing periods of flooding in order to make available a more constant water supply for hydropower generation, navigation, and commercial irrigation. Although few detailed studies have been completed on the impacts of such regularization, those that exist have shown them to have a devastating effect on millions of people (Drijver and Marchand, 1985; Horowitz et al., 1990; Adams, 1992; Hollis et al., 1993; Scudder, 1994; and Acreman et al., in preparation).

The topic has been best researched in West Africa in connection with mainstream dams on the Senegal River and a number of dams in Nigeria. After 3 years of study, an Institute for Development Anthropology team showed that the Manantali Dam as managed by the trinational Senegal Valley Development Authority (OMVS) was adversely affecting up to 500,000 people below the dam (Horowitz et al., 1990). In his 1994 analysis of the Kainji Dam project on the Niger, Roder (1994) notes that adverse downstream impacts include an estimated 60 to 70 percent reduction in the riverine fishery and a 30 percent reduction of seasonally flooded (*fadama*) land that has lowered swamp rice production by 18 percent. Although I have seen no confirmation of his figures, in a FAO technical paper (Welcomme, 1979) Awachie stated that Kainji had also been at least partially responsible for a 100,000-t (metric tons) reduction in yam production further downriver in which case we can assume that the dam's impoverishing effects involved hundreds of thousands of people. On a Niger tributary farther upriver, Adams (1993) estimates that the costs of the Bakolori Dam to downstream villagers in terms of reduced crop, livestock, and fisheries production actually exceeds the benefits realized from the project.

Elsewhere in Nigeria dams constructed mainly for irrigation purposes in the Hadejia-Jama'are system have significantly reduced downstream floodplains—once again at the expense of local producers. Indeed, in at least one major case, evidence suggests another project-related loss, with Barbier et al. demonstrating that the net economic benefits of floodplains that have been significantly reduced by river-control mechanisms were at least \$32 per 1000 m^3 of water versus only \$0.0026 for the irrigated Kano Project when all costs (including operational costs) are included [cited in Acreman et al. (in preparation); see also Hollis et al. (1993)].

Those within a project's vicinity: The James Bay Project and the Cree

Introduction. The James Bay Project provides an excellent case history in that it illustrates the strength of the type of coalition advocating large-scale river-basin development projects, their impoverishing impact on an entire culture area (as opposed to just relocatees and hosts), and the difficulties that local people must face in trying to modify, let alone cancel, projects. Just as TVA was a vision of Roosevelt, the Aswan High Dam and the Volta Dam of Nasser and Nkrumah, respectively, and China's Three Gorges Dam of Mao, Chou Enlai, Deng, and Li Peng, so was the James Bay complex the vision of Premier Bourassa of Quebec in 1985 (Bourassa, 1985). An electrical utility, Hydro-Quebec, was responsible for project planning and design. As Hydro-Quebec's name suggests, the James Bay Project was seen primarily as a single purpose one for the generation of hydropower. A separate parastatal organization, the James Bay Development Corporation, was subsequently established to "develop" the area that would be "opened up" by the project.

The James Bay Project. Like the Lesotho Highlands Water Project, Hydro-Quebec's James Bay Project is one of the five largest contemporary river-basin development projects in the world. If completed, the total generating capacity of more than 12 dams would be about 25,000 MW, of which the generating facilities for 10,000 MW have either already been completed or are under construction. Between 1971 and 1991 over $10 billion have been invested, with the remainder of the complex estimated to cost at least another $40 billion.

Hydropower is seen as Quebec's major resource for both provincial consumption and for export, with 95 percent of the province's present installed capacity coming from hydropower (Hydro-Quebec, 1993). Costs to industrial consumers have intentionally been kept low, with the result that Quebec has successfully induced large, energy-intensive aluminum and manganese processing industries to set up installations along the Gulf of Saint Lawrence.

Although subsequent feasibility studies have changed the design of the Grande-Baleine component of the complex, Fig. 14.1 illustrates the project as conceived in the early 1990s. Roughly, the complex can be broken down into three components, of which the La Grande in the middle of the complex area is nearing completion. Feasibility studies for the northernmost component, Grande-Baleine, were completed in 1993. As for the BNR (abbreviated from the three southern rivers) component, Hydro-Quebec's intention has been to bring that on line by the year 2010. In all, five major river systems are involved with the project also incorporating a number of interbasin transfers to augment reservoir storage capacity.

Figure 14.2 shows the homeland of the James Bay Cree, considered to be one of the most viable indigenous populations in North America. One-third of approximately 11,000 Cree, for example, still support themselves primarily as trappers, hunters, and fishers, so that practically the entire homeland is divided into traplines. It does not require superimposing Fig. 14.2 on Fig. 14.1 to see that, coincidentally, the James Bay hydropower complex coincides almost entirely with the boundaries of the Cree homeland.

Advocates of the project point out that virtually no compulsory relocation is required, as the engineering works and reservoirs require the movement of not a single Cree village. Rather, the only relocation—and number of people is probably less than a few hundred—pertains to a relatively small number of trapline camps. Most, if not all, of those can be resited within each family's trapping territory, so that the number of hosts is even smaller. In other words, well over 90

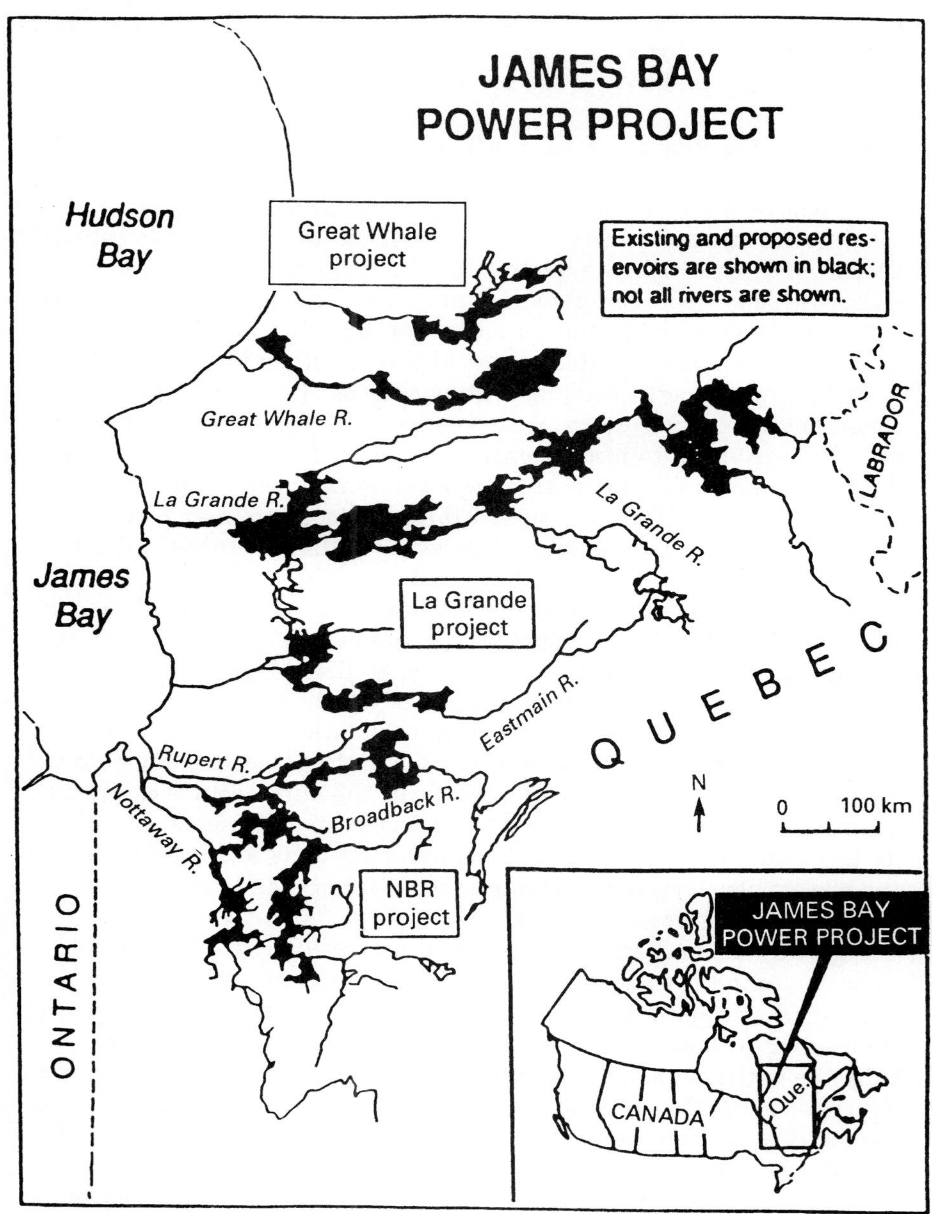

Figure 14.1 James Bay Power Project. (*Source: Probe International.*)

Figure 14.2 Cree Band territories, 1971 (*Salisbury, 1986*).

percent of the people (OPAC) adversely affected by the project are neither relocatees nor hosts.

Social impacts

Introduction. Regardless of complexity and scale, all societies are dynamic open-ended systems involving ongoing processes of change and continuity. It is the dynamics of such a system that make it so difficult to pin down the impacts of specific projects versus what might have happened without the project as a result of other impinging factors, or the proportional importance of such factors with the project. Although more isolated than many societies, the James Bay Cree were in pre-European contact with other ethnic groups for thousands of years. Thereafter, located in between the French- and English-speaking people of southern Quebec and the Inuit to the north, rates of change accelerated with the establishment of various Hudson Bay Company trading posts and the associated fur trade (Feit, 1982) in the latter part of the seventeenth century.

Further contact with the outside world came in the mid-1940s with the initiation of family allowances and other federal government benefits, including the first schools. Construction and occupation of United States Mid Canada Line military bases between 1955 and 1964 brought further contact. Throughout this period, however, isolation continued to characterize the Cree homeland. This changed rapidly once the decision was made in 1971 to proceed with the James Bay Project. The major influences were impacts arising from the 1975 James Bay and Northern Quebec Agreement (JBNQA) and road construction.

The 1975 James Bay and Northern Quebec Agreement (JBNQA). At the time that Quebec's premier announced the James Bay Project in April 1971, there had been no Cree involvement in project planning or design, nor had any form of social-impact analysis been carried out. With construction already under way by the mid-1970s, the emerging Cree leadership believed, probably correctly, that they had little hope of stopping the first phase of the project. By agreeing to its implementation and becoming signatories of the JBNQA, they at least received in return a promise of a degree of self-government (in particular to running their own education and health system and establishing their own police force) and an innovative program for maintaining trapping, hunting, and fishing activities.

Such benefits led anthropologist Richard Salisbury and other friends of the Cree to see JBNQA in a positive light. Certainly it had an impact-mitigating influence. More important, it led to the formation of the Grand Council of the Cree and the Cree Regional

Authority, which provided the Cree with a degree of political self-determination and an institutional structure for facilitating economic development. On the other hand, the land-settlement provisions of the JBNQA split the Cree into eight geographically isolated bands whose exclusive control of surface rights involved only 5 percent of their former lands, with another 15 percent set aside for their exclusive hunting, fishing, and gathering. Aside from preferential access for the Cree to certain mammals, the remaining 80 percent in effect was handed over to Quebec's James Bay Development Corporation.

To summarize, the James Bay Cree agreed to the project in the 1970s in return for a degree of political self-sufficiency at the expense of maintaining control over the development of their homeland. Thereafter neither the Federal Government of Canada nor Quebec Province delivered what they had promised the Cree under the Agreement. Furthermore, the James Bay Development Corporation has largely ignored the Cree in planning and implementing development projects. Tourist and other facilities along the north-south access road, which fell under JBDC jurisdiction, were almost exclusively run by outsiders in 1994 with minimal Cree involvement, including employment.

Roads. While the JBNQA agreement opened up the Cree homeland for hydropower and other development, it was the road system serving the project complex that gave outsiders and Cree alike greater access to the former homeland. During Phase 1 (La Grande) of the project, over 1000 km of roads was constructed north and south through the southern and central areas and east and west through the central portion. Should Phase 2 (Grande-Baleine) eventually proceed, 685 km of roads will be added which will cross 21 of the 37 Cree traplines north of the La Grande river system.

Roads, of course, can bring people out as well as into an area. They also tend to benefit those who can acquire and maintain transport. In the Cree case, however, the overall impact has been negative. Project-facilitated roads, for example, have sped up timber extraction by outsiders in the NBR areas. According to Feit, such logging has proved to be extremely destructive to Cree interests, with the Cree's Grand Chief estimating that clear-cutting was destroying at least one family trapline (averaging 600 km^2 in that area) per year.

Roads have also increased hunting pressure, including poaching, by outsiders and Cree alike. A significant decrease of moose, so important as a Cree food source, in the better-vegetated southern area is one effect. Another is the impact on Cree culture of the sheer number of outsiders entering the area. During 1991, for example, 26,500 people in 10,850 vehicles traveled up the road to La Grande. In regard to other impacts, both Cree and Inuit in the Grande-Baleine contact area blame different perspectives concerning the James Bay Agreement for the widening gulf between members of the two ethnic groups. As for implementation of the La Grande phase, that has had two quite different types of impact. One is the mercury contamination of reservoir fish (with inundation increasing the release of methyl mercury) to the extent that mercury contamination of some Cree, for whom fish are an all important dietary component, significantly exceeds World Health Organization (WHO) standards. Causality there can rather easily be ascertained. Such is not the case with the other type of impact, which includes a high incidence of sexually transmitted diseases (STDs) and such social pathologies as increasing spousal abuse and suicide, especially among young women. Unlike the situation on other reservations in both Canada and the United States, the Cree have been able to avoid dependence on provincial and federal welfare. Unemployment insurance, for example, and aid to dependent children make up a relatively small proportion of Cree income; only 6 percent in 1985, for example. Can the increasing rates of STDs and social pathologies be associated, therefore, with the project-related, accelerated rates of contact with outsiders and outside ideas, which Cree elders and leaders also tend to associate with an erosion of cultural values? Quite possibly they can.

Project-Affected People and Resistance Movements

Planners are apt to attribute local opposition to water-resource development projects to outside agitators. Resistance to Kariba on the north bank of the Zambezi, for example, was attributed by the British colonial administration to the growing influence of external political organizers agitating for national independence. That was in the late 1950s. When approximately 1000 local citizens vilified senior officials at a meeting to protest the arrival of equipment in late 1990 for initiating the first phase of the southern Okavango Integrated Water Development Project (SOIWDP), the Government of Botswana blamed international environmentalists and largely expatriate-owned, safari firm interests.

In both the Kariba and Okavango cases, however, resistance came primarily from project-affected people who believed construction would be at their expense (Colson, 1971; Scudder et al., 1993). Opposition in both cases appeared to be close to universal. Among project-affected people in SOIWDP's impact areas, opposition was based on the people's conviction that their lives had deteriorated following previous attempts to manipulate Okavango flows. The misgivings of those affected by Kariba have been shown to be correct; inundation of the fertile alluvial soils cultivated by villagers before removal were one factor in the deterioration of their living standards in recent decades. While suspension of SOIWDP continues, so that implementation impacts can only be surmised, the IUCN SOIWDP review team concluded that the people's concerns were indeed valid—that project implementation would have adversely affected the living standards of the majority.

To date, resistance movements have involved relocatees more than hosts and other project-affected people (Scudder, 1996). Worldwide, with only a few exceptions, a majority of those involved in thousands of water-resource development projects have resisted removal. Exceptions themselves are illuminating. The construction of the Aswan High Dam in the 1960s appears to have been welcomed by those remaining in Nubia only because the earlier construction of the Aswan Dam, plus two dam heightenings, had so impoverished those who had relocated to the reservoir margins that up to 100 percent of the men in some of the villages closest to the dam had to seek employment in the urban centers of Egypt and the Sudan. Resettlement in the 1960s was welcomed by men and women alike because they believed it would reunite families, in the downstream Kom Ombo resettlement area, that had been split up because of earlier relocation impacts. In the case of TVA's Norris Dam, the majority initially welcomed the project because, as tenant farmers and poor landowners, they believed, erroneously, that completion would bring industry and higher-paying industrial jobs.

Looking to the future, resistance movements can be expected to increase because of the increasing opposition—as currently in Brazil and India—of national and international human rights and environmental nongovernment organizations (NGOs) that are willing to become advocates for project-affected people. Their arguments, in turn, will be supported by increasing emphasis on the rehabilitation and reorientation of existing facilities, by the extent to which benefits have been overestimated and costs—especially to project-affected people and the environment—underestimated, and by increasing evidence that greater benefits in many cases can accrue from the implementation of alternatives that do not require major engineering works. The growing sophistication and recent international networking of indigenous people affected by major projects will also strengthen resistance, with the James Bay Cree, for example, playing a major role in the suspension during 1994 of Hydro-Quebec's Grande-Baleine Project. Other factors are reduced international aid as well as donor guidelines relating to the relocation process, although to date their implementation continues to be unsatisfactory (World Bank, 1994).

On the other hand, for resistance movements to succeed, national or provincial governments must also be amenable to strong economic and political arguments and pressures, and characterized by a judiciary to which project-affected people have access.

Helping Project-Affected People Become Beneficiaries

Introduction

According to the World Bank and other sources, construction on over 200 large dams begins each year. Although consideration of a wider range of alternatives, along with increasing opposition, will lead to alternatives for some future projects, other dams will be constructed. The same applies to water-resource development projects involving other types of structural approaches such as channelization, canals, and embankments. The sections that follow present, first, an example of a nonstructural alternative to a carefully planned but nonetheless flawed water-resource development project, and second, approaches more likely to include project-affected people as beneficiaries where major engineering works continue to be constructed.

An alternative to major engineering works: the IUCN review team's alternative to Botswana's Southern Okavango Integrated Water Development Project (SOIWDP)

Rising in Angola, the Okavango River is one of the largest in central and southern Africa and the eleventh largest on the continent. Shortly after entering Botswana, the river flows into the Okavango Delta, from which less than 5 percent of its annual flow escapes. The largest oasis in the world, the delta includes over 15,000 km^2 of perennial and seasonal swamp and seasonal grassland interspersed with dryland tongues and hundreds of islands. Supporting a rich flora and fauna, it is this mosaic of wetlands and drylands which makes the delta one of the natural wonders of the world.

The Okavango Delta is not a pristine wilderness, however, having been utilized for thousands of years by local people. Concentrated along its southern and western margins, and along the Boteti River, the current population that is dependent on the delta's land and water resources numbers approximately 100,000 people. The Okavango region is also host to a growing tourism industry which soon may pass agriculture (including cattle exports) as the country's second most important source of foreign exchange (diamond mining, which provides approximately 80 percent of foreign exchange earnings, is the most important).

Following up on a joint study with UN agencies in the 1970s on the "Okavango Delta as a Primary Water Resource for Botswana" (UNDP/FAO, 1977), the Government of Botswana formed an interministerial Okavango Water Development Committee in 1982. From the start, attention was focused on one of the 16 possible schemes that the UN-financed study had identified. Subsequently known as the Southern Okavango Integrated Water Development Project (SOIWDP), terms of reference for the necessary feasibility studies were drawn up in the mid-1980s and awarded to a major international engineering corporation which also completed design studies.

The Government's selection of the SOIWDP option was due in part to its being one of the more "environmentally friendly" of the 16 options. The project's major stated goal was to stimulate development. As set forth in the draft terms of reference, project implementation would increase food production and living standards among approximately 50,000 people living in the southern portion of the delta by developing 10,000 ha of commercial irrigation and improving the productivity of 5000 ha of existing flood-recession agriculture. Water supplies for villagers and their livestock would also be improved.

While the emphasis was on raising living standards of the low-income majority, other goals included improving water supplies for the rapidly growing regional center of Maun and for two diamond mines which contributed approximately 50 percent of national production. Such commendable goals would be obtained at the expense of a minimal number of relocatees and hosts, and the project works and associated reservoirs would require the physical relocation of less than 100 people.

Implementing the SOIWDP would involve the completion of two phases of major engineering works (Fig. 14.3). During the first phase channelization of the lower 42 km of the major tributary currently draining the Okavango was designed to increase the flow of water into a major reservoir to be created by a dam approximately 10 km downstream from where the Boteti exits the Okavango Delta. Control structures would be built at either end of a wing-shaped reservoir to achieve improved natural resource management and flood-recession agriculture. The second phase would involve a second dam and major reservoir lower down the Boteti, from which pipelines would distribute water to the mines and a nearby subdistrict center.

The Cabinet approved SOIWDP implementation in late 1988. When the contractor selected to undertake construction began to mobilize equipment in late 1990, strong objections to the project were expressed that December and in January 1991, when perhaps 1000 citizens demanded at a public meeting that the project be stopped. Although convinced that objections were based on misinformation, to

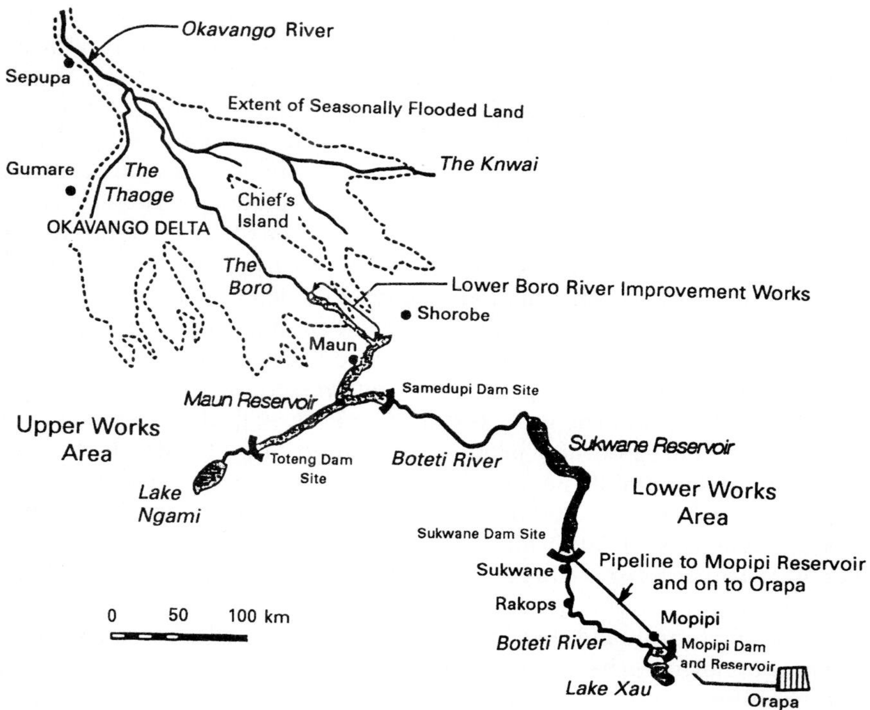

Figure 14.3 Botswana's Okavango River System and the Southern Okavango Integrated Water Development Project (Scudder, 1993c).

its credit, the government agreed to suspend the project while carrying out further review. Selected to carry out that review, the World Conservation Union (IUCN) responded by fielding an interdisciplinary team of 13 scientists who carried out field studies between October 1991 and June 1992, with a final report submitted in October 1992 (Scudder et al., 1993).

In that final report, IUCN recommended that SOIWDP be terminated once and for all. While that recommendation was based on a number of major project flaws detailed in the IUCN reports, a major one was that SOIWDP, contrary to its major goal, would disadvantage rather than benefit a majority of the area's low-income population. The IUCN report also emphasized that local people were correct in their informed opposition to SOIWDP—opposition that the IUCN team found was much stronger than the government had realized. In reaching their conclusions, the IUCN team had divided the project area into nine zones. In six of these, SOIWDP's economic impacts would have been negative on local residents because of reduced flood recession agriculture and grazing below the dams, in the two reservoir basins, and along the channelized Okavango tributary. In two others, impacts would at best be neutral, while they would be positive in only one zone which contained less than 1000 people.

Once again, the way in which a project evolved undermined what were stated to be its primary goals: namely, improving living standards for project-affected people. For example, SOIWDP was not redesigned when the hectarage to be commercially irrigated was dropped from 10,000 to 1200 hectares because of inadequate soils. Nor was it redesigned when the control structure for improving the 5000 ha of flood-recession cultivation was dropped for cost considerations.

Accepting the range of government goals for both the rural and urban sectors of the area, IUCN suggested as an alternative to SOIWDP an approach that, in IUCN's opinion, would better meet those goals at lower financial cost without requiring major engineering works. Falling below the government's poverty-datum line, the majority of low-income people in the Okavango region seek security through a wide range of economic activities. These include flood recession and rainfed agriculture; livestock management; wage labor; a wide range of small-scale commercial activities such as sale of crafts as well as palm wine, thatch, and other plant products; and gathering, hunting, and fishing. The IUCN strategy suggested increasing the productivity of each component in ways that would be sustainable environmentally, economically, and institutionally. A system of decentralized markets was also recommended to better integrate the services and produce needs of the urban sector with rural production and service needs. The IUCN strategy also emphasized greater involvement of local communities in natural-resource management and the evolving tourism industry, including participation with safari companies in joint ventures.

For Maun, the IUCN review suggested a range of industrial enterprises that would make better use of local produce, and conjunctive use of surface and ground water to resolve the town's increasingly serious water crisis. Currently Maun's water supplies are entirely based on boreholes tapping the Shashe wellfield and the adjacent Thamalakane River. Concluding that the Thamalakane could meet Maun's needs 80 percent of the time with a <5 percent reduction in flow until well into the twenty-first century, the IUCN team recommended the construction of an appropriate pipeline and water-treatment plant such as would have been required anyhow if the SOIWDP had been implemented. During years of heavier flooding, recharge of the Shashe wellfield was recommended along with exploratory drilling in two adjacent wellfields with the expectation that they, together with the Shashe, could provide sufficient water during drought periods when the Thamalakane dried up. As for the diamond mines, the IUCN team agreed with their management that groundwater would suffice for the life of the mines with low opportunity costs to surrounding cattle posts and communities.

There are mixed signals as to whether the Government of Botswana will pursue the IUCN team's alternative. Encouraging is the recent contract with an international firm in partnership with local firms to drill 42 exploratory wells in the vicinity of Maun over a 22-month period, including two deep wells searching for freshwater below a saline layer, with the intention of bringing on line 10 new production wells. Two temporary water-treatment plants have also been installed to draw water from the Thamalakane. As for the rural sector, various initiatives are under way which are consistent with the IUCN team's strategy. On the other hand, senior politicians have continued during 1994 and 1995 to make statements in support of implementing the still-suspended SOIWDP.

Increasing local participation

In recent years, all governments, donors, academics, and NGOs have been emphasizing the need for local people to have greater involvement in project planning, implementation, management, and evaluation. That emphasis is welcome and important. As with the implementation of plans that actually make project-affected people beneficiaries, however, the extent to which local people have actually been involved in participatory water-resource development has been disappointing. Aside from lack of political will on the part of governments to actually decentralize decision making, there are several issues that must be dealt with.

One is a difference in definition as to what local participation means. In Botswana, for example, government presents its use of customary meetings (*khotla*), at which local people discuss vital issues with their leaders, as a form of local participation. In connection with the Southern Okavango Integrated Water Development Project, however, such meetings were used more to inform local people of government's intentions, and to solicit their reactions to those intentions, than to actually involve them in the planning process. A second issue relates to government and donor hesitation to link decentralization of decision making with decentralization of financial resources for implementing those decisions; yet the first without the second is meaningless.

A third issue relates to local populations themselves rather than to governments and donors. Ironically, increased emphasis on the need for local participation is occurring at a time when customary participatory institutions are weakening because of increasing incorporation of local communities within wider political economies. Whether by governments, donors, or local people themselves, increasing emphasis is being placed on private ownership of resources as opposed to customary systems based on limited access to communal resources. Within communities, educated individuals are placing increasing emphasis on the household as opposed to extended kin groups and customary institutions of cooperation such as work parties. Moreover, households themselves are becoming increasingly fractionated due to differing interests among members (Dwyer and Bruce, 1988), which is one factor behind a rising proportion of female-headed households.

Such circumstances, also including the increasing differentiation within communities, must be dealt with if local participation is to play the role it should in improving the living standards of project-affected people. A prerequisite will be participatory appraisal (Kumar, 1993) whereby local people and planners work together to determine how best to institutionalize local participation. In some cases, customary institutions such as funeral and other social-welfare groupings and cooperative labor parties may be adjusted to new conditions. More often, however, new institutional forms will probably be required; an effective example is the Grand Council of the Cree which arose during the 1970s in response to the Province of Quebec's James Bay Project. Another institution extending beyond specific communities, which would appear to be especially applicable to project-affected people clustered within a defined area, is a "people's trust" (Reynolds, 1981). Proposed as a "highlands trust" for the Lesotho Highlands Water Project (Maema and Reynolds, in press), the people's trust idea also would be especially applicable to the 11,000 Cree affected by the James Bay Project.

Key features of a people's trust involve placing funds allocated for benefiting local people in a trust under their control. Such funds should be available at the beginning of the project cycle so that local people can be actively involved during feasibility studies. Although not extending beyond such studies, one recent example is where Hydro-Quebec made funds available to the Grand Council of the Cree to enable them to carry out their own analysis during the 1990s of the impacts of the Grande-Baleine component of the James Bay Project.

Should a decision be made to proceed with a project, funds allocated for resettler, host, and other project-affected people's development would be added to the people's trust. Under appropriate safeguards involving overseeing trustees, a minority of whom would be locally selected, project-affected people, rather than the government, would then decide how those funds should be best used.

Although applied successfully to other circumstances in southern Africa by Reynolds and his colleagues, the people's trust idea has yet to be applied to those involved in large-scale water-resource development projects. Modifications would be required, especially where project-affected people are dispersed as individual households over a wide area as along access roads or transmission lines, or where relocation disperses people to many different locales. Under such circumstances it might be best applied, at least initially, to those living within a future reservoir basin. On the other hand, many features of the trust concept look especially attractive if applied to project-affected people. One is its linkage to a hierarchy of periodic markets which link local communities as producers to government administrative centers as providers of services. Government and private-sector institutions (e.g., banks providing credit), for example, would send representatives to the local markets so that local people would not have to make expensive, and often futile, trips to district and other government centers.

Regardless of the type of institutions utilized or developed, effective local participation must involve a much broader cast of characters than just project-affected people. At the national level, commitment must be reflected in the necessary legislative and judicial framework. Where revenue sharing occurs too late in the project cycle, as in China, where revenue for development purposes comes primarily from electricity sales, mulilateral, bilateral, and NGO donors may be asked to provide some funding. The assistance of NGOs in institution building would also be necessary in many cases. So, too, would be the private sector's involvement in various joint ventures with local communities; examples would be linking agricultural outgrowers to processing facilities, or developing tourism or other nonfarm industries. Assistance from universities and research institutions could also be anticipated for helping project-affected people develop appropriate monitoring and evaluation capabilities.

Irrigation

Advantages. A major option for assisting project-affected people to become beneficiaries is to incorporate them within irrigation schemes. One of the best examples of such incorporation was during the earlier phases (Systems H and C) of Sri Lanka's Accelerated Mahaweli Project, when resettlers and downstream hosts were given priority over all other categories of participants.

Well-designed, -implemented, -managed, and -maintained, major irrigation schemes can produce significant increases in both production and living standards in an environmentally sustainable fashion.

There are examples of such successes from all geographic areas. Adjacent host communities (those not within command areas) may also benefit in a major fashion, as Epstein's (1962, 1973) research in Karnataka has shown, while Goldschmidt's (1978) research in California's Central Valley showed that irrigation schemes based on family farms can be expected to generate more multiplier effects in terms of enterprise development and employment generation in market and service centers than large-scale agribusinesses. Probably the most successful irrigation project in Sri Lanka, the medium-scale small-holder irrigation project of Minneriya, has created one of the most dynamic and affluent towns in Sri Lanka with nonfarm permanent employment exceeding farm employment (Wimaladharma, 1981).

Disadvantages

Where project-affected people are excluded. The main disadvantage of irrigation projects for project-affected people is when they not only are unincorporated within a project but also are actually evicted from their land to make way for it. There are innumerable cases in which the land base of unincorporated project-affected people has been reduced to make way for irrigation projects on which government-selected settlers are predominantly immigrants. Although cases in which hosts to a scheme are evicted from their lands without compensation are far fewer, the danger exists that their proportion is increasing as more and more people compete for less and less natural resources. Water-resource development projects are especially vulnerable to political considerations (Waterbury, 1979; Frederiksen, 1992; Ribeiro, 1994), with large-scale projects frequently influenced by the political ideologies of heads of state (Scudder, 1994).

As land and water resources become scarcer, political elites will be increasingly tempted to either access them or use them to achieve political goals. Three examples illustrate the risks to project-affected people. In recent years the governments of both Mauritania and Somalia have passed land-registration acts that favor individual tenure of those with capital over customary systems of land tenure at the village level. While Mauritania's political elite historically showed little interest in irrigation, the construction of the Manantali Dam on the Senegal River, coupled with drought, led to an elite land grab of irrigable land, backed by the military forces of the state, that forced thousands of ethnically distinct riverine citizens into refugee status in neighboring Senegal, where they remain to this day (Horowitz, 1991). The situation was sufficiently serious that retaliatory measures occurred within Senegal against Mauritanian citizens, with a risk over a number of months of a state of warfare occurring between the two nations.

Prior to commencement of the recent civil war in Somalia, the government also passed a land-registration act similar to that of Mauritania. Following the decision to proceed with the World Bank–assisted Baardherre Dam on the Juba River, national elites also participated in a land grab of potentially irrigable downstream land that was not only at the expense of local villagers (Besterman and Roth, 1988) but may have contributed to a widening of the subsequent civil war (oral communication from Peter Little). In the third case, a cabal within the Mahaweli Authority of Sri Lanka, chaired at that time by the Secretary General of the political party in power, plotted to displace Tamil-speaking hosts from now irrigable lands in a later phase (left bank of system B) of the Mahaweli project by a massive land invasion of poor Singala-speaking settlers in order to split in half the Tamil-speaking minority to the north and east (Gunaratna, 1988). Although the land invasion failed, government-stated policies to not mix settlers of different religious affiliations in the same community and, to select settlers from all ethnic groups according to their percentage in the national population, are being ignored. As a result a disproportionate number of Singala-speaking immigrants have been settled in order to convert the Tamil-speaking majority into a minority in their own lands. Hence the government's largest current project, with the potential to play a unifying role within the country's political economy, has been implemented in a way to exacerbate predictably the ongoing civil war between the Tamil separatists and the government (Scudder, 1993b, in preparation).

Notwithstanding major international funding in all three examples, in none did donors (the World Bank was included in two of those cases) try to intervene. Although quite willing to insist on the implementation of guidelines for relocatees and hosts, no systematic effort was made, or has been made, in other downstream locales, to protect the interests of such adversely affected citizens.

Where project-affected people are included. The disadvantages for project-affected people included within irrigation projects are the same as those for all participants when projects are poorly designed and maintained. Environmental problems affecting project life include siltation of reservoirs, waterlogging and salinization, and loss of biodiversity. Public-health problems include increased incidence of malaria, schistosomiasis, and other waterborne diseases as well as dysenteries. Socioeconomic problems arise from insecurity of tenure and poor land preparation and water distribution as well as from poorly designed production and marketing systems that prevent settler households from raising their living standards beyond a subsistence level. A separate problem is the frequency with which focus on a single cash crop lowers the status of women, especially in cases where they had their own preproject fields or crops. Notwithstanding such problems, incorporation within an irrigation project is still a better option for project-affected people than exclusion. Technical and participatory (e.g., water-use associations) solutions exist, and are well known, to all the preceding disadvantages. What is lacking is their incorporation within plans followed by their implementation.

Reservoir fisheries

Critics of river-basin development projects have tended to underestimate the importance for project-affected people of reservoir fisheries. To benefit, however, training and technical assistance are required, as is protection of the entry of project-affected people during the early years of a new fishery. Otherwise, more competitive fishers from existing reservoirs and natural water bodies can be expected to dominate the new fishery, as has been the case with Ghana's Volta reservoir (still the largest artificial reservoir in the world) and Mali's Manantali Reservoir.

The development of the Lake Kariba reservoir fishery illustrates effective ways for incorporating project-affected people. For an initial 5-year period the fishery was closed to immigrants. During that time a training center was built that offered short courses to small-scale commercial fishers using hand-paddled boats and gill nets with subsequent follow-up in the various fishing camps. Improved boats for dealing with conditions on a large water body were designed with local carpenters trained in their manufacture. Credit for their purchase and that of other gear was made available. Lakeside markets were provided with accessible feeder roads for exporting fresh and smoked sun-dried fish.

Although previously they did not have the technology for fishing mainstream Zambezi waters, the response from local people was rapid with over 2000 resettlers, hosts, and other project-affected people catching over 3000 t annually within a 4-year period. Not only were loan repayments over 90 percent, but savings were invested in ways that enabled a majority of fishers to shift to other activities when the reservoir's initial productivity declined. Especially important was investment in cattle (and especially plow oxen), which played a major role in the rapid development of mixed farming, including small-holder cash cropping of cotton and cereals and sale of livestock. Small village stores and other commercial enterprises were also established. Education of children was emphasized, which allowed many children to proceed on to a secondary-school education, and higher earning opportunities, which their families otherwise could not have afforded. The fishery also provided a major mechanism for further incorporating village women within a market economy by providing them with an outlet for the sale of village produce and the manufacture, within fish camps, of beer.

The major policy failing was not anticipating the decline in productivity that characterizes the formation of new water bodies. Anticipatory planning can at least partially compensate for such a decline by expanding the fishery to capture a wider range of species and to use a wider range of techniques. Use of cages, as in Indonesia and China, has the potential of significantly increasing production. Other means include placing barriers across inlets to create small water bodies when reservoir levels rise which then can be stocked with fingerlings and fertilized. Appropriate introductions can also significantly raise productivity. While landings during the height of the Kariba gillnet fishery during 1963–64 probably did not exceed 7000 t annually, stocking of the reservoir's open waters during the late 1960s with a small sardinelike fish (*Limnothrissa miodon*) currently produces sustainable annual yields of approximately 20,000 t.

The *Limnothrissa* fishery, however, is capital-intensive, and all equipment has been owned by immigrant entrepreneurs until recently. Nonetheless, employment of over 1000 local people on the fishing rigs, and in processing and marketing activities, has provided an important source of income in an otherwise suffering economy (Scudder, 1993a). An incentive system based on nightly catches has increased yields as well as income for both owners and employees. And in the past few years purchase of rigs by local councils and nongovernment organizations has begun to bring a larger share of the benefits to at least some local communities.

Improved design and management of existing and future engineering works for making controlled releases

Introduction. Where dams are to be constructed, design and operations options should include controlled flood releases at strategic times for the benefit of downstream users and habitats (Scudder, 1991; Acreman and Pirot, in preparation). In exceptional cases such releases may be negotiated after construction as with South Africa's Pongolapoort Dam and the Manantali Dam on the Senegal. However, since dam design may preclude them, the best approach is an attempt to influence policy during the planning and design stage.

Controlled flooding is a relatively new concept primarily initiated by researchers and planners. Where involved, local residents have been highly supportive. Government responses have been mixed, while donors have largely ignored such an option. Controlled flooding is not a panacea, however. It may involve tradeoffs with hydropower generation, for example, and what floods are released may be ill-timed or insufficient to offset dam-induced downstream costs.

Nonetheless, where feasible, the advantages of controlled floodwater releases can be expected to outweigh disadvantages. The best examples are where previously constructed dams were unable to retain flood magnitude with the result that sluices were built to pass, for the benefit of downstream habitats and users, silt and nutrient-laden waters during the initial flood. Those sluices were then closed to capture in the reservoir what were then relatively silt-free flows; the original Aswan Dam is one example (Scudder, 1994). While some contemporary dams such as the Aswan High Dam and Kariba were not designed to allow controlled flooding, others such as Cahora Bassa farther downstream on the Zambezi could be so operated to the benefit of downstream wetlands, including the critically important delta and riparian communities. According to Gammelsrod (1992), for example, catch per unit effort of shrimp could be increased by 17 percent along the Sofala Bank by altering distribution of runoff. Even where such operations might reduce hydropower generation, developing international grids, as is currently the case in southern Africa, would allow individual dams to serve a wider range of development purposes.

Advantages. The Aswan Dam, as originally built in 1902 and heightened in 1913 and 1933, was designed to pass most of the silt-laden annual flood of the Nile through its sluices. In recent years the importance of such releases have been emphasized for a number of reasons. Williams's (1993) analysis has shown that properly managed releases from the America River's Folsom Dam in California would preclude the need for further dam construction, including that of the Auburn Dam. California is also pioneering controlled releases for the benefit of fisheries and the ecological health of the Sacramento Delta.

Controlled releases for the benefit of project-affected communities is a more recent concept (Scudder, 1980) which has been tested in only a few cases. Theoretically both reservoir-basin communities and communities below a dam should benefit. Where dams are located well upstream, downstream beneficiaries could number in the millions. For them increased flooding by well-timed controlled releases would benefit fisheries and recharge aquifers for the benefit of community wells and riparian forests. Farmers once again could practice some of the flood-recession agriculture that a more regularized regime would have curtailed. They would also benefit from an increase in flood-recessional grazing—a major benefit in semiarid lands when such grazing often becomes available late in the dry season, when it can make the difference between the death and the survival of livestock.

Within reservoir basins, controlled releases would also increase the extent of the drawdown area for both recessional cultivation and grazing. Both have proved to be valuable, and underestimated, benefits for lakeside communities. Drawdown cultivation at Lake Kariba, for example, has provided the most important single source of food for thousands of people during five serious drought years since the early 1980s. Drawdown areas around Ghana's Lake Volta provide an important source of vegetables, while residents in the Kainji Lake basin reap and sell drawdown fodder. In none of those cases, however, were dams designed to expedite controlled downstream flows. If they had been, the benefits to lakeside communities, as well as to downstream residents, would be significantly increased.

Careful research has attempted to quantify the benefits that would accrue from controlled releases from the recently completed Manantali Dam on the Senegal River and dams in Nigeria's Hadejia-Nguru system. On the basis of several years' analysis of the importance of natural flows for over 500,000 riverine inhabitants, the Senegal research has shown that economic benefits accruing from controlled releases from the Manantali Dam would significantly exceed any costs arising from reduced hydropower generation (Horowitz et al., 1990; Salem-Murdock, in preparation).

Similar conclusions have been reached by government officials and academics alike in regard to Nigeria's Hadejia-Nguru river basin when the benefits from controlled releases are compared with use of the same water on government-managed irrigation schemes (Hollis et al., 1993; Acreman et al., in preparation). The discrepancy there was so great that Nigerian officials at a 1993 meeting agreed unanimously that "flooding in the wetlands made possible by artificial releases from the dams in the wet season should be maintained to make possible the production of rice, dry season agriculture, fuelwood, timber, fish, wildlife, as well as biodiversity and groundwater recharge" (HNWCP/NIPSS, 1993). All sustained by natural flooding, the productivity of those resources has been reduced by dam-induced impoverishment of the floodplains habitat and, by extension, of the dependent human population.

A third example comes from South Africa's Pongola River Basin in the province of Natal-KwaZulu. When an intended irrigation project there failed to materialize below the Pongolapoort Dam as planned during the previous government, fishery biologists recommended experimentation with controlled releases for the benefit of downstream fisheries and Thonga/Tembe fisherfolk. Subsequently experimentation was expanded for the benefit of all aspects of the Thonga/Tembe economy, including flood-recession agriculture and pasture. Initially the affected communities did not participate in making decisions about the timing and duration of releases. More recently, however, they have been brought into the decision-making process through the formation of Water Committees and a Water Committee's Executive (Bruwer et al., in preparation).

Disadvantages. While controlled flooding can improve the productivity of wetlands and riverine communities, in most cases the economies of those concerned would be better served by the natural flood regime. As Hughes (1988) has shown for Kenya's Tana River and generalized for other African floodplains, even extreme floods play a vital role in the maintenance of floodplain productivity.

There are also policy, technical, and environmental constraints to the implementation of controlled releases. Current droughts in western, central, and southern Africa's semiarid lands have adversely affected river flows, hence making it more difficult for working out the type of suboptimization strategies that controlled flooding would involve. Since the early 1980s, for example, Zambezi flows have been only approximately 50 percent of what they were during the previous 20 years, while reduced intakes into the Ivory's Coast Kossou Reservoir and Ghana's Volta Reservoir, on occasion, have had adverse economic effects. Even without drought conditions, the magnitude, timing, and duration of controlled releases may not have the desired effect on productivity, as illustrated by releases from Zambia's Iteshiteshi Dam for the benefit of downstream communities using the wetlands resources of the Kafue Flats (Acreman et al., in preparation).

Technical constraints relate to capacities to know when to release flows in what volume for what duration. The necessary knowledge requires operation and maintenance of a sophisticated hydromet network; its utilization requires well-qualified personnel. Policy constraints are of several sorts. One is the lack of political will to make the necessary releases. Notwithstanding the demonstrated economic and environmental benefits, the trinational Senegal River Basin Authority has been unwilling to make controlled releases beneficial to downstream riparian communities since the completion of the Manantali Dam even though operational turbines have yet to be installed. What releases have been made were so poorly timed that they led to decreases rather than increases in local productivity (Horowitz, 1991).

A second policy constraint is tied in with the unfavorable rural-urban terms of trade that continue to characterize so many countries. While realizing longer-term economic returns that would be in the national interest, controlled releases benefit mostly low-income rural households at the short-term expense of either the urban sector or farmers on commercial irrigation schemes. For example, the hydropower that would have to be forgone under some circumstances would be more for the benefit of urban industrial and residential users than of rural communities. Donors and private investors also might favor maximizing power generation in order to facilitate debt repayment as there is no easy way to siphon off such funds from the rising disposal incomes of hundreds of thousands of low-income rural households.

Conclusions and Lessons Learned

An obvious and major conclusion is that the adverse social and environmental impacts of large-scale water resource development projects have been seriously underestimated. Not only is a much larger number of people involved than acknowledged by researchers, government planners, and donors, but there has also been a failure to acknowledge the range and magnitude of impacts on different categories of people, effects on relocatees, for example, differ from those on hosts as well as on other project-affected people living below dams. This situation is due partially to a paucity of research, and especially of research designed to assess long-term impacts. But lack of knowledge may be as much a symptom as a cause of the problem. Granted the extent of impoverishment that often results along with other costs, the persistence of such projects remains to be explained, as does the inadequacy of environmental- and social-impact assessments and the failure of the World Bank, as the major multilateral donor, to complete ex post audits of costs and benefits.

This chapter has focused on large-scale projects. For that reason, it has not emphasized the positive and negative social impacts of thousands of small-scale water-resource development projects. Little has been said, for example, about the ability of environmentally sustainable irrigation, including individually and communally owned tube wells, to increase productivity and raise living standards. Rural and urban projects that have improved both the quality and the quantity of community water supplies, along with successful efforts to institutionalize community water-user associations to operate and maintain such projects, have also been ignored. Ironically, large-scale river-basin development projects have tended to reduce both water quality and quantity for project-affected people. Quantity has been reduced by increased evapotranspiration and seepage from reservoirs. Where no longer resident on the Zambezi or the Volta Rivers, Kariba and Volta relocatees who were shifted to inland locales dependent on boreholes and shallow wells have had to cope with reduced water availability, coupled with an increase in waterborne disease. Because of reduced flooding, millions of downstream residents throughout the tropics and subtropics have been adversely affected by lowered aquifers.

As economic solutions to international, national, and regional problems, some large-scale projects look good as planned. The Lesotho Highlands Water Project, for example, if implemented as intended, would provide the high-quality water necessary for the development of south Africa's major urban industrial area and the electricity and revenue for Lesotho without major environmental and social impacts. Designed to be one of the world's higher dams within China's Pearl River System, the Longtan Dam could play a major role regulating flows, including controlled releases, to a cascade of downstream dams, some of which already exist. The problem, however, is that the potential of such large-scale projects is apt to be undermined by a number of factors influencing their implementation.

Lack of knowledge of adverse long-term effects is one such factor that has already been emphasized. Another argument against large-scale projects, and one that has been insufficiently documented, is that they are more apt to be subverted during implementation because of political and broader ideological considerations. Sri Lanka's Accelerated Mahaweli Project is a good example. As planned, this multi-billion-dollar project had the potential not only to significantly raise the living standards of settler households, but also to catalyze a process of regional development through increased enterprise development and nonfarm employment. It also had the potential to play a constructive role in finding solutions to ongoing communal strife between Sri Lanka's Singala-speaking Buddhist majority and Tamil-speaking Hindu minority. That potential has not been realized. Certainly, wrong assumptions, lack of knowledge, and technical and economic problems were involved. But there were also major political and broader ideological constraints.

A devout Buddhist, the implementing agency's powerful chairman refused for a number of years to allow the project to improve its economic potential by incorporating livestock and fisheries within the household and project production systems. In regard to political factors, not only were Tamil-speakers not incorporated, as stipulated by project policy, according to their proportion in the national population, but the project, as implemented, became a mechanism for significantly reducing their numbers in areas where Tamil-speaking villagers had been in the majority. Contrary to another stated policy, poor Singala-speakers were even settled in Tamil-speaking communities within known areas of communal strife. Hence a project which might have helped solve communal problems exacerbated them since, as predicted (Scudder, in preparation), mixing of settlers led to attacks and counterattacks on them by the warring factions.

Large-scale river-basin development projects are sponsored by a powerful coalition including heads of state; multinational corporations of consulting engineers, contractors, and suppliers; and multilateral and bilateral donors. Whether as personal monuments or as means for implementing visions of the future, the names of heads of state are intricately associated with such projects. Examples have already been given. As Ribeiro (1994, p. 50) explains, multinational corporations to remain competitive must move smoothly from one project to another; indeed, "they stimulate the market for them by indicating and proposing new works." Their involvement is facilitated by bilateral donors linking favorable credit arrangements (as through export-import banks) to contracts for nationally based firms and by the World Bank's insistence on international competitive bidding.

Heads of state such as Franklin Roosevelt and Nkrumah at least had visions, based on political ideologies and capital-intensive technology, of "their" projects stimulating regional development. It is this political component that Waterbury (1979) has labeled "hydropolitics." According to the World Bank's Frederiksen (1992, p. 4), "the conditions encountered in a country's water sector reflect the political demands and the wisdom and leadership in these matters more than any other factor." In regard to specific cases, Waterbury's assessment of the Aswan High Dam concludes that "the history of this project is testimony to the primacy of political considerations determining virtually all technical choices with the predicted result that a host of unanticipated technical and ecological crises have emerged that now entail more political decisions" (ibid., p. 4).

Also using the term *hydropolitics* is the title of his analysis of the binational (Argentina-Paraguay) Yacyreta High Dam, Ribeiro notes statements by Argentine specialists of better alternatives to Yacyreta in the form of smaller dams, better sites for high dams, and use of natural gas. He emphasizes several other factors supporting the decision to proceed with Yacyreta. One is project specific, involving competition between Argentina and Brazil (with its binational Itaipu High Dam) for "regional hegemony" (ibid., p. 45). The other can be generalized to other national political economies. Rejecting the word *development,* Ribeiro sees large-scale projects, such as Yacyreta, as "a form of production linked to economic expansion" into "outpost" areas (ibid., p. 163).

Ribeiro's interpretation would also appear to be applicable to both the James Bay Project and the Southern Okavango Integrated Water Development Project (SOIWDP). In addition to economic expansion, the former project gives the Province of Quebec increased control over indigenous lands of ambiguous legal status as far as that province is concerned. SOIWDP would not only channel more water to the diamond mines (and hence to the advantage of both the government and the DeBeers/Anglo-American multinational corporation) but also provide increased access for Botswanan elite from the more densely populated eastern regions to Okavangan resources in the form of water for irrigation and grazing and water for cattle. Rhetoric aside, intended beneficiaries of neither project are supposed to be local people. Impacts on such people are ignored or played down— to the fullest extent possible—throughout the project cycle. Appropriate assessment of those impacts—because they are widespread—is expensive, which is another reason why they are underassessed.

If more local people are to become beneficiaries, not only must World Bank–type guidelines for relocatees be extended to all project-affected people, but the horizon of environment- and social-impact assessments must be expanded. The same guidelines, however, must also be applied to other options, including, for example, increased reliance on coal-generated electricity. Should that be done, in many cases other alternatives to major project works would be seen as preferable. And where major works continue to be the "least-cost alternative," such major changes in design and operation as controlled flooding would be called for.

Acknowledgments

This chapter draws heavily on several previous publications of the author, including, especially "Development-induced impoverishment, resistance and river basin development," to be published in *Resisting Impoverishment—Tackling the Consequences of Development-Induced Displacement,* Chris McDowell, ed., Berghahn Books, Oxford. The author is also indebted to Michael M. Cernea, Elizabeth Colson, Scott G. Ferguson, Robert Goodland, Michael M. Horowitz, Patrick McCully, and Martin ter Woort for their comments on an earlier version.

References

Acreman, M., and J.-Y. Pirot, ed. (in preparation). *Hydrological Management & Wetland Conservation in Africa.* Gland, Switzerland: International Union for the Conservation of Nature and Natural Resources (IUCN/World Conservation Union).

Acreman, M., J.-Y. Pirot, and T. Scudder (in preparation). "The role of artificial flooding in the integrated development of river basins in Africa." In *Hydrological Management & Wetland Conservation in Africa,* M. Acreman and J.-Y. Pirot, eds. Gland, Switzerland: International Union for the Conservation of Nature and Natural Resources (IUCN/World Conservation Union).

Adams, W. M. (1992). *Wasting the Rain: Rivers, People and Planning in Africa.* London: Earthscan Publications.

Adams, W. M. (1993). "Development's Deaf Ear: Downstream Users and Water Releases from the Bakolori Dam, Nigeria." *World Development* **21**(9): 1405–1416.

Bartolome, M., and A. Barabas (1990). *La presa Cerro de Oro y el Ingeniero Gran Dios: Relocalizacion y Etnocidio Chinanteco en Mexico.* Col. Presencias nos. 19, 20. Mexico City: Consejo National para la Cutura y las Artes, INI.

Besterman, C., and M. Roth (1988). *Report on Land Tenure in the Middle Jubba Valley: Issues and Policy Recommendations.* JESS Report no. 34. Burlington, Vt.: Associates in Rural Development.

Biswas, A. K. (1987). "Monitoring and evaluation of irrigated agriculture." *Food Policy* (Feb.): 47–61.

Bourassa, R. (1985). *Power from the North.* Ottawa: Prentice-Hall, Canada.

Bruwer, C., C. Poultney, and Z. Nyathi (in preparation). "Community based hydrological management of the Pongola Floodplain." In *Hydrological Management & Wetland Conservation in Africa,* M. Acreman and J.-Y. Pirot, ed. Gland, Switzerland: International Union for the Conservation of Nature and Natural Resources (IUCN/World Conservation Union).

Cernea, M. M. (1988). *Involuntary Resettlement in Development Projects: Policy Guidelines in the World Bank-Financed Projects.* World Bank Technical Paper no. 180. Washington, D.C.: World Bank.

Cernea, M. M., and S. E. Guggenheim, ed. (1993). *Anthropological Approaches to Resettlement: Policy, Practice, and Theory.* Boulder, Colo.: Westview Press.

Chandler, W. U. (1984). *The Myth of TVA: Conservation and Development in the Tennessee Valley, 1933–1983.* Cambridge, Mass.: Ballinger.

Colson, E. F. (1971). *The Social Consequences of Resettlement: The Impact of Kariba Resettlement upon the Gwembe Tonga.* Manchester University Press for Institute for African Studies, University of Zambia.

Dickson, I. J., R. Goodland, P. H. Leblond, F. L. Leistriz, C. Limoges, and T. Scudder (1994). *Hydro-Quebec's Grande-Baleine Environmental Impact Study: An Assessment Report.* Report prepared for Hydro-Quebec.

Drijver, C. A., and M. Marchand (1985). *Taming the Floods: Environmental Aspects of Floodplain Development in Africa.* Leiden: Centre for Environmental Studies, State University of Leiden.

Dwyer, D., and J. Bruce (1988). *A Home Divided: Women and Income in the Third World.* Stanford, Calif.: Stanford University Press.

Epstein, T. S. (1962). *Economic Development and Social Change in South India.* Manchester, U. K.: Manchester University Press.

Epstein, T. S. (1973). *South India: Yesterday, Today and Tomorrow; Mysore Villages Revisited.* New York: Holmes & Meier.

Feit, H. A. (1982). "The future of hunters within nation-states: anthropology and the James Bay Cree." In *Politics and History in Band Societies,* E. Leacock and R. Lee, ed. Cambridge University Press.

Frederiksen, H. D. (1992). *Water Resources Institutions: Some Principles and Practices.* World Bank Technical Paper no. 191. Washington, D. C.: The World Bank.

Freeman, S. D. (1985). "TVA and the national power policy," *Tennessee Law Review* **49** (4): 679–685.

Gammelsrod, T. (1992). "Variation in shrimp abundance on the Sofala Bank, Mozambique, and its relation to the Zambezi River runoff," *Estuarine, Coastal and Shelf Science* **35:** 91–103.

Gaventa, J. (1982). "Book review: TVA and the dispossessed: the resettlement of population in the Norris Dam area," *Tennessee Law Review* **49**(4): 979–983.

Gleick, P. H., ed. (1993). *Water in Crisis: A Guide to the World's Fresh Water Resources.* New York and Oxford: Oxford University Press.

Goldschmidt, W. R. (1978). *As You Sow: Three Studies in the Social Consequences of Agribusiness.* Montclair, N. J.: Allanheld, Osmun.

Goldsmith, E., and N. Hildyard, eds. (1984). *The Social and Environmental Effects of Large Dams:* vol. 1, *Overview.* Camelford, Cornwall (U.K.): Wadebridge Ecological Centre.

Goldsmith, E., and N. Hildyard, eds. (1986). *The Social and Environmental Effects of Large Dams:* vol. 2, *Case Studies.* Camelford, Cornwall: Wadebridge Ecological Centre.

Goldsmith, E., N. Hildyard, and D. Trussel, eds. (1992). *The Social and Environmental Effects of Large Dams:* vol. 3, *A Review of the Literature.* Camelford, Cornwall: Wadebridge Ecological Centre.

Grant, N. L. (1990). *TVA and Black Americans: Planning for the Status Quo.* Philadelphia: Temple University Press.

Gunaratna, M. H. (1988). *For a Sovereign State.* Ratmalana, Sri Lanka: Sarvodaya Book Publishing Services.

HNWCP/NIPSS (Hadejia-Nguru Wetlands Conservation Project and the National Institute for Policy and Strategic Studies) (1993). *Proceedings of the Workshop on the Management of the Water Resources of the Komadugu-Yobe Basin* (Kuru, April 1–2, 1993). Kuru, Nigeria: National Institute Press.

Hollis, G. E., W. M. Adams, and M. Aminu-Kano (1993). *The Hadejia-Nguru Wetlands: Environment, Economy and Sustainable Development of a Sahelian Floodplain Wetland.* Gland, Switzerland: IUCN.

Horowitz, M. M. (1991). "Victims upstream and down." *Journal of Refugee Studies* **2:** 164–181.

Horowitz, M. M., M. Salem-Murdock, et al. (1990). *The Senegal River Basin Monitoring Activity Synthesis Report.* Binghamton, N. Y.: Institute for Development Anthropology.

Hughes, F. M. R. (1988). "The ecology of African floodplain forests in semi-arid and arid zones: a review." *Journal of Biogeography* **15:**127–140.

Hunter, J. M., L. Rey, K. Y., Chu, E. O. Adekolu-John, and K. E. Mott (1993). *Parasitic Diseases in Water Resources Development: The Need for Intersectoral Negotiation.* Geneva: World Health Organization.

Hydro-Quebec (1993). *Grande-Baleine Complex Feasibility Study.* Quebec: Hydro-Quebec.

IUCN (The World Conservation Union) (ongoing). *IUCN—The World Conservation Union Newsletter: Wetlands Programme.* Gland, Switzerland: IUCN.

Kumar, K. ed. (1993). *Rapid Appraisal Methods.* Regional and Sectoral Studies, the World Bank. Washington, D.C.: The World Bank.

Lahmeyer-MacDonald Consortium (1986). *Lesotho Highlands Water Project. Feasibility Study Report.* Maseru.

Maema, M., and N. Reynolds (in press). "Lesotho Highlands Water Project-induced displacement: context, impacts, rehabilitation strategies, implementation experience, and future options. In *Resisting Impoverishment: Tackling the Consequences of Development-Induced Displacement,* C. McDowell, ed. Oxford: Berghahn Books.

McDonald, M. J., and J. Muldowny (1982). *TVA and the Disposed: The Resettlement of Population in the Norris Dam Area.* Knoxville: University of Tennessee Press.

Morse, B., and T. Berger (1992). *Sardar Sarovar: Report of the Independent Review.* Ottawa: Resource Futures International.

Quebec Government (1991 ed.). *The James Bay and Northern Quebec Agreement and Complementary Agreements.* Quebec: Les Publications of Quebec.

Reynolds, N. (1981). *The Design of Rural Development: Proposals for the Evolution of a Social Contract Suited to Conditions in Southern Africa.* Parts I and II, Working Papers 40 and 41. Capetown: Southern Africa Labour and Development Research Unit.

Ribeiro, G. L. (1994). *Transnational Capitalism and Hydropolitics in Argentina: The Yacyreta High Dam.* Gainesville: University Press of Florida.

Roder, W. (1994). *Human Adjustment to Kainji Reservoir in Nigeria: An Assessment of the Economic and Environmental Consequences of a Major Man-made Lake in Africa.* Lanham, Md.: University Press of America.

Salem-Murdock, M. (in preparation). "The sociology of wetland conservation: the Senegal River Valley." In *Hydrological Management & Wetland Conservation in Africa,* M. Acreman and J.-Y. Pirot, eds. Gland, Switzerland: International Union for the Conservation of Nature and Natural Resources (IUCN/World Conservation Union).

Salisbury, R. F. (1986). *A Homeland for the Cree: Regional Development in James Bay 1971–1981.* Quebec: McGill-Queen's University Press.

SCOPE (Scientific Committee on Problems of the Environment) (1972). *Man-made Lakes as Modified Ecosystems.* Scope Report 2. International Council of Scientific Unions (ICSU). Paris: ICSU.

Scudder, T. (1962). *The Ecology of the Gwembe Tonga.* Manchester University Press for Rhodes-Livingstone Institute.

Scudder, T. (1980). "River basin development and local initiatives in African savanna environments." In *Human Ecology in Savanna Environments,* D. R. Harris, ed. London: Academic Press.

Scudder, T. (1991). "The need and justification for maintaining transboundary flood regimes: the Africa case," *Special Issue of Natural Resources Journal on The International Law of the Hydrologic Cycle* **31**(1): 75–107.

Scudder, T. (1993a). "Development-induced relocation and refugee studies: 37 years of change and continuity among Zambia's Gwembe Tonga," *Journal of Refugee Studies.* **6**(2): 123–152.

Scudder, T. (1993b). "Monitoring a large-scale resettlement program with repeated household interviews." In *Rapid Appraisal Methods,* K. Kumar, ed. Regional and Sectoral Studies. Washington, D.C.: The World Bank.

Scudder, T. (1993c). "Development Strategies for Botswana's Okavango Delta." In *Hydraulic Engineering '93,* vol. 1, H. W. Shen, S. T. Su, and F. Wen, eds., p. 419.

Scudder, T. (1994). "Recent experiences with river basin development in the tropics and subtropics," *Natural Resources Forum* **18**(2): 101–113.

Scudder, T. (1996). "Development-induced impoverishment, resistance, and river basin development." In *Resisting Impoverishment—Tackling the Consequences of Development-Induced Displacement,* C. McDowell, ed. Oxford: Berghahn Books.

Scudder, T. (in preparation). "Constraints to the development of settler incomes and production-oriented participatory organizations in large-scale government-sponsored projects: the Mahaweli case." In *Proceedings of the Symposium on Irrigation and Society in the Context of Mahaweli, Sri Lanka,* S. Hettige, ed. Sri Lanka: Vishva Lekha Publishers.

Scudder, T., and E. Colson (1982). "From welfare to development: a conceptual framework for the analysis of dislocated people." In *Involuntary Migration and Resettlement: The Problems and Responses of Dislocated People,* A. Hansen and A. Oliver-Smith, eds. Boulder, Colo.: Westview Press.

Scudder, T., R. E. Manley, R. W. Coley, R. K. Davis, J. Green, G. W. Howard, S. W. Lawry, D. Martz, Peter P. Rogers, A. R. D. Taylor, S. D. Turner, G. F. White, and E. P. Wright (1993). *IUCN Review of the Southern Okavango Integrated Water Development Project.* The IUCN Wetlands Programme. Gland, Switzerland: Samara Publishing Company for IUCN (World Conservation Union).

Selznick, P. (1949). *TVA and the Grass Roots: A Study in the Sociology of Formal Organization.* Berkeley: University of California Press.

Stanley, D. J., and A. G. Warne (1993). "Nile Delta: recent geological evolution and human impact," *Science.* **260:** 628–634.

Tshabalala, M. (1994). *Phase 1B Socioeconomic Census Report: Mohale, 1993,* vol. 1 (main report). Maseru, Lesotho: Environment Department, Lesotho Highlands Development Authority.

United Nations Development Programme (UNDP) and Food and Agriculture Organization (FAO) (1977). *Investigation of the Okavango Delta as a Primary Water Source: Botswana, Project Findings and Recommendations.* Rome: FAO.

Waterbury, J. (1979). *Hydropolitics of the Nile Valley.* Syracuse, N.Y.: Syracuse University Press.

Welcomme, R. ed. (1979). *Fisheries management in Large Rivers.* FAO Technical Report no. 194. Rome, United Nations Food and Agriculture Organization.

Wheeler, W. B., and M. J. McDonald (1986). *TVA and the Tellico Dam, 1936–1979: A Bureaucratic Crisis in Post-industrial America.* Knoxville: University of Tennessee Press.

White, G. F. (1985). *Environment* **27**(3) (editorial).

White, G. F. (1988). "The environmental effects of the High Dam at Aswan," *Environment* **30**(7): 5–11, 34–40.

Williams, P. B. (1993). "Assessing the true value of flood control reservoirs: the experience of Folsom Dam in the February 1986 flood." Paper presented at the 1993 American Society of Civil Engineers National Conference on Hydraulic Engineering

and International Symposium on Engineering Hydrology, San Francisco, July 25–30.
Wimaladharma, K. (1981). Unpublished manuscript and field notes on Sri Lanka's Minneriya project.
World Bank (1985). *The Experience of the World Bank with Government-Sponsored Land Settlement*. Report no. 5625. Washington, D.C.: Operations Evaluation Department, the World Bank.
World Bank (1994). *Resettlement and Development. The Bankwide Review of Projects Involving Resettlement 1986–1993*. Washington, D.C.: The World Bank.

Chapter

15

Resettlement

Thayer Scudder
California Institute of Technology
Pasadena, California
and Institute for Development Anthropology
Binghamton, New York

Introduction

Writers on development tend to refer to a range of resettlement types which include spontaneous resettlement, facilitated spontaneous resettlement, sponsored voluntary resettlement, and involuntary resettlement. This chapter is concerned only with involuntary resettlement in connection with water-resource development. Emphasis is on large-scale river-basin development projects. In good part because of the unsatisfactory implementation record and the millions of people annually impoverished, in recent years resettlement from such projects has become a major topic of interest to the extent that resettlement studies have become a separate subfield.

In 1994 the World Bank published an annotated reference bibliography on development-induced involuntary resettlement that lists over 800 sources (Guggenheim, 1994). Those sources, and others, are drawn from many social-science and public-health disciplines. Especially relevant are writings in economics (Nelson, 1973), political science (Chambers, 1969), and anthropology (Scudder and Colson, 1982; Scudder, 1991) dealing with resettlement as a dynamic process which can best be understood as a succession of stages through which those relocated must move if resettlement is to be successfully implemented. Still-earlier writings in psychology (Fried, 1963) and sociology (Gans, 1962, 1968; Hartman, 1974) dealing with involuntary resettlement caused by urban renewal have provided valuable insights as to how a majority respond during the early stages of the resettlement process.

Since 1980, the World Bank has pioneered the application of what was largely academic research to Bank-financed projects involving involuntary resettlement (World Bank, 1980, 1994a; Cernea, 1988), of which water-resource development projects involve the most people. Because the World Bank is also the largest single donor funding such projects, and completed during 1994 what is the most detailed analysis of experiences with resettlement, this chapter focuses on Bank-financed projects. While Bank officials most responsible for resettlement believe that plans can be implemented that will protect—indeed, benefit—both resettlers and host populations, unfortunately, the record to date indicates that it is virtually impossible to restore the living standards of the majority. The potential is there to make such people project beneficiaries because of their proven capacity to adapt to new conditions and because governments could use such projects to create new opportunities for project-affected people. Nonetheless, a large number of constraints counter that potential. Two are of particular importance. The first, most important, is the lack of will on the part of governments to implement resettlement with development. Largely because resettlement is merely a project by-product involving predominantly poor people, often ethnic minorities, who have little political clout, engineers, decision makers, and politicians show little concern for their welfare. A second major constraint is the complexity of the resettlement process, including the difficulty of sustaining initial successes in improving living standards.

Although there may be individual success stories, looking to the future there is little cause to believe that the overall record will improve. While World Bank–financed projects are the most closely supervised today, current attempts within the Bank to establish guidelines dealing with environmental and cultural issues, including involuntary resettlement, while less prominent, are worrisome. Also worrisome is the increasing bilateral and multilateral donor emphasis on getting the private sector to build and operate water-resource development projects.

This chapter starts off with an appropriate goal for involuntary resettlement. Analysis of scale and impact of resettlement is followed by discussion of donor involvement and why the resettlement record has been so unsatisfactory. While optimism is hardly warranted, a wide range of development strategies is presented next; these strategies could improve the record if backed by appropriate policies and political will on the part of governments and donors.

The Goal of Involuntary Resettlement

The goal of involuntary resettlement must be for those removed, and the host populations among whom they are resettled, to become project beneficiaries. This means that the income and living standards of the large majority *must* improve to the extent that such improvement is obvious both to themselves and to external evaluators. Such a goal is justifiable in terms of both human rights and economics. Inadequate resettlement creates dependence and impoverishment, which lowers the stream of project benefits through its failure, on one hand, to incorporate whatever contribution those relocated might make, and, on the other hand, by creating an increased dependence on safety nets. It also jeopardizes the life of the project by increasing siltation and decreasing water quality since poorly relocated and impoverished people within a reservoir basin have little alternative to overutilizing their environment.

Analysis of the global record with resettlement also suggests that the goal of raising the majority's living standards is possible but very difficult. The main resource are those relocated themselves. Although far more research is needed on the later stages of the resettlement process, what little evidence exists suggests that within a few years of removal, the majority are more receptive to development than their neighbors who were not displaced. That is partly because resettlement is apt to remove a range of cultural constraints to future entrepreneurial activities and initiative, including land tenural, political, and economic constraints. That hypothesis, however, should never be used as a reason for resettlement for two major reasons. One is the multidimensional stress associated with resettlement's initial years. The other is the difficulty, in terms of the availability of land and other natural resources, and of employment opportunities, of sustaining a process of development once commenced.

The Scale and Impact of Involuntary Resettlement

Scale

The number of people annually displaced by development-induced involuntary resettlement actually exceeds the much more publicized number of refugees fleeing across international borders because of political repression and warfare (Cernea, 1996). Dam construction has been the largest single cause, although its proportional importance currently is dropping as increasing numbers of people are moved to make way for construction and renewal of urban and transportation infrastructure. According to Cernea, for example, approximately 4 million people are displaced by the annual construction of "some 300 large dams (above 15 m high) typically begun each year" (Cernea, 1994b, p. 47). Most such removal has occurred since the end of World War II.

Construction of dams and irrigation projects in China and India are responsible for the largest number on a country-by-country basis. In China over 10 million people were relocated in connection with water-development projects between 1960 and 1990 (*Beijing Review,* 1992), while Fernandes et al. (1989) have estimated that a still-larger number has been relocated in India over a 40-year period. Table 15.1 lists 28 completed or under-construction projects that have been responsible for the largest number of relocated people since the mid-1950s. As projects under construction indicate, large future relocations can be anticipated. In response to the world's increasing demand for energy, for both national development purposes and poverty alleviation, a

TABLE 15.1 Reservoir Resettlement (1)

Project	Country	Number of resettlers
Three Gorges	China	1,250,000
Danjiangkou	China	383,000
Sanmenxia	China	319,000
Xinjiang	China	306,000
Dongpinghu	India	278,000
Upper Krishna II	India	220,000
Xiaolangdi	China	180,000
MCIP III Irrigation	India	168,000
Andhra Pradesh Irrigation II	India	150,000
Gujarat Med. Irrigation II	India	140,000
Sardar Sarovar	India	127,000
Tehri	India	105,000
Aswan High Dam	Egypt	100,000
Subarnarekha Group	India	100,000
Kossou	Ivory Coast	85,000
Akosombo	Ghana	84,000
Longtan	China	73,000
Shuikou I and II	China	67,000
Mahaweli I–IV	Sri Lanka	60,000
Saguling	Indonesia	60,000
Kariba	Zambia and Zimbabwe	57,000
Cirata	Indonesia	56,000
Sobradinho	Brazil	55,000
Paulo Af. IV	Brazil	52,000
Yacyreta	Argentina and Paraguay	50,000
Itaparica	Brazil	50,000
Kainji	Nigeria	44,000
Yantan	China	40,000

SOURCE: Adapted from Goodland (1994, p. 425) and World Bank (1994a, p. 2/8) except for Kariba and Mahaweli figures, which are from the author's records.

conservative forecast is for hydropower production to double between 1990 and 2020 (Goldemberg et al., 1987). Irrigation can also be expected to increase to meet future demands for food and with it, the need for further resettlement.

Only a small minority of water-resource development projects have involved, and benefited from, donor supervision. The larger ones have also benefited from the advice of internationally recruited engineering panels of experts. Only rarely, however, have such panels been complemented by similar panels dealing with environmental and resettlement issues. Although the largest single donor, the World Bank, estimates that "Bank-funded projects account, conservatively, for some 3 percent of the resettlement caused by dam construction worldwide" (World Bank, 1994a, p. i), people displaced by dams and irrigation projects have made up 63 percent of those displaced by the Bank's total portfolio of projects that involve involuntary resettlement. Lessons learned have come disproportionately from such projects. One of those lessons concerns the difficulty of realizing successful outcomes.

Impacts

According to the World Bank's Senior Adviser for Social Policy and Sociology, "forced population displacement caused by dam construction is the single most serious counter-developmental social consequence of water resource development" (Cernea, 1990, p. 1). The Bank's Environment Adviser concurs: "Involuntary resettlement is arguably the most serious issue of hydro-projects nowadays." He goes on to add, "it may not be improving, and is numerically vast" (Cernea, 1994, p. 149). Add the adverse effects that most resettlement to date has also had on incorporating host populations and on habitats surrounding resettlement sites, and the impact magnitude increases still further.

For the majority of resettlers, multidimensional stress is associated with the earlier stages of the resettlement process. Although synergistically interrelated, for purposes of analysis, its components can be broken down into physiological, psychological, and sociocultural stress. Physiological stress is caused by increased morbidity and mortality rates (Scudder, 1975; PEEM, 1985; Ault, 1989). It has also been shown to have an effect on fertility (Clark et al., 1995).

Aside from the frequent increase of schistosomiasis and malaria that is associated with reservoir formation and irrigation, one of the ironies of resettlement has been the inability of governments to provide adequate water supplies. As at Kariba, waterborne diseases increase where a riverine population with access to an ample supply of good-quality water is moved inland to sites where people are dependent on poorly maintained and inadequate wells and boreholes on poor aquifers. For purposes of administration and because of insufficient replacement land, and also to provide better social infrastructure, governments frequently aggregate dispersed communities into smaller numbers of settlements where increased population densities, inadequate water supplies, and poor sanitation increase the incidence of such killers as measles and a wide range of health-threatening parasites. Although primarily an urban-based disease, during the 1990s cholera also has become a serious problem in the two most densely settled, and degraded, Kariba north-bank resettlement areas.

Drawing on research among low-income people displaced by urban redevelopment, Fried (1963) summed up psychological stress as the "grieving for a lost home" syndrome, which is even more appropriate for dam-removed rural populations with strong ties to the land that have had continuity over centuries. Anxiety for the future, especially among the elderly, women, and those whose status and well-being are dependent on immovable resources, is another component. Nontransferability of knowledge and cessation, at least temporarily, of a wide range of behavioral patterns, statuses, and institutions cause sociocultural stress. Invariably some economic activities, and knowledge associated with those activities, have less relevance after removal. That is especially the case where people are moved inland or away from a riverine habitat as in the case of the Aswan High Dam. It is also the case where farmers practicing rain-fed agriculture are subsequently incorporated within irrigation projects. Furthermore, some members of all communities are involved in activities which the larger society is apt to see as illegal or otherwise inappropriate. Resettlement tends to increase not only population densities but also government authority and control. That can be very stressful for ethnic minorities and other communities that formerly had little outside contact.

Land-based and other rituals may never be reestablished, especially where ridiculed or seen as culturally threatening by host populations. Resettlement is especially stressful for people in leadership positions since their impotence in defending the interests of regions, communities, and households is revealed when relocation occurs. Political leaders, in particular, may be caught in a "no win" situation. If they support removal, they lose their constituency since people do not wish to be forced to move. If they oppose removal, their effectiveness is adversely affected when displacement occurs. Either way, leadership has been undermined during the initial years following removal.

It is one of the paradoxes of resettlement, however, that the initially stressful cessation or inapplicability of important statuses and institutions may subsequently foster a more dynamic process of economic development and community formation. Less inhibited by previously restricting customs (relating, for example, to land tenural patterns and community rituals) and by entrenched leaders, aspiring entrepreneurs and leaders are apt to find themselves in a more flexible environment.

While incorporation within a wider political economy, with greater contact with government departments, also has the potential for creating new opportunities for displaced communities, the general record to date is for resettlement to create impoverishment. In recent writings, Cernea (1990, 1996) has formulated an eight-component model of impoverishment risks associated with resettlement. Four components with serious implications for improving living standards are landlessness, loss of access to community or kin-based common property, joblessness, and marginalization (in which households lose a portion of their productive resources). If not problem-prone in terms of an inadequate or poor natural-resource base, most resettlement areas contain host populations with whom those relocated find themselves in competition for jobs; land for crop production and grazing; a wide variety of natural products such as fuel, thatch, and timber; and markets. Market access is a problem especially for small nonfarm enterprises since weavers, carpenters, and masons and small shopkeepers find themselves in competition with already-established host operators.

Social disintegration, Cernea's fifth component, is characterized by the type of stress-inducing reduction and loss of cultural inventory previously noted, including loss of leadership and capacity for group and community action. Recent donor research at Kariba about 40 years after removal relates the low capacity for self-help to resettlement-induced dependency (GTZ, 1993). While other factors are involved, that observation is disconcerting since it suggests that the adverse social effects of resettlement can last several generations, including among a generation born and brought up in resettlement areas. In the Kariba case, another destabilizing factor has been economic downturn. Although due partially to a bankrupt political economy at the national level and drought, downturn is also resettlement-related because of serious environmental degradation in too densely populated resettlement areas. Associated with widespread alcohol abuse, intracommunity violence and theft, and an increasing predisposition to attribute poverty and misfortune to the witchcraft of neighbors and kin, community unraveling—a major, global problem today—has been intensified by resettlement (Scudder, 1993).

Synergistically interrelated with the previous five components are food insecurity due to reduced production and productivity and increased morbidity and mortality. Although less common today, since most resettlement projects provide at least some housing, is homelessness—Cernea's final component.

Donor Involvement in Involuntary Resettlement

The scale, complexity, and unsatisfactory record to date of the resettlement process requires, if improvement is to occur, international cooperation, facilitation including training and supervision, and finance. Although pioneering such a role since 1980, World Bank documents, as well as those of many researchers, continue to document that, generally speaking, resettlement outcomes remain unsatisfactory even where governments have the political will to do "a good job." Doing "a good job" requires resettlement expertise that is unavailable in most countries.

It also requires funds. The appraised costs, for example, of the resettlement component of 20 World Bank–financed projects between 1986 and 1993 was estimated at 9 percent of total costs (Cernea, 1994). That percentage rises as the number of people to be relocated increases, so that resettlement in connection with China's Three Gorges Project will require over one-third of project funds if implemented in accordance with Chinese policy guidelines. For large river-basin development projects costing over a billion dollars, 9 percent requires the availability of $100 million for the resettlement component alone. Few countries have access to such funds. When they do, they may be allocated to other purposes not just because of greater priority given to the construction of physical infrastructure but also because of frequent cost overruns. Another factor is that funding for resettlement with development is required for years *after* the completion of infrastructure. Unless some form of project-related revenue sharing—as from generation of electricity—is institutionalized, by then previously available funds may be gone.

Various World Bank publications have shown how donor policy guidelines initiated since 1980 have had positive effects on resettlement planning and implementation that extends far beyond specific projects. The World Bank has been the key agency pushing such guidelines with the result that currently similar guidelines have been accepted by OECD (1992) and are accepted or are under consideration in several of the regional development banks. At the national level, several countries, including China and Turkey, have general policies for development-induced resettlement, while others, including Brazil and Columbia, have sectoral projects dealing with hydroprojects. Now run through the Bank's Economic Development Institute (EDI), such countries have benefited from a Bank-initiated training program in which country representatives are sent to attend courses held in the various regions. China, in particular, has also benefited from recently introduced "standalone" Bank-funded resettlement projects; an example is Xiaolangdi, with the Bank providing a $110 million credit for the resettlement component. Such a standalone project not only guarantees the necessary funding but also provides supervision that is intended to ensure compliance with agreed-on policies and implementation plans.

Prior to 1980, bilateral and multilateral donors alike considered development-induced resettlement to be strictly an internal affair. Some governments, including the United States, made little effort to relocate people as communities or to help them improve their living standards. Cash compensation or replacement housing for individual families was the best that could be expected in most such cases. Other governments, both colonial and national, however, took resettlement more seriously. African examples in the 1950s and 1960s include Kariba (northern Rhodesia only), Akosombo (Ghana), Aswan High Dam (Egypt and the Sudan), and Kainji (Nigeria). In those cases governments tried to relocate people as social units of their choice as well as complementing physical removal with some development. Two of those projects, Kainji and Aswan (in regard to Egyptian Nubians), continue to be two of the more successful projects implemented to date, although in both cases "success" has been due more to special circumstances than to implementing a well-planned resettlement with development project.

In the mid-1970s, Michael Cernea initiated within the Bank a series of sociological seminars. One (Scudder, 1977), dealing primarily with involuntary resettlement, provided case material that documented how the large majority of those relocated had been impoverished by Bank-financed water-resource development projects. Under Cernea's leadership, and with the assistance of David Butcher, an effort was initiated within the Bank to take a more responsible position toward those relocated. A milestone was the issuing in 1980 of the Bank's first Operational Manual Statement (Number 2.33) entitled *Social Issues of Involuntary Resettlement in Bank-Financed Projects*. According to a subsequent review (World Bank, 1985), that policy statement had an immediately beneficial impact on Bank-financed projects appraised during 1980–1982. Although less attention was paid to resettlement issues during the next 2 years, strengthened policy and operational procedures were issued in 1986 and 1990 (World Bank, 1986, 1990).

During 1993/94 the Bank's Environmental Department coordinated a major review of all Bank-financed projects with resettlement components that were in its portfolio between 1986 and 1992. That included field visits by Environment Department staff and consultants, as well as assessment of some cases that did not involve Bank assistance. The final report (World Bank, 1994a) pulled together reviews, based largely on reports from supervisory missions, undertaken by each regional bureau, as well as by the Bank's environment, legal, and operations evaluation (World Bank, 1993d) departments.

Without question, the Bank's review is the most detailed assessment of the record to date with development-induced involuntary resettlement. That assessment is especially relevant for water-resource development since 63 percent of those displaced were resettled as a result of dam construction. Furthermore, the Bank's policy emphasis on resettlement projects as development projects has had a beneficial impact on resettlement planning and implementation that has extended beyond Bank-financed projects.

Especially influential has been the Bank's emphasis on alternatives that either reduce or eliminate the need for resettlement. However, a careful examination of the evidence shows very few cases after 15 years of Bank activities where relocation can be expected to improve the living standards of the majority, in a sustainable fashion, over the longer term. While such a conclusion is not surprising, granted the complexity and dynamics of the resettlement process, the fact that further large-scale relocation is inevitable requires analysis to go beyond the reasons for failure to a consideration of ways in which a majority of those relocated might become project beneficiaries who also contribute significantly to the stream of project benefits. Again, however, it is important to emphasize the constraints to achieving successful outcomes. Looking to the future, the increasing emphasis within bilateral and multilateral donor agencies on development initiated by the private sector as opposed to governments may impose a new constraint characterized by less interest in and supervision of the resettlement component.

The Record to Date

With only a few exceptions, the more successful water-resource development resettlement has been carried out in connection with World Bank–financed projects. Few, however, have met the Bank's minimal goal of restoring for the majority their prerelocation living standards. Because dam sites often are located in rugged and remote areas, future reservoir basins are apt to contain ethnic minorities. That has often been the case with indigenous people in Canada, Mexico, and the United States. It is also often characteristic of South America and India. Even in China, where over 95 percent of the population is of Han derivation, vulnerable ethnic minorities are involved as with the

Longtan project, where site preparations are under way for one of the world's higher dams. For such people "it has rarely, if ever, been possible to achieve adequate resettlement" (Goodland, 1994, p. 149).

The social-science literature (Colson, 1971; Scudder, 1973; Hansen and Oliver-Smith, 1982; Cernea and Guggenheim, 1993; Refugee Studies Programme, in press), and that of environmentalists and journalists (Goldsmith and Hildyard, 1984, 1986; Pearce, 1992; McCully, 1996), is replete with case material suggesting that the same is true for other categories of people. There are few exceptions. Two apparent ones are Nigeria's Kainji Dam project and Egypt's relocation of approximately 50,000 Nubians in connection with the construction of the Aswan High Dam. Both are characterized by special features which are not transferable. In the Egyptian case, the majority of those relocated had already been marginalized as a result of relocation in connection with the original construction of the Aswan Dam at the turn of the century, or subsequent heightening in 1913 and 1933.

In the Kainji case, annual drawdown tends to expose about half of the shallow reservoir basin, including a large, highly productive former island in the middle of the Niger channel. That vast area is immensely productive for drawdown cultivation, grazing, and gathering (especially harvesting, as a cash crop, an indigenous grass for livestock production). Local utilization of that potential is probably the major reason why Roder can state a quarter century after removal that "when all is considered the resettlement was remarkably successful" (Roder, 1994, p. 10).

While the Bank's 1994 review is rather optimistic about the possibility of restoring incomes, few supporting examples are included. Reasons include inadequate preresettlement baseline data against which future living standards can be measured, as well as difficulties in acquiring such data. As reported by the Bank's regional bureaus in 1993 and 1994, the inadequacy of baseline data for sub-Saharan Africa renders speculation about income restoration meaningless, while critical baseline data for northern Africa and the Middle East, East Asia and the Pacific, South Asia, and Latin America and the Caribbean were either not available or were "often non-existent" (World Bank, 1994a, p. 4/4).

What data are available tend to document inadequate implementation. According to the regional report on Latin America and the Caribbean, "few of the resettled people in the earlier period can be said to have improved their living standards" (ibid.), while the report for East Asia and the Pacific noted that "the available evidence suggests that most of the affected in Indonesia have not been able to regain their former standard of living" (ibid.). As for India, "Projects approved after 1990 take social considerations into account in the design of physical components of resettlement. Still, economic rehabilitation has apparently been slow in all projects" (World Bank, 1994b).

As for more successful cases of implementation, the Bank's review discusses only four. Each case provides important examples of strategies which can raise living standards if properly planned and implemented. None, however, can be considered an unqualified success either because those considered successful constitute only a minority of those relocated, because a proportion of the population on whom there is too little or no data is omitted from the analysis, or because the case refers to too early a stage in the resettlement process. One example discusses cage aquaculture systems on Indonesia's Saguling Reservoir. Such systems are important, with untapped potential for other reservoirs. While those involved have raised their living standards, they nonetheless remain a minority, with living standards of those relocated as farmers around the reservoir margin are worse than before removal.

Improved living standards also involve only a minority of those relocated in the second case, which is the Krishna subproject of Maharashtra Irrigation II. If the entire project of six schemes were assessed, the record would have been even less satisfactory. Success in that case is attributed to receipt of irrigated land in the project's command area. Of 6,517 families that lost all their arable land, or if landless, their livelihood, only 2,138 (one-third of the total) received irrigated land with government assistance. That in itself was rather exceptional, with local leaders in the command area encouraging farmers to sell land to those resettlers, many of whom belonged to the same Maratha caste (World Bank, 1993b, p. v). Nonetheless, they have done well, with the large majority not only having higher living standards than before displacement but also higher standards than hosting villages.

What about the remaining two-thirds? A minority preferred to resettle within the catchment area above the dam; "the perception in these villages was that they were not better off than before." Indeed, they were "decidedly worse off" (ibid., p. 16) in comparison to those incorporated within the command area. As for those who accepted cash compensation (almost always inadequate in India as elsewhere) and moved farther away, or received no compensation at all, nothing is known. That includes 45 percent of the total families who were ineligible to receive either land or a house plot within the command area because they were either uncompensated joint owners of land or landless. Also unknown is the fate of another 273 families in 8 villages who lost over 50 percent (as opposed to all) of their land and of 1,056 families in 11 villages who lost some land (but less than 50 percent). For joint families, only one family would have received cash compensation while the unknown number of landless households dependent on families that moved received nothing.

The other two cases, Shuikou in China and Khao Laem Hydroelectric in Thailand, are encouraging in that living standards of a majority appear to be rising. However, what success has been achieved has occurred during the initial years of resettlement. It remains to be seen whether it is sustainable. The point is an important one. There are other cases where initial success unraveled thereafter. One such example pertains to the 35,000 people resettled in the late 1950s on the Zambian side of the Lake Kariba reservoir. For a 10-year period in the 1960s and 1970s, the majority of those relocated were able to raise their living standards to the extent that an evaluator 15 years after removal would have labeled Kariba North Bank resettlement "a success." Such is not the case today, where environmental degradation caused by dam-induced increases of population densities in resettlement areas is a major factor in the current impoverishment of both relocated and host populations (Scudder, 1993).

According to McCully (1996), the Bank's review has overestimated Khao Laem's benefits in claiming that "incomes for all households rose after resettlement." That statement is based on a different reading of a 1993 report from the Bank's Operations Evaluation Department (OED) (World Bank, 1993a). According to that report, preresettlement resources were never evaluated properly (ibid., p. 16), while postresettlement figures refer only to average incomes (ibid., p. iii). Averages do not show the range of incomes among relocated families. The OED review also notes that they do not include those [perhaps one family in five and certainly 15 percent (ibid., pp. 8–9)] who, because they did not have identity cards, were denied entry to the resettlement sites.

In the Shuikou case the resettlement of 67,000 people was completed only in 1992, with impoundment occurring during 1993. Although relocation commenced in 1987, most of those relocated in the early years would have lost access to their land only in 1993. The Government of China probably has the best relocation policies in the world. From careful monitoring during the first 5 years, reports on Shuikou also indicate that an exceptional effort is being made there to diversify farm and nonfarm opportunities. Nonetheless, it is too early to assess the extent to which current efforts will be successful. Especially problematic are the 30,000 resettlers (and especially the older ones) who must shift from agriculture to nonfarm enterprises. The competitiveness of those enterprises over the longer terms also remains to be seen, as does the cost-effectiveness of citrus as a cash crop granted the importance placed on new citrus orchards at Three

Gorges and elsewhere in China. As in Sri Lanka's Mahaweli project, income levels for families with members employed during the project's construction phase may drop after the completion of physical infrastructure. Yet another problem is inflation, especially for the increased proportion of people who must now buy more of their food or purchase inputs for agricultural intensification.

This critique is not intended to denigrate the World Bank's efforts to improve resettlement outcomes. Because of those efforts, policies and planning have improved. Even where political will on the part of borrowers is there (and all too frequently it is not), what has been more difficult to achieve is implementation that actually raises living standards for the majority above their prerelocation levels. Reasons for that failure need careful analysis in order to point up the unfavorable odds that must be overcome if successful implementation is to occur. Overcoming those odds will be very difficult. It will require new goals and new approaches. It will also require careful analysis of how best to incorporate into resettler production systems such activities as reservoir fisheries and drawdown utilization, and irrigation, which have benefited resettlers in some projects.

Reasons That Unsatisfactory Resettlement Programs Continue

Introduction

Reasons are many as to why the record with resettlement continues to be unsatisfactory. For governments and donors, these include lack of awareness of the complexity and dynamics of resettlement, inadequate and inappropriate goals, and an inadequate perception of the impacts of resettlement on displaced people and hosts. They also include inadequate commitment and implementation at the government level. Relating more to those relocated and their hosts, other reasons are lack of empowerment, loss of resiliency and increased dependence through incorporation within a wider political economy, host-relocate conflicts, and—increasingly—the problem of sustainability. Many of these characteristics are synergistically interrelated. Even with the best intentions, in most cases it is unlikely that they can be overcome.

The complexity and dynamics of involuntary resettlement

Complexity. It is hard to imagine a more daunting task for planners than designing and implementing a program that requires, on one hand, the humane shifting of an unwilling, traumatized population and, on the other hand, the creation in a problem-prone habitat of economically and socially viable and sustainable communities. Only one case is known in which a majority appeared to welcome compulsory resettlement. That was the relocation in the mid-1960s of over 50,000 Egyptian Nubians in connection with the construction of the Aswan High Dam. There it was the women who welcomed the move. That was because three previous relocations in connection with the construction and heightening of the original Aswan Dam had left those relocated to the reservoir margin more and more isolated geographically. It had also cut women and children off from family heads who were forced, because of a lack of economic resources and opportunities, to seek wage employment in the cities of Egypt and Sudan. In other words, they welcomed moving because of the failure of past resettlement to provide them with a viable existence.

Involuntary resettlement requires everyone to move—the willing and unwilling, the young and the old, the strong and the weak, and the healthy and the sick. Because of a general preference of resettlers to move short distances in order to remain as close as possible to familiar habitats and neighbors, governments usually allow resettlement within the vicinity of former homes. All too often, however, such areas are either marginal (as when people move inland from flooded river valleys), densely populated, or both.

As global populations rise, resettlement areas are increasingly problem-prone in terms of availability of arable land, water supplies, and other essential natural resources; market access; and social and physical infrastructure. Such resource constraints increase the complexity of resettlement. Because of land scarcity, more and more farm households will have to switch to nonfarm employment or, at the very least, increase the proportion of income obtained from nonfarm employment. Little is known about how to accomplish that goal, especially for couples middle-aged and older. Its accomplishment is a major policy issue which will require much experimentation accompanied by careful monitoring and evaluation.

Approximately 50 percent of farmers to be relocated in connection with China's Three Gorges Dam (at full storage level of 175 m, its construction will require the relocation of approximately 1.25 million people) must shift to such nonfarm employment as village industries, while others must intensify agriculture and shift to new crops. Similar shifts are becoming increasingly necessary in India and other densely populated countries. In Lesotho, the absence of replacement land will require major changes in the household economies of previously isolated rural villagers currently affected by the first of several phases of the Highlands Water Project. In all such cases, many of those undergoing relocation will have to change production techniques and occupations if they are to maintain, let alone improve, their preproject living standards. The global experience is that job training alone has not worked in the past because the necessary jobs have not materialized. As for the creation of village industries, a high failure rate is to be expected for a wide range of reasons, including management constraints, quality of products (noncompetitiveness), and lack of markets.

Resettlement is also complex because of what the resettlers have to do to adapt and prosper in their new habitat. In addition to coping with the trauma and sense of powerlessness which characterizes their initial removal, they also have to adapt to a daunting range of new conditions in a new habitat. They have to come to terms with being strangers in a land where the host population has already established political and ritual authority. In most cases they also have to adjust to tighter administrative control, denser populations, and new production techniques. Available data also indicate that they have to cope with temporarily heightened morbidity and mortality rates. Finally, unlike some political refugees, the majority of those displaced from future reservoir basins do not have the option of returning home.

The dynamics of the resettlement process. Successful resettlement takes time. Scudder has formulated a four-stage framework that covers the generation that moves and the first generation born in the resettlement areas (Scudder, 1981; Scudder and Colson, 1982). Following an initial planning-and-recruitment stage (stage 1), the second stage documents the struggle to adjust to loss of homeland and to new surroundings. As described earlier, it is characterized by multidimensional stress. Throughout, the majority respond by clinging to familiar routines and relying on kin and coethnics as much as possible. They are risk-averse, behaving as if a sociocultural system were a closed system until they gain a degree of sufficiency and begin to feel at home.

While stage 2, adjustment and coping, often continues for a minimum of 2 years following removal, in many settlement areas the majority never arrive at stage 3, which must be reached if the potential of the resettlement process is to be realized. This is because the third stage is when economic development and community formation are most likely to occur; when, almost paradoxically, a majority of settler households become risk takers within a relatively short period of time. A number of indicators characterize the transition between stages 2 and 3. These include the naming of physical features and increased emphasis on community, as opposed to household development as reflected in the establishment of funeral and other social-welfare associations and places of worship, including churches, temples, and mosques in different cases.

Institutional development then continues throughout stage 3, which may also be characterized by a resurgence of cultural symbols—almost a renaissance—as community members reaffirm control over their lives. Economic development is fostered as households increasingly pursue dynamic investment strategies. Farmers, for instance, initially begin shifting from a reliance on consumption crops to higher-value cash crops. Increased emphasis is also placed on education of children. Production systems at the household level also begin to diversify, not so much as a risk-avoidance strategy but as a means for reallocating family labor into more lucrative enterprises, including livestock management and small-scale nonfarm enterprises. Initially small businesses are run from the household's homestead allotment with subsequent expansion to service centers within the settlement area and, if especially successful, to urban centers including national capitols where real estate investments may also be made.

Stage 3 development must be sustainable into the next generation for a resettlement project to be considered successful. *Sustainability* here refers to institutional sustainability (as required for managing irrigation systems, or various types of cooperatives) as well as to environmental and economic sustainability. Stage 4 commences when the next generation of settlers takes over from the pioneers and when that generation is able to compete successfully with other areas of a country for national resources. It is also characterized by the devolution of what management responsibilities are held by specialized resettlement agencies to the community of resettlers and to the various line ministries.

Social scientists have found this analytical framework of four stages lasting over two generations useful in predicting how a majority of households will respond at various times during the resettlement process. While accepting the concept of dynamic stages, some, including de Wet (1995), have pointed out that the high level of generalization obscures the importance of environmental and societal variation, which is apt to influence how different categories of people behave during the various stages. Partridge et al. (1982) emphasize, for example, that too little is known about how different income and social groups experience resettlement. In relatively stratified societies, wealthier households may adjust much more rapidly by using resettlement compensation as an opportunity to capitalize new ventures. Where they decide to avoid resettlement sites by shifting to towns or other rural areas, their departure also provides new opportunities for aspiring households to develop more rapidly in government-sponsored resettlement areas.

Inadequate and inappropriate goals

Inadequate goals. Governments and donors, including the World Bank, continue to have inadequate goals for those undergoing resettlement. Although the Bank states the desirability of those being displaced increasing their incomes and improving their living standards, the impression given by Bank documents is that it is sufficient to merely restore preresettlement incomes. That impression even permeates the World Bank (1994a) *Bankwide Review of Project.* While the first section of that review states that "the Bank's policy is to improve the former living standards and earning capacities of displaced people—or at least restore them," thereafter the emphasis is on restoration (ibid., pp. 3/8, 4/1, 4/3, 4/15; especially pp. 4/20, 8/1).

Attempts to merely restore incomes will end up further impoverishing what are, with few exceptions, already poor people. Several reasons support that conclusion. Planning for river-basin development projects is a lengthy procedure. A 10-year planning horizon is common; 20 years, as with India's Sardar Sarovar Project, or even 30 years, as in the case of China's Three Gorges, may also occur. Once the possibility of a future dam arises, governments are much less likely to make investments in what may become a future reservoir basin. In other words, by the time a decision has been made to build a dam, or donors get involved in project appraisals, the living standards of those to be relocated have already been adversely affected.

A second reason why attempts to merely restore incomes can be expected to further impoverish displaced people pertains to the resettlement process itself. During the years immediately following removal, and in some cases during the years immediately preceding removal, income levels tend to drop. Once people realize that removal is imminent, new investments which are tied to preresettlement sites are no longer made. While such investments may not have been major in the area without the project, they would have occurred nonetheless. As for the period immediately following removal, adjusting to new habitats, neighbors, and government programs reduces time and energy for restoring whatever living standards those displaced previously had. During that time period (which seldom lasts for less than a year), people's living standards drop [see also Cernea (1996)]. Furthermore, expenses following resettlement are apt to be greater than before removal, which is another reason why merely restoring incomes is insufficient. Increased costs are a problem especially for resettlers who have to purchase food supplies that they were able to produce previously.

Another weakness of policies that only attempt to restore income levels is when a gap, which may continue for years, occurs between people's physical removal and implementation of development plans. That constraint remains a very serious problem today in connection with the Lesotho Highlands Water Project. Elsewhere, as with Sri Lanka's Accelerated Mahaweli Project, incomes for workers and small contractors may improve during the construction phase during the initial years of resettlement, but drop thereafter because of poor synchronization between physical removal and the intended development plan for resettlement areas. That problem may also affect a number of Chinese projects where sale of hydropower, which is to generate funds for resettlement but is postponed until after reservoir filling commences, is poorly phased with the physical removal of resettlers.

If donor and government policies do not improve living standards, then the circumstances described will, in fact, further impoverish those involved. The only solution to this problem is to insist from the start not only that resettlers should be project beneficiaries but also to define specifically what that means in terms of income and living standards expected from proposed occupations. If that is deemed impossible, development alternatives should be sought. So that living standards will improve, all resettlement projects should be planned and implemented as development projects, with the definition of development incorporating the needs and expectations of resettlers and hosts. While the World Bank recognizes such a goal as desirable, the wording of its policies, and the realities of the river-basin development and resettlement processes, impede the realization of that goal. The way the policies are worded and described also allows governments like China and Uganda (Government of Uganda, 1995) to settle for less than their own policies stipulate.

Inappropriate goals: the problem of commitment to equitable resettlement at the national level. Large dams are built by government agencies, sometimes supported by donor agencies. While both have a responsibility for planning and implementing resettlement programs that are sustainable environmentally, economically, and institutionally, national governments have tended to be more deficient than donors. Part of the reason is lack of political will, due in part because most communities undergoing involuntary resettlement have little political clout. The United States is a case in point, where the benefits of the Uniform Relocation Act of 1975 relate primarily to "decent, safe and sanitary housing" for individual families, as there is no provision for community relocation or for a development component—which is by far the most important single feature of any resettlement plan. The result is that in the United States, as elsewhere, most dam resettlement continues to be poorly planned, poorly financed, and poorly implemented.

Among ongoing projects, India's Sardar Sarovar project is an example. There the author's discussions with some high-level project officials brought out their surprise as to why the World Bank was so concerned about relocation issues—after all, those requiring movement were only low-caste peasants or tribal people. Equal disdain on the part of some officials was evident during the initial years of the Lesotho Highlands Water Project. More recently, arguments over financing resettlement by both Lesotho and South African officials have seriously delayed implementation of the project's Environmental Action Plan.

Another example of inappropriate goals is where resettlement is used to achieve political goals. As discussed in Chap. 14, a cabal within the Mahaweli Authority of Sri Lanka planned to use the resettlement of poor Singala-speaking Buddhists as a means for driving a wedge through the Tamil-speaking areas of northern and eastern Sri Lanka (Gunaratna, 1988). Although government policy was not to combine people of different linguistic, ethnic, and religious communities in the same settlement, that policy was ignored in System B, where Tamil-speakers were in the majority. In addition to using the security forces to harass the host population, the Mahaweli Authority of Sri Lanka shifted defenseless Singala-speaking settlers into Tamil communities, with retaliatory killings used as an excuse for further "ethnic cleansing" by the government. At one point the government of China aired, but then retracted, a similar political goal when suggesting that several hundred thousand resettlers from the future Three Gorges Project be resettled in western China among Turkic-speakers.

Inadequate perception of the impact of involuntary resettlement on the part of policymakers and government officials

Even behavioral and social scientists, no doubt because they are highly mobile themselves, are periodically surprised by the extent of trauma caused by different types of involuntary resettlement, not just of low-income people with strong ties to homes and surroundings, but also of highly educated individuals and families. In her New Haven psychiatric practice, Weissman initially was puzzled by the depression of a number of her women clients (Weissman and Paykel, 1972). Eventually she attributed it to the fact that their husbands had recently moved from previous homes to take up positions at Yale University. For husbands, and their perception as it related to their families, the move was an upward one. For their wives (and for their children), who had to leave familiar routines, places, and friends, the move had the characteristics of involuntary resettlement. Scudder and Scudder (1981) also noted how physiological, psychological, and social stress reactions characterized wealthy women in the United States who saw their move from beneath the flight pattern of an expanding airport as involuntary.

It is even easier for well-educated and mobile river-basin development policymakers and government officials to underestimate the difficulties that a majority of people will undergo because of involuntary resettlement. Colson's *The Social Consequences of Resettlement* (1971) vividly reproduces the recollections of Kariba North Bank resettlers of the day they were moved from their riverine gardens and homes along the Zambezi back into the bush. Summarizing her notes, Colson wrote that the people "got out of the lorries in the midst of a howling wilderness, and looked around and were downright scared. Something that continued when they heard the elephants and hyenas that night." Tremmel (1994) has reproduced in still-greater detail the recollections of the Kariba South Bank resettlers who hoped that "the water would follow them."

It is all too easy to gloss over with a few sentences the trauma associated with a people's inability to protect their old homes followed immediately by the need to adapt to new areas where they now are the strangers. Although a major research gap is the impact of involuntary resettlement on children, underestimated in particular are impacts on women (Koenig, undated manuscript), many of whom previously have not traveled much beyond their homes.

Two recent experiences have especially impressed the author of this chapter. One was in India—in a village of tribal people from Madhya Pradesh who had been relocated to Gujarat in the early 1980s in connection with the Sardar Sarovar Project. There a reputable social-science research institution reported high rates of infant mortality during 1986/87 and 1987/88, with the number of infants dying during one of those years exceeding the number of new births. At the time it was hard to imagine what the impact of such deaths could be on families already stressed by their recent removal and the unsatisfactory land base and employment opportunities in their new "home."

The other experience was during November 1989 while the author was reinterviewing, with Kapila P. Vimaladharma, a small sample of Sri Lankan households whose fortunes had been followed for 5 to 10 years. All were Mahaweli Project settlers, some of whom had been involuntarily relocated because of dam and canal construction in connection with one of the world's largest river-basin development projects. At the time conflict was accelerating between an ultranationlist movement (the JVP) and the government security forces. On one hand, the JVP had killed one household head as well as the sister-in-law of another; on the other hand, it had recently killed five police officers with a land mine. In retaliation for those deaths, the security forces had burned down the houses of over 20 Mahaweli settlers who had the misfortune of living in the vicinity. They also killed and burned the bodies of at least 50 youths. Caught in the middle, Mahaweli women resettlers told how they lived in fear for themselves and their families (especially their sons). Yet, when asked which year had been the most difficult for them since their resettlement, the majority (including those households that had been involuntarily relocated in connection with project activities) picked the 12-month period following their arrival during which no communal strife had occurred!

There is no way that social cost/benefit analyses can accurately reflect the hardships involved, which is why it is the responsibility of planners and informed social scientists to emphasize the need to consider alternative development strategies which do not involve compulsory relocation. Where alternatives are rejected, it is their responsibility, on one hand, to emphasize the importance of keeping the number of resettlers as low as possible, and, on the other hand, to assist in efforts to implement equitable and sustainable plans that will improve the living standards of the majority.

Inadequate implementation

There are many implementation constraints in addition to the complexity and difficulty of implementing an adequate dam resettlement development program even where the best of intentions exist. One relates to the timing of resettlement activities in regard to donor disbursements in connection with dam and other infrastructure construction. Another is financial. Still another relates to low priorities placed on resettlement by river-basin development agencies as well as ambivalence within the World Bank concerning enforcement of its own guidelines through suspension of disbursements and withdrawal from project support when borrowers do not observe contractual obligations. Changes in government policies can also negate a good start. Such constraints are analyzed in the sections that follow so that officials, researchers, nongovernment organizations (NGOs), and others with resettlement responsibilities and concerns can have a better idea of the difficult context in which resettlement with development is carried out, and the problems that must be overcome.

Timing, financial, and institutional constraints. Dam resettlement is a by-product of a much larger project. It also tends to extend well beyond the completion of the dam itself. Because of cost overruns during construction, funds may no longer be available for completing resettlement activities. This is especially the case with the development phase of resettlement, which must continue after reservoir filling if living standards in relocated communities are to rise.

If resettlement is planned and implemented according to the criteria laid down by the World Bank, the financial costs of a resettlement with a development program can be expected to cost a minimum of $3000 per capita at current prices. As numbers of potential resettlers increase, such costs can exceed 10 percent of total project costs, especially where funds are made available—as they should be—for incorporating the host population.

While governments may agree to meet such costs during the planning phase, finding the necessary funds is another matter. There is also the risk that monies set aside for relocation with development will be diverted to complete the construction of physical infrastructure. This is especially the case where cost overruns occur, as they frequently do with large-scale infrastructure projects, since such costs tend to fall primarily on the borrower, with donor credits and loans being for fixed amounts. Since a major resettlement program is apt to extend for 10 years or more, there is also the problem of national emergencies and changes in government priorities diverting funds elsewhere.

Lack of funds at Kpong was responsible for the inadequate implementation of a relatively good plan (even though too much emphasis was placed on housing as opposed to productive activities). At Mahaweli, reduced financial support from both the government and donors today threatens the viability of the entire, well-planned settlement component. In the Kpong case "the lending agencies had assigned the highest priority to resettlement" (Futa, 1983, p. 106). They, as well as the Ghanaian officials involved, realized that "the establishment of the economic base was critical, as the Akosombo experience had vividly demonstrated" (ibid., p. 105). Development plans that were carefully drawn up included improved cultivation and livestock-management practices for that majority of the resettlers who were farmers. Because it was recognized that grazing after resettlement would be insufficient, special emphasis was placed on a livestock program which would involve irrigated fodder crops. Emphasis was also placed on improved housing and social services.

During a July 1989 visit, Futa (whose previous visit to Kpong had been in 1981) and the author could see that a major effort had been made to implement the housing and social-services aspects of the plan. Funds, however, had not been allocated for the agricultural development program or for other productive activities. Although conditions varied among the six resettlement villages, the majority of the resettlers were still subsistence farmers. Many considered themselves worse off as farmers because they had been allocated only 1000 ha of arable land in return for their previous 1500 ha, some of which was of better quality. A World Bank evaluation reached a similar conclusion: "the resettlement outcome was unsatisfactory: incomes have not been restored to the levels which prevailed prior to moving" (World Bank, 1993c, p. iii).

Another problem is the unwillingness of donors to fund resettlement costs. Although the World Bank today has, without question, the best set of guidelines, their implementation would significantly increase project costs. Yet, with only a few exceptions, the Bank itself has been unwilling to fund that implementation. Such funding is important not just because of the finance provided but also because its provision results in closer supervision by the donor. A case in point is funding by the U.S. Agency for International Development (AID) for the resettlement of approximately 12,000 people in connection with Mali's Manantali Dam. There funding for physical resettlement was linked to careful monitoring which played a role in reaching a satisfactory outcome. Unfortunately, AID reduced the impact of its involvement by not providing additional funds for development activities. Those have been unsatisfactory, especially for the majority relocated below the dam for whom inadequate land was allocated (Horowitz et al., 1993).

In regard to institutional constraints, even with World Bank–funded projects appraisal missions involving knowledgeable social scientists seldom find a satisfactory institutional capacity for planning and implementing resettlement. Institutional constraints increase dramatically with the number of resettlers. It is one thing to plan and implement a pilot project involving one or two communities; it is quite a different thing to resettle tens of thousands of resettlers per year during the height of resettlement programs involving over 100,000 resettlers (as with India's Sardar Sarovar Project) or over 1 million (as would be the case if China's Three Gorges Dam is built).

Solutions to financial and institutional constraints are difficult even where the political will is present to plan and implement resettlement with development. In China, a new policy since the early 1980s for national projects is to allocate a proportion of the revenue from hydropower generation for the resettlement and development of relocated people. This policy has had impressive results in correcting "remaining problems" in the reservoir basin of the Danjiangkou Dam over 10 years after relocation and is now applied to all new reservoirs. It is not sufficient, however, for providing funds for physical removal and initial development of the majority of resettlers who must move before turbines for generating hydropower are operational.

Just as donor funding is necessary for the construction of physical infrastructure, so, too, is donor funding needed for the resettlement component of large-scale development projects such as major dams. While both AID and the World Bank have provided funds for resettlement components associated with dam construction, few projects have so benefited. Such funding needs to be increased. In addition to being tied—along with borrower counterpart funds—to development activities, donor funds can also play an important role in reducing institutional and other constraints.

Inadequate implementation due to low-priority placed on resettlement component by both national and donor agencies. The principal player in resettlement activities remains the national borrower. Where the political will is not present, implementation will suffer unless strong pressure is brought to bear on government agencies by organized resettlers, NGOs, and concerned individuals and donors, and unless donors are willing to tie compliance with resettlement guidelines to the disbursement of project funds. While the sanction of stopping disbursements may be explicit in project documents, it is rare for donors to withdraw funding once major contracts have been let. This is especially the case, as in World Bank–financed projects, where international competitive bidding is required, and where construction by large international and national firms is proceeding according to technical specifications. Donor leverage also is reduced if their funds are allocated only for major construction activities since disbursement is apt to be completed before completion of resettlement and development activities. In addition to tying assistance to adequate resettlement planning and implementation, the best approach for donors is to provide separate funding for the resettlement component, backed up by adequate monitoring throughout the resettlement process and by cooperative interaction not just with government agencies and local people but also with concerned NGOs.

Changes in government policies and donor funding priorities negating a good start. Resettlement programs are especially vulnerable to changes in government and donor policies because they take much time to plan and implement. During that time period, the government in power may shift priorities or new governments may come to power with new priorities following elections or coups. Unexpected events, such as natural disasters, and civil disturbances and wars as well as changes in donor policies, may also divert funds with adverse effects on resettlement programs. The current situation in Sri Lanka illustrates a number of these constraints as they relate to the settlement components of the Accelerated Mahaweli Programme. In terms of poverty alleviation, at risk is not just the country's largest single development project, and first major attempt at regional planning, but also one of the major projects in the world.

The settlement-resettlement component of the Mahaweli Project is in jeopardy for both resettlers and voluntary settlers for several reasons. First, a serious drain on the finances of a poor country, is the civil war between the government and Tamil separatists on one hand and the government and the militantly Buddhist JVP on the other hand. Second was a change of attitude within the governing party toward Mahaweli after its return to power as a result of the 1988 elections. Between 1977 and the 1988 elections, United Nationalist Party (UNP) President Jayewardene had provided crucial political backing to the Mahaweli Project. During that time, one of the rising stars within the UNP had been the Minister for Mahaweli Development, and he had appointed, with President Jayewardene's approval, the Secretary General of the UNP as the first chairman of the Mahaweli Authority of Sri Lanka. Together the three made a formidable team which was able to attract over a billion dollars in donor funding.

Although the UNP was returned to power in the 1988 elections, the new president had different priorities. Thereafter Mahaweli was deemphasized by both the UNP government and donors (some of whom switched funding to the new president's priorities), hence jeopardizing the entire settlement component (Scudder and Vimaladharma, 1989).

While little can be done to alter the political process when national leaders switch priorities, it should be possible to educate donors about the need for longer-term funding for projects that involve the settlement of people in problem-prone new lands. As actions regarding structural reforms by the IMF, the World Bank, and bilateral donors such as AID illustrate, donors do have leverage when they choose to use it. The pity is that leverage is not used more often for the benefit of the low-income majority.

The lack of empowerment of relocated and host populations

Most dam projects are constructed in relatively isolated areas that are inhabited by low-income populations. Not infrequently, ethnic minorities and rural communities with low social status are involved; native people in the United States and Canada and tribal and lower-caste peoples in India are examples. Such people have little political clout. This is true both in the local political arena and at the national level. As a result, people at risk of being forcibly relocated are at a disadvantage in attempting to pressure governments to plan and implement for them resettlement with development projects. Without the assistance of outside agencies, there is little that they can do to offset this constraint.

There are two ways to go here. One is to encourage more donors and governments to formulate, and implement, resettlement guidelines which involve all projects with resettlement and development components and resettlers as project beneficiaries. The other is through the advocacy of NGOs and individuals on behalf of resettlers. That advocacy has more potential for achieving results where NGOs can bring class action suits against governments as in India, where NGOs have accused state and national government agencies of violating constitutional rights as they relate to project-affected people. While the best solution is where well-intentioned governments work closely with resettlers and hosts, NGOs, and donors in resettlement planning and implementation, advocacy (including an initial adversary relationship with government) may be necessary to achieve such a relationship.

Loss of resilience and increased dependence on government through incorporation within a wider political economy

Prior to resettlement, riverine inhabitants tend to be relatively self-sufficient, with resilience provided by a range of strategies for making a living. Their dependence often relates more to nature than to government. When such people are moved, population densities tend to be increased, while quality and quantity of arable land is often reduced. More people on less land makes it unlikely that those relocated can revert back to their preremoval economy on a sustainable basis. As for those who must shift from farm to nonfarm employment, or from rural to urban employment, reverting back to an agricultural mode of production is virtually impossible hence increasing their dependency on a wider political economy.

A further problem is that resettlement areas tend to be problem-prone or more complex or both, hence increasing the resettlers' need for government assistance, as well as that of the now more-crowded host population. Areas selected for rural relocation are available usually *because* they are problem-prone, especially in terms of availability of arable land. Although reservoir formation causes their removal, ironically people are also frequently relocated in areas with inadequate supplies of potable water. That is one reason why such areas are available for resettlement in the first place. To provide water, resettlement authorities utilize a range of technologies, all of which present, sooner or later, new maintenance and financial problems for relocated communities.

Incorporation of resettlers who previously practiced rain-fed and flood-recession agriculture into irrigation projects illustrates the type of greater complexity in which those displaced may find themselves. In such cases they become dependent on governments to provide irrigation water, at least at the project level, and to operate and maintain the irrigation system at that level. When insufficient water is delivered as is the case with Sudanese Nubians relocated at Khashm el Girba (Salem-Murdock, 1989), or Mahaweli settlers in System H (Scudder and Vimaladharma, 1989), the resettlers suffer. Even when incomes increase, such increases may be offset, as in the Mahaweli case, by the higher cost of inputs and by indebtedness from acquiring those inputs.

The logical approach here is to realize from the start that resettlement usually incorporates both resettlers and hosts within a wider political economy. Following that realization, the next step is to provide the people with diversified and income-generating opportunities that are sustainable, and with the training and extension required for realizing and enhancing those opportunities. Also essential is institutional development, whereby resettlers can gradually take over managerial responsibilities from resettlement and other agencies. At Mahaweli, for example, the Mahaweli Economic Agency has become part of the problem because of its inability to correct land preparation and irrigation-system deficiencies, and its increasingly adverse relationship with the settlers. In Scudder and Vimaladharma's 1989 report, the first recommendation stated that "the phased handing over of increasing management responsibilities to settler organizations" should share the highest priority with diversification of the settlers' production systems (Scudder and Vimaladharma, 1989, pp. 6–7).

Host-resettler conflicts

Even where political leaders among the host population agree to the movement of resettlers into their midst (as was the case at Kariba), sooner or later conflicts between the two can be expected. They arise because of competition of a larger population over a diminished (in terms of per capita) land base, as well as over access to job opportunities, social services, and political power. Conflicts over grazing can be expected from the start (as illustrated by the Indian Krishna subproject); provision of water, forage, and sustenance for livestock is a chronic weakness of resettlement programs. Conflicts over arable land can be expected to intensify as the children of resettlers and hosts seek fields to support newly established families. Since political leaders are apt to welcome resettlers as a means of enlarging their constituencies, their resentment can be expected to grow when initially hesitant resettlers appoint their own leaders, some of whom may take over the position of host leaders, especially where resettlers are in the majority.

Few resettlement projects in which host-resettler conflict has not appeared have been documented. In the Lusitu resettlement area for the Kariba project, where resettlers outnumbered the host population, initially relationships between the two categories were good. That was in the late 1950s and early 1960s, when land was plentiful, with hosts providing fields to resettler "friends" in return for various kinds of assistance. Today those hosts increasingly are requesting the return of that land, while their chiefs have accused the resettler chiefs of usurping their custodianship of the land. Jonathan Habarad believes that the increased amount of strife within the Lusitu area below the dam, which involved armed robbery of resettler stores, may have a host-resettler dimension (verbal communication to the author).

In Mahaweli's System H, nervous resettlers and voluntary settlers told Vimaladharma and the author of threats by the host population to drive them off project lands which originally belonged to the hosts. In the catchment surrounding the Kotmale reservoir basin, where over 40 percent of the Kotmale Dam resettlers chose to stay, Ben-Ami (verbal communication, 1989) reported that the combination of increased population densities and environmental degradation was creating a critical situation not just for the habitat but also in regard to conflicts between the resettlers and the host population.

About 10 years after resettlement in connection with Ghana's Kpong Dam, fighting broke out in early 1989 between resettlers in one of the six villages and the host population. At least 10 people died. In that situation, conflict was triggered by the decision of the resettlers to fill a chieftaincy vacancy with one of their own people. Pointing out that their current land had been given to them by the government, they insisted on their right to appoint their own chief. The hosts disagreed, and fighting—no doubt exacerbated by tensions over arable land and grazing—broke out.

It is best to anticipate from the outset the inevitability of host-resettler conflicts. The best approach for minimizing them is to provide the host population with access to at least some of the improved social services and economic-development opportunities intended for the resettlers. While such an approach will increase the financial costs of resettlement in the short run, in the long run it will enhance the possibility of multiplier effects as well as reduce the intensity of conflict. Access to new schools and medical facilities, as well as to extension services, is probably the cheapest approach. Necessary, in most cases it is not sufficient, simply because increased population densities sooner or later will require some intensification of production among both hosts and resettlers. Unfortunately, such incorporation of the host population within resettlement programs is rare. One of the few examples, and one that relates to planning rather than implementation, comes from China, where Three Gorges planners have included the hosts within their agricultural development program.

Immigrants

Few reports comment on the extent to which immigrants tend to benefit disproportionately from whatever benefits may occur within the reservoir basin (see Chap. 14, this volume). Resettlers, for example, seldom have the skills or education to join the labor-force-building physical infrastructure or operating that infrastructure. Less skilled to exploit a large open body of water, resettlers and hosts are even apt to find reservoir fisheries eventually dominated by immigrants, as occurred at Volta in Ghana and at Selingue and Manantali in Mali. That problem was partially alleviated at Kariba by closing the new fishery to outsiders between 1959 and 1964, during which time training was provided to local people. Nonetheless, today they represent less than 50 percent of the fishers. They are also losing land today to more powerful agribusinesses and other commercially minded corporations and individuals who wish to privatize lands which belong to resettler and host communities under customary systems of land tenure.

The problem of sustainability

While sustainability is a problem with all types of rural development, it is a particularly difficult problem where relocation is involved simply because increased population densities in receiving areas put further pressure on the natural resource base. Kariba resettlement, for example, is a major factor responsible for the accelerating degradation of soils, removal of vegetative cover, and erosion of tributary upper catchment basins throughout the Middle Zambezi Valley. Kariba was the earliest major, dam-induced resettlement project in the tropics, and current events there may well forecast the future for more recent removals.

By the late 1970s, what formerly had been a well-forested woodland savanna habitat was increasingly taking on a Sahelian character. Formerly buried lateral roots of the few remaining trees, including baobabs, now might be as much as a foot above the ground as a result of gully, sheet, splatter, and wind erosion. Such environmental degradation is a major factor involved in the economic downturn and community unraveling that began during the 1970s (Scudder and Habarad, 1991). By 1995, grazing and browse were seldom available for livestock toward the end of the dry season, with cattle frequently dying from hunger or driven into adjacent areas, including a wildlife-management area, hence exporting degradation to a wider area. By

1995, over 50 percent of the population of some villages had emigrated from one major resettlement area. Host communities were also adversely affected and involved in outmigration.

Environmental degradation is also accelerating in the upper catchment of Sri Lanka's Mahaweli River basin. While such degradation preceded dam construction, it is particularly serious in the new reservoir basins. This is also due to increased population densities, since relocatees who did not wish to resettle in irrigation command areas downstream chose to relocate on the slopes surrounding the reservoirs that cover their former homes. As noted previously, such settlement also increases the risk of conflict between the resettlers and the host population.

Raising Living Standards and Making Resettlers Project Beneficiaries

Introduction

Since large dams will continue to be built, what can be done to improve the record with involuntary resettlement? The paragraphs that follow first examine strategies endorsed by the World Bank as the major international financier of large-scale river-basin development projects. Necessary but not sufficient, they need be complemented by a range of other approaches, including more knowledge of what works and doesn't work on the basis of long-term research. Future strategies also must require greater local participation and the improvement of living standards, while feasibility studies must focus more on hosts and other project-affected people. That in itself will attract more attention to alternatives. Also necessary are additional development initiatives which have yet to receive sufficient attention. Only then can the odds of further resettler impoverishment be reduced.

Actions taken or intended by the World Bank

In the final section of their 1994 review, the World Bank lists a series of "actions to improve performance." All are important; they warrant summarizing here. Most important is the recommendation to improve project design in ways that "avoid or reduce displacement." Some success has already been achieved there with previous Bank-financed projects, such as Saguling, where lowering the height of the dam 5 m during the design stage reduced resettlement by nearly 50 percent.

To improve government capabilities, recommended strategic priorities were to "enhance the borrower's commitment" by financing only those projects with acceptable policies and legal frameworks; "enhance the borrower's institutional capacity"; "provide adequate

Bank financing"; and "diversify project vehicles," whereby the Bank complements the financing of physical infrastructure with standalone resettlement projects. Although too few have been funded to date, financing standalone resettlement projects, such as the $110 million credit for the Xiaolangdi Project on China's Yellow River, is especially important. Two reasons are the increasing financial cost of adequate resettlement and the fact that resettlement with development takes more time than constructing physical infrastructure; hence the risk, as with Ghana's Kpong, that funds will be exhausted by the time they are needed unless they are from a protected source.

Other recommended strategic priorities include the need to "strengthen the Bank's institutional capacity" so as to improve the Bank's ability to deal with the different stages of the project cycle, and to improve "the content and frequency of resettlement supervision." Where possible, "remedial and retrofitting actions" are also emphasized in connection with previously funded, but inadequately implemented, World Bank–funded projects such as Kenya's Kiambere. One future candidate for such action would be Kariba North Bank, where Zambia's Electricity Supply Corporation (ZESCO) has indicated an interest in helping with resettlement area rehabilitation. Finally, the Bank emphasizes that more attention will be paid to promoting "people's participation" and NGO facilitation of "local institutional development" (1994a, pp. 8/1–8/4).

Greater local participation and NGO involvement

If the living standards of project-affected people are to improve, significantly increased local participation is absolutely essential. Equally essential is increased involvement of local and international NGOs. The local people–NGO linkage is necessary because of several characteristics of involuntary resettlement. As argued in Chap. 14, people's participation should commence during the earliest stages of project planning. That means involvement in the appraisal process so that local people not only become more aware of project implications for their lives but also can use that knowledge to plan for their becoming project beneficiaries.

A major constraint to such early local participation is the fact that planning for large projects is apt to precede a decision to proceed or not to proceed by one or more decades. During that time period local people are apt to be unaware of what is going on, hence the need for NGO involvement. Increasingly effective in monitoring proposed water-resource development schemes, there now is a need for NGOs such as the Environmental Defence Fund (EDF) and the International Rivers Network (IRN) to play a broader role than is currently the case. That role should include alerting local people early on to a possible project. It should also include helping those people increase their visibility and especially their political clout. And it requires their assistance in helping project-affected people develop the institutional capacity to participate in the planning process.

Various constraints to making such a linkage effective have been outlined in Chap. 14. At the local level, they include a global erosion of local participatory institutions. At the government level they include unwillingness to allow decentralized decision making and finance as well as inadequate laws and judicial institutions. But there are also NGO constraints. These may be financial in the case of local NGOs, in which case assistance from international donors will be essential as when OXFAM (Oxford Committee for Famine Relief) provided assistance to local NGOs resisting India's Sardar Sarovar Project or when OXFAM and Christian Aid provided assistance to the Highlands Church Action Group for improving its capacity to work with local people affected by the Lesotho Highlands Water Project.

NGO constraints may also be ideological, including, for example, an agenda that opposes all water-resource development projects that involve major infrastructure. In such cases, early linkage with project-affected people will at least help such NGOs to familiarize themselves with what local people want. Worldwide, the large majority of river-basin inhabitants want to raise their living standards; in other words, they want development. Among other prerequisites, that will require some electrification. If river basin development is a reasonable alternative for providing energy and other benefits at a reasonable cost (as ascertained by local people through their participation in the planning process), then NGOs will have to decide whose interests come first—theirs or project-affected people's.

A case in point is the Lesotho Highlands Water Project, which has the potential to play a critically important role in South Africa's development. If successful, that development will provide employment and raise living standards for millions of low-income South Africans, increasing stability throughout southern Africa, including Lesotho. In the process, however, thousands of project-affected people in the highlands of Lesotho may become worse off if implementation of the Lesotho Highlands Development Authority's Rural Development Plan continues to be inadequate.

There are no easy answers for NGOs, as the example of India's Sardar Sarovar illustrates. A local NGO, ARCH-Vahini, was carrying on social welfare work in some of the 19 SSP-impacted villages in the state of Gujarat prior to the commencement of dam construction. Early on, ARCH-Vahini used the court system and other means to ensure a wider range of benefits accruing to those villagers. When its efforts were successful, it joined with government agencies in implementing resettlement through its involvement in committees acquiring land for resettler occupation.

Becoming part of the resettlement-development process in Gujarat was at the expense of ARCH-Vahini's relationship with NGOs working with resettlers in the other two project-affected states. While pressured to improve conditions for the relatively small minority of SSP resettlers who came from within its own boundaries, Gujarat did not honor required obligations to facilitate resettlement with development for resettlers from the other two states, who had the option of relocating to irrigable land in SSP command areas within Gujarat. Nor were those two states willing to plan adequately for the in-state resettlement of that majority of their citizens who wished to remain close to their former homes. In those two states local NGOs continue efforts to stop a project, the implementation of which will worsen the living standards of well over 100,000 tribal and peasant people if canal resettlers are added to reservoir resettlers. While understandable—because of differing circumstances and the difficult decisions involved—the existing split between the various NGOs is not in the interests of local people. Far better would have been maintenance of a unified front based on agreement as when to complement adversarial positions with cooperative positions.

The various other means for facilitating local participation outlined in Chap. 14 are all equally applicable to resettlement, especially where residents in a future reservoir basin are not dispersed over a wide area. In such cases, "people's trusts," combined with marketing networks (Reynolds, 1981), present the opportunity for resettlers to actively decide how to use revenue allocated for their resettlement and development. Joint ventures with private-sector firms involved in agricultural, industrial, and tourism activities arising from the project also warrant careful assessment. Then, too, there are new opportunities provided by the reservoir to the resettlers.

Higher goals: raising living standards

If the goal of resettlement projects is to maintain living standards, it is essential that they be improved. This is for the reasons already outlined. First, because of the long planning process involved in major water-resource development projects, living standards for the majority in future reservoir basins have been adversely affected in most cases even before resettlement begins. Second, their rate of "without project" improvement will also exceed that which occurs during the years immediately preceding and following resettlement. Third, living standards tend to drop immediately following removal except for that minority of families benefiting from project employment. Furthermore, following removal, costs are apt to increase because of the need to purchase food and other commodities such as fuel which were grown or otherwise available locally.

To become project beneficiaries as well as to contribute to the stream of project benefits, the total of all water-resource development projects for resettlers should be to improve living standards still further. Because of increased expenses that accompany resettlement with development, inflation, and the risk that what baseline data are obtained will underestimate income (hence making it easier for borrowers and donors to state that incomes have "improved" following removal), income alone is an insufficient index for measuring improvement. For ascertaining impacts on living standards, other indices should be added for assessing housing and household furnishings, lighting and fuel, water and sanitation, and production equipment as well as access to such social services as schools and health facilities.

Development strategies

Land for land. For resettlers who wish to remain agriculturists, there is no substitute for a resettlement-development strategy that provides new arable land for land lost. Land is a heritable resource that can be passed on to future generations. That is not the case with such other development options as employment in new industries or other nonfarm occupations. Although pensions and people's trusts could be heritable, the former present problems in the case of divorce or other family unraveling, while experience with people's trusts has yet to encompass a second generation.

Government officials, as in India's Sardar Sarovar Project and Lesotho's Highland Water Project, may conclude that no further land is available. In both these cases, however, evidence to the contrary was produced by NGOs as well as by owners themselves. To access such land, a range of options exist in addition to government-facilitated "willing seller–willing buyer" transactions, use of state lands, or government acquisition based on eminent domain. One example includes land sharing with host communities. Practiced in China, host communities are asked to sign a land-sharing agreement to "take care" of relocated families in return for a share of project-development funds that are used for intensifying production on host lands that are not shared with resettlers (written communication from Martin ter Woort, 1996). Such an approach is especially applicable to cases in which irrigation replaces rain-fed cultivation as in India or where high-value crops replace low-value ones, as with tree crops in China.

Forecasting reservoir levels and land-use planning within the drawdown area. One of the few benefits arising from reservoir formation is the productivity of the drawdown area for flood-recession (and perhaps flood-rise) agriculture, grazing, and aquaculture. At Kariba, where recurrent droughts have devastated the lake basin since the early 1980s, crop production from thousands of hectares of land in the drawdown area that are cultivated in maize interplanted with legumes and cucurbits has been the main source of food during 4 out of 6 years during the 1990s. Colonization of extensive areas by two indigenous grasses (*Panicum repens* and *Vossia* sp.)—which provide some of the best grazing in central Africa—has enabled many cattle, driven down to the lakeside from overutilized inland pastures, to survive. At Volta, flood-recession agriculture and use of small pumps has supported a flourishing vegetable industry, while at Kainji another indigenous grass (*Echinochloa stagnina*) is harvested as an important cash crop for the livestock industry. In each case, other productive uses are possible, including building weirs across flooded inlets which become fish ponds when water recedes, or perhaps even planting floating rice when reservoirs commence rising.

In each case, land-use planning which includes forecasts on a fortnightly basis about falls and rises in reservoir levels would not only make better use of available resources but also increase the area that could be utilized by different crops (including fodder crops in lower areas exposed for shorter time periods) as well as the reliability of such utilization. Zoning the drawdown area for different uses other than the cultivation of crops and grazing, including uses for forest reserves, wildlife management, and tourism, would also reduce conflicts among users. For example, at Kariba current conflicts occur within villages between those who wish to use the drawdown area for grazing and those who wish to use it for crop production. Conflicts also involve villagers with educated, wealthy, and politically well-connected emigrants and outsiders who would like to privatize village lands for commercial activities.

On the basis of sophisticated hydromet systems for flood controland hydropower-generation purposes, forecasts of reservoir behavior could be made for the benefit of farmers by whatever agencies have dam management responsibilities. At least for Africa, however, no such forecasts are made in spite of the appreciable benefits that could follow quickly at low financial cost. Benefits for a still greater number of project-affected people could follow if reservoir drawdown was synchronized with controlled releases downstream (Scudder, 1980). As discussed in Chap. 14, Institute for Development Anthropology research on the Senegal River has shown that economic benefits for downstream users from controlled releases from the Manantali Reservoir would exceed whatever costs might accrue from reduced electricity generation (Horowitz et al., 1990). If such releases were synchronized with reservoir drawdown, the accuracy of forecasts could be further improved for the benefit of project-affected people within the reservoir basin.

Joint ventures between resettlers and the private sector. Although increasing emphasis is placed by donors on private-sector involvement in development, scant attention has been paid to ways in which private-sector initiative, capital, and management might be tapped to facilitate resettler development. Although the same is true with government-sponsored settlement schemes involving settler volunteers, at least some joint ventures involving cotton, sugar cane, and tree crops have been implemented in Africa and Asia with encouraging results. Ideally, the private sector would provide processing facilities for crops, fish, and other animal products, hence providing employment while eliminating marketing constraints. Should they require their own nuclear estate for experimental work and some production, that should be kept as small as possible, with most production provided by resettler and host outgrowers.

In the Zambian portion of the Middle Zambezi Valley, the author discussed such joint ventures with three private-sector business enterprises during summer 1995. One had already funded, on its own initiative, 800 ha of outgrowers cotton for its ginnery during the 1994/95 rainy season. Although that experiment proved costly, because of drought-caused low production, the firm intends a similar experiment during the 1995/96 rainy season. Further expansion, however, would require additional funding. A second enterprise has provided a clinic for local resettlers and hosts as well as improving, maintaining, and extending a portion of the road system. Their hope is to connect one road to their own lake transport system to serve fishing villages currently being implemented by the Department of Fisheries and cross-reservoir transport to Zimbabwe. They are also interested in joint ventures with adjacent villages that deal with wildlife management and tourism. As for a third private-sector business, in addition to providing employment for up to 2500 workers during the agricultural off-season, they are interested in a possible outgrowers scheme for such crops as paprika, cotton, sesame (for oil production), and papaya (for extracting papain).

Such joint ventures should be established only after careful planning to ensure that they are actually in the interest of resettlers, hosts, and other project-affected people. Other private-sector initiatives at Kariba have actually penalized local villagers by coopting already scarce land which is no longer available to them. On the other hand, potential benefits that could arise from private sector–resettler joint ventures warrant taking some risks on at least a pilot basis. For example, in Sri Lanka's Mahaweli System H, an experiment in which the Ceylon Tobacco Corporation assumed responsibility for distributary canal system maintenance, agricultural extension, initiation of water-user associations, and marketing activities worked well until taken over by the government's Mahaweli Authority (World Bank, 1985). Similarly, a private bank in a portion of that system has done a credible job in providing agricultural credit and advice to individual farmers and farm associations.

Periurban enterprise development. Increasingly, lack of suitable arable land is requiring resettlement programs to attempt shifting rural farm families into nonfarm occupations. While educated youth may welcome such an opportunity, it is far more difficult for less-well-educated older men and women. For them, an intriguing compromise would be resettlement as periurban farmers. In Lesotho, where the government states that there is no replacement land, preliminary surveys in the Mohale Reservoir basin (Phase 1B) indicate that approximately 50 percent of project-affected people would prefer resettlement to lowland towns if they must shift from their present mountain villages (Tshabalala, 1994). For those households headed by older men and women, periurban horticulture, fruit growing, and small livestock management presumably would have considerable appeal.

According to Sechaba Consultants (1995, p. 116), "home gardens are the single largest source of domestic supply of vegetables." Although a relatively small percentage of such gardens, 121 ha were planted in urban areas during the 1991/92 summer season. While 92 percent of sales in rural areas went directly to consumers as opposed to vendors, the proportion sold to vendors in urban areas, where home plot growers presumably have a comparative advantage, rose to 38 percent. Lesotho is not the only south African country where periurban horticulture is of increased importance. Other examples include the greenbelt surrounding Mozambique's capital, where women's initiative has created a band of small irrigated gardens that serve the urban market, and small-scale producers close to Kenya's Nairobi. Combining horticulture with fruit production and small-livestock management could further diversify production and increase income levels. The potential for that increasing proportion of resettlers who are willing to shift to towns either within or without reservoir basins would appear to be considerable, provided households have access to sufficiently large plots (e.g., 500 m^2) and water. Yet, with the major exception of China, which places considerable emphasis on what is called "sideline" production at the household level, that potential is seldom integrated into resettlement planning.

Research and monitoring

Major river-basin development projects are apt to be the largest single investment in national development plans. Although the World Bank has been the major donor financing such projects, neither the Bank, other donors, nor governments have completed internally or funded externally long-term studies of the resettlement process. That deficiency should be corrected. While the Bank has pioneered the funding through borrower agencies of environment and resettlement panels that draw on international expertise to complement dam-safety panels, such panels at best play a supervisory role, based on one or two short visits per year. As for independent monitoring by research institutions that the Bank requires for major projects, they, too, are no substitute for detailed research in the field.

Especially important are long-term studies of projects where planning is for success and where results during the initial years of implementation are encouraging. Comparative analysis of four major projects in China—Shuikou, Xiaolangdi, Longtan, and Three Gorges—would be especially appropriate. Shuikou is already considered by the World Bank to be a success story, while resettlement experts at Acres International are impressed with construction-site resettlement at Xiaolangdi. Well aware that human-rights and environmental advocates are watching carefully, the Government of China has already commenced applying its national resettlement policies to Longtan and Three Gorges. But resettlement results must be carefully assessed through time. Does "success" continue after 5, 10, and more years? Does it extend to the next generation? If, yes, why? If no, what constraints were involved? Far more knowledge is needed of what activities bring success, and whether they are sustainable and transferable.

More knowledge, as provided by long-term studies, is also needed on the impacts of increased population densities that accompany most resettlement on the physical and biotic environment if the type of degradation which has plagued Kariba resettlement is to be avoided in the future. Also needed is more research on how different categories of people respond during the various stages of the resettlement process and on the policy implications of those responses. Such studies must continue during the later stages; our knowledge is weakest about the outcomes of resettlement programs a generation or more after their initiation. Especially valuable would be a comparative analysis after, say, 20 years of a number of resettlement projects which earlier research had suggested were better than most. Such research would help us learn the extent to which initial benefits were passed on to the second generation and were sustainable.

Conclusion

The record to date with resettlement in connection with the large majority of water-resource development projects is unsatisfactory. Few countries have the national or sectoral policies required for planning and implementing resettlement with development. Even where the political will exists to implement such policies, such as currently appears to be the case in China, the odds are not encouraging because of the complexity and difficulty of the resettlement process and of other constraints. To the fullest extent possible, other alternatives that do not require resettlement should be more carefully assessed.

Where resettlement is inevitable, the first step should be to examine the extent to which different designs will alter the magnitude of resettlement, by, for example, different siting and height of dams, or construction of protective works around densely populated areas. For what resettlement remains, the basic goal and commitment *must* be not merely to restore incomes but also to improve living standards. Whether on human-rights, socioeconomic, or environmental grounds, if a project cannot make beneficiaries of those who are most affected, it should not be implemented. Once appropriate goals are set, the widest possible range of strategies should be examined for improving the living standards of resettlers, hosts, and, as argued in Chap. 14, other project-affected people. The record to date indicates that the complexities involved require the cooperation of the implementing agencies with all project-affected people, NGOs, the private sector where appropriate, and donors.

Acknowledgments

This chapter draws heavily on several previous publications of the author, including "Development refugees: reflections on involuntary relocation in connection with dam construction," to be published in *Barrages Internationaux et Cooperation,* Francoise Conac, ed., Pantheon-Sorbonne, University of Paris I. The author is also indebted to Daniel Aronson, Elizabeth Colson, Scott G. Ferguson, Michael M. Horowitz, and Martin ter Woort for commenting on an earlier version. References are supplemented by the author's field notes dealing with the following projects listed according to number of resettlers or potential resettlers: Three Gorges (China), Danjiangkou (China), Sardar Sarovar (India), Aswan High Dam (Egypt and Sudan), Akosombo (Ghana), Longtan (China), Mahaweli (Sri Lanka), Kariba (Zambia and Zmbabwe), Kainji (Nigeria), Yantan (China), Kpong (Ghana), and Highlands Water (Lesotho).

References

Ault, S. K. (1989). "The health effect of involuntary resettlement." Paper prepared for the Task Force on Involuntary Resettlement of the American Anthropological Association.

Beijing Review (April 6–12, 1992). "Trial resettlement of residents." Vol. 35, no. 14, pp. 28–32.

Cernea, M. M. (1988). *Involuntary Resettlement in Development Projects: Policy Guidelines in World Bank-Financed Projects.* World Bank Technical Paper no. 80. Washington, D.C.: World Bank.

Cernea, M. M. (1990). *Poverty Risks from Population Displacement in Water Resources Development.* Development Discussion Paper 355. Cambridge, Mass: Harvard Institute for International Development.

Cernea, M. M. (1994a). "Hydropower dams and social impacts: a sociological perspective." Paper presented at the International Hydropower and Environment Conference, Oslo-Hofmenkollen, Sept. 12–14.

Cernea, M. M. (1994b). "Population resettlement and development," *Finance & Development* (Sept.): 46–49.

Cernea, M. M. (1996). "Understanding and preventing impoverishment from displacement: reflections on the state of the knowledge." In *Resisting Impoverishment—Tackling the Consequences of Development-Induced Displacement,* C. McDowell, ed. Oxford: Berghahn Books.

Cernea, M. M., and S. E. Guggenheim, eds. (1993). *Anthropological Approaches to Resettlement: Policy, Practice and Theory.* Boulder, Colo.: Westview Press.

Chambers, R. (1969). *Settlement Schemes in Tropical Africa: A Study of Organizations and Development.* New York: Praeger.

Clark, S., E. Colson, J. Lee, and T. Scudder (1995). "Ten thousand Tonga: a longitudinal anthropological study from southern Zambia, 1956–1991," *Population Studies* **49:** 91–109.

Colson, E. (1971). *The Social Consequences of Resettlement: The Impact of the Kariba Resettlement upon the Gwembe Tonga.* Manchester University Press for Institute for African Studies, University of Zambia.

de Wet, C. (1995). *Moving Together, Drifting Apart: Villagisation in a South African Homeland.* Johannesburg: Witwatersrand University Press.

Fernandes, W., J. C. Das, and S. Rao (1989). "Displacement and rehabilitation: extent and prospects." In *Development, Displacement, and Rehabilitation: Issues for a National Debate,* W. Fernandes and E. G. Thukral, eds. New Delhi: Indian Social Institute.

Fried, M. (1963). "Grieving for a lost home." In *The Urban Condition,* L. Duhl, ed. New York: Basic Books.

Futa, A. B. (1983). "Water resources development—organization of a resettlement programme (a case study of the Kpong Resettlement Programme in Ghana)," *Water International* **8:** 98–108.

Gans, H. J. (1962). *The Urban Villagers,* New York: Free Press of Glencoe.

Gans, H. J. (1968). *People and Plans: Essays on Urban Problems and Solutions.* New York: Basic Books.

Goldemberg. J., T. B. Johansson, A. K. N. Reddy, and R. H. Williams (1987). *Energy for a Sustainable World.* Washington, D.C.: World Resources Institute.

Goldsmith, E., and N. Hildyard (1984, 1986). *Social and Environmental Effects of Large Dams,* Vols. 1 (*Overview*) and 2 (*Case Studies*). Camelford, Cornwall (U. K.): Wadebridge Ecological Centre.

Goodland, R. (1994). "Ethical priorities in environmentally sustainable energy systems: the case of tropical hydropower." In *Energy Needs in the Year 2000: Ethical and Environmental Perspectives,* W. R. Shea, ed. Canton, Mass.: Watson Publishing International.

Government of Uganda (1995). *Resettlement Policy and Institutional Capacity for Resettlement Planning in Uganda.* Kampala: Office of the Prime Minister.

GTZ (Deutsche Gesellschaft fur Technische Zusammenarbeit) (1993). *Report of a Project Progress Review of Siavonga Agricultural Development Project* (SADP)—*Zambia: Summary Report.* GmbH, Eschborn: GTZ.

Guggenheim, S. (1994). *Involuntary Resettlement: An Annotated Reference Bibliography for Development Research.* Environment Working Paper no. 64. Washington, D.C.: The World Bank.

Gunaratna, M. H. (1988). *For a Sovereign State.* Ramalana, Sri Lanka: Sarvodya Vishva Lekha.

Hansen, A., and A. Oliver-Smith, eds. (1982). *Involuntary Migration and Resettlement: The Problems and Responses of Dislocated People.* Boulder, Colo.: Westview Press.

Hartman, C. (1974). *Yerba Buena: Land Grab and Community Resistance in San Francisco.* San Francisco: Glide Publications for University of California, Berkeley.

Horowitz, M. M., M. Salem-Murdock, et al. (1990). *The Senegal River Basin Monitoring Activity Synthesis Report.* Binghamton, N.Y.: Institute for Development Anthropology.

Horowitz, M. M., D. Koenig, C. Grimm, and Y. Conate (1993). "Resettlement at Manantali, Mali: short-term success, long-term problems." In *Anthropological Approaches to Resettlement: Policy, Practice and Theory,* M. M. Cernea and S. Guggenheim, eds. Boulder, Colo.: Westview Press.

Koenig. D. (undated manuscript). "Women and resettlement."

McCully, P. (1996). *Silenced Rivers: The Ecology and Politics of Large Dams.* London: Zed Books.

Morse, B., and T. Berger (1992). *Sardar Sarovar: The Report of the Independent Review.* Ottawa: Resource Futures International.

Nelson, M. (1973). *The Development of Tropical Lands: Policy Issues in Latin America.* Baltimore: Johns Hopkins Press for Resources for the Future.

OECD (Organization for Economic Cooperation and Development) (1992). *Guidelines for Aid Agencies on Involuntary Displacement and Resettlement in Development Projects.* Development Assistance Committee. Paris: OECD.

Partridge, W. L., A. B. Brown, and J. B. Nugent, eds. (1982). "The Papaloapan Dam and Resettlement Project: human ecology and health impacts." In *Involuntary Migration and Resettlement: The Problems and Responses of Dislocated People,* A. Hansen and A. Oliver-Smith, eds. Boulder, Colo.: Westview Press.

Pearce, F. (1992). *The Dammed: Rivers, Dams, and the Coming World Water Crisis.* London: Bodley Head.

PEEM (Panel of Experts on Environmental Management for Vector Control) (1985). "The environmental impact of population resettlement and its effect on vector-borne diseases." In *Report of the Fifth Meeting.* Geneva: WHO/PEEM Secretariat.

Refugee Studies Programme (1996). *Resisting Impoverishment—Tackling the Consequences of Development-Induced Displacement,* C. McDowell, ed. Oxford: Berghahn Books.

Reynolds, N. (1981). *The Design of Rural Development: Proposals for the Evolution of a Social Contract Suited to Conditions in Southern Africa.* Parts I and II. Saldru Working Papers no. 41 and 42. Capetown: Southern Africa Labour and Development Research Unit.

Roder, W. (1994). *Human Adjustment to the Kainji Reservoir in Nigeria: An Assessment of the Economic and Environmental Consequences of a Major Man-made Lake in Africa.* Lanham, Md.: University Press of America.

Salem-Murdock, M. (1989). *Arabs and Nubians in New Halfa: A Study of Settlement and Irrigation.* Salt Lake City: University of Utah Press.

Scudder, T. (1973). "The human ecology of big projects: river basin development and resettlement." In *Annual Review of Anthropology,* B. Siegel, ed. Palo Alto, Calif.: Annual Reviews.

Scudder, T. (1975). "Resettlement." In *Man-Made Lakes and Human Health,* N. F. Stanley and M. P. Alpers, eds. London: Academic Press for Institute of Biology.

Scudder, T. (May 1977). "Some policy implications of compulsory relocation in connection with river basin development and other projects impacting upon low income populations" and subsequent memo, "Points referring to all categories of settlement projects." Unpublished papers prepared for the World Bank's In-House Seminar in Sociology.

Scudder, T. (1980). River basin development and local initiatives in African savanna environments. In *Human Ecology in Savanna Environments,* D. R. Harris, ed. London: Academic Press.

Scudder, T. (1981). "What it means to be dammed: the anthropology of large-scale development projects in the tropics and subtropics," *Engineering and Science* **54**(4): 9–15.

Scudder, T. (1991). A new sociological framework for the analysis of new land settlements. In *Putting People First: Sociological Variables in Rural Development,* 2d ed., M. M. Cernea, ed. New York: Oxford University Press for the World Bank.

Scudder, T. (1993). "Development-induced relocation and refugee studies: 37 years of change and continuity among Zambia's Gwembe Tonga," *Journal of Refugee Studies.* **6**(2): 123–152.

Scudder, T., and E. Colson (1982). "From welfare to development: a conceptual framework for the analysis of dislocated people." In *Involuntary Migration and Resettlement: The Problems and Responses of Dislocated People,* A. Hansen and A. Oliver-Smith, eds. Boulder, Colo.: Westview Press.

Scudder, T., and J. Habarad (1991). "Local responses to involuntary relocation and development in the Zambian portion of the Middle Zambezi Valley." In *Migrants in Agricultural Development,* J. A. Mollett, ed. London: Macmillan.

Scudder, T., and M. E. D. Scudder (1981). *The Social Impacts of the John Wayne Airport of Orange County.* Report commissioned by the Mariners Community Association.

Scudder, T., and K. P. Vimaladharma (1989). "*The Accelerated Mahaweli Programme (AMP) and Dry Zone Development: Report Number 7.*" Binghamton, N.Y.: Institute for Development Anthropology, sponsored by the United States Agency for International Development.

Sechaba Consultants (1995). *Lesotho's Long Journey: Hard Choices at the Crossroads.* Maseru: Sechaba Consultants for Irish Aid.

Tremmel, M. (1994). *The People of the Great River: The Tonga Hoped the Water Would Follow Them.* Silveira House Social Series no. 9. Gweru. Zimbabwe: Mambo Press.

Tshabalala, M. (1994). *Phase 1B Socioeconomic Census Report: Mohale, 1993,* Vol. 1 (main report). Maseru, Lesotho: Environmental Department. Lesotho Highlands Development Authority.

Weissman, M. M., and E. S. Paykel (1972). "Moving," *Yale Magazine.*

World Bank (1980). *Social Issues Associated with Involuntary Resettlement in Bank-Financed Projects.* Operational Manual Statement 2.33. Washington, D.C.: The World Bank.

World Bank (1985). *The Experience of the World Bank with Government-Sponsored Land Settlement.* Report no. 5625. Operations Evaluation Department. Washington, D.C.: The World Bank.

World Bank (1986). *Operations Policy Issues in the Treatment of Involuntary Resettlement.* Operational Manual Statement no. 10.08. Washington, D.C.: The World Bank.

World Bank (1990). *Involuntary Resettlement.* Operational Directive no. 4.30. Washington, D.C.: The World Bank.

World Bank (1993a). *Early Experience with the Involuntary Resettlement: Impact Evaluation on Thailand Khao Laem Hydroelectric (Loan 1770-TH).* Report no. 12131. Washington, D.C.: The World Bank, Operations Evaluation Department.

World Bank (1993b) *Early Experience with Involuntary Resettlement: Impact Evaluation on India Maharashtra Irrigation II Project (Credit 954-IN).* Report no. 12133. Washington, D.C.: The World Bank, Operations Evaluation Department.

World Bank (1993c). *Early Experience with Involuntary Resettlement: Impact Evaluation on Ghana Kpong Hydroelectric Project (Loan 1380-GH).* Report no. 12141. Washington, D.C.: The World Bank, Operations Evaluation Department.

World Bank (1993d). *Early Experience with Involuntary Resettlement: Overview*. Report no. 12142. Washington, D.C.: The World Bank, Operations Evaluation Department.

World Bank (1994a). *Resettlement and Development: The Bankwide Review of Projects Involving Involuntary Resettlement 1986–1993*. Washington, D.C.: The World Bank.

World Bank (1994b). *Resettlement and Rehabilitation in India: A Status Update of Projects Involving Involuntary Resettlement* (in two vols). Washington, D.C.: The World Bank.

Index

Absolute subsidence, 134
Abstraction points, 118
Acid deposition:
  African lakes, 344
  fishery resource threats, 443, 446, 447
Acidification:
  water quality modeling, 250
  water quality monitoring, 217–218
A/C index, 306
Acrolein, 493
Administration (*see* Institutional principles for resource management)
Adsorption, 123, 124
Advection, 123, 257
Advection-dispersion equations, 109, 258–259
Advocacy NGOs, 564–565
Aeration zone, shallow aquifers, 122
Africa:
  drinking water sources, 125
  erosion in, 40, 43
  eutrophication in, 323, 325, 343–345
  fishery resources, 451–452
    management and enhancement of, 466
    stocking practices, 464
  potable water reuse, 249–430
  social impacts of development projects (*see* Resettlement; Social impacts of development projects)
Agasicles hygrophila, 502
Agenda 21, 210–211
Aggregate dead zone (ADZ) model, 264–266, 291
Agriculture, 2, 702, 703
  contaminants of groundwater, 139, 141, 142
Agriculture (*Cont.*):
  erosion, 39–42
    conservation planting, 45, 46, 47
    control techniques, 59–61
    costs of, 54–56
    mulching, 51
    universal soil-loss equation, 92
  and eutrophication, 308, 358–359, 362
  fishery resource threats, 443, 446, 451, 456
  integrated approach to groundwater management, 200, 201
  social impacts of development projects, 647, 655
  wastewater reuse, 385, 386
  water quality modeling, 251
  water quality monitoring, 228
  water use in, 38–39
    global trends, 5, 6, 8, 126
    irrigation (*see* Irrigation; Waterlogging and salinity)
  weeds and, 481
  (*See also* Erosion; Irrigation)
Airborne contaminants (*see* Acid deposition; Atmospheric deposition)
Algae, 482–483
  eutrophication (*see* Eutrophication)
  fish and, 510
  herbicides, 493, 496–500
  manual removal of, 485
  modeling of phosphorus effects, 276–277
  toxins, 481
  water quality monitoring, 229
  weeds and, 481
Allelochemicals, 501
Alligator weed (*Alternanthera philoxeroides*), 502

Allocation of resources, 542–546
  economic mechanisms (*see* Economic mechanisms for resource management)
  institutional deficiencies, 538–539
  opportunity costs and, 572
Alluvial aquifer, 120, 121
Alluvial channel boundaries, 76
Alluvial deposits, land subsidence, 131
Alluvial plains:
  conjunctive use of surface water and groundwater, 186
  surface and groundwater interdependence, 128–129
Alpine lake eutrophication, 324, 350
Alternanthera philoxeroides (alligator weed), 502
Altitude, and eutrophication, 309
Aluminum, fishery resource threats, 446
Amazonian tropical lake eutrophication, 309
Amazon River fish resources, 454
Ambient permits, 596
Amelioration, 530
Anabranching stream, 74, 76
Analysis, chemical (*see* Water quality monitoring)
Angle of repose, 81
Anopheles gambiae, 32
Aquaculture, 440, 466, 702
  in Asia, 454
  environmental impact of, 448–450
  fishery resource threats, 448–449
  status of, 442
  wastewater reuse in, 408
  weeds and, 481
Aquashade, 509
Aquatic communities:
  fishery enhancement, 460–461
  water quality bioassays, 241–242
Aquatic weeds (*see* Weeds, aquatic)
Aquifers:
  assessment approach, 167
  confined storage coefficients, 121
  contamination of, 135
  flow system characterization, 118
  hydrogeologic characteristics, 120
  mechanics of land subsidence, 131
  regulation and legislation, 197
  resource management and planning approaches, 174–175
  role in water-resource systems, 127–128
Aquifers (*Cont.*):
  saltwater intrusion, 139, 143–147
  transport processes, 122
Aral Sea, 454
Araphinidineae (A/C index), 306
Argentina:
  eutrophication in, 345
  fishery resources, 455
  population and per capita water availability in, 9
  social impacts of development projects, 660
Arid and semiarid regions:
  environmental impact assessment, 169
  increasing demand for water, 7
Arizona:
  land subsidence in, 132–133
  wastewater reuse, 387
Arizona Nuclear Power Project, Palo Verde Nuclear Generating Station, 414–415
Asia:
  contaminants of groundwater, 141
  drinking water sources, 125
  erosion in, 40
  eutrophication in, 320, 322, 324, 333, 338–341
  fishery resources, 452–454, 463, 464
  land subsidence in, 132–133
  social impacts of development projects, 635
  water quality in, 142
  water use trends, 5, 6, 126
Asulam, 492
Aswan Dam, 25–26, 34, 93, 655
  resettlement of dislocated populations, 677, 681
  social impacts of development projects, 627, 634, 636, 643, 660
Atmospheric deposition:
  African lakes, 344
  and eutrophication, 358
  fishery resource threats, 446
  water quality modeling, 251
  water quality monitoring, 228
Attenuation capacity, pollution risk assessment, 173
Australia:
  erosion in, 40
  eutrophication in, 320, 341–342, 354–355
  fishery resources, 456–457, 466
  global water use trends, 6

Australian swamp stonecrop, 525
Azolla, 487

Bangladesh:
  contaminants of groundwater, 141
  fishpasses in, 461
  groundwater use in, 127
  population and per capita water availability in, 9
  social impacts of development projects, 627, 635
Bankside weeds, 483
Bank storage, 127
Baseflow, 127
Basin management, assignment of, 61–562
Bed load, 88
Bed material, stream properties, 76
Benthos:
  and eutrophication, 300, 307
  fishery resource threats, 444
  modification for weed control, 504
  water quality models, 257
  weed control, 516
Bentonite, 86
Bhima project, 28, 32
Bioassays, water quality monitoring, 234, 241–242
Biochemical oxygen demand:
  mass balance modeling, 260
  water quality modeling, 250–251, 252
  water quality model predictive equations, 287–288
  water quality monitoring, 213, 227, 241, 243
Biocoenotic models of water quality, 255–256
Biodiversity, soil, 55–56
Biological contaminants of groundwater, 139, 140
Biological control methods for aquatic weeds, 500–507
Biological indicators of eutrophication, 306–308
Biological processes:
  eutrophication (*see* Eutrophication)
  reactions of contaminants, 123, 124
  water quality models, 256–257
  water quality monitoring, 255–256
Biomanipulation:
  eutrophic lake restoration, 360–361
  fishery resource protection, 458
Biomanipulation (*Cont.*):
  weed control, 504–505
Biomass fuels, 29
Biota:
  soil erosion and, 50–51, 52, 53
  water quality monitoring, 224
  (*See also* Microbiology)
Black-box models of water quality, 253–254
BLOOM II model, 356
Boats, weed-cutting, 485–487
Boreholes, 221, 223
Boron, 402, 405
Boulders, 75, 76
Brahmaputra basin fishery, 453, 464
Braided stream, 74, 75, 76
Brazil:
  eutrophication in, 344–345
  fishery resources, 455, 456, 462
  population and per capita water availability in, 9
  social impacts of development projects, 660
  water quality monitoring, 241
Brine-injection wells, 136
Buckets, weed, 487–488
Bulrushes, 523

Cadmium, 455
Calcium carbonate, 513
Calcium ions, 101, 102, 103
Calibration of water quality prediction models, 255, 256
California:
  land subsidence in, 131, 132–133
  Wastewater Reclamation Criteria, 431, 435
  wastewater reuse, 417–419, 408–413, 423–426, 431–435
Cambarus, 521
Canada:
  eutrophication in, 312, 313, 314, 318, 320, 321, 337–338
  James Bay Project, 627, 633, 636–642, 643, 660
  population and per capita water availability in, 9
  water quality monitoring standards, 213
Capital investment, 534, 581
  financing, 573–574
  pricing issues, 603

Capture fisheries status, 441–442
Carbon:
  and eutrophication, 312
  fishery resource threats, 447
Carbonaceous biochemical oxygen demand (CBOD), 273
Carbonates, prediction of soil salinity changes, 108
Carbon cycle, water quality models, 290
Carbon dioxide, weed control, 512–513
Carp, 462, 466, 505, 506–507, 521
Caspian Sea, 454
Catla catla, 452
Centrales (A/C index), 306
Ceratophyllum, 482
Cercospora rodmanii, 503
Chanute, Kansas episode, 427–429
Chemical contaminants of groundwater, 139, 140
Chemical control methods, aquatic weeds, 489–497
Chemical oxygen demand, water quality monitoring, 227
Chemical processes:
  and eutrophication, 300, 301
  reactivity of contaminants, 123, 124
  solute transport models, 154–158
  water quality models, 255–256, 256–257
Chemical contaminants of groundwater (*see* Water quality monitoring)
Chemistry:
  academic disciplines, 199
  water quality monitoring (*see* Water quality monitoring)
Chile, contingent water rights in, 606
China:
  erosion in, 40, 43
  eutrophication in, 305–306, 320, 340–341, 361–362
  fishery resource, 452–453, 464
  global water use trends, 5, 6
  groundwater use in, 127
  irrigation in, 99
  population and per capita water availability in, 9
  resettlement of dislocated populations, 670
  social impacts of development projects, 627, 633, 635, 636, 650
  water use in, 38
Chinese carp, 462
Chinese grass carp, 505, 506–507, 521
Chironomus anthracinus, 307
Chloride:
  irrigation with recycled water, 402, 404
  water quality monitoring, 213, 245
Chlorine:
  irrigation with recycled water, 402
  wastewater pretreatment for groundwater recharge, 416
Chlorophyll:
  eutrophication models, 348–351
  trophic state monitoring, 303, 304
Cirrhinus mrigala, 452
Classic biological control, 501, 502–503
Classification of groundwater resources, 181, 182
Clay, 120
  and eutrophication, 340
  hydraulic conductivity, 120, 121
  sediment size classes, 75, 78
  and stream bed forms, 83
  stream properties, 76
  and subsidence of land, 131
  types of, 81
Climate:
  environmental assessment, 148–149
  and eutrophication, 309
  groundwater recharge area locations, 118–119
  prediction of soil salinity changes, 107
  regional groundwater use, 124–127
  and transport of contaminants, 122–123
  and waterlogging and salinity effects, 100, 101
  water quality effects, 217
  water quality monitoring, 229
  weed control, 514–515
Climate changes, 19
Cobbles, 75, 76, 78
Coefficient of drag, 79
Cohesion of particulates, 81
Colby's method, 89, 90, 91–92
Colorado, wastewater reuse, 387
Colorado River, 93
Common carp, 521
Competition for water resources, 12
Compost, aquatic weed uses as, 523
Compression index, 161
Conductivity, electrical, 245, 402, 403–405
Conductivity, hydraulic, 120–121, 173, 183

Confined aquifers, land subsidence, 131, 132–133
Confined storage coefficients, 121
Confining layers, assessment approach, 167
Conflict resolution, 604–605, 608–610
Conflicts, disputes, wars:
resettled populations, 694–696
social impacts of development projects, 651–652, 659
water rights, 604–605, 608–610
Conglomerate, 120
Conjunctive use of surface and groundwater, 183–189
Connate water, 145
Conservation of water
erosion control and, 59–61
pricing and, 602
Conservation planting, 45, 46, 47, 59–61
Conservative transport models, 154–158
Consolidated formations, 120, 121
Consolidation model, 159–162
Consolidation of soil, 130–131
Consolidometer test, 172
Constrained optimization methods, 288
Contact load, 88
Contaminants (*see* Pollution)
Contingent water rights, 606
Continuously stirred tank reactor (CSTR) method, 260–262, 264–266, 268–270, 278–286
computer programs, 290
detention time, 281–286
segmentation, 286
Controlled releases, 702–703
Convection, 123
Cooling towers, wastewater reuse, 413–414
Copper salts, 493, 497
Costs, 12
environmental control of weeds, 517
reclamation of salt-affected soils, 114–115
of services, 571
treatment of wastewater for groundwater recharge, 420–421
water-loss, erosion and, 54–59
(*See also* Economics)
Crassula helmsii, 525
Crayfish, 521
Crisis in demand for water, 7–14
Critical shear stress, particulates, 87, 88
Cropping practices:
conservation planting, 45, 46, 47, 59–61
Cropping practices (*Cont.*):
mulching, 51
Crop productivity (*see* Erosion)
Crop yield:
on eroded soil, 45, 47–52
irrigation with recycled water, 401, 412
with waterlogging and salinity, 109–113
Ctenopharyngodon idella, 505, 506–507, 521
Cuba:
fishery resource, 462
fishery resources, 455
Culex quinquefasciatus, 32
Culture-enhanced fisheries, 463
Cyanobacteria blooms, 316–319
Cyperus antiquorum, 523
Cyperus papyrus, 344, 523
Cyprinus carpio, 521

Dalapon, 492
2,4-D-Amine, 492
Dams:
fishery resource enhancement, 458, 460, 461
fishery resource threats, 443, 444, 456, 457
infrastructure, 606
institutional foundations of resource management, 548
sedimentation behind, 93–94
social impacts of (*see* Resettlement; Social impacts of development projects)
Danube River fishery resource, 457
Daphnia, 351, 504
Darcy equation, 152–153
Data:
assessment approach, 167, 168, 169, 170
environmental assessment, 147–148
management models, 192–195
water quality monitoring, 220, 221, 237
Data collection:
assignment of functions, 558–59
water quality monitoring (*see* Water quality monitoring)
Data validation, 164
Dead zones, 259, 264–266, 291
Deforestation, 232, 443
Demand for water, 581, 584, 585, 586–591
increases in, 7–14
wastewater reuse, 394–395

Demand management, 174
Demand-oriented approach to groundwater management, 177
Denver Water Department Potable Water Reuse Demonstration Plant, 430–431
Deposition of sediment, 73
Depth:
  and eutrophication, 309
  and stream bedforms, 83–84
  water quality monitoring, 229
  weed growth, 513–514
Design:
  government function, 560–561
  irrigation and drainage channels, 511, 514
  program for water quality monitoring, 220, 221
  water quality monitoring, error sources, 236
Desorption, 123, 124
Deterministic models:
  flow regime, 151–152
  water quality, 253, 254
Developing countries:
  trends in water use, 4, 5
  water quality monitoring, 239–243
  (*See also specific regions*)
Development of water resources, 13–14
  environmental impact assessment, 20–35
  planning, 147
  social impacts of (*see* Resettlement; Social impacts of development projects)
Dichlobenil, 493, 494
2,6-Dichlorobenzonitrile, 493
2,2-Dichloropropionic acid, 492
Diffusion, 123, 257, 258
Diquat, 492, 493
Diquat alginate, 494
Direct injection groundwater recharge, 416, 419–421, 423
Direct reuse of wastewater, 382
Discharges of waste (*see* Pollution; Waste discharges)
Discharge zone, 119, 138, 174–175
Discontinuous rating form, 84
Dispersion, 123, 257, 258
Disputes (*see* Conflicts, disputes, wars)
Dissolved oxygen (*see* Oxygen)
Diuron, 493
Domestic water use:
  comparison with agricultural water use, 38
  global water use trends, 6
  population and per capita water availability, 8, 9, 10, 11
Drainage, 533–534, 568
  conjunctive use of surface water and groundwater, 184
  and crop productivity, 88
  disposal of saline effluents, 115
  financing, 571
  Indus basin, 130
  soil erosion and, 56
  utilities concept, 562–563
  weed control, 504, 514–515
DRAINMOD, 109
Dredging:
  management of algae blooms and eutrophication, 329
  weed control, 488, 516, 519–520
Drinking water:
  classification of groundwater resources, 181
  global patterns of groundwater use, 125–127
  water quality monitoring, 213, 226–227
  (*See also* Potable water)
Drought, 186, 530, 605–606
  conjunctive use of surface water and groundwater, 186
  fishery resource threats, 452
Dynamic models:
  eutrophication, 355–357
  of water quality, 254–255

Earth Summit, 210
Ecological models of water quality, 255–256
Ecological modification (*see* Physical modifications of environment)
Economic and Social Commission for Asia and the Pacific (ESCAP), 125–127
Economic development:
  and environment, 14–15
  sustainable, 15–19
Economic impact, environmental assessment, 22–24
Economic mechanisms for resource management:
  efficient allocation of water, 582–593

Economic mechanisms for resource management, efficient allocation of water (*Cont.*):
demand estimation, 586–591
supply estimation, 591–593
efficient pricing and water permits, 593–603
government involvement in water allocation, 579–582
implementing pricing, 598–599
irrigation charges, 599–601
issues in, 602–603
municipal and industrial charges, 601–602
tradeable permits, 597–598
water quality management, 595–597
intertemporal allocation, 615–619
scarcity and, 615–617
uncertainty and, 617–619
market approaches to water management, 603–612
alternatives to sale of water rights, 611
conflict resolution, 608–610
distribution of water rights, 610–611
government action and, 611–612
infrastructure, 607
management of systems, 607
third-party impacts, 608
water rights, 605–606
open-access problem, 613–615
Economics:
conjunctive use systems, 188
costs of water, 12
erosion and water-loss costs, 54–59
financial aspects of resource management, 569–575
integrated approach to groundwater management, 200, 201
reclamation of salt-affected soils, 114–115
resource management and planning approaches, 178
supply and demand for water, 8–11
transport of water, 7–8
wastewater reuse, 387
economic and financial analysis, 392–394
market assessment, 389–392
miscellaneous factors, 394–395
reports, 396–397
user contracts, 297–298
water quality monitoring, 220–221
Ecosystems:
in-stream water uses, 610
(*See also* Aquatic communities)
Effluent discharges (*see* Pollution)
Egypt, 2, 39
Aswan Dam, 25–26, 34
population and per capita water availability in, 9
schistosomiasis in, 32
social impacts of development projects, 627, 634, 636, 643
Eichhornia crassipes (water hyacinth), 449–450, 451, 478, 479, 515, 525
biogas production, 523
biological control of, 502–503
environmental control, 509
in Indonesia, 343
Eifelmaare lakes, 300
Einstein's method, 89, 90
Electrical conductivity:
irrigation with recycled water, 402, 403–405
water quality monitoring, 245
Elkhorn River, 84
Elodea canadensis, 478, 525
Emergency use of reclaimed municipal wastewater for potable supply, 427–429
Emergent weeds, 481–482, 491–493
Empirical models of water quality, 253–254
Endothal, 493
Energy:
conjunctive use of surface water and groundwater, 187
global water use trends, 8
Energy costs, erosion and, 54
Energy generation:
fishery resource threats, 443, 444
James Bay Project, 627, 633, 636–642, 643
Energy-grade line, stream bedform, 84–85
Enforcement:
government functions, 563–564
water rights, 606
England, 8, 383
Enteric viruses (*see* Pathogens)
Environment:
environmental assessment, 147–174
framework of, 167–174
mathematical models, 151–162
methodologies, 148–151

Environment, environmental assessment (*Cont.*):
modeling process, 162–165
planning model, 165–167
groundwater problems, 128–147
contamination, 135–143
land subsidence, 130–135
overcharging, 128–130
overexploitation, 130
saltwater intrusion, 143–147
hydrologic system, 117–128
flow system characterization, 118–124
groundwater use, 124–127
interaction with surface water, 127
role of groundwater in water resource systems, 127–128
management, 174–201
conjunctive use of surface and groundwater, 183–189
data requirement and monitoring role, 192–195
key issues and problems, 176–178
models in, 189–192
national strategy, 195–201
protection management, 181–183
protection policy, 180–181
protection strategy, 179–180
water quality issues, 178–183
Environmental assessment, 14–15, 20–35, 169–170
absence of integrated approach, 27–29
framework of, 167–174
lack of adequate knowledge, 29–33
limited framework for, 21–27
mathematical models, 151–162
methodologies, 148–151
modeling process, 162–165
monitoring and evaluation, 33–35
planning model, 165–167
remediation of contaminated water, 179
Environmental control of weeds, 507–518
advantages and disadvantages of, 517–518
carbon dioxide, 512–513
integrated methods, 519
light limitation, 508–511, 517
nutrients, 511–512, 517
sediments, 515–517
water, 513–515, 517
Epiphytes, 493
Erosion:
Aswan Dam and, 34
Erosion (*Cont.*):
control of, and water conservation, 59–61
in croplands and pasturelands, 39–42
and eutrophication, 340
factors affecting, 43–45, 46, 47
model of erosion and water-loss effects on crop productivity, 52–53
and reduced productivity, 45, 47–52
transport of salts, 101
universal soil-loss equation, 92
and water-loss costs, 54–59
and water-loss processes, 42–43
water use in agriculture and forestry, 38–39
(*See also* Sedimentation)
Error sources in water quality monitoring, 236–238
Europe:
domestic water sources, 13, 125
erosion in, 40
eutrophication in, 320, 323, 324, 326–331, 350, 356
Alpine project of OECD, 330–331
Eastern Europe, 332–333
Nordic project of OECD, 326–330
shallow lakes and reservoirs, 331–332
fishery resources, 456–457, 466
groundwater protection zones, 183
wastewater reuse, 383
water use, 5–6, 125
European Inland Fisheries Advisory Commission (EIFAC), 227
Eutrophication:
in Africa, 343–345, 451
aquaculture impacts, 448–449
in Asia and Pacific, 142, 338–341, 453
in Eastern Europe, 332–333
fishery resource threats, 446, 447, 452, 453
modeling, 260, 276
in North America, 333–338
in Oceania, 341–343
overall trends, 317–326
research approaches, features of, 298–300
in South America, 345–347
trophic state and, 300–317
control of, factors in, 308–317
criteria and indicators, 301–308
definition, 300–301, 302
lake types, 301

Eutrophication, trophic state and (*Cont.*):
nutrients, 309–316
water quality modeling, 250, 251
in western Europe, 326–331
Alpine project of OECD, 330–331
Nordic project of OECD, 326–330
shallow lakes and reservoirs, 331–332
Evapotranspiration:
water and salt balance in irrigated land, 105
water loss as, 38
secondary salinization, 101
Evolution of institutions, 534–535
Existence value, 591
Expansion costs, 572
Exportation of water, 7–8
Externalities, sustainable development, 18
External quality control, 235

Fairness doctrine, 572
Fallout (*see* Atmospheric deposition)
Fall velocity, 79, 80, 86
Farm crop budget analysis, 586
Fault zones, 119
Fecal coliforms (*see* Pathogens; Public health)
Feedlots, 136
Fertilizers, 139, 140, 141, 142
Field data:
assessment approach, 168
biocoenotic models, 256
groundwater flow distribution, 122
water quality monitoring, 236
(*See also* Water quality monitoring)
Filariasis, 32
Financing:
government function, 542, 564
institutional basis of resource management, 569–575
local agency limitations, 539
Fish:
and eutrophication, 309
herbicide toxicity, 493
James Bay Project and, 642
weed control, 503, 504, 505–507, 510
integrated methods, 518–519
utilization of weeds by, 521
Fisheries, 30, 702, 703
conservation and enhancement options, 457–465
Fisheries, conservation and enhancement options (*Cont.*):
aquatic environment modification, 460–461
future developments, 457–460
management, 465
stocking, 461–464
definition, 440
global and regional issues, 450–457
Africa, 451–452
Asia, 452–454
Australia, Europe, and North America, 456–457
Latin America, 455–456
in-stream water uses, 610
integrated management of resources, 467
prospects, 465–467
social impacts of development projects, 653–654, 656
status of, 440–442
stress, fish response to, 450
threats to resources, 442–450
aquaculture, 448–449
introductions, 449–450
overexploitation, 447–448
physical modifications, 443–445
pollution, 445–447
water quality monitoring, 213, 227
weeds and, 480, 481
Flood control, 533, 568
financing, 571
social impacts of development projects, 635–636
utilities concept, 562–563
Flooding, erosion and, 57
Floodplains, 76, 635, 636, 657
Florida, wastewater reuse, 387
Flow characteristics, 118–124
academic disciplines, 199
assessment approach, 167, 170
channel design, 514
fishery resource enhancement, 460
fishery resource threats, 443, 445
and herbicide performance, 495
hydrogeology, 119–120
physical properties, 120–122
storage area development, 185, 186
stream properties, 76, 82–84
transport processes, 122–124
water quality models, 256–259
water quality monitoring, 223, 229–230, 231, 232

Flow characteristics (*Cont.*):
  weed control, 516
  and weed problems, 478
  weeds and, 450, 479
  (*See also* Hydrology)
Flow model, 152–154
Fluid drag, 79
Flume studies, 85
Fluoride, 142
Fluoridone, 493
Fluvial geomorphology, 74–75, 76, 77
Flux monitoring, 215–216
Food and Agriculture Organization of UN, 227, 243, 466, 467
Food web:
  eutrophication, 307
  eutrophic lake restoration, 361
  fishery resource enhancement, 460
  fishery resource threats, 446
  pesticide bioaccumulation, 454
  water quality modeling, 255–256
Food-web models of eutrophication, 357
Forestry, water use in, 39
Fosamine ammonium, 492
Fuelwoods, 29

Ganges River, 453, 464
Genetic algorithms, 288
Genetically modified fish, 464
Geochemistry, 135, 199
Geology (*see* Hydrogeology)
Geomorphology:
  conjunctive use of surface water and groundwater, 187
  fishery resource threats, 443
  fluvial, 74–75, 76, 77
Germany:
  domestic water sources, 13
  eutrophication in, 320
  wastewater reuse, 383
Global Environment Monitoring System (GEMS), 215–216, 243–245, 445
Global resources:
  erosion rates, 39–41
  eutrophication (*see* Eutrophication)
  fishery, 450–457
    Africa, 451–452
    Asia, 452–454
    Australia, Europe, and North America, 456–457
    Latin America, 455–456
  topsoil, 51

Global resources (*Cont.*):
  water quality monitoring, 210, 243–245
  water supply, 39
  water use, 2–7, 124–127
    factors in water crisis, 7–14
    in irrigation, 99–100
Glyphosate, 492, 509
Goals of groundwater protection strategy, 189–180
Government:
  allocation of water resources, 589–582
  integrated approach to groundwater management, 200
  (*See also* Institutional principles for resource management; Regulation)
Grass carp, 505, 506–507, 518–519, 521
Gravel:
  consolidated and semiconsolidated formations, 120
  hydraulic conductivity, 120, 121
  sediment size classes, 75, 78
  stream properties, 76
Great Britain, 13
Great Lakes of Africa, 343–344
Great Lakes of North America, 320, 336–337
Green manure, aquatic weeds as, 523
Groundwater:
  global water crisis, 13
  interaction with surface water, 127
  role in water resource systems, 127–128
  use of, 124–127
Groundwater problems:
  contamination, 135–143
  land subsidence, 130–135
  overcharging, 128–130
  overexploitation, 130
  saltwater intrusion, 143–147
Groundwater recharge (*see* Recharge)
Guidelines, water quality monitoring, 212, 213
Gypsum, 108

Habitat:
  fishery resource threats, 442, 443–445
  weeds and, 481
Hardness of water, 513
Health (*see* Pathogens; Public health)
Heat fluxes, water quality reaction model extensions, 270–271

Heavy metals, 139, 141, 142
  mining wastes, 455
  wastewater reuse, 412
  water quality modeling, 251
  water quality monitoring standards, 213
Herbicides:
  with biological control approaches, 503
  integrated weed control, 518–519
  weed removal, 490–496
Herbivores, weed control, 522
Hilsa ilisha, 464
Historical data validation, 164
Horticulture, 704
Hungary, eutrophication in, 361
Hydraulic conductivity, 120–121, 173, 183
Hydraulic models, 254
Hydrilla verticillata, 478, 482, 503, 509, 513
Hydrogen peroxide, 490
Hydrogeology, 119–120
  academic training needs, 198–199
  assessment approach, 167
  and contaminant transport, 196–197
  environmental assessment, 148, 149, 151
  flow system characteristics, 119–120
  groundwater recharge area locations, 118–119
  management model data requirements, 192–193
  prediction of groundwater flow, 122
  water quality effects, 217
  water quality monitoring, 221, 223, 229, 230, 231, 232
Hydrograph analysis, 149, 151
Hydrologic cycle:
  conjunctive use of surface water and groundwater, 184–185
  environmental assessment, 149–150
  storage area development, 185, 186
Hydrology, 117–128, 210
  conjunctive use of surface water and groundwater, 187
  flow system characterization, 118–124
  and groundwater movement, 120–121
  groundwater use, 124–127
  interaction with surface water, 127
  prediction of groundwater flow, 122
  resource management and planning approaches, 174–175
  role of groundwater in water resource systems, 127–128
Hydrology (*Cont.*):
  water quality effects, 217
  water quality monitoring, 220, 221, 229, 230, 231, 232
  water quality prediction and management, 256–259
  (*See also* Flow characteristics)
Hydropolitics, 660
Hydropower, 589–590
Hypothalmichthys molitrix (silver carp), 521

Impact monitoring, 214
Impoundments:
  fishery resource threats, 442–450
  infrastructure, 606
  pollution risk assessment methods, 173
  (*See also* Dams; Reservoirs; Surface water)
India, 2
  Bhima project, 28
  contaminants of groundwater, 140–141
  erosion in, 40
  fishery resources, 453, 464
  irrigation in, 99
  population and per capita water availability in, 9, 10
  RAJAD project, 113
  resettlement of dislocated populations, 670
  social impacts of development projects, 627
  water use in, 5, 6, 126
Indonesia:
  erosion in, 40
  eutrophication in, 342–343
  global water use trends, 6
  population and per capita water availability in, 9
Indus River:
  fishery resource threats, 445
  salinity control and reclamation projects, 129–130
  Tarbella Dam, 93
Industrial effluents:
  contaminants of groundwater, 139, 140, 141
  water quality modeling, 251
  water quality monitoring, 228
Industrial sector:
  estimating demand for water, 589
  fishery resource threats, 446, 456

Industrial sector (*Cont.*):
global water use trends, 5, 6, 8, 126
integrated approach to groundwater management, 200
pricing water, 601–602
wastewater reuse, 385, 386, 413–415
Industry NGOs, 565–566
Infiltration galleries, 145
Infiltration rates, wastewater reuse, 412, 418
Inflation, 602
Inflow:
assessment approach, 167
conjunctive use of surface water and groundwater, 185
Infrastructure, 604, 607
Infrastructure financing, 573–574
Inland fisheries (*see* Fisheries)
Insects:
biological weed control, 502, 503
disease vectors, 32, 450
Institutional principles for resource management:
financial aspects, 569–575
nature of, 529–542
common deficiencies, 538–539
evolution of, 531–534
general principles, 530–531
modification of, 539–542
political and institutional change, 534–535
public participation, 535–538
reform and modification of, 539–542, 575–576
responsibilities and organization of public and private sector, 549–564
functional independence, 553–557
functional linkages, 550–553
government functions, 557–564
role of nongovernment entities and public, 564–569
Integrated approach:
fishery resources management, 467
groundwater management, 199–201
water quality monitoring and water quality management, 210–211
Integrated (conjunctive) use of surface and groundwater, 184–189
Integrated weed control methods, 518–520
Intensification strategies, fishery resource, 459–460
Interdisciplinary approach to groundwater management, 199–201
Interest groups:
conjunctive use systems, 188–189
stakeholder participation, 535–538
(*See also* Local participation; Public)
Intermediate flow path, 119
International Co-operative Programme on Assessment and Monitoring of Acidification of Rivers and Lakes, 217–218
International trade in water, 7–8
Introductions:
aquatic weeds, 449–450, 525
fishery enhancement, 461–464
fishery resource threats, 443, 449–450
Inundative biological control, 501–502, 503–505
Invertebrates:
eutrophication, 307
herbicide toxicity, 493
soil biota, erosion and, 43, 50–51
weed control, 521
Investor-owned private companies, 568–569
Iron, 141, 142, 512
Irrigation, 15, 18, 532, 604
in alluvial plains, 129
channel design, 511, 514
conjunctive use of surface water and groundwater, 184
contamination of groundwater, 135–138
environmental assessment, 27–29
estimating demand for water, 586–587
fishery resource threats, 453, 454
institutional foundations of resource management, 548
pricing water, 599–601
private sector projects, 568
public health risks, 29–33
surface and groundwater interaction, 127
wastewater reuse, 385, 386, 387, 399–413
case study: *Monterey Wastewater Reclamation Study for Agriculture*, 408–413
comparison of regulatory criteria, 408, 409
health risks, 406–408
Israel project, 421
pretreatment processes, 399, 400, 401
water quality, 401–406
water and salt balance in irrigated land, 104–107

Irrigation (*Cont.*):
waterlogging and salinity effects (*see* Waterlogging and salinity)
water quality monitoring standards, 213
and water table levels, 101, 102
water use trends, 6, 125
weeds and, 480
Israel, wastewater reuse projects, 421–422
Italy, eutrophication in, 320, 331

James Bay Project, 627, 633, 636–642, 643, 660
Japan:
erosion in, 40
eutrophication in, 314–315, 319, 320, 338
per capita water use in, 8, 9, 10, 11
wastewater reuse, 383
Java, 40, 343
Joint ventures, 581

Kaolinite, 81, 131
Kapenta, 462
Kinematic dispersion, 123
Kinematic viscosity, 85, 86, 87
Kinetic models of water quality, 255–256

Labeo rohita, 452
Laboratory tests:
assessment approach, 168
water quality monitoring, error sources, 236
Lake Mead, 93
Lake Nassar, 93
Lakes:
erosion and, 57
eutrophication (*see* Eutrophication)
flow system characterization, 118, 119
sediments (*see* Sedimentation)
water quality monitoring, 222, 245
Landfills, 136, 138, 228
Landscaping, wastewater reuse, 385, 386, 387, 401, 407, 409
Land subsidence, 130–135
assessment approach, 168
conjunctive use of surface water and groundwater, 184
modeling susceptibility to, 171–172
Land-subsidence model, 158–162
Land use:
fishery resource threats, 443
regulation and legislation, 197, 198
resource management and planning approaches, 174–175
water management linkages, 546–547, 550
water quality monitoring, 232
(*See also* Erosion)
La Plata River, 454
Lates niloticus (Nile perch), 462, 463
Latin America:
drinking water sources, 125
eutrophication in, 319, 345
fishery resources, 455–456, 462, 466
Latitude and eutrophication, 317, 318
Leaching:
irrigation before sowing for salt removal, 113
irrigation with recycled water, 404
reclamation of salt-affected soils, 114
water and salt balance in irrigated lands, 107
Lead:
contaminants of groundwater, 142
water quality monitoring standards, 213
Leasing of water rights, 611
Legal issues:
conjunctive use systems, 188
resource management and planning approaches, 178, 180
(*See also* Institutional principles for resource management)
Legislation, 142, 196–198
environmental impact assessment requirements, 21
water rights and allocation mechanisms, 544, 545
Lemna, 485, 488, 509, 515
Light:
and eutrophication, 309, 316
modeling of phosphorus effects, 277
and trophic state, 303
water clarity, 513
weed control methods, 508–511, 517
Limestone, 121
Limnology (*see* Eutrophication)
Limnothrissa miodon, 462
Linear programming approach, 586
Line functions, separation of, 553–554, 555

Livestock, 29, 30, 702
aquatic weeds as silage, 522–523
contamination of groundwater, 136
erosion on croplands and pasturelands, 39–42
fishery resource threats, 444
groundwater use patterns, 125
social impacts of development projects, 647, 655
water quality monitoring standards, 213
Local autonomy, 541
Local flow, 119
Local government, 533, 548, 557, 563
Local participation, 535–538
relocation planning, 698–700
social impacts of development projects, 648–650
Local water resource agencies, 539

Magdalena River, 454, 455
Magnesium ions and salts, 101, 102, 103, 108
Malaria, 27, 30, 31–32, 450
Malawi, 39
Malaysia, 141
Management, 142, 174–201
conjunctive use of surface and groundwater, 183–189
data requirement and monitoring role, 192–195
economic mechanisms (*see* Economic mechanisms for resource management)
eutrophication models, 347
fishery resources, 452, 457–465, 467
institutional principles (*see* Institutional principles for resource management)
key issues and problems, 176–178
models in, 189–192
national strategy, 195–201
protection management, 181–183
protection policy, 180–181
protection strategy, 179–180
water quality (*see* Water quality prediction and management)
water quality issues, 178–183
water quality monitoring (*see* Water quality monitoring)
Management goals, 542–543
Management model, 165–166
Mansonia, 32
Manual weed control, 484–485
Mapping, stratigraphic, 167
Marginal cost curve of water, 593
Marginal costs, 572
Marginal product, value of, 582–584
Markets (*see* Economic mechanisms for resource management)
Mass-balance models:
dissolved salt concentrations, 109
eutrophication, 351–355
water quality modeling, 260–266, 267, 278–286
water quality prediction and management, 260–266, 267
Mass-transport models, 154–158
Mathematical models:
environmental assessment, 151–162
(*See also* Models)
Meandering stream, 74, 75
Meandering streams, 76
Mechanical control:
of algae, 496
aquatic weeds, 484–489, 516, 518–519
Mechanical dispersion, 123
Melosira, 329
Melosira islandica ssp *helvetica*, 330
Mercury, 455, 642
Metals:
water quality bioassays, 242
water quality monitoring standards, 213
(*See also* Heavy metals)
Meteorology, environmental assessment, 148–14, 1519
Mexico:
fishery resources, 455, 462
irrigation in, 99
land subsidence in, 131, 132–133
population and per capita water availability in, 9
Michaelis-Menton equation, 273–274
Microbial control of weeds, 503
Microbiology:
aquaculture impacts, 448
contaminants of groundwater, 139, 140
irrigation with recycled water, 405–406, 407, 408, 409
nitrogen cycle modeling, 274–275
soil, erosion and, 50–51, 52, 53
wastewater reuse, 385, 386–387
health risk estimetion, 431–435
industrial cooling tower operations, 414

Microbiology (*Cont.*):
waterlogging and salinity effects, 100
water quality monitoring, 213, 224, 241
(*See also* Pathogens; Public health)
Microcystis, 341, 343
Microcystis aeruginosa, 316, 317
MIKE 11, 291
Mining wastes, 455
Mixing processes, 258
Models:
assessment approach, 168, 169, 170
DRAINMOD, 109
environmental assessment, 150–165
flow model, 152–154
land-subsidence model, 158–162
solute-transport model, 154–158
erosion and water-loss effects on crop productivity, 52–53
eutrophication, 347–357
dynamic models, 355–357
for environmental management, 347
mass-balance models, 351–355
nitrogen and phosphorus loading, 357–359
phosphorus versus chlorophyll, 348–351
recovery strategy, 357–362
remediation and restoration, 359–362
management, 189–192
performance of, 169
water quality (*see* Water quality prediction and management)
Modification of institutions, 539–542
Molecular diffusion, 123, 257, 258
Monitoring:
defined, 211
environmental impact assessment, 33–35
eutrophication, 303
evaluation of adequacy of, 169
government functions, 563–564
institutional foundations of resource management, 547
management, 192–195
management models, 193–194
water quality (*see* Water quality monitoring)
Monitoring wells, 192–193
Monopolies, 568, 580
Monterey Wastewater Reclamation Study for Agriculture, 408–413
Montmorillonite, 81, 131
Mosquitoes, 32
Mougeotia, 330
Mulching, 51
Multipurpose monitoring, 216
Municipalities:
estimating demand for water, 587–589
global water use trends, 125, 126
pricing water, 601–602
Municipal waste:
contamination of groundwater, 135–138
discharge of, 18
and eutrophication, 358
(*See also* Pollution; Sewage)
Municipal wastewater:
reuse of
groundwater recharge, 418, 419–421
as potable water, 427–429
(*See also* Wastewater reuse)
water quality monitoring, 240
Mylossoma argenteum (silver-dollar fish), 521
Myriophyllum spicatum, 503, 525

Namibia, potable water reuse, 249–430
National Environmental Protection Act of 1970, 21
National strategy, groundwater management, 195–201
Natural control methods, aquatic weeds, 497–500
Neochetina eichhorniae, 502–503
Neohydronomus affinis, 503
Neohydronomus pulchellus, 503
Netherlands, 13, 356
Neural networks, 253–254
Nigeria, 8
erosion in, 43, 51
no-till agriculture in, 60
population and per capita water availability in, 9
Nile perch (*Lates niloticus*), 462, 463
Nile river:
Aswan Dam (*see* Aswan Dam)
fishery resource threats, 445
Nitrates, 13
contaminants of groundwater, 139, 141
and weed growth, 511
Nitrobacter, 274, 275
Nitrogen:
eutrophication (*see* Eutrophication)
irrigation with recycled water, 405
water quality models, 290

Nitrogen (*Cont.*):
water quality monitoring, 241, 243
GEMS/WATER, 245
reaction model extensions, 274–276
standards, 213
(*See also* Nutrients)
Nitrogen-fixing organisms, 313
Nitrosomonas, 274, 275
Nonconservative transport models, 154–158
Nongovernment organizations (NGOs), 564–569, 693, 698–700
Nonpoint-source pollutants:
eutrophication, 358
wetlands for remediation, 362
Nonresponding-type of lake, 306
North America:
eutrophication in, 322, 324, 333–338
fishery resources, 456–457, 466
(*See also* Canada; Mexico; United States)
Norway, 327
Not-for-profit NGOs, 566–568
Nuclear contaminants, 139
Nuclear power plants, wastewater reuse, 414–415
Nuphar, 482
Nutrients:
aquaculture impacts, 448–449
and eutrophication, 309–317
eutrophication (*see* Eutrophication)
fishery resource threats, 445, 447
irrigation with recycled water, 405
soil, erosion and, 49, 53
water and salt balance in irrigated lands, 107
water quality model extensions, 274–277
water quality models, 255, 290
water quality monitoring, 213, 241, 243
weed control, 511–512, 517
Nymphaea, 482

Objectives, water quality monitoring, 220, 221
Oceania:
eutrophication in, 322, 324, 341–343
fishery resource threats, 453–454
global water use trends, 6
Oligotrophic:
defined, 300
qualitative criteria, 301–302
One-dimensional advection-dispersion equation, 259
One-dimensional plug-flow reactor modeling, 262–264, 267–268
One-dimensional water quality simulation throughout system, 267–270
Open-access problem in water resource management, 613–615
Operating costs, 571–572
Operational studies, assessment approach, 168
Operations, 553–554
groundwater management, 174
Operations and maintenance, 534, 541, 553, 555
financing, 574–575
pricing issues, 603
Opportunity costs, 572, 604
Option value, 591
Orconectes, 521
Organic chemicals, 142
contaminants of groundwater, 139
fishery resource threats, 457
wastewater reuse, 424–426
Organic matter:
aquaculture impacts, 448–449
and eutrophication, 300
fishery resource threats, 447
soil erosion and, 49–50, 53
Orinoco River, 454
Oscillatoria agardhii, 328–329
Oscillatoria bornettii, 327
Oscillatoria rubescens, 330
Overcharging, 128–130
Overexploitation, 130
assessment approach, 168
fishery resource threats, 447–448, 452
Ownership of resources, 542–543
Oxygen:
aquaculture impacts, 448–449
and eutrophication, 300
eutrophication, 307
fishery resource threats, 447
herbicide effects, 496
mass balance modeling, 260
straw antialgal factor, 498, 499
water quality modeling, 250–251, 252, 287–288, 289–290
water quality monitoring, 227, 241, 243
GEMS/WATER, 245
standards, 213

Russia (*Cont.*):
water quality monitoring standards, 213

Saline intrusion (*see* Saltwater intrusion)
Saline sodic soils, defined, 103–104
Saline soils, defined, 103
Salinity (*see* Waterlogging and salinity)
Salinity control and reclamation projects (SCARPS), 129–130
Salinization:
in Asia and Pacific, 142
conjunctive use of surface water and groundwater, 184
definitions, 103–104
estimating supply of water, 593
fishery resource threats, 446, 452
irrigation with recycled water, 402, 403–405
saltwater intrusion, 143–147
water quality modeling, 250, 251
water quality monitoring, 245
(*See also* Waterlogging and salinity)
SALMO model, 356
Saltwater intrusion, 139, 143–147, 408
contaminants of groundwater, 139, 140, 141
modeling susceptibility to, 171–172
Salvinia, 343, 515
Salvinia molesta, 450, 502, 509
Sampling:
water quality monitoring, 222–223, 224–225
(*See also* Water quality monitoring)
water quality monitoring, error sources, 236
Sand:
consolidated and semiconsolidated formations, 120
hydraulic conductivity, 120, 121
sediment size classes, 75, 78
stream properties, 76
Sand and gravel stream bedforms, 81–87
Sandstone, 120, 121
Sanitary landfills, 136, 138–139
San Joaquin Valley, land subsidence in, 131, 132–133
Santa Clara Valley, land subsidence in, 131
Saturated zone, 118, 173
Scandinavia, eutrophication in, 325, 326–330
Scarcity, 593, 615–617
Scarcity rent, 584, 585
Schistosomiasis, 27, 30, 32, 33, 34, 450, 652
Seasonal cycles:
water quality monitoring, 229
water quality reaction model extensions, 270–271
Secchi disk transparency, 304, 305
Secondary production in food chain, 255–256
Secondary salinization, 101
Sediment:
in Asia and Pacific, 142
fishery resource threats, 442–450
inflow of, 94
and life of reservoir, 185
modeling of susceptibility to subsidence, 172
phosphate release from, 512
water quality models, 290
water quality monitoring, 222
weed control, 515–517
Sedimentation, 73, 73–97
bed forms, 81–87
concepts, 73–74
and eutrophication, 300
fishery resource threats, 447, 451
glossary, 96–97
particle size and properties of sediment, 75, 78, 79–81
reservoirs, 92–95
river forms (fluvial geomorphology), 74–75, 76, 77
transport of sediment, 87–92
water quality modeling, 250
yield, sediment, 92
Seepage, flow system characterization, 119
Segmentation, water quality models, 286
Semiconsolidated formations, 120
Service charges, 571–573
Service operation NGOs, 566–568
Service responsibilities, 562–563
Sewage:
administration of, 533
contamination of groundwater, 135–138, 139, 140, 141
and eutrophication, 358
water quality monitoring, 240
Sewage treatment:
fishery resource protection, 458
recycling of water, 187

Shale, 121
Shape factor, 79
Shear stress, particulates, 79, 87, 88
Shields parameter, 87, 88
Silt, 120
  and eutrophication, 300
  fishery resource threats, 447
  light reduction, 510
  sediment size classes, 75, 78
  stream properties, 76
Siltation:
  erosion and, 56, 57
  fishery resource threats, 453
    in Asia, 454
    in South America, 456
  (*See also* Sedimentation)
Silver carp (*Hypothalmichthys molitrix*), 521
Silver-dollar fish (*Mylossoma argenteum*), 521
Single-objective water quality monitoring, 217
Sinuosity of streams, 76
Sive method of sediment size distribution, 75
Slope, and erosion, 43
Sludge disposal, contamination of groundwater, 135–138
Social benefits, in-stream water uses, 610–611
Social factors:
  conjunctive use of surface water and groundwater, 187
  water resource management, 176, 466
  (*See also* Institutional principles for resource management)
Social goals, 581
Social impacts of development projects, 15
  benchmark study data, 631–632
  benefits to project-affected people, 644–658
    alternative plans, 644–648
    controlled releases, 654–658
    disadvantages, 651–653
    irrigation, 650–651
    local participation, 648–650
    reservoir fisheries, 653–654
  case studies, 633–642
    downstream affected people, 634–636
    James Bay project, 636–640
    social impacts, 640–642
  definitions, 625–627
Social impacts of development projects (*Cont.*):
  environmental assessment, 22–24
  immigrants, 633
  numbers of people affected, 627–630
  political and ideological factors, 659–660
  relocation (*see* Resettlement)
  resistance movements, 642–644
  unforeseen consequences, 659
Sodium ions:
  irrigation with recycled water, 402, 404–405
  (*See also* Salinization; Saltwater intrusion; Waterlogging and salinity)
Soil:
  and erosion, 44–45
  irrigation with recycled water, 402, 404–405
  sediments (*see* Sedimentation)
  waterlogging and salinity (*see* Waterlogging and salinity)
Soil chemistry, 199
Soil erosion (*see* Erosion)
Soil formation rates, 51, 52
Soil-loss equation, universal (USLE), 92
Soil nutrients, erosion and, 49
Soil water, flow system characterization, 119
Solute-transport model, 154–158
South America:
  erosion in, 40
  eutrophication in, 319, 322–323, 324, 345–347
  fishery resources, 455
  global water use trends, 6
  social impacts of development projects, 660
Spain, eutrophication in, 332
Special interests, 534
Species diversity, fishery resource threats, 447
Specific weight of particulates, 80, 95
Spicata molesta, 525
Sri Lanka, 632, 650–651
Stakeholder participation, 535–538
Standards:
  institutional foundations of resource management, 547–549
  water quality monitoring, 212, 213
State autonomy, 541
Static models of water quality, 254–255

Statistical models of water quality, 253–254
Stephanodisculs hantzschii, 330
Stochastic models of water quality, 253, 254
Stockholm Action Plan, 24
Stocking, fish, 452, 461–464
Stoke's law, 80
Storage:
  development of, 185–186
  role of groundwater in water-resource systems, 127–128
Storage-coefficient values, 120–121, 122
Storm events:
  surface and groundwater interaction, 127
  water quality models, 256
Straight stream, 74
Stratification of aquifer, 172
Stratigraphy, 119, 167
Straw:
  algae control with, 497–500, 519
  mulching with, 51
Streams:
  bedforms, 81–87
  erosion and, 56
  flow system characterization, 118, 119
  planform classification, 74–75, 76, 77
  water quality monitoring, 245
Stress, fish response to, 450
Submerged plants, 482
Subsidence of land (*see* Land subsidence)
Subsidies, 563
Subsoil conditions and land subsidence, 131, 132–133
Subtraction, 581
Sum of excessive water table rises above 30-cm depth, 110
Supply of water, 581, 584, 585, 591–593
Surface evaporation, secondary salinization, 101
Surface spreading, groundwater recharge, 416, 418, 423
Surface water:
  conjunctive use with groundwater, 183–189
  erosion and, 56
  eutrophication (*see* Eutrophication)
  flow system characterization, 118, 119
  groundwater interactions with, 127
  institutional foundations of resource management, 552
Surface water (*Cont.*):
  water quality monitoring, 215–216, 217–218, 221, 222
    GEMS/WATER, 245
  water rights establishment, 606
  (*See also* Dams; Impoundments)
Suspended sediments, 88–89
Sustainability, 13–14, 15–19, 553
  environmental impact studies in past 25 years, 23–24
  global water shortages, 39
  resettled populations and environmental degradation, 696–697
Sweden, eutrophication in, 314, 328–329
Switzerland, eutrophication in, 330
Synedra, 330

Tarbella Dam, Indus River, 93
Taxation, water resources as basis, 2
Taylor, one-dimensional advection-dispersion equation of, 259
Technological change, 556
Temperature:
  and eutrophication, 316–317
  fishery resource threats, 444, 451
  sedimentation, Colby's correction curves, 91
  and waterlogging and salinity effects, 100
  water quality monitoring, 245
  water quality reaction model extensions, 270–271
  weed control, 514–515
Tennessee Valley Authority, 624–625
Terbutryn, 493, 494, 497
Texas:
  erosion in, 47
  wastewater reuse projects, 387, 419–421
Thailand:
  contaminants of groundwater, 141
  land subsidence in, 132–133, 135
Third-party impacts, 606
Tilapia, 462, 521
Till, 121
Time representations in water quality modeling, 254–255
Toll goods, 581
Topography:
  groundwater recharge area locations, 118–119
  universal soil-loss equation, 92

Topsoil resources, global, 51
Total sediment discharge, 89
Toxic chemicals:
  contaminants of groundwater, 142
  water quality modeling, 250
Toxicity:
  irrigation with recycled water, 404
  water quality bioassays, 242
Toxins, weeds and, 481
Transferable water permit, 594, 597–598
Transmissive properties of geologic formations, 119
Transport:
  academic disciplines, 199
  of contaminants, 122–124
  models of, 154–158
  of salts, ions, and minerals, 101
  water quality models, 256, 257–259, 290
  water quality monitoring, 215–216
Transportation of water, 7–8
Transport of samples, 236
Trap efficiency, 94–95
Treatment processes, 533
  aquatic plant uses, 524
  contamination of groundwater, 135–138
  fishery resource protection, 458
  wastewater reuse, 382
    groundwater recharge, 416–417, 420, 421–422, 423, 424
    for irrigation, 399, 400, 401
    in nuclear generating plant, 415
Trend monitoring, 214–215
Triazine herbicides, 493, 497
Triploid fish, 464
Trophic state index (TSI) system, 303
Turbidity, 513
  trophic state monitoring, 303, 304, 305
  water quality monitoring standards, 213
  weed control, 510–511
  (*See also* Sediment; Silt)
Turbulence, 257
Turkey, 9
Two-stage channels, 520
Typha, 523

Uncertainty:
  estimating supply of water, 592
  and resource allocation, 617–619
  resource management and planning approaches, 178
  sustainable development, 19
Unconsolidated formations, 120, 172
United Nations Conference on Environment and Development (UNCED), 14–15, 210
United Nations Conference on the Human Environment, 24
United Nations Educational, Scientific and Cultural Organization (UNESCO), 243
United Nations Environment Programme (UNEP), 21, 242
United Nations Food and Agriculture Organization, 227, 243, 466, 467
United States:
  classification of groundwater resources, 181–182
  domestic water sources, 13
  erosion in, 40, 41–42, 57, 58, 59
  eutrophication in, 318–319, 320, 350
  groundwater use patterns, 125
  irrigation in, 99–100
  land subsidence in, 131, 132–133
  population and per capita water availability in, 9
  potable water reuse, 427–429, 430–431
  wastewater reuse, 383, 387
  water quality monitoring, 218
  water rights in, 605
United States Environmental Protection Agency trophic state classification, 302
Universal soil-loss equation (USLE), 92
Unplanned reuse of wastewater, 382
Unsaturated zones, 118, 183
Upconing, 145
Urban areas:
  Africa, water quality deterioration, 344
  contaminants of groundwater, 139, 140
  land subsidence in coastal cities, 134
  population and per capita water availability, 8, 9, 10
  water quality monitoring, 239–240, 241
Urban horticulture, 704
Urbanization, 232, 531
  and eutrophication, 358–359
  fishery resource threats, 443, 451, 452, 453
Uroglena, 339
Uroglena americana, 339–340
USDA universal soil-loss equation, 92
Use of resources:
  monitoring in relation to, 212–216

Use of resources (*Cont.*):
water quality monitoring baseline values, 223–226
Users:
conjunctive use systems, 188–189
resource management and planning approaches, 177, 178
Utilities concept, 562–563

Vadose zone, 199
Value of marginal product, 582–584
Vegetation:
African lakes, 344
crop yield effects of salinity and waterlogging, 109–113
and erosion, 43–44
erosion control techniques, 59–61
eutrophication process, 305–306
eutrophic lake restoration, 361
fishery resource protection, 458
fishery resource threats, 443
salt removal, 101
stream properties, 76
(*See also* Weeds, aquatic)
Verification of model, 164, 255
Viruses (*see* Pathogens; Public health)
Viscosity, and stream bedforms, 83–84, 85, 86, 87
Visual accumulation tube analysis, 75
Voids:
geologic formations, 119, 120
(*See also* Porosity; Porous media)
Volga River, 457

Wales, 8
Wash load, 88
Waste discharges, 18
contamination of groundwater, 135–138, 139
and eutrophication, 358
regulation and legislation, 197
water quality monitoring, 227–228
(*See also* Contamination)
Waste disposal, 209
contaminants of groundwater, 139, 140–141, 141
institutional deficiencies, 538
integrated approach to groundwater management, 200, 201
water quality monitoring, 228
Waste management:
administration of, 533
institutional foundations of resource management, 548
local government entities, 563
utilities concept, 562–563
Waste treatment (*see* Treatment processes)
Wastewater, 210
water quality monitoring, 240
(*See also* Pollution)
Wastewater reuse:
categories of wastewater, 383–385
definitions, 382–383
groundwater recharge, 415–425
guidelines for, 422–423
Israel projects, 421–422
pretreatment requirements, 416–417
projects in Southern California, 417–419
proposed California regulations, 423–426
techniques, 416
Texas projects, 419–421
for industrial processes, 413–415
for irrigation, 399–413
case study: *Monterey Wastewater Reclamation Study for Agriculture*, 408–413
comparison of regulatory criteria, 408, 409
health risks, 406–408
pretreatment processes, 399, 400, 401
water quality, 401–406
microbial health risk estimation, 431–435
planning, 387–398
basic function, 388–389
economic and financial analysis, 392–394
market assessment, 389–392
miscellaneous factors, 394–395
reports, 396–397
user contracts, 297–298
for potable water, 426–431
potential and current status, 385–387
Water chemistry, 199
Water crisis, 7–14
Water cycle:
environmental assessment, 149–150
erosion and (*see* Erosion)
irrigation and, 104–107
weeds and, 480

Water erosion, 42, 43, 44, 47, 58
Waterfowl, 481, 505, 522
Water holding capacity of soil, 53
Water hyacinth (see *Eichhornia crassipes*)
Water lettuce (*Pistia stratiotes*), 503, 525
Water level:
  fishery resource enhancement, 460
  fishery resource threats, 453, 454
  planned releases from impoundments, 654–658
  in soil (*see* Water table)
  weed control, 514–515
Water lilies, 482, 509
Waterlogging and salinity, 15, 41, 117
  characteristics of saline and alkali soils, 103–104
  conjunctive use of surface water and groundwater, 186
  and crop yield, 109–113
  environmental assessment, 23
  occurrence, 100–103
  prediction of salinity changes, 107–109, 110
  reclamation and remediation, 113–115
  water and salt balance in irrigated lands, 104–107
Water quality, 533
  aquaculture impacts, 448–449
  aquatic plant uses in water treatment, 524
  aquatic weeds and, 450
  assessment approach, 168
  baseline values, 167
  conjunctive use of surface water and groundwater, 184
  economic mechanisms (*see* Economic mechanisms for resource management)
  estimating supply of water, 592
  eutrophication (*see* Eutrophication)
  and fishery resources, 440
    enhancement, 460
    threats, 456
  and herbicide performance, 495
  institutional foundations of resource management
    institutional deficiencies, 538
    linkages with management, 550–552
    standards, regulations, and administrative rules, 547–549
  management and planning issues, 178–183
  pollution risk assessment method, 173
Water quality (*Cont.*):
  pricing and permits, 595–596
  regulation and legislation, 197
  wastewater reuse, 390, 395, 398–399
    California's *Wastewater Reclamation Criteria*, 431, 435
    industrial cooling tower operations, 414
    for irrigation, 401–406, 407, 408, 409
  weeds and, 481
  (*See also* Pollution)
Water Quality Analysis Simulation Program (WASP), 290
Water quality degradation as operating cost, 572
Water quality monitoring:
  defining objectives, 216–219
  definitions, 211–216
    flux monitoring, 215–216
    impact monitoring, 214
    monitoring in relation to use, 212–214
    multipurpose monitoring, 216
    trend monitoring, 214–215
  in developing countries, 239–243
  global scale, 243–245
  program components, 219–235
    data collection methods, 230–231
    data integration, 231–232
    data interpretation, 232–234
    management recommendations and program evaluation, 234–235
    preliminary surveys, 219–222
    sampling, 229–230
    sampling media, 222–223, 224–225
    selection of variables, 224, 226–229
  quality assurance and control, 235–239
  recommendations, 246
Water quality prediction and management:
  calibrating and assessing model applicability and precision, 286–289
  classification of models, 253–256
  currently available models, 289–291
  hydraulic considerations, 256–259
  mass balance modeling, 260–266, 267
  model development and solution procedure, 278–286
  model selection and complexity, 252–253
  one-dimensional water quality simulation throughout system, 267–270

Water quality prediction and management (*Cont.*):
reactions, extensions of models, 270–277
Water resources:
allocation of (*see* Allocation of resources; Economic mechanisms for resource management)
groundwater role, 127–128
(*See also* Water supply)
Water rights, 542–546
economic mechanisms (*see* Economic mechanisms for resource management)
institutional deficiencies, 538
Watershed, surface and groundwater interaction, 127
Water shortages, 39
Water supply:
conjunctive use of surface water and groundwater, 184
local government entities, 563
utilities concept, 562–563
Water table:
in alluvial plains, 129
conjunctive use of surface water and groundwater, 186
and crop yield, 110
irrigation and, 101, 102
modeling susceptibility to subsidence or saline intrusion, 171–172
pollutants entering recharge and discharge zones, 138
prediction of groundwater flow, 122
and subsidence of land, 131
upward movement, salinization effects, 101
Water treatment (*see* Treatment processes)
Wave energy, weed control, 515
Weathering, 101
Weeds, aquatic, 524–526
control methods, 483–520
biological methods, 500–507
chemical, 489–497
environmental, 507–518
integrated, 518–520
mechanical, 484–489
natural methods, straw, 497–500
definitions, 477–478
extent of problem, 478–479
fishery resource threats, 449–450, 451
fish introductions and, 463
in Indonesia, 343
problems caused by, 479–481
types of, 481–483
uses of, 520–524
Weirs, 463
Wellhead protection areas, 181–182
Wells, water quality monitoring, 221
Wetlands, 360, 362
White amur, 505, 506–507
Whittier Narrows Groundwater Recharge Project, 417–419
Wildlife habitat, erosion and, 56
Wind erosion, 42–43, 44, 47
costs of, 58
off-site damage, 57
World Bank, 21
(*See also* Resettlement; Social impacts of development projects)
World Health Organization:
schistosomiasis estimates, 33
wastewater reuse guidelines, 408, 409
water quality monitoring, 213, 215–216, 244
World Meterological Organization, 243
WQRRSQ model, 290

Zinc, 405
Zones of benefit, 571
Zooplankton, 307

ABOUT THE EDITOR-IN-CHIEF

Professor Asit K. Biswas (Oxford, England) is the past president of the International Water Resources Association and chairs the Technical Programme Committee of the World Water Council. Associated with major universities in Canada, England, Mexico, and Sweden, he has been senior adviser to the heads of five United Nations agencies, all major international aid organizations, and 17 governments on water-environment issues. He is the author of 53 books and more than 500 technical papers. Acknowledged as one of the world's leading water experts, his work has been translated into 31 languages.